Water Resources Engineering

First Edition

Larry W. Mays, Ph.D., P.E., P.H.
Professor of Civil and Environmental Engineering
Arizona State University
Tempe, Arizona

John Wiley & Sons, Inc.
New York/Chichester/Weinheim/Brisbane/Toronto/Singapore

ACQUISITIONS EDITOR	Wayne Anderson
MARKETING MANAGER	Katherine Hepburn
SENIOR PRODUCTION EDITOR	Patricia McFadden
DESIGN DIRECTOR	Maddy Lesure
ILLUSTRATION EDITOR	Anna Melhorn

This book was set in Times by Argosy and printed and bound by Hamilton Printing. The cover was printed by Lehigh Press, Inc.

Cover: Department of the Interior Bureau of Reclamation Phoenix Area Office; Joe Madrigal, Jr.

This book is printed on acid-free paper.

Library of Congress Cataloging in Publication Data:

Mays, Larry W.
  Water Resources Engineering/Larry W. Mays. — 1st ed.

  ISBN 0-471-29783-6

Printed in the United States of America

10 9 8 7 6 5 4 3 2 1

Acknowledgments

Water Resources Engineering is the result of teaching classes over the past 24 years at the University of Texas at Austin and Arizona State University. So first and foremost, I would like to thank the many students that I have taught over the years. Several of my past Ph.D. students have helped me in many ways through their review of the material and help in development of the solutions manual. These former students include Drs. Aihua Tang, Guihua Li, John Nicklow, Burcu Sakarya, Kaan Tuncok, Carlos Carriaqa, Bing Zhao, and Messele Ejeta. I would like to acknowledge Arizona State University, especially the time afforded me to pursue this book.

I would like to thank Wayne Anderson of John Wiley & Sons, Inc. for having faith in me through his willingness to publish the book. Reviewers of the book manuscript included Professors Howard H. Chang (San Diego State University), Neil S. Grigg (Colorado Sate University), G. V. Loganathan (Virginia Tech), and Jerome A. Westphal (University of Missouri at Rolla). They all provided insight and many suggestions.

During my academic career as a professor I have received help and encouragement from so many people that it is not possible to name them all. These people represent a wide range of universities, research institutions, government agencies, and professions. To all of you I express my deepest thanks.

Water Resources Engineering has been a part of a personal journey that began years ago when I was a young boy with a love of water. This love of water resources has continued throughout my life, even in my spare time, being an avid snow skier and fly-fisherman. Books are companions along the journey of learning and I hope that you will be able to use this book in your own exploration of the field of water resources. Have a wonderful journey.

Larry W. Mays
Scottsdale, Arizona

I would like to dedicate this book to humanity and human welfare.

Preface

Water Resources Engineering can be used for the first undergraduate courses in hydraulics, hydrology, or water resources engineering and for upper level undergraduate and graduate courses in water resources engineering design. This book is also intended as a reference for practicing hydraulic engineers, civil engineers, mechanical engineers, environmental engineers, and hydrologists.

Water resources engineering, as defined for the purposes of this book, includes both water use and water excess management. The fundamental water resources engineering processes are the hydrologic processes and the hydraulic processes. The common threads that relate to the explanation of these processes are the fundamentals of fluid mechanics using the control volume approach. The hydraulic processes include pressurized pipe flow, open-channel flow, and groundwater flow. Each of these in turn can be subdivided into various processes and types of flow. The hydrologic processes include rainfall, evaporation, infiltration, rainfall-runoff, and routing, all of which can be further subdivided into other processes. Knowledge of the hydrologic and hydraulic processes is extended to the design and analysis aspects. This book, however, does not cover the water quality management aspects of water resources engineering.

Water resources development has had a long history, basically beginning when humans changed from being hunters and food gatherers to developing of agriculture and settlements. This change resulted in humans harnessing water for irrigation. As humans developed, they began to invent and develop technologies, and to transport and manage water for irrigation. The first successful efforts to control the flow of water were in Egypt and Mesopotamia. Since that time humans have continuously built on the knowledge of water resources engineering. This book builds on that knowledge to present state-of-the-art concepts and practices in water resources engineering.

Water Resources Engineering is divided into four parts: Part I – Hydraulics; Part II – Hydrology; Part III – Engineering Analysis and Design for Water Use; and Part IV – Engineering Analysis and Design for Water Excess Management. Part I consists of six chapters that introduce the basic processes of hydraulics. Chapter 1 is a very brief introduction to water resources. Chapter 2 is a review of basic fluid mechanics principles. Chapter 3 presents the control volume approach for continuity, energy, and momentum. Chapters 4, 5, and 6 cover pressurized pipe flow, open-channel flow, and groundwater flow. Part II presents four chapters that cover the basics of hydrology: Chapter 7 on hydrologic processes; Chapter 8 on rainfall-runoff analysis; Chapter 9 on routing and Chapter 10 on and probability and frequency analysis. Part III, on engineering analysis and design for water use, consists of three chapters: Chapter 11 on water withdrawals and uses; Chapter 12 on water distribution systems; and Chapter 13 on water for hydroelectric generation. Part IV, on engineering analysis and design for water excess management, includes four chapters: Chapter 14 on water excess management; Chapter 15 on stormwater control: storm sewers and detention; Chapter 16 on stormwater control: street and highway drainage and culverts; and Chapter 17 on the design of hydraulic structures for flood control storage systems.

Several first courses could be taught from this book: a first course on hydraulics, a first course on hydrology, a first course on water resources engineering analysis and design, and a first course on hydraulic design. The flowcharts on the following pages illustrate the topics and chapters that could be covered in these courses.

This is a comprehensive book covering a large number of topics that would be impossible to cover in any single course. This was done purposely because of the wide variation in the manner in which faculty teach these courses or variations of these courses. Also, to make this book more valuable to the practicing engineer or hydrologist, the selection of these topics and the extent of coverage in each chapter were considered carefully. I have attempted to include enough example problems to make the theory more applicable, more understandable, and most of all more enjoyable to the student and engineer.

Students using this book will most likely have had an introductory fluid mechanics course based on the control volume approach. Chapter 2 should serve as a review of basic fluid concepts and Chapter 3 should serve as a review of the control volume concepts. Control volume concepts are then used in the succeeding chapters to introduce the hydrologic and hydraulic processes. Even if the student or engineer has not had an introductory course in fluid mechanics, this book can still be used, because the concepts of fluid mechanics and the control volume approach are covered.

I sincerely hope that this book will be a contribution toward the goal of better engineering in the field of water resources. I constantly remind myself of the following quote from Baba Diodum: "In the end we will conserve only what we love, we will love only what we understand, and we will understand only what we are taught."

This book has been another part of a personal journey of mine that began as a young boy with an inquisitive interest and love of water, in the streams, creeks, ponds, lakes, rivers, and oceans, and water as rain and snow. Coming from a small Illinois town situated between the Mississippi and Illinois Rivers near Mark Twain's country, I began to see and appreciate at an early age the beauty, the useful power, and the extreme destructiveness that rivers can create. I hope that this book will be of value in your journey of learning about water resources.

First Undergraduate Hydraulics Course

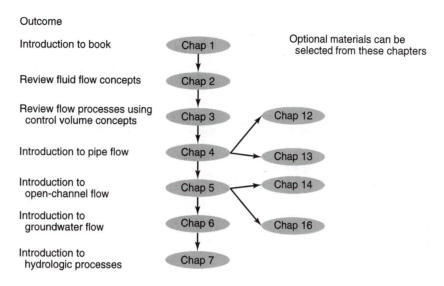

Outcome

Introduction to book

Review fluid flow concepts

Review flow processes using
control volume concepts

Introduction to pipe flow

Introduction to
open-channel flow

Introduction to
groundwater flow

Introduction to
hydrologic processes

Optional materials can be
selected from these chapters

First Undergraduate Hydrology Course

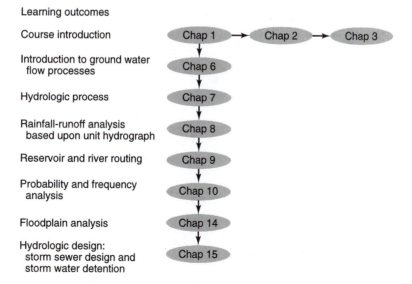

Learning outcomes

Course introduction

Introduction to ground water
flow processes

Hydrologic process

Rainfall-runoff analysis
based upon unit hydrograph

Reservoir and river routing

Probability and frequency
analysis

Floodplain analysis

Hydrologic design:
storm sewer design and
storm water detention

Undergraduate Hydraulic Design Course

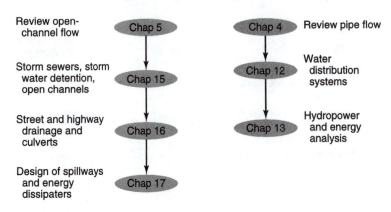

Water Resources Engineering

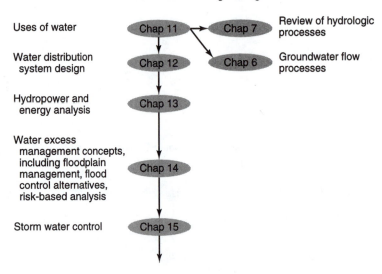

Contents

Chapter 1

Introduction

1.1 BACKGROUND

Water resources engineering (and management) as defined for the purposes of this book includes engineering for both *water supply management* and *water excess management* (see Figure 1.1.1). This book does not cover the *water quality management (or environmental restoration)* aspect of water resources engineering. The two major processes that are engineered are the *hydrologic processes* and the *hydraulic processes*. The common threads that relate to the explanation of the hydrologic and hydraulic processes are the fundamentals of fluid mechanics. The hydraulic processes include three types of flow: pipe (pressurized) flow, open-channel flow, and groundwater flow.

The broad topic of *water resources* includes areas of study in the biological sciences, engineering, physical sciences, and social sciences, as illustrated in Figure 1.1.1. The areas in biological sciences range from ecology to zoology, those in the physical sciences range from chemistry to meteorology to physics, and those in the social sciences range from economics to sociology. Water resources engineering as used in this book focuses on the engineering aspects of hydrology and hydraulics for water supply management and water excess management.

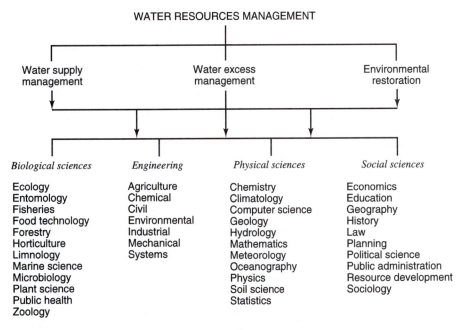

Figure 1.1.1 Ingredients of water resources management (from Mays (1996)).

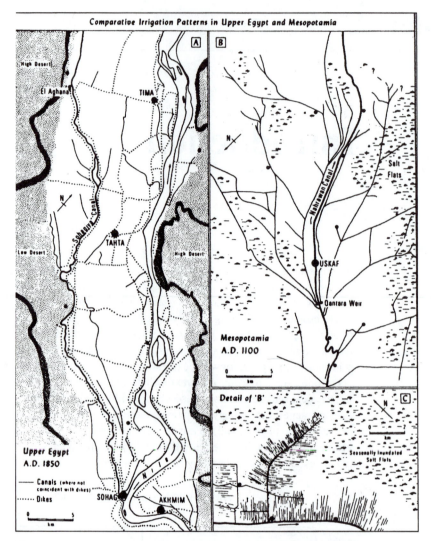

Figure 1.1.2 Comparative irrigation networks in Upper Egypt and Mesopotamia. (*a*) Example of linear, basin irrigation in Sohag province, ca. AD 1850; (*b*) Example of radial canalization system in the lower Nasharawan region southeast of Baghdad, Abbasid (A.D. 883–1150). (Modified from R. M. Adams (1965), Fig. 9. Same scale as Egyptian counterpart). (*c*) Detail of field canal layout in *b*. (Simplified from R. M. Adams (1965), Fig. 10. Figure as presented in Butzer (1976)).

Water resources engineering not only includes the analysis and synthesis of various water problems through the use of the many analytical tools in hydrologic engineering and hydraulic engineering but also extends to the design aspects.

Water resources engineering has evolved over the past 9,000 to 10,000 years as humans have developed the knowledge and techniques for building hydraulic structures to convey and store water. Early examples include irrigation networks built by the Egyptians and Mesopotamians (see Figure 1.1.2) and by the Hohokam in North America (see Figure 1.1.3). The world's oldest large dam was the Sadd-el-kafara dam built in Egypt between 2950 and 2690 B.C. The oldest known pressurized water distribution (approximately 2000 B.C.) was in the ancient city of Knossos on Crete (see Mays, 1999, 2000, for further details). There are many examples of ancient water systems throughout the world.

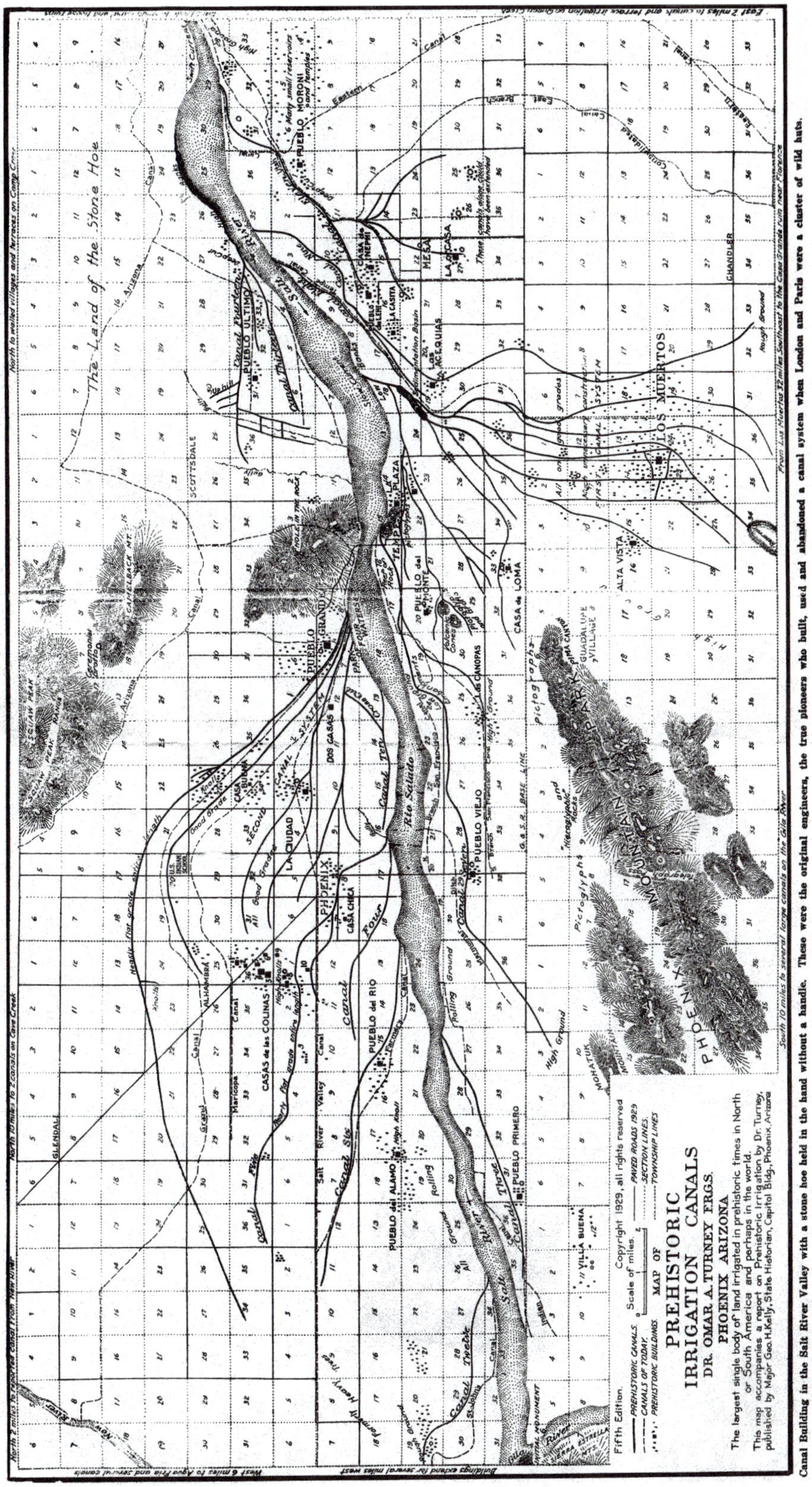

Figure 1.1.3 Canal building in the Salt River Valley with a stone hoe held in the hand without a handle. These were the original engineers, the true pioneers who built, used, and abandoned a canal system when London and Paris were clusters of wild huts (from Turney (1922)). (Courtesy of Salt River Project, Phoenix, Arizona.)

1.2 THE WORLD'S FRESHWATER RESOURCES

Among today's most acute and complex problems are water problems related to the rational use and protection of water resources (see Gleick, 1993). Associated with water problems is the need to supply humankind with adequate clean freshwater. Data collected on global water resources by Soviet scientists are listed in Table 1.2.1. These obviously are only approximations and should not be considered as accurate (Shiklomanov, 1993). Table 1.2.2 presents the dynamics of actual water availability in different regions of the world. Table 1.2.3 presents the dynamics of water use in the world by human activity. Table 1.2.4 presents the annual runoff and water consumption by continents and by physiographic and economic regions of the world.

Table 1.2.1 Water Reserves on the Earth

	Distribution area (10^3 km^2)	Volume (10^3 km^3)	Layer (m)	Percentage of global reserves Of total water	Percentage of global reserves Of fresh-water
World ocean	361,300	1,338,000	3,700	96.5	—
Groundwater	134,800	23,400	174	1.7	—
Freshwater		10,530	78	0.76	30.1
Soil moisture		16.5	0.2	0.001	0.05
Glaciers and permanent snow cover	16,227	24,064	1,463	1.74	68.7
Antarctic	13,980	21,600	1,546	1.56	61.7
Greenland	1,802	2,340	1,298	0.17	6.68
Arctic islands	226	83.5	369	0.006	0.24
Mountainous regions	224	40.6	181	0.003	0.12
Ground ice/permafrost	21,000	300	14	0.022	0.86
Water reserves in lakes	2,058.7	176.4	85.7	0.013	—
Fresh	1,236.4	91	73.6	0.007	0.26
Saline	822.3	85.4	103.8	0.006	—
Swamp water	2,682.6	11.47	4.28	0.0008	0.03
River flows	148,800	2.12	0.014	0.0002	0.006
Biological water	510,000	1.12	0.002	0.0001	0.003
Atmospheric water	510,000	12.9	0.025	0.001	0.04
Total water reserves	510,000	1,385,984	2,718	100	—
Total freshwater reserves	148,800	35,029	235	2.53	100

Source: Shiklomanov (1993).

Table 1.2.2 Dynamics of Actual Water Availability in Different Regions of the World

Continent and region	Area (10^6 km^2)	Actual water availability (10^3 m^3 per year per capita) 1950	1960	1970	1980	2000
Europe	10.28	5.9	5.4	4.9	4.6	4.1
North	1.32	39.2	36.5	33.9	32.7	30.9
Central	1.86	3.0	2.8	2.6	2.4	2.3
South	1.76	3.8	3.5	3.1	2.8	2.5
European USSR (North)	1.82	33.8	29.2	26.3	24.1	20.9
European USSR (South)	3.52	4.4	4	3.6	3.2	2.4
North America	24.16	37.2	30.2	25.2	21.3	17.5
Canada and Alaska	13.67	384	294	246	219	189
United States	7.83	10.6	8.8	7.6	6.8	5.6
Central America	2.67	22.7	17.2	12.5	9.4	7.1

Table 1.2.2   Dynamics of Actual Water Availability in Different Regions of the World *(continued)*

Continent and region	Area (10⁶ km²)	Actual water availability (10³ m³ per year per capita)				
		1950	1960	1970	1980	2000
Africa	30.10	20.6	16.5	12.7	9.4	5.1
North	8.78	2.3	1.6	1.1	0.69	0.21
South	5.11	12.2	10.3	7.6	5.7	3.0
East	5.17	15.0	12	9.2	6.9	3.7
West	6.96	20.5	16.2	12.4	9.2	4.9
Central	4.08	92.7	79.5	59.1	46.0	25.4
Asia	44.56	9.6	7.9	6.1	5.1	3.3
North China and Mongolia	9.14	3.8	3.0	2.3	1.9	1.2
South	4.49	4.1	3.4	2.5	2.1	1.1
West	6.82	6.3	4.2	3.3	2.3	1.3
South-east	7.17	13.2	11.1	8.6	7.1	4.9
Central Asia and Kazakhstan	2.43	7.5	5.5	3.3	2.0	0.7
Siberia and Far East	14.32	124	112	102	96.2	95.3
Trans-Caucasus	0.19	8.8	6.9	5.4	4.5	3.0
South America	17.85	105	80.2	61.7	48.8	28.3
North	2.55	179	128	94.8	72.9	37.4
Brazil	8.51	115	86	64.5	50.3	32.2
West	2.33	97.9	77.1	58.6	45.8	25.7
Central	4.46	34	27	23.9	20.5	10.4
Australia and Oceania	8.59	112	91.3	74.6	64.0	50.0
Australia	7.62	35.7	28.4	23	19.8	15.0
Oceania	1.34	161	132	108	92.4	73.5

Source: Shiklomanov (1993).

Table 1.2.3   Dynamics of Water Use in the World by Human Activity

Water users[a]	1900 (km³ per year)	1940 (km³ per year)	1950 (km³ per year)	1960 (km³ per year)	1970 (km³ per year)	1975 (km³ per year)	1980 (km³ per year)	1980 (%)	1990[b] (km³ per year)	1990[b] (%)	2000[b] (km³ per year)	2000[b] (%)
Agriculture												
Withdrawal	525	893	1,130	1,550	1,850	2,050	2,290	69.0	2,680	64.9	3,250	62.6
Consumption	409	679	859	1,180	1,400	1,570	1,730	88.7	2,050	86.9	2,500	86.2
Industry												
Withdrawal	37.2	124	178	330	540	612	710	21.4	973	23.6	1,280	24.7
Consumption	3.5	9.7	14.5	24.9	38.0	47.2	61.9	3.2	88.5	3.8	117	4.0
Municipal supply												
Withdrawal	16.1	36.3	52.0	82.0	130	161	200	6.0	300	7.3	441	8.5
Consumption	4.0	9.0	14	20.3	29.2	34.3	41.1	2.1	52.4	2.2	64.5	2.2
Reservoirs												
Withdrawal	0.3	3.7	6.5	23.0	66.0	103	120	3.6	170	4.1	220	4.2
Consumption	0.3	3.7	6.5	23.0	66.0	103	120	6.2	170	7.2	220	7.6
Total (rounded off)												
Withdrawal	579	1,060	1,360	1,990	2,590	2,930	3,320	100	4,130	100	5,190	100
Consumption	417	701	894	1,250	1,540	1,760	1,950	100	2,360	100	2,900	100

[a] Total water withdrawal is shown in the first line of each category, consumptive use (irretrievable water loss) is shown in the second line.
[b] Estimated.

Source: Shiklomanov (1993).

Table 1.2.4 Annual Runoff and Water Consumption by Continents and by Physiographic and Economic Regions of the World

Continent and region	Mean annual runoff (mm)	Mean annual runoff (km^3 per year)	Aridity index (R/LP)	Water consumption (km^3 per year) 1980 Total	1980 Irretrievable	1990 Total	1990 Irretrievable	2000 Total	2000 Irretrievable
Europe	310	3,210	—	435	127	555	178	673	222
North	480	737	0.6	9.9	1.6	12	2.0	13	2.3
Central	380	705	0.7	141	22	176	28	205	33
South	320	564	1.4	132	51	184	64	226	73
European USSR (North)	330	601	0.7	18	2.1	24	3.4	29	5.2
European USSR (South)	150	525	1.5	134	50	159	81	200	108
North America	340	8,200	—	663	224	724	255	796	302
Canada and Alaska	390	5,300	0.8	41	8	57	11	97	15
United States	220	1,700	1.5	527	155	546	171	531	194
Central America	450	1,200	1.2	95	61	120	73	168	93
Africa	150	4,570	—	168	129	232	165	317	211
North	17	154	8.1	100	79	125	97	150	112
South	68	349	2.5	23	16	36	20	63	34
East	160	809	2.2	23	18	32	23	45	28
West	190	1,350	2.5	19	14	33	23	51	34
Central	470	1,909	0.8	2.8	1.3	4.8	2.1	8.4	3.4
Asia	330	14,410	—	1,910	1,380	2,440	1,660	3,140	2,020
North China and Mongolia	160	1,470	2.2	395	270	527	314	677	360
South	490	2,200	1.3	668	518	857	638	1,200	865
West	72	490	2.7	192	147	220	165	262	190
South-east	1,090	6,650	0.7	461	337	609	399	741	435
Central Asia and Kazakhstan	70	170	3.1	135	87	157	109	174	128
Siberia and Far East	230	3,350	0.9	34	11	40	17	49	25
Trans-Caucasus	410	77	1.2	24	14	26	18	33	21
South America	660	11,760	—	111	71	150	86	216	116
Northern area	1,230	3,126	0.6	15	11	23	16	33	20
Brazil	720	6,148	0.7	23	10	33	14	48	21
West	740	1,714	1.3	40	30	45	32	64	44
Central	170	812	2.0	33	20	48	24	70	31
Australia and Oceania	270	2,390		29	15	38	17	47	22
Australia	39	301	4.0	27	13	34	16	42	20
Oceania	1,560	2,090	0.6	2.4	1.5	3.3	1.8	4.5	2.3
Land area (rounded off)	—	44,500	—	3,320	1,450	4,130	2,360	5,190	2,900

Source: Shiklomanov (1993).

1.3 WATER USE IN THE UNITED STATES

Dziegielewski et al. (1996) define *water use* from a hydrologic perspective as all water flows that are a result of human intervention in the hydrologic cycle. The National Water Use Information Program (NWUI Program), conducted by the United States Geological Survey (USGS), used this perspective on water use in establishing a national system of water-use accounting. This accounting system distinguishes the following water-use flows: (1) water withdrawals for off-stream purposes, (2) water deliveries at point of use or quantities released after use, (3) consumptive use, (4) conveyance loss, (5) reclaimed wastewater, (6) return flow, and (7) in-stream flow (Solley et al., 1993). The relationships among these human-made flows at various points of measurement are illustrated in Figure 1.3.1. Figure 1.3.2 illustrates the estimated water use by tracking the sources,

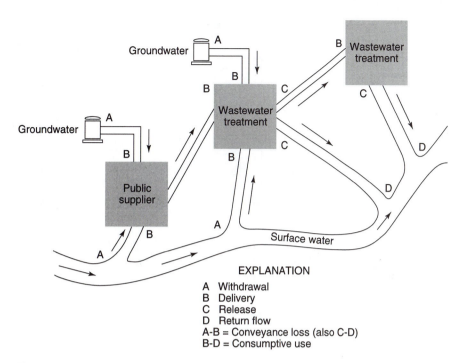

Figure 1.3.1 Definition of water-use flows and losses (from Solley et al. (1993)).

uses, and disposition of freshwater using the hydrologic accounting system given in Figure 1.3.1. Table 1.3.1 defines the major purposes of water use.

Table 1.3.1   Major Purposes of Water Use

Water-use purpose	Definition
Domestic use	Water for household needs such as drinking, food preparation, bathing, washing clothes and dishes, flushing toilets, and watering lawns and gardens (also called residential water use).
Commercial use	Water for motels, hotels, restaurants, office buildings, and other commercial facilities and institutions.
Irrigation use	Artificial application of water on lands to assist in the growing of crops and pastures or to maintain vegetative growth in recreational lands such as parks and golf courses.
Industrial use	Water for industrial purposes such as fabrication, processing, washing, and cooling.
Livestock use	Water for livestock watering, feed lots, dairy operations, fish farming, and other on-farm needs.
Mining use	Water for the extraction of minerals occurring naturally and associated with quarrying, well operations, milling, and other preparations customarily done at the mine site or as part of a mining activity.
Public use	Water supplied from a public water supply and used for such purposes as firefighting, street washing, municipal parks, and swimming pools.
Rural use	Water for suburban or farm areas for domestic and livestock needs, which is generally self-supplied.
Thermoelectric power use	Water for the process of the generation of thermoelectric power.

Source: Solley et al. (1993).

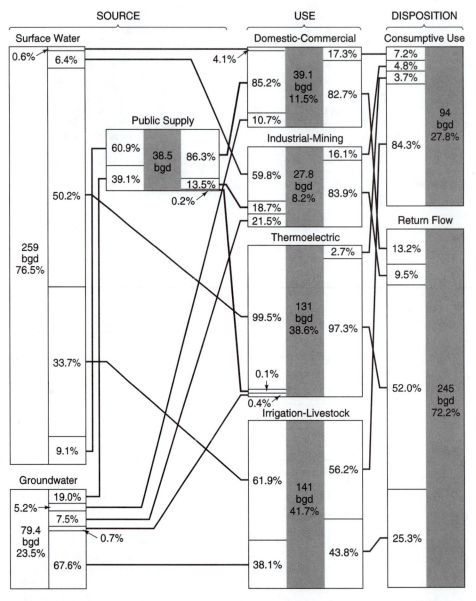

Figure 1.3.2 Estimated water use in the United States, 1990. Freshwater withdrawals and disposition of water in billion gallons per day (bgd). For each water use category, this diagram shows the relative proportion of water source and disposition and the general distribution of water from source to disposition. The lines and arrows indicate the distribution of water from source to disposition for each category; for example, surface water was 76.5 percent of total freshwater withdrawn, and, going from "Source" to "Use" columns, the line from the surface water block to the domestic and commercial block indicates that 0.6 percent of all surface water withdrawn was the source for 4.1 percent of total water (self-supplied withdrawals, public supply deliveries) for domestic and commercial purposes (from Solley, Pierce, and Perlman, (1993)).

1.4   SYSTEMS OF UNITS

The analysis of pressurized (conduit) flow, open-channel flow, and groundwater flows requires an understanding of the elements of fluid mechanics (presented in Chapter 2). A review of the mechanics of materials is a prerequisite to the examination of fluid mechanics principles. Table

1.4.1 lists of the basic mechanical properties of matter with their dimensions and units in the SI system. In the United States much of the technology related to water resources engineering is still based upon the foot-pound-second (FPS) system of units, or what are referred to in this book as U.S. Customary Units. Table 1.4.2 provides a set of correction factors for converting U.S. customary units to SI units.

Table 1.4.1   Definitions, Dimensions, and SI Units for Basic Mechanical Properties

Property	Symbol	Definition	SI Unit	SI symbol	Derived	Basic
Mass	M		kilogram	kg		kg
Length	l		meter	m		m
Time	t		second	s		s
Area	A	$A = l^2$				m^2
Volume	V	$V = l^3$				m^3
Velocity	v	$v = l/t$				m/s
Acceleration	a	$a = l/t^2$				m/s^2
Force	F	$F = Ma$	newton	N	N	$kg \cdot m/s^2$
Weight	w	$w = Mg$	newton	N	N	$kg \cdot m/s^2$
Pressure	p	$p = F/A$	pascal	Pa	N/m^2	$kg/m \cdot s^2$
Work	W	$W = Fl$	joule	J	$N \cdot m$	$kg \cdot m^2/s^2$
Energy		Work done	joule	J	$N \cdot m$	$kg \cdot m^2/s^2$
Mass density	p	$p = M/V$				kg/m^2
Weight density	γ	$\gamma = w/V$			N/m^3	$kg/m^2 \cdot s^2$
Stress	σ, τ	Internal response to external p	pascal	Pa	N/m^2	$kg/m \cdot s^2$
Strain	ϵ	$\epsilon = \Delta V/V$				Dimensionless
Young's modulus	E	Hooke's law			N/m^2	$kg/m \cdot s^2$

Source: Freeze and Cherry (1979).

Table 1.4.2   Conversion Factors FPS (Foot-Pound-Second) System of Units to SI Units

	Multiply	By	To obtain
Length	ft	3.048×10^{-1}	m
	ft	3.048×10	cm
	ft	3.048×10^{-4}	km
	mile	1.609×103	m
	mile	1.609	km
Area	ft^2	9.290×10^{-2}	m^2
	mi^2	2.590	km^2
	acre	4.047×10^3	m^2
	acre	4.047×10^{-3}	km^2
Volume	ft^3	2.832×10^{-2}	m^3
	U.S. gal	3.785×10^{-3}	m^3
	U.K. gal	4.546×10^{-3}	m^3
	ft^3	2.832×10	ℓ
	U.S. gal	3.785	ℓ
	U.K. gal	4.546	ℓ
Velocity	ft/s	3.048×10^{-1}	m/s
	ft/s	3.048×10	cm/s
	mi/h	4.470×10^{-1}	m/s
	mi/h	1.609	km/h
Acceleration	ft/s^2	3.048×10^{-1}	m/s^2

Table 1.4.2 Conversion Factors FPS (Foot-Pound-Second) System of Units to SI Units *(continued)*

	Multiply	By	To obtain
Mass	lb_m*	4.536×10^{-1}	kg
	slug*	1.459×10	kg
	ton	1.016×10^3	kg
Force and weight	lb_f*	4.448	N
	poundal	1.383×10^{-1}	N
Pressure and stress	psi	6.895×10^3	Pa or N/m^2
	lb_f/ft^2	4.788×10	Pa
	poundal/ft^2	1.488	Pa
	atm	1.013×10^5	Pa
	in Hg	3.386×10^3	Pa
	mb	1.000×10^2	Pa
Work and energy	ft-lbf	1.356	J
	ft-poundal	4.214×10^{-2}	J
	Btu	1.055×10^{-3}	J
	calorie	4.187	J
Mass density	lbm/ft^3	1.602×10	kg/m^3
	slug/ft^3	5.154×10^2	kg/m^3
Weight density	lb_f/ft^3	1.571×10^2	N/m^3
Discharge	ft^3/s	2.832×10^{-2}	m^3/s
	ft^3/s	2.832×10	ℓ/s
	U.S. gal/min	6.309×10^{-5}	m^3/s
	U.K. gal/min	7.576×10^{-5}	m^3/s
	U.S. gal/min	6.309×10^{-2}	ℓ/s
	U.K. gal/min	7.576×10^{-2}	ℓ/s
Hydraulic conductivity	ft/s	3.048×10^{-1}	m/s
(see also Table 2.3)	U.S. gal/day/ft^2	4.720×10^{-7}	m/s
Transmissivity	ft^2/s	9.290×10^{-2}	m^2/s
	U.S. gal/day/ft	1.438×10^{-7}	m^2/s

*A body whose mass is 1 lb mass (lb_m) has a weight of 1 lb force (lb_f), 1 lb_f is the force required to accelerate a body of 1 lb_m to an acceleration of g = 32.2 ft/s^2. A slug is the unit of mass which, when acted upon by a force of 1 lb_f, acquires an acceleration of 1 ft/s^2.

Source: Freeze and Cherry (1979).

1.5 WHAT IS WATER?

The water molecule is a unique combination of hydrogen and oxygen atoms, with electrons being shared between them as shown in Figure 1.5.1. The symmetry of the distribution of electrons leaves one side of each molecule with a positive charge, resulting in an electrostatic attraction between molecules. Water molecules can form four such relatively weak hydrogen bonds. The hydrogen, or polar, bonds of water molecules are much weaker than the covalent bonds between hydrogen and oxygen within the molecule. These polar bonds cause water molecules to cluster in tetrahedral patterns, as shown in Figure 1.5.2 for ice. In the solid state, the tetrahedral arrangement of the bonding produces a tetrahedral crystalline structure. In the fluid state, increases in temperature weaken the hydrogen bonding.

Ice processes heat energy from the vibration of atoms and molecules in the fixed structure. As ice warms, the vibrations increase to the point where the tetrahedral structure breaks down and the ice melts. The molecules of the liquid phase are closer than in the solid state, as illustrated in Figure 1.5.2, making water slightly more dense than ice at its melting point. Molecules of water in the liquid phase vibrate faster as temperature rises. Once the vibrations are great enough, some molecules are thrown from (or escape) the liquid surface in a process called evaporation, forming

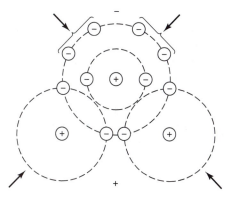

Figure 1.5.1 The water molecule (after Sutcliffe (1968)).

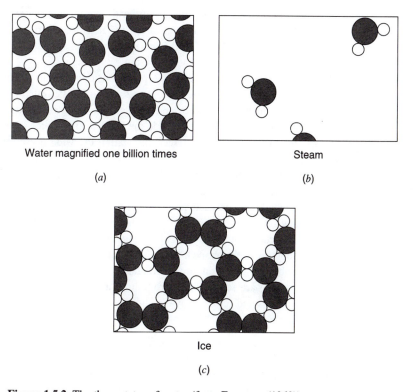

Figure 1.5.2 The three states of water (from Feynman (1963)).

a gaseous or vapor phase. This evaporation consumes a large amount of energy, called *latent heat of vaporization.* The phase changes for water are: (1) *evaporation*—liquid to vapor, (2) *condensation*—vapor to liquid, (3) *sublimation*—vapor to solid or solid to vapor, (4) *melting*—solid to liquid, and (5) *freezing*—liquid to solid.

The physical properties of water are unique among substances with similar molecular mass. Water has the highest specific heat of any known substance, which means that temperature change within it occurs very slowly. Compared to most common liquids, water has a high viscosity and a high surface tension, which are caused by the hydrogen bonding. This produces capillary rise of water in soils and causes rain to form into spherical droplets. Physical properties of water in the solid and liquid phases vary with temperature. In these states the variation in density differs more significantly than in most liquids. Water in the gaseous phase (water vapor) exerts a partial pressure in the atmosphere, referred to as its *vapor pressure*. In the atmosphere above a liquid water surface, water molecules are constantly being exchanged between the air and the water. In a drier atmosphere, the rate of uptake of molecules is greater than the rate of return to the surface. At a state of equilibrium, when the number of molecules leaving the surface is equal to the number arriving, saturation of the vapor pressure of air has been reached. Additional water molecules to the air are balanced by deposition on the water surface. The latent heat of vaporization is about eight times larger than is necessary to melt ice, and about 600 times larger than its heat capacity (the energy necessary to raise water temperature by 1°C).

1.6   THE FUTURE OF WATER RESOURCES

The management of water resources can be subdivided into three broad categories: (1) *water-supply management*, (2) *water-excess management*, and (3) *environmental restoration*. All modern multipurpose water resources projects are designed and built for water-supply management and/or water-excess management. In fact, throughout human history all water resources projects have been designed and built for one or both of these categories. A *water resources system* is a system for redistribution, in space and time, of the water available to a region to meet societal needs (Plate, 1993). Water can be utilized from surface water systems, from groundwater systems, or from conjunctive/ground surface water systems.

When discussing water resources, we must consider both the quantity and the quality aspects. The hydrologic cycle must be defined in terms of both water quantity and water quality. Because of the very complex water issues and problems that we face today, many fields of study are involved in their solution. These include the biological sciences, engineering, physical sciences, and social sciences (see Figure 1.1.1), illustrating the wide diversity of disciplines involved in water resources.

As the twenty-first century approaches, we are questioning the viability of our patterns of development, industrialization, and resources usage. We are now beginning to discuss the goals of attaining an equitable and sustainable society in the international community. Looking into the future, a new set of problems face us, including the rapidly growing population in developing countries; uncertain impacts of global climate change; possible conflicts over shared freshwater resources; thinning of the ozone layer; destruction of rain forests; threats to wetland, farmland, and other renewable resources; and many others.

These problems are very different from those that humans have faced before. The fact that there are so many things undiscovered by the human race leads me to the statement by Sir Isaac Newton, shortly before his death in 1727:

> *I do not know what I may appear to the world, but to myself I seem to have been only like a boy playing on the sea shore, and diverting myself in now and then finding a smoother pebble or a prettier shell than ordinary, while the great ocean of truth lay all undiscovered before me.*

REFERENCES

Adams, R. M., *Heartland of Cities, Surveys of Ancient Settlement and Land Use in the Central Floodplain of the Euphrates*, University of Chicago Press, Chicago, 1965.

Butzer, K. W., *Early Hydraulic Civilization in Egypt*, University of Chicago Press, Chicago, 1976.

Dziegielewski, B., E. M. Opitz, and D. R. Maidment, "Water Demand Analysis," Chapter 23 in *Water Resources Handbook* (edited by L.W. Mays), McGraw-Hill, New York, 1996.

Feynman, R. P., *Six Easy Pieces*, Perseus Books, Reading, MA, 1995.

Feynman, R. P., R. B. Leighton, and M. Sands, *The Feynman Lecture Notes on Physics,* Vol. I, Addison-Wesley, Reading, MA, 1963.

Freeze, R. A. and J. A. Cherry, *Groundwater*, Prentice-Hall Inc., Englewood Cliffs, NJ, 1979.

Gleick, P. H., *Water in Crisis*, Oxford University Press, Oxford, 1993.

Mays, L. W., "Water Resources: An Introduction," in *Water Resources Handbook* (edited by L.W. Mays), McGraw-Hill, New York, 1996.

Mays, L. W., "Introduction," in *Hydraulic Design Handbook* (edited by L. W. Mays), McGraw-Hill, New York, 1999.

Mays, L. W., "Introduction," in *Water Distribution Systems Handbook* (edited by L. W. Mays), McGraw-Hill, New York, 2000.

Plate, E. J., "Sustainable Development of Water Resources: A Challenge to Science and Engineering," *Water International*, International Water Resources Association, 18(2):84–94, June 1993.

Shiklomanov, I., "World Fresh Water Resources," in *Water in Crisis* (edited by P. H. Gleick), Oxford University Press, New York, 1993.

Solley, W. B., R. R. Pierce, and H. A. Perlman, "Estimated Use of Water in the United States in 1990," U.S. Geological Survey Circular 1081, Washington, DC, 1993.

Sutcliffe, J., *Plants and Water*, Edward Arnold, London, 1968.

Turney, O. S., Map of Prehistoric Irrigation Canals, Map No. 002004, Archaeological Site Records Office, Arizona State Museum, University of Arizona, Tucson, 1922.

Chapter 2

Principles of Flow in Hydrosystems

The purpose of this chapter is to present some of the fundamental principles of fluid mechanics including fluid properties. Much greater detail can be found in fluid mechanics texts such as the excellent books by Frazini and Finnemore (1997), Munson, et al. (1998), and Roberson and Crowe (1993).

2.1 PROPERTIES INVOLVING MASS OR WEIGHT OF WATER

Mass density, often called *density*, is the mass per unit volume, with units of kilograms (kg) per cubic meter (m^3) or $N \cdot s^2/m^4$ in SI units. The Greek symbol ρ (rho) is used to denote density. The mass density of water at 4°C is 1000 kg/m^3 or 1.94 $slugs/ft^3$. For most applications in hydrologic and hydraulic processes, the density is assumed to be constant so that water is assumed incompressible. Incompressibility does not always mean constant density because salt in water changes the density of water without changing its volume.

Specific weight is the gravitational force (weight) per unit volume of water, denoted by the Greek symbol γ (gamma). The specific weight of water at 4°C is 9810 N/m^3 or 62.4 lb/ft^3. The relationship between density and specific weight is

$$\rho = \frac{\gamma}{g} \tag{2.1.1}$$

Specific gravity of a fluid refers to the ratio of the specific weight of a given liquid to the specific weight of water. Tables 2.1.1 and 2.1.2 list the various physical properties of water in English units and SI units, respectively. The relationship between temperature scales is

$$°C = \frac{5}{9}(°F - 32) \text{ or } °F = \frac{9}{5}°C + 32.$$

Table 2.1.1   Physical Properties of Water in English Units

Temp. (°F)	Specific weight, γ (lb/ft^3)	Density, ρ (slugs/ft^3)	Viscosity, $10^5\mu$ (lb · sec/ft^2)	Kinematic viscosity $10^5\nu$ (ft^2/sec)	Surface tension, 100σ (lb/ft)	Vapor-pressure head, p_v/γ (ft)	Bulk modulus of elasticity, $10^{-3}\beta$ (lb/in^2)
32	62.42	1.940	3.746	1.931	0.518	0.20	293
40	62.43	1.940	3.229	1.664	0.514	0.28	294
50	62.41	1.940	2.735	1.410	0.509	0.41	305
60	62.37	1.938	2.359	1.217	0.504	0.59	311
70	62.30	1.936	2.050	1.059	0.500	0.84	320
80	62.22	1.934	1.799	0.930	0.492	1.17	322
90	62.11	1.931	1.595	0.826	0.486	1.61	323
100	62.00	1.927	1.424	0.739	0.480	2.19	327
110	61.86	1.923	1.284	0.667	0.473	2.95	331
120	61.71	1.918	1.168	0.609	0.465	3.91	333
130	61.55	1.913	1.069	0.558	0.460	5.13	334
140	61.38	1.908	0.981	0.514	0.454	6.67	330
150	61.20	1.902	0.905	0.476	0.447	8.58	328
160	61.00	1.896	0.838	0.442	0.441	10.95	326
170	60.80	1.890	0.780	0.413	0.433	13.83	322
180	60.58	1.883	0.726	0.385	0.426	17.33	313
190	60.36	1.876	0.678	0.362	0.419	21.55	313
200	60.12	1.868	0.637	0.341	0.412	26.59	308
212	59.83	1.860	0.593	0.319	0.404	33.90	300

Table 2.1.2 Physical Properties of Water in SI Units

Temp. (°C)	Specific weight, γ (N/m^3)	Density, ρ (kg/m^3)	Viscosity, $10^3\mu$ (n · s/m^2)	Kinematic viscosity, $10^6\nu$ (m^2/s)	Surface tension, 100σ (N/m)	Vapor-pressure head, p_v/γ (m)	Bulk modulus of elasticity, $10^{-7}\beta$ (N/m^2)
0	9805	999.9	1.792	1.792	7.62	0.06	204
5	9806	1000.0	1.519	1.519	7.54	0.09	206
10	9803	999.7	1.308	1.308	7.48	0.12	211
15	9798	999.1	1.140	1.141	7.41	0.17	214
20	9789	998.2	1.005	1.007	7.36	0.25	220
25	9779	997.1	0.894	0.897	7.26	0.33	222
30	9767	995.7	0.801	0.804	7.18	0.44	223
35	9752	994.1	0.723	0.727	7.10	0.58	224
40	9737	992.2	0.656	0.661	7.01	0.76	227
45	9720	990.2	0.599	0.605	6.92	0.98	229
50	9697	988.1	0.549	0.556	6.82	1.26	230
55	9679	985.7	0.506	0.513	6.74	1.61	231
60	9658	983.2	0.469	0.477	6.68	2.03	228
65	9635	980.6	0.436	0.444	6.58	2.56	226
70	9600	977.8	0.406	0.415	6.50	3.20	225
75	9589	974.9	0.380	0.390	6.40	3.96	223
80	9557	971.8	0.357	0.367	6.30	4.86	221
85	9529	968.6	0.336	0.347	6.20	5.93	217
90	9499	965.3	0.317	0.328	6.12	7.18	216
95	9469	961.9	0.299	0.311	6.02	8.62	211
100	9438	958.4	0.284	0.296	5.94	10.33	207

EXAMPLE 2.1.1   According to Table 2.1.2, the specific weight of water at 20°C is 9789 N/m³; what is its density?

SOLUTION   Using equation (2.1) and (N = kg · m/s²), we get

$$\rho = \frac{\gamma}{g} = \frac{9789 \text{ N/m}^3}{9.81 \text{ m/s}^2} = 998 \text{ N} \cdot \text{s}^2/\text{m}^4 = 998 \text{ kg/m}^3$$

EXAMPLE 2.1.2   According to Table 2.1.1, the specific weight of water at 50°F is 62.41 lb/ft³; what is its density?

SOLUTION   Using equation (2.1) and (slugs = lb · s²/ft), we get

$$\rho = \frac{\gamma}{g} = \frac{62.41 \text{ lb/ft}^3}{32.2 \text{ ft/s}^2} = 1.94 \text{ slugs/ft}^3$$

2.2 VISCOSITY

In the flow of water shear force exists, producing fluid friction. *Viscosity* is the measure of its resistance to shear or angular deformation. For a velocity gradient, dv/dy, the shear stress τ (tau) between any two thin sheets of fluid is

$$\tau = \mu \frac{dv}{dy} \qquad (2.2.1)$$

where μ (mu) is the dynamic viscosity and v is the velocity. The velocity gradient is the time rate of strain. Thus the definition of dynamic viscosity is the ratio of shear stress to the velocity gradient,

$$\mu = \frac{\tau}{dv/dy} \qquad (2.2.2)$$

Kinematic viscosity is the ratio of the dynamic viscosity to the density in which the gradient force dimension cancels out in μ/ρ. The Greek symbol v (nu) is used to identify the kinematic viscosity

$$v = \mu/\rho \qquad (2.2.3)$$

Kinematic viscosity has been defined because many equations include μ/ρ. Refer to Tables 2.1.1 and 2.1.2 for values of viscosity as a function of temperature.

The shear stress in fluids is involved with the cohesion forces between molecules. Stress applied to fluids causes motion, whereas solids can resist shear stress in a static condition. Considering flow of water in a pipe, the water near the center of the pipe has a greater velocity than the water near the wall. Shear force is increased or decreased in direct proportion to increases or decreases in relative velocity.

Shear stress has units of N/m². Dynamic viscosity has units of

$$\mu = \frac{\tau}{\dfrac{dv}{dy}} \equiv \frac{\text{N/m}^2}{(\text{m/s})/\text{m}} = \frac{\text{N} \cdot \text{s}}{\text{m}^2}$$

Kinematic viscosity has units of

$$v = \frac{\mu}{\rho} \equiv \frac{\text{N} \cdot \text{s/m}^2}{\text{N} \cdot \text{s}^2/\text{m}^4} = \frac{\text{m}^2}{\text{s}}$$

The SI unit for dynamic viscosity is centipoise (cP), in which $1 \text{ cP} = 1 \text{ N} \cdot \text{s/m}^2 \times 10^{-3}$. The SI unit for kinematic viscosity is centistoke (cst), in which $1 \text{ cst} = 1 \text{ m}^2/\text{s} \times 10^{-6}$.

Ideal fluids are defined as the ones in which viscosity is zero, i.e., there is no friction. Such fluids do not exist in reality but the concept is useful in many types of fluid analysis. *Real fluids* do consider viscosity effects so that shear force exists whenever motion takes place, thus producing fluid friction.

EXAMPLE 2.2.1

Water has a kinematic viscosity of 10 poises. What is the kinematic viscosity in ft^2/s?

SOLUTION

A poise is measured in dyne-seconds per cm^2. Since $1 \text{ lb} = 444800$ dynes and $1 \text{ ft} = 30.48$ cm, then $1 \text{ lb} \cdot \text{s/ft}^2 = 444800 \text{ dyne} \cdot \text{s}/(30.48 \text{ cm})^2 = 478.8$ poises. The conversion of poise to $\text{lb} \cdot \text{s/ft}^2$ (for dynamic viscosity) is then $1 \text{ lb} \cdot \text{s/ft}^2 = 478.8$ poises, so that

$$\mu = 10/478.8 = 0.0209 \text{ lb} \cdot \text{s/ft}^2$$

The kinematic viscosity is then computed using equation (2.2.3), $v = \mu/\rho$, as

$$v = \frac{(0.0209)(32.2)}{62.4} = 0.0108 \text{ ft}^2/\text{s}.$$

2.3 ELASTICITY

Elasticity (or *compressibility*) is important when we talk about water hammer in the hydraulics of pipe flow. Elasticity of water is related to the amount of deformation (expansion or contraction) induced by a pressure change. Elasticity is characterized by the *bulk modulus of elasticity*, *E*, which is defined as the ratio of relative change in volume, $d\forall/\forall$, due to a differential change in pressure, dp, so that

$$E = -\frac{dp}{d\forall/\forall} \tag{2.3.1}$$

Also $d\rho/\rho = d\forall/\forall$, so that

$$E = \frac{dp}{d\rho/\rho} \tag{2.3.2}$$

Refer to Tables 2.1.1 and 2.1.2 for values of the bulk modulus of elasticity as a function of temperature.

EXAMPLE 2.3.1

By approximately how much should the pressure on water be increased or what pressure must be applied to water at 60°F to reduce its volume 1 percent?

SOLUTION

Using equation (2.3.1), $E = -dp/(d\forall/\forall)$, where $E = 311,000 \text{ lb/in}^2$ (Table 2.1.1), we get

$$dp = p_2 - 0, \text{ and } d\forall/\forall = -0.01; \text{ then}$$

$$311,000 = -\frac{p_2 - 0}{-0.01}$$

$$p_2 = 3,110 \text{ lb/in}^2 \text{ (psi)}$$

EXAMPLE 2.3.2	Considering a bulk modulus of elasticity of water at 10°C as 2.1 GPa (GPa = gigapascal, 10^9 Pa), what pressure is required to reduce its volume by 1 percent?
SOLUTION	Using equation (2.3.1), $E = -dp/(d\forall/\forall)$, where E = 2.1 GPa, $dp = p_2 - 0$, and $d\forall/\forall = -0.01$, then we get

$$2.1 = -\frac{p_2 - 0}{-0.01}$$

$$p_2 = 0.021 \text{ GPa}$$

$$= 21 \text{ MPa (megapascal, } 10^6 \text{ Pa)}$$

2.4 PRESSURE AND PRESSURE VARIATION

Pressure, p, is the force F acting over an area A, denoted as

$$p = \lim_{\Delta A \to 0} \frac{\Delta F}{\Delta A} = \frac{dF}{dA} \tag{2.4.1}$$

Pressure at a point is equal in all directions. The pressure at a depth y (neglecting pressure on the surface of water) is

$$p = \gamma y \tag{2.4.2}$$

Units of pressure are N/m² (Pascal), lb/in² (psi), lb/ft², feet of water, and inches of mercury.

Gauge pressure uses atmospheric pressure as the datum. *Absolute pressure* is the pressure above absolute zero. *Vacuum* refers to pressure less than atmospheric pressure. At absolute zero pressure is a perfect vacuum. Figure 2.4.1 illustrates the relationship among various pressures.

The *pressure force F* exerted by water on a plane area A is the product of the area and the pressure at its centroid, expressed as

$$F = pA = \gamma y_c A \tag{2.4.3}$$

where y_c is the vertical depth of the water over the centroid.

For static water, the only variation in pressure is with the elevation in the fluid, i.e.

$$\frac{dp}{dz} = -\gamma \tag{2.4.4}$$

where z refers to elevation. Equation (2.4.4) is the basic equation for hydraulic pressure variation with elevation. For water on a horizontal plane, the pressure everywhere on this plane is constant. The greatest possible change in hydrostatic pressure occurs along a vertical path through water.

Considering the specific weight to be constant, equation (2.4.4) can be integrated to obtain

$$p = -\gamma z + \text{constant} \tag{2.4.5}$$

or

$$\left(\frac{p}{\gamma} + z \right) = \text{constant} \tag{2.4.6}$$

The term $\left(p/\gamma + z \right)$ is called the *piezometric head*, which is then constant through any incompressible static fluid. The pressure and elevation at two different points in a static incompressible fluid are then

$$\frac{p_1}{\gamma} + z_1 = \frac{p_2}{\gamma} + z_2 \tag{2.4.7}$$

<table>
<tr><td>**EXAMPLE 2.4.1**</td><td>What is the horizontal pressure acting at the face of a dam at 100 ft? Consider a water temperature of approximately 50°F.</td></tr>
</table>

SOLUTION

Using equation (2.4.2), $p = \gamma y$ where $\gamma = 62.4$ lb/ft³, we get

$$p = \gamma y = (62.4 \text{ lb/ft}^3)(100 \text{ ft}) = 6{,}240 \text{ lb/ft}^2$$

$$= 43.3 \text{ lb/in}^2$$

or alternatively using SI units, $\gamma = 9.81$ kN/m³, $y = 100$ ft/(3.281 ft/m) = 30.48 m, we get

$$p = (9.81 \text{ kN/m}^3)(30.48 \text{ m})$$

$$= 299 \text{ kN/m}^2$$

$$= 299{,}000 \text{ N/m}^2.$$

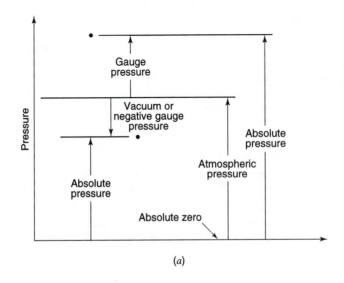

(a)

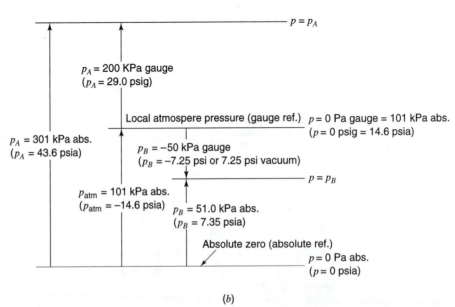

(b)

Figure 2.4.1 (a) Relationship between various pressures (from Chaudhry (1996)); (b) Example of pressure relation (from Roberson & Crowe (1993)).

EXAMPLE 2.4.2

A tank contains water under pressure at 10 kPa. The depth of water in the tank is 6 m. What is the pressure at the bottom of the tank?

SOLUTION

The total pressure is the pressure at the top of the water (10 kPa) plus the pressure at the depth of 6 m.

$$p = 10 + \gamma y$$

$$= 10 \text{ kN/m}^2 + (9.81 \text{ kN/m}^3)(6 \text{ m})$$

$$= 58.86 \text{ kN/m}^2 \text{ or } 58.86 \text{ kPa}$$

Also because 1 lb/in^2 = 6.894 kN/m^2, the pressure is

$$p = 58.86 \text{ kN/m}^2 \left(\frac{1 \text{ lb/in}^2}{6.894 \text{ kN/m}^2} \right) = 8.538 \text{ lb/in}^2.$$

EXAMPLE 2.4.3

An open tank contains 5 ft of water covered with 1 ft of oil (specific gravity = 0.86). What is the pressure at the interface and at the bottom of the tank?

SOLUTION

At the interface the pressure is

$$p = \gamma y = [0.86 \times 62.4 \text{ lb/ft}^3] \, (1.0 \text{ ft}) \times (1 \text{ ft}^2/144 \text{ in}^2)$$

$$= 0.373 \text{ lb/in}^2$$

or

$$p = 0.373 \text{ lb/in}^2 \times \left(\frac{6.894 \text{ kN/m}^2}{1 \text{ lb/in}^2} \right) = 2.571 \text{ kN/m}^2. \text{ At the bottom of the tank, the pressure is}$$

$$p = 0.373 \text{ lb/in}^2 + (62.4 \text{ lb/ft}^3) \, (5 \text{ ft}) \times (1 \text{ ft}^2/144 \text{ in}^2)$$

$$= 2.54 \text{ lb/in}^2 \text{ (or } 17.51 \text{ kN/m}^2)$$

EXAMPLE 2.4.4

At an elevation of 5000 ft above mean sea level the absolute pressure is 12.24 psia at a temperature of 40°F. A gauge attached to a tank reads 4.0 in Hg vacuum. What is the absolute pressure in the tank?

SOLUTION

First determine the gauge pressure assuming the specific gravity of mercury (Hg) as 13.6:

$$p_{gauge} = [(13.6)(62.4 \text{ lb/ft}^3)][4.0 \text{ in} \times (1 \text{ ft}/12 \text{ in})](1 \text{ ft}^2/144 \text{ in}^2)$$

$$= 1.964 \text{ lb/in}^2 \text{ vacuum}$$

$$= -1.964 \text{ lb/in}^2$$

The absolute pressure is

$$p_{abs} = 12.24 + (-1.964)$$

$$= 10.28 \text{ psia}$$

2.5   SURFACE TENSION

Molecules of water below the surface act on each other by forces that are equal in all directions. Molecules near the surface have a greater attraction for each other. Molecules on the surface are not able to bond in all directions and consequently form stronger bonds with adjacent molecules. The water surface acts like a stretched membrane seeking a minimum possible area by exerting a tension on the adjacent portion of the surface or an object in contact with the water surface. This *surface tension* acts in the plane of the surface as illustrated in Figure 2.5.1 for capillary action. Refer to Tables 2.1.1 and 2.1.2 for values of surface tension as a function of temperature.

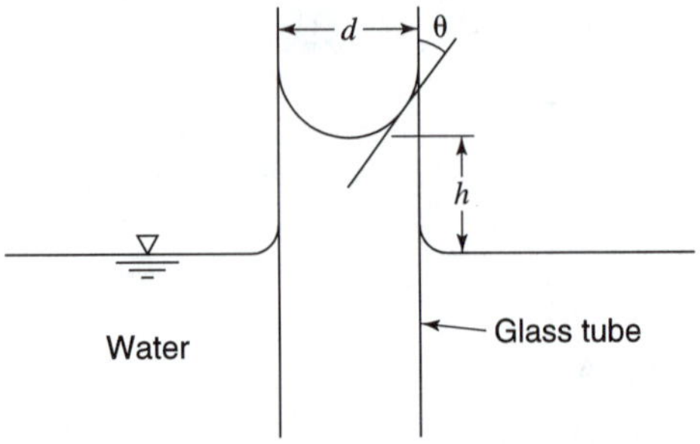

Figure 2.5.1 Capillary action. The effect of surface tension is illustrated for the capillary rise in a small glass tube. θ is the angle of the tangent to the meniscus where it contacts the wall of the tube. Surface tension force acts around the circumference of the tube.

EXAMPLE 2.5.1

A small droplet of water of diameter 0.015 inch is in contact with the air. If the pressure inside the droplet is 0.03 psi greater than the atmosphere, what is the surface tension?

SOLUTION

The pressure force F is the product of the pressure and the area $pA = p\,(\pi\,d^2/4)$. The force exerted by the surface tension is the product of the circumference of the droplet (πd) and the surface tension σ, $(\pi d)\sigma$. The pressure force and the surface tension force are in balance, so

$$p\,(\pi d^2/4) = (\pi d)\sigma$$

$$\sigma = pd/4$$

$$= \frac{\left[\left(0.03\,\dfrac{\text{lb}}{\text{in}^2}\right)\left(144\,\dfrac{\text{in}^2}{\text{ft}^2}\right)\right]\left[\dfrac{0.015\ \text{in}}{12\ \text{in/ft}}\right]}{4}$$

$$= 0.00135\ \text{lb/ft}$$

EXAMPLE 2.5.2

Two clean parallel glass plates separated by a distance of 1 mm are placed in water that is at 20°C. How far does the water rise above the water surface due to capillary action?

SOLUTION

From Table 2.1.2, the surface tension is $\sigma = 0.0736$ N/m. Consider the free-body diagram in Figure 2.5.2.

The angle of contact between the water and glass is assumed to be zero. The objective is to sum the forces in the vertical direction; i.e., the force due to surface tension (F_σ) equals the weight of the fluid F_w, considering unit width of the plates:

$$F_\sigma - F_w = 0$$

$$2[\sigma\,(1)] - (1/1000)\,(1)\,(h)\,\gamma = 0$$

$$2\,[0.0736\,(1)] - (0.001)\,(1)\,(h)\,(9.79 \times 10^3) = 0$$

$$h = 0.015\ \text{m}$$

$$= 15\ \text{mm}$$

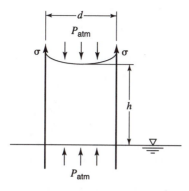

Figure 2.5.2 Free-body diagram of two parallel glass plates for example 2.5.2.

2.6 FLOW VISUALIZATION

There are two viewpoints on the motion of fluids, the *Eulerian* viewpoint and the *Lagrangian* viewpoint. The Lagrangian viewpoint focuses on the motion of individual fluid particles and follows these particles for all time. It is more common, however, in hydrologic and hydraulic processes to consider that fluids form a continuum wherein the motion of particles is not traced. This Eulerian viewpoint then focuses on a particular point or control volume in space and considers the motion of fluid that passes through as a function of time.

Streamlines are lines drawn through a fluid field so that the velocity vectors of the fluid at all points on the streamlines are tangent to the streamline at any instant in time. The tangent of the curve at any point along the streamline is the direction of the velocity vector at that point in the flow field. Examples of streamlines are shown in Figure 2.6.1. In the Eulerian viewpoint, the total velocity is expressed as a function of position along a streamline, x, and time, t.

$$\mathbf{V} = \mathbf{V}(x,t) \tag{2.6.1}$$

A *uniform flow* is defined as one in which the velocity does not change from point to point along any of the streamlines in the flow field. Thus the streamlines are straight and parallel, so that

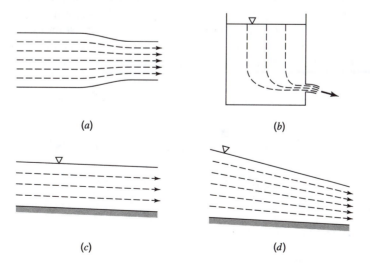

(a)   (b)

(c)   (d)

Figure 2.6.1 Streamlines. (*a*) Flow in a conduit; (*b*) Flow from a slot; (*c*) Open-channel flow (uniform); (*d*) Open-channel flow (nonuniform).

$$\frac{\partial \mathbf{V}}{\partial x} = 0 \text{ (uniform flow)} \qquad (2.6.2)$$

When streamlines are not straight there is a directional change in velocity. If they are not parallel there is a change in speed along the streamlines. Under such circumstance the flow is *nonuniform flow* and

$$\frac{\partial \mathbf{V}}{\partial x} \neq 0 \text{ (nonuniform flow)} \qquad (2.6.3)$$

So this flow pattern has streamlines that are curved in space (converging or diverging) as shown in Figure 2.6.1(d) for nonuniform open-channel flow.

The variation in velocity with respect to time at a given point in a flow field is also used to classify flow. *Steady flow* occurs when the velocity at a point in the flow field does not vary in magnitude or direction with respect to time:

$$\frac{\partial \mathbf{V}}{\partial t} = 0 \text{ (steady flow)} \qquad (2.6.4)$$

Unsteady flow occurs when the velocity does vary in magnitude or direction at a point in the flow field with respect to time.

2.7   LAMINAR AND TURBULENT FLOW

Turbulent flow is caused by eddies of varying size within the flow that create a mixing action. The fluid particles follow irregular and erratic paths and no two particles have similar motion. Turbulent flow then is irregular with no definite flow patterns. Flow in rivers is a good example of turbulent flow. The index used to relate to turbulence is the *Reynolds number*

$$R_e = \frac{VD\rho}{\mu} = \frac{VD}{v} \qquad (2.7.1)$$

where D is a characteristic length such as the diameter of a pipe. For pipe flow the flow is generally turbulent for $R_e > 2000$.

Laminar flow does not have the eddies that cause the intense mixing and therefore the flow is very smooth. The fluid particles move in definite paths and the fluid appears to move by the sliding of laminations of infinitesimal thickness relative to the adjacent layers. The viscous shear of the fluid particles produces the resistance to flow. The resistance to flow varies with the first power of the velocity. For pipe flow the flow is generally laminar for $R_e < 2000$.

Flow can be *one-*, *two-*, or *three-dimensional flow*, for which one, two, or three coordinate directions respectively, are required to describe the velocity and property changes in a flow field.

| EXAMPLE 2.7.1 | Water flows full in a 5-ft diameter pipe at a velocity of 10 ft/s. What is the Reynolds number? The temperature is 50°F. |

SOLUTION

The Reynolds number is computed using equation (2.7.1), $R_e = VD/v$, where $v = 1.41 \times 10^{-5}$ ft²/s from Table 2.1.1.

$$R_e = \frac{(10 \text{ ft/s})(5 \text{ ft})}{1.41 \times 10^{-5} \text{ ft}^2/\text{s}} = 3.55 \times 10^6$$

This flow is turbulent flow.

EXAMPLE 2.7.2

Water flows full in a 1.5 m diameter pipe at a velocity of 3.0 m/s. What is the Reynolds number? The temperature of the water is 20°C.

SOLUTION

The Reynolds number is computed using equation (2.7.1) where $v = 1.007 \times 10^{-6}$ m²/s, from Table 2.1.2:

$$R_e = \frac{3.0 \text{ m/s } (1.5 \text{ m})}{1.007 \times 10^{-6} \text{ m}^2/\text{s}} = 4.47 \times 10^6$$

This flow is turbulent flow.

2.8 DISCHARGE

Discharge, or *flow rate*, is the volume rate of flow that passes a given section in a flow stream. The flow velocity v varies across a flow field, as for the example of pipe flow in Figure 2.8.1. The rate of flow through a differential area dA is $v dA$ so that the total volume can be expressed by integrating over the entire flow section as

$$Q = \int_A v \, dA \tag{2.8.1}$$

Using the mean velocity V, the discharge is defined as

$$Q = AV \tag{2.8.2}$$

By defining an area vector as one that has the magnitude of the area and is oriented normal to the area, then $V\cos\theta dA = \mathbf{V} \cdot d\mathbf{A}$. The discharge is then

$$Q = \int_A \mathbf{V} \cdot d\mathbf{A} \tag{2.8.3}$$

and for a constant (mean) velocity over the cross-sectional area of flow the discharge is

$$Q = \mathbf{V} \cdot \mathbf{A} \tag{2.8.4}$$

The *mass rate of flow* past a flow section is

$$\dot{m} = \int_A \rho v \, dA \tag{2.8.5a}$$

$$= \rho \int_A v \, dA \tag{2.8.5b}$$

$$= \rho Q \tag{2.8.5c}$$

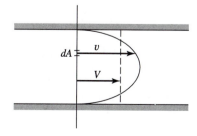

Figure 2.8.1 Velocity distribution in a pipe flow.

EXAMPLE 2.8.1   Water flows in a 4-in diameter pipe at a mean velocity of 5 ft/s. What are the discharge and mass rate of flow at a temperature of 50°F?

SOLUTION   Using the continuity equation (2.8.2), the discharge is

$$Q = AV = \left[\frac{\pi(4/12)^2}{4} \, \text{ft}^2 \right](5 \text{ ft/s })$$

$$= 0.436 \text{ ft}^3/\text{s or cfs}$$

$$= 0.436 \text{ ft}^3/\text{s} \times \frac{1 \text{ gpm}}{0.002228 \text{ ft}^3/\text{s}}$$

$$= 195.7 \text{ gpm}$$

The mass rate of flow is computed using equation (2.8.5c), $\dot{m} = \rho Q$, so for $\rho = 1.94$ slugs/ft^3

$$\dot{m} = (1.94 \text{ slugs/ft}^3)(0.436 \text{ ft}^3/\text{s})$$

$$= 0.846 \text{ slugs/s}$$

EXAMPLE 2.8.2   Water flows in a 10 cm diameter pipe at a mean velocity of 1.5 m/s. What are the discharge and mass rate of flow at a temperature of 5°F?

SOLUTION   Using the continuity equation (2.8.2), the discharge is

$$Q = AV = \left[\frac{\pi \, (10 \text{ cm}/100 \text{ cm}/\text{m})^2}{4} \right](1.5 \text{ m/s})$$

$$= 0.0118 \text{ m}^3/\text{s}$$

The mass rate of flow is computed using equation (2.8.5c), $\dot{m} = \rho Q$, where $\rho = 1000$ kg/m^3:

$$\dot{m} = (1000 \text{ kg/m}^3)(0.0118 \text{ m}^3/\text{s})$$

$$= 11.8 \text{ kg/s}$$

Also, because 14.59 kg = 1 slug, $\dot{m} = 11.8/14.59 = 0.809$ slugs/s.

EXAMPLE 2.8.3   Laminar flow in a circular pipe has a parabolic velocity profile, so that the rate of discharge is represented by the volume of the paraboloid. Determine the ratio of the mean velocity to the maximum velocity.

SOLUTION   For a paraboloid the velocity v can be represented as

$$v = v_{max}\left[1 - \left(\frac{r}{r_o} \right)^2 \right]$$

where v is the velocity at a radius r from the centerline of the pipe, v_{max} is the maximum velocity, and r_o is the radius of the pipe. Using equation (2.8.1), we get

$$Q = \int_A v \, dA = \int_0^{r_0} v(2\pi \, rdr)$$

$$= \int_0^{r_0} v_{max} \left[1 - \left(\frac{r}{r_0} \right)^2 \right] (2\pi \, r dr)$$

$$= 2\pi \, v_{max} \left[\frac{r^2}{2} - \frac{r^4}{4r_0^2} \right]_0^{r_0}$$

$$= 2\pi \, v_{max} \left[\frac{r_0^2}{2} - \frac{r_0^4}{4r_0^2} \right]$$

$$Q = v_{max} \left(\frac{\pi r_0^2}{2} \right)$$

The mean velocity is computed using equation (2.8.2) as

$$V = Q/A = \frac{v_{max} \left(\dfrac{\pi r_0^2}{2} \right)}{\left(\pi r_0^2 \right)}$$

$$V = \frac{v_{max}}{2}$$

The ratio of the mean velocity V to the maximum velocity v_{max} is then $V/v_{max} = 0.5$.

PROBLEMS

2.1.1 If 10 m³ of a liquid weighs 60 kN, calculate its specific weight, density, and specific gravity.

2.1.2 The specific gravity of a certain oil is 0.7. Calculate its specific weight and density in SI and English units.

2.1.3 What is the unit of force in SI units? What is it identical to?

2.1.4 What is the unit of mass in English units? What is it identical to?

2.1.5 A rocket weighing 9810 N on earth is landed on the moon where the acceleration due to gravity is approximately one-sixth that at the earth's surface. What is the mass of the rocket on the earth and the moon, and the weight of the rocket on the moon?

2.1.6 The summer temperature in Arizona often exceeds 110°F. What is the Celsius equivalent temperature?

2.1.7 What will be the percentage change in the density of water if temperature changes from 20°C to (a) 50°C and (b) 80°C?

2.2.1 The specific gravity and kinematic viscosity of a certain liquid are 1.5 and 4×10^{-4} m²/s, respectively. What is its dynamic viscosity?

2.2.2 If the dynamic viscosity of a liquid is 2.09×10^{-5} lb · s/ft², what is its dynamic viscosity in N · s/m² and centipoises?

2.2.3 If the kinematic viscosity of an oil is 800 centistokes, what is its kinematic viscosity in m²/s and ft²/s?

2.2.4 Derive the conversion of stokes to ft²/s.

2.2.5 A liquid has a density of 700 kg/m³ and a dynamic viscosity of 0.0042 kg/m · s. What is its kinematic viscosity in m²/s, ft²/s, centistokes, and stokes?

2.3.1 What is the change in volume of 10 ft³ of water at 50°F when it is subjected to a pressure increase of 200 psi?

2.3.2 What is the bulk modulus of elasticity of a liquid if its volume is reduced by 0.05% by the application of a pressure of 160 psi?

2.3.3 Determine the bulk modulus of elasticity of a liquid by using the following data: The volume is 3000 cm² (= 3 L) at 2 MPa, and the volume is 2750 cm³ at 3 MPa.

2.3.4 The density of sea water is 1025 kg/m³ at the ocean surface. The bulk modulus of elasticity of sea water is 234×10^7 Pa. What is the pressure at a depth, if the change in density between the surface and that depth is 1.5?

2.3.5 Suppose it is desired to reduce a given volume of water by 1% by increasing the pressures at two different temperatures, 50°F and 100°F. Compare the changes in the pressures applied at these temperatures. Calculate the percentage of the pressure change at 100°F with reference to that at 50°F.

2.4.1 A fluid pressure is 10 kPa above standard atmospheric pressure (101.3 kPa). Express the pressure as gauge and absolute.

2.4.2 If the pressure 15 ft below the free surface of a liquid is 30 psi, calculate its specific weight and specific gravity.

2.4.3 If the absolute pressure at a bottom of an open tank filled with oil (specific gravity = 0.8) is 200 kPa, what is the depth of oil in the tank?

2.4.4 A fluid pressure is 5 psi below the standard atmospheric pressure (14.7 psi). Express the pressure as vacuum and absolute.

2.4.5 Express the standard atmospheric pressure (14.7 psi) as feet of water and inches of mercury.

2.4.6 A Bourdon gauge reads a vacuum of 10 in of mercury (specific gravity =13.6) when the atmospheric pressure is 14.3 psia. Calculate the corresponding absolute pressure.

2.4.7 What is the height of the column of a water barometer for an atmospheric pressure of 100 kPa, if the water is at 10°C, 50°C, and 100°C?

2.4.8 Compare the pressure forces exerted on the face of two dams that hold a fresh water of 100 kg/m^3 density and a salty water of 1030 kg/m^3. Assume the faces of both dams are vertical.

2.5.1 Determine the equation for calculating the capillary rise h in the tube shown in Figure 2.5.1 below in terms of θ, σ, γ, and r.

2.5.2 Derive the relation between the gauge pressure p inside a spherical droplet of a liquid and the surface tension.

2.5.3 Derive the relation between the gauge pressure p inside the spherical bubble and the surface tension.

2.5.4 Calculate the pressure inside the spherical droplet of water having a diameter of 0.5 mm at 25°C, if the pressure outside the droplet is standard atmospheric pressure (101.3 kPa).

2.5.5 What is the capillary rise of water in a glass tube having a diameter of 0.02 in at 70°F (take $\theta = 0$)?

2.5.6 Calculate the capillary depression of mercury in a glass tube having a diameter of 0.03 in at 68°F (take $\theta = 0$). The surface tension of mercury is 0.03562 lb/ft, and its specific gravity is 13.57. Take $\gamma_w = 62.3$ lb/ft^3 at 68°F.

2.5.7 Calculate the force necessary to lift a thin wire ring of 10-mm diameter from a water surface at 10°C.

2.7.1 Heavy fuel oil flows in a pipe that has a 6-in diameter with a velocity of 7 ft/s. The kinematic viscosity is 0.00444 ft^2/s. What is the Reynolds number?

2.8.1 The discharge of water in a 20-cm diameter pipe is 0.02 m^3/s. Calculate the velocity, Reynolds number, and mass rate of flow. Assume the temperature is 15°C.

2.8.2 Water flows in a channel having a slope of 25° at a mean velocity of 3 ft/s. The depth of flow that is measured along a vertical line is 2 ft. Calculate the discharge if the width of the channel is 1 ft.

2.8.3 Water flows in a 10-cm diameter pipe at 500 kg/min. Calculate the discharge and mean velocity if the temperature is 20°C.

2.8.4 Water flows into a weigh tank for 30 min. Calculate the increase in the weight of the tank, if the discharge is 1.5 ft^3/s. The temperature is 50°F.

2.8.5 The hypothetical velocity distribution in a horizontal, rectangular, open-channel is $v = 0.3$ y (m/s), where v is the velocity at a distance y (m) above the bottom of the channel. If the vertical depth of flow is 2 m and the width of the channel is 4 m, what are the maximum velocity, the discharge, and the mean velocity?

2.8.6 Assume water flows full in a 10-mm piezometer pipe at a temperature of 20°C. What is the approximate maximum discharge for which a laminar flow may be expected?

REFERENCES

Chaudhry, H., "Principles of Flow of Water," Chapter 2 in *Water Resources Handbook* (edited by L. W. Mays), McGraw-Hill, New York, 1996.

Franzini, J. B. and E. J. Finnemore, *Fluid Mechanics*, Ninth Edition, McGraw-Hill, New York, 1997.

Munson, B. R., D. F. Young, T. H. Okiishi, *Fundamentals of Fluid Mechanics*, Third Edition Update, John Wiley and Sons, Inc., New York, 1998.

Roberson, J. A. and C. T. Crowe, *Engineering Fluid Mechanics*, Fifth Edition, Houghton Mifflin Company, Boston, MA, 1993.

Chapter 3

Flow Processes and Hydrostatic Forces

3.1 CONTROL VOLUME APPROACH FOR HYDROSYSTEMS

Hydrosystem processes transform the space and time distribution of water in hydrologic systems throughout the hydrologic cycle, in natural and human-made hydraulic systems, and in water resources systems that include both hydrologic and hydraulic systems. The commonality of all hydrosystems is the physical laws that define the flow of fluid in these systems. A consistent mechanism for developing these physical laws is called the *control volume approach*.

The simplified concept of a system is very important in the control volume approach because of the extreme complexity of hydrosystems. Typically a system is defined from the fluids viewpoint is defined as a given quantity of mass. A *system* is also a set of connected parts that form a whole. For the present discussion the fluids viewpoint will be used, in which the system has a *system boundary* or *control surface* (CS) as shown in Figure 3.1.1. A control surface is the surface that surrounds the control volume. The control surface can coincide with physical boundaries such as the wall of a pipe or the boundary of a watershed. Part of the control surface may be a hypothetical surface through which fluid flows.

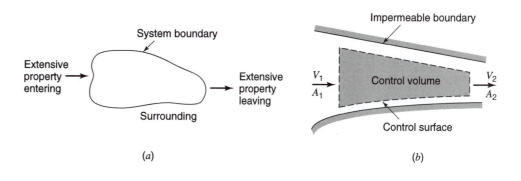

Figure 3.1.1 Control volume approach. (*a*) System and surrounding; (*b*) Control volume as a system.

Two properties, *extensive properties* and *intensive properties*, are used in the control volume approach to apply physical properties for discrete masses to a fluid flowing continuously through a control volume. Extensive properties are related to the total mass of the system (control volume), whereas intensive properties are independent of the amount of fluid. The extensive properties are mass m, momentum mV, and energy E. Corresponding intensive properties are mass per unit mass, momentum per unit mass, which is velocity v, and energy per unit mass e. In other words, for an extensive property B, the corresponding intensive property β is defined as the quantity of B per unit mass, $\beta = dB/dm$. Both the extensive and intensive properties can be scalar or vector quantities.

The relationship between intensive and extensive properties for a given system is defined by the following integral over the system:

$$B = \int_{system} \beta \, dm = \int \beta \rho \, d\forall \tag{3.1.1}$$

where dm and $d\forall$ are the differential mass and differential volume, respectively, and ρ is the fluid density.

The volume rate of flow past a given area A is expressed as

$$Q = \mathbf{V} \cdot \mathbf{A} \tag{3.1.2}$$

where $\mathbf{V}$ is the velocity vector of flow and $\mathbf{A}$ is the area vector, which is directed normal to the area and points outward from the control volume.

For the control volume in Figure 3.1.1, the net flowrate $\dot{Q}$ is

$$\begin{aligned} \dot{Q} &= Q_{out} - Q_{in} \\ &= \mathbf{V}_2 \cdot \mathbf{A}_2 - \mathbf{V}_1 \cdot \mathbf{A}_1 \\ &= \sum_{CS} \mathbf{V} \cdot \mathbf{A} \end{aligned} \tag{3.1.3}$$

In other words, the dot product $\mathbf{V} \cdot \mathbf{A}$ for all flows in and out of a control volume is the net rate of outflow.

The mass rate of flow out of the control volume is

$$\frac{dm}{dt} = \dot{m} = \sum_{CS} \rho \mathbf{V} \cdot \mathbf{A} \tag{3.1.4}$$

The rate of flow of extensive property B is the product of the mass rate and the intensive property:

$$\frac{dB}{dt} = \dot{B} = \sum_{CS} \beta \rho \mathbf{V} \cdot \mathbf{A} \tag{3.1.5}$$

If the velocity varies across the flow section, then it must be integrated across the section, so that the above equation for the rate of flow of extensive property $\dot{B}$ from the control volume becomes

$$\dot{B} = \int_{CS} \beta \rho \mathbf{V} \cdot d\mathbf{A} \tag{3.1.6}$$

Considering the system in Figure 3.1.2, the control volume is defined by the control surface at time $t(\text{I} + \text{II})$ with extensive property B_t. At time $t + \Delta t$ the control volume, defined by the control surface, $(\text{II} + \text{III})$ has moved and has extensive property $B_{t+\Delta t}$. The rate of change of extensive property B is

$$\frac{dB}{dt} = \lim_{\Delta t \to 0} \left[\frac{B_{t+\Delta t} - B_t}{\Delta t} \right] \tag{3.1.7}$$

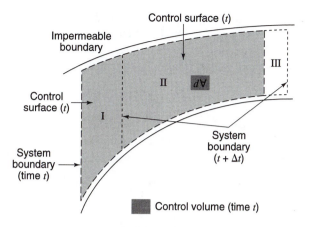

Figure 3.1.2 Control volume at times t and $t + \Delta t$.

The mass of the system at time $t + \Delta t$, $m_{\text{sys},t+\Delta t}$, is

$$m_{\text{sys},t+\Delta t} = m_{t+\Delta t} + \Delta m_{\text{out}} - \Delta m_{\text{in}} \tag{3.1.8}$$

where
$m_{t+\Delta t}$ = mass of fluid within the control volume at time $t + \Delta t$
Δm_{out} = mass of fluid that has moved out of the control volume in time Δt
Δm_{in} = mass of fluid that has moved into the control volume in time Δt

The extensive property of the system at time $t + \Delta t$ is

$$B_{\text{sys}} = B_{\text{CV},t+\Delta t} + \Delta B_{\text{out}} - \Delta B_{\text{in}} \tag{3.1.9}$$

where $B_{\text{CV},t+\Delta t}$ = amount of extensive property in the control volume at time $t + \Delta t$
ΔB_{out} = amount of extensive property of the system that has moved out of the control volume in time Δt
ΔB_{in} = amount of extensive property of the system that has moved into the control volume in time Δt

The time rate of change of extensive property of the system is

$$\frac{dB_{\text{sys}}}{dt} = \lim_{\Delta t \to 0} \left[\frac{\left(B_{\text{CV},t+\Delta t} + \Delta B_{\text{out}} - \Delta B_{\text{in}} \right) - B_{\text{CV},t}}{\Delta t} \right] \tag{3.1.10}$$

The expression can be rearranged to yield

$$\frac{dB_{\text{sys}}}{dt} = \lim_{\Delta t \to 0} \left[\frac{B_{\text{CV},t+\Delta t} - B_{\text{CV},t}}{\Delta t} \right] + \lim_{\Delta t \to 0} \left[\frac{\Delta B_{\text{out}} - \Delta B_{\text{in}}}{\Delta t} \right]$$

$$= \begin{Bmatrix} \text{Rate of change with} \\ \text{respect to time of} \\ \text{extensive property} \\ \text{in the control volume} \end{Bmatrix} + \begin{Bmatrix} \text{Net flow of} \\ \text{extensive property} \\ \text{from the control} \\ \text{volume} \end{Bmatrix} \tag{3.1.11}$$

$$= \frac{dB_{\text{CV}}}{dt} + \frac{dB}{dt}$$

The derivative $\dfrac{dB_{\text{CV}}}{dt} = \dfrac{d}{dt} \displaystyle\int_{\text{CV}} \beta\rho \; d\forall$ and $\dfrac{dB}{dt}$ is defined by equation (3.1.5), so that the *control volume equation for one-dimensional flow* becomes

$$\frac{dB_{\text{sys}}}{dt} = \frac{d}{dt} \int_{\text{CV}} \beta\rho \, d\forall + \sum_{\text{CS}} \beta\rho \, \mathbf{V} \cdot \mathbf{A} \tag{3.1.12}$$

The above equation for the general control volume equation was derived for one-dimensional flow so that the rate of flow of B at each section is $\beta\rho\mathbf{V}\cdot\mathbf{A}$. A more general form for rate of flow of an extensive property considers the velocity as variable across a section. Using equation (3.1.6), then, the *general control volume equation* is expressed as

$$\frac{dB_{sys}}{dt} = \frac{d}{dt}\int_{CV}\beta\rho\,d\forall + \int_{CS}\beta\rho\,\mathbf{V}\cdot d\mathbf{A} \tag{3.1.13}$$

This general control volume equation (also referred to as the *Reynolds transport theorem*) states that the total rate of change of extensive property of a flow is equal to the rate of change of extensive property stored in the control volume, $\dfrac{d}{dt}\int_{CV}\beta\rho\;d\forall$, plus the net rate of outflow of extensive property through the control surface, $\int_{CS}\beta\rho\;\mathbf{V}\cdot d\mathbf{A}$.

Throughout this book the general control volume equation (approach) is applied to develop continuity, energy, and momentum equations for hydrosystem (hydrologic and hydraulic) processes.

3.2   CONTINUITY

In order to write the continuity equation, the extensive property is mass ($B = m$) and the intensive property $\beta = dB/dm = 1$. By the law of conservation of mass, the mass of a system is constant, therefore $dB/dt = dm/dt = 0$. The general form of the continuity equation is then

$$0 = \frac{d}{dt}\int_{CV}\rho\,d\forall + \int_{CS}\rho\,\mathbf{V}\cdot d\mathbf{A} \tag{3.2.1}$$

which is the *integral equation of continuity for an unsteady, variable-density flow*. Equation (3.2.1) can be rewritten as

$$\int_{CS}\rho\,\mathbf{V}\cdot d\mathbf{A} = -\frac{d}{dt}\int_{CV}\rho\;d\forall \tag{3.2.2}$$

which states that the net rate of outflow of mass from the control volume is equal to the rate of decrease of mass within the control volume.

For flow with constant density, equation (3.2.2) can be expressed as

$$\int_{CS}\mathbf{V}\cdot d\mathbf{A} = -\frac{d}{dt}\int_{CV}d\forall \tag{3.2.3}$$

The continuity equation for flow with a uniform velocity across the flow section and constant density is expressed as

$$\sum_{CS}\mathbf{V}\cdot\mathbf{A} = -\frac{d}{dt}\int_{CV}d\forall \tag{3.2.4}$$

For a *constant-density, steady one-dimensional flow,* such as water flowing in a conduit, the velocity is the mean velocity, then

$$\sum_{CS}\mathbf{V}\cdot\mathbf{A} = 0 \tag{3.2.5}$$

For pipe conduit flow we consider a control volume between two locations of the pipe, at sections 1 and 2, then the continuity equation is

$$-V_1 A_1 + V_2 A_2 = 0 \tag{3.2.6a}$$

or

$$V_1 A_1 = V_2 A_2 \tag{3.2.6b}$$

or

$$Q_1 = Q_2 \tag{3.2.6c}$$

For a *constant-density unsteady flow*, consider the integral $\int_{CV} d\forall$ as the volume of fluid stored in a control volume denoted by S, so that

$$\frac{d}{dt}\int_{CV} d\forall = \frac{dS}{dt} \tag{3.2.7}$$

The net outflow is defined as

$$\int_{CS} \mathbf{V} \cdot d\mathbf{A} = \int_{outlet} \mathbf{V} \cdot d\mathbf{A} + \int_{inlet} \mathbf{V} \cdot d\mathbf{A}$$

$$= Q(t) - I(t) \tag{3.2.8}$$

Then the integral equation of continuity is determined by substituting equations (3.2.7) and (3.2.8) into equation (3.2.2) to obtain

$$Q(t) - I(t) = -\frac{dS}{dt} \tag{3.2.9}$$

or more commonly expressed as

$$\frac{dS}{dt} = I(t) - Q(t) \tag{3.2.10}$$

This continuity expression is used extensively in describing hydrologic processes.

EXAMPLE 3.2.1

A river section is defined by two bridges. At a particular time the flow at the upstream bridge is 100 m^3/s and at the same time the flow at the downstream bridge is 75 m^3/s. At this particular time, what is the rate at which water is being stored in the river section, assuming no losses?

SOLUTION

Using the continuity equation (3.2.10) yields

$$\frac{dS}{dt} = Q_{up}(t) - Q_{down}(t)$$

$$= 100 \ m^3/s - 75 \ m^3/s$$

$$= 25 \ m^3/s$$

EXAMPLE 3.2.2

A reservoir has the following monthly inflows and outflows in relative units:

Month	J	F	M	A
Inflows	10	5	0	5
Outflows	5	5	10	0

If the reservoir contains 30 units of water in storage at the beginning of the year, how many units of water in storage are there at the end of April?

SOLUTION

The integral equation (3.2.10) of continuity is used to perform a routing of flows into and out of the reservoir. Because the inflow and outflows are for discrete time intervals, the continuity equation (3.2.10) can be reformulated as

$$dS = I(t)dt - Q(t)dt$$

and integrated over time intervals $j = 1, 2, \ldots J$ of each length Δt:

$$\int_{S_{j-1}}^{S_j} dS = \int_{(j-1)\Delta t}^{j\Delta t} I(t)dt - \int_{(j-1)\Delta t}^{j\Delta t} Q(t)dt$$

or

$$S_j - S_{j-1} = I_j - Q_j \text{ for } j = 1, 2, \ldots$$

$$\Delta S_j = I_j - Q_j$$

where I_j and Q_j are the volumes of inflow and outflow for the jth time interval. The cumulative storage is $S_{j+1} = S_j + \Delta S_j$. For the first interval of time,

$$\Delta S_1 = I_1 - Q_1 = 10 - 5 = 5$$

Then $S_2 = S_1 + \Delta S_1 = 30 + 5 = 35$. The remaining computations are:

Time	I_j	Q_j	ΔS_j	S_j
1	10	5	5	30
2	5	5	0	35
3	0	10	−10	25
4	5	0	5	30

EXAMPLE 3.2.3

Consider the steady flow of water through a nozzle in which the upstream diameter of $D_2 = 30$ cm reduces to a downstream diameter of $D_2 = 20$ cm. For a flowrate of 0.08 m³/s, compute the mean velocities for the upstream and downstream diameters.

SOLUTION

Using the continuity equations (3.2.6), $Q = V_1 A_1 = V_2 A_2$, we get

$$V_1 = Q/A_1 = \frac{0.08}{\left[\pi (30/100)^2 / 4\right]} = 1.13 \text{ m/s}$$

$$V_2 = Q/A_2 = \frac{0.08}{\left[\pi (20/100)^2 / 4\right]} = 2.55 \text{ m/s}$$

3.3 ENERGY

This section uses the first law of thermodynamics along with the control volume approach to develop the energy equation for fluid flow in hydrologic and hydraulic processes. An energy balance for hydrologic and hydraulic processes considers an accounting of all inputs and outputs of energy to and from a system. By the *first law of thermodynamics*, the rate of change of energy, E, with time is the rate at which heat is transferred into the fluid, dH/dt, minus the rate at which the fluid does work on the surroundings, dW/dt, expressed as

$$\frac{dE}{dt} = \frac{dH}{dt} - \frac{dW}{dt} \tag{3.3.1}$$

The total energy of a fluid system is the sum of the internal energy E_u the kinetic energy E_k, and the potential energy E_p; thus

$$E = E_u + E_k + E_p \tag{3.3.2}$$

The extensive property is the amount of energy in the system, $B = E$:

$$B = E_u + E_k + E_p \tag{3.3.3}$$

and the intensive property is

$$\beta = \frac{dB}{dm} = e = e_u + e_k + e_p \tag{3.3.4}$$

where e represents the energy per unit mass. Also, the rate of change of extensive property with respect to time is

$$\frac{dB}{dt} = \frac{dE}{dt} = \frac{dH}{dt} - \frac{dW}{dt} \tag{3.3.5}$$

The *energy balance equation* is now derived by substituting β (equation (3.3.4)) and dB/dt (equation (3.3.5)) into the general control volume equation (3.1.12),

$$\frac{dE}{dt} = \frac{dH}{dt} - \frac{dW}{dt} = \frac{d}{dt}\int_{CV} e\rho\, d\forall + \sum_{CS} e\rho \mathbf{V} \cdot \mathbf{A} \tag{3.3.6}$$

Next we can replace e by equation (3.3.4):

$$\frac{dH}{dt} - \frac{dW}{dt} = \frac{d}{dt}\int_{CV} \left(e_u + e_k + e_p\right)\rho\, d\forall + \sum_{CS} \left(e_u + e_k + e_p\right)\rho\, \mathbf{V} \cdot \mathbf{A} \tag{3.3.7}$$

The kinetic energy per unit mass e_k is the total kinetic energy of mass with velocity V divided by the mass m:

$$e_k = \frac{mV^2/2}{m} = \frac{V^2}{2} \tag{3.3.8}$$

The potential energy per unit mass e_p is the weight of the fluid $\gamma\forall$ times the centroid elevation z of the mass divided by the mass:

$$e_p = \frac{\gamma\,\forall z}{m} = \frac{\gamma\,\forall z}{\rho\forall} = gz \tag{3.3.9}$$

because $\gamma/\rho = g$.

Now the *general energy equation for unsteady variable density flow* can be written as

$$\frac{dH}{dt} - \frac{dW}{dt} = \frac{d}{dt}\int_{CV}\left(e_u + \frac{1}{2}V^2 + gz\right)\rho\, d\forall + \sum_{CS}\left(e_u + \frac{1}{2}V^2 + gz\right)\rho\mathbf{V} \cdot \mathbf{A} \tag{3.3.10}$$

For steady flow, equation (3.3.10) reduces to

$$\frac{dH}{dt} - \frac{dW}{dt} = \sum_{CS}\left(e_u + \frac{1}{2}V^2 + gz\right)\rho\mathbf{V} \cdot \mathbf{A} \tag{3.3.11}$$

The work done by a system on its surroundings can be divided into *shaft work*, W_s, and *flow work*, W_f. Flow work is the result of pressure force as the system moves through space and shaft work is any other work besides the flow work. In the control volume in Figure 3.1.2, the force on the upstream end of the fluid is p_1A_1 and the distance traveled over time Δt is $l_1 = V_1\Delta t$. Work done on the surrounding fluid as a result of this force is then the product of the force p_1A_1 in the direction of motion and the distance traveled, $V_1\Delta t$. The work force on the upstream end is then

$$W_{f_1} = -V_1 p_1 A_1 \Delta t \tag{3.3.12a}$$

and on the downstream end is

$$W_{f_2} = V_2 p_2 A_2 \Delta t \qquad (3.3.12b)$$

At the upstream end a negative sign must be used because the pressure force on the surrounding fluid acts in the opposite direction to the motion of the system boundary. The rate of work at the upstream and downstream ends are respectively

$$\frac{dW_{f_1}}{dt} = -V_1 p_1 A_1 \qquad (3.3.13)$$

and

$$\frac{dW_{f_2}}{dt} = V_2 p_2 A_2 \qquad (3.3.14)$$

The rate of flow work can then be expressed in general terms as

$$\frac{dW_f}{dt} = p\,\mathbf{V} \cdot \mathbf{A} \qquad (3.3.15)$$

or for all streams passing through the control volume as

$$\frac{dW_f}{dt} = \sum_{CS} p\mathbf{V} \cdot \mathbf{A} = \sum_{CS} \frac{p}{\rho}\,\rho\mathbf{V} \cdot \mathbf{A} \qquad (3.3.16)$$

The net rate of work on the system can now be expressed as

$$\frac{dW}{dt} = \frac{dW_s}{dt} + \sum_{CS} \frac{p}{\rho}\,\rho\,\mathbf{V} \cdot \mathbf{A} \qquad (3.3.17)$$

Using equation (3.3.17), the *general energy equation* (3.3.10) *for unsteady variable density flow* can be expressed as

$$\frac{dH}{dt} - \frac{dW_s}{dt} - \sum_{CS} \frac{p}{\rho}\,\rho\,\mathbf{V} \cdot \mathbf{A} = \frac{d}{dt} \int_{CV} \left(e_u + \frac{1}{2}V^2 + gz \right)\rho\,d\forall + \sum_{CS} \left(e_u + \frac{1}{2}V^2 + gz \right)\rho\mathbf{V} \cdot \mathbf{A} \qquad (3.3.18)$$

which can be written as

$$\frac{dH}{dt} - \frac{dW_s}{dt} = \frac{d}{dt} \int_{CV} \left(e_u + \frac{1}{2}V^2 + gz \right)\rho\,d\forall + \sum_{CS} \left(\frac{p}{\rho} + e_u + \frac{1}{2}V^2 + gz \right)\rho\,\mathbf{V} \cdot \mathbf{A} \qquad (3.3.19)$$

For steady flow, equation (3.3.19) reduces to

$$\frac{dH}{dt} - \frac{dW_s}{dt} = \sum_{CS} \left(\frac{p}{\rho} + e_u + \frac{1}{2}V^2 + gz \right)\rho\,\mathbf{V} \cdot \mathbf{A} \qquad (3.3.20)$$

EXAMPLE 3.3.1

Determine an expression based upon the energy concept that relates the pressures at the upstream and downstream ends of the nozzle in example 3.2.3, assuming steady flow, neglecting change in internal energy, and assuming $dH/dt = 0$ and $dW_s/dt = 0$.

SOLUTION

Using the energy equation (3.3.20) for steady flow yields

$$\frac{dH}{dt} - \frac{dW_s}{dt} = \sum_{CS} \left(\frac{p}{\rho} + e_u + \frac{1}{2}V^2 + gz \right)\rho\,\mathbf{V} \cdot \mathbf{A}$$

Neglecting dH/dt and dW_s/dt the above energy equation can be expressed as

$$\int_{A_2} \left(\frac{p_2}{\rho} + e_{u_2} + \frac{1}{2} V_2^2 + gz_2 \right) \rho\, V_2\, dA_2 - \int_{A_1} \left(\frac{p_1}{\rho} + e_{u_1} + \frac{1}{2} V_1^2 + gz_1 \right) \rho\, V_1\, dA_1 = 0$$

which can be modified to

$$\int_{A_2} \left(\frac{p_2}{\rho} + e_{u_2} + gz_2 \right) \rho\, V_2\, dA_2 + \int_{A_2} \frac{\rho\, V_2^3}{2}\, dA_2 - \int_{A_1} \left(\frac{p_1}{\rho} + e_{u_1} + gz_1 \right) \rho\, V_1\, dA_1 - \int_{A_1} \frac{\rho\, V_1^3}{2}\, dA_1 = 0$$

For hydrostatic conditions, $\left(\dfrac{p}{\rho} + e_u + gz \right)$ is constant across the system, which allows these term to be taken outside the integral:

$$\left(\frac{p_2}{\rho} + e_{u_2} + gz_2 \right) \int_{A_2} \rho\, V_2 dA_2 + \int_{A_2} \frac{\rho\, V_2^3}{2}\, dA_2 - \left(\frac{p_1}{\rho} + e_{u_1} + gz_1 \right) \int_{A_1} \rho\, V_1 dA_1 - \int_{A_1} \frac{\rho\, V_1^3}{2}\, dA_1 = 0$$

The term $\int \rho V dA$ is the mass rate of flow, $\dot{m}$, and the term $\int \dfrac{\rho V^3}{2} dA = \dot{m} \dfrac{V^2}{2}$, so

$$\left(\frac{p_2}{\rho} + e_{u_2} + gz_2 \right) \dot{m} + \dot{m} \frac{V_2^2}{2} - \left(\frac{p_1}{\rho} + e_{u_1} + gz_1 \right) \dot{m} - \dot{m} \frac{V_1^2}{2} = 0$$

Dividing through by $\dot{m}g$ and rearranging yields

$$\frac{p_1}{\rho g} + \frac{e_{u_1}}{g} + z_1 + \frac{V_1^2}{2g} = \frac{p_2}{\rho g} + \frac{e_{u_2}}{g} + z_2 + \frac{V_2^2}{2g}$$

$\gamma = \rho g$ and rearranging yields

$$\frac{p_1}{\gamma} + \frac{V_1^2}{2g} + z_1 = \frac{p_2}{\gamma} + \frac{V_2^2}{2g} + z_2 + \frac{e_{u_2} - e_{u_1}}{g}$$

Neglecting changes in internal energy, $(e_{u_2} - e_{u_1})/g = 0$

$$\frac{p_1}{\gamma} + \frac{V_1^2}{2g} + z_1 = \frac{p_2}{\gamma} + \frac{V_2^2}{2g} + z_2$$

Assuming the control volume is horizontal, $z_1 = z_2$, then

$$\frac{p_1}{\gamma} + \frac{V_1^2}{2g} = \frac{p_2}{\gamma} + \frac{V_2^2}{2g}$$

This energy equation relates the pressures assuming steady flow, $z_1 = z_2$, neglecting change of internal energy in the fluid and assuming $dH/dt = 0$ and $dW_s/dt = 0$.

EXAMPLE 3.3.2

For the nozzle in example 3.2.3, determine the pressure change through the nozzle between the upstream and downstream end of the nozzle. Assume steady flow, neglect changes in internal energy of the fluid, assume $dH/dt = 0$ and $dW_s/dt = 0$, and say that the nozzle is horizontal. Assume the temperature is 20°C.

SOLUTION

For example 3.2.3, the velocities determined were $V_1 = 1.13$ m/s and $V_2 = 2.55$ m/s. Using the energy equation derived in example 3.3.1 yields

$$\frac{p_1}{\gamma} + \frac{V_1^2}{2g} = \frac{p_2}{\gamma} + \frac{V_2^2}{2g}$$

$$p_1 - p_2 = \left(V_2^2 - V_1^2\right)\frac{\gamma}{2g}$$

$$= \left[(2.55)^2 - (1.13)^2\right] \times \frac{9.79 \text{ kN/m}^3}{2 \times 9.81 \text{ m/s}^2}$$

$$= (5.226 \text{ m}^2/\text{s}^2)\,(0.499 \text{ kN s}^2/\text{m}^4)$$

$$= 2.608 \text{ kN/m}^2 = 2.608 \text{ kPa} = 2608 \text{ Pa}$$

The pressure change is a pressure decrease of 2608 Pa.

3.4   MOMENTUM

In order to derive the general momentum equation for fluid flow in a hydrologic or hydraulic system, we use the control volume approach along with Newton's second law. *Newton's second law* states that the summation of all external forces on a system is equal to the rate of change of momentum of the system

$$\sum \mathbf{F} = \frac{d(\text{momentum})}{dt} \tag{3.4.1}$$

To apply the control volume approach the extensive property is momentum, $B = m\mathbf{v}$ and the intensive property is the momentum per unit mass, $\beta = d(m\mathbf{v})/dt$, so

$$\sum \mathbf{F} = \frac{d(m\mathbf{v})}{dt} \tag{3.4.2}$$

A lowercase $\mathbf{v}$ is used to denote that this velocity is referenced to the inertial reference frame and to distinguish it from $\mathbf{V}$.

Using the general control volume equation (3.1.13),

$$\frac{dB_{sys}}{dt} = \frac{d}{dt} \int_{CV} \beta\rho \, d\forall + \int_{CS} \beta\rho \, \mathbf{V} \cdot d\mathbf{A} \tag{3.1.13}$$

and from equation (3.4.2) then

$$\sum \mathbf{F} = \frac{d}{dt} \int_{CV} \mathbf{v}\rho \, d\forall + \int_{CS} \mathbf{v}\rho\mathbf{V} \cdot d\mathbf{A} \tag{3.4.3}$$

which is the *integral momentum equation for fluid flow*. For steady flow, equation (3.4.3) reduces to

$$\sum \mathbf{F} = \int_{CS} \mathbf{v}\rho \, \mathbf{V} \cdot d\mathbf{A} \tag{3.4.4}$$

When a uniform velocity occurs in the stream crossing the control surface, the integral momentum equation is

$$\sum \mathbf{F} = \frac{d}{dt} \int_{CV} \mathbf{v}\rho \, d\forall + \sum_{CS} \mathbf{v}\rho\mathbf{V} \cdot \mathbf{A} \tag{3.4.5}$$

The momentum can be written for the coordinate directions x, y, and z in the Cartesian coordinate system as

$$\sum F_x = \frac{d}{dt}\int_{CV} v_x \rho \, d\forall + \sum_{CS} v_x \left(\rho \mathbf{V} \cdot \mathbf{A}\right) \tag{3.4.6}$$

$$\sum F_y = \frac{d}{dt}\int_{CV} v_y \rho \, d\forall + \sum_{CS} v_y \left(\rho \mathbf{V} \cdot \mathbf{A}\right) \tag{3.4.7}$$

$$\sum F_z = \frac{d}{dt}\int_{CV} v_z \rho \, d\forall + \sum_{CS} v_z \left(\rho \mathbf{V} \cdot \mathbf{A}\right) \tag{3.4.8}$$

For a steady flow the time derivative in equation (3.4.5) drops out, yielding

$$\sum \mathbf{F} = \sum_{CS} v\rho \, \mathbf{V} \cdot \mathbf{A} \tag{3.4.9}$$

For a steady flow in which the cross-sectional area of flow does not change along the length of the flow, $\displaystyle\sum_{CS} v\rho \, \mathbf{V} \cdot \mathbf{A} = 0$, (referred to as uniform flow), equation (3.4.9) reduces to

$$\sum \mathbf{F} = 0 \tag{3.4.10}$$

3.5   PRESSURE AND PRESSURE FORCES IN STATIC FLUIDS

In section 2.4, pressure, absolute pressure, gauge pressure, piezometric head, and pressure force were defined. This section extends that conversation to hydrostatic forces on submerged surfaces and buoyancy.

3.5.1   Hydrostatic Forces

Hydraulic engineers have many engineering applications in which they have to compute the force being exerted on submerged surfaces. The hydrostatic force on any submerged plane surface is equal to the product of the surface area and the pressure acting at the centroid of the plane surface. Consider the force on the plane surface shown in Figure 3.5.1. This plane surface can be divided into an infinite number of differential horizontal planes with width dy and area dA. The distance to the incremental area from the axis O–O is y. The pressure on dA is $p = \gamma y \sin \theta$ so that the force dF is $dF = p\,dA = \gamma y \sin \theta \, dA$. The force on the entire submerged plane is obtained by integrating the differential force on the differential area:

$$F = \int_A \gamma y \sin \theta \, dA \tag{3.5.1a}$$

$$= \gamma \sin \theta \int_A y \, dA \tag{3.5.1b}$$

$$= \gamma \sin \theta y_c A \tag{3.5.1c}$$

where γ and $\sin \theta$ are constants. The integral $\displaystyle\int_A y \, dA$ is by definition the first moment of the area

and $\displaystyle\int_A y(dA/A) = y_c$ is the distance from the O–O axis to the centroid (center of gravity) of the

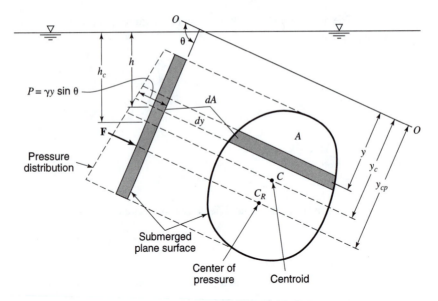

Figure 3.5.1 Hydrostatic pressure on a plane surface.

submerged plane. The vertical distance to the centroid can be defined as $h_c = y_c \sin \theta$, so that the force on the submerged plane is

$$F = \gamma h_c A \qquad (3.5.2)$$

Engineers are normally interested in the forces that are in excess of the ambient atmospheric pressures. Keep in mind that atmospheric pressure, for most applications, acts on both sides of the submerged surface so that gauge pressure is of importance.

Even though pressure forces acting on a submerged surface are distributed throughout the surface, engineers are interested in the location of the *center of pressure*, which is the point on the submerged surface where the resultant force acts. The moment equation is

$$y_{cp} F = \int y \, dF \qquad (3.5.3)$$

where $dF = p \, dA$, so

$$y_{cp} F = \int_A yp \, dA \qquad (3.5.4)$$

and $p = \gamma y \sin \theta$, so

$$y_{cp} F = \int_A \gamma y^2 \sin \theta \, dA = \gamma \sin \theta \int_A y^2 \, dA \qquad (3.5.5)$$

The integral $\int_A y^2 \, dA = I_0$ is the *moment of inertia (moment of the area)*, with respect to an axis formed by the interaction of the plane containing the surface and the free surface. This can also be expressed with respect to the horizontal centroidal axis of the area by the *parallel axis theorem* as

$$I_0 = \bar{I} + y_c^2 A \qquad (3.5.6)$$

Equation (3.5.5) can now be expressed as

$$y_{cp} F = \gamma \sin \theta \, I_0 = \gamma \sin \theta \left(\bar{I} + y_c^2 A \right) \qquad (3.5.7)$$

Substituting equation (3.5.1c) and solving for y_{cp} yields

$$y_{cp} = y_c + \frac{\bar{I}}{y_c A}$$ (3.5.8)

The vertical distance to the center of pressure h_{cp} is then

$$h_{cp} = y_{cp} \sin \theta$$ (3.5.9)

EXAMPLE 3.5.1

Derive the expression for the depth to the center of pressure y_{cp} for a rectangular area $(b \times h)$ vertically submerged with the long side (h) at the liquid surface.

SOLUTION

Using equation (3.5.8),

$$y_{cp} = y_c + \frac{\bar{I}}{y_c A}$$

where $y_c = h/2$, $\bar{I} = bh^3/12$, and $A = bh$, we get

$$y_{cp} = \frac{h}{2} + \frac{bh^3/12}{\left(\frac{h}{2}\right)(bh)} = \frac{h}{2} + \frac{h}{6} = \frac{4}{6}h = \frac{2}{3}h$$

EXAMPLE 3.5.2

Determine the hydrostatic force and the location of the center of pressure on the 25 m long dam shown in Figure 3.5.2. The face of the dam is at an angle of 60°. Assume 20°C.

SOLUTION

The diagram in Figure 3.5.2 shows the pressure distribution. Using equation (3.5.2), $h_c = 2.5$ m and $A = (25$ m $\times$ 5$)/\sin 60°$, so the hydrostatic force is

$$F = \gamma h_c A = \left(9.79 \frac{kN}{m^3}\right)(2.5 \text{ m})\left(25 \text{ m} \times \frac{5}{\sin 60°} \text{ m}\right)$$

$$= 3.532 \text{ kN}$$

The center of pressure is at 2/3 of the total water depth, $(2/3) \times 5 = 3.33$ m.

EXAMPLE 3.5.3

Consider a vertical rectangular gate $(b = 4$ m and $h = 2$ m) that is vertically submerged in water so that the top of the gate is 4 m below the surface of the water (as shown in Figure 3.5.3). Determine the total resultant force on the gate and the location of the center of pressure.

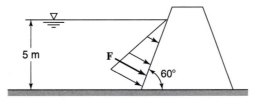

Figure 3.5.2 Hydrostatic force on dam for example 3.5.2.

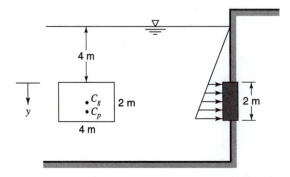

Figure 3.5.3 Vertical rectangular gate for example 3.5.3.

SOLUTION

Use the free-body diagram in Figure 3.5.3. The total resultant force is computed using equation (3.5.2), $F = \gamma h_c A$, where $h_c = 4 + (2/2) = 5$ m:

$$F = (9.79 \text{ kN/m}^3)(5 \text{ m})(4 \times 2 \text{ m}^2) = 396.1 \text{ kN}$$

The location of the center of pressure is computed using equation (3.5.8):

$$y_{cp} = y_c + \frac{\bar{I}}{y_c A} = 5 + \frac{\dfrac{4 \times 2^3}{12}}{5 \times (4 \times 2)}$$

$$= 5.067 \text{ m}$$

Alternatively, this problem can be solved by the simple integration $F = \displaystyle\int_0^2 \gamma\, h\, dA$, where $dA = 4\, dy$:

$$F = \int_0^2 \gamma h\, dA = \int_0^2 (9.79)(4 + y)(4 dy)$$

$$= 39.16 \left[4y + \frac{y^2}{2} \right]_0^2 = 39.16 \left[4 \times 2 + \frac{2^2}{2} \right]$$

$$= 391.6 \text{ kN}$$

EXAMPLE 3.5.4

Consider an inclined rectangular gate with water on one side as shown in Figure 3.5.4. Determine the total resultant force acting on the gate and the location of the center of pressure.

SOLUTION

To determine the total resultant force, $F = \gamma h_c A$, where $h_c = 5 + 1\backslash2(4\cos 60°)$, so that

$$F = (62.4) \left[5 + \frac{1}{2}(4\cos 60°) \right] (4 \times 6) = 8{,}986 \text{ lb}$$

The location of the center of pressure is

$$y_{cp} = y_c + \frac{\bar{I}}{y_c A}$$

where $y_c = \dfrac{5}{\cos 60°} + \dfrac{4}{2} = 12$ ft, $\bar{I} = bh^3/12 = 6 \times 4^3/12 = 32$ ft^4 and $A = 6 \times 4 = 24$ ft^2:

$$y_{cp} = 12 + \frac{32}{12 \times 24} = 12.11 \text{ ft}$$

Using equation (3.5.9), $h_{cp} = y_{cp} \sin \theta = 12.11 (\sin 30°) = 6.06$ ft

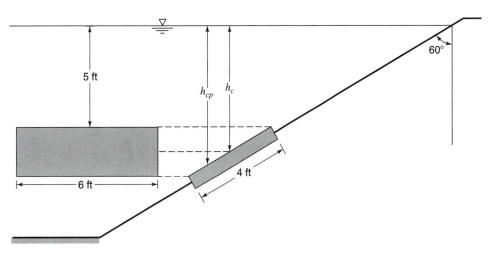

Figure 3.5.4 Inclined rectangular gate for example 3.5.4.

3.5.2   Buoyancy

The submerged body in Figure 3.5.5 is acted upon by gravity and the pressure of the surrounding fluid. On the upper surface of the submerged body, the vertical force is F_y and is equal to the weight of the volume ABCD above the surface. The vertical component of force F'_y on the bottom is the weight of the volume of fluid ABCED. The difference between the two volumes ABCD and ABCED is the volume of the submerged body. Applying the momentum principle, from equation (3.4.7) we get

$$\sum F_y = 0 \tag{3.5.10}$$

The buoyant force F_b is the weight of the volume of fluid DCE and is equal to the weight of the volume of fluid displaced, so that

$$F_b - F'_y + F_y = 0 \tag{3.5.11a}$$

or

$$F_b = F'_y - F_y \tag{3.5.11b}$$

Archimedes' principle (about 250 B.C.) states that the weight of a submerged body is reduced by an amount equal to the weight of liquid displaced by the body. This principle may be viewed as the difference of vertical pressure forces on the two surfaces DC and DEC. Floating bodies are partially submerged due to the balance of the body weight and buoyancy force.

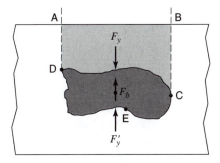

Figure 3.5.5 Forces on a submerged body. Buoyant force, F_b, passes through the centroid of the displaced volume and acts through a point called the center of buoyancy.

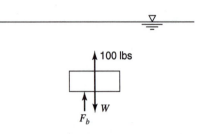

Figure 3.5.6 Free-body diagram for example 3.5.6.

EXAMPLE 3.5.5

A metal block weighs 400 N in air, but when completely submerged in water it weighs 250 N. What is the volume of the metal block?

SOLUTION

Essentially the buoyant force F_b is equal to the weight of water displaced by the metal block, i.e.

$$F_b = 400 \text{ N} - 250 \text{ N} = 150 \text{ N}$$

The weight $W = (9.79)(1000) \, \forall$, where $\forall$ is the volume.

$$150 = (9.79)(1000) \, \forall$$

$$\forall = 0.0153 \text{ m}^3$$

EXAMPLE 3.5.6

An object is 1 ft thick by 1 ft wide by 2 ft long. It weighs 100 lbs at a depth of 10 ft. What is the weight of the object in air and what is its specific gravity?

SOLUTION

Use the free-body diagram in Figure 3.5.6. The summation of forces acting on the object in the vertical direction is

$$\sum F_y = 100 + F_b - W = 0$$

where F_b is the buoyant force and W is the weight of the object.

$$F_b = (62.4 \frac{\text{lb}}{\text{ft}^3})(1 \text{ ft})(1 \text{ ft})(2 \text{ ft})$$

$$= 124.8 \text{ lbs}$$

$$100 + 124.8 - W = 0$$

$$W = 224.8 \text{ lb}$$

The specific gravity is $224.8/124.8 = 1.8$.

3.6 VELOCITY DISTRIBUTION

We discussed in section 2.8 that the actual velocity varies throughout a flow section (see Figure 2.8.1 for pipe flow as an illustration). Figure 3.6.1 illustrates velocity profiles in various open-channel flow sections. As a result of these nonuniform velocity distributions in pipe flow and open-channel flow, the velocity head is generally greater than the value computed according to $V^2/2g$ where V is the mean velocity. When using the energy principle, the true velocity head is expressed as $\alpha V^2/2g$, where α is a *kinetic energy correction factor*. Chow (1959) also referred to α as an energy coefficient or *Coriolis coefficient*.

Consider the velocity distribution shown in Figure 2.8.1. The mass of fluid flowing through an area dA per unit time is $(\gamma/g)vdA$, where v is the velocity through area dA. The flow of kinetic energy per unit time through this area is $(\gamma/g) \, v \, dA \, (v^2/2) = (\gamma/2g)v^3dA$. For a known velocity distribution, the total kinetic energy flowing through the section per unit time is

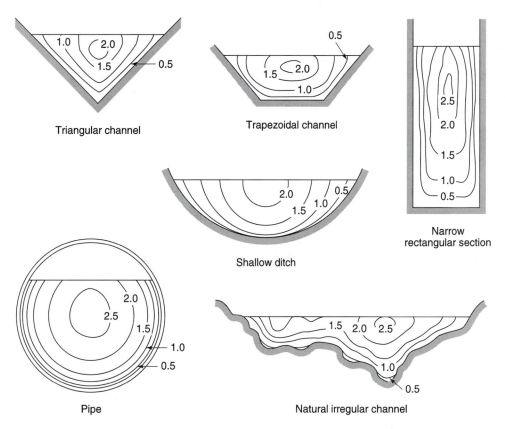

Figure 3.6.1 Typical curves of equal velocity in various channel sections (from Chow (1959)).

$$\text{Total kinetic energy} = \frac{\gamma}{2g}\int_A v^3 dA \qquad (3.6.1)$$
$$\text{(per unit time)}$$

Using the mean flow velocity V and the coefficient α, the total energy per unit weight is $\alpha V^2/2g$; because the flow across the entire section is γAV. The total kinetic energy transmitted is

$$\text{Total kinetic energy} = (\gamma AV)\left(\alpha\frac{V^2}{2g}\right)$$
$$\text{(per unit time)}$$

$$= \gamma\alpha A\frac{V^3}{2g} \qquad (3.6.2)$$

From equations (3.6.1) and (3.6.2) we get

$$\frac{\gamma}{2g}\int_A v^3 dA = \gamma\alpha A\frac{V^3}{2g} \qquad (3.6.3)$$

and we then solve for the kinetic energy correction factor:

$$\alpha = \frac{1}{AV^3}\int_A v^3 dA \qquad (3.6.4)$$

The value of α for flow in circular pipes flowing full with a parabolic velocity distribution is equal to 2 for laminar flow and normally ranges from 1.03 to 1.06 for turbulent flow. Because α is not known precisely, it is not commonly used in pipe flow calculations, and the kinetic energy of fluid per unit weight is $V^2/2g$. The values of α for open-channel flow varies by the type of channel flow. For example, in regular channels, flumes, and spillways, α ranges between 1.10 and 1.20, and for river valleys and areas α ranges between 1.5 and 2.0 with an average of 1.75 (see Chow, 1959).

The nonuniform distribution of velocity also affects the computation of momentum in open-channel flow. The corrected momentum of water passing through a channel section per unit time is

$$\underset{\text{(per unit time)}}{\text{Total momentum}} = \beta_m \frac{\gamma}{g} QV$$

$$= \beta_m \frac{\gamma}{g} AV^2 \qquad (3.6.5)$$

where β_m is the *momentum correction factor*, also called the *momentum coefficient* or *Boussinesq coefficient* by Chow (1959). The momentum of water passing through an elemental area dA per unit time is the product of the mass per unit time $(\gamma/g)\, vdA$ and the velocity v, which is $(\gamma/g)\, v^2 dA$. The total momentum of fluid per unit time is

$$\underset{\text{(per unit time)}}{\text{Total momentum}} = \frac{\gamma}{g} \int_A v^2 dA \qquad (3.6.6)$$

From equations (3.6.5) and (3.6.6), we get

$$\beta_m \frac{\gamma}{g} AV^2 = \frac{\gamma}{g} \int_A v^2 dA \qquad (3.6.7)$$

Solving for the momentum correction factor β_m yields

$$\beta_m = \frac{1}{AV^2} = \int_A v^2 dA \qquad (3.6.8)$$

According to Chow (1959), the value of β_m for fairly straight prismatic channels varies between 1.01 to 1.12 and for river valleys β_m varies between 1.17 and 1.33.

| EXAMPLE 3.6.1 | Show that the kinetic energy correction factor is $\alpha = 2$ for laminar flow in a circular pipe. |

SOLUTION In example 2.8.3, the parabolic velocity distribution is expressed as

$$v = v_{max}\left[1 - \left(\frac{r}{r_0}\right)^2\right]$$

The rate of $V/v_{max} = 0.5$ was derived and $A = \pi r^2$. Using equation (3.6.4) yields

$$\alpha = \frac{1}{AV^3} \int_A v^3 dA = \frac{1}{AV^3} \int_0^{r_0} v^3 (2\pi r)dr$$

$$= \frac{1}{\pi r_0^2 (0.5 v_{max})^3} \int_0^{r_0} \left\{ v_{max}\left[1 - \left(\frac{r}{r_0}\right)^2\right]\right\}^3 (2\pi r)dr$$

$$= \frac{16}{r_0^2} \int_0^{r_0} \left[1 - \left(\frac{r}{r_0}\right)^2\right]^3 r\, dr = 2$$

PROBLEMS

3.2.1 It is required to reduce a pipe of diameter 8 in to a minimum-diameter pipe that allows the downstream velocity not to exceed twice the upstream velocity. Determine the diameter of the pipe. Assume smooth transition.

3.3.1 Water flows through a pipe of diameter 3 inches. If it is desired to use another pipe for the same flow rate such that the velocity head in the second pipe is four times the velocity head in the first pipe, determine the diameter of the pipe.

3.4.1 If the pipeline in Problem 3.2.1 is horizontal, what is the proportion of the potential energy head at the upstream cross-section that is changed to kinetic energy head at the downstream cross-section? Determine the answer in terms of the discharge.

3.5.1 Derive an expression for the depth to the center of pressure for a triangle of height h and base b that is vertically submerged in water with the vertex at the water surface.

3.5.2 Derive an expression for the depth to the center of pressure for a triangle of height h and base b that is vertically submerged in water with the vertex a distance x below the water surface.

3.5.3 Determine the magnitude and the location of the hydrostatic force on the 2-m by 4-m vertical rectangular gate shown in Figure 3.5.3 if the top of the gate is 6 m below the water surface. Also determine the total hydrostatic force on the gate and the location of the center of pressure.

3.5.4 Suppose a vertical flat plate supports water on one side and oil of specific gravity 0.86 on the other side, as shown in Figure P3.5.4. How deep should the oil be so that there is no net horizontal force on the plate? Calculate the moments of the pressure forces about the base of the plate. Are the magnitudes of the moments equal? Why?

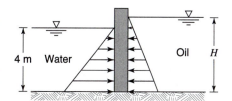

Figure P3.5.4 Vertical flat plate for problem 3.5.4.

3.5.5 Suppose a steel material of specific gravity of 7.8 is attached to a wood of specific gravity 0.8 as shown in Figure P3.5.5 If it is required that the material does not sink or rise when left in static water, what should be the proportion of the volume of the steel to that of the wood in Figure P3.5.5?

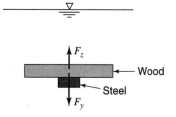

Figure P3.5.5 Problem 3.5.5 System

3.5.6 Rework example 3.5.4 if the top of the inclined rectangular gate is 3 ft below the water surface.

3.6.1 Figure P3.6.1 shows a compound open-channel cross-section. Determine the energy correction factor α. Assume uniform velocities within the subsections.

3.6.2 Determine the momentum correction factor β for Problem 3.6.1.

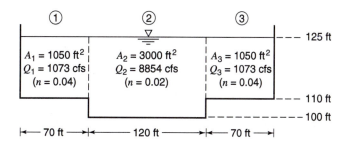

Figure P3.6.1 Compound open-channel cross-section for Problem 3.6.1

REFERENCES

Chow, V. T., *Open-Channel Hydraulics*, McGraw-Hill, New York, 1959.

Roberson, J. A., and C. T. Crowe, *Engineering Fluid Mechanics*, Houghton Mifflin Company, Boston, MA, 1990.

Chapter 4

Hydraulic Processes: Pressurized Pipe Flow

4.1 CLASSIFICATION OF FLOW

Hydraulics is typically defined as the study of liquid (water) flow in pipes and open channels, referred to as pipe flow and open-channel flow, respectively. Pipe flow and open-channel flow are similar in many ways but have one major difference. *Open-channel flow* occurs when there is a free surface, whereas pipe flow does not have a free surface. *Pipe flow* here refers to pressurized flow in pipes as long as there is not a free surface. Open-channel flow can occur in pipes. Analogies can be made between pipe flow and groundwater flow in confined aquifers. Also, groundwater flow in unconfined (water table) conditions is analogous to open-channel flow. The major difference is the geometry of the flow paths in groundwater flow as compared to pipe or open-channel flow. Because of the many varied-flow paths that occur in groundwater flow, the macroscopic average of the liquid (water) and medium properties is used. The control volumes for these three types of flow are illustrated in Figure 4.1.1.

Both pipe flow and open-channel flow are expressed in terms of the discharge Q, cross-sectional averaged velocity V, and cross-sectional area of flow A. They are related mathematically as

$$Q = AV \tag{4.1.1}$$

and

$$A = \int_0^y B \, dy \tag{4.1.2}$$

where y is the depth of flow and B is the channel width, so that

$$Q = \int_A v \, dA \tag{4.1.3}$$

v is the local point velocity along the direction normal to A. For groundwater flow, the discharge is expressed as

$$Q = Aq \tag{4.1.4}$$

where A is the total cross-sectional area including the space occupied by the porous medium and q is the *Darcy flux* or volumetric flow rate per unit area of porous medium Q/A, also sometimes

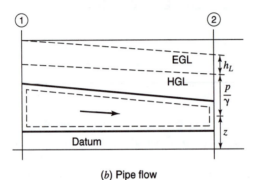

(a) Open-channel flow

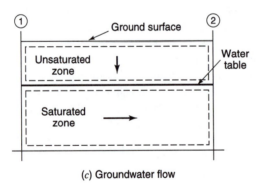

(b) Pipe flow

(c) Groundwater flow

Figure 4.1.1 Control volumes for open-channel flow, pipe flow, and groundwater flow.

called as the *specific discharge* or *Darcy velocity*. This velocity assumes that flow occurs through the entire cross-section of the porous medium without regard to solids and pores. *Darcy's law* states that flow through a porous medium is proportional to the headloss and inversely proportional to the length of the flow path, $Q{\sim}dh/dL$. Darcy's flux is expressed as

$$q = -K\frac{dh}{dL}$$

(4.1.5)

and

$$Q = -KA\frac{dh}{dL}$$

(4.1.6)

where K is the saturated hydraulic conductivity (L/T) or units of velocity.

The criterion to differentiate viscosity effects, i.e. turbulent or laminar flow, is the *Reynolds number, R_e*:

$$R_e = \frac{VD}{\nu} = \frac{V4R}{\nu} \tag{4.1.7}$$

where D is a characteristic length such as pipe diameter, R is the *hydraulic radius* defined as the cross-sectional area of flow A divided by the wetted perimeter, P, of A, and ν is the kinematic viscosity of the fluid. For full pipe flow, $R = A/P = \pi D^2/4/\pi D = D/4$. The critical number is around 2000 for pressurized pipe flow and 500 for open-channel flow. Below this critical value is laminar flow and above this number is turbulent flow. For groundwater flow the Reynolds number can be expressed with the velocity being the Darcy flux, q, and the effective grain size (d_{10}) is used for D. The *effective grain size* (d_{10}) is used to indicate that 10 percent (by weight) of a sample has a diameter smaller than d_{10}. Darcy's law is typically valid for Reynolds number $R_e < 1$ and does not depart seriously until $R_e = 10$, which can be thought of as the upper limit for Darcy's law.

The criterion to differentiate gravity effects, i.e., subcritical or supercritical flow, is the *Froude number*:

$$F_r = \frac{V}{\sqrt{gD_h}} \tag{4.1.8}$$

where D_h is the *hydraulic depth* defined as the cross-sectional area of flow divided by the top width of flow. The critical value of the Froude number is 1, $F_r = 1$ for critical flow, i.e., when $F_r < 1$ the flow is subcritical and when $F_r > 1$ the flow is supercritical. Table 4.1.1 lists the flow classification for the criteria of time, space, viscosity, and gravity.

Table 4.1.1   Flow Classification

Classification Criteria	Flow Classification	Book Section Defined
Time	Steady flow	2.6
	Unsteady flow	2.6
Space	Uniform flow	2.6
	Nonuniform flow	2.6
	Gradually varied	5.3
	Rapidly varied	5.3
Viscosity (Reynolds number)	Laminar	2.7
	Turbulent	2.7
Gravity (Froude number)	Subcritical	5.2
	Critical	5.2
	Supercritical	5.2

EXAMPLE 4.1.2

Water flows in a trapezoidal channel having a bottom width of 15 ft and side slopes of 1 vertical to 1.5 horizontal at a rate of 200 cfs. Is the flow a subcritical or a supercritical flow if the depth of the flow is 2 ft?

SOLUTION

To find out whether the flow is subcritical or supercritical, the Froude number should be calculated by equation (4.1.8):

$$F_r = \frac{V}{\sqrt{gD_h}}$$

The top width of the channel T is

$$T = 15 + 2 \times (2 \times 1.5) = 21$$

To determine the mean velocity, cross-sectional area must be computed:

$A = 2 \times (15 + 21)/2 = 36 \text{ ft}^2$

and from equation (4.1.1)

$V = Q/A$

$V = 200/36 = 5.56 \text{ ft/s}$

the hydraulic depth is

$D_h = A/T = 36/21 = 1.71 \text{ ft}$

Then

$$F_r = \frac{5.56 \,\text{ft/s}}{\sqrt{32.2 \,\text{ft/s}^2 \times 1.71 \,\text{ft}}} = 0.75$$

F_r is < 1, so the flow is subcritical flow.

4.2 PRESSURIZED (PIPE) FLOW

4.2.1 Energy Equation

In section 3.3, the general energy equation for steady fluid flow was derived as equation (3.3.20). Using equation (3.3.20) and considering pipe flow between sections 1 and 2 (Figure 4.1.1b), the energy equation for pipe flow is expressed as

$$\frac{dH}{dt} - \frac{dW_s}{dt} = \int_{A_2}\left(\frac{P_2}{\rho} + e_{u_2} + \frac{1}{2}V_2^2 + gz_2\right)\rho V_2 dA_2 - \int_{A_1}\left(\frac{P_1}{\rho} + e_{u_1} + \frac{1}{2}V_1^2 + gz_1\right)\rho V_1 dA_1 \qquad (4.2.1)$$

which can be modified to

$$\frac{dH}{dt} - \frac{dW_s}{dt} = \int_{A_2}\left(\frac{P_2}{\rho} + e_{u_2} + gz_2\right)\rho V_2 dA_2 + \int_{A_2}\frac{\rho V_2^3}{2}dA_2 - \int_{A_1}\left(\frac{P_1}{\rho} + e_{u_1} + gz_1\right)\rho V_1 dA_1 - \int_{A_1}\frac{\rho V_1^3}{2}dA_1$$

$$(4.2.2)$$

Flow is uniform at sections 1 and 2 and therefore hydrostatic conditions prevail across the section. For hydrostatic conditions, $p/\rho + e_u + gz$ is constant across the system. This allows the term $(p/\rho + e_u + gz)$ to be taken outside the integral, so that (4.2.2) can be expressed as

$$\frac{dH}{dt} - \frac{dW_s}{dt} = \left(\frac{P_2}{\rho} + e_{u_2} + gz_2\right)\int_{A_2}\rho V_2 dA_2 + \int_{A_2}\frac{\rho V_2^3 dA_2}{2}$$

$$- \left(\frac{P_1}{\rho} + e_{u_1} + gz_1\right)\int_{A_1}\rho V_1 dA_1 - \int_{A_1}\frac{\rho V_1^3}{2}dA_1$$

$$(4.2.3)$$

The term $\int \rho V \, dA = \dot{m}$ is the mass rate of flow and $\int_A (\rho V^3/2)dA = (\rho V^3/2)A = \dot{m}(\rho V^2/2)$.

Substituting these definitions, equation (4.2.3) now becomes

$$\frac{dH}{dt} - \frac{dW_s}{dt} = \left(\frac{P_2}{\rho} + e_{u_2} + gz_2\right)\dot{m} + \dot{m}\frac{V_2^2}{2} - \left(\frac{P_1}{\rho} + e_{u_1} + gz_1\right)\dot{m} - \dot{m}\frac{V_1^2}{2} \qquad (4.2.4)$$

Dividing through by $\dot{m}$ and rearranging then yields

$$\frac{1}{\dot{m}}\left(\frac{dH}{dt} - \frac{dW_s}{dt}\right) + \frac{P_1}{\rho} + e_{u_1} + gz_1 + \frac{V_1^2}{2} = \frac{P_2}{\rho} + e_{u_2} + gz_2 + \frac{V_2^2}{2} \qquad (4.2.5)$$

The shaft-work term (dW_s/dt) can be the result of a turbine (W_T) or a pump (W_P) in the system, so that the shaft work term can be expressed as

$$\frac{dW_S}{dt} = \frac{dW_T}{dt} - \frac{dW_P}{dt} \qquad (4.2.6)$$

The minus sign occurs because a pump does work on the fluid and conversely a turbine does shaft work on the surroundings. The term dW/dt has the units of power, which is work per unit time. Substituting equation (4.2.6) into (4.2.5) and dividing through by g results in

$$\left(\frac{1}{\dot{m}g}\right)\frac{dW_P}{dt} + \frac{P_1}{g} + z_1 + \frac{V_1^2}{2g} = \left(\frac{1}{\dot{m}g}\right)\frac{dW_T}{dt} + \frac{P_2}{g} + z_2 + \frac{V_2^2}{2g} + \left[\frac{e_{u_2} - e_{u_1}}{g} - \frac{1}{\dot{m}g}\frac{dH}{dt}\right] \qquad (4.2.7)$$

Each term in the above equation has the dimension of length. The following are defined:

Head supplied by pumps:
$$h_P = \left(\frac{1}{\dot{m}g}\right)\frac{dW_P}{dt} \qquad (4.2.8)$$

Head supplied by turbines:
$$h_T = \left(\frac{1}{\dot{m}g}\right)\frac{dW_T}{dt} \qquad (4.2.9)$$

Head loss (loss of mechanical energy due to viscous stress):

$$h_L = \left[\left(\frac{e_{u_2} - e_{u_1}}{g}\right) - \left(\frac{1}{\dot{m}g}\right)\frac{dH}{dt}\right] \qquad (4.2.10)$$

The term $\left((e_{u_2} - e_{u_1})/g\right)$ represents finite increases in internal energy of the flow system because some of the mechanical energy is converted to thermal energy through viscous action between fluid particles. The term $-(1/\dot{m}g)dH/dt$ represents heat generated through energy dissipation that escapes the flow system.

Using h_P, h_T, and h_L, the energy equation (4.2.7) is now expressed as

$$\frac{P_1}{\gamma} + z_1 + \frac{V_1^2}{2g} + h_P = \frac{P_2}{\gamma} + z_2 + \frac{V_2^2}{2g} + h_T + h_L \qquad (4.2.11)$$

This energy equation is expressed with the velocity representing the mean velocity; p/γ is referred to as the *pressure head* and $V^2/2g$ is referred as the *velocity head.*

Because the velocity is not actually uniform over a cross-section, an *energy correction factor* is typically introduced, which is defined as (see section 3.6)

$$\alpha = \frac{\displaystyle\int_A v^3 dA}{V^3 A} \qquad (4.2.12)$$

where v is the velocity at any point in the section. Using α_1 and α_2 for sections 1 and 2, the energy equation can be expressed as

$$\frac{P_1}{\gamma} + z_1 + \alpha_1\frac{V_1^2}{2g} + h_P = \frac{P_2}{\gamma} + z_2 + \alpha_2\frac{V_2^2}{2g} + h_T + h_L \qquad (4.2.13)$$

Because α has a value very near 1 for pipe flow, it is typically omitted.

EXAMPLE 4.2.1	The flow of water in a horizontal pipe of constant cross-section has a pressure gauge at location 1 with a pressure of 100 psig and a pressure gauge at location 2 with a pressure of 75 psig. What is the head-loss between the two pressure gauges?
SOLUTION	For the energy equation (4.2.11), $h_P = 0$, $h_T = 0$, $z_1 = z_2$, and because the pipe size is constant $V_1^2/2g = V_2^2/2g$; then (4.2.11) reduces to

$$h_L = \frac{P_1 - P_2}{\gamma} = \frac{(100 \text{ lb/in}^2 - 75 \text{ lb/in}^2) \times 144 \text{ in}^2/\text{ft}^2)}{62.4 \text{ lb/ft}^3} = \frac{25 \times 144}{62.4}$$

$$h_L = 57.7 \text{ ft}$$

EXAMPLE 4.2.2	For the simple pipe system shown in Figure 4.2.1, the pressures are $p_1 = 14$ kPa, $p_2 = 12.5$ kPa, and $p_3 = 10$ kPa. Determine the headloss between 1 and 2 and the headloss between 1 and 3. The discharge is 7 L/s.
SOLUTION	The energy equation between 1 and 2 is

$$\frac{P_1}{\gamma} + \frac{V_1^2}{2g} + z_1 = \frac{P_2}{\gamma} + \frac{V_2^2}{2g} + z_2 + h_{L_{1-2}}$$

where $\dfrac{V_1^2}{2g} = \dfrac{V_2^2}{2g}$ and $z_1 = z_2$, so

$$\frac{P_1}{\gamma} = \frac{P_2}{\gamma} + h_{L_{1-2}} \quad \text{and} \quad h_{L_{1-2}} = \frac{P_1 - P_2}{\gamma} = \frac{14 - 12.5}{9.79} = 0.153 \text{ m}$$

The energy equation between 1 and 3 is

$$\frac{P_1}{\gamma} + \frac{V_1^2}{2g} + z_1 = \frac{P_3}{\gamma} + \frac{V_3^2}{2g} + z_3 + h_{L_{1-3}}$$

$$V_1 = Q/A_1 = \frac{(7/1000)}{\left[\dfrac{\pi \, (60/1000)^2}{4}\right]} = \frac{0.007}{0.0028} = 2.477 \text{ m/s}$$

$$V_3 = Q/A_3 = \frac{(7/1000)}{\left[\dfrac{\pi \, (40/1000)^2}{4}\right]} = \frac{0.007}{0.0013} = 5.385 \text{ m/s}$$

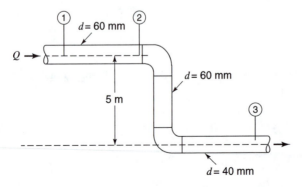

Figure 4.2.1 Pipe system for example 4.2.2.

Now using the above energy equation, we get

$$\frac{14}{9.79} + \frac{(2.477)^2}{2 \times 9.81} + 5 = \frac{10}{9.79} + \frac{(5.385)^2}{2 \times 9.81} + 0 + h_{L_{1-3}}$$

$$1.43 + 0.313 + 5 = 1.022 + 1.478 + 0 + h_{L_{1-3}}$$

$$h_{L_{1-3}} = 4.24 \text{ m}$$

EXAMPLE 4.2.3

A *Venturi meter* is a device, as shown in Figure 4.2.2, that is inserted into a pipeline to measure the incompressible flow rate. These meters consist of a convergent section that reduces the diameter to between one-half and one-fourth the pipe diameter, followed by a divergent section. Pressure differences between the position just before the Venturi and at the throat are measured by a differential manometer. Using the energy equation, develop an equation for the discharge in terms of the pressure difference $p_1 - p_2$. Use a coefficient of discharge that takes into account frictional effects.

SOLUTION

Write the energy equation between 1 and 2:

$$\frac{p_1}{\gamma} + \frac{V_1^2}{2g} + z_1 = \frac{p_2}{\gamma} + \frac{V_2^2}{2g} + z_2 + h_L$$

For $z_1 = z_2$ and assuming $h_L = 0$, the energy equation reduces to

$$\frac{p_1}{\gamma} + \frac{V_1^2}{2g} = \frac{p_2}{\gamma} + \frac{V_2^2}{2g}$$

$$V_1^2 - V_2^2 = 2g\left(\frac{p_2 - p_1}{\gamma}\right)$$

From continuity, $A_1 V_1 = A_2 V_2$, so $V_1 = V_2(A_2/A_1)$. Substituting V_1 into the above energy equation yields

$$\left[V_2 \frac{A_2}{A_1}\right]^2 - V_2^2 = 2g\left(\frac{p_2 - p_1}{\gamma}\right)$$

$$\left[\left(\frac{A_2}{A_1}\right)^2 - 1\right]V_2^2 = 2g\left(\frac{p_2 - p_1}{\gamma}\right)$$

Multiplying through by (−1) and solving for V_2, we get

$$V_2 = \sqrt{\frac{1}{1-(A_2/A_1)^2}} \sqrt{2g\left(\frac{p_1 - p_2}{\gamma}\right)}$$

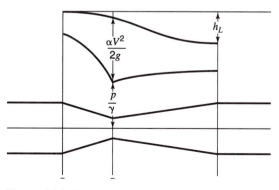

Figure 4.2.2 Venturi tube.

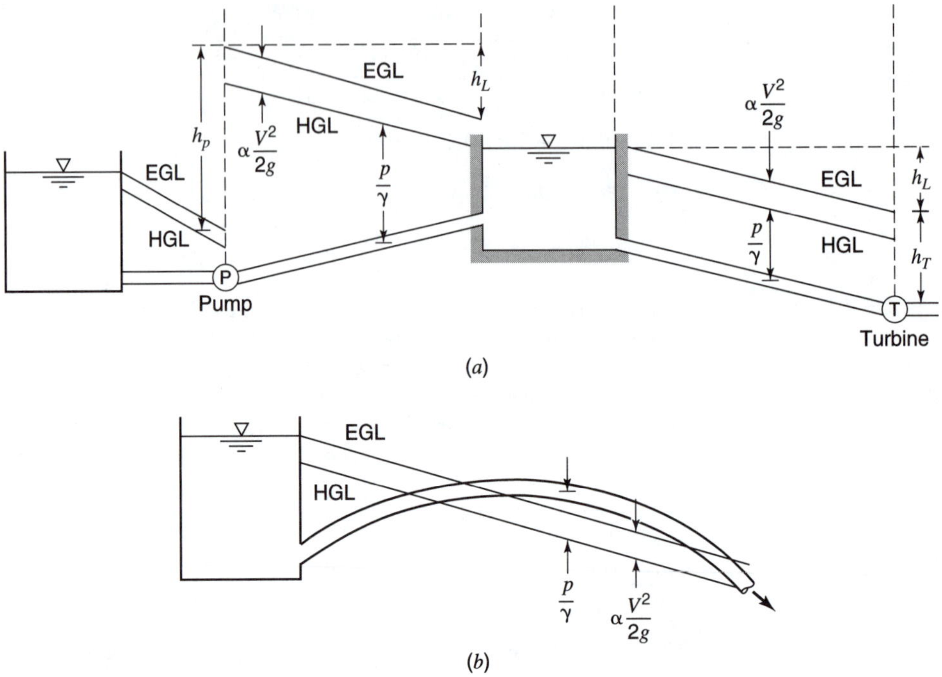

(a)

(b)

Figure 4.2.3 Energy and hydraulic grade lines. (a) EGL and HGL for system with pump and turbine; (b) System with subatmospheric pressure (pipe is above the HGL).

which neglects friction effects. A coefficient of discharge C_d is used in the continuity equation to account for friction effect, so

$$Q = C_d A_2 V_2 = C_d A_2 \left[\frac{1}{1-(A_2/A_1)^2}\right]^{1/2} \left[2g\frac{P_1-P_2}{\gamma}\right]^{1/2} = C_d \frac{A_2}{\sqrt{1-(A_2/A_1)^2}} \sqrt{2g\frac{P_1-P_2}{\gamma}}$$

The value of C_d must be determined experimentally.

4.2.2   Hydraulic and Energy Grade Lines

The terms in equation (4.2.13) have dimensions of length and units of feet or meters. The concept of the *hydraulic grade line (HGL)* and the *energy grade line (EGL)* can be used to give a physical relationship to these terms. The HGL is essentially the line p/γ above the center line of a pipe, which is the distance water would rise in a piezometer tube attached to the pipe. The EGL is a distance of $\alpha V^2/2g$ above the HGL. Figure 4.2.3 illustrates the HGL and EGL.

4.3   HEADLOSSES

The energy equation (4.2.13) for pipe flow was derived in the previous section. This equation has a headloss term h_L that was defined as the loss of mechanical energy due to viscous stress. The objective of this section is to discuss the headlosses that occur in pressurized pipe flow.

4.3.1   Shear-Stress Distribution of Flow in Pipes

Consider steady flow (laminar or turbulent) in a pipe (a cylindrical element of fluid) of uniform cross-section as shown in Figure 4.3.1. For a uniform flow the general form of the integral momentum equation in the x-direction is expressed by

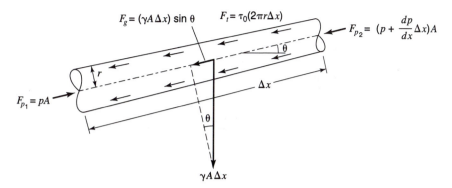

Figure 4.3.1 Cylindrical element of water.

$$\sum F_x = \frac{d}{dt}\int_{CV} v_x \rho\, d\forall + \sum_{CS} v_x(\rho \mathbf{V}\cdot\mathbf{A}) \qquad (3.4.6)$$

where $\left(\dfrac{d}{dt}\right)\displaystyle\int_{CV} v_x\rho\, d\forall = 0$ because the flow is steady and $\displaystyle\sum_{CS} v_x(\rho\mathbf{V}\cdot\mathbf{A}) = 0$ because there is no

flow of extensive property through the control surface so that

$$\sum F_x = 0 \qquad (4.3.1)$$

The forces are: (1) the *pressure forces*, $F_{p_1} = pA$ and $F_{p_2} = \left(p + dp/dx\,\Delta x\right)A$; (2) the *gravity force* due to the weight of the water, $F_g = (\gamma A\Delta x)\sin\theta$; and (3) the *shearing force* $F_\tau = \tau(2\pi r\Delta x)$ where τ is the shearstress. The sum of forces is

$$F_{p_1} - F_{p_2} - F_g - F_\tau = 0 \qquad (4.3.2)$$

or

$$pA - \left(p + \frac{dp}{dx}\Delta x\right)A - (\gamma A\Delta x)\sin\theta - t(2\pi r\Delta x) = 0 \qquad (4.3.3)$$

Equation (4.3.3) reduces to

$$-\frac{dp}{dx}\Delta x A - (\gamma A\Delta x)\sin\theta - \tau(2\pi r\Delta x) = 0 \qquad (4.3.4)$$

Solving for the shear stress by using $dz = \sin\theta\, dx$ gives

$$\tau = \frac{r}{2}\left[-\frac{d}{dx}(p + \gamma z)\right] \qquad (4.3.5)$$

Equation (4.3.5) indicates that τ is zero at the center of the pipe where $r = 0$ and increases linearly to a maximum at the pipe wall (see Figure 4.3.2). Keep in mind that $p + \gamma z$ is constant across the section because the streamlines are straight and parallel in a uniform flow, so that there is no acceleration of fluid normal to the streamline. In other words, hydrostatic conditions prevail across the flow section, resulting in $p + \gamma z$ being constant across the flow section. The gradient $d(p + \gamma z)/dx$ is therefore negative and constant across the flow section for uniform flow.

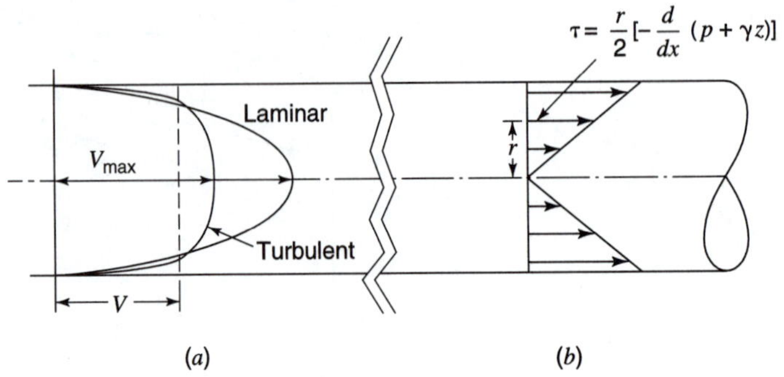

Figure 4.3.2 Distribution of velocity and shear stress for pipe flow. (*a*) Velocity distribution; (*b*) shear stress.

4.3.2   Velocity Distribution of Flow in Pipes

Laminar Flow

In Chapter 2 the following shear stress equation (2.2.1) was introduced:

$$\tau = \mu \frac{dv}{dy} \tag{2.2.1}$$

where μ is the dynamic viscosity and dv/dy is the velocity gradient. For our purposes here, the gradient is actually $dv/dy = -dv/dr$, so that using equation (4.3.5) we get

$$\tau = \mu \frac{dv}{dr} = \frac{r}{2}\left[-\frac{d}{d_x}(p + \gamma z)\right] \tag{4.3.6}$$

Integrating this equation by the separation of variables and boundary condition ($r = r_0$, $v = 0$) yields

$$v = \frac{r_0^2 - r^2}{4\mu}\left[-\frac{d}{dx}(p + \gamma z)\right] \tag{4.3.7}$$

This equation indicates that the velocity distribution for laminar pipe flow is parabolic across the section and has the maximum velocity at the pipe center (refer to Figure 4.3.2). By integrating the velocity across the section using $Q = \int v\, dA$ and using the energy equation, the headloss for laminar flow is

$$h_{L_f} = \frac{32\mu\, LV}{\gamma D^2} \tag{4.3.8}$$

where D is the diameter of the pipe.

Turbulent Flow—Smooth Pipes

For a smooth pipe, the following velocity distribution equations are based upon experiment (Roberson and Crowe, 1990):

$$\frac{u}{u_*} = \frac{u_* y}{v} \quad \text{for } 0 < \frac{u_* y}{v} < 5 \tag{4.3.9}$$

and

$$\frac{u}{u_*} = 5.75\log\frac{u_* y}{v} + 5.5 \quad \text{for } 20 < \frac{u_* y}{v} < 10^5 \tag{4.3.10}$$

where u is the velocity, y is the distance from the pipe wall, v is the kinematic viscosity, u_* is the shear velocity $(\sqrt{\tau_0/\rho})$, and τ_0 is the shear stress at the pipe wall.

The velocity distribution for turbulent flow can also be approximated using power law formulas of the form $u/u_{max} = (y/r_0)^m$, where u_{max} is the velocity at the center of the pipe, r_0 is the pipe radius, and m is an exponent that increases with Reynolds number (see Roberson and Crowe, 1990, for values of m).

Turbulent Flow—Rough Pipes

Experimental results on flow in rough pipes indicate the following form of equation for velocity distribution of turbulent flow in rough pipes:

$$\frac{u}{u_*} = 5.75\log\frac{y}{k} + B \tag{4.3.11}$$

where y is the distance from the wall, k is a measure of the height of the roughness elements, and B is a function of the roughness characteristics. Nikuradse (1933) determined the value of $B = 8.5$, y was measured from the geometric mean of the wall surface, and $k_s = k$ was the sand grain size, so that

$$\frac{u}{u_*} = 5.75\log\frac{y}{k_s} + 8.5 \tag{4.3.12}$$

Nikuradse's work revealed that (1) for low Reynolds numbers and small sand grains the flow resistance is basically the same as for smooth pipes (roughness elements became submerged in the viscous sublayer) and (2) for high Reynolds numbers the resistance coefficient is a function of only the relative roughness k_s/D, where D is the diameter (the viscous sublayer is very thin so that the roughness elements project into the flow, causing flow resistance from drag of the individual roughness elements). Figure 4.3.3 presents values of relative roughness for different kinds of pipe as a function of pipe diameter.

EXAMPLE 4.3.1

For the system shown in Figure 4.3.4, determine the headloss per unit length of pipe and the discharge, assuming laminar flow. ($\mu = 1.002 \times 10^{-3}$ N-s/m^2).

SOLUTION

The headloss equation (4.3.8) for laminar pipe flow is used to determine the velocity of flow in the pipe, where $h_L = 10$ m, $D = 5$ mm, $\gamma = 9.79$ kN/m^3, $\mu = 1.002 \times 10^{-3}$ N-s/m^2:

$$V = \frac{\gamma h_L D^2}{32\mu L} = \frac{(9.79 \text{ kN/m}^3 \times 1000)(10\text{m})(5/1000\text{m})^2}{32(1.002 \times 10^{-3} \text{ N} - \text{s/m}^2)(600\text{m})} = 0.125 \text{m/s}$$

Check the Reynolds number:

$$R_e = \frac{VD}{v} = \frac{VD\rho}{\mu} = \frac{0.125 \times (5/1000) \times 1000}{1.002 \times 10^{-3}} = 6.24 \times 10^2 = 624$$

Flow is laminar. The flowrate is then

$$Q = AV = \left[\frac{\pi(5/1000)^2}{4}\right] \times 0.125 \text{ m/s} = 2.45 \times 10^{-6} \text{m}^3/\text{s}$$

$$= 2.45 \times 10^{-6}(1000)(60)$$

$$= 0.147 \text{ L/min}$$

The headloss per unit length of pipe is (10 m/600 m) = 0.0167 m/m.

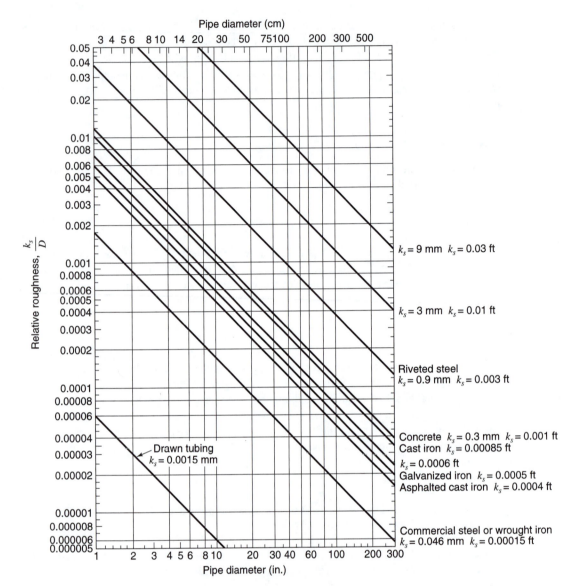

Figure 4.3.3 Relative roughness for various kinds of pipe (from Moody (1944)).

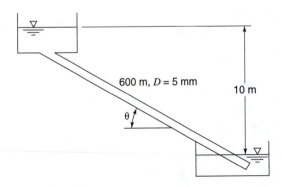

Figure 4.3.4 Pipe system for example 4.3.1.

4.3.3   Headlosses from Pipe Friction

Various equations have been proposed to determine the headlosses due to friction, including the Darcy–Weisbach, Chezy, Manning, Hazen–Williams, and Scobey formulas. These equations relate the friction losses to physical characteristics of the pipe and various flow parameters. The Darcy–Weisbach equation is scientifically based and applies to both laminar and turbulent flows. The *Darcy–Weisbach equation* is

$$h_{L_f} = f\frac{L}{D}\frac{V^2}{2g} \tag{4.3.13}$$

where h_{L_f} is the headloss due to pipe friction, f is the dimensionless friction factor, L is the length of the conduit, D is the inside diameter of the pipe, V is the mean flow velocity, and g is the acceleration due to gravity.

The friction factor is a function of the Reynolds number (R_e) and the relative roughness k_s/D, where k_s is the average nonuniform roughness of the pipe. For laminar flow ($R_e < 2000$) the friction factor is

$$f = \frac{64}{R_e} \tag{4.3.14}$$

where

$$R_e = \frac{VD}{\nu} \tag{4.3.15}$$

and ν is the kinematic viscosity. For turbulent flow in

$$\textit{Smooth pipe} \quad \frac{1}{\sqrt{f}} = 2\log_{10}\left(R_e\sqrt{f}\right) - 0.8 \text{ for } R_e > 3000 \tag{4.3.16}$$

$$\textit{Rough pipe} \quad \frac{1}{\sqrt{f}} = 2\log_{10}\frac{D}{(k_s)} + 1.14 = 1.14 - 2\log_{10}\left(\frac{k_s}{D}\right) \tag{4.3.17}$$

where r is the pipe radius. Equations (4.3.16) and (4.3.17) were proposed by von Karman and Prandtl based upon experiments by Nikuradse (1932).

Colebrook and White (1939) proposed the following semi-empirical formula:

$$\frac{1}{\sqrt{f}} = -2.0\log_{10}\left(\frac{k_s/D}{3.7} + \frac{2.51}{R_e\sqrt{f}}\right) \tag{4.3.18}$$

The above equation is asymptotic to both the smooth and rough pipe equations (4.3.16) and (4.3.17) and is valid for the entire nonlaminar range of the Moody diagram.

Moody (1944) developed the *Moody diagram*, shown in Figure 4.3.5 using experimental data on commercial pipes, the Colebrook–White equation, and the Prandtl–Karman experimental data. Knowing the pipe roughness (relative roughness) and the Reynolds number, the friction factor can be obtained from the Moody diagram. The use of Manning's equation and the Hazen–Williams equation is discussed in Chapter 12.

EXAMPLE 4.3.2

Water flows in a 1000-m long pipeline of diameter 200 mm at a velocity of 5 m/s. The pipeline is new cast iron pipe with $k_s = 0.00026$ m. Determine the headloss in the pipeline. Use $\nu = 1.007 \times 10^{-6}$ m²/s.

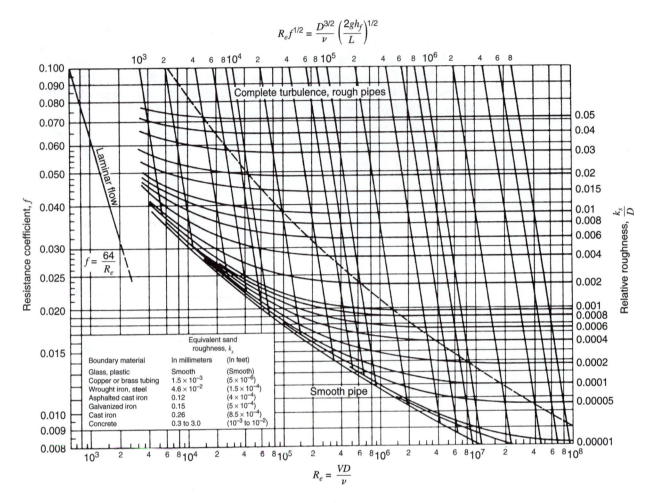

$$R_e f^{1/2} = \frac{D^{3/2}}{\nu}\left(\frac{2gh_f}{L}\right)^{1/2}$$

Figure 4.3.5   Resistance coefficient f versus R_e (from Moody (1994)).

SOLUTION

The headloss is computed using the Darcy–Weisbach equation (4.3.13):

$$h_{L_f} = f\frac{L}{D}\frac{V^2}{2g}$$

To determine the friction factor from Moody diagram the Reynolds number must be computed:

$$R_e = \frac{VD}{\nu} = \frac{5\times(200/1000)}{1.007\times10^{-6}} = 9.93\times10^5$$

The relative roughness is $\dfrac{k_s}{D} = \dfrac{0.00026}{0.200} = 0.0013$. From Figure 4.3.5, $f = 0.021$, so

$$h_{L_f} = 0.021\frac{1000}{(200/1000)}\frac{(5)^2}{2(9.81)} = 133.8\text{m}.$$

EXAMPLE 4.3.3

A 10-cm diameter 2000 m long pipeline connects two reservoirs open to the atmosphere. What is the discharge in the pipeline if the water surface elevation difference of the reservoirs is 50 m? Assume a smooth pipe and $\nu = 1.02 \times 10^{-6}\,\text{m}^2/\text{s}$.

SOLUTION

The energy equation between the reservoir surfaces is used to determine the headloss due to friction, which is then used to determine the velocity of flow in the pipeline:

$$\frac{P_1}{\gamma} + \frac{V_1^2}{2g} + z_1 = \frac{P_2}{\gamma} + \frac{V_2^2}{2g} + z_2 + h_{L_f}$$

$$0 + 0 + z_1 = 0 + 0 + z_2 + h_{L_f}$$

$$h_{L_f} = z_1 - z_2 = 50 \text{ m}$$

$$h_{L_f} = f \frac{L}{D} \frac{V^2}{2g}$$

$$f\left(\frac{2000}{10/100}\right) \frac{V^2}{2 \times 9.81} = 50$$

$$f V^2 = 0.0491 \text{ (or } V = 0.2215/\sqrt{f} \text{)}$$

This must be solved for V using a trial-and-error procedure by assuming f, computing V, then computing the Reynolds number. Assume $f = 0.02$, $V = 0.2215/\sqrt{0.02} = 1.566 \text{ m/s}$:

$$R_e = \frac{VD}{\nu} = \frac{1.566(10/100)}{1.02 \times 10^{-6}} = 1.54 \times 10^5$$

From Figure 4.3.5, $f = 0.016$, $V = 0.2215/\sqrt{0.016} = 1.751 \text{ m/s}$, so

$$R_e = \frac{VD}{\nu} = \frac{1.751(10/100)}{1.02 \times 10^{-6}} = 1.72 \times 10^5$$

From Figure 4.3.5, $f = 0.016$, which is close enough. The discharge is

$$Q = AV = \frac{\pi(10/100)^2}{4}(1.751) = 0.0137 \text{m}^3\text{/s}.$$

EXAMPLE 4.3.4

Compute the friction factor for flow having a Reynolds number of 1.37×10^4 and relative roughness $k_s/D = 0.000375$ using the Colebrook–White formula.

SOLUTION

Use equation (4.3.18) $\frac{1}{\sqrt{f}} = -2.0 \log_{10}\left(\frac{0.000375}{3.7} + \frac{2.51}{1.37 \times 10^4 \sqrt{f}}\right)$ and solve using trial and error

$f = 0.0291$. Referring to the Moody diagram (Figure 4.3.5), we would read approximately 0.028 or 0.029.

4.3.4 Form (Minor) Losses

Headlosses are also caused by inlets, outlets, bends, and other appurtenances such as fittings, valves, expansions, and contractions. These losses, referred to as *minor losses, form losses,* or *secondary losses*, are caused by flow separation and the generation of turbulence. Headlosses produced, in general, can be expressed by

$$h_{L_m} = K \frac{V^2}{2g} \tag{4.3.19}$$

where K is the loss coefficient (see Table 4.3.1) and V is the mean velocity. Table 4.3.2 lists loss coefficients for various transitions and fittings. Table 4.3.3 lists loss coefficients for common hydraulic valves.

Table 4.3.1 Minor Loss Coefficients for Pipe Flow

Type of Minor Loss	K Loss in Terms of $V^2/2g$
Pipe fittings:	
90° elbow, regular	0.21–0.30
90° elbow, long radius	0.14–0.23
45° elbow, regular	0.2
Return bend, regular	0.4
Return bend, long radius	0.3
AWWA tee, flow through side outlet	0.5–1.80
AWWA tee, flow through run	0.1–0.6
AWWA tee, flow split side inlet to run	0.5–1.8
Valves:	
Butterfly valve ($\theta = 90°$ for closed valve)*	
$\theta = 0°$	0.3–1.3
$\theta = 10°$	0.46–0.52
$\theta = 20°$	1.38–1.54
$\theta = 30°$	3.6–3.9
$\theta = 40°$	10–11
$\theta = 50°$	31–33
$\theta = 60°$	90–120
Check valves (swing check) fully open	0.6–2.5
Gate valves (4 to 12 in) fully open	0.07–0.14
1/4 closed	0.47–0.55
1/2 closed	2.2–2.6
3/4 closed	12–16
Sluice gates:	
As submerged port in 12 in wall	0.8
As contraction in conduit	0.5
Width equal to conduit width and without top submergence	0.2
Entrance and exit losses:	
Entrance, bellmouthed	0.04
Entrance, slightly taunted	0.23
Entrance, square edged	0.5
Entrance, projecting	1.0
Exit, bellmouthed	$0.1\left(\dfrac{V_1^2}{2g} - \dfrac{V_2^2}{2g}\right)$
Exit, submerged pipe to still water	1.0

*Loss coefficients for partially open conditions may vary widely. Individual manufacturers should be consulted for specific conditions.

Source: Adapted from Velon and Johnson (1993).

Table 4.3.2 Loss Coefficients for Various Transitions and Fittings

Description	Sketch	Additional Data		K	Source
Pipe entrance $h_{l_m} = K_e \, V^2/2g$		r/d 0.0 0.1 >0.2		K_e 0.50 0.12 0.03	(a)
Contraction $h_{l_m} = K_C \, V_2^2/2g$		D_2/D_1 0.0 0.20 0.40 0.60 0.80 0.90	K_C $\theta = 60°$ 0.08 0.08 0.07 0.06 0.05 0.04	K_C $\theta = 180°$ 0.50 0.49 0.42 0.32 0.18 0.10	(a)
Expansion $h_{l_m} = K_E \, V_1^2/2g$		D_1/D_2 0.0 0.20 0.40 0.60 0.80	K_E $\theta = 10°$ 0.13 0.11 0.06 0.03	K_E $\theta = 180°$ 1.00 0.92 0.72 0.42 0.16	(a)
90° miter bend	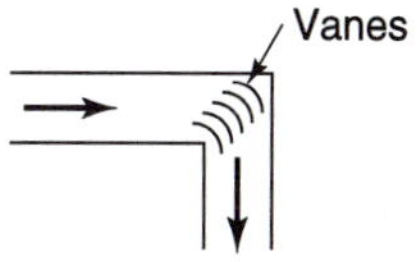	Without vanes	$K_b = 1.1$		(b)
		With vanes	$K_b = 0.2$		(b)
Smooth bend		r/d 1 2 4 6	K_b $\theta = 45°$ 0.10 0.09 0.10 0.12	K_b $\theta = 90°$ 0.35 0.19 0.16 0.21	(c) and (d)
Threaded pipe fittings	Globe valve—wide open Angle valve—wide open Gate valve—wide open Gate valve—half open Return bend Tee 90° elbow 45° elbow		$K_v = 10.0$ $K_v = 5.0$ $K_v = 0.2$ $K_v = 5.6$ $K_b = 2.2$ $K_t = 1.8$ $K_b = 0.9$ $K_b = 0.4$		(b)

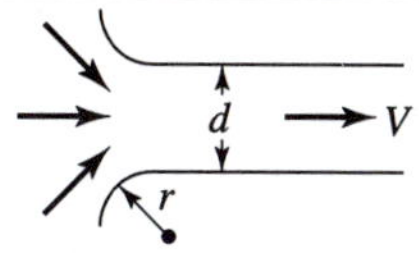

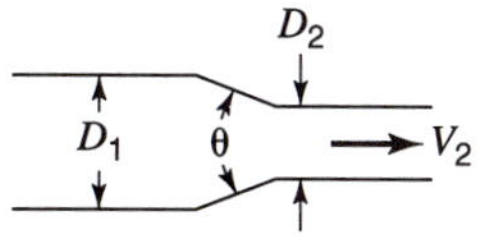

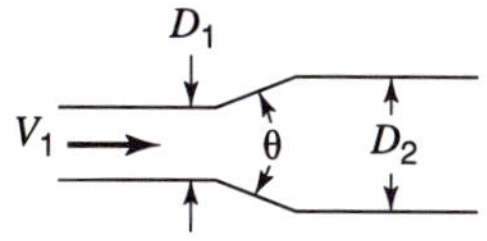

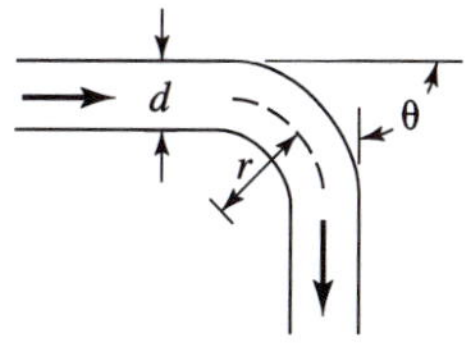

(a) ASHRAE (1977)
(b) Streeter (1961)
(c) Bei (1938)
(d) Idel'chik (1966)

Source: after Roberson et al. (1988).

Table 4.3.3 Values of K_v, for Certain Common Hydraulic Valves

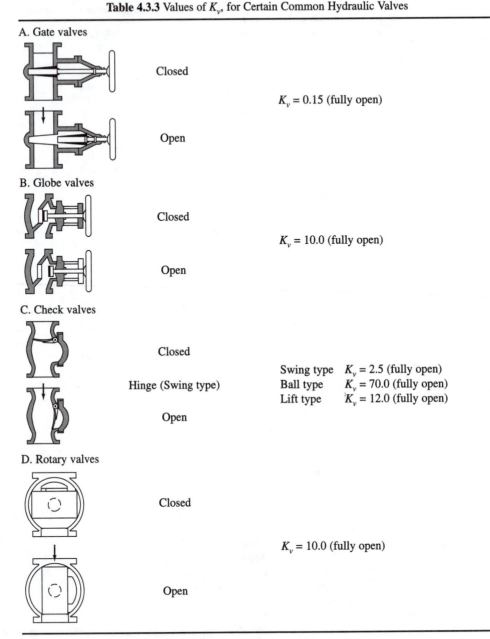

Source: from Hwang and Houghtalen (1996).

For sudden expansions or enlargements (see Figure 4.3.6), the headlosses can be expressed as

$$h_{L_m} = \frac{(V_1 - V_2)^2}{2g} \tag{4.3.20}$$

For *gradual enlargements* (see Figure 4.3.7), such as *conical diffusers*, the headloss is expressed as

$$h_{L_m} = K' \frac{(V_1 - V_2)^2}{2g} \tag{4.3.21}$$

The values of K' as a function of cone angle are given in Figure 4.3.8.

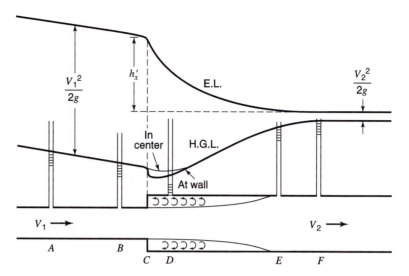

Figure 4.3.6 Loss due to sudden enlargement (from Daugherty and Franzini (1997)).

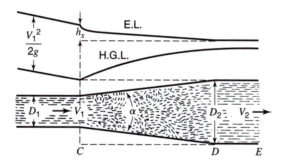

Figure 4.3.7 Loss due to gradual enlargement (from Daugherty and Franzini (1977)).

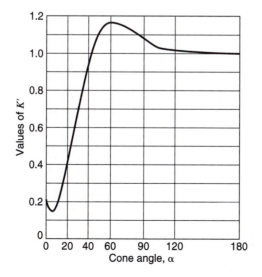

Figure 4.3.8 Loss coefficient for conical diffusers (from Daugherty and Franzini (1977)).

The loss of head due to a *sudden contraction* (see Figure 4.3.9) may be expressed as

$$h_{L_m} = K_c \frac{V_2^2}{2g} \tag{4.3.22}$$

where the values of K_c are a function of the diameter ratios D_2/D_1.

Entrance losses (see Figure 4.3.10) are computed using

$$h_{L_m} = K_e \frac{V^2}{2g} \tag{4.3.23}$$

where values of K_e are found in Table 4.3.1.

Exit (or discharge) losses (see Figure 4.3.11 and Table 4.3.1) from the end of a pipe into a reservoir that has a negligible velocity are expressed as

$$h_{L_m} = \frac{V^2}{2g} \tag{4.3.24}$$

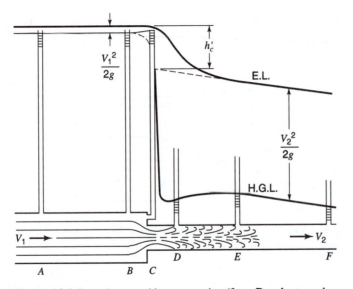

D_2/D_1	K_c
0.0	0.50
0.1	0.45
0.2	0.42
0.3	0.39
0.4	0.36
0.5	0.33
0.6	0.28
0.7	0.22
0.8	0.15
0.9	0.06

Figure 4.3.9 Loss due to sudden contraction (from Daugherty and Franzini (1977)).

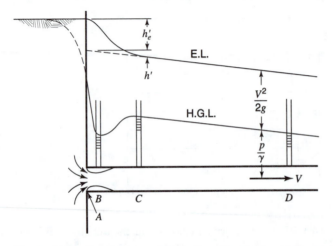

Figure 4.3.10 Conditions at entrance (from Daugherty and Franzini (1977)).

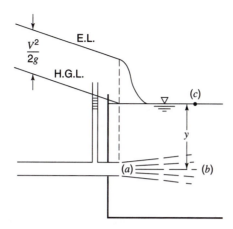

Figure 4.3.11   Discharge loss (from Daugherty and Franzini (1977)).

EXAMPLE 4.3.5   Water flows from reservoir 1 to reservoir 2 through a 12-in diameter, 600-ft long pipe line as shown in Figure 4.3.12. The reservoir 1 surface elevation is 1000 ft and reservoir 2 surface elevation is 950 ft. Consider the minor losses due to the sharp-edged entrance, the globe valve, the two bends (90° elbow), and the sharp-edged exit. The pipe is galvanized iron with $k_s = 0.0005$ ft. Determine the discharge from reservoir 1 to reservoir 2. For 120°F, $\nu = 0.609 \times 10^{-5}$ ft²/s.

SOLUTION   The energy equation between 1 and 2 is

$$\frac{P_1}{\gamma} + \frac{V_1^2}{2g} + z_1 = \frac{P_2}{\gamma} + \frac{V_2^2}{2g} + z_2 + h_{L_f} + \sum h_{L_m}$$

For the reservoir surface $p_1/\gamma = 0$ and $V^2/2g = 0$, so

$$0 + 0 + 1000 = 0 + 0 + 950 + h_{L_f} + \sum h_{L_m}$$

$$50 = h_{L_f} + \sum h_{L_m}$$

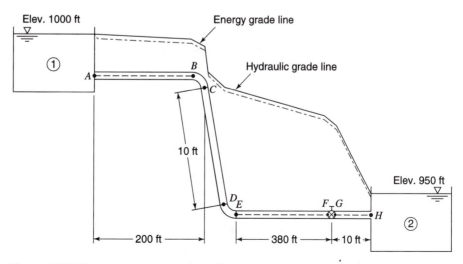

Figure 4.3.12 Pipe system for example 4.3.5.

where

$$h_{L_f} = \frac{fLV^2}{D2g} = f\frac{600}{(12/12)}\frac{V^2}{2 \times 32.2} = 9.32fV^2$$

The minor losses are

$$\sum h_{L_m} = h_{L_{entrance}} + h_{L_{elbow}} + h_{L_{elbow}} + h_{L_{globevalve}} + h_{L_{exit}}$$

$$= K_e\frac{V^2}{2g} + 2K_{elbow}\frac{V^2}{2g} + K_{valve}\frac{V^2}{2g} + \frac{V^2}{2g}$$

$$= \left(K_e + 2K_{elbow} + K_{valve} + 1\right)\frac{V^2}{2g}$$

where $K_e = 0.5$, $K_{elbow} = 0.25$, and $K_{valve} = 1.5$. The energy equation is now expressed as

$$50 = (0.5 + 2 \times 0.25 + 1.5 + 1.0)\frac{V^2}{2g} + 9.32fV^2$$

$$50 = 3.5\frac{V^2}{2g} + 9.32fV^2$$

$$50 = (0.054 + 9.32f)V^2$$

$$V = \sqrt{50/(0.054 + 9.32f)}$$

Assuming fully turbulent flow and using $k_s/D = 0.0005/1 = 0.0005$, we get $f = 0.0165$ from Figure 4.3.5, then

$$V = \sqrt{50/(0.054 + 9.32 \times 0.0165)}$$

$$V = 15.51 \text{ ft/s}$$

Compute $R_e = \dfrac{VD}{\nu} = \dfrac{15.51 \times 1}{0.609 \times 10^{-5}} = 2.55 \times 10^6$. Referring to Figure 4.3.5 (Moody diagram), we see that the value of f is OK. Now use the continuity equation to determine Q:

$$Q = AV = \left[\pi(12/12)^2/4\right](15.51 \text{ ft/s}) = 12.18 \text{ ft}^3/\text{s}$$

4.4 FORCES IN PIPE FLOW

Changes in direction or magnitude of flow velocity of a fluid causes changes in the momentum of the fluid (see Figure 4.4.1). The forces that are required to produce the change in momentum come from the pressure variation within the fluid and from forces transmitted to the fluid from the pipe walls. Applying the momentum principle (equation 3.4.6) to the control volumes in Figure 4.4.1, we get

$$\sum F_x = \frac{d}{dt}\int_{CV} v_x\rho \, d\forall + \sum_{CS} v_x\rho\mathbf{V}\cdot\mathbf{A} \tag{3.4.6}$$

where

$$\frac{d}{dt}\int_{CV} v_x\rho \, d\forall = 0$$

because flow is steady and

$$\sum_{CS} v_x\rho\mathbf{V}\cdot\mathbf{A} = \rho V_{x_2}Q - \rho V_{x_1}Q \tag{4.4.1}$$

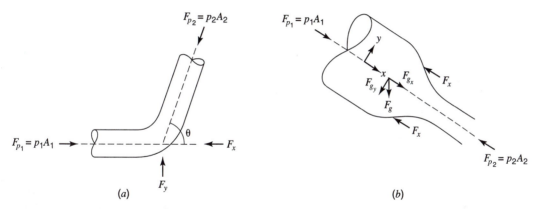

Figure 4.4.1 Forces in pipe flow. (*a*) Pipe bend; (*b*) Contraction.

The momentum entering from the upstream is $-\rho V_{x_1} Q$ and the momentum leaving the control volume is $\rho V_{x_2} Q$. The summation of forces in the x direction for the contraction in Figure 4.4.1b is

$$\sum F_x = p_1 A_1 - p_2 A_2 - F_x + F_{g_x} \tag{4.4.2}$$

where F_{g_x} is the weight of the fluid in the x-direction in the contraction. The momentum principle for the x-direction can be stated using equation (3.4.6) as

$$\sum F_x = \rho V_{x_2} Q - \rho V_{x_1} Q = \rho Q \left(V_{x_2} - V_{x_1} \right) \tag{4.4.3}$$

Similar equations can be developed for other directions:

$$\sum F_y = \rho Q \left(V_{y_2} - V_{y_1} \right) \tag{4.4.4}$$

EXAMPLE 4.4.1

The nozzle in example 3.2.3 has the dimensions shown in Figure 4.4.2. What is the force exerted on the flange bolts for a flow rate of 0.08 m³/s? The upstream pipe pressure is 75 kPa.

SOLUTION

The velocities were computed as $V_1 = 1.13$ m/s and $V_2 = 2.55$ m/s in the solution to example 3.2.3. Using equation (4.4.3) yields

$$\sum F_x = \rho Q(V_{x_2} - V_{x_1})$$

$$F_1 - F_2 - F_{\text{bolt}} = \rho \, Q(V_{x_2} - V_{x_1})$$

where $F_1 = p_1 A_1$ and $F_2 = p_2 A_2 = 0$ (because this end of the nozzle is open to the atmosphere and $p_2 = 0$):

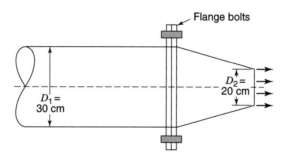

Figure 4.4.2 Nozzle for example 4.4.1.

$$p_1 A_1 - F_{\text{bolt}} = \rho Q (V_{x_2} - V_{x_1})$$

$$(75,000 \text{ N/m}^2) \left(\frac{\pi (30/100)^2 \text{ m}^2}{4} \right) - F_{\text{bolt}} = \left(1000 \text{ kg/m}^3 \right) \left(0.08 \text{ m}^3/\text{s} \right) \left(2.55 \text{ m/s} - 1.13 \text{ m/s} \right)$$

where 1 Pa = 1 N/m^2 and 1 kg/m^3 = 1 N-s^2/m^4, so

$5301.4 \text{ N} - F_{\text{bolt}} = 113.6 \text{ N}$

$F_{\text{bolt}} = 5415 \text{ N (or 1217 lb)}$

EXAMPLE 4.4.2

Water flows through a horizontal 45° reducing bend shown in Figure 4.4.3, with a 36-in diameter upstream and a 24-in diameter downstream, at the rate of 20 cfs under a pressure of 15 psi at the upstream end of the bend. Neglecting the head loss in the bend, calculate the force exerted by the water on the bend.

SOLUTION

The free-body diagram shown in Figure 4.4.3 is used. To solve this problem, equations (4.4.3) and (4.4.4) will be used:

$$\sum F_x = \rho Q (V_{x_2} - V_{x_1}) \tag{4.4.3}$$

$$\sum F_y = \rho Q (V_{y_2} - V_{y_1}) \tag{4.4.4}$$

The first objective is to determine the velocities in order to apply the energy equation to determine p_2:

$V_{x_1} = V_1 = Q/A_1 = 20/\pi(3)^2/4 = 2.83 \text{ ft/s}$

$V_{y_1} = 0$

$V_2 = Q/A = 20/\pi(2)^2/4 = 6.37 \text{ ft/s}$

$V_{x_2} = V_{y_2} = V_2 \cos 45° = 6.37(0.707) + 4.50 \text{ ft/s}$

Next the energy equation is applied horizontally, $z_1 = z_2 = 0$ and $h_L = 0$:

$$\frac{p_1}{\gamma} + \frac{V_1^2}{2g} = \frac{p_2}{\gamma} + \frac{V_2^2}{2g}$$

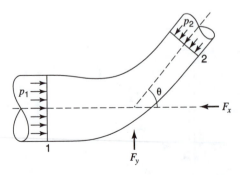

Figure 4.4.3 Reducing bend for example 4.4.2.

$$\frac{15 \text{ lb/in}^2 \times 144 \text{ in}^2/\text{ft}^2}{62.4 \text{ lb/ft}^3} + \frac{(2.83 \text{ ft/s})^2}{2(32.2) \text{ ft/s}^2} = \frac{p_2}{\gamma} + \frac{(6.37 \text{ ft/s})^2}{2(32.2) \text{ ft/s}^2}$$

$$34.615 + 0.124 = \frac{p_2}{\gamma} + 0.630$$

$$\frac{p_2}{\gamma} = 34.11 \text{ ft}$$

$$p_2 = 34.11 \times \frac{62.4}{144} = 14.78 \text{ lb/in}^2$$

Using equation (4.4.3) yields

$$\sum F_x = p_1 A_1 - p_2 A_2 \cos 45^\circ - F_x = \rho Q(V_{x_2} - V_{x_1})$$

$$F_x = p_1 A_1 - p_2 A_2 \cos 45^\circ - \rho Q(V_{x_2} - V_{x_1})$$

$$F_x = (15 \text{ lb/in}^2)(144 \text{ in}^2/\text{ft}^2)\left(\pi \frac{3^2}{4}\text{ft}^2\right) - (14.78 \text{ lb/in}^2)(144 \text{ in}^2/\text{ft}^2)\left(\pi \frac{2^2}{4}\text{ft}^2\right)$$

$$-\left[1.94 \frac{\text{slugs}}{\text{ft}^3}\right](20 \text{ ft}^3)(4.50 \text{ ft/s} - 2.83 \text{ ft/s})$$

$$F_x = 15{,}268 - 6686 - 65$$

$$= 8517 \text{ lb}$$

Using equation (4.4.4), we get

$$\sum F_y = F_{y_1} - F_{y_2} + F_y = \rho Q(V_{y_2} - V_{y_1})$$

where $F_{y_1} = 0$ because there is no pressure component in the y direction at 1.

$$-p_2 A_2 \sin 45^\circ + F_y = \rho Q(V_{y_2} - V_{y_1})$$

$$F_y = p_2 A_2 \sin 45^\circ + \rho Q(V_{y_2} - V_{y_1})$$

$$F_y = (14.78)(144)(\pi \frac{2^2}{4})(0.707) + (1.94)(20)(4.50 - 0)$$

$$F_y = 4727 + 174$$

$$F_y = 4901 \text{ lb}$$

The resultant force is $F = \sqrt{(8517)^2 + (4901)^2} = 9826$ lb at an angle of $\theta = \tan^{-1}(4901/8517) = 30^\circ$.

4.5 PIPE FLOW IN SIMPLE NETWORKS

4.5.1 Series Pipe Systems

Consider the simple series pipe system in Figure 4.5.1a. Through continuity the discharge is equal in each pipe:

$$Q = Q_1 = Q_2 = Q_3 \tag{4.5.1}$$

and through energy the total headloss is the sum of headlosses in each pipe:

$$h_L = h_{L_1} + h_{L_2} + h_{L_3} \tag{4.5.2}$$

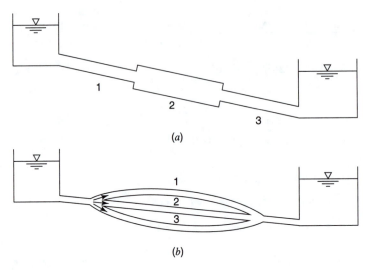

Figure 4.5.1 Pipes in (a) series and (b) parallel.

EXAMPLE 4.5.1

Water flows at a rate of 0.030 m^3/s from reservoir 1 to reservoir 2 through three pipes connected in series ($f = 0.025$) as shown in Figure 4.5.2. Neglecting minor losses, determine the difference in water surface elevation.

SOLUTION

Write the energy equation from 1 to 2:

$$z_1 = z_2 + h_{L_f} \qquad (\text{or } h_{L_f} = z_1 - z_2)$$

$$h_{L_f} = h_{L_A} + h_{L_B} + h_{L_C} = z_1 - z_2$$

Using the Darcy-Weisbach equation (4.3.13)

$$h_{L_f} = f \frac{L}{D} \frac{V^2}{2g}$$

the energy equation is

$$z_1 - z_2 = 0.025 \frac{1000}{[20/1000]} \frac{V_A^2}{2(9.81)} + 0.025 \frac{1500}{[180/1000]} \frac{V_B^2}{2(9.81)} + 0.025 \frac{2000}{[220/1000]} \frac{V_C^2}{2(9.81)}$$

$$z_1 - z_2 = 6.37 V_A^2 + 10.62 V_B^2 + 11.58 V_C^2$$

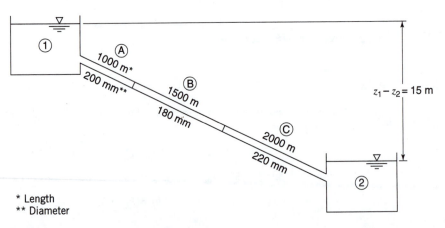

* Length
** Diameter

Figure 4.5.2 Pipe system for example 4.5.1.

Use continuity to determine the velocities:

$$V_A = Q/A_A = 0.03 \bigg/ \frac{\pi(20/1000)^2}{4} = 0.955 \text{ m/s}$$

$$V_B = Q/A_B = 0.03 \bigg/ \frac{\pi(20/1000)^2}{4} = 1.180 \text{ m/s}$$

$$V_C = Q/A_C = 0.03 \bigg/ \frac{\pi(20/1000)^2}{4} = 0.790 \text{ m/s}$$

$$z_1 - z_2 = 6.37\,(0.955)^2 + 10.62\,(1.180)^2 + 11.58\,(0.790)^2$$

$$= 5.81 + 14.79 + 7.22$$

$$= 27.82 \text{ m}$$

4.5.2   Parallel Pipe Systems

Consider the simple parallel pipe system in Figure 4.5.1b. Through continuity the total flow is the sum of flow in each of the pipes:

$$Q = Q_1 + Q_2 + Q_3 \tag{4.5.3}$$

Through energy the flow distribution in the parallel pipes is such that the headloss in each pipe is equal:

$$h_L = h_{L_1} = h_{L_2} = h_{L_3} \tag{4.5.4}$$

EXAMPLE 4.5.2

The three-pipe system shown in Figure 4.5.3 has the following characteristics:

Pipe	D (in)	L (ft)	f
A	8	1500	0.020
B	6	2000	0.025
C	10	3000	0.030

Find the flowrate of water in each pipe and the pressure at point 3. Neglect minor losses.

SOLUTION

Write the energy equation from 1 to 2:

$$\frac{p_1}{\gamma} + \frac{V_1^2}{2g} + z_1 = \frac{p_2}{\gamma} + \frac{V_2^2}{2g} + z_2 + h_{L_f}$$

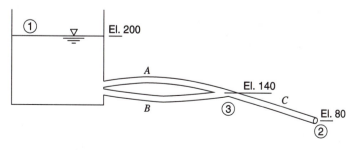

Figure 4.5.3 Pipe system for example 4.5.2.

$$0 + 0 + 200 = 0 + \frac{V_2^2}{2g} + 80 + h_L$$

where V_2 is V_C; $h_{L_f} = h_{L_A} + h_{L_C}$, and using the Darcy-Weisbach equation (4.3.13), we get

$$120 = \frac{V_C^2}{2g} + f_A \frac{L_A}{D_A} \frac{V_A^2}{2g} + f_C \frac{L_C}{D_C} \frac{V_C^2}{2g}$$

Because pipes A and B are parallel, the headloss in A is equal to the headloss in B, so the headloss for B can also be used in the above energy equation instead of for A. This energy equation has two unknowns, V_A and V_C, so that continuity can be used as a second equation:

$$Q_A + Q_B = Q_C$$

$$A_A V_A + A_B V_B = A_C V_C$$

which introduces a third unknown V_A. Because $h_{L_A} = h_{L_B}$, the third equation is

$$f_A \frac{L_A}{D_A} \frac{V_A^2}{2g} = f_B \frac{L_B}{D_B} \frac{V_B^2}{2g}$$

$$0.020 \left(\frac{1500}{8/12} \right) \frac{V_A^2}{2(32.2)} = 0.025 \left(\frac{2000}{6/12} \right) \frac{V_B^2}{2(32.2)}$$

$$0.699 V_A^2 = 1.553 V_B^2$$

$$V_A^2 = 2.221 V_B^2$$

$$V_B = 0.671 V_A$$

So now we have three equations and three unknowns. Using the continuity equation, we get

$$\left[\pi \frac{(8/12)^2}{4} \right] V_A + \left[\pi \frac{(6/12)^2}{4} \right] V_B = \left[\pi \frac{(10/12)^2}{4} \right] V_C$$

$$8^2 V_A + 6^2 V_B = 10^2 V_C$$

substituting

$$V_B = 0.671 V_A$$

$$64 V_A + 36(0.671) V_A = 100 V_C$$

$$88.16 V_A = 100 V_C$$

$$V_A = 1.134 V_C$$

Substitute $V_A = 1.134 V_C$ into the energy equation

$$120 = \frac{V_C^2}{2g} + 0.020 \left(\frac{1500}{8/12} \right) \frac{(1.134 V_C)^2}{2g} + 0.030 \left(\frac{3000}{10/12} \right) \frac{V_C^2}{2g}$$

and solve for V_C:

$$120 = (1 + 57.82 + 108) \frac{V_C^2}{2g} = 166.82 \frac{V_C^2}{2(32.2)}$$

$$V_C = 6.805 \text{ ft/s}$$

The flow rate is then

$$Q = A_C V_C = \left[\pi \frac{(10/12)^2}{4} \right] \times 6.805 = 3.712 \text{ ft}^3/\text{s}$$

The pressure at 3 can be computed using the energy equation from 1 to 3 or from 3 to 2. Using

$$\frac{p_3}{\gamma} + \frac{V_3^2}{2g} + z_3 = \frac{p_2}{\gamma} + \frac{V_2^2}{2g} + z_2 + h_{L_{3-2}}$$

Because $V_3 = V_2$, the velocity head terms cancel out:

$$\frac{p_3}{\gamma} + 140 = 0 + 80 + h_{L_{3-2}}$$

$$\frac{p_3}{\gamma} = -60 + f_C \frac{L_C V_C^2}{D_C 2g}$$

$$= -60 + 0.030 \frac{3000}{10/12} \frac{6.805^2}{2(32.2)}$$

$$= -60 + 77.66$$

$$= 17.66$$

and

$$p_3 = (62.4)(17.66) = 1102 \ \text{lb/ft}^2$$

$$= 7.65 \ \text{lb/in}^2$$

EXAMPLE 4.5.3

The pipe system shown in Figure 4.5.4 connects two reservoirs that have an elevation difference of 20 m. This pipe system consists of 200 m of 50-cm concrete pipe (pipe A), that branches into 400 m of 20-cm pipe (pipe B) and 400 m of 40-cm pipe (pipe C) in parallel. Pipes B and C join into a single 50-cm pipe that is 500 m long (pipe D). For $f = 0.030$ in all the pipes, what is the flow rate in each pipe of the system?

SOLUTION

The objective is to compute the velocity in each pipe. We know that $V_A = V_D$ because they are the same diameter pipe, $h_{L_B} = h_{L_C}$ because pipes B and C are in parallel, and $Q_A = Q_D = Q_B + Q_C$. Express $h_{L_B} = h_{L_C}$ in term of the velocities

$$f_B \frac{L_B}{D_B} \frac{V_B^2}{2g} = f_C \frac{L_C}{D_C} \frac{V_C^2}{2g}$$

Since $f_B = f_C$ and $L_B = L_C$,

$$\frac{V_B^2}{D_B} = \frac{V_C^2}{D_C}$$

$$\frac{V_B^2}{20/100} = \frac{V_C^2}{40/100} \quad \text{or} \quad V_B^2 = \frac{1}{2} V_C^2 \quad \text{or} \quad V_B = \frac{V_C}{\sqrt{2}} \quad \text{or} \quad V_C = \sqrt{2} \ V_B$$

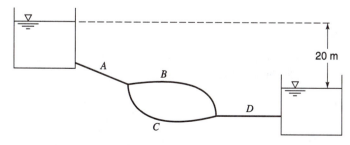

Figure 4.5.4 Pipe system for example 4.5.3.

Using $Q_A = Q_B + Q_C$, we get

$$\left[\frac{\pi(50/100)^2}{4}\right]V_A = \left[\frac{\pi(20/100)^2}{4}\right]V_B + \left[\frac{\pi(40/100)^2}{4}\right]V_C$$

$$50^2 V_A = 20^2 V_B + 40^2 V_C$$

Substituting $V_C = \sqrt{2}\,V_B$ yields

$$2500 V_A = 400 V_B + 1600(\sqrt{2}\,V_B)$$

$$V_A = 1.065 V_B \text{ or } V_B = 0.939\,V_A$$

Next convert the parallel pipes to a single equivalent $D_A = D_D = 50$-cm diameter pipe, with a length of L_E:

$$f\frac{L_B}{D_B}\frac{V_B^2}{2g} = f\frac{L_E}{D_A}\frac{V_A^2}{2g}$$

$$\frac{L_B}{D_B}V_B^2 = \frac{L_E}{D_A}V_A^2$$

$$\frac{400}{(20/100)}V_B^2 = \frac{L_E}{(50/100)}V_A^2$$

$$1000 V_B^2 = L_E V_A^2$$

Substitute $V_B = 0.939 V_A$:

$$1000(0.939 V_A)^2 = L_E V_A^2$$

$$L_E = 882 \text{ m}$$

Write the energy equation from reservoir surface to reservoir surface $\Sigma h = 20$ m

$$20 = \frac{f(L_A + L_E + L_D))}{(50/100)}\frac{V_A^2}{2g}$$

$$20 = \frac{0.030(200 + 882 + 500)}{(50/100)}\frac{V_A^2}{2(9.81)}$$

$$V_A = 2.033 \text{ m/s}, \quad Q_A = \frac{\pi(50/100)^2}{4}(2.033) = 0.399 \text{ m}^3/\text{s}$$

Also, $V_A = V_D$, so $Q_D = Q_A$:

$$V_A = 0.939\ (2.033) = 1.909 \text{ m/s}, \quad Q_B = \frac{\pi(20/100)^2}{4}(1.909) = 0.060 \text{ m}^3/\text{s}$$

$$Q_C = Q_A - Q_B = 0.399 - 0.060 = 0.339 \text{ m}^3/\text{s}.$$

4.5.3 Branching Pipe Flow

Consider the branching pipe system shown in Figure 4.5.5. The following energy equations can be written (neglecting the velocity heads):

$$z_A = z_D + \frac{p_D}{\gamma} + h_{L_{AD}} \tag{4.5.5}$$

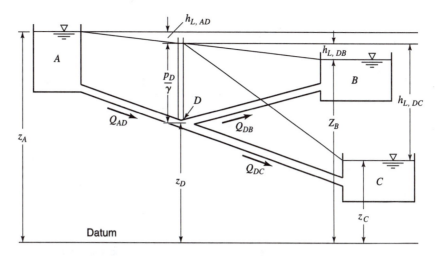

Figure 4.5.5 Branching pipe system.

$$z_B = z_D + \frac{p_D}{\gamma} - h_{L_{DB}} \tag{4.5.6}$$

$$z_C = z_D + \frac{p_D}{\gamma} - h_{L_{DC}} \tag{4.5.7}$$

where the headlosses are defined using the Darcy–Weisbach equation (4.3.13)

$$h_{L_{AD}} = f_{AD} \frac{L_{AD}}{D_{AD}} \frac{V_{AB}^2}{2g} \tag{4.5.8}$$

$$h_{L_{DB}} = f_{DB} \frac{L_{DB}}{D_{DB}} \frac{V_{DB}^2}{2g} \tag{4.5.9}$$

$$h_{L_{DC}} = f_{DC} \frac{L_{DC}}{D_{DC}} \frac{V_{DC}^2}{2g} \tag{4.5.10}$$

The continuity equation is

$$Q_{AD} = Q_{DB} + Q_{DC} \tag{4.5.11}$$

or

$$A_{AD}V_{AD} = A_{DB}V_{DB} + A_{DC}V_{DC} \tag{4.5.12}$$

By substituting the headloss expressions (equations (4.5.8) – (4.5.10)) respectively into equations (4.5.5)–(4.5.7), the three energy equations have four unknowns, p_D/γ, V_{AD}, V_{DB}, and V_{DC}. The continuity equation (4.5.12) provides the fourth equation to solve for the four unknowns.

4.6   MEASUREMENT OF FLOWING FLUIDS IN PRESSURE CONDUITS

4.6.1   Measurement of Static Pressure

Measuring the static pressure in a flowing fluid requires that the measuring device fit the streamlines as closely as possible. This is required so that no disturbance in the flow will occur. For straight reaches of pipe conduit, the static pressure is usually measured by using a piezometer,

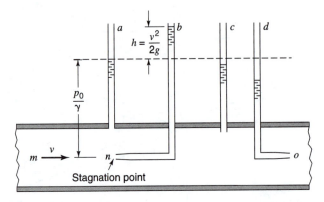

Figure 4.6.1 Piezometer (from Daugherty and Franzini (1977)).

shown in Figure 4.6.1, a pressure gauge or U-tube manometer. The piezometer opening in the side of the pipe conduit should be normal to and flush with the surface.

Measuring the static pressure in a flow field requires the use of a static tube (Figure 4.6.2). For this device, the pressure is transmitted to a gauge or a manometer through piezometric holes that are evenly spaced around the circumference of the tube. The device must be perfectly aligned with the flow.

4.6.2 Measurement of Velocity

Using a pitot tube, the pressure at the forward stagnation point, p_s, in a flowing fluid is $p_s = p_0 + 1/2\rho v^2$, where p_0 and v are the pressure and velocity, respectively, in the undisturbed flow upstream from the tube. The stagnation pressure can be measured by a tube facing upstream. So for a closed conduit it is necessary also to measure the static pressure and to subtract this from the total pitot reading to obtain the differential head h. The differential pressure may be measured with any suitable manometer arrangement.

$$\frac{p_0}{\gamma} + \frac{v^2}{2g} = \frac{p_s}{\gamma} \tag{4.6.1}$$

$$v^2 = 2g\left(\frac{p_s}{\gamma} - \frac{p_0}{\gamma}\right) \tag{4.6.2}$$

$$v = \sqrt{2g\left(\frac{p_s}{\gamma} - \frac{p_0}{\gamma}\right)} \tag{4.6.3}$$

This equation gives the ideal velocity in the stream where the pitot tube is located. To obtain the true velocity the right-hand side of the equation must be multiplied by a factor varying from 0.98 to 0.995.

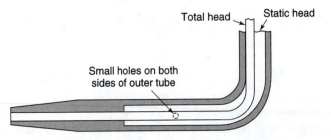

Figure 4.6.2 Pitot-static tube (from Daugherty and Franzini (1977)).

A combined pitot-static tube, shown in Figure 4.6.2, can also be used to measure static pressure. This pressure is measured through two or more holes drilled through the outer tube into an annular space. The velocity is found using:

$$v = C_I \sqrt{2g\left(\frac{p_s}{\gamma} - \frac{p_0}{\gamma}\right)}$$
(4.6.4)

where C_I is a coefficient that must be found by calibrating the instrument in the laboratory. These tubes have been shown to record a pressure slightly less than the true static pressure because of the increase in velocity past the tube, so the C_I is used.

4.6.3 Measurement of Discharge

Several techniques (devices) exist for measuring discharge in pressure conduits. Among the devices used for the measurement of discharge are constriction meters, shown in Figure 4.6.3.

Flow of fluid through a constriction in a pressure conduit results in a lowering of pressure at the constriction. The drop in piezometric head between the undisturbed flow and the constriction is a function of flowrate. This is the basis for using constriction meters to measure discharge. The types of constriction meters include orifices, nozzles, and tubes.

Venturi Tube (or Meter): A converging tube is an efficient device for converting pressure head to velocity head and a diverging tube converts velocity head to pressure head. The two are combined to form a Venturi tube, shown in Figures 4.2.2 and 4.6.3a. It consists of a tube with a constricted throat, which produces an increased velocity accompanied by a reduction in pressure, followed by a gradually diverging portion where the velocity is transformed back into pressure with slight friction losses.

Flow Nozzle: If the diverging discharge cone of a Venturi tube is omitted, the result is a flow nozzle, as shown in Figure 4.6.3b. This nozzle is usually installed between the flanges of a pipeline. The Venturi equation can be employed for the flow nozzle; however, it is more convenient and customary to include the correction for velocity of approach with the coefficient of discharge, so that

$$Q = KA_2 \sqrt{2g\left[\left(\frac{p_1}{\gamma} + z_1\right) - \left(\frac{p_2}{\gamma} + z_2\right)\right]}$$
(4.6.5)

where A_2 is the area of the nozzle throat and K is the *flow coefficient*, defined as

$$K = \frac{C}{\sqrt{1 - \left(D_2/D_1\right)^4}}$$
(4.6.6)

in which C is a discharge coefficient. Values of K range from 0.92 to 1.12.

Orifice Meters: An orifice (see Figure 4.6.3c) is an opening in the wall of a tank or in a plate normal to the axis of a pipe. The plate is either at the end of the pipe or in some intermediate location. The thickness of the wall or plate is very small relative to the size of opening. An orifice meter, as shown in Figure 4.6.3c, may also be used in a pipeline in the same manner as the Venturi tube or the flow nozzle. Flow rate through an orifice meter can be expressed as

$$Q = KA_0 \sqrt{2g\left[\left(\frac{p_1}{\gamma} + z_1\right) - \left(\frac{p_2}{\gamma} + z_2\right)\right]}$$
(4.6.7)

which has the same form as the flow nozzle equation except that A_2 is replaced by A_0, the cross-sectional area of the orifice opening.

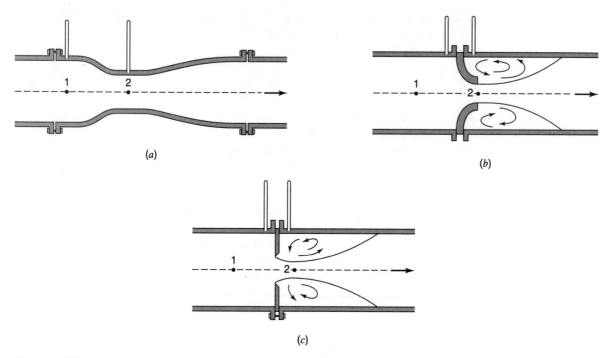

(a)

(b)

(c)

Figure 4.6.3 Three types of constriction meters for pipe flow. (*a*) Venturi meter; (*b*) Flow nozzle; (*c*) Orifice meter (from Linsley et al. (1992)).

PROBLEMS

4.1.1 Water flows in a pipe of 3/8-in diameter at a temperature of 70°F. The pressures at 1 and 2 (see Figure P4.1.1) are found to be 15 psi and 20 psi, respectively. Determine the direction of flow. What is the minimum discharge above which the flow will not be laminar?

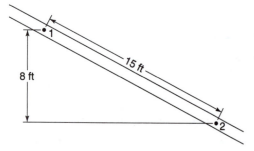

Figure P4.1.1

4.1.2 An experiment was done to determine the hydraulic conductivity K of an aquifer. A dye was injected into the aquifer at point 1 (see Figure P4.1.2) and was 36 hours later observed at point 2. The piezometric head difference between points 1 and 2 was observed to be 0.5 m. Determine the hydraulic conductivity of the aquifer. Take the porosity of the aquifer as 0.24.

Figure P4.1.2

4.1.3 The piezometric heads at points 1 and 2 in Figure P4.1.3 are found to be 75 ft and 72.5 ft. If the hydraulic conductivity of the aquifer is 50 ft/day, what is the Darcy flux? Determine the discharge by taking the average thickness of the aquifer as 100 ft.

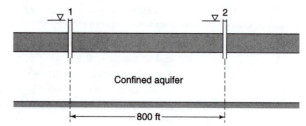

Figure P4.1.3

4.2.1 A Venturi meter with a throat diameter of 150 mm is connected to a pipe of diameter 250 mm to measure the discharge in the pipe. The pressures just upstream of the connection and at the throat were found to be 140 kPa and 80 kPa, respectively. Determine the flow rate in the pipe. Take the coefficient of discharge for the Venturi meter as 0.98.

4.2.2 The pressure difference between the upstream end section and the throat section of a Venturi meter connected to a pipe flow is found be 12 psi. The diameters of the pipe and the throat of the Venturi meter are 1-7/8-in and 1-1/8-in, respectively. The actual flow in the pipe is 0.353 ft³/s. Calibrate the Venturi meter for Reynolds numbers at the throat greater than 2×10^6.

4.2.3 Draw (to scale) the hydraulic grade line (HGL) and the energy grade line (EGL) of the system in Figure 4.3.12 (example 4.3.5). Take the length of each pipe as given and neglect the height of the elbows.

4.2.4 Suppose the water fountain in Fountain Hills, Arizona (see Figure P4.2.4), rises vertically to 150 ft above the lake (when operated). Neglecting wind effects and minor losses, determine the velocity at which the water is ejected.

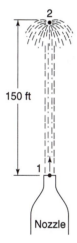

Figure P4.2.4

4.3.1 Develop the expression for the headloss in a pipe for steady, laminar flow of an incompressible fluid (equation (4.3.8)).

4.3.2 Suppose a globe valve ($K = 10$) is present in a pipe line of diameter 300 mm that has a friction factor f of 0.020. What is the equivalent length of this pipe that would cause equal headloss as the globe valve for the same discharge? Repeat this problem for a pipe of 150-mm diameter that has the same friction factor.

4.3.3 Two materials (wrought iron, $k_s = 0.046$ mm, and galvanized iron, $k_s = 0.15$ mm) are being considered for a new pipe line. The expected discharge is 0.15 m³/s. Both the headloss and cost are sought for. Wrought iron pipe costs 5 cents more than the galvanized iron pipe for every meter length of the pipe. Determine the tradeoff between the cost and the energy head for pipe diameters of 200 mm and 150 mm. Take the temperature as 15°C.

4.3.4 For the pipe system shown in Figure P4.3.4, determine the proportion of each pipe so that the pipe friction loss in each pipe is the same. Assume the same friction factor in all pipes.

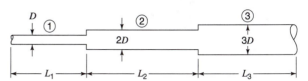

Figure P4.3.4

4.3.5 For the pipe system in problem 4.3.4, compare the expansion losses. Suppose the diameter of the third pipe was twice that of the second. What can you infer by comparing the headlosses at the expansion?

4.4.1 A plate is held against a horizontal water jet in the horizontal plane as shown in Figure P4.4.1. The jet has a diameter of 30 mm and an unknown velocity. A force of 200 N is required to hold the plate in the position. Determine the velocity of flow of the jet just before it hits the plate. What is the discharge?

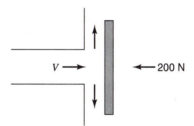

Figure P4.4.1

4.4.2 Suppose the nozzle in example 4.4.1 is connected to the pipe by flange bolts of 1-cm diameter. If the allowable tensile stress in the bolts is 330 N/cm², how many bolts are required for a safe connection?

4.4.3 Suppose the horizontal reducing bend in example 4.4.2 has an unknown bend angle (45° in example 4.4.2). What should be this bend angle so that the horizontal component force F_x is three times the vertical component force, F_y, in magnitude?

4.5.1 If the headloss in Example 4.5.1 were 15 m, what would be the discharge? Also, determine the velocity in each pipe.

4.5.2 The rate of flow in the pipe system in Figure P4.5.2 is 0.05 m³/s. The pressure at point 1 is measured to be 260 kPa. All the pipes are galvanized iron with roughness value of 0.15 mm. Determine the pressure at point 2. Take the loss coefficient for the sudden contraction as 0.05 and $\nu = 1.141 \times 10^{-6}$ m²/s.

Figure P4.5.2

4.5.3 The pressure difference between points 1 and 2 in the series pipe system in Figure P4.5.3 is 15 psi. All the pipes are galvanized iron with roughness value of 0.0005 ft. The loss coefficient at the sudden contraction is 0.05. Determine the flow rate in the system. The prevailing temperature is 70°F.

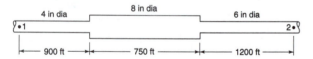

Figure P4.5.3

4.5.4 The pressure at point 1 in the parallel pipe system shown in Figure P4.5.4 is 750 kPa. If the flow rate through the system is 0.50 m³/s, what is the pressure at point 2? Neglect minor losses. All the pipes are steel with roughness value of 0.046 mm. Also, determine the fraction of the flow in each of the parallel pipes and check your solution. Take $\nu = 1.141 \times 10^{-6}$ m²/s.

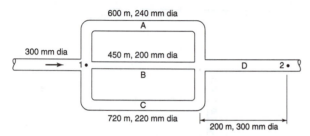

Figure P4.5.4

4.5.5 In problem 4.5.4 above, how far will the water flow before all its energy head is exhausted? Assume pipe D continues horizontally downstream without any other structure.

4.5.6 If the pressure difference between points 1 and 2 in Figure P4.5.6 is 30 psi, what will be the flow rate? The pipes are galvanized iron with $k_s = 0.0005$ ft. Take $\nu = 1.06 \times 10^{-5}$ ft²/s and neglect minor losses.

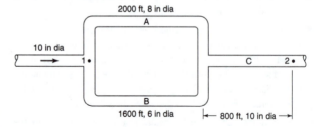

Figure P4.5.6

4.5.7 For the branching pipe system given in Figure P4.5.7, determine the flow to and the elevation of the third reservoir. Neglect minor losses and the velocity heads. The pipes are galvanized iron with $k_s = 0.0005$ ft and $\nu = 1.06 \times 10^{-5}$ ft²/s.

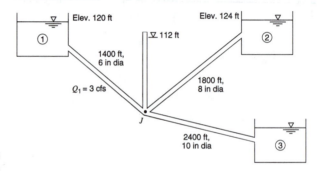

Figure P4.5.7

REFERENCES

ASHRAE, *ASHRAE Handbook, 1977 Fundamentals*, Am. Soc. of Heating, Refrigerating and Air Conditioning Engineers, New York, 1977.

Beij, K. H., "Pressure Losses for Fluid Flow in 90° Pipe Bends," *J. Res. Nat. Bur. Std.*, 21, 1938.

Chaudhry, M. H., *Open-Channel Flow*, Prentice Hall, Englewood Cliffs, NJ, 1993.

Chow, V. T., *Open-Channel Hydraulics*, McGraw-Hill, New York, 1959.

Colebrook, C. F., and C. M. White, "Turbulent Flow in Pipes with Particular Reference to the Transition Region Between Smooth and Rough Pipe Laws." *Institute of Civil Engineers*, London, vol. 11, p. 133, 1939.

Daugherty, R. L., and J. B. Franzini, *Fluid Mechanics with Engineering Application*, 7th edition, McGraw-Hill, New York, 1977.

French, R. H., *Open-Channel Hydraulics*, McGraw-Hill, New York, 1985.

Henderson, F. M., *Open Channel Flow*, Macmillan, New York, 1966.

Hwang, N. H. C., and R. J. Houghtalen, *Fundamentals of Hydraulic Engineering Systems*, 3rd edition, Prentice-Hall, Upper Saddle River, NJ, 1996.

Idel'chik, I. E., *Handbook of Hydraulic Resistance Coefficients of Local Resistance and of Friction*, Trans. A. Barouch, Israel Program for Scientific Translation, 1966.

Linsley, R. K., J. B. Franzini, D. L. Freyberg, and G. Tchobanoglous, *Water-Resources Engineering*, 4th edition, McGraw-Hill, New York, 1992.

Manning, R., "On the Flow of Water in Open Channels and Pipes," *Transactions Institute of Civil Engineers of Ireland*, vol. 20, pp. 161–209, Dublin, 1891; Supplement, vol. 24, pp. 179–207, 1895.

Moody, L. F., "Friction Factors for Pipe Flow," *Trans., Amer. Soc. Mech. Engrs.*, 66, Nov., 1944.

Nikuradse, J., "Gesetzmassigkeiten der turbulenten Stromung in glatten Rohren," *VDI Forschungsheft* 356, 1932.

Nikuradse, J. "Stromungsgesetze in rauhen Rohren." *VDI-Forschungsh*, 361, 1933. Also translated in *NACA Tech*. Memo 1292.

Roberson, J. A., and C. T. Crowe, *Engineering Fluid Mechanics*, Houghton Mifflin, Boston, MA, 1990.

Roberson, J. A., J. J. Cassidy, and M. H. Chaudhry, *Hydraulic Engineering*, Houghton Mifflin, Boston, 1988.

Streeter, V. L. (ed.) *Handbook of Fluid Dynamics*. McGraw-Hill, New York, 1961.

Velon, J. P., and T. J. Johnson, "Water Distribution and Treatment," *Davis' Handbook of Applied Hydraulics*, 4th edition, edited by V. I. Zippano and H. Hasen, McGraw-Hill, New York, 1993.

Chapter **5**

Hydraulic Processes: Open-Channel Flow

Open-channel flow refers to that flow whose top surface is exposed to atmospheric pressure. The topic of open-channel flow is covered in detail in textbooks such as Chow (1959), Henderson (1966), French (1985), Townson (1991), and Chaudhry (1993).

5.1 STEADY UNIFORM FLOW

This section describes the continuity, energy, and momentum equations for steady uniform flow in open-channels. Consider the control volume shown in Figure 5.1.1 in which the channel cross-section slope and boundary roughness are constant along the length of the control volume. For *uniform flow* the velocity is uniform throughout the control volume, so that $V_1 = V_2$ for the control volume in Figure 5.1.1. Hence for a uniform flow, $Q_1 = Q_2$, $A_1 = A_2$, $V_1 = V_2$, and $y_1 = y_2$. The depth of flow in uniform open-channel flow is also referred to as the *normal depth*. Figure 5.1.2 shows an open-channel flow, in an aqueduct of the Central Arizona Project.

5.1.1 Energy

The energy equation for open-channel flow can be derived in a similar manner as the energy equation for pipe flow (equation 4.2.13) using the control volume approach. In section 3.3, the general energy equation for steady fluid flow was derived as equation (3.3.20). Considering open-channel flow in the control volume in Figure 5.1.1, the energy equation can be expressed as

$$
\begin{aligned}
\frac{dH}{dt} = &\int_{A_2} \left(\frac{p_2}{\rho} + e_{u_2} + \frac{1}{2}V_2^2 + gz_2 \right) \rho V_2 dA_2 \\
&- \int_{A_1} \left(\frac{p_1}{\rho} + e_{u_1} + \frac{1}{2}V_1^2 + gz_1 \right) \rho V_1 dA_1
\end{aligned}
\tag{5.1.1}
$$

Assume the energy correction factor (section 3.6) is $\alpha = 1.0$. Refer to equation (3.6.4) for the definition of α. The shaft-work term is $dW_s/dt = 0$ because no pump or turbine exists. Because hydrostatic conditions prevail, the terms $(p/\rho + e_u + gz)$ can be taken outside the integral in equation (5.1.1):

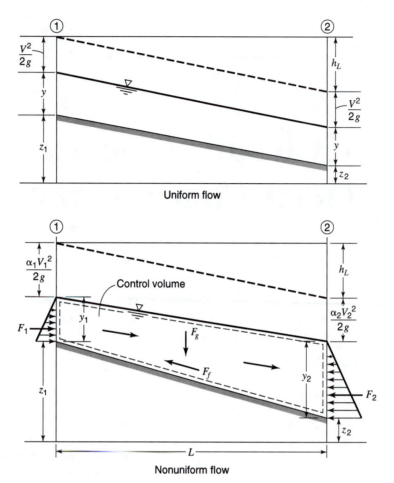

Figure 5.1.1 Open-channel flow: uniform and nonuniform flow.

$$\frac{dH}{dt} = \left(\frac{p_2}{\rho} + e_{u_2} + gz_2\right)\int_{A_2} \rho V_2 dA_2 + \int_{A_2}\frac{\rho V_2^2 dA_2}{2} - \left(\frac{p_1}{\rho} + e_{u_1} + gz_1\right)\int_{A_1} \rho V_1 dA_1 + \int_{A_1}\frac{\rho V_1^2}{2}dA_1 \quad (5.1.2)$$

The terms $\int \rho V dA = \dot{m}$ are the mass rate of flow at sections 1 and 2 and the terms

$\int_A (\rho V^3/2) dA = (\rho V^3/2)A = \dot{m}\dfrac{V^2}{2}$, so that equation (5.1.2) becomes

$$\frac{dH}{dt} = \left(\frac{p_2}{\rho} + e_{u_2} + gz_2\right)\dot{m} + \dot{m}\frac{V_2^2}{2} - \left(\frac{p_1}{\rho} + e_{u_1} + gz_1\right)\dot{m} - \dot{m}\frac{V_1^2}{2} \quad (5.1.3)$$

Dividing through by $\dot{m}g$ and rearranging yields

$$\frac{p_1}{\gamma} + z_1 + \frac{V_1^2}{2g} = \frac{p_2}{\gamma} + z_2 + \frac{V_2^2}{2g} + \left[\frac{e_{u_2} - e_{u_1}}{g} - \frac{1}{\dot{m}g}\frac{dH}{dt}\right] \quad (5.1.4)$$

Similar to equation (4.2.10), the terms in square brackets represent the headloss h_L due to viscous stress (friction). This energy lost due to friction effects per unit weight of fluid is denoted as h_L.

The energy equation for one-dimensional flow in an open-channel is

$$\frac{p_1}{\gamma} + z_1 + \alpha_1\frac{V_1^2}{2g} = \frac{p_2}{\gamma} + z_2 + \alpha_2\frac{V_2^2}{2g} + h_L \quad (5.1.5)$$

Figure 5.1.2 Hayden-Rhodes Aqueduct, Central Arizona Project. (Courtesy of the U.S. Bureau of Reclamation, (1985), photograph by Joe Madrigal Jr.)

where we have put back in the energy correction factor (see section 3.6). Pressure is hydrostatically distributed, and thus $p/\gamma + z$ is constant at each section in the control volume, so that $p_1/\gamma = y_1$ and $p_2/\gamma = y_2$. The energy equation for *nonuniform open-channel flow* is expressed as

$$y_1 + z_1 + \alpha_1 \frac{V_1^2}{2g} = y_2 + z_2 + \alpha_2 \frac{V_2^2}{2g} + h_L \qquad (5.1.6)$$

For *uniform flow*, $V_1 = V_2$ and $y_1 = y_2$, so

$$h_L = z_1 - z_2 \qquad (5.1.7)$$

By dividing both sides by L, the length of the control volume (channel), the following headloss per unit length of channel, S_f, is obtained as

$$S_f = \frac{h_L}{L} = \frac{z_1 - z_2}{L} \qquad (5.1.8)$$

so that the friction slope equals the channel bottom slope. The channel bottom slope $S_0 = \tan \theta$, where θ is the angle of inclination. If θ is small ($< 10°$), then $\tan \theta \approx \sin \theta = (z_1 - z_2)/L$.

5.1.2 Momentum

The forces acting upon the fluid control volume in Figure 5.1.1 are friction, gravity, and hydrostatic pressure. The friction force, F_f, is the product of the wall shear stress τ_0 and the area over which it acts, PL, where P is the wetted perimeter of the cross-section, thus

$$F_f = -\tau_0 PL \qquad (5.1.9)$$

where the negative sign indicates that the friction force acts opposite to the direction of flow. The gravity force F_g relates to the weight of the fluid γAL, where γ is the specific weight of the fluid (weight per unit volume). The gravity force on the fluid is the component of the weight acting in the direction of flow, that is,

$$F_g = \gamma AL \sin \theta \qquad (5.1.10)$$

The hydrostatic forces are denoted as F_1 and F_2, and are identical for uniform flow so that $F_1 - F_2 = 0$.

For a steady uniform flow, the general form of the integral momentum equation (3.4.6) in the x direction is

$$\sum F = \sum_{cs} v_x (\rho \mathbf{V} \cdot \mathbf{A}) \qquad (5.1.11)$$

or

$$F_1 + F_g + F_f - F_2 = 0 \qquad (5.1.12)$$

where $\sum_{cs} v_x (\rho \mathbf{V} \cdot \mathbf{A}) = 0$. Because $F_1 = F_2$, then by equation (5.1.12) $F_g + F_f = 0$, or

$$\gamma AL \sin \theta - \tau_0 PL = 0 \qquad (5.1.13)$$

For θ small, $S_0 \approx \sin \theta$ so

$$\gamma AL S_0 = \tau_0 PL \qquad (5.1.14)$$

which states that for steady uniform flow the friction and gravity forces are in balance and $S_0 = S_f$. Solving equation (5.1.14) for the wall shear stress (for steady uniform flow) yields

$$\tau_0 = \frac{\gamma AL S_0}{PL} \qquad (5.1.15)$$

or

$$\tau_0 = \gamma R S_0 = \gamma R S_f \tag{5.1.16}$$

where $R = A/P$ is the hydraulic radius. Equation (5.1.16) expresses the effects of friction through the wall shear stress τ_0 as represented from a momentum viewpoint and through the rate of energy dissipation S_f represented from an energy viewpoint. Consequently equation (5.1.16) expresses a linkage between the momentum and energy principles.

The shear stress τ_0 for fully turbulent flow can be expressed as a function of density, velocity, and resistance coefficient C_f as

$$\tau_0 = C_f \rho \left(\frac{V^2}{2} \right) \tag{5.1.17}$$

Equating (5.1.16) and (5.1.17) yields

$$C_f \, \rho \frac{V^2}{2} = \gamma R S_0 \tag{5.1.18}$$

and solving for the velocity gives

$$V = \sqrt{\frac{2g}{C_f}} \sqrt{R S_0} \tag{5.1.19}$$

Defining $C = \sqrt{2g/C_f}$, then equation (5.1.19) can be simplified to the well known *Chezy equation*

$$V = C\sqrt{R S_0} \tag{5.1.20}$$

where C is referred to as the *Chezy coefficient*.

Robert Manning (1891, 1895) derived the following empirical relation for C based upon experiments:

$$C = \frac{1}{n} R^{1/6} \tag{5.1.21}$$

where n is the Manning's roughness coefficient. Values of n are listed in Table 5.1.1. Values of n for natural channels have been also published by the U.S. Geological Survey (Barnes, 1967). Substituting C from equation (5.1.21) into equation (5.1.20) results in the *Manning equation*

$$V = \frac{1}{n} R^{2/3} S_0^{1/2} \tag{5.1.22}$$

which is valid for SI units and $S_0 = S_f$.

Table 5.1.1   Values of the Roughness Coefficient n
(Boldface figures are values generally recommended in design)

Type of channel and description	Minimum	Normal	Maximum
A. Closed conduits flowing partly full			
A-1. Metal			
a. Brass, smooth	0.009	**0.010**	0.013
b. Steel			
1. Lockbar and welded	0.010	0.012	0.014
2. Riveted and spiral	0.013	0.016	0.017
c. Cast iron			
1. Coated	0.010	0.013	0.014
2. Uncoated	0.011	0.014	0.016

Table 5.1.1   Values of the Roughness Coefficient *n* *(continued)*
(Boldface figures are values generally recommended in design)

Type of channel and description	Minimum	Normal	Maximum
d. Wrought iron			
1. Black	0.012	0.014	0.015
2. Galvanized	0.013	0.016	0.017
e. Corrugated metal			
1. Subdrain	0.017	0.019	0.021
2. Storm drain	0.021	**0.024**	0.030
A-2. Nonmetal			
a. Lucite	0.008	0.009	0.010
b. Glass	0.009	**0.010**	0.013
c. Cement			
1. Neat, surface	0.010	0.011	0.013
2. Mortar	0.011	0.013	0.015
d. Concrete			
1. Culvert, straight and free of debris	0.010	0.011	0.013
2. Culvert with bends, connections, and some debris	0.011	**0.013**	0.014
3. Finished	0.011	0.012	0.014
4. Sewer with manholes, inlet, etc., straight	0.013	0.015	0.017
5. Unfinished, steel form	0.012	0.013	0.014
6. Unfinished, smooth wood form	0.012	**0.014**	0.016
7. Unfinished, rough wood form	0.015	0.017	0.020
e. Wood			
1. Stave	0.010	0.012	0.014
2. Laminated, treated	0.015	0.017	0.020
f. Clay			
1. Common drainage title	0.011	0.013	0.017
2. Vitrified sewer	0.011	0.014	0.017
3. Vitrified sewer with manholes, inlet, etc.	0.013	0.015	0.017
4. Vitrified subdrain with open joint	0.014	0.016	0.018
g. Brickwork			
1. Glazed	0.011	0.013	0.015
2. Lined with cement mortar	0.012	0.015	0.017
h. Sanitary sewers coated with sewage slimes, with bends and connections	0.012	0.013	0.016
i. Paved invert, sewer, smooth bottom	0.016	0.019	0.020
j. Rubble masonry, cemented	0.018	0.025	0.030
B. Lined or built-up channels			
B-1. Metal			
a. Smooth steel surface			
1. Unpainted	0.011	**0.012**	0.014
2. Painted	0.012	0.013	0.017
b. Corrugated	0.021	0.025	0.030
B-2. Nonmetal			
a. Cement			
1. Neat, surface	0.010	0.011	0.013
2. Mortar	0.011	0.013	0.015
b. Wood			
1. Planed, untreated	0.010	0.012	0.014
2. Planed, creosoted	0.011	0.012	0.015
3. Unplaned	0.011	0.013	0.015
4. Plank with battens	0.012	0.015	0.018
5. Lined with roofing paper	0.010	0.014	0.017

Table 5.1.1 Values of the Roughness Coefficient *n* (continued)
(*Boldface figures are values generally recommended in design*)

Type of channel and description	Minimum	Normal	Maximum
c. Concrete			
1. Trowel finish	0.011	**0.013**	0.015
2. Float finish	0.013	0.015	0.016
3. Finished, with gravel on bottom	0.015	0.017	0.020
4. Unfinished	0.014	0.017	0.020
5. Gunite, good section	0.016	0.019	0.023
6. Gunite, wavy section	0.018	0.022	0.025
7. On good excavated rock	0.017	0.020	—
8. On irregular excavated rock	0.022	0.027	—
d. Concrete bottom float finished with sides of			
1. Dressed stone in mortar	0.015	0.017	0.020
2. Random stone in mortar	0.017	0.020	0.024
3. Cement rubble masonry, plastered	0.016	0.020	0.024
4. Cement rubble masonry	0.020	0.025	0.030
5. Dry rubble or riprap	0.020	0.030	0.035
e. Gravel bottom with sides of			
1. Formed concrete	0.017	0.020	0.025
2. Random stone in mortar	0.020	0.023	0.026
3. Dry rubble or riprap	0.023	0.033	0.036
f. Brick			
1. Glazed	0.011	**0.013**	0.015
2. In cement mortar	0.012	**0.015**	0.018
g. Masonry			
1. Cemented rubble	0.017	0.025	0.030
2. Dry rubble	0.023	0.032	0.035
h. Dressed ashlar	0.013	0.015	0.017
i. Asphalt			
1. Smooth	0.013	0.013	—
2. Rough	0.016	0.016	—
j. Vegetal lining	0.030	—	0.500
C. Excavated or dredged			
a. Earth, straight and uniform			
1. Clean, recently completed	0.016	0.018	0.020
2. Clean, after weathering	0.018	**0.022**	0.025
3. Gravel, uniform section, clean	0.022	0.025	0.030
4. With short grass, few weeds	0.022	0.027	0.033
b. Earth, winding and sluggish			
1. No vegetation	0.023	0.025	0.030
2. Grass, some weeds	0.025	0.030	0.033
3. Dense weeds or aquatic plants in deep channels	0.030	0.035	0.040
4. Earth bottom and rubble sides	0.028	0.030	0.035
5. Stony bottom and weedy banks	0.025	0.035	0.040
6. Cobble bottom and clean sides	0.030	0.040	0.050
c. Dragline-excavated or dredged			
1. No vegetation	0.025	0.028	0.033
2. Light brush on banks	0.035	0.050	0.060
d. Rock cuts			
1. Smooth and uniform	0.025	0.035	0.040
2. Jagged and irregular	0.035	0.040	0.050
c. Channels not maintained, weeds and brush uncut			
1. Dense weeds, high as flow depth	0.050	0.080	0.120
2. Clean bottom, brush on sides	0.040	0.050	0.080
3. Same, highest stage of flow	0.045	0.070	0.110
4. Dense brush, high stage	0.080	0.100	0.140

Table 5.1.1   Values of the Roughness Coefficient *n (continued)*
(Boldface figures are values generally recommended in design)

Type of channel and description	Minimum	Normal	Maximum
D. Natural streams			
D-1. Minor streams (top width at flood stage <100 ft)			
a. Streams on plain			
1. Clean, straight, full stage, no rifts or deep pools	0.025	0.030	0.033
2. Same as above, but more stones and weeds	0.030	0.035	0.040
3. Clean, winding, some pools and shoals	0.033	0.040	0.045
4. Same as above, but some weeds and stones	0.035	0.045	0.050
5. Same as above, lower stages, more ineffective slopes and sections	0.040	0.048	0.055
6. Same as 4, but more stones	0.045	0.050	0.060
7. Sluggish reaches, weedy, deep pools	0.050	0.070	0.080
8. Very weedy reaches, deep pools, or floodways with heavy stand of timber and underbrush	0.075	0.100	0.150
b. Mountain streams, no vegetation in channel, banks usually steep, trees and brush along banks submerged at high stages			
1. Bottom: gravels, cobbles, and few boulders	0.030	0.040	0.050
2. Bottom: cobbles with large boulders	0.040	0.050	0.070
D-2. Flood plains			
a. Pasture, no brush			
1. Short grass	0.025	0.030	0.035
2. High grass	0.030	0.035	0.050
b. Cultivated areas			
1. No crop	0.020	0.030	0.040
2. Mature row crops	0.025	0.035	0.045
3. Mature field crops	0.030	0.040	0.050
c. Brush			
1. Scattered brush, heavy weeds	0.035	0.050	0.070
2. Light brush and trees, in winter	0.035	0.050	0.060
3. Light brush and trees, in summer	0.040	0.060	0.080
4. Medium to dense brush, in winter	0.045	0.070	0.110
5. Medium to dense brush, in summer	0.070	0.100	0.160
d. Trees			
1. Dense willows, summer, straight	0.110	0.150	0.200
2. Cleared land with tree stumps, no sprouts	0.030	0.040	0.050
3. Same as above, but with heavy growth of sprouts	0.050	0.060	0.080
4. Heavy stand of timber, a few down trees, little undergrowth, flood stage below branches	0.080	0.100	0.120
5. Same as above, but with flood stage reaching branches	0.100	0.120	0.100
D-3. Major streams (top width at flood stage > 100 ft). The *n* value is less than that for minor streams of similar description, because banks offer less effective resistance.			
a. Regular section with no boulders or brush	0.025	—	0.060
b. Irregular and rough section	0.035	—	0.100

Source: Chow (1959).

Manning's equation in SI units can also be expressed as

$$Q = \frac{1}{n} A R^{2/3} S_0^{1/2} \tag{5.1.23}$$

For V in ft/sec and R in feet (U.S. Customary units), equation (5.1.22) can be rewritten as

$$V = \frac{1.49}{n} R^{2/3} S_0^{1/2} \tag{5.1.24}$$

and equation (5.1.23) can be written as

$$Q = \frac{1.49}{n} A R^{2/3} S_0^{1/2} \tag{5.1.25}$$

where A is in ft^2 and $S_0 = S_f$. Table 5.1.2 lists the geometric function for channel elements.

To determine the normal depth (for uniform flow), equation (5.1.23) or (5.1.25) can be solved with a specified discharge. Because the original shear stress τ_0 in equation (5.1.17) is for fully turbulent flow, Manning's equation is valid only for fully turbulent flow. Henderson (1966) presented the following criterion for fully turbulent flow in an open-channel:

$$n^6 \sqrt{RS_f} \geq 1.9 \times 10^{-3} \qquad (R \text{ in feet}) \tag{5.1.26a}$$

$$n^6 \sqrt{RS_f} \geq 1.1 \times 10^{-3} \qquad (R \text{ in meters}) \tag{5.1.26b}$$

Table 5.1.2   Geometric Functions for Channel Elements

Section:	Rectangle	Trapezoid	Triangle	Circle
Area A	$B_w y$	$(B_w + zy)y$	zy^2	$\frac{1}{8}(\theta - \sin\theta)d_o^2$
Wetted perimeter P	$B_w + 2y$	$B_w + 2y\sqrt{1+z^2}$	$2y\sqrt{1+z^2}$	$\frac{1}{2}\theta d_o$
Hydraulic radius R	$\dfrac{B_w y}{B_w + 2y}$	$\dfrac{(B_w + zy)y}{B_w + 2y\sqrt{1+z^2}}$	$\dfrac{zy}{2y\sqrt{1+z^2}}$	$\dfrac{1}{4}\left(1 - \dfrac{\sin\theta}{\theta}\right)d_o$
Top width B	B_w	$B_w + 2zy$	$2zy$	$\left[\sin\left(\dfrac{\theta}{2}\right)\right]d_o$ or $2\sqrt{y(d_o - y)}$
$\dfrac{2dR}{3Rdy} + \dfrac{1}{A}\dfrac{dA}{dy}$	$\dfrac{5B_w + 6y}{3y(B_w + 2y)}$	$\dfrac{(B_w + 2zy)(5B_w + 6y\sqrt{1+z^2}) + 4zy^2\sqrt{1+z^2}}{3y(B_w + zy)(B_w + 2y\sqrt{1+z^2})}$	$\dfrac{8}{3y}$	$\dfrac{4(2\sin\theta + 3\theta - 5\theta\cos\theta)}{3d_o\theta(\theta - \sin\theta)\sin(\theta/2)}$
				where $\theta = 2\cos^{-1}\left(1 - \dfrac{2y}{d_o}\right)$

Source: Chow (1959) (with additions).

EXAMPLE 5.1.1   An 8-ft wide rectangular channel with a bed slope of 0.0004 ft/ft has a depth of flow of 2 ft. Assuming steady uniform flow, determine the discharge in the channel. The Manning roughness coefficient is $n = 0.015$.

SOLUTION   From equation (5.1.25), the discharge is

$$Q = \frac{1.49}{n} AR^{2/3} S_0^{1/2}$$

$$= \frac{1.49}{0.015}(8)(2)\left[\frac{(8)(2)}{8+2(2)}\right]^{2/3}(0.0004)^{1/2}$$

$$= 38.5 \text{ ft}^3/\text{s}$$

EXAMPLE 5.1.2   Solve example 5.1.1 using SI units.

SOLUTION   The channel width is 2.438 m, with a depth of flow of 0.610 m. Using equation (5.1.23), the discharge is

$$Q = \frac{1}{n} AR^{2/3} S_0^{1/2}$$

$$= \frac{1}{0.015}(2.438)(0.610)\left[\frac{(2.438)(0.610)}{2.438+2(0.610)}\right]^{2/3}(0.0004)^{1/2}$$

$$= 1.09 \text{ m}^3/\text{s}$$

EXAMPLE 5.1.3   Determine the normal depth (for uniform flow) if the channel described in example 5.1.1 has a flow rate of 100 cfs.

SOLUTION   This problem is solved using equation (5.1.33) with Q_j defined by equation (5.1.25):

$$Q_j = \frac{1.49}{n} S_0^{1/2} \frac{(B_w y_j)^{5/2}}{(B_w + 2y_j)^{2/3}}$$

$$Q_j = \frac{1.49}{0.015}(0.0004)^{1/2}\frac{(8y_j)^{5/3}}{(8+2y_j)^{2/3}} = 1.987\frac{(8y_j)^{5/3}}{(8+2y_j)^{2/3}}$$

Using a numerical method such as Newton's method (see Appendix A), the normal depth is 3.98 ft.

5.1.3   Best Hydraulic Sections for Uniform Flow in Nonerodible Channels

The conveyance of a channel section increases with an increase in the hydraulic radius or with a decrease in the wetted perimeter. Consequently the channel section with the smallest wetted perimeter for a given channel section area will have maximum conveyance, referred to as the *best hydraulic section* or the cross-section of greatest hydraulic efficiency. Table 5.1.3 presents the geometric elements of the best hydraulic sections for six cross-section shapes. These sections may not always be practical because of difficulties in construction and use of material. The concept of best hydraulic section is only for nonerodible channels. Even though the best hydraulic section gives the minimum area for a given discharge, it may not necessarily have the minimum excavation.

Table 5.1.3   Best Hydraulic Sections

Cross-section	Area A	Wetted perimeter P	Hydraulic radius R	Top width T	Hydraulic depth D	Section factor Z
Trapezoid, half of a hexagon	$\sqrt{3}y^2$	$2\sqrt{3}y$	$^1/_2 y$	$\frac{4}{3}\sqrt{3}y$	$^3/_4 y$	$\frac{3}{2}y^{2.5}$
Rectangle, half of a square	$2y^2$	$4y$	$^1/_2 y$	$2y$	y	$2y^{2.5}$
Triangle, half of a square	y^2	$2\sqrt{2}y$	$\frac{1}{4}\sqrt{2}y$	$2y$	$^1/_2 y$	$\frac{\sqrt{2}}{2}y^{2.5}$
Semicircle	$\frac{\pi}{2}y^2$	πy	$^1/_2 y$	$2y$	$\frac{\pi}{4}y$	$\frac{\pi}{4}y^{2.5}$
Parabola, $T = 2\sqrt{2}\,y$	$^4/_3\sqrt{2}y^2$	$^8/_3\sqrt{2}y$	$^1/_2 y$	$2\sqrt{2}y$	$^2/_3 y$	$^8/_9\sqrt{3}y^{2.5}$
Hydrostatic catenary	$1.39586y^2$	$2.9836y$	$0.46784y$	$1.917532y$	$0.72795y$	$1.19093y^{2.5}$

Source: Chow (1959).

EXAMPLE 5.1.4

Determine the cross-section of greatest hydraulic efficiency for a trapezoidal channel if the design discharge is 10.0 m³/sec, the channel slope is 0.00052 and Manning's $n = 0.025$.

SOLUTION

From Table 5.1.3, the hydraulic radius should be $R = y/2$, so that the width B and area A are

$$B = \frac{2\sqrt{3}y}{3} = 1.155y \quad \text{(Because } B = \frac{1}{3}P \text{ for half of a hexagon)}$$

$$A = \sqrt{3}y^2 = 1.732y^2$$

Manning's equation (5.1.23) is used to determine the depth:

$$Q = \frac{1}{n}AR^{2/3}S_0^{1/2} = \frac{1}{0.025}(1.732y^2)\left(\frac{y}{2}\right)^{2/3}(0.00052)^{1/2} = 10$$

so

$$\frac{10 \times 0.025 \times 2^{2/3}}{1.732(0.00052)^{1/2}} = y^{8/2}$$

Thus, $y = 2.38$ m, so that $B = 2.75$ m and $A = 9.81$ m².

5.2   SPECIFIC ENERGY, MOMENTUM, AND SPECIFIC FORCE

5.2.1   Specific Energy

The *total head* or *energy head*, H, at any location in an open-channel flow can be expressed as

$$H = y + z + \frac{V^2}{2g} \tag{5.1.6}$$

which assumes that the velocity distribution is uniform (i.e., $\alpha = 1$) and the pressure distribution is hydrostatic (i.e., $p = \gamma y$). Using the channel bottom as the datum (i.e., $z = 0$) then define the total head above the channel bottom as the *specific energy*,

$$E = y + \frac{V^2}{2g} \tag{5.2.1}$$

Using continuity ($V = Q/A$), the specific energy can be expressed in terms of the discharge as

$$E = y + \frac{Q^2}{2gA^2} \tag{5.2.2}$$

Specific energy curves, such as are shown in Figures 5.2.1 and 5.2.2, can be derived using equation (5.2.2).

   Critical flow occurs when the specific energy is minimum for a given discharge (i.e., $dE/dy = 0$), so that

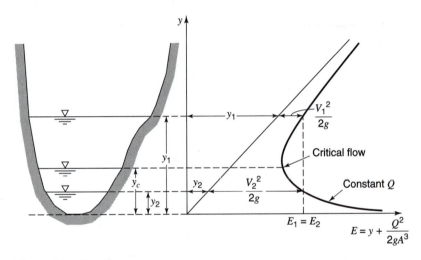

Figure 5.2.1 Specific energy.

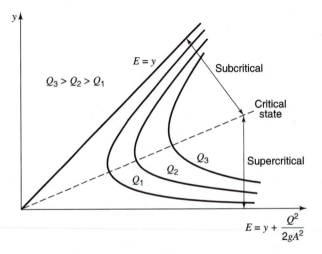

Figure 5.2.2 Specific energy showing subcritical and super-critical flow ranges.

$$\frac{dE}{dy} = 1 - \frac{Q^2}{gA^3}\frac{dA}{dy} = 0 \tag{5.2.3}$$

Referring to Figure 5.2.1, the top width is defined as $T = dA/dy$ so equation (5.2.3) can be expressed as

$$1 - \frac{TQ^2}{gA^3} = 0 \tag{5.2.4}$$

or

$$\frac{TQ^2}{gA^3} = 1 \tag{5.2.5}$$

To denote critical conditions use T_c, A_c, V_c, and y_c, so

$$\frac{T_c Q^2}{gA_c^{\ 3}} = 1 \tag{5.2.6}$$

or

$$\frac{V_c^{\ 2}}{g} = \frac{A_c}{T_c} \tag{5.2.7}$$

Equation (5.2.6) or (5.2.7) can be used to determine the critical depth and/or the critical velocity.
 Rearranging equation (5.2.7) yields

$$\frac{V_c^{\ 2}}{g(A_c/T_c)} = 1 \tag{5.2.8}$$

The *hydraulic depth* is defined as $D = A/T$ so equation (5.2.7) becomes

$$\frac{V_c^{\ 2}}{gD_c} = 1 \tag{5.2.9}$$

or

$$\frac{V_c}{\sqrt{gD_c}} = 1 \tag{5.2.10}$$

This is basically the *Froude number*, F_r, which is 1 at critical flow:

$$F_r = \frac{V}{\sqrt{gD}} \begin{cases} <1 & \text{subcritical flow} \\ =1 & \text{critical flow} \\ >1 & \text{supercritical flow} \end{cases} \tag{5.2.11}$$

 Figure 5.2.2 illustrates the range of subcritical flow and the range of supercritical flow along with the location of the critical states. Note the relationship of the specific energy curves and the fact that $Q_3 > Q_2 > Q_1$. Figure 5.2.1 illustrates the alternate depths y_1 and y_2 for which $E_1 = E_2$ or

$$y_1 + \frac{V_1^2}{2g} = y_2 + \frac{V_2^2}{2g} \tag{5.2.12}$$

For a rectangular channel $D_c = A_c/T_c = y_c$, so equation (5.2.10) for critical flow becomes

$$\frac{V_c}{\sqrt{gy_c}} = 1 \tag{5.2.13}$$

If we let q be the flow rate per unit width of channel for a rectangular channel, i.e. $q = Q/B$ where $T = B$, the width of the channel (or $q = Q/T$), then equation (5.2.6) can be rearranged, $T_c Q^2/(gT_c{}^3 y_c{}^3) = q^2/(gy_c) = 1$, and solved for y_c to yield

$$y_c = \left(\frac{q^2}{g}\right)^{1/3}$$

(5.2.14)

EXAMPLE 5.2.1

Compute the critical depth for the channel in example 5.1.1 using a discharge of 100 cfs.

SOLUTION

Using equation (5.2.13), $V_c = \sqrt{gy_c} = Q/A = 100/8y_c$, so

$$y_c^{3/2} = \frac{100}{8\sqrt{g}} \quad \text{or} \quad y_c = \left(\frac{100}{8\sqrt{g}}\right)^{2/3} = 1.69 \text{ ft}$$

Alternatively, using equation (5.2.14) yields

$$y_c = \left(\frac{(100/8)^2}{g}\right)^{1/3} = 1.69 \text{ ft}$$

EXAMPLE 5.2.2

For a rectangular channel of 20 ft width, construct a family of specific energy curves for $Q = 0, 50, 100$, and 300 cfs. Draw the locus of the critical depth points on these curves. For each flow rate, what is the minimum specific energy found from these curves?

SOLUTION

The specific energy is computed using equation (5.2.1):

$$E = y + \frac{V^2}{2g} = y + \frac{1}{2g}\left(\frac{Q}{A}\right)^2 = y + \frac{1}{2g}\frac{Q^2}{(20y)^2} = y + \frac{Q^2}{25760y^2}$$

Computing critical depths for the flow rates using equation (5.2.14) with $q = Q/B$ yields

$$Q = 0: \qquad y_c = \sqrt[3]{\frac{Q^2}{B^2 g}} = 0$$

$$Q = 50 \text{ cfs}: \qquad y_c = \sqrt[3]{\frac{Q^2}{B^2 g}} = \sqrt[3]{\frac{50^2}{20^2(32.2)}} = 0.58 \text{ ft}$$

$$Q = 100 \text{ cfs}: \qquad y_c = \sqrt[3]{\frac{Q^2}{B^2 g}} = \sqrt[3]{\frac{100^2}{20^2(32.2)}} = 0.92 \text{ ft}$$

$$Q = 300 \text{ cfs}: \qquad y_c = \sqrt[3]{\frac{Q^2}{B^2 g}} = \sqrt[3]{\frac{300^2}{20^2(32.2)}} = 1.9 \text{ ft}$$

Computed specific energies are listed in Table 5.2.1.

Table 5.2.1 Computed Specific Energy Values for Example 5.2.2

Depth, y (ft)	Specific energy, E (ft-lb/lb)			
	$Q = 0$	$Q = 50$	$Q = 100$	$Q = 300$
0.5	0.50	0.89	2.05	14.86
0.6	0.60	0.87	1.68	10.57
0.8	0.80	0.95	1.41	6.41
1.0	1.00	1.10	1.39	4.59
1.2	1.20	1.27	1.47	3.69
1.4	1.40	1.45	1.60	3.23
1.6	1.60	1.64	1.75	3.00
1.8	1.80	1.83	1.92	2.91
2.0	2.00	2.02	2.10	2.90
2.2	2.20	2.22	2.28	2.94
2.4	2.40	2.42	2.47	3.02
2.6	2.60	2.61	2.66	3.13
2.8	2.80	2.81	2.85	3.26
3.0	3.00	3.01	3.04	3.40
3.5	3.50	3.51	3.53	3.79
4.0	4.00	4.01	4.02	4.22
4.5	4.50	4.50	4.52	4.68
5.0	5.00	5.00	5.02	5.14

The specific energy curves are shown in Figure 5.2.3. The minimum specific energies are:

$Q = 50$ cfs:　　　$E_{min} = 0.868$

$Q = 100$ cfs:　　$E_{min} = 1.379$

$Q = 300$ cfs:　　$E_{min} = 2.868$

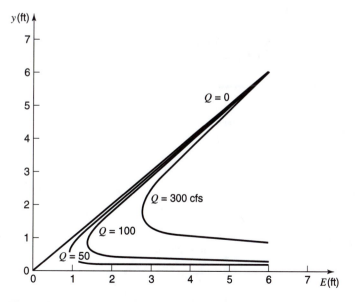

Figure 5.2.3 Specific energy curves for example 5.2.2.

EXAMPLE 5.2.3

A rectangular channel 2 m wide has a flow of 2.4 m³/s at a depth of 1.0 m. Determine if critical depth occurs at (a) a section where a hump of $\Delta z = 20$ cm high is installed across the channel bed, (b) a side wall constriction (with no humps) reducing the channel width to 1.7 m, and (c) both the hump and side wall constrictions combined. Neglect head losses of the hump and constriction caused by friction, expansion, and contraction.

SOLUTION

(a) The computation is focused on determining the critical elevation change in the channel bottom (hump) Δz_{crit} that causes a critical depth at the hump. The energy equation is $E = E_{min} + \Delta z_{crit}$ or $\Delta z_{crit} = E - E_{min}$, where E is the specific energy of the channel flow and E_{min} is the minimum specific energy, which is at critical depth by definition. If $\Delta z_{crit} \leq \Delta z$ then critical depth will occur. Using equation (5.2.2) yields

$$E = y + \frac{Q^2}{2gA^2} = y + \frac{q^2}{2gy^2}$$

which can be solved for q:

$$q = \sqrt{2g(y^2 E - y^3)}$$

Differentiating this equation with respect to y because maximum q and minimum E are equivalent (see Figure 5.2.4) yields

$$\frac{dq}{dy} = \frac{d}{dy}\left[\sqrt{2g(y^2 E - y^3)}\right] = 0$$

$$y_c = \frac{2}{3}E_{min} \quad \text{or} \quad E_{min} = \frac{3}{2}y_c$$

To compute specific energy, use

$$E = y + \frac{q^2}{2gy^2} = 1.0 + \frac{(2.4/2.0)^2}{2(9.81)(1)^2} = 1.073 \text{ m}$$

Next compute E_{min} using $E_{min} = 3/2y_c$, where $y_c = \left(q^2/g\right)^{1/3}$ (equation (5.2.14)):

$$y_c = \left[\frac{(2.4/2.0)^2}{9.81}\right]^{1/3} = 0.528 \text{ m}$$

So $E_{min} = 3/2(0.528 \text{ m}) = 0.792$ m. Then $\Delta z_{crit} = E - E_{min} = 1.073 - 0.792 = 0.281$ m. In this case $\Delta z = 20$ cm $= 20/100$m $= 0.2$ m $< \Delta z_{crit} = 0.281$ m. Therefore, y_c does not occur at the hump.

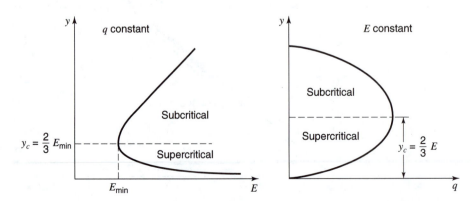

Figure 5.2.4 Specific energy curve and y versus q for constant E.

(b) The critical depth at the side wall constriction is

$$y_c = \left[\frac{(2.4/1.7)^2}{9.81}\right]^{1/3} = 0.588 \text{ m}$$

Thus $E_{min} = (3/2)y_c = (3/2)(0.588) = 0.882$ m. E is computed above as $E = 1.073$ m. Because $E_{min} = 0.882$ m $< E = 1.073$ m, critical depth does not occur at the constriction. Remember that energy losses are negligible so that the specific energy in the constriction and upstream of the constriction must be equal. For critical flow to occur, the constriction width can be computed as follows: $E_{min} = E =$

1.073 m $= (3/2)y_c$, so that $y_c = 0.715$ m. Then using equation (5.2.14), $0.715 = \left[(2.4/B_c)^2/9.81\right]^{1/3}$ and $B_c = 1.267$ m.

(c) With both the hump and the side wall constriction, y_c is 0.588 m, so $E_{min} = 0.882$ m. Then $\Delta z_{crit} = E - E_{min} = 1.073 - 0.882 = 0.191$ m.

Because $\Delta z = 20/100$ m $= 0.20$ m $> \Delta z_{crit} = 0.191$ m, critical depth will occur at the hump with a constriction.

5.2.2 Momentum

Applying the momentum principle (equation 3.4.6) to a short horizontal reach of channel with steady flow (Figure 5.2.5), we get

$$\sum F_x = \frac{d}{dt}\int_{cv} v_x \rho \, d\forall + \sum_{cs} v_x \rho \mathbf{V} \cdot \mathbf{A} \tag{2.5.6}$$

where

$$\frac{d}{dt}\int_{cv} v_x \rho \, d\forall = 0 \tag{5.2.15}$$

The momentum entering from the upstream is $-\rho\beta_1 V_1 Q$ and the momentum leaving the control volume is $\rho\beta_2 V_2 Q$, where β is called the *momentum correction factor* that accounts for the nonuniformity of velocity (equation 3.6.8), so that

$$\sum v_x \rho \mathbf{V} \cdot \mathbf{A} = -\rho\beta_1 V_1 Q + \rho\beta_2 V_2 Q \tag{5.2.16}$$

The forces are

$$\sum F_x = F_1 - F_2 + F_g - F_f' \tag{5.2.17}$$

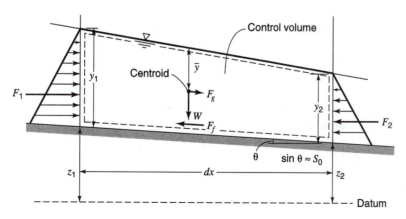

Figure 5.2.5 Application of momentum principle.

The hydrostatic forces are

$$F_1 = \gamma A \bar{y}_1 \tag{5.2.18}$$

and

$$F_2 = \gamma A \bar{y}_2 \tag{5.2.19}$$

where $\bar{y}_1$ and $\bar{y}_2$ are the distances to the centroid. The gravity force F_g due to the weight W of the water is $W \sin \theta = \rho g A dx \sin \theta$, where $W = \rho g A dx$. Because the channel slope is small, $S_0 \approx \sin \theta$, and the force due to gravity is

$$F_g = \rho g \bar{A} dx S_0 \tag{5.2.20}$$

where $\bar{A} = (A_1 + A_2)/2$ is the average cross-sectional area of flow. The external force due to friction created by shear between the channel bottom and sides of the control volume is $-\tau_0 P dx$ where τ_0 is the bed shear stress and P is the wetted perimeter. From equation (5.1.6), $\tau_0 = \gamma R S_f = \rho g (A/P) S_f$. So the friction force is then

$$F_f' = -\rho g A S_f dx \tag{5.2.21}$$

For our purposes here we will continue to use F_g and F_f'.

Substituting equations (5.2.15) through (5.2.21) into (3.4.6) gives

$$\gamma A_1 \bar{y}_1 - \gamma A_2 \bar{y}_2 + W \sin \theta - F_f' = -\rho \beta_1 V_1 Q + \rho \beta_2 V_2 Q \tag{5.2.22}$$

which is the *momentum equation for steady state open-channel flow*.

It should be emphasized that in the energy equation the F_f (loss due to friction) is a measure of the internal energy dissipated in the entire mass of water in the control volume, whereas F_f' in the momentum equation measures the losses due to external forces exerted on the water by the wetted perimeter of the control volume. Ignoring the small difference between the energy coefficient α and the momentum coefficient β in gradually varied flow, the internal energy losses are practically identical with the losses due to external forces (Chow, 1959). For uniform flow, $F_g = F_f'$.

Application of the energy and momentum principles in open-channel flow can be confusing at first. It is important to understand the basic differences, even though the two principles may produce identical or very similar results. Keep in mind that *energy is a scalar quantity* and *momentum is a vector quantity* and that *energy considers internal losses* in the energy equation and *momentum considers external resistance* in the momentum equation. The energy principle is simpler and clearer than the momentum principle; however, the momentum principle has certain advantages in application to problems involving high internal-energy changes, such as the hydraulic jump (Chow, 1959), which is discussed in section 5.5 on rapidly varied flow.

5.2.3 Specific Force

For a short horizontal reach (control volume) with $\theta = 0$ and the gravity force $F_g = W \sin \theta = 0$, the external force of friction F_f', can be neglected so $F_f' = 0$ and $F_g = 0$. Also assuming $\beta_1 = \beta_2$, the momentum equation (5.2.22) reduces to

$$\gamma A_1 \bar{y}_1 - \gamma A_2 \bar{y}_2 = -\rho V_1 Q + \rho V_2 Q \tag{5.2.23}$$

Substituting $V_1 = Q/A_1$ and $V_2 = Q/A_2$, dividing through by γ and substituting $1/g = \rho/\gamma$, and then rearranging yields

$$\frac{Q^2}{g A_1} + A_1 \bar{y}_1 = \frac{Q^2}{g A_2} + A_2 \bar{y}_2 \tag{5.2.24}$$

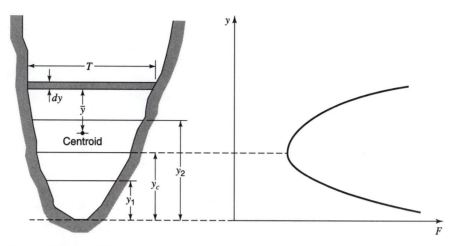

Figure 5.2.6 Specific force curves.

The *specific force F* (Figure 5.2.6) is defined as

$$F = \frac{Q^2}{gA} + A\bar{y} \tag{5.2.25}$$

which has units of ft^3 or m^3. The minimum value of the specific force with respect to the depth is determined using

$$\frac{dF}{dy} = \frac{d\left(\dfrac{Q^2}{gA}\right)}{dy} + \frac{d(A\bar{y})}{dy} = 0 \tag{5.2.26}$$

which results in

$$\frac{dF}{dy} = -\frac{TQ^2}{gA^2} + A = 0 \tag{5.2.27}$$

Refer to Chow (1959) or Chaudhry (1993) for the proof and further explanation of $d(A\bar{y})/dy = A$. Equation (5.2.27) reduces to $-V^2/g + A/T = 0$ where the hydraulic depth $D = A/T$, so

$$\frac{V^2}{g} = D \text{ or } \frac{V^2}{gD} = 1 \tag{5.2.28}$$

which we have already shown is the criterion for critical flow (equation (5.2.9) or (5.2.10)). Therefore, at critical flow the specific force is a minimum for a given discharge.

Summarizing, critical flow is characterized by the following conditions:

- Specific energy is minimum for a given discharge.
- Specific force is minimum for a given discharge.
- Velocity head is equal to half the hydraulic depth.
- Froude number is equal to unity.

Two additional conditions that are not proven here are (Chow, 1959):

- The discharge is maximum for a given specific energy
- The velocity of flow in a channel of small slope with uniform velocity distribution is equal to the celerity of small gravity waves in shallow water caused by local disturbances.

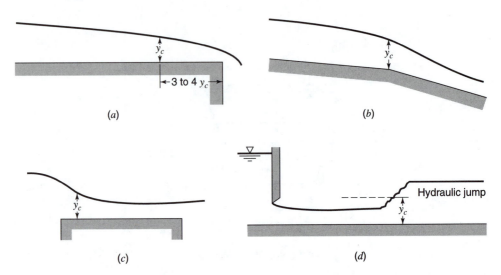

Figure 5.2.7 Example locations of critical flow. (*a*) Critical depth near free overfall; (*b*) Change in grade of channel bottom. (*c*) Flow over a broad-crested weir. (*d*) Flow through hydraulic jump.

When flow is at or near the critical state, minor changes in specific energy near critical flow causes major changes in depth (see Figures 5.2.1 or 5.2.2), causing the flow to be unstable. Figure 5.2.7 illustrates examples of locations of critical flow.

EXAMPLE 5.2.4   Compute the specific force curves for the channel and flow rates used in example 5.2.2.

SOLUTION   The specific force values are computed using equation (5.2.25) with the values presented in Table 5.2.2. The curves are plotted in Figure 5.2.8.

Table 5.2.2   Computed Specific Force Curve Values for Example 5.2.4

Depth, y (ft)	Specific force, F (ft^3)			
	$Q = 0$	$Q = 50$	$Q = 100$	$Q = 300$
0.1	0.10	38.92	155.38	1397.62
0.2	0.40	19.81	78.04	699.16
0.4	1.60	11.30	40.42	350.98
0.6	3.60	10.07	29.48	236.52
0.8	6.40	11.25	25.81	181.09
1.0	10.00	13.88	25.53	149.75
1.2	14.40	17.63	27.34	130.86
1.4	19.60	22.37	30.69	119.42
1.6	25.60	28.03	35.30	112.94
1.8	32.40	34.56	41.03	110.04
2.0	40.00	41.94	47.76	109.88
2.2	48.40	50.16	55.46	111.92
2.4	57.60	59.22	64.07	115.83
2.6	67.60	69.09	73.57	121.35
2.8	78.40	79.79	83.95	128.31

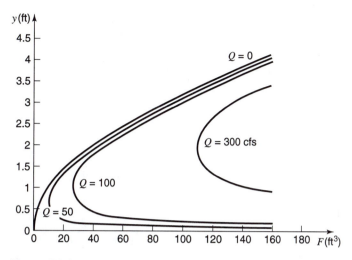

Figure 5.2.8 Specific force curves.

Table 5.2.2 Computed Specific Force Curve Values for Example 5.2.4 *(continued)*

Depth, y (ft)	Specific force, F (ft³)			
	$Q = 0$	$Q = 50$	$Q = 100$	$Q = 300$
3.0	90.00	91.29	95.18	136.58
3.5	122.50	123.61	126.94	162.43
4.0	160.00	160.97	163.88	194.94
4.5	202.50	203.36	205.95	233.56
5.0	250.00	250.78	253.11	277.95

5.3 STEADY, GRADUALLY VARIED FLOW

5.3.1 Gradually Varied Flow Equations

Several types of open-channel flow problems can be solved in hydraulic engineering practice using the concepts of nonuniform flow. The first to be discussed are *gradually varied flow problems* in which the change in the water surface profile is small enough that it is possible to integrate the relevant differential equation from one section to an adjacent section for the change in depth or change in water surface elevation. Consider the energy equation (5.1.6) previously derived for non-uniform flow using the control volume approach (with $\alpha_1 = \alpha_2 = 1$):

$$y_1 + z_1 + \frac{V_1^2}{2g} = y_2 + z_2 + \frac{V_2^2}{2g} + h_L \tag{5.1.6}$$

Because $h_L = S_f L = S_f \Delta x$ letting $\Delta y = y_2 - y_1$ and $\Delta z = z_1 - z_2 = S_0 \Delta x$, then equation (5.1.6) can be expressed as

$$S_0 \Delta x + \frac{V_1^2}{2g} = \Delta y + \frac{V_2^2}{2g} + S_f \Delta x \tag{5.3.1}$$

Rearranging yields

$$\Delta y = S_0 \Delta x - S_f \Delta x - \left(\frac{V_2^2}{2g} - \frac{V_1^2}{2g} \right) \tag{5.3.2}$$

and then dividing through by Δx results in

$$\frac{\Delta y}{\Delta x} = S_0 - S_f - \left(\frac{V_2^2}{2g} - \frac{V_1^2}{2g} \right) \frac{1}{\Delta x} \tag{5.3.3}$$

Taking the limit as $\Delta x \to 0$, we get

$$\lim_{\Delta x \to 0} \left(\frac{\Delta y}{\Delta x} \right) = \frac{dy}{dx} \tag{5.3.4}$$

and

$$\lim_{\Delta x \to 0} \left(\frac{V_2^2}{2g} - \frac{V_1^2}{2g} \right) \left(\frac{1}{\Delta x} \right) = \frac{d}{dx} \left(\frac{V^2}{2g} \right) \tag{5.3.5}$$

Substituting these into equation (5.3.3) and rearranging yields

$$\frac{dy}{dx} + \frac{d}{dx} \left(\frac{V^2}{2g} \right) = S_0 - S_f \tag{5.3.6}$$

The second term $\dfrac{d}{dx} \left(\dfrac{V^2}{2g} \right)$ can be expressed as $\left[\dfrac{d\left(\dfrac{V^2}{2g} \right)}{dy} \right] \dfrac{dy}{dx}$, so that equation (5.3.6) can be simplified to

$$\frac{dy}{dx} \left[1 + \frac{d\left(\dfrac{V^2}{2g} \right)}{dy} \right] = S_0 - S_f \tag{5.3.7}$$

or

$$\frac{dy}{dx} = \frac{S_0 - S_f}{\left[1 + \dfrac{d\left(\dfrac{V^2}{2g} \right)}{dy} \right]} \tag{5.3.8}$$

Equations (5.3.7) and (5.3.8) are two expressions of *the differential equation for gradually varied flow*. Equation (5.3.8) can also be expressed in terms of the Froude number. First observe that

$$\frac{d}{dy} \left(\frac{V^2}{2g} \right) = \frac{d}{dy} \left(\frac{Q^2}{2gA^2} \right) = -\left(\frac{Q^2}{gA^3} \right) \frac{dA}{dy} \tag{5.3.9}$$

By definition, the incremental increase in cross-sectional area of flow dA, due to an incremental increase in the depth dy, is $dA = Tdy$, where T is the top width of flow (see Figure 5.3.1). Also $A/T = D$, which is the hydraulic depth. Equation (5.3.9) can now be expressed as

$$\frac{d}{dy} \left(\frac{V^2}{2g} \right) = -\left(\frac{Q^2}{gA^3} \right) \frac{Tdy}{dy} = -\frac{Q^2}{gA^2} \left(\frac{T}{A} \right) = -\frac{Q^2}{gA^2 D} \tag{5.3.10a}$$

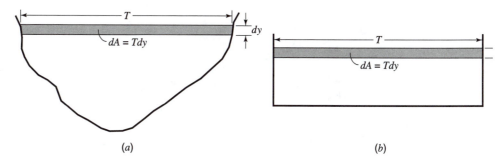

Figure 5.3.1 Definition of top width ($T = dA/dy$). (*a*) Natural channel; (*b*) Rectangular channel.

$$= -F_r^2 \qquad (5.3.10b)$$

where $F_r = \dfrac{V}{\sqrt{gD}} = \dfrac{Q}{A\sqrt{gD}}$. Substituting equation (5.3.10b) into (5.3.8) and simplifying, we find that the gradually varied flow equation in terms of the Froude number is

$$\frac{dy}{dx} = \frac{S_0 - S_f}{1 - F_r^2} \qquad (5.3.11)$$

EXAMPLE 5.3.1

Consider a vertical sluice gate in a wide rectangular channel ($R = A/P = By/(B + 2Y) \approx y$ because $B \gg 2y$). The flow downstream of a sluice gate is basically a jet that possesses a vena contracta (see Figure 5.3.2). The distance, L, from the sluice gate to the vena contracta as a rule is approximated as the same as the sluice gate opening (Chow, 1959). The coefficients of contraction for vertical sluice gates are approximately 0.6, ranging from 0.598 to 0.611 (Henderson, 1966). The objective of this problem is to determine the distance from the vena contracta to a point b downstream where the depth of flow is known to be 0.5 m deep. The depth of flow at the vena contracta is 0.457 m for a flow rate of 4.646 m³/s per meter of width. The channel bed slope is 0.0003 and Manning's roughness factor is $n = 0.020$.

SOLUTION

To compute the distance, Δx from y_a to y_b, the gradually varied flow equation (5.3.1) can be used,

$$S_0 \Delta x + \frac{V_1^2}{2g} = \Delta y + \frac{V_2^2}{2g} + S_f \Delta x$$

where $\Delta y = y_2 - y_1$. Solving for Δx, we get

$$(S_0 - S_f)\Delta x = \Delta y + \left(\frac{V_2^2}{2g} - \frac{V_1^2}{2g} \right)$$

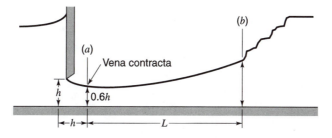

Figure 5.3.2 Flow downstream of a sluice gate in a wide rectangular channel.

$$\Delta x = \frac{\Delta y + \left(\dfrac{V_2^2}{2g} - \dfrac{V_1^2}{2g}\right)}{S_0 - S_f}$$

The friction slope is computed using Manning's equation (5.1.22) with average values of the hydraulic radius

$$V_{ave} = \frac{1}{n} R_{ave}^{2/3} S_f^{1/2}$$

so

$$S_f = \left[\frac{n V_{ave}}{R_{ave}^{2/3}}\right]^2$$

Let us use the following values for this example:

Location	y (m)	$R = y$ (ft)	V (m/s)	$V^2/2g$ (m)
a	0.457	0.457	10.17	5.27
b	0.500	0.500	9.292	4.40

Now we get

$$V_{ave} = \frac{10.17 + 9.292}{2} = 9.73 \text{ m/s}$$

$$R_{ave} = \frac{0.457 + 0.500}{2} = 0.479 \text{ m}$$

$$S_f = \left[\frac{0.020(9.73)}{0.479^{2/3}}\right]^2 = 0.101 \text{ m/m}$$

The distance from a to b, Δx, is

$$\Delta x = \frac{(0.500 - 0.457) + (4.40 - 5.27)}{0.0003 - 0.101} = \frac{-0.827}{-0.101} = 8.21 \text{ m}$$

The distance from the sluice gate to b is $\left(\dfrac{0.457}{0.6}\right) + 8.21 = 8.97$ m.

5.3.2   Water Surface Profile Classification

Channel bed slopes may be classified as mild (M), steep (S), critical (C), horizontal (H) ($S_0 = 0$), and adverse (A) ($S_0 < 0$). To define the various types of slopes for the mild, steep, and critical slopes, the normal depth y_n and critical depth y_c are used:

$$\text{Mild: } y_n > y_c \text{ or } \frac{y_n}{y_c} > 1 \tag{5.3.12a}$$

$$\text{Steep: } y_n < y_c \text{ or } \frac{y_n}{y_c} < 1 \tag{5.3.12b}$$

$$\text{Critical: } y_n = y_c \text{ or } \frac{y_n}{y_c} = 1 \tag{5.3.12c}$$

The horizontal and adverse slopes are special cases because the normal depth does not exist for them. Table 5.3.1 lists the types and characteristics of the various types of profiles and Figure 5.3.3 shows the classification of gradually varied flow profiles.

Table 5.3.1   Types of Flow Profiles in Prismatic Channels

Channel slope	Designation			Relation of y to y_n and y_c			General type of curve	Type of flow
	Zone 1	Zone 2	Zone 3	Zone 1	Zone 2	Zone 3		
Horizontal $S_o = 0$	None			$y > \; y_n$	$> \; y_c$		None	None
		H2		y_n	$> y > y_c$		Drawdown	Subcritical
			H3	y_n	$> \; y_c > y$		Backwater	Supercritical
Mild $0 < S_o < S_c$	M1			$y > \; y_n$	$> \; y_c$		Backwater	Subcritical
		M2		y_n	$> y > y_c$		Drawdown	Subcritical
			M3	y_n	$> \; y_c > y$		Backwater	Supercritical
Critical $S_o = S_c > 0$	C1			$y > \; y_c$	$= \; y_n$		Backwater	Subcritical
		C2		y_n	$= y = y_c$		Parallel to channel bottom	Uniform-critical
			C3	y_c	$= \; y_n > y$		Backwater	Supercritical
Steep $S_o > S_e > 0$	S1			$y \; > \; y_c$	$> \; y_n$		Backwater	Supercritical
		S2		y_c	$> y > y_n$		Drawdown	Supercritical
			S3	y_c	$> \; y_n > y$		Backwater	Supercritical
Adverse $S_o < 0$	None			$y > (y_n)^*$	$> \; y_c$		None	None
		A2		$(y_n)^*$	$> y > y_c$		Drawdown	Subcritical
			A3	$(y_n)^*$	$> \; y_c > y$		Backwater	Supercritcal

* y_n in parentheses is assumed a positive value.

Source: Chow (1959).

The three zones for mild slopes are defined as

Zone 1: $y > y_n > y_c$

Zone 2: $y_n > y > y_c$

Zone 3: $y_n > y_c > y$

The energy grade line, water surface, and channel bottom are all parallel for uniform flow, i.e., $S_f = S_0 =$ slope of water surface when $y = y_n$. From Manning's equation for a given discharge, $S_f < S_0$ if $y > y_n$.

Now consider the qualitative characteristics using the three zones.

Zone 1 (M1 profile): $y > y_n$; then $S_f < S_0$ or $S_0 - S_f = +$

$$F_r < 1 \text{ since } y > y_c, \text{ so } 1 - F_r^2 = +$$

by equation (5.3.11), $\dfrac{dy}{dx} = \dfrac{S_0 - S_f}{1 - F_r^2} = \dfrac{+}{+} = +$

then y increases with x so that $y \rightarrow y_n$

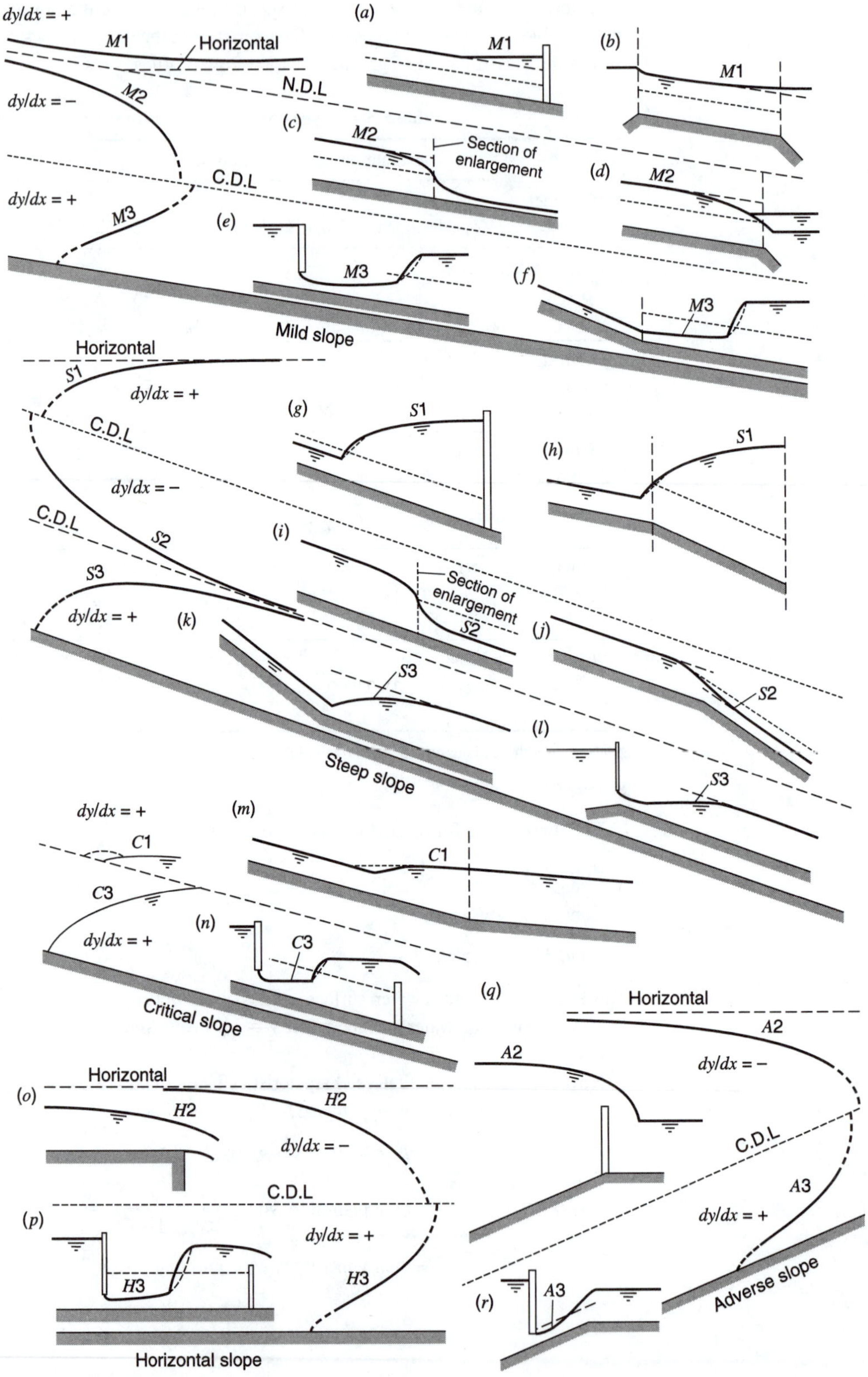

Figure 5.3.3 Flow profiles (from Chow (1959)).

Zone 2 (M2 profile): $y < y_n$; then $S_f > S_0$ or $S_0 - S_f = -$

$$F_r < 1 \text{ since } y > y_c, \text{ so } 1 - F_r^2 = +$$

by equation (5.3.11), $\dfrac{dy}{dx} = \dfrac{S_0 - S_f}{1 - F_r^2} = \dfrac{-}{+} = -$

then y decreases with x so that $y \rightarrow y_c$

Zone 3 (M3 profile): $y < y_n$; then $S_f > S_0$ or $S_0 - S_f = -$

$$F_r > 1 \text{ since } y < y_c \text{ so } 1 - F_r^2 = -$$

by equation (5.3.11), $\dfrac{dy}{dx} = \dfrac{S_0 - S_f}{1 - F_r^2} = \dfrac{-}{-} = +$

then y increases with x so that $y \rightarrow y_c$

This analysis can be made of the other profiles. The results are summarized in Table 5.3.1.

EXAMPLE 5.3.2

For the rectangular channel described in examples 5.1.1, 5.1.3 and 5.2.1, classify the type of slope. Determine the types of profiles that exist for depths of 5.0 ft, 2.0 ft, and 1.0 ft with a discharge of 100 ft³/s.

SOLUTION

In example 5.1.3, the normal depth is computed as $y_n = 3.97$ ft, and in example 5.2.1, the critical depth is computed as $y_c = 1.69$ ft. Because $y_n > y_c$, this is a mild channel bed slope.

For a flow depth of 5.0 ft, $5.0 > y_n > y_c$, so that an M1 profile with a backwater curve exists (refer to Table 5.3.1 and Figure 5.3.3). The flow is subcritical.

For a flow depth of 2.0 ft, $y_n > 2.0 > y_c$, so that an M2 profile with a drawdown curve exists. The flow is subcritical.

For a flow depth of 1.0 ft, $y_n > y_c > 1.0$, so that an M3 profile with a backwater curve exists. The flow is supercritical.

5.4 GRADUALLY VARIED FLOW FOR NATURAL CHANNELS

5.4.1 Development of Equations

As an alternate to the procedure presented above, the gradually varied flow equation can be expressed in terms of the water surface elevation for application to natural channels by considering $w = z + y$ where w is the water surface elevation above a datum such as mean sea level. The total energy H at a section is

$$H = z + y + \alpha \frac{V^2}{2g} = w + \alpha \frac{V^2}{2g} \tag{5.4.1}$$

including the energy correction factor α. The change in total energy head with respect to location along a channel is

$$\frac{dH}{dx} = \frac{dw}{dx} + \frac{d}{dx}\left(\alpha \frac{V^2}{2g}\right) \tag{5.4.2}$$

The total energy loss is due to friction losses (S_f) and contraction-expansion losses (S_e):

$$\frac{dH}{dx} = -S_f - S_e \tag{5.4.3}$$

S_e is the slope term for the contraction-expansion loss. Substituting (5.4.3) into (5.4.2) results in

$$-S_f - S_e = \frac{dw}{dx} + \frac{d}{dx}\left(\alpha\frac{V^2}{2g}\right) \tag{5.4.4}$$

The friction slope S_f can be expressed using Manning's equation (5.1.23) or (5.1.25):

$$Q = KS_f^{1/2} \tag{5.4.5}$$

where K is defined as the *conveyance* in SI units

$$K = \frac{1}{n}AR^{2/3} \tag{5.4.6a}$$

or in U.S. Customary units as

$$K = \frac{1.486}{n}AR^{2/3} \tag{5.4.6b}$$

for equations (5.1.23) or (5.1.25), respectively. The friction slope (from equation 5.4.5) is then

$$S_f = \frac{Q^2}{K^2} = \frac{Q^2}{2}\left[\frac{1}{K_1^2} + \frac{1}{K_2^2}\right] \tag{5.4.7}$$

with the conveyance effect $\frac{1}{K^2} = \frac{1}{2}\left[\frac{1}{K_1^2} + \frac{1}{K_2^2}\right]$, where K_1 and K_2 are the conveyances,

respectively, at the upstream and the downstream ends of the reach. Alternatively, the friction slope can be determined using an average conveyance, i.e., $\bar{K} = (K_1 + K_2)/2$ and $Q = \bar{K}S_f^{1/2}$; then

$$S_f = \frac{Q^2}{\bar{K}^2} \tag{5.4.8}$$

The *contraction-expansion loss term* S_e can be expressed for a *contraction loss* as

$$S_e = \frac{C_c}{dx}\left[\alpha_2\frac{V_2^2}{2g} - \alpha_1\frac{V_1^2}{2g}\right] \text{ for } d\left(\alpha\frac{V^2}{2g}\right) = \left(\alpha_2\frac{V_2^2}{2g} - \alpha_1\frac{V_1^2}{2g}\right) > 0 \tag{5.4.9}$$

and for an *expansion loss* as

$$S_e = \frac{C_e}{dx}\left[\alpha_2\frac{V_2^2}{2g} - \alpha_1\frac{V_1^2}{2g}\right] \text{ for } d\left(\alpha\frac{V^2}{2g}\right) < 0 \tag{5.4.10}$$

The *gradually varied flow equation for a natural channel* is defined by substituting equation (5.4.7) into equation (5.4.4):

$$-\frac{Q^2}{2}\left[\frac{1}{K_1^2} + \frac{1}{K_2^2}\right] - S_e = \frac{dw}{dx} + \frac{d}{dx}\left(\alpha\frac{V^2}{2g}\right) \tag{5.4.11a}$$

or

$$-\frac{Q^2}{2}\left[\frac{1}{K_1^2} + \frac{1}{K_2^2}\right] - S_e = \frac{w_2 - w_1}{\Delta x} + \frac{1}{\Delta x}\left[\alpha_2\frac{V_2^2}{2g} - \alpha_1\frac{V_1^2}{2g}\right] \tag{5.4.11b}$$

Rearranging yields

$$w_1 + \alpha_1\frac{V_1^2}{2g} = w_2 + \alpha_2\frac{V_2^2}{2g} + \frac{Q^2}{2}\left[\frac{1}{K_1^2} + \frac{1}{K_2^2}\right]\Delta x + S_e\Delta x \tag{5.4.12}$$

EXAMPLE 5.4.1

Derive an expression for the change in water depth as a function of distance along a prismatic channel (i.e., constant alignment and slope) for a gradually varied flow.

SOLUTION

We start with equation (5.3.11):

$$\frac{dy}{dx} = \frac{S_o - S_f}{1 - F_r^2}$$

with

$$F_r = \left(V/\sqrt{gD}\right)^2 = \left(Q/A\sqrt{gD}\right)^2 = \left(Q/A\sqrt{gA/T}\right)^2 = \frac{Q^2 T}{gA^3} \, . \text{ Then}$$

$$\frac{dy}{dx} = \frac{S_0 - S_f}{\left(1 - \frac{Q^2 T}{gA^3}\right)}$$

To determine a gradually varied flow profile, this equation is integrated.

EXAMPLE 5.4.2

For a river section with a subcritical discharge of 6,500 ft³/s, the water surface elevation at the downstream section is 5710.5 ft with a velocity head of 3.72 ft. The next section is 500 ft upstream with a velocity head of 1.95 ft. The conveyances for the downstream and upstream sections are 76,140 and 104,300, respectively. Using expansion and contraction coefficients of 0.3 and 0.1, respectively, determine the water surface elevation at the upstream section.

SOLUTION

Using equation (5.4.12), the objective is to solve for the upstream water surface elevation w_1:

$$w_1 = w_2 + a_2 \frac{V_2^2}{2g} - a_1 \frac{V_1^2}{2g} + \frac{Q^2}{2}\left[\frac{1}{K_1^2} + \frac{1}{K_2^2}\right]\Delta x + S_e \Delta x$$

$$= w_2 + \Delta\left(a\frac{V^2}{2g}\right) + \frac{Q^2}{2}\left[\frac{1}{K_1^2} + \frac{1}{K_2^2}\right]\Delta x + S_e \Delta x$$

The friction slope term is

$$\frac{Q^2}{2}\left[\frac{1}{K_1^2} + \frac{1}{K_2^2}\right]\Delta x = \frac{6500^2}{2}\left[\frac{1}{76,140^2} + \frac{1}{104,300^2}\right]500 = 2.79 \text{ ft}$$

$$\Delta\left(\alpha\frac{V^2}{2g}\right) = 3.72 - 1.95 = +1.77 \text{ ft}$$

Because $\Delta[\alpha(V^2/2g)] > 0$, a contraction exists for flow from cross-section 1 to cross-section 2. Then

$$S_e \Delta x = C_c\left[\alpha_2 \frac{V_2^2}{2g} - \alpha_1 \frac{V_1^2}{2g}\right] = 0.1(1.77) = 0.177$$

The water surface elevation at the upstream cross-section is then

$$w_1 = 5710.5 + 1.77 + 2.79 + 0.177 = 5715.2 \text{ ft}$$

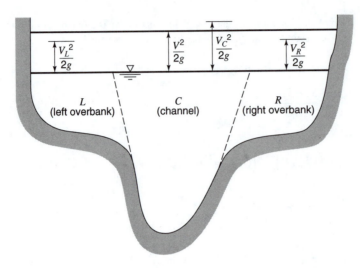

Figure 5.4.1 Compound channel section.

5.4.2 Energy Correction Factor

In section 3.6 the formula for the kinetic energy correction factor is derived as

$$\alpha = \frac{1}{AV^3} \int_A v^3 \, dA \qquad (3.6.4)$$

which can be approximated as

$$\alpha \approx \frac{\sum v^3 \Delta A}{V^3 A} \qquad (5.4.13)$$

where V is the mean velocity.

Consider a compound channel section as shown in Figure 5.4.1 that has three flow sections. The objective is to derive an expression for the energy coefficient in terms of the conveyance for a compound channel, so that the velocity head for the entire channel is $\alpha(V^2/2g)$ where V is the mean velocity in the compound channel. Equation (5.4.13) can be expressed as

$$a = \frac{\sum_{i=1}^{N} V_i^3 A_i}{V^3 \sum_{i=1}^{N} A_i} \qquad (5.4.14)$$

where N is the number of sections (subareas) of the channel (e.g., in Figure 5.4.1, $N = 3$), V_i is the mean velocity in each section (subarea), and A_i is the cross-sectional area of flow in each section (subarea).

The mean velocity can be expressed as

$$V = \frac{\sum_{i=1}^{N} V_i A_i}{\sum_{i=1}^{N} A_i} \qquad (5.4.15)$$

Substituting $V_i = Q_i/A_i$ and equation (5.4.15) for V into (5.4.14) and simplifying yields

$$\alpha = \frac{\displaystyle\sum_{i=1}^{N}\left(\frac{Q_i}{A_i}\right)^3 A_i}{\left[\dfrac{\displaystyle\sum_{i=1}^{N}\left(\dfrac{Q_i}{A_i}\right)A_i}{\displaystyle\sum_{i=1}^{N}A_i}\right]^3 \left(\displaystyle\sum_{i=1}^{N}A_i\right)}$$

$$= \frac{\displaystyle\sum_{i=1}^{N}\left(\frac{Q_i^3}{A_i^2}\right)\left(\displaystyle\sum_{i=1}^{N}A_i\right)^2}{\left(\displaystyle\sum_{i=1}^{N}Q_i\right)^3}$$

(5.4.16)

Now using equation (5.4.5) for each section, we get

$$Q_i = K_i S_{f_i}^{1/2}$$

(5.4.17)

and solving for $S_{f_i}^{1/2}$ yields

$$S_{f_i}^{1/2} = \frac{Q_i}{K_i}$$

(5.4.18)

Assuming that the friction slope is the same for all sections, $S_{f_i} = S_f$ $(i = 1,...,N)$, then according to equation (5.4.18)

$$\frac{Q_1}{K_1} = \frac{Q_2}{K_2} = \frac{Q_3}{K_3} = \cdots = \frac{Q_N}{K_N}$$

(5.4.19)

This leads to

$$Q_1 = K_1 \frac{Q_N}{K_N}$$

$$Q_2 = K_2 \frac{Q_N}{K_N}$$

$$\vdots$$

$$Q_N = K_N \frac{Q_N}{K_N}$$

(5.4.20)

and the total discharge is

$$Q = \sum Q_i = \frac{Q_N}{K_N} \sum K_i$$

(5.4.21)

Substituting the above expression for $\sum Q_i$ and

$$Q_i = K_i \frac{Q_N}{K_N}$$

(5.4.22)

into equation (5.4.16) and simplifying, we get

$$\alpha = \frac{\sum\limits_{i=1}^{N}\left(\dfrac{K_i^3}{A_i^2}\right)\left(\sum\limits_{i=1}^{N}A_i\right)^2}{\left(\sum\limits_{i=1}^{N}K_i\right)^3} \tag{5.4.23}$$

or

$$\alpha = \frac{\sum\limits_{i=1}^{N}\left(\dfrac{K_i^3}{A_i^2}\right)(A_t)^2}{(K_t)^3} \tag{5.4.24}$$

where A_t and K_t are the totals.
The friction slope for the reach is

$$S_{f_i} = \left(\frac{\sum Q_i}{\sum K_i}\right)^2 = \frac{Q^2}{\left(\sum K_i\right)^2} \tag{5.4.25}$$

by eliminating Q_N/K_N from equations (5.4.19) through (5.4.21).

EXAMPLE 5.4.3

For the compound cross-section at river mile 1.0 shown in Figure 5.4.2, determine the energy correction factor α. The discharge is $Q = 11,000$ cfs and the water surface elevation is 125 ft.

SOLUTION

Step 1   Compute the cross-sectional areas of flow for the left overbank (L), channel (C), and right overbank (R):

$A_L = 1050$ ft^2, $A_c = 3000$ ft^2, $A_R = 1050$ ft^2

Step 2   Compute the hydraulic radius for L, C, and R:

$R_L = 1050/85 = 12.35$ ft; $R_c = 3000/140 = 21.40$ ft; $R_R = 1050/85 = 12.35$ ft

Step 3   Compute the conveyance factor for L, C, and R:

$$K_L = \frac{1.486A_L R_L^{2/3}}{n} = \frac{1.486(1050)(12.35)^{2/3}}{0.04} = 208,700$$

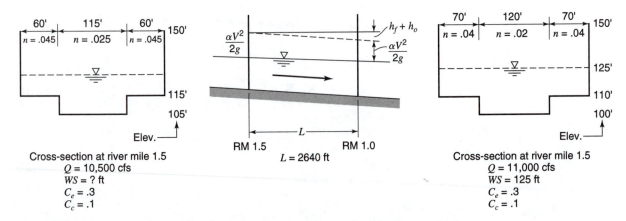

Cross-section at river mile 1.5
$Q = 10,500$ cfs
$WS = ?$ ft
$C_e = .3$
$C_c = .1$

$L = 2640$ ft

Cross-section at river mile 1.5
$Q = 11,000$ cfs
$WS = 125$ ft
$C_e = .3$
$C_c = .1$

Figure 5.4.2 Cross-section and reach length data for example 5.4.3 (from Hoggan (1997)).

$$K_C = \frac{1.486(3000)(21.4)^{2/3}}{0.02} = 1,716,300$$

$$K_R = 208,700 \quad (K_R = K_L)$$

Step 4 Compute totals A_t and K_t:

$$K_t = K_L + K_C + K_R = 2,133,700$$

$$A_t = A_L + A_C + A_R = 5100 \text{ ft}^2$$

Step 5 Compute K^3/A^2 and $\Sigma\, K^3/A^2$:

$$K_L{}^3/A_L{}^2 = 8.25 \times 10^9$$

$$K_C{}^3/A_C{}^2 = 561.8 \times 10^9$$

$$K_R{}^3/A_R{}^2 = 8.25 \times 10^9$$

$$\Sigma\, K^3/A^2 = 578.3 \times 10^9$$

Step 6 Use equation (5.4.24) to compute α:

$$\alpha = \frac{\sum_{i=1}^{N}\left(\dfrac{K_i^3}{A_i^2}\right)(A_t)^2}{(K_t)^3} = \frac{(578.3 \times 10^9)(5100)^2}{(2,133,700)^3} = 1.55$$

EXAMPLE 5.4.4

For the data in Figure 5.4.2, start with the known water surface elevation at river mile 1.0 and determine the water surface at river mile 1.5 (adapted from Hoggan, 1997).

SOLUTION

Computations are presented in Table 5.4.1.

Table 5.4.1 Standard Step Backwater Computation

(1) River mile	(2) Water surface elevation W_K (ft)	(3) Area A_2 (ft)	(4) Hydraulic radius R	(5) Manning roughness n	(6) Convey-ance K	(7) Average conveyance $\bar{K}$	(8) Average friction slope $\bar{S}_f$	(9) Friction loss h_L (ft)	(10) $\dfrac{K^3}{A^2}$ (10^9)	(11) Energy correction factor α	(12) Velocity V (ft/sec)	(13) $\alpha\dfrac{V^2}{2g}$ (ft/sec)	(14) $\Delta\left(\dfrac{V^2}{2g}\right)$ (ft/sec)	(15) h_o (ft)	(16) W_K (ft)
1.0	125.0	1050.	12.35	0.040	208,700.				8.25						
		3000.	21.43	0.020	1,716,300.				561.80						
		1050.	12.35	0.040	208,700.				8.25						
		5100.			2,133,700.				578.30	1.55	2.16	0.11			
1.5	126.1	666.0	9.37	0.045	97,650.				2.10						
		2426.5	17.97	0.025	989,400.				164.50						
		666.0	9.37	0.045	97,650.				2.10						
		3758.5			1,184,700.	1,659,200	0.00004	0.106	168.70	1.43	2.79	0.17	−0.06	0.02	125.07
1.5	125.0	600.	8.57	0.045	83,000.				1.59						
		2300.	17.04	0.025	905,300.				140.22						
		600.	8.57	0.045	83,000.				1.59						
		3500			1,071,300.	1,602,350	0.000043	0.113	143.30	1.43	3.00	0.20	−0.09	0.03	125.05

*$\alpha = \dfrac{(A_t)^2 \Sigma\, K_i^3/A_i^2}{(K_t)^3}$

**$h_o = C_c \,|\, \Delta\,(\alpha V^2/2g)\,|$ for $\Delta\,(\alpha V^2/2g) < 0$ (loss due to channel expansion); $h_o = C_c \,|\, \Delta\,(\alpha V^2/2g)\,|$ for $\Delta\,(\alpha V^2/2g) > 0$ (loss due to channel contraction).

***$W_2 = W_1 + \Delta\,(\alpha\, V^2/2g) + h_L + h_o = 125.0 + (−0.06) + 0.106 + 0.02 = 125.066 \cong 125.07$.

Source: Hoggan (1997).

5.4.3   Application for Water Surface Profile

The change in head with respect to distance x along the channel has been expressed in equation (5.4.2) as

$$\frac{dH}{dx} = \frac{dw}{dx} + \frac{d}{dx}\left(\alpha\frac{V^2}{2g}\right)$$ (5.4.2)

The total energy loss term is $dH/dx = -S_f - S_e$, where S_f is the friction slope defined by equation (5.4.7) or equation (5.4.8) and S_e is the slope of the contraction or expansion loss. The differentials dw and $d[\alpha(V^2/2g)]$ are defined over the channel reach as $dw = w_k - w_{k+1}$ and

$d\left[\alpha(V^2/2g)\right] = \alpha_{k+1}\dfrac{V_{k+1}^2}{2g} - \alpha_k\dfrac{V_k^2}{2g}$, where we define $k+1$ at the downstream and k at the

upstream. River cross-sections are normally defined from downstream to upstream for gradually varied flow. Equation (5.4.2) is now expressed as

$$w_k + \alpha_k\frac{V_k^2}{2g} = w_{k+1} + \alpha_{k+1}\frac{V_{k+1}^2}{2g} + S_f dx + S_e dx$$ (5.4.26)

The *standard step procedure* for water surface computations is described in the following steps:

a. Start at a point in the channel where the water surface is known or can be approximated. This is the *downstream boundary condition* for subcritical flow and the *upstream boundary condition* for supercritical flow. Computation proceeds upstream for subcritical flow and downstream for supercritical flow. Why?

b. Choose a water surface elevation w_k at the upstream end of the reach for subcritical flow or w_{k+1} at the downstream end of the reach for supercritical flow. This water surface elevation will be slightly lower or higher depending upon the type of profile (see Chow, 1959; Henderson, 1966; French, 1985; or Chaudhry, 1993).

c. Next compute the conveyance, corresponding friction slope and expansion and contraction loss terms in equation (5.4.26) using the assumed water surface elevation.

d. Solve equation (5.4.26) for w_{k+1} (supercritical flow) or w_k (subcritical flow).

e. Compare the calculated water surface elevation w with the assumed water surface elevation w'. If the calculated and assumed elevations do not agree within an acceptable tolerance (e.g., 0.01 ft), then set $w_{k+1}' = w_{k+1}$ (for supercritical flow) and $w_k' = w_k$ (for subcritical flow) and return to step (c).

Computer models for determining water surface profiles using the standard step procedure include the HEC-2 model and the newer HEC-RAS model. HEC-RAS River Analysis System (developed by the U.S. Army Corps of Engineers (USACE) Hydrologic Engineering Center) computes water surface profiles for one-dimensional steady, gradually varied flow in rivers of any cross-section (HEC, 1997a–c). HEC-RAS can simulate flow through a single channel, a dendritic system of channels, or a full network of open-channels (sometimes called a fully looped system).

HEC-RAS can model sub- or supercritical flow, or a mixture of each within the same analysis. A graphical user interface provides input data entry, data modifications, and plots of stream cross-sections, profiles, and other data. Program options include inserting trapezoidal excavations on cross-sections, and analyzing the potential for bridge scour. The water surface profile through structures such as bridges, culverts, weirs, and gates can be computed. The World Wide Web address to obtain the HEC-RAS model is www.hec.usace.army.mil.

EXAMPLE 5.4.5

A plan view of the Red Fox River in California is shown in Figure 5.4.3, along with the location of four cross-sections. Perform the standard step calculations to determine the water surface elevation at cross-section 3 for a discharge of 6500 ft^3/s. Figures 5.4.4a, b, and c are plots of cross-sections at 1, 2, and 3, respectively. Figures 5.4.5a, b, and c are the area and hydraulic radius curves for cross-sections 1, 2, and 3 respectively. Use expansion and contraction coefficients of 0.3 and 0.1, respectively. Manning's roughness factors are presented in Figure 5.4.4. The downstream starting water surface elevation at cross-section 1 is 5710.5 ft above mean sea level. This example was originally adapted by the U.S. Army Corps of Engineers from material developed by the U.S. Bureau of Reclamation (1957).

SOLUTION

The computations for this example are illustrated in Table 5.4.2.

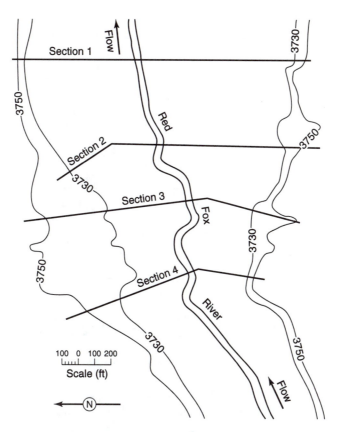

Figure 5.4.3 Map of the Red Fox River indicating cross-sections for water surface profile analysis (from U.S. Bureau of Reclamation (1957)).

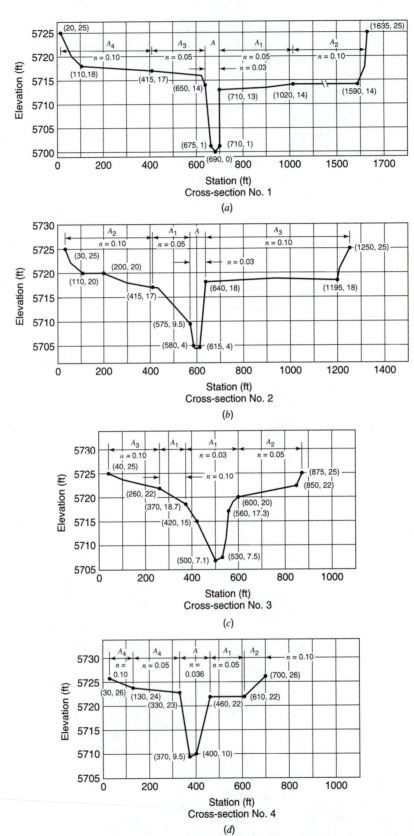

Figure 5.4.4 Cross-sections of the Red Fox River (from U.S. Bureau of Reclamation (1957)).

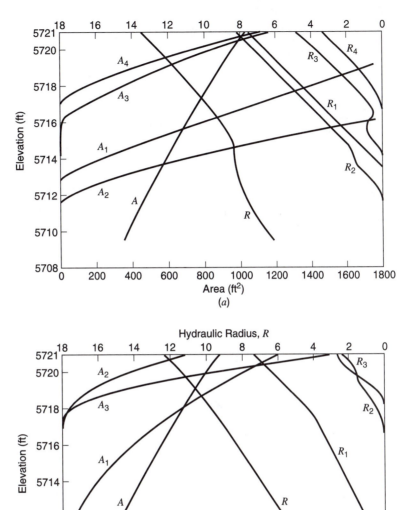

Figure 5.4.5 Area elevation and hydraulic radius—elevation curves for cross-sections 1 to 4. (*a*) Cross-section 1; (*b*) Cross-section 2 (from U.S. Bureau of Reclamation (1957)).

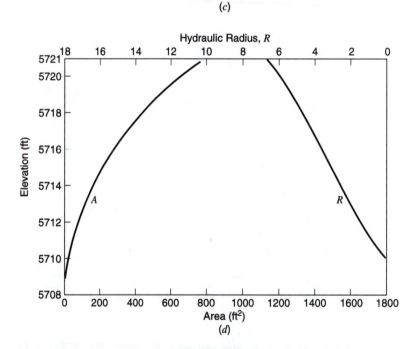

Figure 5.4.5 (*continued*) Area elevation and hydraulic radius—elevation curves for cross-sections 1 to 4. (*c*) Cross-section 3; (*d*) Cross-section 4. (from U.S. Bureau of Reclamation (1957)).

Table 5.4.2   Standard Step Backwater Computation for Red Fox River

(1)	(2)	(3)	(4)	(5)	(6)	(7)	(8)	(9)	(10)	(11)	(12)	(13)	(14)	(15)	(16)
Cross-section	Water surface elevation W_{k+1} (ft)	Area A_2 (ft)	Hydraulic radius R (ft)	Manning roughness n	Convey-ance K	Average conveyance $\bar{K}$	Average friction slope $S_f(10^{-3})$ (ft/ft)	Friction loss h_L (ft)	$\dfrac{K^3}{A^2}$ (10^6)	Energy correction factor α	Velocity V (ft/sec)	$\alpha\dfrac{V^2}{2g}$ (ft)	$\Delta\left(\alpha\dfrac{V^2}{2g}\right)$ (ft)	h_o (ft)	W_k (ft)
1	5710.5*	420	7.0	0.03	76,100					1.0	15.5	3.72			
2	5714.7	470	7.6	0.03	90,100				3311.1						
		260	2.5	0.05	14,200				42.0						
		730			104,300	90,200	5.19	2.60	3353.5	1.58	8.90	1.95	+1.77	0.18	5715.07
	5715.0														
		500	7.85	0.03	97,800				3741.8						
		300	2.7	0.05	17,300				57.5						
		800			115,100	95,600	4.62	2.31	3799.3	1.59	8.13	1.63	+2.09	0.21	5715.1
3	5718	1145	5.85	0.03	184,100	149,600	1.89	0.76		1.0	5.68	0.50	+1.13	0.11	5717.1
	5717.1	970	5.6	0.03	151,500	133,300	2.38	0.95		1.0	6.70	0.70	+.93	0.09	5717.1

*Known starting water surface elevation.

**$W_{k+1} = 5710.5 + 1.77 + 2.60 + 0.18 = 5715.05 = 5715.1$; $\alpha = (A_t)^2 \sum K_i^3/A_i^2/(K_t)^3$; $h_o = C_e \mid = (\alpha V^2/2g) \mid$ for $\Delta (\alpha V^2 2g) < 0$ (loss due to channel expansion; $h_o = C_e \mid \Delta (\alpha V^2/2g) \mid$ for $\Delta (\alpha V^2/2g) > 0$ (loss due to channel contraction); $W_{K+1} = W_K + \Delta (\alpha V^2/2g) + h_L + h_o$.

Source: Hoggan (1997).

5.5   RAPIDLY VARIED FLOW

Rapidly varied flow occurs when a water flow depth changes abruptly over a very short distance. The following are characteristic features of rapidly varied flow (Chow, 1959):

- Curvature of the flow is pronounced, so that pressure distribution cannot be assumed to be hydrostatic.
- The rapid variation occurs over a relatively short distance so that boundary friction is comparatively small and usually insignificant.
- Rapid changes of water area occur in rapidly varied flow, causing the velocity distribution coefficients α and β to be much greater than 1.0.

Examples of rapidly varied flow are hydraulic jumps, transitions in channels, flow over spillways, flow in channels of nonlinear alignment, and flow through nonprismatic channel sections such as flow in channel junctions, flow through trash racks, and flow between bridge piers.

The discussion presented in this chapter is limited to the hydraulic jump. The *hydraulic jump* occurs when a rapid change in flow depth occurs from a small depth to a large depth such that there is an abrupt rise in water surface. A hydraulic jump occurs wherever supercritical flow changes to subcritical flow. Hydraulic jumps can occur in canals downstream of regulating sluices, at the foot of spillways, or where a steep channel slope suddenly becomes flat.

Figure 5.5.1 illustrates a hydraulic jump along with the specific energy and specific force curves. The depths of flow upstream and downstream of the jump are called *sequent depths* or *conjugate depths*. Because hydraulic jumps are typically short in length, the losses due to shear along the wetted perimeter are small compared to the pressure forces. Neglecting these forces and assuming a horizontal channel ($F_g = 0$), the momentum principle can be applied as in section 5.2.3 to derive equation (5.2.24):

$$\frac{Q^2}{gA_1} + A_1\bar{y}_1 = \frac{Q^2}{gA_2} + A_2\bar{y}_2 \qquad (5.2.24)$$

Consider a rectangular channel of width $B > 0$, so $Q = A_1V_1 = A_2V_2$, $A_1 = By_1$, $A_2 = By_2$, $\bar{y}_1 = y_1/2$ and $\bar{y}_2 = y_2/2$.

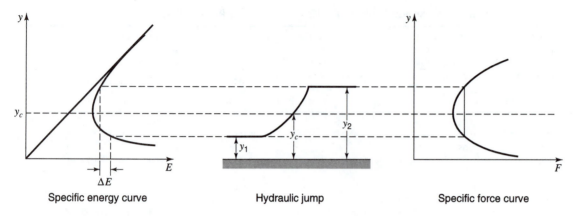

Figure 5.5.1 Hydraulic jump.

$$\frac{Q^2}{gBy_1} + By_1\left(\frac{y_1}{2}\right) = \frac{Q^2}{gBy_2} + By_2\left(\frac{y_2}{2}\right)$$ (5.5.1)

Simplifying yields

$$\frac{Q^2}{g}\left(\frac{1}{y_1} - \frac{1}{y_2}\right) = \frac{1}{2}B^2\left(y_2^2 - y_1^2\right)$$

$$\frac{Q^2}{g}(y_2 - y_1) = \frac{1}{2}B^2 y_1 y_2\left(y_2^2 - y_1^2\right)$$

$$\frac{B^2 y_1^2 V_1^2}{g}(y_2 - y_1) = \frac{1}{2}B^2 y_1 y_2(y_2 + y_1)(y_2 - y_1)$$

$$\frac{y_1 V_1^2}{g} = \frac{1}{2}y_2(y_2 + y_1)$$

Dividing by y_{12}, we get

$$\frac{2V_1^2}{gy_1} = \frac{y_2}{y_1}\left(\frac{y_2}{y_1} + 1\right)$$ (5.5.2)

The Froude number for a rectangular channel is $F_{r_1} = V_1/\sqrt{gD_1} = V_1/\sqrt{gy_1}$; therefore equation (5.5.2) reduces to

$$\left(\frac{y_2}{y_1}\right)^2 + \frac{y_2}{y_1} - 2F_{r_1}^2 = 0$$ (5.5.3)

or

$$\frac{y_2}{y_1} = \frac{1}{2}\left(-1 + \sqrt{1 + 8F_{r_1}^2}\right)$$ (5.5.4)

Alternatively, $Q = B^2 y_2^2 V_2^2$ and $F_{r_2} = V_2/\sqrt{gy_2}$ could have been used to derive

$$\frac{y_1}{y_2} = \frac{1}{2}\left(-1 + \sqrt{1 + 8F_{r_2}^2}\right)$$ (5.5.5)

Equations (5.5.4) and (5.5.5) can be used to find the sequent depths of a hydraulic jump. The use of hydraulic jumps as energy dissipaters is further discussed in Chapters 15 and 17.

EXAMPLE 5.5.1

Consider the 8-ft wide rectangular channel used in examples 5.1.1, 5.1.3, and 5.3.1 with a discharge of 100 cfs. If a weir were placed in the channel and the depth upstream of the weir were 5 ft, would a hydraulic jump form upstream of the weir?

SOLUTION

For the discharge of 100 cfs, the normal depth is $y_n = 3.97$ ft from example 5.1.3, and the critical depth is $y_c = 1.69$. Because $y_c < y_n < 5$ ft, a hydraulic jump would not form. As a result of $y_n > y_c$, a mild slope exists. For a jump to form, $y_n < y_c$, which is a steep slope.

EXAMPLE 5.5.2

For example 5.3.1, determine whether a hydraulic jump will occur.

SOLUTION

The normal depth and critical depth must be computed and compared. Using equation (5.2.14) with $q = 4.646$ m³/s per meter of width, we get

$$y_c = \left(\frac{q^2}{g}\right)^{1/3} = \left(\frac{4.646^2}{9.81}\right)^{1/3} = 1.30 \text{ m.}$$

The depths of flow at $y_a = 0.457$ m and at $y_b = 0.5$ m, so this flow is supercritical flow.

Next, the normal depth is computed using Manning's equation (5.1.23):

$$Q = \frac{1}{n} AR^{2/3} S_0^{1/2}$$

or

$$q = \frac{1}{n}(y_n)(y_n)^{2/3} S_0^{1/2} = \frac{1}{n} y_n^{5/3} S_0^{1/2}$$

$$4.646 = \frac{1}{0.020} y_n^{5/3} (0.0003)^{1/2}$$

Thus $y_n = (5.365)^{3/5} = 2.74$ m. Under these conditions $y_n > y_c > 0.5$ m, an M3 water surface profile exists and a hydraulic jump occurs. If normal depth occurs downstream of the jump, what is the depth before the jump? Using equation (5.5.5) with $y_2 = 2.74$ m, we find that y_1 is 0.5 m.

EXAMPLE 5.5.3

A rectangular channel is 10.0 ft wide and carries a flow of 400 cfs at a normal depth of 3.00 feet. Manning's $n = 0.017$. An obstruction causes the depth just upstream of the obstruction to be 8.00 ft deep. Will a jump form upstream from the obstruction? If so, how far upstream? What type of curve will be present?

SOLUTION

First a determination must be made if a jump will form by comparing the normal depth and critical depth. Using equation (5.2.14), we find

$$y_c = \left(\frac{q^2}{g}\right)^{1/3} = \left(\frac{40^2}{32.2}\right)^{1/3} = 3.68 \text{ ft}$$

Because $y_n < y_c < 8$ ft, a supercritical flow exists and a hydraulic jump will form. If the depth before the jump is considered to be normal depth, $y_n = y_1 = 3$, then the conjugate depth y_2 can be computed using equation (5.5.4), where the Froude number is

$$F_{r_1}^2 = \left(\frac{V_1}{\sqrt{gy_1}}\right)^2 = \frac{V_1^2}{gy_1} = \frac{q^2}{gy_1^3}$$

so

$$y_2 = \frac{y_1}{2}\left(-1+\sqrt{1+\frac{8q^2}{gy_1^3}}\right)$$

$$= \frac{3}{2}\left(-1+\sqrt{1+\frac{8(40)^2}{g(3)^3}}\right)$$

$$= 4.45 \text{ ft}$$

Next the distance Δx from the depth of 4.45 ft to the depth of 8 ft is determined using equation (5.3.1):

$$S_0\Delta x + \frac{V_2^2}{2g} = \Delta y + \frac{V_3^2}{2g} + S_f\Delta x$$

which can be rearranged to yield

$$(S_0 - S_f)\Delta x = y_3 + \frac{V_3^2}{2g} - y_2 - \frac{V_2^2}{2g}$$

$$= E_3 - E_2$$

so

$$\Delta x = \frac{E_3 - E_2}{S_0 - S_f} = \frac{\Delta E}{S_0 - S_f}$$

To solve for Δx, first compute E_2 and E_3:

Depth (ft)	A (ft^2)	R (ft)	V (ft/s)	$V^2/2g$ (ft)	E (ft)	R_{ave} (ft)	V_{ave} (ft/s)
8	80	3.08	5.00	0.388	8.388		
4.45	44.5	2.35	8.99	1.25	5.70	2.72	7.00

Now compute S_f from Manning's equation using equation (5.1.24) with V_{ave} and R_{ave}, and rearrange to yield

$$S_f = \frac{n^2V_{ave}^2}{2.22R_{ave}^{4/3}} = \frac{0.017^2 \times 7^2}{2.22 \times 2.72^{4/3}} = 0.00168$$

Compute S_0 using Manning's equation with the normal depth:

$$Q = \frac{1.49}{n}AR^{2/3}S_0^{1/2}$$

$$400 = \frac{1.49}{0.017} \times 3 \times 10 \times \left(\frac{30}{16}\right)^{2/3}\sqrt{S_0}$$

Thus, $S_0 = 0.0100$. Now, using $\Delta x = \dfrac{\Delta E}{S_0 - S_f} = \dfrac{8.388 - 5.71}{0.0100 - 0.00168} = 322$ ft, the distance from the conjugate depth of the jump $y_2 = 4.45$ ft downstream to the depth y_3 of 8 ft (location of the obstruction) is 322 ft. In other words, the hydraulic jump occurs approximately 322 ft upstream of the obstruction. The type of water surface profile after the jump is an S1 profile.

EXAMPLE 5.5.4

A hydraulic jump occurs in a rectangular channel 3.0 m wide. The water depth before the jump is 0.6 m, and after the jump is 1.6 m. Compute (a) the flow rate in the channel (b) the critical depth (c) the head loss in the jump.

SOLUTION

(a) To compute the flow rate knowing $y_1 = 0.6$ m and $y_2 = 1.6$ m, equation (5.5.4) can be used:

$$y_2 = \frac{y_1}{2}\left[-1+\sqrt{1+\frac{8q^2}{gy_1^3}}\right]$$

in which $F_{r_1} = q/\sqrt{gy_1^3}$ has been substituted:

$$1.6 = \frac{0.6}{2}\left[-1+\sqrt{1+\frac{8q^2}{9.81(0.6)^3}}\right]$$

$$6.33 = \sqrt{1+3.775q^2}$$

$$40.07 = 3.775q^2$$

$$q = 3.26 \text{ m}^3/\text{s per meter width of channel}$$

and

$$Q = 3q = 9.78 \text{ m}^3/\text{s}.$$

(b) Critical depth is computed using equation (5.2.14):

$$y_c = \left(\frac{q^2}{g}\right)^{1/3} = \left(\frac{3.26^2}{9.81}\right)^{1/3} = 1.03 \text{ m}$$

(c) The headloss in the jump is the change in specific energy before and after the jump

$$h_L = \Delta E = E_1 - E_2$$

so

$$h_L = y_1 + \frac{V_1^2}{2g} - y_2 - \frac{V_2^2}{2g}$$

so

$$V_1 = Q/A_1 = 9.78/(3 \times 0.6) = 5.43 \text{ m/s}$$

$$V_2 = Q/A_2 = 9.78/(3 \times 1.6) = 2.04 \text{ m/s}$$

$$h_L = 0.6 + \frac{5.43^2}{2(9.81)} - 1.6 - \frac{2.04^2}{2(9.81)}$$

$$= 0.6 + 1.5 - 1.6 - 0.21$$

$$h_L = 0.29 \text{ m}.$$

EXAMPLE 5.5.5

Derive an equation to approximate the headloss (energy loss) of a hydraulic jump in a horizontal rectangular channel in terms of the depths before and after the jump, y_1 and y_2.

SOLUTION

The energy loss can be approximated by

$$h_L = E_1 - E_2$$

$$= \left(y_1 + \frac{V_1^2}{2g} \right) - \left(y_2 + \frac{V_2^2}{2g} \right)$$

The velocities can be expressed as $V_1 = Q/A_1 = q/y_1$ and $V_2 = q/y_2$, so

$$h_L = y_1 + \frac{1}{2g} \frac{q^2}{y_1^2} - y_2 - \frac{1}{2g} \frac{q^2}{y_2^2}$$

$$= \frac{q^2}{2g} \left(\frac{1}{y_1^2} - \frac{1}{y_2^2} \right) + (y_1 - y_2)$$

The balance between hydrostatic forces and the momentum flux per unit width of the channel can be expressed using equation (5.2.23):

$$\sum F = F_1 - F_2 = \rho q (V_2 - V_1) = \gamma A_1 \bar{y}_1 - \gamma A_2 \bar{y}_2 = \rho q V_2 - \rho q V_1$$

where $A = (1)(y_1)$ and $\bar{y}_1 = y_1/2$, so

$$\frac{\gamma}{2} y_1^2 - \frac{\gamma}{2} y_2^2 = \rho q \left(\frac{q}{y_2} - \frac{q}{y_1} \right)$$

Solving, we get

$$\frac{q^2}{g} = y_1 y_2 \left(\frac{y_1 + y_2}{2} \right)$$

Substituting this equation into the above equation for the headloss and simplifying gives

$$\Delta E = \frac{(y_2 - y_1)^3}{4 y_1 y_2}$$

5.6   DISCHARGE MEASUREMENT

5.6.1   Weir

A *weir* is a device (or overflow structure) that is placed normal to the direction of flow. The weir essentially backs up water so that in flowing over the weir the water goes through critical depth. Weirs have been used for the measurement of water flow in open-channels for many years. Weirs can generally be classified as *sharp-crested weirs* and *broad-crested weirs*. Weirs are discussed in detail in Bos et al. (1984), Brater et al. (1996), and Replogle et al. (1999).

A *sharp-crested weir* is basically a thin plate mounted perpendicular to the flow with the top of the plate having a beveled, sharp edge, which makes the nappe spring clear from the plate (see Figure 5.6.1). The rate of flow is determined by measuring the head, typically in a stilling well (see Figure 5.6.2) at a distance upstream from the crest. The head H is measured using a gauge.

Suppressed rectangular weir

These sharp-crested weirs are as wide as the channel and the width of the nappe is the same length as the crest. Referring to Figure 5.6.1, consider an elemental area $dA = Bdh$ and assume the velocity is $\sqrt{2gh}$; then the elemental flow is

$$dQ = Bdh\sqrt{2gh} = B\sqrt{2g}h^{1/2}\,dh \tag{5.6.1}$$

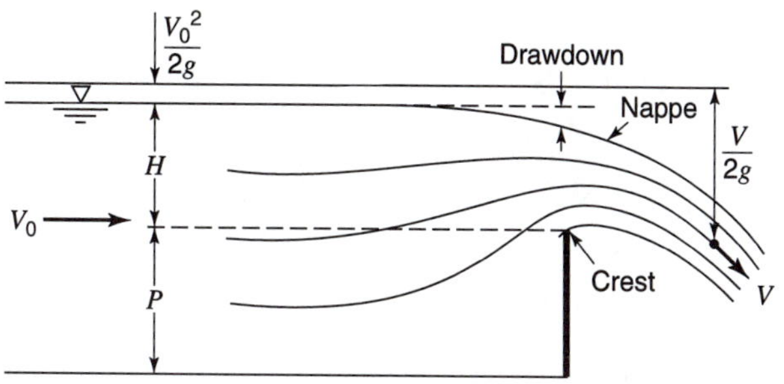

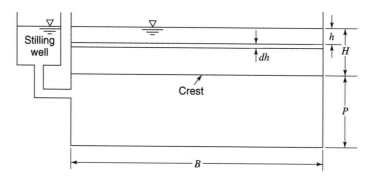

Figure 5.6.1 Flow over sharp-crested weir.

Figure 5.6.2 Rectangular sharp-crested weir without end contraction.

The discharge is expressed by integrating equation (5.6.1) over the area above the top of the weir crest:

$$Q = \int_0^H dQ = \sqrt{2g} B \int_0^H h^{1/2} dh = \frac{2}{3}\sqrt{2g} BH^{3/2} \tag{5.6.2}$$

Friction effects have been neglected in derivation of equation (5.6.2). The drawdown effect shown in Figure 5.6.1 and the crest contraction indicate that the streamlines are not parallel or normal to the area in the plane. To account for these effects a coefficient of discharge C_d is used, so that

$$Q = C_d \frac{2}{3}\sqrt{2g} BH^{3/2} \tag{5.6.3}$$

where C_d is approximately 0.62. This is the basic equation for a suppressed rectangular weir, which can be expressed more generally as

$$Q = C_w BH^{3/2} \tag{5.6.4}$$

where the C_w is the weir coefficient, $C_w = C_d \frac{2}{3}\sqrt{2g}$. For U.S. customary units, $C_w \approx 3.33$, and for SI units $C_w \approx 1.84$.

If the velocity of approach V_a where H is measured is appreciable, then the integration limits are

$$Q = \sqrt{2g} B \int_{\frac{V_a^2}{2g}}^{H+\frac{V_a^2}{2g}} h^{1/2} dh = C_w B \left[\left(H + \frac{V_a^2}{2g} \right)^{3/2} - \left(\frac{V_a^2}{2g} \right)^{3/2} \right] \tag{5.6.5a}$$

When $\left(\dfrac{V_a^2}{2g}\right)^{3/2} \approx 0$, equation (5.6.5a) can be simplified to

$$Q = C_w B \left(H + \frac{V_a^2}{2g} \right)^{3/2}$$

(5.6.5b)

Contracted rectangular weirs

A *contracted horizontal weir* is another sharp crested weir with a crest that is shorter than the width of the channel and one or two beveled end sections so that water contracts both horizontally and vertically. This forces the nappe width to be less than B. The effective crest length is

$$B' = B - 0.1\,nH$$

(5.6.6)

where $n = 1$ if the weir is placed against one side wall of the channel so that the contraction on one side is suppressed and $n = 2$ if the weir is positioned so that it is not placed against a side wall.

Triangular weir

Triangular or *V-notch weirs* are sharp-crested weirs that are used for relatively small flows, but have the advantage that they can also function for reasonably large flows as well. Referring to Figure 5.6.3, the rate of discharge through an elemental area, dA, is

$$dQ = C_d \sqrt{2gh}\,dA$$

(5.6.7)

where $dA = 2xdh$, and, $x = (H - h)\tan\dfrac{\theta}{2}$ so $dA = 2(H - h)\tan\left(\dfrac{\theta}{2}\right)dh$. Then

$$dQ = C_d \sqrt{2gh}\left[2(H - h)\tan\left(\frac{\theta}{2}\right)dh \right]$$

(5.6.8)

and

$$Q = C_d\, 2\sqrt{2g}\, \tan\left(\frac{\theta}{2}\right)\int_0^H (H - h)h^{1/2}dh$$

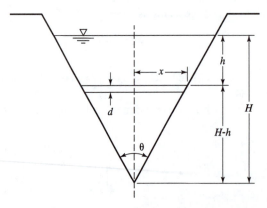

Figure 5.6.3 Triangular sharp-crested weir.

$$= C_d 2\sqrt{2g} \tan\left(\frac{\theta}{2}\right) \int_0^H (H-h)h^{1/2}\,dh$$

$$= C_d\left(\frac{8}{15}\right)\sqrt{2g} \tan\left(\frac{\theta}{2}\right) H^{5/2}$$

$$= C_w H^{5/2} \tag{5.6.9}$$

The value of C_w for a value of $\theta = 90°$ (the most common) is $C_w = 2.50$ for U.S. customary units and $C_w = 1.38$ for SI units.

Broad-crested weir

Broad-crested weirs (refer to Figure 5.6.4) are essentially critical-depth weirs in that if the weirs are high enough, critical depth occurs on the crest of the weir. For critical flow conditions $y_c = \left(q^2/g\right)^{1/3}$ and $E = \frac{3}{2}y_c$ for rectangular channels:

$$Q = B \cdot q = B\sqrt{gy_c^3} = B\sqrt{g\left(\frac{2}{3}E\right)^3} = B\left(\frac{2}{3}\right)^{3/2}\sqrt{g}E^{3/2}$$

or, assuming the approach velocity is negligible:

$$Q = B\left(\frac{2}{3}\right)^{3/2}\sqrt{g}H^{3/2}$$

$$Q = C_w BH^{3/2} \tag{5.6.10}$$

Figure 5.6.5 illustrates a broad-crested weir installation in a concrete-lined canal.

EXAMPLE 5.6.1   A rectangular, sharp-crested suppressed weir 3 m long is 1.0 m high. Determine the discharge when the head is 150 mm.

SOLUTION   Using equation (5.6.4), $Q = 1.84\,BH^{1.5}$, the discharge is

$$Q = 1.84(3)\left(\frac{150}{1000}\right)^{1.5} = 0.321 \text{ m}^3/\text{s}$$

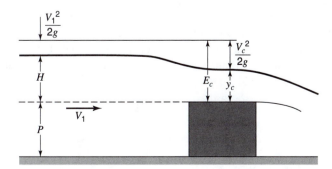

Figure 5.6.4 Broad-crested weir.

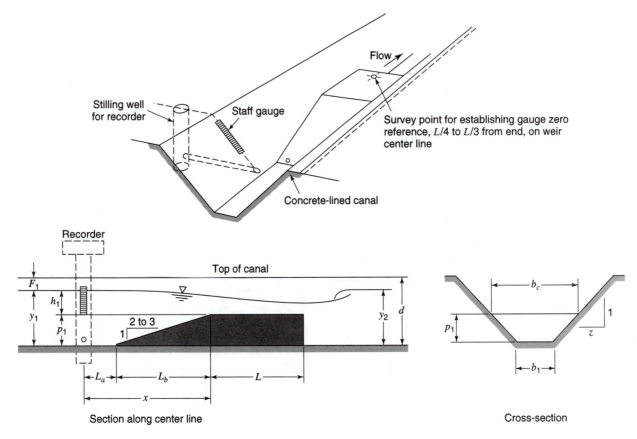

Figure 5.6.5 Broad-crested weir in concrete-lined canal (from Bos et al. (1984)).

5.6.2   Flumes

Bos et al. (1984) provide an excellent discussion of flumes. A weir is a control section that is formed by raising the channel bottom, whereas a *flume* is formed by narrowing a channel. When a control section is formed by raising both the channel bottom and narrowing it, the structure is usually called a *flume*. Figure 5.6.6 shows a distinction between weirs and flumes.

Weirs can result in relatively large headlosses and, if the water has suspended sediment, can cause deposition upstream of the weir, resulting in a gradual change of the weir coefficient . These disadvantages can be overcome in many situations by the use of flumes.

Figure 5.6.7 illustrates the general layout of a flow measuring structure. Most flow measurement and flow regulating structures consist of (a) a *converging transition*, where subcritical flow is accelerated and guided into the throat without flow separation; (b) a *throat* where the water accelerates to supercritical flow so that the discharge is controlled; and (c) a *diverging transition* where flow velocity is gradually reduced to subcritical flow and the potential energy is recovered.

One of the more widely used flumes is the *Parshall flume* (U.S. Bureau of Reclamation, 1981), which is a Venturi-type flume illustrated in Figure 5.6.8. The discharge equation for these flumes with widths of 1 ft (0.31 m) to 8 ft (2.4 m) is

$$Q = 4WH_a^{1.522W^{0.026}}$$

(5.6.11)

where Q is the discharge in ft^3/s, W is the width of the flume throat and H_a is the upstream head in ft. For smaller flumes, e.g. 6-inch flumes,

$$Q = 2.06H_a^{1.58} \qquad (5.6.12)$$

and for 9-inch flumes

$$Q = 3.07H_a^{1.53} \qquad (5.6.13)$$

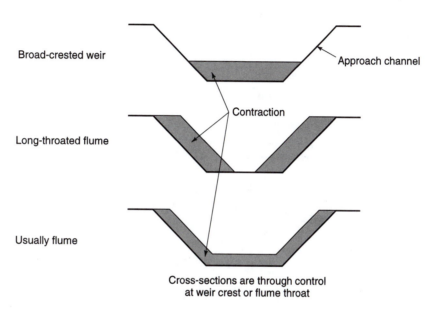

Figure 5.6.6 Distinction between a weir and a flume (from Bos et al. (1984)).

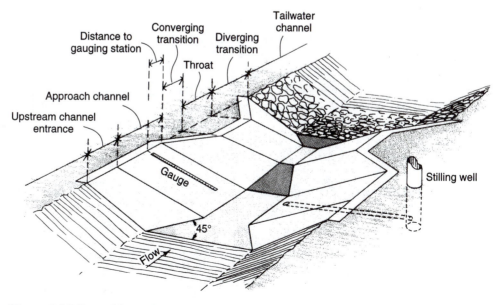

Figure 5.6.7 General layout of a flow-measuring structure (from Bos et al. (1984)).

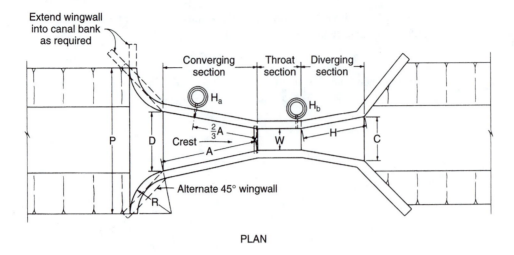

PLAN

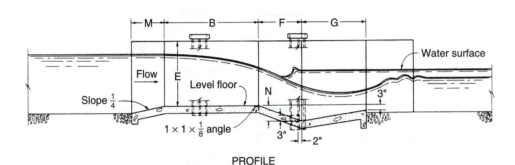

PROFILE

W		A		$\frac{2}{3}$A		B		C		D		E		F		G		M		N		P		R		FREE-FLOW CAPACITY	
																										MINIMUM	MAXIMUM
FT.	IN.	FT.	IN.	FT.	IN.	FT.	IN.	FT.	IN.	FT.	IN.	FT.	IN.	FT.	IN.	FT.	IN.	FT.	IN.	FT.	IN.	FT.	IN.	FT.	IN.	CFS	CFS
0	6	2	$\frac{7}{16}$	1	$4\frac{5}{16}$	2	0	1	$3\frac{1}{2}$	1	$3\frac{5}{8}$	2	0	1	0	2	0	1	0	0	$4\frac{1}{2}$	2	$11\frac{1}{2}$	1	4	.05	3.9
	9	2	$10\frac{5}{8}$	1	$11\frac{1}{8}$	2	10	1	3	1	$10\frac{5}{8}$	2	6	1	0	1	6	1	0		$4\frac{1}{2}$	3	$6\frac{1}{2}$	1	4	.09	8.9
1	0	4	6	3	0	4	$4\frac{7}{8}$	2	0	2	$9\frac{1}{4}$	3	0	2	0	3	0	1	3		9	4	$10\frac{3}{4}$	1	8	.11	16.1
1	6	4	9	3	2	4	$7\frac{7}{8}$	2	6	3	$4\frac{3}{8}$	3	0	2	0	3	0	1	3		9	5	6	1	8	.15	24.6
2	0	5	0	3	4	4	$10\frac{7}{8}$	3	0	3	$11\frac{1}{2}$	3	0	2	0	3	0	1	3		9	6	1	1	8	.42	33.1
3	0	5	6	3	8	5	$4\frac{3}{4}$	4	0	5	$1\frac{7}{8}$	3	0	2	0	3	0	1	3		9	7	$3\frac{1}{2}$	1	8	.61	50.4
4	0	6	0	4	0	5	$10\frac{5}{8}$	5	0	6	$4\frac{1}{4}$	3	0	2	0	3	0	1	6		9	8	$10\frac{3}{4}$	2	0	1.3	67.9
5	0	6	6	4	4	6	$4\frac{1}{2}$	6	0	7	$6\frac{5}{8}$	3	0	2	0	3	0	1	6		9	10	$1\frac{1}{4}$	2	0	1.6	85.6
6	0	7	0	4	8	6	$10\frac{3}{8}$	7	0	8	9	3	0	2	0	3	0	1	6		9	11	$3\frac{1}{2}$	2	0	2.6	103.5
7	0	7	6	5	0	7	$4\frac{1}{4}$	8	0	9	$11\frac{3}{8}$	3	0	2	0	3	0	1	6		9	12	6	2	0	3.0	121.4
8	0	8	0	5	4	7	$10\frac{1}{8}$	9	0	11	$1\frac{3}{4}$	3	0	2	0	3	0	1	6		9	13	$8\frac{1}{4}$	2	0	3.5	139.5

Figure 5.6.8 Standard Parshall flume dimensions. 103-D-1225 (from U.S. Bureau of Reclamation (1978)).

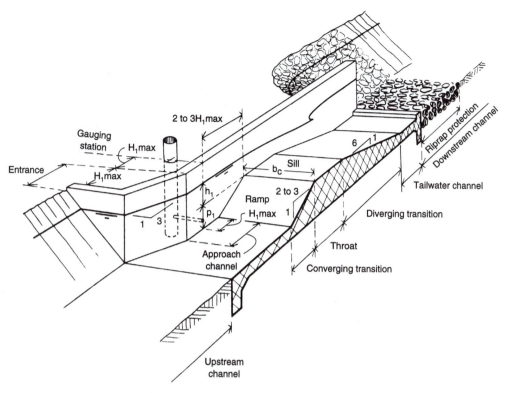

Figure 5.6.9 Flow-measuring structure for earthen channel with rectangular control section (from Bos et al. (1984)).

EXAMPLE 5.6.2

Water flows through a Parshall flume with a throat width of 4.0 ft at a depth of 2.0 ft. What is the flow rate?

SOLUTION

Using equation (5.6.11),

$$Q = 4.0(4.0)(2.0)^{(1.522)(4.0)^{0.026}} = 47.8 \text{ ft}^3/\text{s}$$

Figure 5.6.9 illustrates a flow-measuring structure for unlined (earthen) channels which are longer, and consequently more expensive, than structures for concrete-lined channels. For concrete-lined channels, the approach channel and sides of the control section are already available.

5.6.3   Stream Flow Measurement: Velocity-Area-Integration Method

As for weirs and flumes, stream flow is not directly measured. Instead, water level is measured and stream flow is determined from a *rating curve*, which is the relationship between water surface elevation and discharge.

Water surface level (elevation) can be measured manually or automatically. *Crest stage gauges* are used to measure flood crests. They consist of a wooden staff gauge placed inside a pipe with small holes for water entry. Cork in the pipe floats as the water rises and adheres to the staff (scale) at the highest water level. *Bubble gauges* (shown in Figure 5.6.10) sense the water surface level by bubbling a continuous stream of gas (usually carbon dioxide) into the water. The pressure required to continuously force the gas stream out beneath the water surface is a measure of the depth of water over the nozzle of the bubble stream. The pressure is measured with a manometer assembly to provide a continuous record of water level in the stream (gauge height).

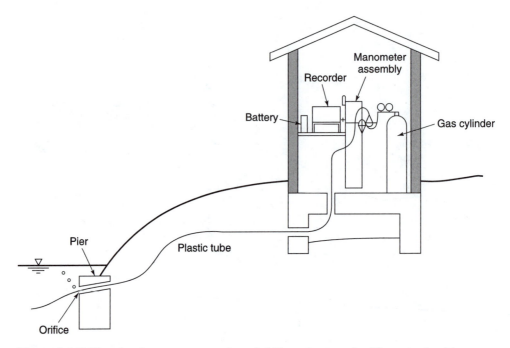

Figure 5.6.10 Water level measurement using a bubble gauge recorder. The water level is measured as the back pressure on the bubbling stream of gas by using a mercury manometer (from Rantz et al. (1982)).

Rating curves are developed using a set of measurements of discharge and gauge height in the stream. The discharge is $Q = AV$ where V is the mean velocity normal to the cross-sectional area of flow A, which is a function of the gauge height. So in order to measure discharge the velocity and the gauge height must be determined. In a stream or river, the velocity varies with depth, as discussed in Chapter 3 (Figure 3.6.1). Therefore, the velocity must be recorded at various locations and depths across the stream.

Referring to Figure 5.6.11, the total discharge is computed by summing the incremental discharge calculated from each measurement i, $i = 1, 2, ..., n$, of velocity V_i and depth y_i. These measurements represent average values over the width Δw_i of the stream. The total discharge is computed using

$$Q = \sum_{i=1}^{n} V_i y_i \Delta w_i \qquad (5.6.14)$$

Both theory and experimental evidence indicate that the mean velocity in a vertical section can be closely approximated by the average of the velocities at 0.2 depth and 0.8 depth below the water surface, as shown in Figure 5.6.11. If the stream is shallow, it may be possible to take a single measurement of velocity at a 0.6 depth.

To measure the velocity in a stream, a *current meter*, which is an impellor device, can be used. The speed at which the impellor rotates is proportional to the flow velocity. Figure 5.6.12a shows a propeller type current meter on a wading rod and Figure 5.6.12b shows a Price current meter, which is the most commonly used velocity meter in the United States. Refer to Wahl, et al. (1995) for detailed descriptions on the stream-gauging program of the U.S. Geological Survey.

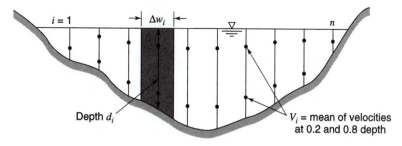

Figure 5.6.11 Computation of discharge from stream gauging data.

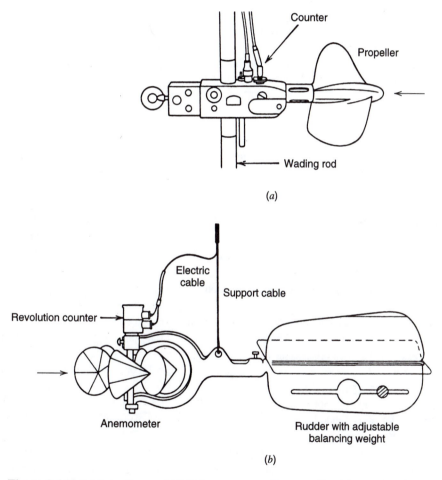

Figure 5.6.12 (*a*) Propeller- and (*b*) Price-type current meters (from James (1988)).

PROBLEMS

5.1.1 A 2-m wide rectangular channel with a bed slope of 0.0005 has a depth of flow of 1.5 m. Manning's roughness coefficient is 0.015. Determine the steady uniform discharge in the channel.

5.1.2 Determine the uniform flow depth in a rectangular channel 2.5 m wide with a discharge of 3 m^3/s. The slope is 0.0004 and Manning's roughness factor is 0.015.

5.1.3 Determine the uniform flow depth in a trapezoidal channel with a bottom width of 8 ft and side slopes of 1 vertical to 2 horizontal. The discharge is 100 ft^3/s. Manning's roughness factor is 0.015 and the channel bottom slope is 0.0004.

5.1.4 Determine the uniform flow depth in a trapezoidal channel with a bottom width of 2.5 m and side slopes of 1 vertical to 2 horizontal with a discharge of 3 m^3/s. The slope is 0.0004 and Manning's roughness factor is 0.015.

5.1.5 Determine the cross-section of the greatest hydraulic efficiency for a trapezoidal channel with side slope of 1 vertical to 2 horizontal if the design discharge is 10 m^3/s. The channel slope is 0.001 and Manning's roughness factor is 0.020.

5.1.6 For a trapezoidal shaped channel ($n = 0.014$ and slope S_o = 0.0002 with a 20-ft bottom width and side slopes of 1 vertical to 1.5 horizontal), determine the normal depth for a discharge of 1000 cfs.

5.1.7 Show that the best hydraulic trapezoidal section is one-half of a hexagon.

5.2.1 Solve example 5.2.2 for discharges of 0, 25, 75, 125, and 200 ft^3/s.

5.2.2 Rework example 5.2.3 for a 30-cm high hump and a side wall constriction that reduces the channel width to 1.6 m.

5.2.3 Compute the critical depth for the channel in problem 5.1.1.

5.2.4 Compute the critical depth for the channel in problem 5.1.2.

5.2.5 Rework example 5.2.4 with discharges of 0, 25, 75, 125, and 200 cfs.

5.2.6 Rework example 5.2.5 using a discharge at river mile 1.0 of 8000 cfs and a discharge of 7500 cfs at river mile 1.5. The water surface elevation at river mile 1.0 is 123.5 ft. All other data is the same.

5.3.1 Resolve example 5.3.1 for a channel bed slope of 0.003.

5.3.2 A 2.45-m wide rectangular channel has a bed slope of 0.0004 and Manning's roughness factor of 0.015. For a discharge of 2.83 m^3/sec, determine the type of water surface profile for depths of 1.52 m, 0.61 m, and 0.30 m.

5.3.3 Rework problem 5.3.2 with a bed slope of 0.004.

5.3.4 If the channel of problem 5.1.6 is preceded by a steep slope and followed by a mild slope and a sluice gate as shown in Figure P5.3.4, sketch a possible water surface profile with the elevations to a scale of 1 inch to 10 ft. Consider a discharge of 1500 cfs. For this discharge, the normal depth for a slope of 0.0003 is 8.18 ft and for a slope of 0.0002 is 9.13 ft.

5.3.5 Sketch possible water surface profiles for the channel in Figure P5.3.4. First locate and mark the control points, then sketch the profiles, marking each profile with the appropriate designation. Show any hydraulic jumps that occur.

5.4.1 Rework example 5.4.2 using equation (5.4.8), $S_f = Q^2/\bar{K}^2$.

5.4.2 Resolve example 5.4.3 with a discharge of 10000 cfs and a downstream water surface elevation of 123.5 ft.

5.4.3 Rework example 5.4.4 using a discharge at river mile 1.0 of 8000 cfs and a discharge of 7500 cfs at river mile 1.5. The water surface elevation at river mile 1.0 is 123.5 ft. All other data are the same.

5.4.4 Consider a starting (assumed) water surface elevation of 5719.5 ft at cross-section 4 for example 5.4.5 and determine water surface elevation at cross-section number 4.

5.4.5 Consider a starting (assumed) water surface elevation of 5717.6 ft at cross-section 4 for the example 5.4.5 and determine the computed water surface elevation at cross-section 4.

5.5.1 Consider a 2.45-m wide rectangular channel with a bed slope of 0.0004 and a Manning's roughness factor of 0.015. A weir is placed in the channel and the depth upstream of the weir is 1.52 m for a discharge of 5.66 m^3/s. Determine if a hydraulic jump forms upstream of the weir.

5.5.2 A hydraulic jump occurs in a rectangular channel 4.0 m wide. The water depth before the jump is 0.4 m and after the jump is 1.7 m. Compute the flow rate in the channel, the critical depth and the headloss in the jump.

5.5.3 Rework example 5.5.3 with a flow rate of 450 cfs at a normal depth of 3.2 ft. All other data remain the same.

5.5.4 Rework example 5.5.4 if the depth before the jump is 0.8 m and all other data remain the same.

5.6.1 A rectangular, sharp crested weir with end contraction is 1.6 m long. How high should it be placed in a channel to maintain an upstream depth of 2.5 m for 0.5 m^3/s flow rate?

5.6.2 For a sharp-crested suppressed weir ($C_w = 3.33$) of length $B = 8.0$ ft, $P = 2.0$ ft, and $H = 1.0$ ft, determine the discharge over the weir. Neglect the velocity of approach head.

5.6.3 Rework problem 5.6.2 incorporating the velocity of approach head (equation (5.6.5a)).

5.6.4 Rework example 5.6.2 using equation (5.6.5b).

5.6.5 A rectangular sharp-crested weir with end contractions is 1.5 m long. How high should the weir crest be placed in a channel to maintain an upstream depth of 2.5 m for 0.5 m^3/s flow rate?

5.6.6 Determine the head on a 60° V-notch weir for a discharge of 150 l/s. Take $C_d = 0.58$.

5.6.7 The head on a 90° V-notch weir is 1.5 ft. Determine the discharge.

5.6.8 Determine the weir coefficient of a 90° V-notch weir for a head of 180 mm for a flow rate of 20 l/s.

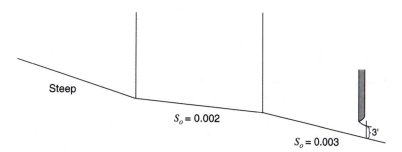

Figure P5.3.4

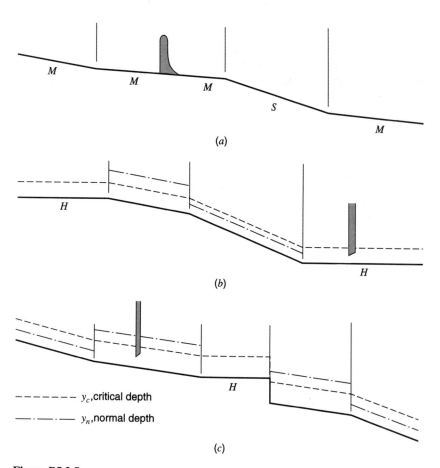

(a)

(b)

(c)

-------- y_c,critical depth

—·—·— y_n,normal depth

Figure P5.3.5

5.6.9 Determine the required head for a flow of 3.0 m3/s over a broad-crested weir 1.5 m high and 3 m long with a well-rounded upstream corner ($C_w = 1.67$).

5.6.10 Water flows through a Parshall flume with a throat width of 4.0 ft at a depth of 7.5 ft. Determine the flow rate.

5.6.11 Water flows through a Parshall flume with a throat width of 5.0 ft at a depth of 3.4 ft. Determine the flow rate.

5.6.12 The following information was obtained from a discharge measurement on a stream. Determine the discharge.

Distance from bank (ft)	Depth (ft)	Mean velocity (ft)
0	0.0	0.00
12	.1	0.37
32	4.4	0.87
52	4.6	1.09
72	5.7	1.34
92	4.3	0.71
100	0.0	0.00

REFERENCES

Barnes, H. H., Jr., *Roughness Characteristics of Natural Channels*, U.S. Geological Survey, Water Supply Paper 1849, U.S. Government Printing Office, Washington, DC, 1962.

Bos, M. G., J. A. Replogle, and A. J. Clemmens, *Flow Measuring Flumes for Open Channel System*, John Wiley and Sons, New York, 1984.

Brater, E. F., H. W. King, J. E. Lindell, and C.Y. Wei, *Handbook of Hydraulics*, 7th edition, McGraw-Hill, New York, 1996.

Chaudhry, M. H., *Open-Channel Flow*, Prentice Hall, Englewood Cliffs, NJ, 1993.

Chow, V. T., *Open-Channel Hydraulics*, McGraw-Hill, New York, 1959.

Chow, V. T., D. R. Maidment, and L. W. Mays, *Applied Hydrology*, McGraw-Hill, New York, 1988.

French, R. H., *Open-Channel Hydraulics*, McGraw-Hill, New York, 1985.

Henderson, F. M., *Open-Channel Flow*, Macmillan, New York, 1966.

Hoggan, D. H., *Computer-Assisted Floodplain Hydrology and Hydraulics*, 2nd edition, McGraw-Hill, New York, 1997.

Hydrologic Engineering Center (HEC), *HEC-RAS River System Analysis System, User's Manual, Version 2.2*, U.S. Army Corps of Engineers Water Resources Support Center, Davis, CA, 1997a.

Hydrologic Engineering Center (HEC), *HEC-RAS River Analysis System, Hydraulic Reference Manual, Version 2.0*, U.S. Army Corps of Engineers Water Resources Support Center, Davis, CA, 1997b.

Hydrologic Engineering Center (HEC), *HEC-RAS River Analysis System, Applications Guide, Version 2.0*, U.S. Army Corps of Engineers Water Resources Support Center, Davis, CA, 1997c.

James, L. G., *Principles of Farm Irrigation System Design*, John Wiley and Sons, Inc., New York, 1988.

Leopold, W. B., *Water, Rivers, and Creeks*, University Science Books, Sausalito, CA, 1997.

Manning, R., "On the Flow of Water in Open Channels and Pipes," *Transactions Institute of Civil Engineers of Ireland*, vol. 20, pp. 161−209, Dublin, 1891; Supplement, vol. 24, pp. 179−207, 1895.

Rantz, S. E., et al., *Measurement and Computation of Stream Flow*, vol. 1, *Measurement of Stage and Discharge*, Water Supply Paper 2175, U.S. Geological Survey, 1982.

Replogle, J. A., A. J. Clemmens, and C. A. Pugh, "Hydraulic Design of Flow Measuring Structures," *Hydraulic Design Handbook*, edited by L. W. Mays, McGraw-Hill, New York, 1999.

Townson, J. M., *Free-Surface Hydraulics*, Unwin Hyman, London, 1991.

U.S. Bureau of Reclamation, *Guide for Computing Water Surface Profiles*, 1957.

U.S. Bureau of Reclamation, *Design of Small Canal Structures*, U.S. Government Printing Office, Denver, CO, 1978.

U.S. Bureau of Reclamation, *Water Measurement Manual*, U.S. Government Printing Office, Denver, CO, 1981.

Wahl, K. L., W. O. Thomas, Jr., and R. M. Hirsch, "Stream-Gauging Program of the U.S. Geological Survey, *Circular 11123*, U.S. Geological Survey, Reston, VA, 1995.

Chapter **6**

Hydraulic Processes: Groundwater Flow

Groundwater hydrology is the science that considers the occurrence, distribution, and movement of water below the surface of the earth (Todd, 1980). It is concerned with both the quantity and quality aspects of this water (Charbeneau, 2000). This chapter only considers the quantity aspects, in particular the hydraulic flow processes, emphasizing *aquifer hydraulics* (also see Batu, 1988; Charbeneau, 2000; and Delleur, 1999).

6.1 GROUNDWATER CONCEPTS

Groundwater flows through *porous media, fractured media,* and *large passages (Karst)*. Porous media consist of solid material and *voids* or *pore space*. Porous media contains relatively small openings in the solid and is a permeable medium allowing the flow of water. The porous media that we typically are interested in for groundwater flow include natural soils, unconsolidated sediments, and sedimentary rocks. The size range of particles in a soil is referred to as the *soil texture*. Grain size determines the particle size classification. The fraction of clay, slit, and sand in a soil texture has been described by the *soil texture triangle* shown in Figure 6.1.1. Each point on the triangle corresponds to different percentages by mass (weight) of clay, sand, and silt.

The subsurface occurrence of groundwater, as shown in Figure 6.1.2, can be divided into the *vadose zone (zone operation)* and the *zone of saturation*. The vadose zone, also called the *unsaturated* or *partially saturated zone*, is the subsurface media above the water table. The term vadose is derived from the Latin *vadosus*, meaning shallow. Flow in the unsaturated or vadose zone is discussed further in Section 7.4. This chapter is focused on saturated flow, which is referred to as *groundwater flow*.

Groundwater originates through infiltration, influent streams, seepage from reservoirs, artificial recharge, condensation, seepage from oceans, water trapped in the sedimentary rock (*connote water*), and *juvenile* water (volcanic, magmatic, and cosmic). Any significant quantity of subsurface water is stored in subsurface formations defined as *aquifers*. An aquifer may be defined as a formation that contains sufficient saturated permeable material to yield significant quantities of water to wells (Lohman et al., 1972). Aquifers are usually of large areal extent and are essentially underground storage reservoirs. They may be overlain or underlain by a *confining bed*, which is a relatively impermeable material adjacent to the aquifer.

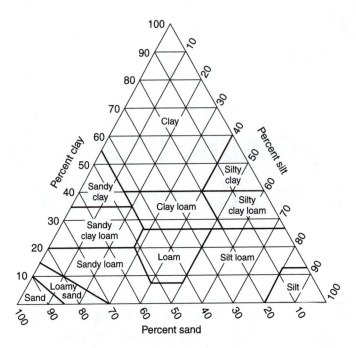

Figure 6.1.1 Triangle of soil textures for describing various combinations of sand, silt, and clay (from U.S. Soil Conservation Service (1951)).

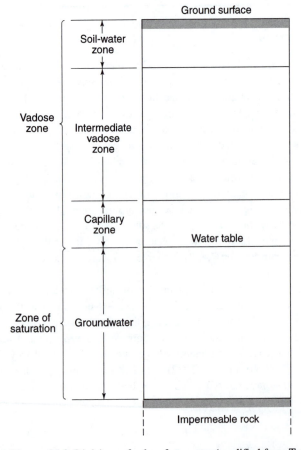

Figure 6.1.2 Divisions of subsurface water (modified from Todd (1980)).

The following are different types of confining beds (Todd, 1980):

- *Aquilude*—A saturated but relatively impermeable material that does not yield appreciable quantities of water to wells; clay is an example.
- *Aquifuge*—A relatively impermeable formation neither containing nor transmitting water; solid granite belongs in this category.
- *Aquitard*—A saturated but poorly permeable stratum that impedes groundwater movement and does not yield water freely to wells, but that may transmit appreciable water to or from adjacent aquifers and, where sufficiently thick, may constitute an important groundwater storage zone; sandy clay is an example.

Table 6.1.1 lists the various aquifer types and their characteristics.

Table 6.1.1   Summary of the Characteristics of Principal Aquifer Types

Aquifer Type	Lithology	Groundwater Flow Regime	Aquifer properties				Residence Times
			Porosity (percent)	Permeability (m/day)	Specific Yield (percent)	Natural Flow Rates (m/day)	
Shallow alluvium	Gravel	Intergranular	25–35	100–1000	12–25	2–10	Could be very short;
	Sand	Intergranular	30–42	1–50	10–25	0.05–1	a few months to years,
	Silt	Intergranular	40–45	0.0005–0.1	5–10	0.001–0.1	depending on volume
Deep sedimentary formations	Sand and silts	Intergranular	30–40	0.1–5	2–10	0.001–0.01	Many thousands of years
Sandstone	Cemented quartz grains	Intergranular and fissure	10–30	0.1–10	8–20	0.001–0.1	Tens to hundreds of years
Limestone	Cemented carbonate	Mainly fissure	5–30	0.1–5.0	5–15	0.001–1	Tens to hundreds of years
Karstic limestone	Cemented carbonate	Fissures and channels	5–25	100–10,000	5–15	10–2,000	A few hours to days
Volcanic rock							
Basalt	Fine grained crystalline	Fissure	2–15	0.1–1,000	1–5	1–500	Very wide range; can be very short
Tuff	Cemented grains	Intergranular and fissure	15–30	0.1–5	10–20	0.001–1	Wide range
Igneous and metamorphic rocks (granites and gneisses):							
Fresh	Crystalline	Fissure	0.1–2	10^{-7}–5×10^{-5}		10^{-6}–10^{-2}	Thousands of years, but can be rapid where fractured
Weathered	Disaggregated crystalline	Intergranular and fissure	10–20	0.1–2	1–5	0.001–0.1	Tens to hundreds of years

Source: P. J. Chilton, as presented in Gleick (1993).

Aquifers are classified as *unconfined* or *confined* depending upon the presence or absence of a water table (Figure 6.1.3). An *unconfined aquifer* is one in which a water table serves as the upper surface of the *zone of saturation*, also known as a *free, phreatic,* or *non-artesian aquifer.* Changes in the water table (rising or falling) correspond to changes in the volume of water in storage within an aquifer. A *confined aquifer* is one in which the groundwater is confined under pressure greater than atmospheric by overlying, relatively impermeable strata. Confined aquifers are also known as *artesian* or *pressure aquifers.* Water enters such aquifers in an area where the confining bed rises to the surface or ends underground and such an area is known as a *recharge area.* Changes of the water levels in wells penetrating confined aquifers result primarily from changes in pressure rather than changes in storage volumes.

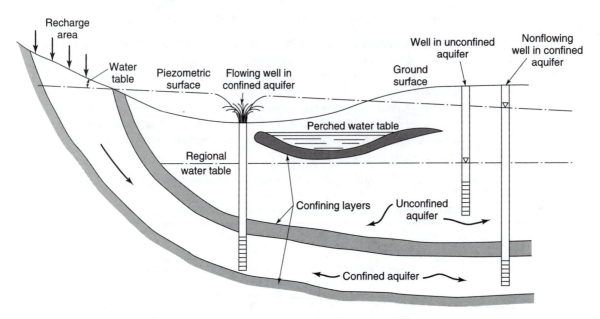

Figure 6.1.3 Types of aquifers (from U.S. Bureau of Reclamation (1981)).

Aquifer Properties

Aquifers perform two important functions—a *storage function* and a *conduit function*. In other words, aquifers store water and also function as a pipeline. When water is drained from a saturated material under the influence of gravity, only a portion of the total saturated volume in the pores is released. Part of the water is retained in interstices due to the losses of the molecular attraction, adhesion, and cohesion. The *specific yield S_y*, which is the storage term for unconfined aquifers, is the volume of water drained from a saturated sample of unit volume (1 ft^3 or 1 m^3) with a unit decrease in the water table. *Specific retention S_r* is the quantity of water that is retained in the unit volume after gravity drainage. The sum of the specific yield and the specific retention for saturated aquifers is the porosity, $\alpha = S_y + S_r$. *Porosity* is the pore volume divided by the total volume, expressed as a percent. Porosity represents the potential storage of an aquifer, but does not indicate the amount of water a porous material will yield. Figure 6.1.4 illustrates the various types of porosities.

The *storativity (or storage coefficient)* of an aquifer is the volume of water the aquifer releases from or takes into storage per unit surface area of the aquifer per unit decline or rise of head. This is illustrated in Figure 6.1.5, considering a vertical column of unit area extending through a confined aquifer and an unconfined aquifer. In both cases the storage coefficient S equals the volume of water (ft^3 or m^3) released from the aquifer when the piezometric surface or water table declines one unit of distance (1 ft or 1 m). The storativity then has dimensions of ft^3/ft^3 or m^3/m^3. In the case of unconfined aquifers, the storativity corresponds to the specific yield. Confined aquifers have storativities in the range of $10^{-5} \le S \le 10^{-3}$. These small values indicate that large pressure changes are required to produce substantial water yields. Storativity can be determined in the field by pump tests. The *specific storage S_s* of a saturated aquifer is the volume of water that a unit volume of aquifer releases from storage under a unit decline in hydraulic head, i.e., $S = S_s b$ where b is the thickness of a confined aquifer.

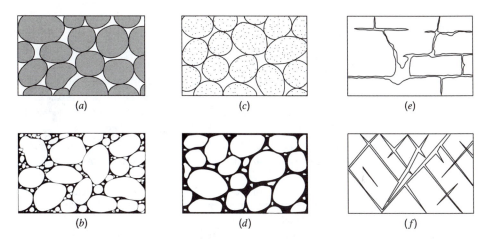

Figure 6.1.4 The various types of porosity: (*a*) Well-sorted sedimentary deposit possessing high porosity; (*b*) Poorly sorted sedimentary deposit of low porosity; (*c*) Similar to *a* but consisting of pebbles that are themselves porous, so the porosity of the deposit is very high; (*d*) Also similar to *a* but the porosity has been diminished by the deposition of mineral matter in the interstices; (*e*) Rocks rendered porous by solution; and (*f*) Rocks rendered porous by fracturing (from O. E. Meinzer (1923)).

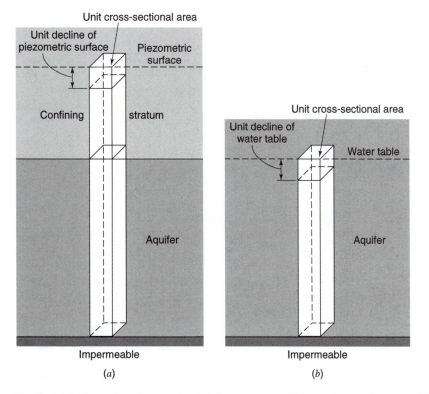

Figure 6.1.5 Illustrative sketches for defining storage coefficient of (*a*) confined and (*b*) unconfined aquifers (from Todd (1980)).

Hydraulic conductivity (also referred to as *coefficient of permeability*) is the property related to the conduit function of an aquifer. It is the measure of ease of moving groundwater through aquifers, with dimensions of (L/T). The hydraulic conductivity K is the rate of flow of water through a cross-section of unit area of the aquifer under a unit hydraulic gradient. The *hydraulic gradient* is the head loss divided by the distance between the two points. The hydraulic conductivity is commonly expressed in gallons per day/ft² or in ft/day in U.S. customary units or in cm/d or m/d in SI units. Table 6.1.2 presents a range of hydraulic conductivity. Table 6.1.3 lists, in both the SI and the FPS systems, the dimensions and common units for some of the basic groundwater parameters. The hydraulic conductivity is a function of both the porous medium and the fluid properties. Table 6.1.1 summarizes of the characteristics of principal aquifer types.

Table 6.1.2 Range of Values of Hydraulic Conductivity and Permeability

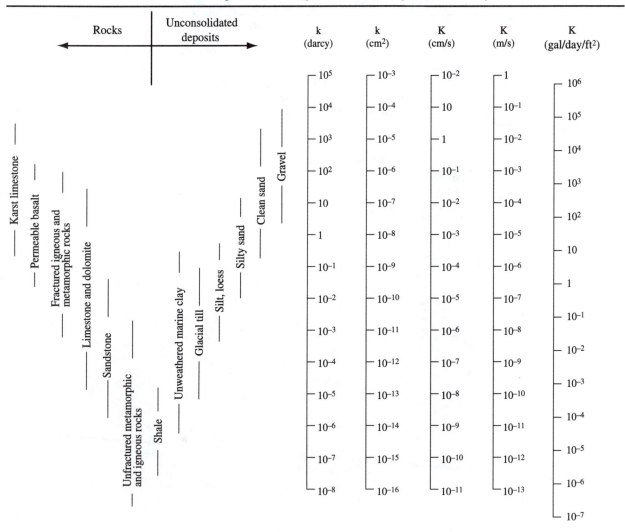

Source: Freeze and Cherry (1979).

Table 6.1.3   Dimensions and Common Units for Basic Groundwater Parameters

| Parameter | Symbol | Systeme International* SI | | Foot-Pound-Second System** FPS | |
		Dimension	Units	Dimension	Units
Hydraulic head	h	$[L]$	m	$[L]$	ft
Pressure head	ψ	$[L]$	m	$[L]$	ft
Elevation head	z	$[L]$	m	$[L]$	ft
Fluid pressure	p	$[M/LT^2]$	N/m^2 or Pa	$[F/LT^2]$	lb/ft^2
Fluid potential	Φ	$[L^2/T^2]$	m^2/s^2	$[L^2/T^2]$	ft^2/s^2
Mass density	ρ	$[M/L^3]$	kg/m^3	—	—
Weight density	γ	—	—	$[F/L^3]$	lb/ft^3
Specific discharge	v	$[L/T]$	m/s	$[L/T]$	ft/s
Hydraulic conductivity	K	$[L/T]$	m/s	$[L/T]$	ft/s

*Basic dimensions are length L, mass M, and time T.
**Basic dimensions are length L, force F, and time T.

Source: Freeze and Cherry (1979).

A *specific* or *intrinsic* permeability, which is a function of the medium alone and not the fluid properties, is $k = Cd^2$ where C is a constant of proportionality and d is the grain size diameter. This k has simply been referred to as permeability (Freeze and Cherry, 1979). Table 6.1.2 also presents ranges of k values, which have dimensions of $[L^2]$. The *darcy* is also a unit of permeability, where 1 darcy is the permeability that leads to a specific discharge of 1 cm/s for a fluid with a viscosity of 1 cP (1 centipoise, cP = N = s/m^2 × 10^{-3}) under a hydraulic gradient of 1 cm/cm. One darcy is approximately equal to 10^{-8} cm^2 (Freeze and Cherry, 1979). Refer to Table 6.1.2 for the range of values of k.

Closely related to the hydraulic conductivity is the *transmissivity* (or *transmissibility*), which indicates the capacity of an aquifer to transmit water through its entire thickness. The transmissivity, T, is the flow rate (ft^2/s or m^2/s) through a vertical strip of the aquifer 1 unit wide (1 ft or 1 m) and extending through the saturated thickness under a unit hydraulic gradient. The transmissivity is equal to the hydraulic conductivity multiplied by the saturated thickness of the aquifer: $T = Kb$, in which b is the saturated thickness of the aquifer.

Heterogeneity and Anisotropy of Hydraulic Conductivity

In geologic formations the hydraulic conductivity usually varies through space, referred to as *heterogeneity*. A geologic formation is *homogeneous* if the hydraulic conductivity is independent of position in the formation, i.e., $K(x,y,z)$ = constant. A geologic formation is *heterogeneous* if the hydraulic conductivity is dependent on position in the formation, $K(x,y,z) \neq$ constant.

Hydraulic conductivity may also show variations with the direction of measurement at a given point in the formation. A geologic formation is *isotropic* at a point if the hydraulic conductivity is independent of the direction of measurement at the point, $K_x = K_y = K_z$. A geologic formation is *anisotropic* at a point if the hydraulic conductivity varies with the direction of measurement at that point, $K_x \neq K_y \neq K_z$.

Groundwater Basins

Aquifers exist in both *consolidated* (mainly bedrock) and *unconsolidated* (mainly surface deposits) *materials*. Figure 6.1.6 illustrates the distribution of major aquifers of different types. *Groundwater basins* are a group of interrelated bodies of groundwater linked together in a larger flow system (Marsh, 1987). These basins are typically complex, three-dimensional systems. The spatial configurations of groundwater basins are determined by regional geology.

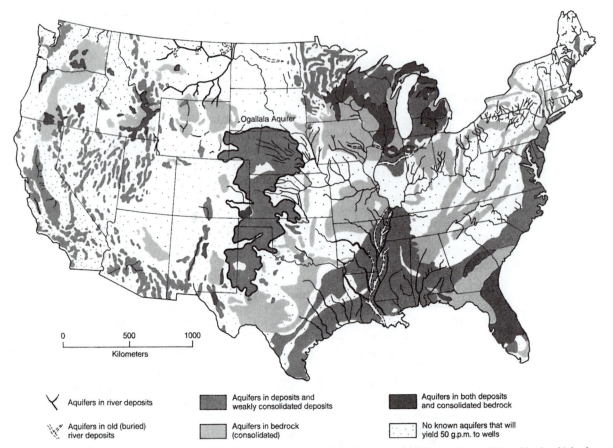

Figure 6.1.6 Distribution of major aquifers of different types (river deposits, other deposits, and consolidated bedrock) in the coterminous United States. A major aquifer is defined as one composed of material capable of yielding 50 gallons per minute or more to an individual well and having water quality generally not containing more than 2000 parts per million of dissolved solids (from Marsh (1987)).

Groundwater Movement

Groundwater in its natural state is invariably moving and this movement is governed by hydraulic principles. The flow through an aquifer is expressed by *Darcy's law*, which is the foundation of groundwater hydraulics. This law states that the flow rate through porous media is proportional to the headloss and inversely proportional to the length of the flow path, expressed mathematically as

$$Q = -KA\frac{dh}{dL} \tag{6.1.1}$$

Figure 6.1.7 illustrates Darcy's law where $h = dh$ and $L = dL$ so that $Q = KAh/L$.

The Reynold's number for flow in porous media is

$$R_e = \frac{\rho VD}{\mu} \tag{6.1.2}$$

where ρ is fluid density, V is the apparent velocity, D is the average grain diameter (an approximation to average pore diameter, and μ is viscosity of the fluid. The upper limit of validity for Darcy's law is for R_e ranging between 1 and 10, and there is really no lower limit. In almost all natural groundwater motion $R_e < 1$; therefore, Darcy's law is applicable to natural groundwater problems. In summary, this law is valid when the flow is laminar or without turbulence.

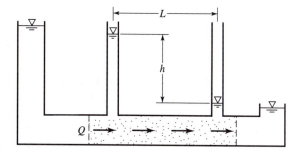

Figure 6.1.7 Illustration of Darcy's law (from U.S. Bureau of Reclamation (1981)).

EXAMPLE 6.1.1

Two very large reservoirs are connected by a conduit of two types of porous medium as shown in Figure 6.1.8. Compute the magnitude of flow per unit width between the two reservoirs if $K_1 = 1$ gpd/ft^2 and $K_2 = 2$ gpd/ft^2.

SOLUTION

Darcy's law for sections 1 and 2 where $Q_1 = Q_2$ is

$$K_1 A_1 \frac{dh_1}{dL_1} = K_2 A_2 \frac{dh_2}{dL_2}.$$

Because $A_1 = A_2$,

$$\frac{\Delta h_1}{\Delta L_1} = 2 \frac{\Delta h_2}{\Delta L_2}$$

and

$$\Delta h_1 + \Delta h_2 = 9$$

The two above equations have two unknowns, Δh_1 and Δh_2, so they can be solved to obtain $\Delta h_2 = 4.69$ ft and $\Delta h_1 = 4.36$ ft. The flow rate is computed using Darcy's equation,

$$Q_1 = 1 \cdot 10 \cdot \frac{4.36}{16} = 2.725 \text{ gpd}$$

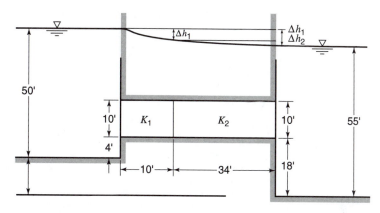

Figure 6.1.8 Two reservoirs connected by conduit of two porous medium for Example 6.1.1.

6.2   SATURATED FLOW

6.2.1   Governing Equations

The control volume for a saturated flow is shown in Figure 6.2.1. The sides have lengths dx, dy, and dz in the coordinate directions. The total volume of the control volume is $dxdydz$ and the volume of water in control volume is $\theta dxdydz$ where θ is the moisture content. With the control volume approach, the extensive property B is the mass of the groundwater, the intensive property is $\beta = dB/dm = 1$, and $dB/dt = 0$, and because no phase change occurs. The general control volume equation for continuity, equation (3.2.1), is applicable:

$$0 = \frac{d}{dt}\int_{CV}\rho\,d\forall + \int_{CS}\rho\mathbf{V}\cdot d\mathbf{A} \tag{3.2.1}$$

The time rate of change of mass stored in the control volume is defined as the time rate of change of fluid mass in storage, expressed as

$$\frac{d}{dt}\int_{CV}\rho\,d\forall = \rho S_S\frac{\partial h}{\partial t}(dxdydz) + \rho W(dxdydz) \tag{6.2.1}$$

where S_s is the specific storage and W is the flow out of the control volume, $W = Q/(dxdydz)$. The term $\rho S_s(\partial h/\partial t)(dxdydz)$ includes two parts: (1) the mass rate of water produced by an expansion of the water under change in density and (2) the mass rate of water produced by the compaction of the porous medium due to change in porosity (see Freeze and Cherry (1979) for more details).

The inflow of water through the control surface at the bottom of the control volume is $qdxdy$ and the outflow at the top is $[q + (\partial q/\partial z)dz]dxdy$, so that the net outflow in the vertical direction z is

$$\int_{CS(z)}\rho\mathbf{V}\cdot d\mathbf{A} = \rho\left(q + \frac{\partial q}{\partial z}dz\right)dxdy - \rho qdxdy = \rho dxdydz\frac{\partial q}{\partial z} \tag{6.2.2}$$

where q in $\partial q/\partial z$ is q_z, i.e., in the z direction.

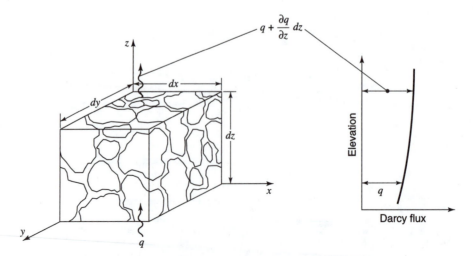

Figure 6.2.1 Control volume for development of the continuity equation in porous medium (from Chow et al. (1988)).

Considering all three directions, we get

$$\int_{CS} \rho \mathbf{V} \cdot d\mathbf{A} = \rho dxdydz \frac{\partial q}{\partial x} + \rho dxdydz \frac{\partial q}{\partial y} + \rho dxdydz \frac{\partial q}{\partial z} \tag{6.2.3}$$

Substituting (6.2.1) and (6.2.3) into (3.2.1) results in

$$0 = \rho S_s \frac{\partial h}{\partial t} dxdydz + W(dxdydz) + \rho dxdydz \frac{\partial q}{\partial x} + \rho dxdydz \frac{\partial q}{\partial y} + \rho dxdydz \frac{\partial q}{\partial z} \tag{6.2.4}$$

Dividing through by $\rho dxdydz$ gives

$$S_s \frac{\partial h}{\partial t} + \frac{\partial q}{\partial x} + \frac{\partial q}{\partial y} + \frac{\partial q}{\partial z} + W = 0 \tag{6.2.5}$$

Using Darcy's law, the Darcy flux in each direction is

$$q_x = -K_x \frac{\partial h}{\partial x} \tag{6.2.6a}$$

$$q_y = -K_y \frac{\partial h}{\partial y} \tag{6.2.6b}$$

$$q_z = -K_z \frac{\partial h}{\partial z} \tag{6.2.6c}$$

Substituting these definitions of q_x, q_y, and q_z into (6.2.5) and rearranging results in

$$\frac{\partial}{\partial x}\left(K_x \frac{\partial h}{\partial x} \right) + \frac{\partial}{\partial y}\left(K_y \frac{\partial h}{\partial y} \right) + \frac{\partial}{\partial z}\left(K_z \frac{\partial h}{\partial z} \right) = S_s \frac{\partial h}{\partial t} + W \tag{6.2.7}$$

This is the equation for *three-dimensional transient flow through a saturated anisotropic porous medium*. For a *homogenous, isotropic medium* ($K_x = K_y = K_z$), equation (6.2.7) becomes

$$\frac{\partial^2 h}{\partial x^2} + \frac{\partial^2 h}{\partial y^2} + \frac{\partial^2 h}{\partial z^2} = \frac{S_s}{K} \frac{\partial h}{\partial t} + \frac{W}{K} \tag{6.2.8}$$

For a *homogenous, isotropic medium and steady-state flow* equation (6.2.8) becomes

$$\frac{\partial^2 h}{\partial x^2} + \frac{\partial^2 h}{\partial y^2} + \frac{\partial^2 h}{\partial z^2} = \frac{W}{K} \tag{6.2.9}$$

For a horizontal confined aquifer of thickness b, $S = S_s b$ and transmissivity $T = Kb$, the two-dimensional form of (6.2.8) with $W = 0$ becomes

$$\frac{\partial^2 h}{\partial x^2} + \frac{\partial^2 h}{\partial y^2} = \frac{S}{T} \frac{\partial h}{\partial t} \tag{6.2.10}$$

The governing equation for radial flow can also be derived using the control volume approach. Alternatively, equation (6.2.10) can be converted into radial coordinates by the relation $r = \sqrt{x^2 + y^2}$. This is known as the *diffusion equation*, expressed as

$$\frac{1}{r} \frac{\partial}{\partial r}\left(r \frac{\partial h}{\partial r} \right) = \frac{\partial^2 h}{\partial r^2} + \frac{1}{r} \frac{\partial h}{\partial r} = \frac{S}{T} \frac{\partial h}{\partial t} \tag{6.2.11}$$

where r is the radial distance from a pumped well and t is the time since the beginning of pumping. For steady-state conditions, $\partial h/\partial t = 0$, so equation (6.2.11) reduces to

$$\frac{1}{r} \frac{\partial}{\partial r}\left(r \frac{\partial h}{\partial r} \right) = 0 \tag{6.2.12}$$

6.2.2   Flow Nets

Flow nets provide a graphical means to illustrate two-dimensional groundwater flow problems. For steady-state conditions, the two-dimensional flow equation (6.2.10) for a homogeneous, isotropic medium becomes

$$\frac{\partial^2 h}{\partial x^2} + \frac{\partial^2 h}{\partial y^2} = 0 \tag{6.2.13}$$

This is the classic partial differential equation form known as the *Laplace equation*. Equation (6.2.13) is linear in terms of the piezometric head h, and its solution depends entirely on the values of h on the boundaries of a flow field in the x–y plane. In other words, h at any point in a flow field can be determined uniquely in terms of h on the boundaries. Laplace's equation arises in other areas such as hydrodynamics and heat flow.

Velocity of flow is normal to lines of constant piezometric heads. This can be seen through examination of Darcy's equation $v_x = K_x(\partial h/\partial x)$ and $v_y = -K_y(\partial h/\partial y)$ and continuity equation $(\partial v_x/\partial x) + (\partial v_y/\partial y) = 0$. Figure 6.2.2 is a hypothetical flow net. The velocity vector **V** has components v_x and v_y with the resultant velocity vector in the direction of decreasing head h. The *streamlines* are a set of lines that are drawn tangent to the velocity vector and normal to the line of constant piezometric head. The family of streamlines is called the *stream function* ψ. In steady flow the streamlines define the paths of flowing particles of fluid. Because the streamlines are normal to the line of constant head, the velocities v_x and v_y can be expressed as

$$v_x = \frac{\partial \psi}{\partial y} \tag{6.2.14}$$

$$v_y = \frac{\partial \psi}{\partial x} \tag{6.2.15}$$

which can be substituted into the differential form of the two-dimensional equation of continuity:

$$\frac{\partial v_x}{\partial x} + \frac{\partial v_y}{\partial y} = 0 \tag{6.2.16}$$

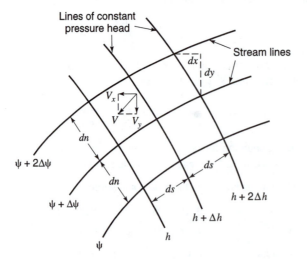

Figure 6.2.2 Hypothetical flow net.

resulting in

$$\frac{\partial^2 \psi}{\partial x^2} + \frac{\partial^2 \psi}{\partial y^2} = 0 \tag{6.2.17}$$

which is also the *Laplace equation* (see equation (6.2.13)).

Because the streamlines are everywhere tangent to velocity vectors, no flow exists across them. The rate of flow is constant between any two streamlines. The flow lines in Figure 6.2.2 have been constructed so that $ds = dh$. The discharge through a cross-sectional area of unit depth perpendicular to the flow net is

$$dq = K\frac{dh}{ds}dh \tag{6.2.18}$$

and since $ds = dh$ then

$$dq = Kdh \tag{6.2.19}$$

and with m sections

$$q = mKdh \tag{6.2.20}$$

If the total head drop across the region of flow is H and there are n divisions of head in the flow net ($H = ndh$), then

$$q = \frac{mKh}{n} \tag{6.2.21}$$

This equation is applicable only to simple flow systems with one recharge boundary and one discharge boundary.

The rate of flow between streamlines can also be expressed as

$$dq = v_x dx - v_y dy \tag{6.2.22}$$

Substituting (6.2.14) and (6.2.15) for v_x and v_y, we get

$$\frac{\partial \psi}{\partial x}dx + \frac{\partial \psi}{\partial y}dy = dq \tag{6.2.23}$$

which implies that

$$dq = d\psi \tag{6.2.31}$$

The value of the stream function is numerically equal to the unit discharge. Also, the increment in unit discharge between two streamlines is equal to the change in the value of the stream function between two streamlines.

Figure 6.2.3 illustrates the development of the flow distribution through the use of *equipotential lines* and *flow lines*. Flow lines represent the paths followed by molecules of water as they move through an aquifer in the direction of decreasing head. Equipotential lines intersect the flow lines at right angles, representing the piezometric-surface or water-table contours. The two families of lines or curves together are referred to as a *flow net*.

Figure 6.2.3 illustrates two example applications showing the flow lines and equipotential lines that form the orthogonal network of cells making up the flow net. Theoretically, the flow through any one cell equals the flow through any other cell. This figure shows contrasting flow nets for channel seepage through layered anisotropic media.

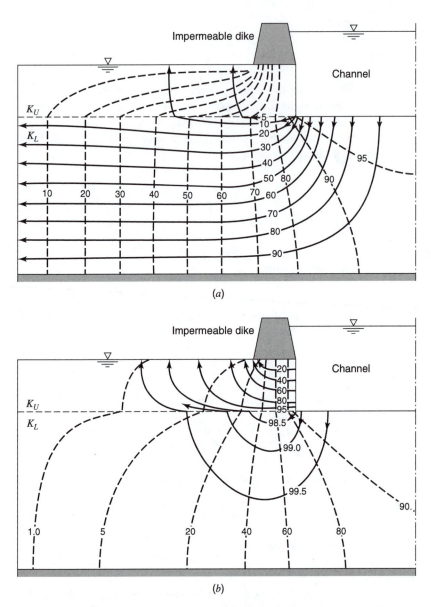

Figure 6.2.3 Flow nets for seepage from one side of a channel through two different anisotropic two-layer systems. (a) $K_u/K_L = 1/50$; (b) $K_u/K_L = 50$. The anisotropy ratio for all layers is $K_x/K_z = 10$ (after Todd and Bear (1961)).

6.3   STEADY-STATE ONE-DIMENSIONAL FLOW

Confined Aquifer

Consider steady-state groundwater flow in a confined aquifer of uniform thickness, as shown in Figure 6.3.1. For one-dimensional steady-state flow, $\partial^2 h/\partial y^2 = 0$ and $\partial h/\partial t = 0$, so equation (6.2.10) for $W = 0$ reduces to

$$\frac{\partial^2 h}{\partial x^2} = 0 \tag{6.3.1}$$

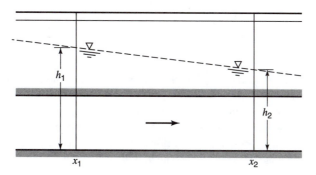

Figure 6.3.1 Flow in a one-dimensional confined aquifer.

which has the solution

$$h = C_1 x + C_2 \tag{6.3.2}$$

where h is the head above a given datum and C_1 and C_2 are constants of integration. For $h = h_1$ at $x = 0$, $h = h_2$ at $x = L$, and $\partial h/\partial x = -(q/K)$ from Darcy's law (equation 6.2.6a), then

$$h = q_x / K \tag{6.3.3}$$

In other words, the head decreases linearly with flow in the x direction, as shown in Figure 6.3.1.

Unconfined Aquifer

For one-dimensional steady-state flow in an unconfined aquifer (Figure 6.3.2), a direct solution of equation (6.2.10) is not possible (see Todd, 1980). To obtain a solution use Darcy's law, $q = -K(\partial h/\partial x)$, and by continuity define the discharge per unit width at any vertical section as

$$\sum_{CS} \rho \mathbf{V} \cdot d\mathbf{A} = Q = -Kh \frac{\partial h}{\partial x} \tag{6.3.4}$$

Equation (6.3.4) can be integrated over the distance from x_1 to x_2 where h_1 and h_2 are the respective heads; then

$$Q \int_{x_1}^{x_2} dx = K \int_{h_1}^{h_2} h \, dh \tag{6.3.5}$$

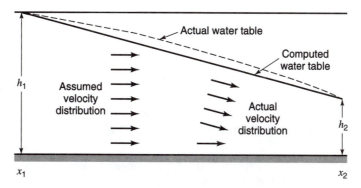

Figure 6.3.2 Flow in a one-dimensional unconfined aquifer.

which becomes

$$Q = K \frac{h_1^2 - h_2^2}{2(x_2 - x_1)} \tag{6.3.6}$$

Equation (6.3.6) indicates a parabolic surface of the saturated portion of the aquifer between x_1 and x_2.

The assumptions in this derivation are (1) the velocity of flow is proportional to the tangent of the hydraulic gradient and (2) the flow is horizontal and uniform every where in a vertical section. The assumption of one-dimensional flow is valid as velocities are small and dh/dx is small. Using the boundary conditions that $x_1 = 0$ and $h_1 = h_0$ and letting $x = X$ and $h = h_2$, then equation (6.3.6) reduces to the well-known *Dupuit equation*

$$Q = \frac{K}{2X}\left(h_0^2 - h^2\right) \tag{6.3.7}$$

Refer to Figure 6.3.3 for an example explanation.

EXAMPLE 6.3.1   A stratum of clean sand and gravel between two channels (see Figure 6.3.3) has a hydraulic conductivity $K = 10^{-1}$ cm/sec and is supplied with water from a ditch ($h_0 = 20$ ft deep) that penetrates to the bottom of the stratum. If the water surface in the second channel is 2 ft above the bottom of the stratum and its distance to the ditch is $x = 30$ ft, which is also the thickness of the stratum, what is the unit flow rate into the gallery?

SOLUTION   The flow is described using the Dupuit equation (6.3.7) for unit flow, where

$$K = 10^{-1} \text{ cm/sec } (60 \text{ sec/min })(1 \text{ in}/2.54 \text{ cm})(1 \text{ ft}/12 \text{ in})(7.48 \text{ gal}/1 \text{ ft}^3) =$$

$$1.54 \text{ gpm/ft}^2 = 2.22 \times 10^3 \text{gpd/ft}^2$$

and

$$Q = \frac{1.5}{2(30)}\left(20^2 - 2^2\right) = 10.16 \text{ gpm/ft} = 1.46 \times 10^4 \text{ gpd/ft}$$

or in SI units with $K = 10^{-1}$ cm/s $= 10^{-3}$ m/s and $h_0 = 20$ ft $= 6.10$ m, $h = 0.61$ m, and $x = 9.14$ m:

$$Q = \frac{1.0^{-3}\left(6.10^2 - 0.61^2\right)}{2(9.14)} = 2.02 \times 10^{-3} \text{ m}^3/\text{s}$$

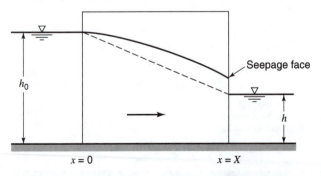

Figure 6.3.3 Steady flow in a one-dimensional unconfined aquifer between two bodies of water with vertical boundaries (application of Dupuit equation (6.3.7)).

6.4   STEADY-STATE WELL HYDRAULICS

6.4.1   Flow to Wells

At initiation of discharge (pumpage) from a well, theoretically the water level or head in the well is lowered relative to the undisturbed condition of the piezometric surface or water table outside the well. In the aquifer surrounding the well, the water flows radially to the lower level in the well. For artesian conditions the actual flow distribution of the flow conforms relatively close to the theoretical shortly after pumping starts. However, in non-artesian (free aquifer) conditions the actual distribution of flow may not conform to the theoretical, as illustrated by the successive stages of development of flow distribution in Figure 6.4.1.

The flow net in Figure 6.4.2 illustrates the distribution of flow in an artesian aquifer for a fully penetrating well and a 100-percent open hole. *Drawdown* is the distance the water level is lowered. When the drawdown falls below the bottom of the upper confining bed, a mixed condition of artesian and non-artesian flow occurs. The flow net in Figure 6.4.3 illustrates the distribution of flow in an artesian aquifer for a well that penetrates through the upper confining bed but not into

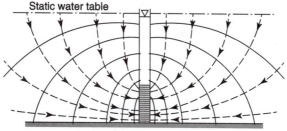

(*a*) Initial stage in pumping a free aquifer. Most water follows a path with a high vertical component from the water table to the screen.

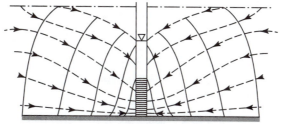

(*b*) Intermediate stage in pumping a free aquifer. Radial component of flow becomes more pronounced but contribution from drawdown cone in immediate vicinity of well is still important.

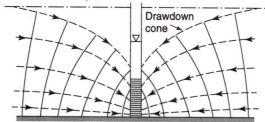

(*c*) Approximate steady state stage in pumping a free aquifer. Profile of cone of depression is established. Nearly all water originating near outer edge of area of influence and stable primarily radial flow pattern established.

Figure 6.4.1 Development of flow distribution about a discharging well in a free aquifer—a fully penetrating and 33-percent open hole (from U.S. Bureau of Reclamation (1981)).

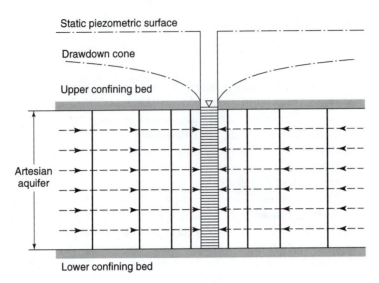

Figure 6.4.2 Distribution of flow to a discharging well in an artesian aquifer—a fully penetrating and 100-percent open hole (from U.S. Bureau of Reclamation (1981)).

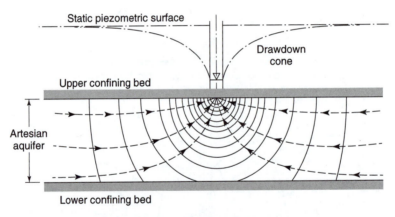

Figure 6.4.3 Distribution of flow to a discharging well—just penetrating to the top of an artesian aquifer (from U.S. Bureau of Reclamation (1981))

the artesian aquifer. A strong vertical component of flow is established out to a distance approximately equal to the thickness of the aquifer. *Drawdown curves (cones)* show the variation of drawdown with distance from the well.

6.4.2 Confined Aquifers

When pumping from a well, water is removed from the aquifer surrounding the well and the piezometric surface is lowered. *Drawdown* is the distance the piezometric surface is lowered (Figure 6.4.4). A radial flow equation can be derived to relate well discharge to drawdown. Consider the confined aquifer with steady-state radial flow to the fully penetrating well that is being pumped, as shown in Figure 6.4.5. For a homogenous, isotropic aquifer, the well discharge at any radial distance r from the pumped well is

$$\sum_{CS} \rho \mathbf{V} \cdot d\mathbf{A} = Q = 2\pi K r b \frac{dh}{dr} \qquad (6.4.1)$$

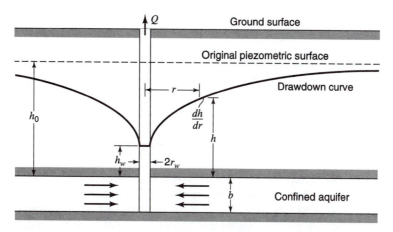

Figure 6.4.4 Well hydraulics for a confined aquifer.

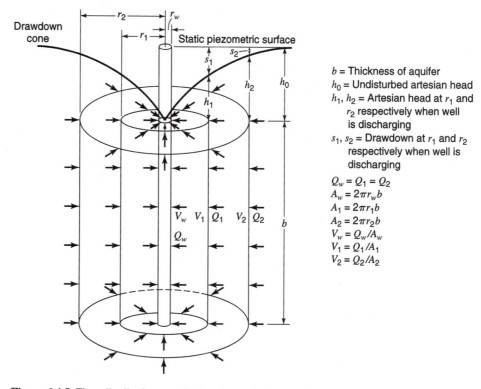

b = Thickness of aquifer
h_0 = Undisturbed artesian head
h_1, h_2 = Artesian head at r_1 and
r_2 respectively when well
is discharging
s_1, s_2 = Drawdown at r_1 and r_2
respectively when well is
discharging

$Q_w = Q_1 = Q_2$
$A_w = 2\pi r_w b$
$A_1 = 2\pi r_1 b$
$A_2 = 2\pi r_2 b$
$V_w = Q_w/A_w$
$V_1 = Q_1/A_1$
$V_2 = Q_2/A_2$

Figure 6.4.5 Flow distribution to a discharging well in an artesian aquifer—a fully penetrating and 100-percent open hole (from U.S. Bureau of Reclamation (1981)).

For boundary conditions of $h = h_1$ at $r = r_1$ and $h = h_2$ at $r = r_2$, then equation (6.4.1) can be integrated:

$$\int_{h_1}^{h_2} dh = \frac{Q}{2\pi Kb} \int_{r_1}^{r_2} \frac{dr}{r}$$

$$h_2 - h_1 = \frac{Q}{2\pi Kb} \ln \frac{r_2}{r_1}$$

(6.4.2)

Solving equation (6.4.2) for Q gives

$$Q = 2\pi Kb \left[\frac{h_2 - h_1}{\ln(r_2/r_1)} \right] \tag{6.4.3}$$

For the more general case of a well penetrating an extensive confined aquifer, there is no limit for r. Referring to Figure 6.4.4, equation (6.4.3) becomes

$$Q = 2\pi Kb \left[\frac{h - h_1}{\ln(r/r_1)} \right] \tag{6.4.4}$$

which shows that the head varies linearly with the logarithm of distance regardless of the rate of discharge.

Equation (6.4.4) is known as the *equilibrium* or *Thiem equation* and enables the aquifer permeability to be determined from a pumped well. Two points define the logarithmic drawdown curve, so drawdown can be measured at two observation wells at different distances from a well that is being pumped at a constant rate. The hydraulic conductivity can be computed using

$$K = \frac{Q \ln(r_2/r_1)}{2\pi b (h_2 - h_1)} \tag{6.4.5}$$

where r_1 and r_2 are the distances and h_1 and h_2 are the heads in the respective observation wells.

EXAMPLE 6.4.1

A well fully penetrates a 25-m thick confined aquifer. After a long period of pumping at a constant rate of 0.05 m³/s, the drawdowns at distances of 50 m and 150 m from the well were observed to be 3 m and 1.2 m, respectively. Determine the hydraulic conductivity and the transmissivity. What type of unconsolidated deposit would you expect this to be?

SOLUTION

Use equation (6.4.5) to determine the hydraulic conductivity with $Q = 0.05$ m³/s, $r_1 = 50$ m, $r_2 = 150$ m, $s_1 = h_0 - h_1$ and $s_2 = h_0 - h_2$, so $s_1 - s_2 = h_2 - h_1 = 3 - 1.2 = 1.8$ m:

$$K = \frac{Q \ln(r_2/r_1)}{2\pi b (s_1 - s_2)} = \frac{0.05 \ln(150/50)}{2\pi (25)(3 - 1.2)} = 1.94 \times 10^{-4} \, \text{m/s}$$

Then the transmissivity is $T = Kb = (1.94 \times 10^{-4})(25) = 4.85 \times 10^{-3}$ m²/s. From Table 6.1.1 for $K = 1.94 \times 10^{-4}$ m/s, this is probably a clean sand or silty sand.

EXAMPLE 6.4.2

A 2-ft diameter well penetrates vertically through a confined aquifer 50 ft thick. When the well is pumped at 500 gpm, the drawdown in a well 50 ft away is 10 ft and in another well 100 ft away is 3 ft. What is the approximate head in the pumped well for steady-state conditions and what is the approximate drawdown in the well? Also compute the transmissivity. Take the initial piezometric level as 100 ft above the datum.

SOLUTION

First determine the hydraulic conductivity using equation (6.4.5):

$$K = \frac{Q \ln(r_2/r_1)}{2\pi b (h_2 - h_1)} = \frac{500 \; \ln(100/50)}{2\pi (50)(97 - 90)} = 0.158 \; \text{gpm/ft}^2$$

Then compute the transmissivity:

$$T = Kb = 0.158 \times 50 = 7.90 \text{ gpm/ft}$$

Now compute the approximate head, h_w, in the pumped well. From

$$h_2 - h_w = \frac{Q\ln(r_2/r_1)}{2\pi Kb}$$

$$h_w = h_2 - \frac{Q\ln(r_2/r_w)}{2\pi Kb}$$

$$h_w = 97 - \frac{500\ln(100/1)}{2\pi(7.90)} = 97 - 46.4 = 50.6 \text{ ft}$$

Drawdown is then $s_w = 100 - 50.6$ ft $= 49.6$ ft.

6.4.3   Unconfined Aquifers

Now consider steady radial flow to a well completely penetrating an unconfined aquifer as shown in Figure 6.4.6. A concentric boundary of constant head surrounds the well. The well discharge is given by

$$\sum_{CS} \rho \mathbf{V} \cdot d\mathbf{A} = Q = 2\pi Khr\frac{dh}{dr} \tag{6.4.6}$$

Integrating equations (6.4.6) with the boundary conditions $h = h_1$ at $r = r_1$ and $h = h_2$ at $r = r_2$ yields

$$\int_{h_1}^{h_2} h\,dh = \frac{Q}{2\pi K}\int_{r_1}^{r_2} \frac{dr}{r} \tag{6.4.7}$$

$$Q = \pi K\left[\frac{h_2^2 - h_1^2}{\ln(r_2/r_1)}\right] \tag{6.4.8}$$

There are large vertical flow components near the well so that this equation fails to describe accurately the drawdown curve near the well, but can be defined for any two distances r_1 and r_2 away from the pumped well.

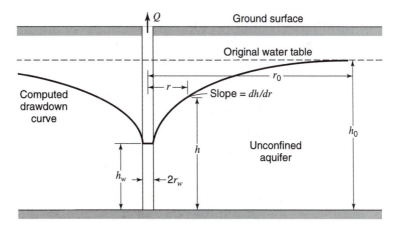

Figure 6.4.6 Well hydraulics for an unconfined aquifer.

EXAMPLE 6.4.3   A well penetrates an unconfined aquifer. Prior to pumping the water level (head) is $h_0 = 25$ m. After a long period of pumping at a constant rate of 0.05 m³/s, the drawdowns at distances of 50 m and 150 m from the well were observed to be 3 m and 1.2 m, respectively. Determine the hydraulic conductivity. What type of deposit would the aquifer material probably be?

SOLUTION   Use equation (6.4.8) and solve for K:

$$K = \frac{Q\ln(r_2/r_1)}{\pi\left(h_2^2 - h_1^2\right)}$$

and solve for K with $Q = 0.05$ m³/s, $r_1 = 50$ m, $r_2 = 150$ m, $h_1 = 25 - 3 = 22$ m, and $h_2 = 25 - 1.2 = 23.8$ m:

$$K = \frac{0.05\ln(150/50)}{\pi\left(23.8^2 - 22^2\right)} = \frac{1.7494\times10^{-2}}{(566.4 - 484)} = 2.12\times10^{-4}\,\text{m/s}$$

The deposit is probably a silty sand or a clean sand.

EXAMPLE 6.4.4   A well 12-in in diameter penetrates 108 ft below the static water table. After a long period of pumping at a rate of 350 gpm, the drawdowns in wells 57 ft and 148 ft from the pumped well were found to be 12 ft and 7.4 ft respectively. What is the transmissivity of the aquifer? What is the approximate drawdown in the pumped well?

SOLUTION   Use equation (6.4.8) for steady-state radial flow to a well in an unconfined aquifer, where $h_1 = 108 - 12 = 96$ ft; $h_2 = 108 - 7.4 = 100.6$ ft; $r_2 = 148$ ft; and $r_1 = 57$ ft. Determine the hydraulic conductivity using equation (6.4.8):

$$K = \frac{Q\ln(r_2/r_1)}{\pi\left(h_2^2 - h_1^2\right)} = \frac{350\cdot\ln(148/57)}{\pi\left[100.6^2 - 96^2\right]} = 0.118 \text{ gpm/ft}^2$$

$$T = Kb = 0.118 \times 108 = 12.74 \text{ gpm/ft}$$

Solve for the approximate head and approximate drawdown at the well:

$$h_w = \sqrt{100.6^2 - \frac{350\cdot\ln(148/0.5)}{\pi\cdot(0.118)}} = 68.90 \text{ ft}$$

$$s_w = 108 - 68.90 = 39.10 \text{ ft}.$$

6.5   TRANSIENT WELL HYDRAULICS—CONFINED CONDITIONS

6.5.1   Nonequilibrium Well Pumping Equation

Consider the confined aquifer with transient flow conditions due to a constant pumping rate Q (L^3/T) at the well shown in Figure 6.4.4. The initial condition is a constant head throughout the aquifer, $h(r, 0) = h_0$ for all r where h_0 is the constant initial head at $t = 0$. The boundary condition assumes (1) no drawdown in hydraulic head at the infinite boundary, $h(\infty, t) = h_0$ for all t with a constant pumping rate and (2) Darcy's law applies:

$$\lim_{r\to0}\left(r\frac{\partial h}{\partial r}\right) = \frac{Q}{2\pi T} \quad \text{for } t > 0 \qquad (6.5.1)$$

The governing differential equation is (6.2.11):

$$\frac{\partial^2 h}{\partial r^2} + \frac{1}{r}\frac{\partial h}{\partial r} = \frac{S}{T}\frac{\partial h}{\partial t} \tag{6.2.11}$$

The confined aquifer is nonleaky, homogenous, isotropic, infinite in areal extent, and the same thickness throughout. The wells fully penetrate the aquifer and are pumped at a constant rate Q. During pumping of such a well, water is withdrawn from storage within the aquifer, causing the cone of depression to progress outward from the well. There is no stabilization of water levels, resulting in a continual decline of head, provided there is no recharge and the aquifer is effectively infinite in areal extent. The rate of decline of the head, however, continuously decreases as the cone of depression spreads. Water is released from storage by compaction of the aquifer and by the expansion of the water. Theis (1935) presented an analytical solution of equation (6.2.11) to solve for the *drawdown s* given as

$$s = h_0 - h = \frac{Q}{4\pi T}\int_u^\infty \left(\frac{e^{-u}}{u}du\right) \tag{6.5.2}$$

where h_0 is the piezometric surface before pumping started, Q is the constant pumping rate (L³/T), T is the transmissivity (L^2/T), and

$$u = \frac{r^2 S}{4Tt} \tag{6.5.3}$$

in which r is in L and t is time in T.

The exponential integral can be expanded into a series expansion as

$$W(u) = \int\left(\frac{e^{-u}}{u}\right)du = -0.5772 - \ln u + u - \frac{u^2}{2\cdot 2!} + \frac{u^3}{3\cdot 3!} + \cdots \tag{6.5.4}$$

where $W(u)$ is called the dimensionless *well function* for nonleaky, isotropic, artesian aquifers fully penetrated by wells having constant discharge conditions. Values of this well function are listed in Table 6.5.1. The well function is also expressed in the form of a type curve, as shown in Figure 6.5.1. Both u and $W(u)$ are dimensionless.

Table 6.5.1 Values of W(u) for Various Values of u

u	1.0	2.0	3.0	4.0	5.0	6.0	7.0	8.0	9.0
$\times 1$	0.219	0.049	0.013	0.0038	0.0011	0.00036	0.00012	0.000038	0.000012
$\times 10^{-1}$	1.82	1.22	0.91	0.70	0.56	0.45	0.37	0.31	0.26
$\times 10^{-2}$	4.04	3.35	2.96	2.68	2.47	2.30	2.15	2.03	1.92
$\times 10^{-3}$	6.33	5.64	5.23	4.95	4.73	4.54	4.39	4.26	4.14
$\times 10^{-4}$	8.63	7.94	7.53	7.25	7.02	6.84	6.69	6.55	6.44
$\times 10^{-5}$	10.94	10.24	9.84	9.55	9.33	9.14	8.99	8.86	8.74
$\times 10^{-6}$	13.24	12.55	12.14	11.85	11.63	11.45	11.29	11.16	11.04
$\times 10^{-7}$	15.54	14.85	14.44	14.15	13.93	13.75	13.60	13.46	13.34
$\times 10^{-8}$	17.84	17.15	16.74	16.46	16.23	16.05	15.90	15.76	15.65
$\times 10^{-9}$	20.15	19.45	19.05	18.76	18.54	18.35	18.20	18.07	17.95
$\times 10^{-10}$	22.45	21.76	21.35	21.06	20.84	20.66	20.50	20.37	20.25
$\times 10^{-11}$	24.75	24.06	23.65	23.36	23.14	22.96	22.81	22.67	22.55
$\times 10^{-12}$	27.05	26.36	25.96	25.67	25.44	25.26	25.11	24.97	24.86
$\times 10^{-13}$	29.36	28.66	28.26	27.97	27.75	27.56	27.41	27.28	27.16
$\times 10^{-14}$	31.66	30.97	30.56	30.27	30.05	29.87	29.71	29.58	29.46
$\times 10^{-15}$	33.96	33.27	32.86	32.58	32.35	32.17	32.02	31.88	31.76

Source: Wenzel (1942).

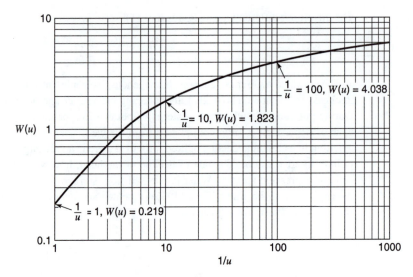

Figure 6.5.1 Type curve for use in solving Theis nonequilibrium equation graphically with values of $W(u)$ (well function of u) corresponding to values of $1/u$. The curve is plotted on a logarithmic scale.

The drawdown is then expressed as

$$s = \frac{Q}{4\pi T} W(u) \tag{6.5.5}$$

This equation is commonly referred to as the *nonequilibrium well pumping equation* or *Theis's equation*.

Equations 6.5.5 and 6.5.3 can be expressed in U.S. customary units (gallon-day-foot system) where s is in ft, Q is in gpm, T is in gpd/ft, r is in ft, and t is in days:

$$s = \frac{114.6Q}{T} W(u) \tag{6.5.6}$$

and

$$u = \frac{1.87r^2 S}{Tt} \tag{6.5.7}$$

or, for t in minutes,

$$u = \frac{2693r^2 S}{Tt} \tag{6.5.8}$$

EXAMPLE 6.5.1

Given that $T = 10,000$ gpd/ft, $S = 10^{-4}$, $t = 2693$ min, $Q = 1000$ gpm, and $r = 1000$ ft, compute the drawdown.

SOLUTION

Step 1. Compute u: (for r in ft, T in gpd/ft, t in minutes)

$$u = \frac{2693r^2 S}{Tt} = \frac{2693 \times 10^6 \times 10^{-4}}{10^4 \times 2693} = 10^{-2}$$

Step 2. Find $W(u)$ from Table 6.5.1: $W(u) = 4.04$

Step 3.   Compute drawdown: (for Q in gal/min and T in gpd/ft)

$$s = \frac{114.6}{T}\frac{Q}{T}W(u) = \frac{114.6 \times 10^3 \times 4.04}{10^4} = 46.30 \text{ ft}$$

6.5.2   Graphical Solution

Theis (1935) developed a graphical procedure for determining T and S by expressing equation (6.5.6) as

$$\log s = \log\left(\frac{114.6Q}{T}\right) + \log W(u) \tag{6.5.9}$$

and equation (6.5.8) as

$$\log t = \log\left(\frac{2693r^2 S}{T}\right) + \log\frac{1}{u} \tag{6.5.10}$$

Because $(114.6Q/T)$ and $(2693r^2 S/Tt)$ are constants for a given distance r from the pumped well, the relation between $\log(s)$ and $\log(t)$ must be similar to the relation between $\log W(u)$ and $\log 1/u$. Therefore, if s is plotted against t and $W(u)$ against $1/u$ on the same double-logarithmic paper, the resulting curves are of the same shape, but horizontally and vertically offset by the constants $(114.6Q/T)$ and $(2693r^2 S/Tt)$. Plot each curve on a separate sheet, then match them by placing one graph on top of the other and moving it horizontally and vertically (keeping the coordinate axis parallel) until the curves are matched. This is further illustrated in Figure 6.5.2.

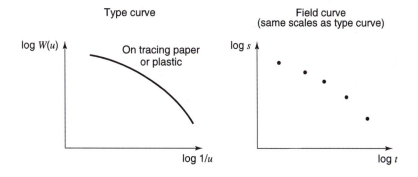

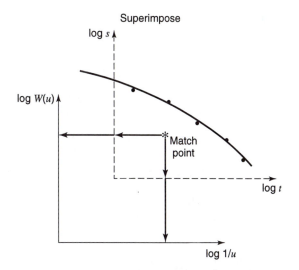

Figure 6.5.2 Graphical procedure to determine T and S from pump test data.

This graphical procedure is used to determine T and S from a field pumping test and requires measurements of a drawdown in at least one observation well. The observation of water levels should be made at proper intervals of time after the instant that pumping starts. Measurements from more than one observation well can also be used.

EXAMPLE 6.5.2

Pump test data (s vs. t) for an observation well has been plotted on log-log paper. The discharge was 150 gpm and the observation well is 300 ft from the pumped well. The match points for the fitted data are: at $W(u) = 1.0$ and $u = 1.0$, $t = 5.5$ min and $s = 0.112$ ft. Compute T and S.

SOLUTION

$$T = \frac{114.6Q}{s}W(u) = \frac{114.6(150)(1)}{0.112} = 153,482 \text{ gpd/ft}$$

and

$$S = \frac{Ttu}{2693r^2} = \frac{(153,482)(5.5)(1)}{2693(300)^2} = 0.00348$$

EXAMPLE 6.5.3

Drawdown was measured during a pumping test at frequent intervals in an observation well 200 ft from a well that was pumped at a constant rate of 500 gpm. The data for this pump test is listed below. These measurements show that the water level is still dropping after 4000 minutes of pumping; therefore, analysis of the test data requires use of the Theis nonequilibrium procedure. Determine T and S for this aquifer.

Pump Test Data

Time (min)	Drawdown (ft)
1	0.05
2	0.22
3	0.40
4	0.56
5	0.70
7	0.94
10	1.2
20	1.8
40	2.5
100	3.4
300	4.5
1000	5.6
4000	7.0

SOLUTION

Step 1. Plot the time-drawdown data on log-log graph paper. The drawdown is plotted on the vertical axis and the time since pumping started on the horizontal axis.

Step 2. Superimpose this plot on the type curve sheet so that the plotted points match the type curve. The axes of both graphs must be kept parallel.

Step 3. Select a match point, which can be any point in the overlap area of the curve sheets. It is usually found most convenient to select a match point where the coordinates on the type curve are known in advance (e.g., $W(u) = 1$ and $1/u = 1$ or $W(u) = 1$ and $1/u = 10$, etc.). Then determine the value of s and t for this match point: $W(u) = 1$, $s = 1$, $1/u = 1$, and $t = 2$.

Step 4. Determine T:

$$T = \frac{114.6}{s}QW(u) = \frac{114.6 \times 500}{1} \times 1 = 57300 \text{ gpd/ft}$$

Step 5. Determine S:

$$S = \frac{Tt}{\dfrac{1}{u} \times 2693 r^2} = \frac{57300 \times 2}{1 \times 2693 \times 200^2} = 1.06 \times 10^{-3}$$

The above procedure can also be used to determine the drawdown in observation wells at rates of pumping different from that used in the pump test. The Theis nonequilibrium analysis can also be used to determine distance-drawdown information, once the T and S are known for a given time after pumping started. Equations (6.5.6) and (6.5.8) can be expressed as

$$\log s = \log\left(\frac{114.6 Q}{T}\right) + \log W(u) \tag{6.5.9}$$

and

$$\log(r^2) = \log\left(\frac{Tt}{2693\ S}\right) - \log\frac{1}{u} \tag{6.5.11}$$

If the values of Q, T, and S are known (determined from a pump test), then the distance-drawdown curve can be determined for any time t. The terms $(114.6Q/T)$ and $(Tt/2693S)$ are known, and for an assumed match point ($W(u)$ and $1/u$), s and r^2 can be determined.

EXAMPLE 6.5.4   For a pumping rate of 100 gpm in a confined aquifer with $T = 10^4$ gpd/ft and $S = 10^{-4}$, determine the distance-drawdown (r^2, s) curve after 269.3 minutes.

SOLUTION   Step 1. Assume $W(u)$ and $1/u$ to compute match points (e.g., $W(u) = 1$ and $1/u = 1$).

Step 2. Compute s:

$$s = \frac{114.6}{T}\ Q\ W(u) = \frac{114.6(100)}{10^4}(1) = 1.146 \text{ ft}$$

Compute r^2:

$$r^2 = \frac{Ttu}{2693\ S} = \frac{10^4 (269.3)(1)}{2693\ S\ (10^{-4})} = 10^7 \text{ ft}^2$$

Step 3. Use the computed match points to align the graph of s on the vertical scale and r^2 on the horizontal scale to the type curve. In other words, match points (r^2, s) with ($1/u$, $W(u)$) by superimposing the graph sheet onto the type curve. Keep all axes parallel. Once the points are matched, trace the type curve on the (r^2, s) graph sheet to form the distance-drawdown curve for t minutes after pumping started.

6.5.3   Cooper–Jacob Method of Solution

Time-Drawdown Analysis

Theis's method for nonequilibrium analysis was simplified by Cooper and Jacob (1946) for the conditions of small values of r and large values of t. From equation (6.5.3), it can be observed that, for small values of r or large values of t, u is small, so that the higher-order terms in equation (6.5.4) become negligible. Then the well function can be expressed as

$$W(u) = -0.5772 - \ln u \tag{6.5.12}$$

which has an error < 3% for $u < 0.1$. The drawdown for this approximation is expressed in U.S. customary units (gallon-day-foot) as

$$s = \frac{114.6\,Q}{T}[-0.5772 - \ln u] \qquad (6.5.13)$$

and in SI units as

$$s = \frac{Q}{4\pi T}[-0.5772 - \ln u] \qquad (6.5.14)$$

Substituting equation (5.6.13) in equation (6.5.8) yields

$$s = \frac{114.6\,Q}{T}\left[-0.5772 - \ln\frac{2693r^2 S}{Tt}\right] \qquad (6.5.15)$$

This shows that the drawdown is a function of log t so that the equation plots as a straight line on semilog paper.

Consider the change in drawdown Δs a distance r from the pumped well, over the time interval t_1 and t_2, which are one log cycle apart. The drawdowns at t_1 and t_2 are, respectively, s_1 and s_2. The change in drawdown Δs is then expressed as

$$\Delta s = s_2 - s_1 = \frac{264\,Q}{T}\left[\log\left(\frac{Tt_2}{2693r^2 S}\right) - \log\left(\frac{Tt_1}{2693r^2 S}\right)\right]$$
$$= \frac{264\,Q}{T}\log\left(\frac{t_2}{t_1}\right) \qquad (6.5.16)$$

If t_2 and t_1 are chosen one log cycle apart, $\log(t_2/t_1) = 1$, the above equation simplifies to

$$\Delta s = \frac{264\,Q}{T} \qquad (6.5.17)$$

in which Q is in gpm and T is in gpd/ft.

Next consider that at time t_0, when pumping begins, the drawdown is 0 in the observation well r ft from the pumped well, so that equation (6.5.15) is expressed as

$$0 = \frac{114.6\,Q}{T}\left[-0.5772 + 2.3\,\log\frac{Tt}{2693r^2 S}\right] \qquad (6.5.18)$$

This equation then reduces to

$$S = \frac{Tt_0}{4790\,r^2} \qquad (6.5.19)$$

The time t_0 is the time intercept on the zero-drawdown axis, keeping in mind that Jacob's method is a straight-line approximation on semilog paper. Refer to Figure 6.5.3.

EXAMPLE 6.5.5   For the time-drawdown data listed below, calculate T and S using Jacob's approximation. After computing T and S, check to see that the basic assumption of this approximation is satisfied. For the values of T and S that you computed, after how many minutes of pumping would Jacob's approximation be valid? The discharge is $Q = 1500$ gpm and the radius $r = 300$ ft.

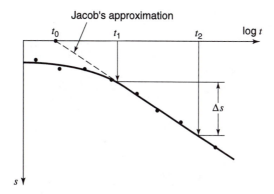

Figure 6.5.3 Illustration of Jacob's method showing the straight-line approximation on semilog paper.

Time after pumping started (min)	Drawdown (ft)
1	0.45
2	0.74
3	0.91
4	1.04
6	1.21
8	1.32
10	1.45
30	2.02
40	2.17
50	2.30
60	2.34
80	2.50
100	2.67
200	2.96
400	3.25
600	3.41
800	3.50
1,000	3.60
1,440	3.81

SOLUTION

Step 1. Plot the field data on semilog paper (s vs. log t) as shown in Figure 6.5.4.

Step 2. Fit a straight line to the data (Figure 6.5.4).

Step 3. Find t_0 from the plot ($t_0 = 0.45$ minutes for $s = 0$).

Step 4. Find Δs for values of t_1 and t_2 one log cycle apart ($t_1 = 10$ min and $t_2 = 100$ min): $\Delta s = s_2 - s_1 = 2.56 - 1.48 = 1.08$ ft.

Step 5. Compute T (from equation 6.5.17):

$$T = \frac{264\ Q}{\Delta s} = \frac{264(1500)}{1.08} = 366{,}700 \text{ gpd/ft}$$

Step 6. Compute S:

$$S = \frac{Tt_0}{4790\ r^2} = \frac{366.700(0.45)}{4790(300^2)} = 3.83 \times 10^{-4}$$

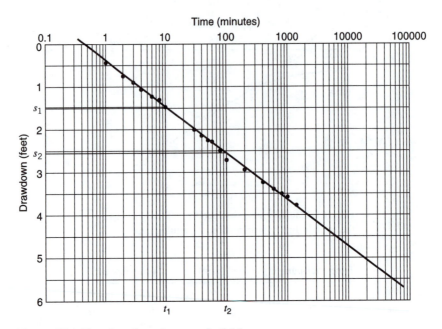

Figure 6.5.4 Time-drawdown for example 6.5.5.

Step 7. Check to see if basic assumption is satisfied:

$$u = \frac{2693(300)^2(3.83 \times 10^{-4})}{366.700(1440)} = 0.00017 < 0.01 \quad \text{OK}$$

Jacob's approximation for this problem is valid for

$$t \geq \frac{2693 r^2 S}{Tu} = \frac{2693(300)^2(3.83 \times 10^{-4})}{366.700(0.01)} \quad \text{where } u = 0.01$$

$$t \geq 12.6 \text{ min}$$

Distance–Drawdown Analysis

A similar analysis using Jacob's approximation can be used to approximate T and S using distance-drawdown field data. The drawdown must be measured simultaneously in three or more observation wells, each at a different distance from the pumped well. The drawdowns in the observation wells plot as a straight line on semilog paper (s vs. log r). By considering the distance r_0 from the pumped well, where the drawdown is 0, then equation (6.5.15) can be expressed as

$$0 = \frac{114.6 \, Q}{T}\left[-0.5772 - \ln\frac{2693 r_0^2 S}{Tt}\right] \tag{6.5.20}$$

which can be simplified to

$$S = Tt/4790 r_0^2 \tag{6.5.21}$$

where r_0 is the intercept at $s = 0$ of the extended straight line fitted to the field distance-drawdown data.

Following the same procedure as for the time-drawdown analysis, consider the change in drawdown $\Delta s = s_2 - s_1$ at a time t over the distance r_2 to r_1, where r_2 and r_1 are chosen to be one log cycle apart on the plot. The drawdowns at r_1 and r_2 are, respectively, s_1 and s_2. The change in drawdown Δs is then expressed as

$$\Delta s = s_2 - s_1 = \frac{264\,Q}{T}\left[\log\frac{2693r_2^2 S}{Tt} - \log\frac{2693r_1^2 S}{Tt}\right] = \frac{528\,Q}{T}\log\left(\frac{r_2}{r_1}\right) \qquad (6.5.22)$$

Because r_2 and r_1 are chosen one log cycle apart, ($\log r_2/r_1 = 1$), the above equation simplifies to

$$\Delta s = \frac{528\,Q}{T} \qquad (6.5.23)$$

EXAMPLE 6.5.6

A well is pumped at 200 gpm for a period of 500 minutes. Three observation wells 1, 2, and 3 (50, 30, and 100 ft from the pumped well, respectively) have drawdowns of 10.6, 13.2, and 7.9 ft respectively. Determine T and S. (from Gehm and Bregman, 1976).

SOLUTION

Step 1. Plot the field data on semilog paper (s vs. log r) as shown in Figure 6.5.5.

Step 2. Fit a straight line to the data (Figure 6.5.5).

Step 3. Find r_0 from the plot ($r_0 = 500$ ft).

Step 4. Find Δs for values of r_1 and r_2 that are selected one log-cycle apart: $\Delta s = s_2 - s_1 = 10.6$ ft.

Step 5. Compute T (from equation 6.5.23):

$$T = \frac{528\,Q}{\Delta s} = \frac{528(200)}{10.6} = 9960 \text{ gpd/ft}$$

Step 6.   Compute S:

$$S = \frac{Tt}{4790r_0^2} = \frac{9960(500)}{4790(500^2)} = 4.16 \times 10^{-3}$$

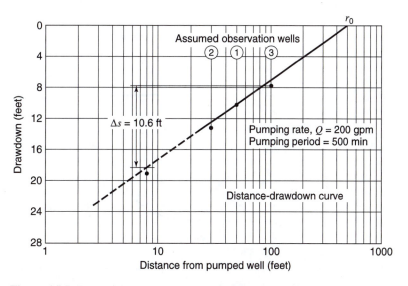

Figure 6.5.5 Trace of the cone of depression plotted on semilog coordinates becomes a straight line. Drawdown in each well was measured at 500 min after starting the pumping test.

6.6 TRANSIENT WELL HYDRAULICS—UNCONFINED CONDITIONS

Unconfined (water table) conditions differ significantly from those of the confined aquifer. For unconfined conditions the pumped water is derived from storage by gravity drainage of the interstices above the cone of depression, by compaction of the aquifer and expansion of the water as pressure is reduced from pumping. For confined conditions the nonequilibrium solution is based on the assumptions that the coefficient of storage is constant and water is released from storage instantaneously with a decline in head. The effects of gravity drainage are not considered in the nonequilibrium solution. Gravity drainage is not immediate and, for unsteady flow of water towards a well for unconfined conditions, is characterized by slow drainage of interstices.

There are three distinct segments of the time-drawdown curve for water table conditions, as shown in Figure 6.6.1. The first segment occurs for a short time after pumping begins: the drawdown reacts in the same manner as an artesian aquifer. In other words, the gravity drainage is not immediate and the water is released instantaneously from storage. It is possible under some conditions to determine the coefficient of transmissivity by applying the nonequilibrium solution to the early time-drawdown data. The coefficient of storage computed using the early time-drawdown is in the artesian range and cannot be used to predict long-term drawdowns. The second segment represents an intermediate stage when the expansion of the cone of depression decreases because of the gravity drainage. The slope of the time-drawdown curve decreases, reflecting recharge. Pump test data deviate significantly from the nonequilibrium theory during the second segment. During the third segment (Figure 6.6.1), the time-drawdown curves conform closely to the nonequilibrium type curves, as shown in Figure 6.6.2. This segment may start from several minutes to several days after pumping starts depending upon the aquifer condition. The coefficient of transmissibility of an aquifer can be determined by applying the nonequilibrium solution to the third segment of time-drawdown data. The coefficient of storage computed from this data will be in the unconfined range, which can be used to predict long-term effects.

Pricket (1965) developed a type curve solution for water table conditions. The following equation for drawdown in an unconfined aquifer with fully penetrating wells and a constant discharge condition Q was presented by Neuman (1975):

$$s = \frac{Q}{4\pi T} W(u_a, u_y, \eta) \tag{6.6.1}$$

where

$$u_a = \frac{r^2 S}{Tt} \quad \text{(applicable for small values of } t) \tag{6.6.2}$$

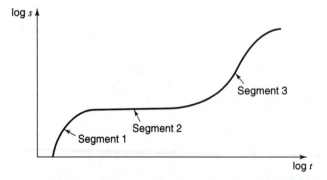

Figure 6.6.1 Three segments of time-drawdown curve for water table conditions.

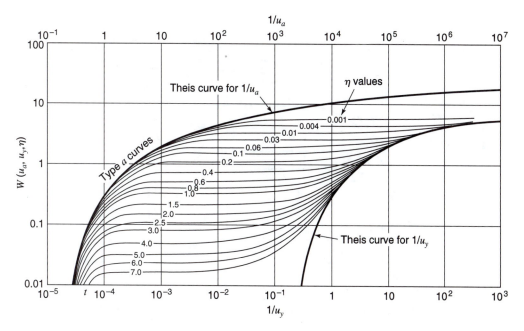

Figure 6.6.2 Theoretical curves of $W(u_a, u_y, \eta)$ versus $1/u_a$ and $1/u_y$ for an unconfined aquifer (after Neuman (1975)).

$$u_y = \frac{r^2 S_y}{Tt} \quad \text{(applicable for large values of } t\text{)} \quad (6.6.3)$$

where $\eta = r^2 k_z/b^2 k_r$.

$W(u_a, u_y, \eta)$ is referred to as the *unconfined well function* and K_r and K_z are the horizontal and vertical hydraulic conductivities, respectively, for an anisotropic aquifer. Neuman (1975) tabulated the well function, as plotted in Figure 6.6.2. For an isotropic aquifer, $K_r = K_z$ and $\eta = r^2/b^2$. The u_a applies only to early-time response when the rate of drawdown is controlled by the elastic storage properties of the aquifer. The u_y applies to the late-time response when the rate of drawdown is controlled by the specific yield.

Distance-drawdown data for water table conditions can be used to compute T and S_y only after the effects of delayed gravity drainage have dissipated in observation wells. After the effects of delayed gravity drainage cease to influence the drawdown in observation wells, the time-drawdown field data conforms closely to the nonequilibrium solution as illustrated above. This length of time was discussed in the previous section.

During the time when delayed gravity drainage is affecting the drawdown in observation wells, the cone of depression is distorted; therefore, the distance-drawdown data for this time cannot be analyzed. However, once the effects of delayed gravity drainage are negligible, then the distance-drawdown data can be analyzed using the nonequilibrium solution technique.

6.7 TRANSIENT WELL HYDRAULICS—LEAKY AQUIFER CONDITIONS

In leaky aquifer systems, water enters the aquifer from adjacent lower-permeability units. These conditions provide an additional source of water and reduce the drawdown predicted by the Theis solution. *Leaky-confined conditions* refer to wells penetrating a confined aquifer overlain by an aquitard, which in turn is overlain by a source bed having a water table. There is vertical leakage through the aquitard (confining layer). The aquifer is assumed to be homogeneous, isotropic, infinite in areal extent, and of constant thickness. Wells fully penetrate the aquifer to which there is radial flow. The confining layer is assumed to be incompressible, neglecting any storage in the

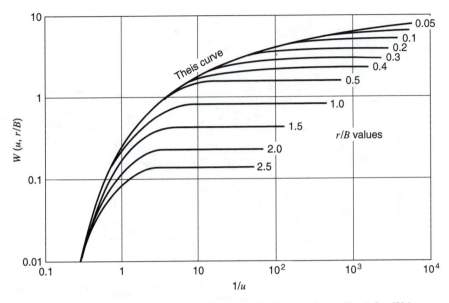

Figure 6.7.1 Theoretical curves of $W(u, r/B)$ versus $1/u$ for a leaky aquifer (after Walton (1960)).

layer. Discharge from a well is supplied from both storage within the aquifer and the vertical leakage. The rate of the vertical leakage is proportional to the difference in head between the water table for the source bed and the piezometric surface of the aquifer.

Hantush and Jacob (1955) developed a solution that describes the drawdown in a radially symmetric leaky confined aquifer separated from an overlying aquifer by a confining bed of lower permeability. Leakage into the pumped aquifer is a function of the vertical hydraulic conductivity of the aquitard, K', the aquitard thickness, b', and the difference in hydraulic head between the overlying aquifer and the aquifer being pumped. The drawdown is expressed as

$$s(r, t) = \frac{Q}{4\pi T} W\left[u, \left(r/B\right)\right] \tag{6.7.1}$$

where the *leaky well function* $W(u, r/B)$ is defined as a function of two dimensionless parameters

$$u = \frac{r^2 S}{4Tt} \tag{6.7.2}$$

and

$$r/B = r\sqrt{\frac{K'}{Kb_b'}} \tag{6.7.3}$$

The well function is plotted in Figure 6.7.1. The key simplification is that the hydraulic head in the upper aquifer remains constant and that water is not released from storage in the aquitard.

6.8 BOUNDARY EFFECTS: IMAGE WELL THEORY

The *nonequilibium Theis solution* for unsteady radial flow to a well is based on the assumption that the aquifer is infinite in areal extent; however, often aquifers are delimited by one or more boundaries, causing the time-drawdown data to deviate under the influence of these boundaries. The boundaries may be walls of impervious soil or rock, which are referred to as *barrier boundaries*, or recharges of water from rivers, streams, or lakes, which are referred to as *recharge boundaries*. The effects of barrier and recharge boundaries on groundwater movement and storage can be described by image well theory.

Walton (1970) has stated the *image well theory* as "the effect of a barrier boundary on the draw-down in a well, as a result of pumping from another well, is the same as though the aquifer were infinite and a like discharging well were located across the real boundary on a perpendicular thereto and at the same distance from the boundary as the real pumping well. For a recharge boundary the principle is the same except that the image well is assumed to be discharging the aquifer instead of pumping from it." These concepts give rise to the use of imaginary wells, referred to as *image wells*, that are introduced to simulate a groundwater system reflecting the effects of known physical boundaries on the system. Essentially, then, through the use of image wells an aquifer of finite extent can be transformed to one of infinite extent. Once this is done the unsteady radial flow equations can be applied to the transformed system. The concept of image wells is further illustrated in Figure 6.8.1 for a barrier boundary and Figure 6.8.2 for a recharge boundary.

6.8.1   Barrier Boundary

In the case of a *barrier boundary*, water cannot flow across the boundary, so that no water is being contributed to the pumped well from the impervious formation. The cone of depression that would exist for a pumped well in an aquifer of infinite areal extent is shown in Figure 6.8.1. Because of the barrier boundary, the cone of depression shown is no longer valid since there can be no flow across the boundary. Placing an image well, discharging in nature, across the barrier boundary creates the effect of no flow across the boundary. The image well must be placed perpendicular to the barrier boundary and at the same distance from the boundary as the real well. The resulting *real cone of depression* is the summation of the components of both the real and image well depression cones, as shown in Figure 6.8.1. Water levels in wells will decline at an initial rate due only to the

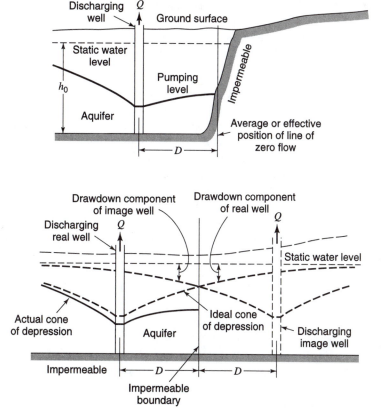

Figure 6.8.1 Relationship of an impermeable boundary and an image well (from U.S. Bureau of Reclamation (1981)).

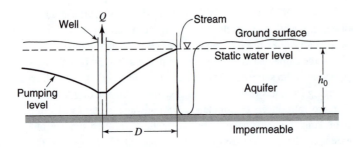

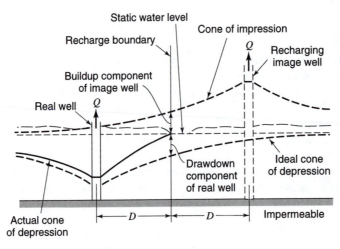

Figure 6.8.2 Relationship of a recharge boundary and an image well (from U.S. Bureau of Reclamation (1981)).

influence of the pumped well. As pumping continues, the barrier boundary effects will begin as simulated by the image well affecting the real well. When the effects of the barrier boundary are realized, the time rate of drawdown will increase (Figure 6.8.3). When this occurs, the total rate of withdrawal from the aquifer is equal to that of the pumped well plus that of the discharging image well, causing the cone of depression of the real well to be deflected downward.

The total drawdown in the real well can be expressed as

$$s_B = s_p + s_i \tag{6.8.1}$$

in which s_B is the total drawdown, s_p is the drawdown in an observation well due to pumping of the production well and s_i is the drawdown due to the discharging image well (barrier boundary).

The total drawdown can be expressed as

$$s_B = \frac{Q}{4\pi T} W(u_p) + \frac{Q}{4\pi T} W(u_i) \tag{6.8.2}$$

where Q is the constant pumping rate (L^3/T), T is the transmissivity (L^2/T), $W(u_p)$, and $W(u_i)$ are dimensionless, and u_p and u_i are

$$u_p = \frac{r_p^2 S}{4Tt_p} \tag{6.8.3}$$

$$u_i = \frac{r_i^2 S}{4Tt_i} \tag{6.8.4}$$

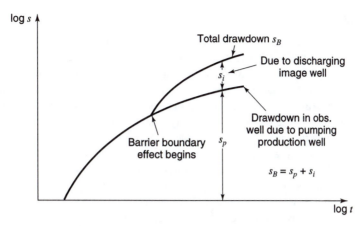

Figure 6.8.3 Barrier boundary effects on time-drawdown curve.

in which r_i and r_p are in L and t is time in T.

The drawdown equation (6.8.1) can also be expressed in U.S. customary units (gal-day-foot) system where S is in ft, Q is in gpm, T is in gpd/ft, r is in ft, and t is in days:

$$s_B = \frac{114.6Q}{T}W(u_p) + \frac{114.6Q}{T}W(u_i) = \frac{114.6Q}{T}\left[W(u_p) + W(u_i)\right] \tag{6.8.5}$$

where

$$u_P = \frac{1.87r_p^2 S}{Tt_p} \tag{6.8.6}$$

$$u_i = \frac{1.87r_i^2 S}{Tt_i} \tag{6.8.7}$$

Now suppose that we choose drawdowns at times t_p and t_i such that $s_p = s_i$, then $W(u_p) = W(u_i)$ and $u_p = u_i$, so that

$$\frac{r_i^2 S}{Tt_i} = \frac{r_p^2 S}{Tt_p} \tag{6.8.8}$$

which reduces to

$$\frac{r_i^2}{t_i} = \frac{r_p^2}{t_p} \tag{6.8.9}$$

Equation (6.8.9) defines the *law of times* (Ingersoll et al., 1948), which states that for a given aquifer, the times of occurrence of equal drawdown vary directly as the squares of distances from an observation well to a production well of equal discharge.

The law of times can be used to determine the distance from an image well to an observation well, using

$$r_i = r_p \sqrt{\frac{t_i}{t_p}} \tag{6.8.10}$$

in which r_i is the distance from the image well to the observation well in ft, r_p is the distance from the pumped well to the observation well in ft, t_p is the time after pumping started and before the barrier boundary is effective, and t_i is the time after pumping started and after the barrier boundary becomes effective, where $s_p = s_i$.

6.8.2   Recharge Boundary

The effect of recharge boundaries can also be simulated using the concepts of image well theory. For an aquifer bounded on one side by a recharge boundary, the cone of depression cannot extend beyond the stream, as shown in Figure 6.8.2. This results in no drawdown along the line of recharge. By placing an image well, recharging in nature, directly opposite and at the same distance from the stream as the real pumped well and also at the same rate, the finite system can be simulated. The resulting cone of depression is the summation of the real well cone of depression without the recharge and the image well cone of depression, as shown in Figure 6.8.2.

The water level in the wells will draw down initially only under the influence of the pumped well. After a time the effects of the recharge boundary will cause the time rate of drawdown to decrease and eventually reach equilibrium conditions. This occurs when recharge equals the pumping rate, as illustrated in Figure 6.8.4.

The drawdown for equilibrium conditions can be expressed as

$$s_r = s_p - s_i \tag{6.8.11}$$

in which s_r is the drawdown in an observation well near a recharge boundary, s_p is the drawdown due to the pumped well, and s_i is the buildup due to the image well (recharge boundary). The drawdown equation can be written as

$$s_r = \frac{4Q}{\pi T}\left[W(u_p) - W(u_i)\right] \tag{6.8.12}$$

or in the gallon-day-foot system as

$$s_r = \frac{114.6Q}{T}\left[W(u_p) - W(u_i)\right] \tag{6.8.13}$$

For large values of time t, the well functions can be expressed as

$$W(u_p) = -0.5772 - \ln u_p \tag{6.7.14}$$

$$W(u_i) = -0.5772 - \ln u_i \tag{6.7.15}$$

This allows equation (6.8.12) to be simplified to

$$s_r = \frac{Q}{4\pi T}\left[-\ln u_p + \ln u_i\right] \tag{6.8.16}$$

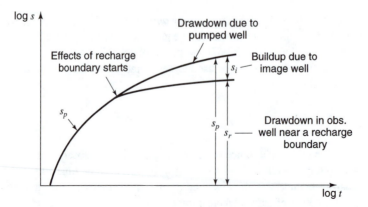

Figure 6.8.4 Recharge boundary effects on time-drawdown curve.

and equation (6.8.13) to

$$s_r = \frac{114.6Q}{T}\left[-\ln u_p + \ln u_i\right]$$

(6.8.17)

Now using the gallon-day-foot system with time in minutes, we get

$$u_p = \frac{2693r_p^2 S}{Tt}$$

(6.8.18)

and

$$u_i = \frac{2693r_i^2 S}{Tt}$$

(6.8.19)

The drawdown in the observation well from equation (6.8.17) is expressed as

$$s_r = \frac{114.6Q}{T}\left[-\ln\left(\frac{2693r_p^2 S}{Tt}\right) + \ln\left(\frac{2693r_i^2 S}{Tt}\right)\right]$$

(6.8.20)

which simplifies to

$$s_r = \frac{528}{T}Q\log\left(\frac{r_i}{r_p}\right)$$

(6.8.21)

Rorabaugh (1956) expressed this equation in terms of the distances between the pumped well and the line of recharges as

$$s_r = \frac{528Q\log\sqrt{(4a^2 + r_p^2 - 4ar_p\cos B_r)/r_p}}{T}$$

(6.8.22)

where a is the distance from the pumped well to the recharge boundary in ft and B_r is the angle between a line connecting the pumped and image wells and a line connecting the pumped and observation wells. Refer to Figure 6.8.5 for an explanation of terms.

6.8.3 Multiple Boundary Systems

Image well theory can also be applied to aquifer systems with multiple boundaries by considering successive reflections on the barrier and recharge boundaries. This is accomplished through a number of image wells. Placing a primary image well across each boundary balances the effect of the pumped well at each boundary. If a pair of converging boundaries is required, each primary image well then produces an unbalanced effect at the opposite boundary. This unbalanced effect is

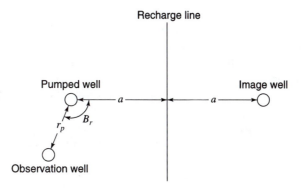

Figure 6.8.5 Definition of terms for equation 6.8.22.

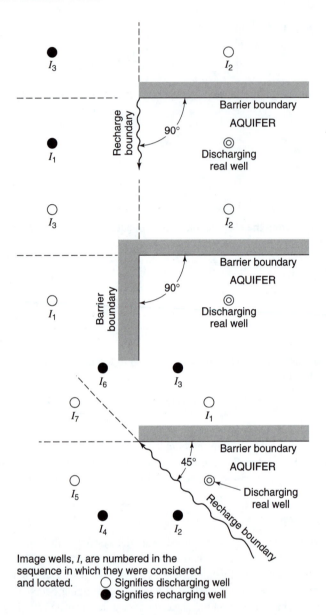

Figure 6.8.6 Plans of image-well systems for several wedge-shaped aquifers (from Ferris et al. (1962)).

corrected by placing secondary image wells until the effects of the real and image wells are balanced at both boundaries. These concepts are illustrated in Figure 6.8.6.

A primary image well placed across a barrier boundary is discharging in character. A primary image well placed across a recharge boundary is recharging in character. A secondary image well placed across a barrier boundary has the same character as its parent image well. A secondary image well placed across a recharge boundary has the opposite character to its parent image well. Figure 6.8.6 shows image well systems for wedge-shaped aquifers. For parallel boundary systems, it is only necessary to add pairs of image wells until the next pair has negligible influence on the sum of all image well effects out to the point.

6.9   SIMULATION OF GROUNDWATER SYSTEMS

6.9.1   Governing Equations

Darcy's law relates the Darcy flux v with dimension L/T to the rate of headloss per unit length of porous medium $\partial h/\partial l$. The negative sign indicates that the total head is decreasing in the direction of flow because of friction. This law applies to a cross-section of porous medium that is large compared to the cross-section of individual pores and grains of the medium. At this scale, Darcy's law describes a steady uniform flow of constant velocity, in which the net force on any fluid element is zero. For unconfined saturated flow, the two forces are gravity and friction. Darcy's law can also be expressed in terms of the transmissivity for confined conditions as

$$v = -\frac{T}{b}\frac{\partial h}{\partial l} \tag{6.9.1}$$

or for unconfined conditions as

$$v = -\frac{T}{h}\frac{\partial h}{\partial l} \tag{6.9.2}$$

Considering two-dimensional (horizontal) flow, a general flow equation can be derived by considering flow through a rectangular element (control volume) shown in Figure 6.9.1. The flow components ($q = Av$) for the four sides of the element are expressed using Darcy's law where $A = \Delta x \cdot h$ for unconfined conditions and $A = \Delta x \cdot b$ for confined conditions, so that

$$q_1 = -T_{x_{i-1,j}}\Delta y_j\left(\frac{\partial h}{\partial x}\right)_1 \tag{6.9.3a}$$

$$q_2 = -T_{x_{i,j}}\Delta y_j\left(\frac{\partial h}{\partial x}\right)_2 \tag{6.9.3b}$$

$$q_3 = -T_{y_{i,j+1}}\Delta x_i\left(\frac{\partial h}{\partial y}\right)_3 \tag{6.9.3c}$$

$$q_4 = -T_{y_{i,j}}\Delta x_i\left(\frac{\partial h}{\partial y}\right)_4 \tag{6.9.3d}$$

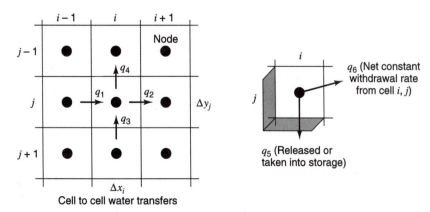

Figure 6.9.1 Finite difference grid.

where $T_{x_{i,j}}$ is the transmissivity in the x flow direction from element (i, j) to element $(i+1, j)$. The terms $(\partial h/\partial x)_1$, $(\partial h/\partial x)_2$, ... define the hydraulic gradients at the element sides 1, 2, ...

The rate at which water is stored or released in the element over time is

$$q_5 = S_{i,j}\Delta x_i \Delta y_i \frac{\partial h}{\partial t} \tag{6.9.4}$$

in which $S_{i,j}$ is the storage coefficient for element (i, j). In addition, the flow rate q_6 for constant net withdrawal or recharge from the element over time interval Δt is considered:

$$q_6 = q_{i,j,t} \tag{6.9.5}$$

in which $q_{i,j,t}$ has a positive value for pumping whereas it has a negative value for recharge.

By continuity, the flow into and out of a grid or cell is

$$q_1 - q_2 + q_3 - q_4 = q_5 + q_6 \tag{6.9.6}$$

Substituting in equations (6.9.3) and (6.9.5) gives

$$-T_{x_{i-1,j}}\Delta y_j\left(\frac{\partial h}{\partial x}\right)_1 + T_{x_{i,j}}\Delta y_i\left(\frac{\partial h}{\partial x}\right)_2 - T_{y_{i,j+1}}\Delta x_i\left(\frac{\partial h}{\partial y}\right)_3 + T_{y_{i,j}}\Delta x_i\left(\frac{\partial h}{\partial y}\right)_4$$

$$= S_{i,j}\Delta x_i \Delta y_i \frac{\partial h}{\partial t} + q_{i,j,t} \tag{6.9.7}$$

Dividing equation (6.9.7) by $\Delta x \Delta y$ and simplifying for constant transmissivities in the x and y directions yields

$$-T_x\left[\frac{\left(\frac{\partial h}{\partial x}\right)_1 - \left(\frac{\partial h}{\partial x}\right)_2}{\Delta x_i}\right] - T_y\left[\frac{\left(\frac{\partial h}{\partial y}\right)_3 - \left(\frac{\partial h}{\partial y}\right)_4}{\Delta y_j}\right] = S_{i,j}\frac{\partial h}{\partial t} + \frac{q_{i,j,t}}{\Delta x_i \Delta y_i} \tag{6.9.8}$$

For Δx and Δy infinitesimally small, the terms in brackets become second derivatives of h; then equation (6.9.8) reduces to

$$T_x\frac{\partial^2 h}{\partial x^2} + T_y\frac{\partial^2 h}{\partial y^2} = S\frac{\partial h}{\partial t} + W \tag{6.9.9}$$

which is the general partial differential equation for unsteady flow in the horizontal direction in which $W = q_{i,j,t}/\Delta x_i \Delta y_j$ is a sink term with dimensions L/T.

In the more general unsteady, two-dimensional heterogeneous anisotropic case, equation (6.9.9) is expressed as

$$\frac{\partial}{\partial x}\left(T_x\frac{\partial h}{\partial x}\right) + \frac{\partial}{\partial y}\left(T_y\frac{\partial h}{\partial y}\right) = S\frac{\partial h}{\partial t} + W \tag{6.9.10a}$$

or more simply

$$\frac{\partial}{\partial x_i}\left(T_{i,j}\frac{\partial h}{\partial x_j}\right) = S\frac{\partial h}{\partial t} + W \qquad i, j = 1, 2 \tag{6.9.10b}$$

6.9.2   Finite Difference Equations

The partial derivative expressions for Darcy's law, equations (6.9.3 a–d), can be expressed in finite difference form for time t in equation (6.9.7) using

$$\left(\frac{\partial h}{\partial x}\right)_1 = \left(\frac{h_{i-1,j,t} - h_{i,j,t}}{\Delta x_i}\right) \tag{6.9.11a}$$

$$\left(\frac{\partial h}{\partial x}\right)_2 = \left(\frac{h_{i,j,t} - h_{i+1,j,t}}{\Delta x_i}\right) \tag{6.9.11b}$$

$$\left(\frac{\partial h}{\partial y}\right)_3 = \left(\frac{h_{i,j+1,t} - h_{i,j,t}}{\Delta y_i}\right) \tag{6.9.11c}$$

$$\left(\frac{\partial h}{\partial y}\right)_4 = \left(\frac{h_{i,j,t} - h_{i,j-1,t}}{\Delta y_i}\right) \tag{6.9.11d}$$

and the time derivative in equation (11.2.7) is

$$\frac{\partial h}{\partial t} = \left(\frac{h_{i,j,t} - h_{i,j,t-1}}{\Delta t}\right) \tag{6.9.12}$$

Substituting equation (6.9.11) and (6.9.12) into equation (6.9.7) yields

$$-T_{x_{i-1,j}} \Delta y_j \left(\frac{h_{i-1,j,t} - h_{i,j,t}}{\Delta x_i}\right) + T_{x_{i,j}} \Delta y_j \left(\frac{h_{i,j,t} - h_{i+1,j,t}}{\Delta x_i}\right)$$

$$-T_{y_{i,j+1}} \Delta x_i \left(\frac{h_{i,j+1,t} - h_{i,j,t}}{\Delta y_i}\right) + T_{y_{i,j}} \Delta x_j \left(\frac{h_{i,j,t} - h_{i,j-1,t}}{\Delta y_j}\right)$$

$$-S_{i,j} \Delta x_i \Delta y_j \left(\frac{h_{i,j,t} - h_{i,j,t-1}}{\Delta t}\right) - q_{i,j,t} = 0 \tag{6.9.13}$$

which can be further simplified to

$$A_{i,j} h_{i,j,t} + B_{i,j} h_{i-1,j,t} + C_{i,j} h_{i+1,j,t} + D_{i,j} h_{i,j+1,t} + E_{i,j} h_{i,j-1,t} + F_{i,j,t} = 0 \tag{6.9.14}$$

where

$$A_{i,j} = \left[T_{x_{i-1,j}} \frac{\Delta y_j}{\Delta x_i} + T_{x_{i,j}} \frac{\Delta y_j}{\Delta x_i} + T_{y_{i,j+1}} \frac{\Delta x_i}{\Delta y_j} + T_{y_{i,j}} \frac{\Delta x_i}{\Delta y_j} - S_{i,j} \frac{\Delta x_i \Delta y_j}{\Delta t}\right] \tag{6.9.15a}$$

$$B_{i,j} = -T_{x_{i-1,j}} \frac{\Delta y_j}{\Delta x_i} \tag{6.9.15b}$$

$$C_{i,j} = -T_{y_{i,j}} \frac{\Delta y_j}{\Delta x_i} \tag{6.9.15c}$$

$$D_{i,j} = -T_{y_{i,j+1}} \frac{\Delta x_j}{\Delta y_j} \tag{6.9.15d}$$

$$E_{i,j} = -T_{y_{i,j}} \frac{\Delta x_i}{\Delta y_j} \tag{6.9.15e}$$

$$F_{i,j,t} = S_{i,j} \frac{\Delta x_j \Delta y_j}{\Delta t} - q_{i,j,t} \tag{6.9.15f}$$

The coefficients $A_{i,j}$, $B_{i,j}$, $C_{i,j}$, and $D_{i,j}$ are linear functions of the thickness of cell (i, j) and the thickness of one of the adjacent cells. For artesian conditions, this thickness is a known constant, so if cell (i, j) and its neighbors are artesian, equation (6.9.14) is linear for all t. For unconfined (water table) conditions, the thickness of cell (i, j) is $h_{i,j,t} - \text{BOT}_{i,j}$, where $BOT_{i,j}$ is the average elevation

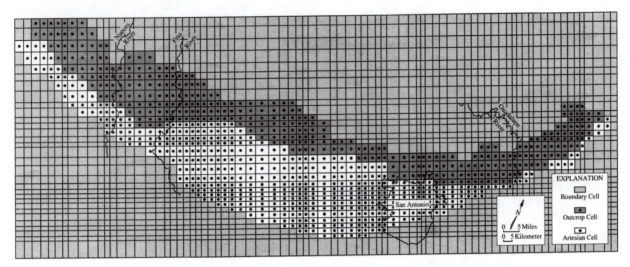

Figure 6.9.2 Cell map used for the digital computer model of the Edwards (Balcones fault zone) aquifer (after Klemt et al. (1979)).

of the bottom of the aquifer for cell (i, j). Then for unconfined conditions, equation (6.9.14) involves products of heads and is nonlinear in terms of the heads.

An *iterative alternating direction implicit (IADI) procedure* can be used to solve the set of equations. The IADI procedure involves reducing a large set of equations to several smaller sets of equations. One such smaller set of equations is generated by writing equation (6.9.14) for each cell or element in a column but assuming that the heads for the nodes on the adjacent columns are known. The unknowns in this set of equations are the heads for the nodes along the column. The head for the nodes along adjoining columns are not considered unknowns. This set of equations is solved by Gauss elimination and the process is repeated until each column is treated. The next step is to develop a set of equations along each row, assuming the heads for the nodes along adjoining rows are known. The set of equations for each row is solved and the process is repeated for each row in the finite difference grid.

Once the sets of equations for the columns and the sets of equations for the row have been solved, one "iteration" has been completed. The iteration process is repeated until the procedure converges. Once convergence is accomplished, the terms $h_{i,j}$ represent the heads at the end of the time step. These heads are used as the beginning heads for the following time step. For a more detailed discussion of IADI procedure, see Peaceman and Rachford (1955), Prickett and Lonnquist (1971), Trescott et al. (1976), or Wang and Anderson (1982).

An example of the application of a two-dimensional finite-difference groundwater model is the Edwards (Balcones Fault Zone) aquifer. This aquifer has been modeled using the GWSIM groundwater simulation model developed by the Texas Water Development Board (1974). GWSIM is a finite-difference simulation model that uses the IADI method, similar to the model by Prickett and Lonnquist (1971). The finite-difference grid for the Edwards aquifer is shown in Figure 6.9.2, which has 856 active cells to describe the aquifer. Wanakule (1989) has modeled only a small portion of the Edwards aquifer called the Barton Springs–Edwards aquifer in Austin, Texas. This application used a finite difference grid system (Figure 6.9.3) containing 330 cells whose dimensions varied from $0.379 \times 0.283 \text{ mi}^2$ to $0.95 \times 1.51 \text{ mi}^2$. Figure 6.9.4 illustrates the 1981 water level contours.

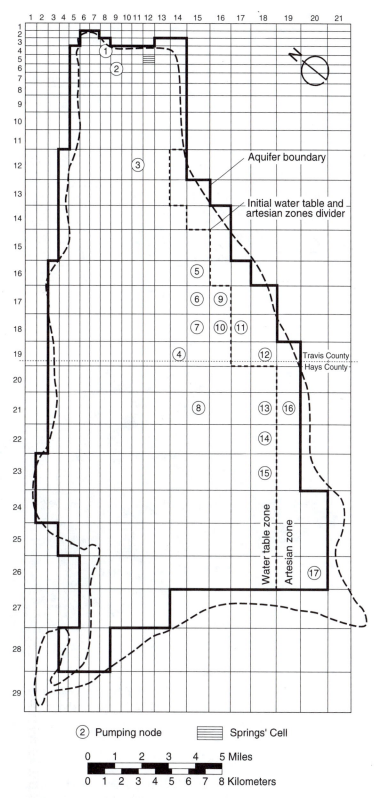

Figure 6.9.3 Pumping locations used in Barton Springs–Edwards Aquifer model (after Wanakule (1989)).

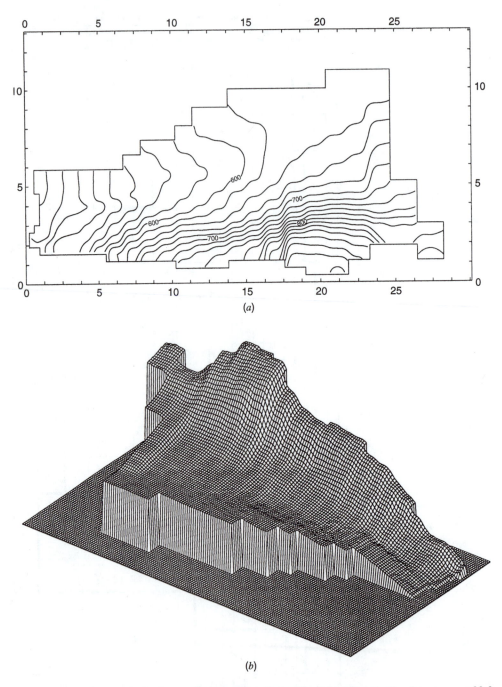

Figure 6.9.4 Water levels for Barton Springs–Edwards Aquifer. (*a*) 1981 water-level contours at 20 ft intervals; (*b*) Perspective block diagram of 1981 water levels viewed from the east side of the aquifer (after Wanakule (1985)).

6.9.3 MODFLOW

One of the most widely used groundwater simulation models in the MODFLOW Model developed by the U.S. Geological Survey (McDonald and Harbaugh, 1988, and Harbaugh and McDonald, 1996). The World Wide Web (WWW) address from which this model can be downloaded is http://water.usgs.gov/software/ground_water.html. MODFLOW is a finite difference model that can be used to solve groundwater flow problems in one-, two-, or three-dimensions. The computer program is divided into a main program and a series of independent subroutines called *modules*. These modules are grouped into "packages," each of which is a group of modules that deals with a single aspect of the simulation. A Basic Package handles tasks that are required for each simulation: the Well Package, which simulates the effects of injection and production wells; the River Package, which simulates the effect of rivers; and the Recharge Package, which simulates the effect of recharge. Others include the Evapotranspiration Package, the Drain Package, the General-Head Boundary Package, and the Solution Procedure Package. Individual packages may or may not be required, depending on the problem being simulated.

PROBLEMS

6.1.1 An experiment was conducted to determine the hydraulic conductivity of an artesian aquifer. The piezometric heads at two points 150 m apart were found to be 55 m and 48.5 m above a datum. A tracer injected into the first piezometer was observed after 32 hours in the second well. A test on porosity of a sample of the aquifer shows that $\alpha = 24\%$. What is the hydraulic conductivity of the aquifer? Suggest what the aquifer material may be and verify that your solution holds true. Take the subsurface temperature as 15°C.

6.2.1 Show that the steady state groundwater flow equation can be expressed in polar coordinate systems as $h = C_1 \ln r + C_2$, where C_1 and C_2 are constants.

6.2.2 Water flows through three confined aquifers in series as shown in Figure P6.2.2. For piezomertic heads in the observation wells of 66.4 m and 60.6 m, determine the flow rate per unit width of the aquifer and the headlosses in each component of the aquifers between the observation wells. If the headlosses in each aquifer between the wells were to be equal, what would be the length of each aquifer?

6.2.3 Suppose an unconfined aquifer lies over a confined aquifer as shown in Figure P6.2.3. Determine the flow out of both aquifers.

6.3.1 Rework example 6.3.1 with $h_0 = 10$ ft.

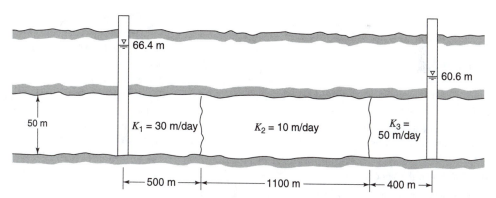

Figure P6.2.2

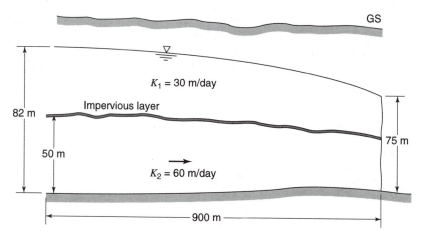

Figure P6.2.3

6.3.2 Consider two strata of the same soil material that lie between two channels. The first stratum is confined and the second one is unconfined, and the water surface elevations in the channels are 24 m and 16 m above the bottom of the unconfined aquifer. What should be the thickness of the confined aquifer for which

1) the discharge through both strata are equal?

2) the discharge through the confined aquifer is half of that through the unconfined aquifer?

6.3.3 Three monitoring wells are used (as shown in Figure P6.3.3) to determine the direction of groundwater flow in a confined aquifer. The piezometric heads in the wells are found to be 52 m in well 1, 49 m in well 2 and 56 m in well 3. Determine the direction of flow.

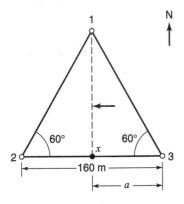

Figure P6.3.3

6.4.1 Rework example 6.4.1 with a constant pumping rate of 0.07 m³/s and the same drawdowns.

6.4.2 Rework example 6.4.2 with a pumping rate of 400 gpm and the same drawdowns.

6.4.3 Rework example 6.4.3 with a constant pumping rate of 0.075 m³/s and the same drawdowns.

6.4.4 Rework example 6.4.4 with the same pumping rate but drawdowns of 10 and 5 ft in the observation wells.

6.4.5 A 50-cm diameter well fully penetrates vertically through a confined aquifer 12 m thick. When the well is pumped at 0.035 m³/s, the heads in the pumped well and the two other observation wells were found to be as shown in Figure P6.4.5. Does this test suggest that the aquifer material is fairly homogeneous in the directions of the observation wells?

6.4.6 Three pumping wells along a straight line are spaced 200 m apart. What should be the steady state pumping rate from each well so that the drawdown in each well will not exceed 2 m: The transmissivity of the confined aquifer which all the wells penetrate fully is 2400 m²/day and all the wells are 40 cm in diameter. Take the thickness of the aquifer $b = 40$ m and the radius of influence of each well to be 800 m.

6.4.7 Reposition the wells in example 6.4.4 such that they form an equilateral triangle (same spacings). For the same restrictions on the drawdown, will the discharge decrease or increase? If so, by how many percent? If not, what difference do you perceive between the two problems?

6.4.8 It is required to dewater a construction site 80 m by 80 m. The bottom of the construction will be 1.5 m below the initial water surface elevation of 90 m. Four pumps are to be used in 0.5 m diameter wells at the four corners of the site. Determine the required steady-state pumping rate. The aquifer has $T = 1600$ m²/day and the wells have each a radius of influence of 600 m.

6.5.1 Work example 6.5.1 with $T = 8,500$ gpd/ft².

6.5.2 A 0.4-m diameter well is pumped continuously at a rate of 5.6 l/s from an aquifer of transmissivity 108 m²/day and storativity 2×10^{-5}. How long will it take before the drawdown in the well reaches 2 m?

6.5.3 In problem 6.5.2, how long will it take before a drawdown at a distance of 400 m from the well becomes 2 m?

6.5.4 In problem 6.5.2, determine the drawdown at a distance 200 m from the well after a) 1 hr, b) 1 day, and c) 1 year of pumping.

6.5.5 A pumping test was performed to determine the transmissivity and the storativity of a confined aquifer. The drawdown vs. time data in an observation well 100 m away from the pumping well is given in the following table The discharge from the pumped well was at a constant rate of 3.6 m³/min throughout the test. Determine the aquifer properties.

Time (min)	Drawdown (m)
0.5	0.12
1	0.18
2	0.25
3	0.28
4	0.31
5	0.33
8	0.38
10	0.41

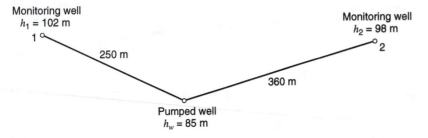

Figure P6.4.5

Time (min)	Drawdown (m)
20	0.47
30	0.51
50	0.56
100	0.62
200	0.69
500	0.79
1000	0.89

6.5.6 The following data was obtained in an observation well 80 m away from the pumped well. The discharge from the pumped well was at 2.5 m³/min. Using Jacob's approximation, determine the aquifer properties. Assume a confined aquifer. Also verify your solution and determine the time after which Jacob's approximation will be valid.

Time (min)	Drawdown (m)
0.5	0.16
1	0.18
2	0.24
3	0.27
4	0.47
5	0.50
7	0.57
10	0.68
20	0.84
30	0.96
50	1.06
100	1.29
200	1.46
500	1.68
1000	1.86

6.8.1 Draw the generalized flow net showing the flow lines and potential lines in the vicinity of a discharging well near a recharge boundary.

6.8.2 Draw the generalized flow net showing the flow lines and potential lines in the vicinity of a discharging well near a barrier boundary.

6.8.3 A well is pumping near a barrier boundary at a rate of 0.03 m³/s from a confined aquifer 20 m thick (see Figure 6.8.3). The hydraulic conductivity of the aquifer is 3.2×10^{-4} m/s and its storativity is 3×10^{-5}. Determine the drawdown in the observation well after 10 hours of continuous pumping. What is the fraction of the drawdown attributable to the barrier boundary?

6.8.4 A 0.5-m diameter well is pumping near a river at an unknown rate from a confined aquifer (see Figure 6.8.4). The aquifer properties are $T = 5.0 \times 10^{-3}$ m²/s and $S = 4.0 \times 10^{-4}$. After eight hours of pumping, the drawdown in the observation well was 0.8 m. Determine the rate of pumping and the drawdown in the pumped well. How great was the effect of the river on the drawdown in the observation well and in the pumped well?

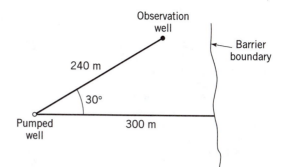

Figure P6.8.3

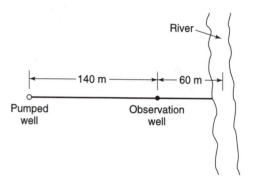

Figure P6.8.4

6.8.5 A production well fully penetrating a nonleaky isotropic artesian aquifer delimited by two barrier boundaries was continuously pumped at a constant rate of 1,485 gpm for a period of four hours. The drawdowns in the table below were observed at a distance of 300 ft in a fully penetrating observation well. Compute the coefficients of transmisivity and storage of the aquifer and the distances to each image well from the observation well.

Time (min)	Drawdown (ft)
2	0.80
3	0.92
4	1.06
5	1.17
6	1.23
7	1.32
8	1.37
9	1.43
10	1.48
20	1.88
30	2.11
40	2.34
50	2.52
60	2.70
70	2.83
80	3.00
90	3.17
100	3.30
200	4.21
300	4.43

REFERENCES

Batu, V., *Aquifer Hydraulics*, Wiley Interscience, New York, 1998.

Bolton, N. S., "Analysis of Data from Non-Equilibrium Pumping Tests Allowing for Delayed Yield from Storage," *Proc. Institute for Civil Engineering* (London), vol. 26, no. 6693, 1963.

Butler, S. S., *Engineering Hydrology*, Prentice-Hall, Englewood Cliffs, NJ, 1957.

Charbeneau, R. J., *Groundwater Hydraulics and Pollutant Transport*, Prentice-Hall, Upper Saddle River, NJ, 2000.

Chow, V. T., D. R. Maidment, and L. W. Mays, *Applied Hydrology*, McGraw-Hill, New York, 1988.

Cooper, H. H., Jr., and C. E. Jacob, "A Generalized Graphical Method for Evaluating Formation Constants and Summarizing Well Field History," *Trans. Amer. Geophys. Union, 27*, pp. 526–534, 1946.

Delleur, J. W. (editor-in chief), *The Handbook of Groundwater Engineering*, CRC Press, Boca Raton, FL, 1999.

Ferris, J. G., D. B. Knowles, R. H. Brown, and R. W. Stallman, "Theory of Aquifer Tests," U.S. Geological Survey Water-Supply Paper 1536-E, 1946.

Freeze, R. A., and J. A. Cherry, *Groundwater*, Prentice-Hall, Englewood Cliffs, NJ, 1979.

Gehm, H. W. and J. I. Bregman, editors, *Handbook of Water Resources and Pollution Control*, Van Nostrand Reinhold Company, New York, 1976.

Gleick, P. H., *Water in Crisis*, Oxford University Press, Oxford, 1993.

Hantush, M. S., "Hydraulics of Wells," in *Advances in Hydroscience*, Academic Press, New York, 1964.

Hantush, M. S., and C. E. Jacob, "Non-Steady Radial Flow in an Infinite Leaky Aquifer," *Trans. Am. Geophys. Union*, vol. 36, no. 1, 1955.

Harbaugh, A. W., and M. G. McDonald, "User's Documentation for MODFLOW—96, an Update to the United States Geological Survey Modular Finite-Difference Groundwater Flow Model," United States Geological Survey, Open-File Report 96-485, 1996.

Ingersoll, L. R., O. J. Zobel, and A. C. Ingersoll, *Heat Conduction with Engineering and Geological Applications*, McGraw-Hill, New York, 1948.

Jacob, C. E., "Flow of Groundwater," Chapter 5 in *Engineering Hydraulics*, edited by H. Rouse, John Wiley & Sons, New York, 1950.

Jacob, C. E., "Radial Flow in a Leaky Artesian Aquifer," *Trans. Am. Geophys. Union*, vol. 27, no. 2, 1946.

Klemt, W. B., T. R. Knowles, G. R. Elder, and T. Sich, "Groundwater Resources and Model Applications for the Edwards (Balcones Fault Zone) Aquifer in the San Antonio Region, Texas," Report 239, Texas Department of Water Resources, Austin, TX, Oct. 1979.

Lohman, S. W., et al., "Definition of Selected Groundwater Terms-Revision and Conceptual Refinements," U.S. Geological Survey Water Supply Paper No. 1988, 1972.

Marsh, W. M., *Earthscape: A Physical Geography*, John Wiley & Sons, New York, 1987.

McDonald, J. M., and A. W. Harbaugh, "A Modular Three-Dimensional Finite-Difference Groundwater Flow Model," *Techniques of Water Resources Investigations of the Unites States Geological Survey, Book 6*, pp. 586, 1988.

Meinzer, O. E., "The Occurrence of Groundwater in the United States," U.S. Geological Survey Water Supply Paper 489, 1923.

Neuman, S. P., "Analysis of Pumping Test Data from Anisotropic Unconfined Aquifers Considering Delayed Gravity Response," *Water Resources Research*, vol. 11, pp. 329–342, 1975.

Peaceman, D. W., and H. H. Rachford, Jr., "The Numerical Solutions of Parabolic and Elliptic Differential Equations," *Journal Soc. Industrial and Applied Mathematics*, vol. 3, pp. 28–41, 1955.

Prickett, T. A., and C. G. Lonnquist, "Selected Digital Computer Techniques for Groundwater Resources Evaluation," *Bulletin No. 55*, Illinois State Water Survey, Urbana, IL, 1971.

Prickett, T. A., "Type-Curve Solution to Aquifer Table Tests Under Water-Table Conditions," *Groundwater*, vol. 3, no.3, 1965.

Roberson, J. A., and C. T. Crowe, *Engineering Fluid Mechanics*, Houghton Mifflin Company, Boston, MA, 1990.

Rorabaugh, M. I., Ground-Water Resources of the Northeastern Part of the Louisville Area, Kentucky, U.S. Geological Survey Water Supply Paper 1360-B, 1956.

Steggewentz, J. H., and J. L. VanNes, "Calculating the Yield of a Well, Taking into Account Replenishment of the Groundwater Above," *Waste Water Engr.*, vol. 41, 1939.

Texas Water Development Board, "GWSIM—Groundwater Simulation Program, Program Document and User's Manual," UM S7405, Austin, TX, 1974.

Theis, C. V., "The Relation Between the Lowering of Piezometric Surface and the Rate and Duration of Discharge of a Well Using Groundwater Storage," *Transactions American Geophysical Union*, vol. 2, pp. 519–524, 1935.

Todd, D. K., *Groundwater Hydrology*, John Wiley and Sons, New York, 1980.

Todd, D. K., and J. Bear, "Seepage Through Layered Anisotropic Porous Media," *Journal of the Hydraulics Division, ASCE*, vol. 87, no. HY 3, pp. 31–57, 1961.

Trescott, P. C., G. F. Pinder, and S. P Larson, Finite-Difference Model for Aquifer Simulation in Two Dimensions with Results of Numerical Experiments, in *U.S. Geological Survey Techniques of Water Resources Investigations*, Book 7, C1, U. S. Geological Survey, Reston, VA, 1976.

U.S. Bureau of Reclamation, *Groundwater Manual*, U.S. Gov. Printing Office, Denver, CO, 1981.

U.S. Soil Conservation Service, *Soil Survey Manual*, Handbook no. 18, U.S. Department of Agriculture, 1951.

Walton, W. C., *Groundwater Resource Evaluation*, McGraw-Hill, New York, 1970.

Walton, W. E., "Leaky Artesian Conditions in Illinois," Illinois State Water Survey, Dept. of Invest. Urbana, IL 39, 1960.

Wanakule, N., "Optimal Groundwater Management Models for the Barton Springs-Edwards Aquifer," Edwards Aquifer Research and Data Center, San Marcos, TX, 1989.

Wang, H. F., and M. P. Anderson, *Introduction to Groundwater Modeling: Finite Difference and Finite Element Models*, W. H. Freeman, San Francisco, CA, 1982.

Wenzel, L. K., Methods for Determining Permeability of Water-Bearing Materials, with Special Reference to Discharging Well Methods, U.S. Geological Survey Water-Supply Paper 887, pp. 192, 1942.

Chapter 7

Hydrologic Processes

7.1 INTRODUCTION TO HYDROLOGY

7.1.1 What Is Hydrology?

The U.S. National Research Council (1991) presented the following definition of hydrology:

> *Hydrology is the science that treats the waters of the Earth, their occurrence, circulation, and distribution, their chemical and physical properties, and their reaction with the environment, including the relation to living things. The domain of hydrology embraces the full life history of water on Earth.*

For purposes of this book we are interested in the engineering aspects of hydrology, or what we might call *engineering hydrology*. From this point of view we are mainly concerned with quantifying amounts of water at various locations (spatially) as a function of time (temporally) for surface water applications. In other words, we are concerned with solving engineering problems using hydrologic principles. This chapter is not concerned with the chemical properties of water and their relation to living things.

Books on hydrology include: Bedient and Huber (1992); Bras (1990); Chow (1964); Chow, Maidment, and Mays (1988); Gupta (1989); Maidment (1993); McCuen (1998); Ponce (1989); Singh (1992); Viessman and Lewis (1996); and Wanielista, Kersten, and, Eaglin (1997).

7.1.2 The Hydrologic Cycle

The central focus of hydrology is the *hydrologic cycle*, consisting of the continuous processes shown in Figure 7.1.1. Water *evaporates* from the oceans and land surfaces to become water vapor that is carried over the earth by atmospheric circulation. The *water vapor* condenses and *precipitates* on the land and oceans. The precipitated water may be *intercepted* by vegetation, become overland flow over the ground surface, *infiltrate* into the ground, flow through the soil as *subsurface flow*, and discharge as *surface runoff*. Evaporation from the land surface comprises evaporation directly from soil and vegetation surfaces, and *transpiration* through plant leaves. Collectively these processes are called *evaportranspiration*. Infiltrated water may percolate deeper to recharge groundwater and later become *springflow* or *seepage* into streams also to become streamflow.

The hydrologic cycle can be viewed on a global scale as shown in Figure 7.1.2. As discussed by the U.S. National Research Council (1991): "As a global average, only about 57 percent of the precipitation that falls on the land P_l returns directly to the atmosphere E_l without reaching the ocean. The remainder is runoff R, which finds it way to the sea primarily by rivers but also through

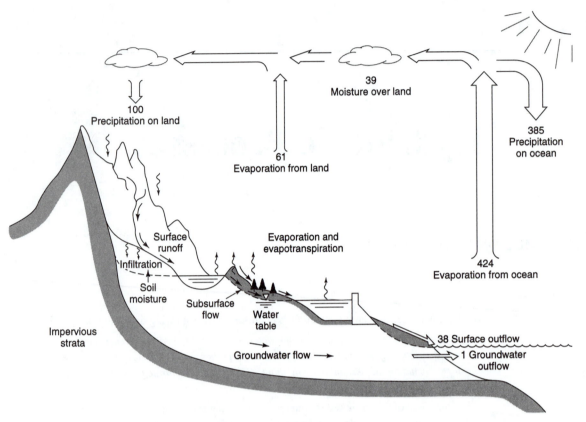

Figure 7.1.1 Hydrologic cycle with global annual average water balance given in units relative to a value of 100 for the rate of precipitation on land (from Chow et al. (1988)).

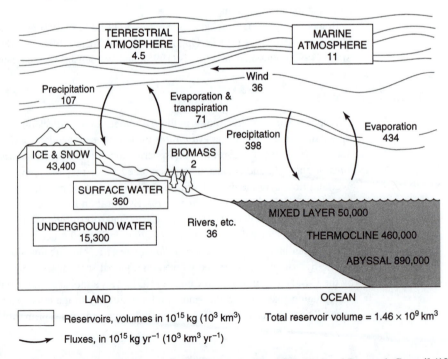

Figure 7.1.2 The hydrologic cycle at global scale (from U.S. National Research Council (1986)).

subsurface (groundwater) movement, and by the calving of icebergs from glaciers and ice shelves *W*. In this gravitationally powered runoff process, the water may spend time in one or more natural storage reservoirs such as snow, glaciers, ice sheet, lakes, streams, soils and sediments, vegetation, and rock. Evaporation from these reservoirs short-circuits the global hydrologic cycle into subcycles with a broad spectrum of scale. The runoff is perhaps the best-known element of the global hydrologic cycle, but even this is subject to significant uncertainty."

7.1.3   Hydrologic Systems

Chow, Maidment, and Mays (1988) defined a *hydrologic system* as a structure or volume in space, surrounded by a boundary, that accepts water and other inputs, operates on them internally, and produces them as outputs. The structure (for surface or subsurface flow) or volume in space (for atmospheric moisture flow) is the totality of the flow paths through which the water may pass as throughput from the point it enters the system to the point it leaves. The boundary is a continuous surface defined in three dimensions enclosing the volume or structure. A *working medium* enters the system as input, interacts with the structure and other media, and leaves as output. Physical, chemical, and biological processes operate on the working media within the system; the most common working media involved in hydrologic analysis are water, air, and heat energy.

The *global hydrologic cycle* can be represented as a system containing three subsystems: the atmospheric water system, the surface water system, and the subsurface water system, as shown in Figure 7.1.3. Another example is the storm-rainfall-runoff process on a watershed, which can be represented as a hydrologic system. The input is rainfall distributed in time and space over the

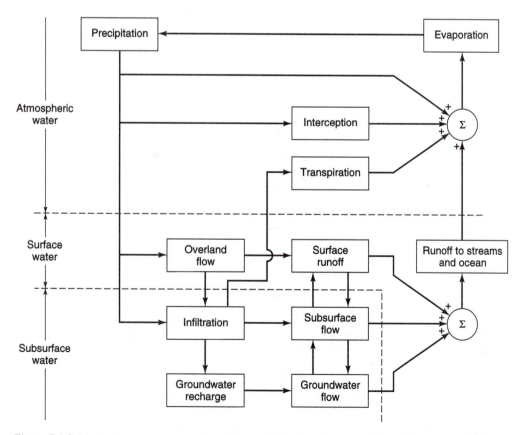

Figure 7.1.3 Block-diagram representation of the global hydrologic system (from Chow et al. (1988)).

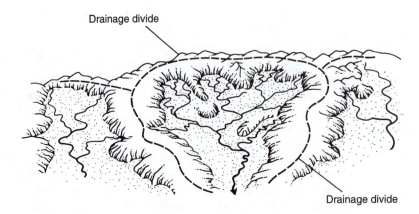

Figure 7.1.4 Schematic diagram of a drainage basin. The high terrain on the perimeter is the drainage divide (from Marsh (1987)).

watershed and the output is streamflow at the watershed outlet. The boundary is defined by the watershed divide and extends vertically upward and downward to horizontal planes.

Drainage basins, *catchments*, and *watersheds* are three synonymous terms that refer to the topographic area that collects and discharges surface streamflow through one outlet or mouth. Catchments are typically referred to as small drainage basins but no specific area limits have been established. Figure 7.1.4 illustrates the drainage basin divide, watershed divide, or catchment divide, which is the line dividing land whose drainage flows toward the given stream from land whose drainage flows away from that stream. Think of drainage basin sizes ranging from the Mississippi River drainage basin to small urban drainage basins in your local community or some small valley in the countryside near you.

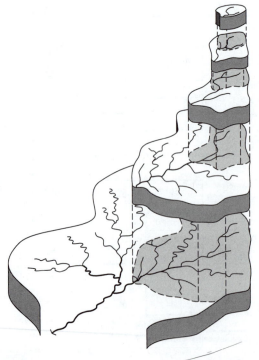

Figure 7.1.5 Illustration of the nested hierarchy of lower-order basins within a large drainage basin (from Marsh (1987)).

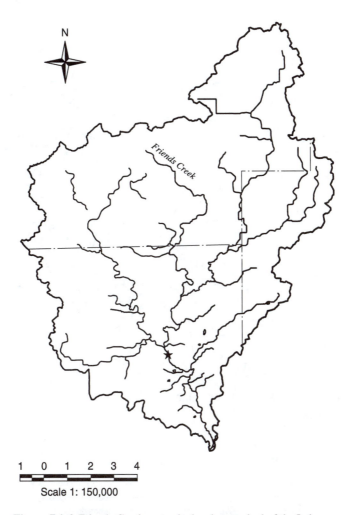

Figure 7.1.6 Friends Creek watershed, subwatershed of the Lake Decatur watershed. (Courtesy of the Illinois State Water Survey, compiled by Erin Hessler Bauer.)

As shown in Figure 7.1.5, drainage basins can be pictured in a pyramidal fashion as the runoff from smaller basins (subsystems) combine to form larger basins (subsystem in system) and the runoffs from these basins in turn combine to form even larger basins, and so on. Marsh (1987) refers to this mode of organization as a *hierarchy* or *nested hierarchy*, as each set of smaller basins is set inside the next layer. A similar concept is that streams that drain small basins combine to form larger streams and so on.

Figures 7.1.6–7.1.10 illustrate the hierarchy of the Friends Creek watershed located in the Lake Decatur watershed (drainage area upstream of Decatur, Illinois, on the Sangamon River with a drainage area of 925 mi^2 or 2,396 km^2). Obviously, the Friends Creek watershed can be subdivided into much smaller watersheds. Figure 7.1.8 illustrates the Illinois River basin (29,000 mi^2) with the Sangamon River. Figure 7.1.9 illustrates the upper Mississippi River basin (excluding the Missouri River) with the Illinois River. Figure 7.1.10 illustrates the entire Mississippi River basin (1.15 million mi^2). This is the largest river basin in the United States, draining about 40 percent of the country. The main stem of the river is about 2400 miles long.

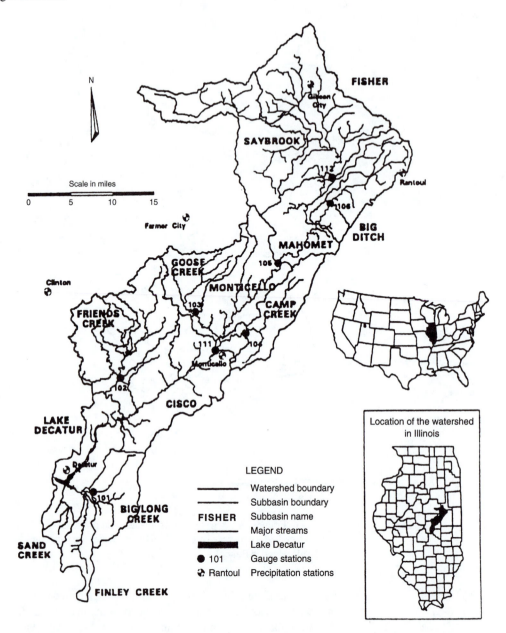

Figure 7.1.7 Lake Decatur watershed in the upper Sangamon River Basin (upstream of Decatur, Illinois). The location of Friends Creek watershed (Figure 7.1.6) is shown (from Demissee and Keefer (1998)).

Figure 7.1.8 The Illinois River Basin showing the Sangamon River (from Bhowmik (1998)).

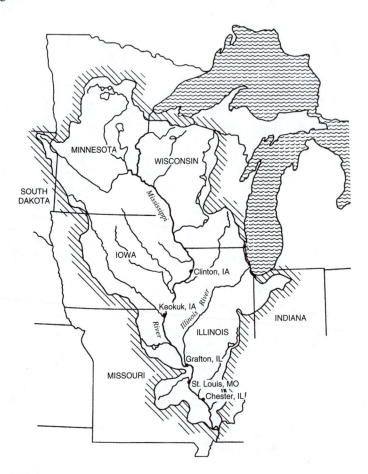

Figure 7.1.9 The Upper Mississippi River excluding the Missouri River. The location of the Illinois River (Figure 7.1.8) is shown (from Bhowmik (1998)).

Figure 7.1.10 The Mississippi River Basin in the United States (from Bhowmik (1998)).

7.1.4 Atmospheric and Ocean Circulation

Atmospheric circulation on Earth is a very complex process that is influenced by many factors. Major influences are differences in heating between low and high altitudes, the earth's rotation, and heat and pressure differences associated with land and water. The general circulation of the atmosphere is due to latitudinal differences in solar heating of the earth's surface and inclination of its axis, distribution of land and water, mechanics of the atmospheric fluid flow, and the Coriolis effect. In general, the atmospheric circulation is thermal in origin and is related to the earth's rotation and global pressure distribution. If the earth were a nonrotating sphere, atmospheric circulation would appear as in Figure 7.1.11. Air would rise near the equator and travel in the upper atmosphere toward the poles, then cool, descend into the lower atmosphere, and return toward the equator. This is called *Hadley circulation*.

The rotation of the earth from west to east changes the circulation pattern. As a ring of air about the earth's axis moves toward the poles, its radius decreases. In order to maintain angular

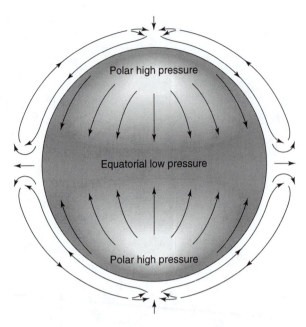

Figure 7.1.11 Atmospheric circulation pattern that would develop on a nonrotating planet. The equatorial belt would heat intensively and would produce low pressure, which would in turn set into motion a gigantic convection system. Each side of the system would span one hemisphere (from Marsh (1987)).

momentum, the velocity of air increases with respect to the land surface, thus producing a westerly air flow. The converse is true for a ring of air moving toward the equator—it forms an easterly air flow. The effect producing these changes in wind direction and velocity is known as the *Coriolis force*.

The idealized pattern of atmospheric circulation has three cells in each hemisphere, as shown in Figure 7.1.12. In the *tropical cell*, heated air ascends at the equator, proceeds toward the poles at upper levels, loses heat, and descends toward the ground at latitude 30°. Near the ground it branches, one branch moving toward the equator and the other toward the pole. In the *polar cell*, air rises at 60° latitude and flows toward the poles at upper levels, then cools, and flows back to 60° near the earth's surface. The *middle cell* is driven frictionally by the other two; its surface flows toward the pole, producing prevailing westerly air flow in the midlatitudes.

The uneven distribution of ocean and land on the earth's surface, coupled with their different thermal properties, creates additional spatial variation in atmospheric circulation. The annual shifting of the thermal equator due to the earth's revolution around the sun causes a corresponding oscillation of the three-cell circulation pattern. With a larger oscillation, exchanges of air between adjacent cells can be more frequent and complete, possibly resulting in many flood years. Also, monsoons may advance deeper into such countries as India and Australia. With a smaller oscillation, intense high pressure may build up around 30° latitude, thus creating extended dry periods, Since the atmospheric circulation is very complicated, only the general pattern can be identified.

The atmosphere is divided vertically into various zones. The atmospheric circulation described above occurs in the *troposphere*, which ranges in height from about 8 km at the poles to 16 km at the equator. The temperature in the troposphere decreases with altitude at a rate varying with the moisture content of the atmosphere. For dry air the rate of decrease is called the *dry adiabatic lapse rate* and is approximately 9.8°C/km. The *saturated adiabatic lapse rate* is less, about 6.5°C/km, because some of the vapor in the air condenses as it rises and cools, releasing heat into the surrounding air. These are average figures for lapse rates that can vary considerably with

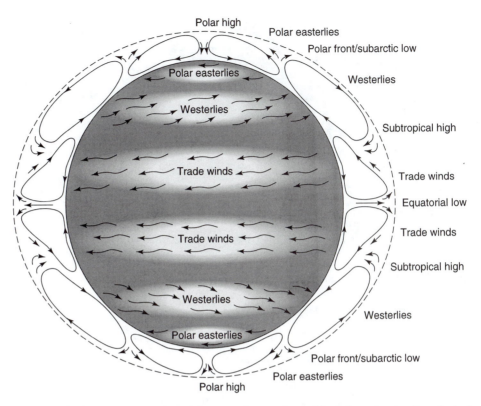

Figure 7.1.12 Idealized circulation of the atmosphere at the earth's surface, showing the principal areas of pressure and belts of winds (from Marsh (1987)).

altitude. The *tropopause* separates the troposphere from the *stratosphere* above. Near the tropopause, sharp changes in temperature and pressure produce strong narrow air currents known as *jet streams* with velocities ranging from 15 to 50 m/s (30 to 100 mi/h). They flow for thousands of kilometers and have an important influence on air-mass movement.

The oceans exert an important control on global climate. Because water bodies have a high volumetric heat capacity, the oceans are able to retain great quantities of heat. Through wave and current circulation, the oceans redistribute heat to considerable depths and even large areas of the oceans. Redistribution is east-west or west-east, and is also across the midaltitudes from the tropics to the subartic, enhancing the overall poleward heat transfer in the atmosphere. Waves are predominately generated by wind. Ocean circulation is illustrated in Figure 7.1.13.

Oceans have a significant effect on the atmosphere; however, an exact understanding of the relationships and mechanisms involved is not known. The correlation between ocean temperatures and weather trends and midlatitude events has not been solved. One trend is the growth and decline of a warm body of water in the equatorial zone of the eastern Pacific Ocean, referred to as El Niño (meaning "The Infant" in Spanish, alluding to the Christ Child, because the effect typically begins around Christmas). The warm body of water develops and expands every five years or so off the coast of Peru, initiated by changes in atmospheric pressure resulting in a decline of the easterly trade winds. This reduction in wind reduces resistance, causing the eastwardly equatorial countercurrent to rise. As El Niño builds up, the warm body of water flows out into the Pacific and along the tropical west coast of the Americas, displacing the colder water of the California and Humboldt currents. One of the interesting effects of this weather variation is the South Oscillation, which changes precipitation patterns—resulting in drier conditions in areas of normally little precipitation.

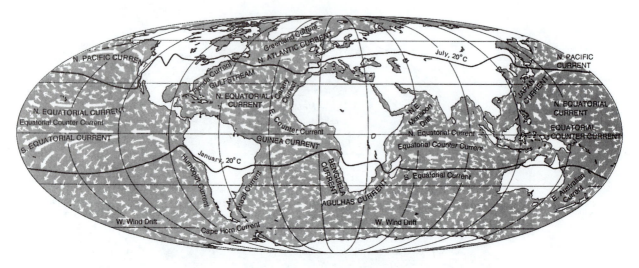

Figure 7.1.13 The actual circulation of the oceans. Major currents are shown with heavy arrows (from Marsh (1987)).

7.2   PRECIPITATION (RAINFALL)

7.2.1   Precipitation Formation and Types

Even though precipitation includes rainfall, snowfall, hail, and sleet, our concern in this book will relate almost entirely to rainfall. The formation of water droplets in clouds is illustrated in Figure 7.2.1. *Condensation* takes place in the atmosphere on *condensation nuclei*, which are very small particles ($10^{-3} - 10\mu$ m) in the atmosphere that are composed of dust or salt. These particles are called *aerosols*. During the initial occurrence of condensation, the droplets or ice particles are very small and are kept aloft by motion of the air molecules. Once droplets are formed they also act as condensation nuclei. These droplets tend to repel one another, but in the presence of an electric field in the atmosphere they attract one another and are heavy enough (~0.1 mm) to fall through

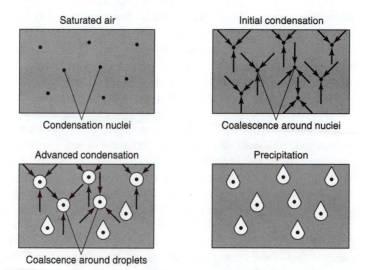

Figure 7.2.1 Precipitation formation. Water droplets in clouds are formed by nucleation of vapor on aerosols, then go through many condensation-evaporation cycles as they circulate in the cloud, until they aggregate into large enough drops to fall through the cloud base (from Marsh (1987)).

the atmosphere. Some of the droplets evaporate in the atmosphere; some of the droplets decrease in size by evaporation; and some of the droplets increase in size by impact and aggregation.

Basically, the formation of precipitation requires lifting of an air mass in the atmosphere; it then cools and some of its moisture condenses. There are three main mechanisms of air mass lifting: *frontal lifting*, *orographic lifting*, and *convective lifting*. Frontal lifting (Figure 7.2.2) is when warm air is lifted over cooler air by frontal passage; orographic lifting (Figure 7.2.3) is when an air mass

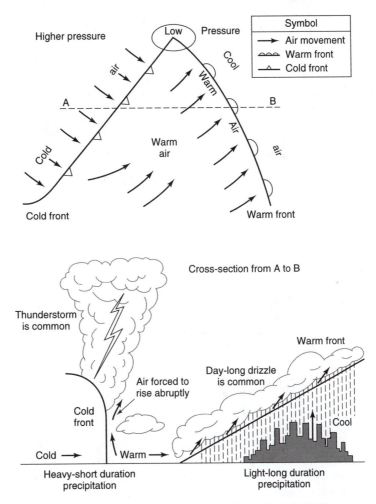

Figure 7.2.2 Cyclonic storms in mid-latitude (from Masch (1984)).

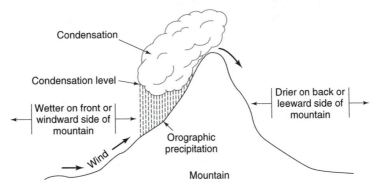

Figure 7.2.3 Orographic storm (from Masch (1984)).

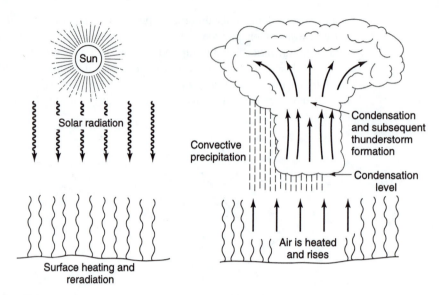

Figure 7.2.4 Convective storm (from Masch (1984)).

rises over a mountain range; and convective lifting (Figure 7.2.4) is when air is drawn upward by convective action such as a thunderstorm cell.

7.2.2 Rainfall Variability

In order to determine the runoff from a watershed and the resulting stream flow, precipitation is one of the primary inputs. Rainfall varies in space and time as a result of the general pattern of atmospheric circulation and local factors. Figure 7.2.5 shows the mean annual precipitation in the United States and Figure 7.2.6 shows the normal monthly distribution of precipitation in the United States Figure 7.2.7 shows the mean annual precipitation of the world.

Rainstorms can vary significantly in space and time. *Rainfall hyetographs* are plots of rainfall depth or intensity as a function of time. Figure 7.2.8a shows an examples of two rainfall hyetographs. Cumulative rainfall hyetographs (rainfall mass curve) can be developed as shown in Figure 7.2.8b.

Isohyets (contours of constant rainfall) can be drawn to develop isohyetal maps of rainfall depth. *Isohyetal maps* are an interpolation of rainfall data recorded at gauged points. An example is shown in Figure 7.2.9 for the Upper Mississippi River Basin storm of January through July 1993. On a much smaller scale, shown in Figure 7.2.10, is the isohyetal map of the May 24–25, 1981, storm in Austin, Texas.

Figure 7.2.11 illustrates the three methods for determining areal average rainfall using rainfall gauge data. These are the *arithmetic–mean method*, the *Thiessen method,* and the *isohyetal method.*

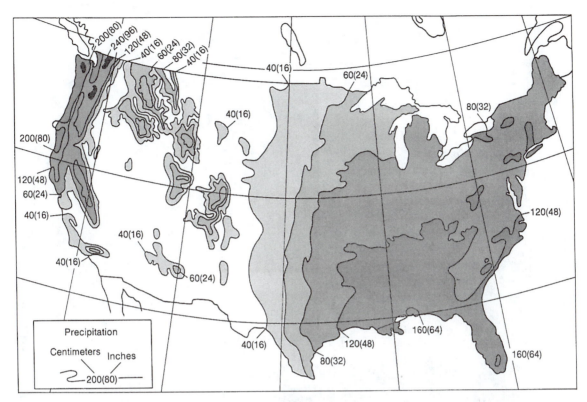

Figure 7.2.5 Mean annual precipitation for the United States in centimeters and inches (from Marsh (1987)).

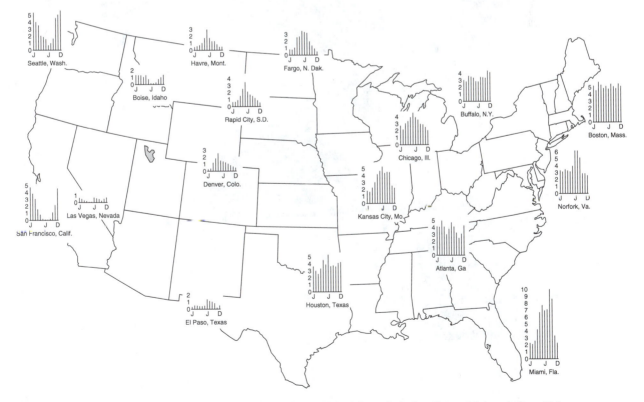

Figure 7.2.6 Normal monthly distribution of precipitation in the United States in inches (1 in = 25.4 mm) (from U.S. Environmental Data Services (1968)).

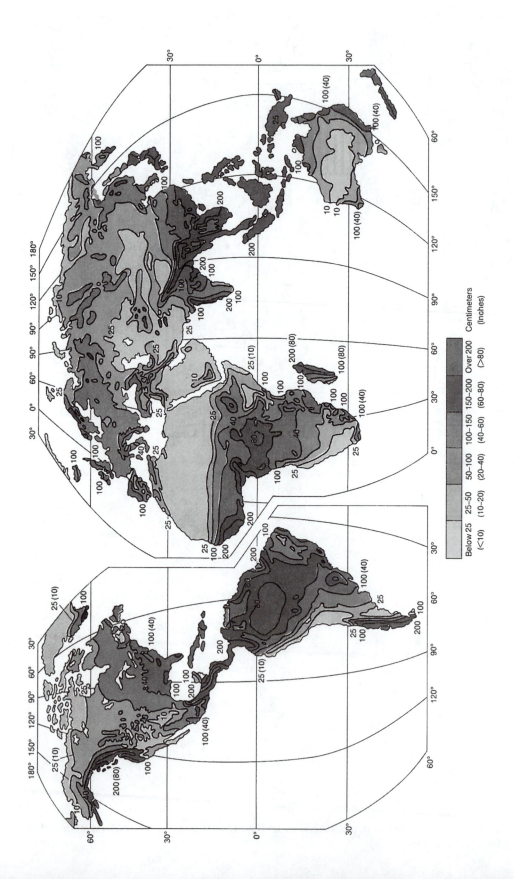

Figure 7.2.7 Average annual precipitation for the world's land areas, excepting Antarctica (from Marsh (1987)).

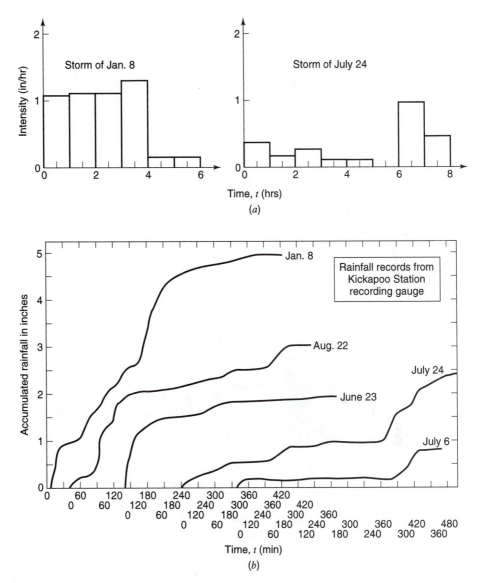

Figure 7.2.8 (*a*) Rainfall hyetographs for Kickapoo Station; (*b*) Mass rainfall curves (from Masch (1984)).

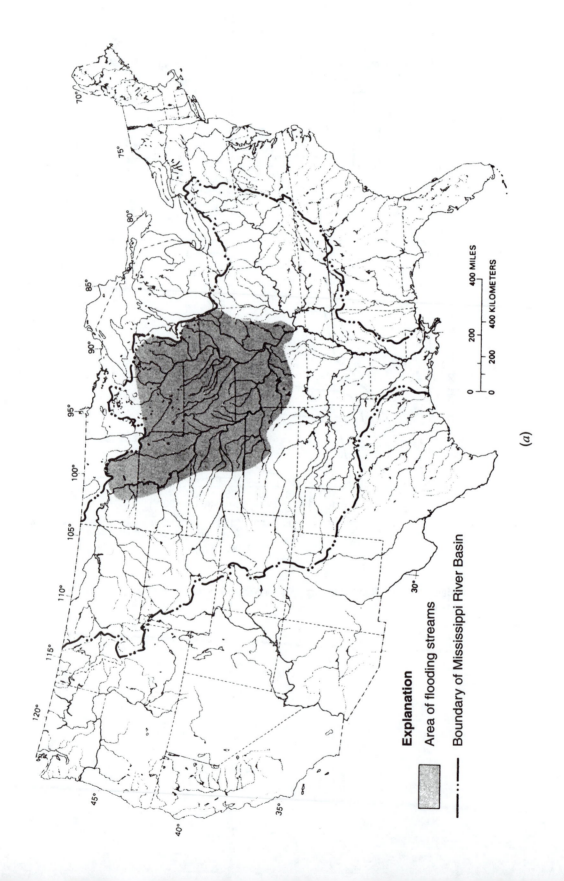

Figure 7.2.9 (*a*) Mississippi River Basin and general area of flooding streams, June through August 1993 (from Parrett et al. (1993)).

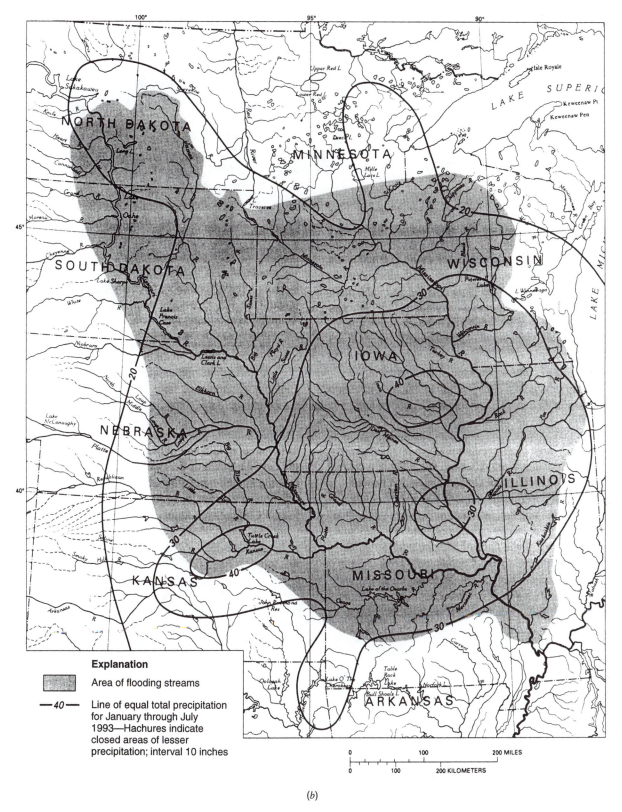

(b)

Figure 7.2.9 (*continued*) (*b*) Areal distribution of total precipitation in the area of flooding in the upper Mississippi River Basin, January through July 1993 (from Parrett et al. (1993)).

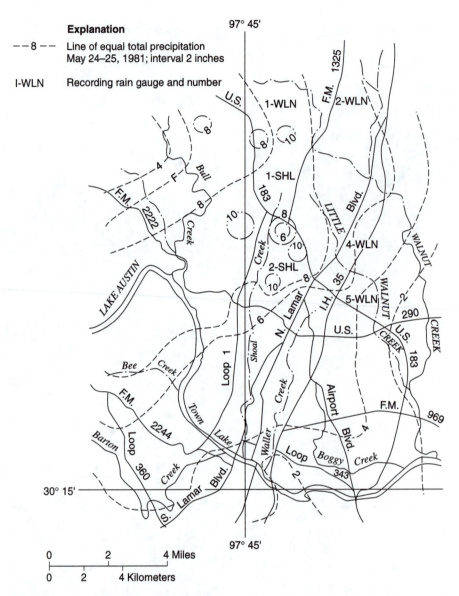

Figure 7.2.10 Isohyetal map of total precipitation (in) on May 24–25, 1981, based on USGS measurements, the City of Austin network, and unofficial precipitation reports (from Moore et al. (1982)).

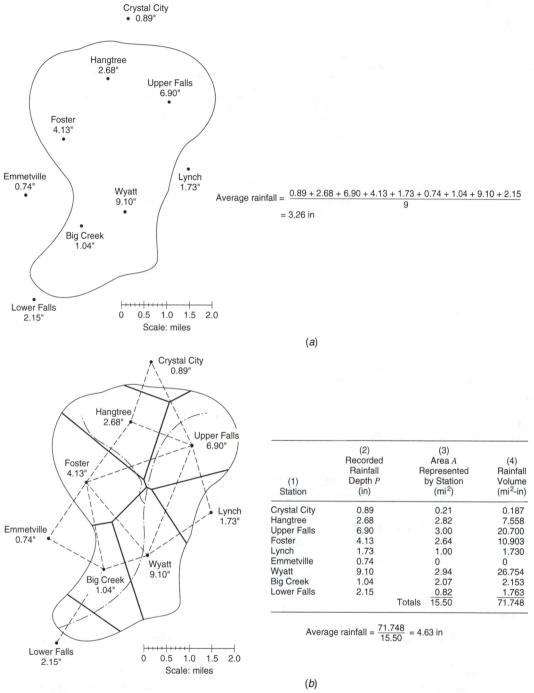

Average rainfall = $\dfrac{0.89 + 2.68 + 6.90 + 4.13 + 1.73 + 0.74 + 1.04 + 9.10 + 2.15}{9}$

= 3.26 in

(a)

(1) Station	(2) Recorded Rainfall Depth P (in)	(3) Area A Represented by Station (mi²)	(4) Rainfall Volume (mi²-in)
Crystal City	0.89	0.21	0.187
Hangtree	2.68	2.82	7.558
Upper Falls	6.90	3.00	20.700
Foster	4.13	2.64	10.903
Lynch	1.73	1.00	1.730
Emmetville	0.74	0	0
Wyatt	9.10	2.94	26.754
Big Creek	1.04	2.07	2.153
Lower Falls	2.15	0.82	1.763
	Totals	15.50	71.748

Average rainfall = $\dfrac{71.748}{15.50}$ = 4.63 in

(b)

Figure 7.2.11 (*a*) Computation of areal average rainfall by the arithmetic-mean method for 24-hour storm. This is the simplest method of determining areal average rainfall. It involves averaging the rainfall depths recorded at a number of gauges. This method is satisfactory if the gauges are uniformly distributed over the area and the individual gauge measurements do not vary greatly about the mean (after Roberson et al. (1998)); (*b*) Computation of areal average rainfall by the Thiessen method for 24-hour storm. This method assumes that at any point in the watershed the rainfall is the same as that at the nearest gauge, so the depth recorded at a given gauge is applied out to a distance halfway to the next station in any direction. The relative weights for each gauge are determined from the corresponding areas of application in a *Thiessen polygon* network, the boundaries of the polygons being formed by the perpendicular bisectors of the lines joining adjacent gauges for J gauges, the area within the watershed assigned to each is A_j and P_j is the rainfall recorded at the jth gauge, the areal average precipitation for the watershed is where the watershed area (after Roberson et al. (1998)).

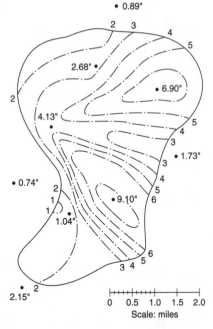

Rainfall Depth on Isohyet (in.)	Average Rainfall Depth (in.)	Area Between Isohyets (mi²)	Rainfall Volume (mi²-in.)
9.1			
8.0	8.55	0.407	3.480
6.0	7.0	1.412	9.884
5.0	5.5	0.841 + 1.375 = 2.216	1.219
4.0	4.5	0.592 + 1.697 = 2.289	10.300
3.0	3.5	3.122	10.927
2.0	2.5	2.599 + 0.431 = 3.030	7.575
	1.5	2.281	3.422
1.0	1.0	0.05	0.050
6.9			
6.0	6.45	0.693	4.470
		Totals 15.500	51.327

Average rainfall = $\frac{51.327}{15.50}$ = 3.31 in.

Scale: miles   0   0.5   1.0   1.5   2.0

Figure 7.2.11 (*continued*) (*c*) Computation of areal average rainfall by the isohyetal method for 24-hour storm. This method constructs isoyets, using observed depths at rain gauges and interpolation between adjacent gauges. Where there is a dense network of rain gauges, isohyetal maps can be constructed using computer programs for automated contouring. Once the isohyetal map is constructed, the area A_j between each pair of isohyets, within the watershed, is measured and multiplied by the average, P_j of the rainfall depths of the two boundary isohyets to compute the areal average precipitation (after Roberson et al. (1998)).

7.2.3   Disposal of Rainfall on a Watershed

A *watershed* is the area of land draining into a stream at a particular location. The various surface water processes in the hydrologic cycle occur on a watershed. Figure 7.2.12 is a schematic illustration of the disposal of rainfall during a storm on a watershed. This figure illustrates the rate (as a function of time) at which water flows or is added to storage for each of the processes. At the beginning of a storm, a large proportion of rainfall contributes to *surface storage* and as water infiltrates the *soil moisture storage* begins. Both *retention storage* and *detention storage* prevail. Retention storage is held for a long period of time and is depleted by evaporation, whereas detention storage is over a short time and is depleted by flow from the storage location.

7.2.4   Design Storms

The determination of flow rates in streams is one of the central tasks of surface water hydrology. For most engineering applications, these flow rates are determined for specified events that are typically extreme events. A major assumption in these analyses is that a certain return period storm results in the same return period flow rates from a watershed. The return period of an event, whether the event is a storm or a flow rate, is the expected value, or the average value measured over a very large number of occurrences. In other words, the return period refers to the time interval for which an event will occur once on the average over a very large number of occurrences.

Hershfield (1961), in a publication often referred to as TP-40, presented isohyetal maps of design rainfall depths for the United States for durations from 30 minutes to 24 hours and return periods from 1 to 100 years. The values of rainfall in these isohyetal maps are point precipitation values, which is precipitation occurring at a single point in space (as opposed to areal precipitation, which is over a larger area). Figure 7.2.13 is the isohyetal map for 100-year 24-hour rainfall. A later publication, U.S. Weather Bureau (1964), included maps for durations for 2 to 10 days, in

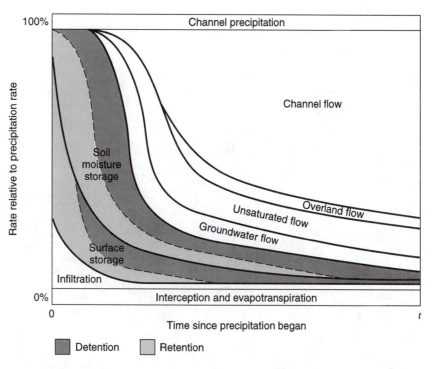

Figure 7.2.12 Schematic illustration of the disposal of precipitation during a storm on a watershed (from Chow et al. (1988)).

what is referred to as TP-49. Miller et al. (1973) presented isohyetal maps for 6- and 24-hour durations for the 11 mountainous states in the western United States, which supersede the corresponding maps in TP-40.

Frederick et al. (1977), in a publication commonly referred to as HYDRO-35, presented isohyetal maps for events having durations from 5 to 60 minutes. The maps of precipitation depths for 5-, 15-, and 60-minute durations and return periods of 2 and 100 years for the 37 eastern states are presented in Figures 7.2.14 (a–f). Depths for a return period are obtained by interpolation from the 5-, 15-, and 60-minute data for the same return period:

$$P_{10min} = 0.41 \, P_{5min} + 0.59 \, P_{15min} \tag{7.2.1a}$$

$$P_{30min} = 0.51 \, P_{15min} + 0.49 \, P_{60min} \tag{7.2.1b}$$

To consider return periods other than 2 or 100 years, the following interpolation equation is used:

$$P_{T\,yr} = aP_{2yr} + bP_{100yr} \tag{7.2.2}$$

where the coefficients a and b are found in Table 7.2.1.

Table 7.2.1 Coefficients for Interpolating Design Precipitation Depths Using Equation (7.2.2)

Return period T years	a	b
5	0.674	0.278
10	0.496	0.449
25	0.293	0.669
50	0.146	0.835

Source: Frederick, Myers, and Auciello (1997).

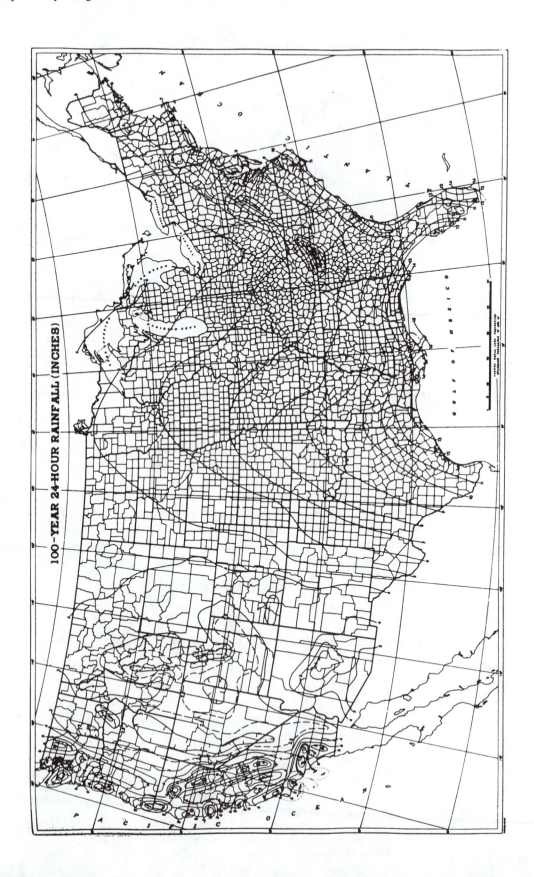

Figure 7.2.13 The 100-year 24-hour rainfall (in) in the United States (from Hershfield (1961)).

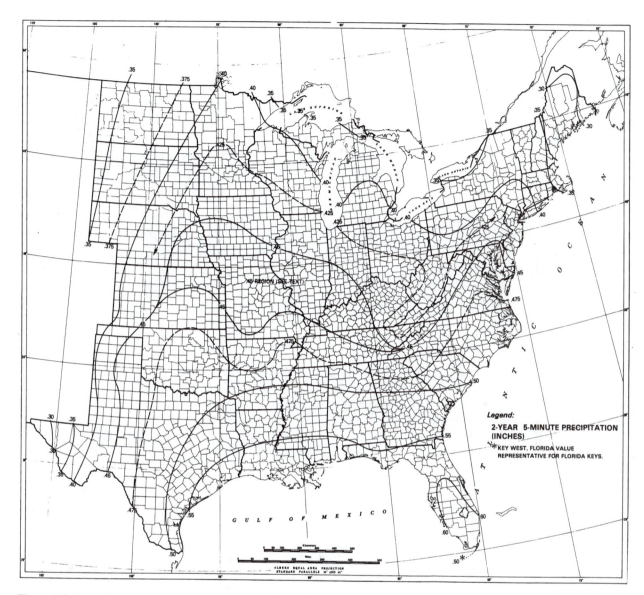

Figure 7.2.14 (*a*) 2-year 5-minute precipitation (in) (from Frederick, Meyers, and Auciello (1977)).

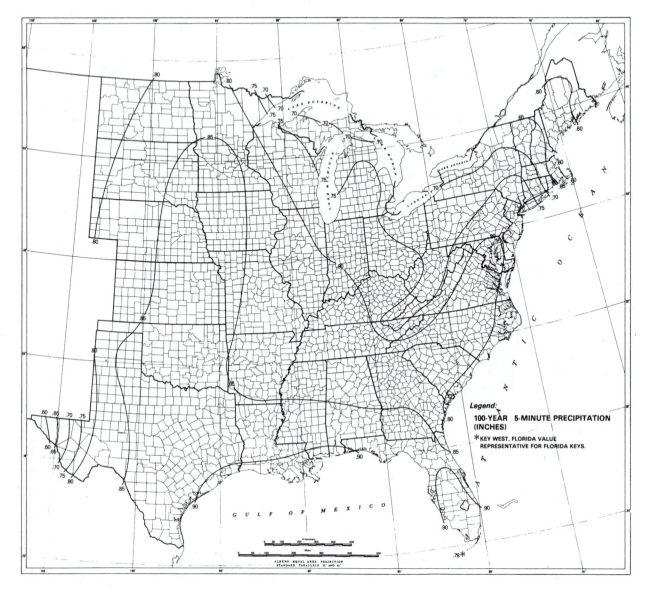

Figure 7.2.14 (*continued*) (*b*) 100-year 5-minute precipitation (in) (from Frederick, Meyers, and Auciello (1977)).

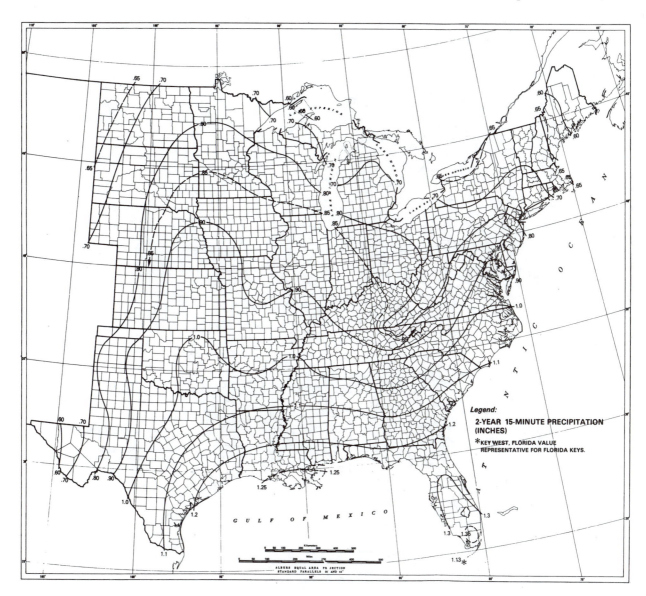

Figure 7.2.14 (*continued*) (*c*) 2-year 15-minute precipitation (in) (from Frederick, Meyers, and Auciello (1977)).

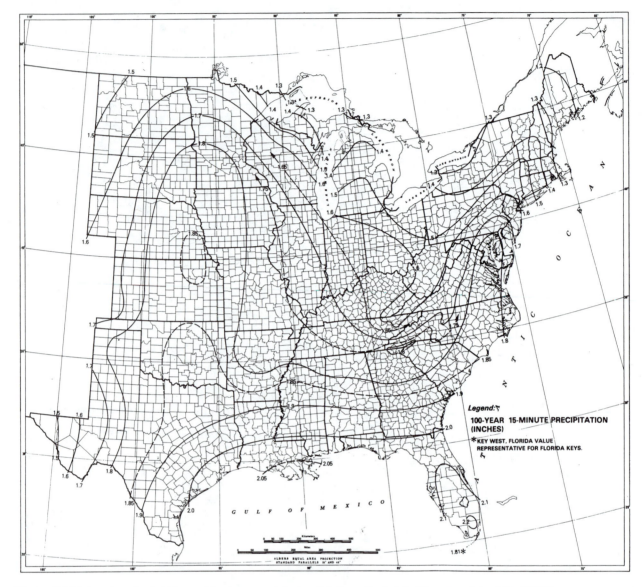

Figure 7.2.14 (*continued*) (*d*) 100-year 15-minute precipitation (in) (from Frederick, Meyers, and Auciello (1977)).

EXAMPLE 7.2.1	Determine the 2-, 10-, 25-, and 100-year rainfall depths for a 15-min duration storm in St. Louis, Missouri.

SOLUTION

From Figure 7.2.14 (HYDRO-35), $P_{2,15} = 0.9$ in and $P_{100,15} = 1.75$ in. $P_{10,15}$ and $P_{25,15}$ are determined using equation (7.2.2). From Table 7.2.1, $a = 0.496$ and $b = 0.449$ for 10 yr and $a = 0.293$ and $b = 0.669$ for 25 years:

$$P_{10yr} = a\,P_{2yr} + b\,P_{100yr}$$

$$P_{10,15} = 0.496 \times 0.9 + 0.449 \times 1.75 = 0.446 + 0.786 = 1.23 \text{ in}$$

$$P_{25,15} = 0.293\,P_{2,15} + 0.669\,P_{100,15} = 0.293 \times 0.9 + 0.669 \times 1.75 = 1.43 \text{ in}$$

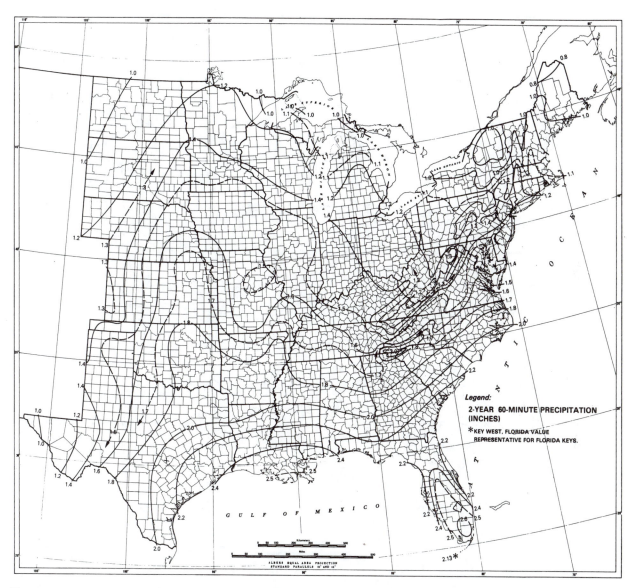

Figure 7.2.14 (*continued*) (*e*) 2-year 60-minute precipitation (in) (from Frederick, Meyers, and Auciello (1977)).

In hydrologic design projects, particularly urban drainage design, the use of *intensity-duration-frequency* relationships is recommended. *Intensity* refers to rainfall intensity (depth per unit time) and in some cases depths are used instead of intensity. *Duration* refers to rainfall duration and *frequency* refers to *return period*, which is the expected value of the *recurrence interval* (time between occurrences). See Chapter 10 for more details. The intensity-duration-frequency (IDF) relationships are also referred to as *IDF curves*. An example of IDF curves are shown in Figure 7.2.15. IDF relationships have also been expressed in equation form, such as

$$i = \frac{c}{T_d^e + f} \tag{7.2.3}$$

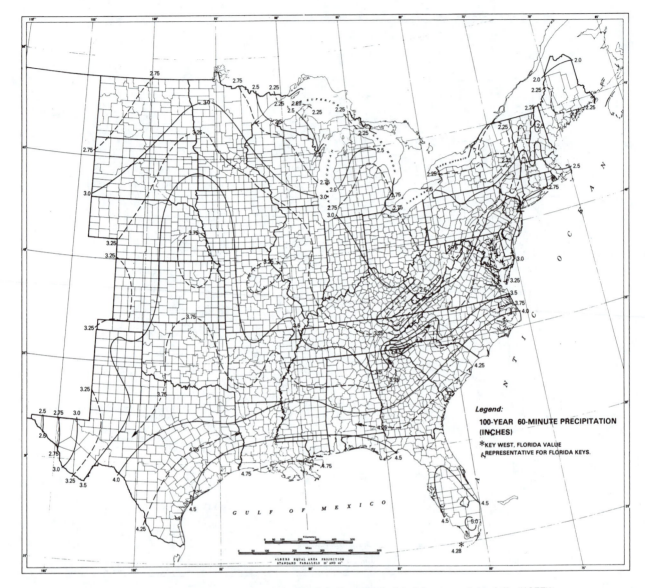

Figure 7.2.14 (*continued*) (*f*) 100-year 60-minute precipitation (in) (from Frederick, Meyers, and Auciello (1977)).

where i is the design rainfall intensity in in/hr, T_d is the duration in minutes, and c, e, and f are coefficients that vary for location and return period. Other forms of these IDF equations include the return period, such as

$$i = \frac{cT^m}{T_d + f} \qquad (7.2.4)$$

and

$$i = \frac{cT^m}{T_d^e + f} \qquad (7.2.5)$$

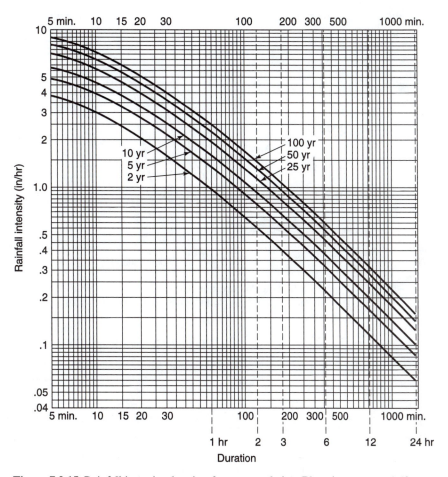

Figure 7.2.15 Rainfall intensity-duration-frequency relation (Phoenix metro area) (from Flood Control District of Maricopa County (1992)).

where T is the return period. In Chapter 15, these IDF equations are used in urban drainage design. Chow et al. (1988) describe in detail how to derive the coefficients for these relationships using rainfall data.

In many types of hydrologic analysis, such as *rainfall-runoff analysis*, to determine the runoff (discharge) from a watershed, the time sequence of rainfall is needed. In such cases it is standard practice to use a *synthetic storm hyetograph*. The United States Department of Agriculture Soil Conservation Service (1973, 1986) developed synthetic storm hyetographs for 6- and 24-hr storms in the United States These are presented in Table 7.2.2 and Figure 7.2.16 as cumulative hyetographs. Four 24-hr duration storms, Type I, IA, II, and III, were developed for different geographic locations in the U.S., as shown in Figure 7.2.17. Types I and IA are for the Pacific maritime climate, which has wet winters and dry summers. Type III is for the Gulf of Mexico and Atlantic coastal areas, which have tropical storms resulting in large 24-hour rainfall amounts. Type II is for the remainder of the United States.

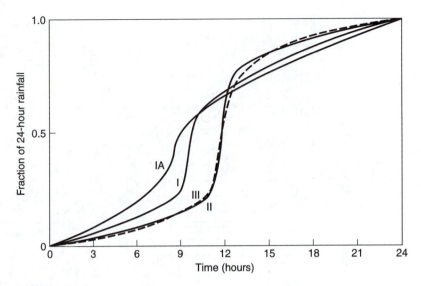

Figure 7.2.16 Soil Conservation Service 24-hour rainfall hyetographs (from U.S. Department of Agriculture Soil Conservation Service (1986)).

Table 7.2.2   SCS Rainfall Distributions

		24-Hour Storm				6-Hour Storm		
			$P_t/P24$					
Hour t	$t/24$	Type I	Type IA	Type II	Type III	Hour t	$t/6$	P_t/P_6
0	0	0	0	0	0	0	0	0
2.0	0.083	0.035	0.050	0.022	0.020	0.60	0.10	0.04
4.0	0.167	0.076	0.116	0.048	0.043	1.20	0.20	0.10
6.0	0.250	0.125	0.206	0.080	0.072	1.50	0.25	0.14
7.0	0.292	0.156	0.268	0.098	0.089	1.80	0.30	0.19
8.0	0.333	0.194	0.425	0.120	0.115	2.10	0.35	0.31
8.5	0.354	0.219	0.480	0.133	0.130	2.28	0.38	0.44
9.0	0.375	0.254	0.520	0.147	0.148	2.40	0.40	0.53
9.5	0.396	0.303	0.550	0.163	0.167	2.52	0.42	0.60
9.75	0.406	0.362	0.564	0.172	0.178	2.64	0.44	0.63
10.0	0.417	0.515	0.577	0.181	0.189	2.76	0.46	0.66
10.5	0.438	0.583	0.601	0.204	0.216	3.00	0.50	0.70
11.0	0.459	0.624	0.624	0.235	0.250	3.30	0.55	0.75
11.5	0.479	0.654	0.645	0.283	0.298	3.60	0.60	0.79
11.75	0.489	0.669	0.655	0.357	0.339	3.90	0.65	0.83
12.0	0.500	0.682	0.664	0.663	0.500	4.20	0.70	0.86
12.5	0.521	0.706	0.683	0.735	0.702	4.50	0.75	0.89
13.0	0.542	0.727	0.701	0.772	0.751	4.80	0.80	0.91
13.5	0.563	0.748	0.719	0.799	0.785	5.40	0.90	0.96
14.0	0.583	0.767	0.736	0.820	0.811	6.00	1.0	1.00
16.0	0.667	0.830	0.800	0.880	0.886			
20.0	0.833	0.926	0.906	0.952	0.957			
24.0	1.000	1.000	1.000	1.000	1.000			

Source: U.S. Dept. of Agriculture Soil Conservation Service (1973, 1986).

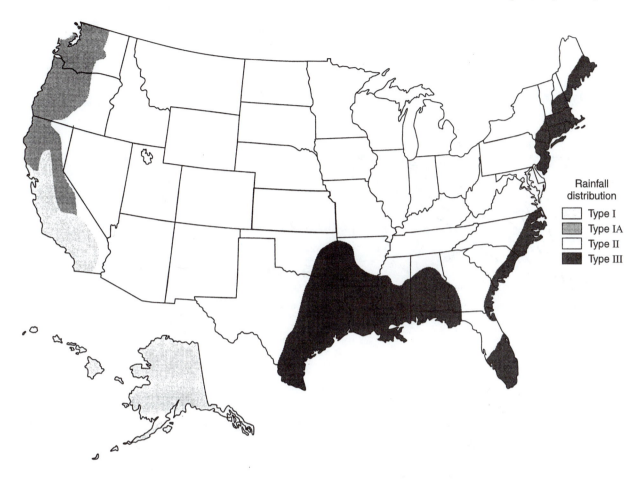

Figure 7.2.17 Location within the United States for application of the SCS 24-hour rainfall hyetographs (from U.S. Department of Agriculture Soil Conservation Service (1986)).

In the midwestern part of the United States the Huff (1967) temporal distribution of storms is widely used for heavy storms on areas ranging up to 400 mi^2. Time distribution patterns were developed for four probability groups, from the most severe (first quartile) to the least severe (fourth quartile). Figure 7.2.18a shows the probability distribution of first-quartile storms. These curves are smooth, reflecting average rainfall distribution with time; they do not exhibit the burst characteristics of observed storms. Figure 7.2.18b shows selected histograms of first-quartile storms for 10-, 50-, and 90-percent cumulative probabilities of occurrence, each illustrating the percentage of total storm rainfall for 10 percent increments of the storm duration. The 50 percent histogram represents a cumulative rainfall pattern that should be exceeded in about half of the storms. The 90 percent histogram can be interpreted as a storm distribution that is equaled or exceeded in 10 percent or less of the storms.

EXAMPLE 7.2.2

Using equation (7.2.3), compute the design rainfall intensities for a 10-year return period, 10-, 20-, and 60-min duration storms for $c = 62.5$, $e = 0.89$, and $f = 9.10$.

SOLUTION

The rainfall intensity duration frequency relationship is then

$$i = \frac{62.5}{T_d^{0.89} + 9.10}$$

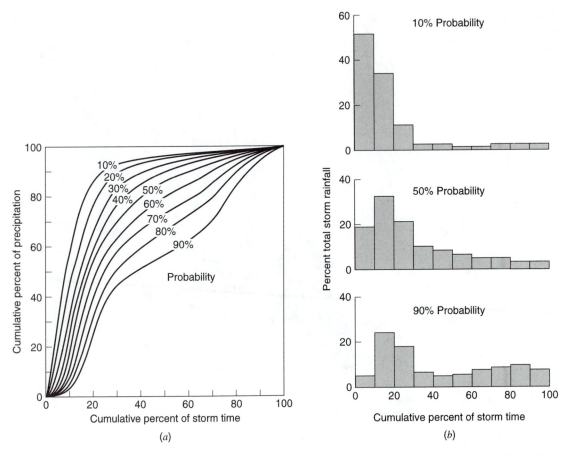

Figure 7.2.18 (*a*) Time distribution of first-quartile storms. The probability shown is the chance that the observed storm pattern will lie to the left of the curve; (*b*) Selected histograms for first-quartile storms (from Huff (1967)).

For $T_d = 10$ min, $i = \dfrac{62.5}{10^{0.89} + 9.10} = 3.71$ in/hr. For $T_d = 20$ min, $i = 2.66$ in/hr, and for $T_d = 60$ min, $i = 1.32$ in/hr.

7.2.5 Estimated Limiting Storms

Estimated limiting values (ELVs) are used for the design of water control structures such as the spillways on large dams. Of particular interest are the *probable maximum precipitation* (PMP) and the *probable maximum storm* (PMS). These are used to derive a *probable maximum flood* (PMF). PMP is a depth of precipitation that is the estimated limiting value of precipitation, defined as the estimated greatest depth of precipitation for a given duration that is physically possible and reasonably characteristic over a particular geographical region at a certain time of year (Chow et al., 1988). Schreiner and Reidel (1978) presented generalized PMP charts for the United States east of the 105th meridian HMR 51. The all-seasons (any time of the year) estimates of PMP are presented in maps as a function of storm area (ranging from 10 to 20,000 mi^2) and storm durations ranging from 6 to 72 hours, as shown in Figure 7.2.19. Regions west of the 105th meridian, the diagram in Figure 7.2.20 shows the appropriate U.S. National Weather Service publication. Hansen et al. (1982) in what is called HMR 52 present the procedure to determine the PMS for areas east of the 105th meridian.

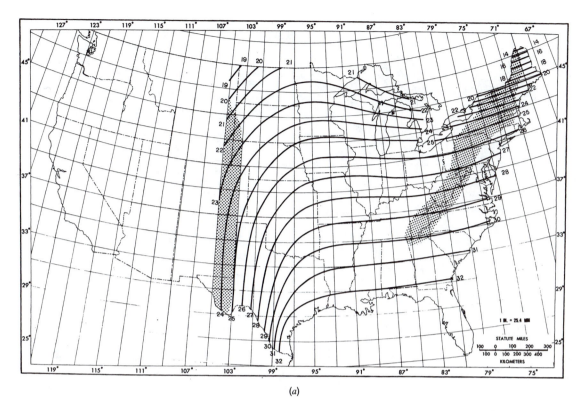

(a)

Figure 7.2.19 (a) Example of all-season PMP (in) for 6 hours, 10 mi² (from Hansen et al. (1982)).

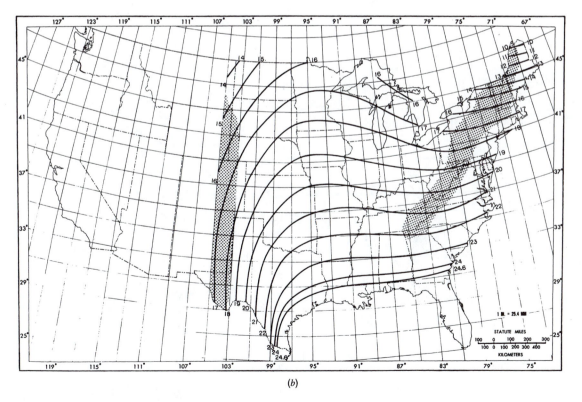

(b)

Figure 7.2.19 (b) Example of all-season PMP (in) for 12 hours, 10 mi² (from Hansen et al. (1982)).

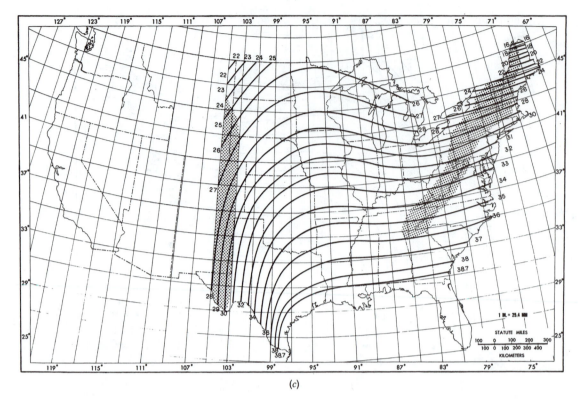

Figure 7.2.19 (*continued*) (*c*) Example of all-season PMP (in) for 6 hours, 200 mi^2 (from Hansen et al. (1982)).

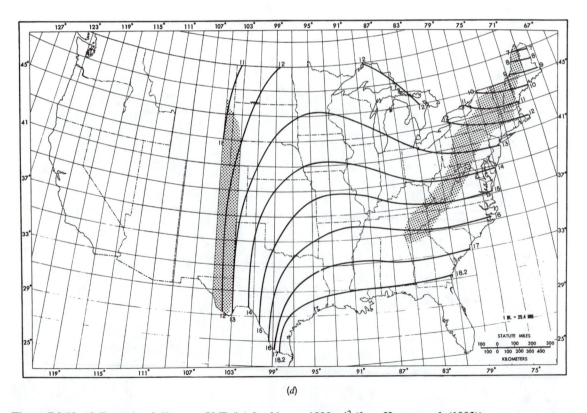

Figure 7.2.19 (*d*) Example of all-season PMP (in) for 6 hours, 1000 mi^2 (from Hansen et al. (1982)).

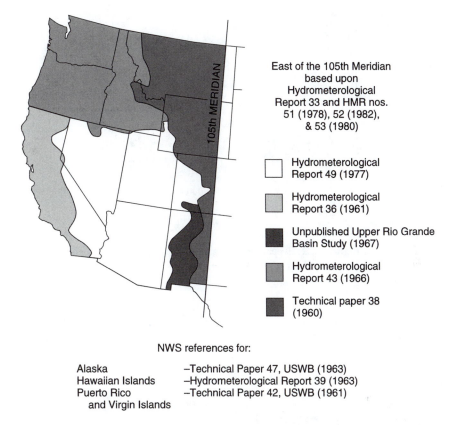

East of the 105th Meridian
based upon
Hydrometerological
Report 33 and HMR nos.
51 (1978), 52 (1982),
& 53 (1980)

☐ Hydrometerological
Report 49 (1977)

▨ Hydrometerological
Report 36 (1961)

■ Unpublished Upper Rio Grande
Basin Study (1967)

▨ Hydrometerological
Report 43 (1966)

■ Technical paper 38
(1960)

NWS references for:

Alaska	–Technical Paper 47, USWB (1963)
Hawaiian Islands	–Hydrometerological Report 39 (1963)
Puerto Rico and Virgin Islands	–Technical Paper 42, USWB (1961)

Figure 7.2.20 Sources on information for probable maximum precipitation computation in the United States (from National Academy of Sciences (1983)).

7.3 EVAPORATION

Evaporation is the process of water changing from its liquid phase to the vapor phase. This process may occur from water bodies, from saturated soils, or from unsaturated surfaces. The evaporation process is illustrated in Figure 7.3.1. Above the water surface a number of things are happening. First of all, there are water molecules in the form of *water vapor*, which are always found above liquid water. In addition, there are some other molecules: (a) two oxygen atoms stuck together by themselves, forming an *oxygen molecule*, and (b) two nitrogen atoms stuck together forming a nitrogen molecule. Above the water surface is the air, a gas, consisting almost entirely of nitrogen, oxygen, some water vapor, and lesser amounts of carbon dioxide, argon, and other things. The molecules in the water are always moving around. From time to time, one on the surface gets knocked away. In other words, molecule by molecule, the water disappears or evaporates.

The computation of evaporation in hydrologic analysis and design is important in water supply design, particularly reservoir design and operation. The supply of energy to provide *latent heat of vaporization* and the *ability to transport water vapor away* from the evaporative surface are the two major factors that influence evaporation. *Latent heat* is the heat that is given up or absorbed when a phase (solid, liquid, or gaseous state) changes. *Latent heat of vaporization* (l_v) refers to the heat given up during vaporization of liquid water to water vapor, and is given as $l_v = 2.501 \times 10^6 - 2370T$, where T is the temperature in °C and l_v is in joules (J) per kilogram.

Three methods are used to determine evaporation: *the energy balance method*, *the aerodynamic method*, and *the combined aerodynamic and energy balance method*. The energy balance method considers the heat energy balance of a hydrologic system and the aerodynamic method considers the ability to transport away from an open surface. Figure 7.3.2 illustrates the radiation and heat

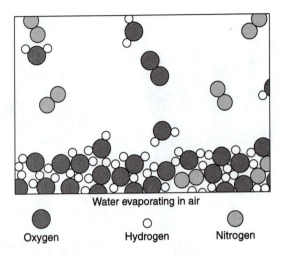

Figure 7.3.1 Evaporation (magnified one billion times)
(from Feynman et al. (1963)).

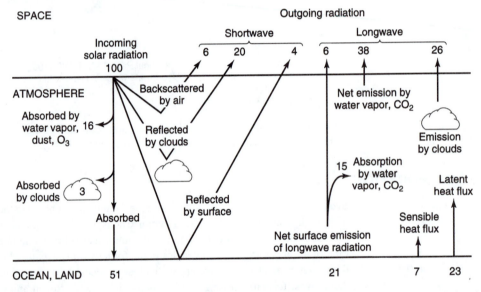

Figure 7.3.2 Radiation and heat balance in the atmosphere and at the earth's surface (from National
Academy of Sciences (1975)).

balance in the atmosphere and at the earth's surface along with relative values for the various
components.

7.3.1 Energy Balance Method

Consider the evaporation pan shown in Figure 7.3.3 with the defined control volume. This control
volume contains water in both the liquid phase and the vapor phase, with densities of ρ_w and ρ_a,
respectively. The continuity equation must be written for both phases. For the liquid phase, the
extensive property $B = m$ (mass of liquid water), the intensive property $\beta = 1$, and $dB/dt = \dot{m}_v$,
which is the mass flow rate of evaporation. Continuity for the liquid phase is then

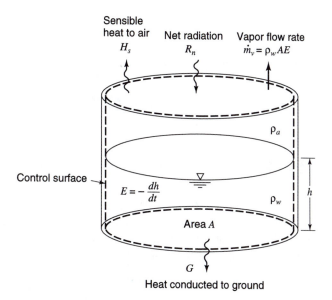

Figure 7.3.3 Control volume defined for continuity and energy
equation development for an evaporation pan (from Chow et al.
(1988)).

$$-\dot{m}_v = \frac{d}{dt} \int_{CV} \rho_w d\forall + \sum_{CS} \rho_w \mathbf{V} \cdot \mathbf{A} \qquad (7.3.1)$$

Because the pan has impermeable sides, there is no flow of liquid water across the control surface,
so $\sum_{CS} \rho_w \mathbf{V} \cdot \mathbf{A} = 0$.

The rate of change of storage is

$$\frac{d}{dt} \int_{CV} \rho_w d\forall = \rho_w A \frac{dh}{dt} \qquad (7.3.2)$$

where A is the cross-sectional area of the pan and h is the depth of water. Substituting equation
(7.3.2) into (7.3.1) gives

$$-\dot{m}_v = \rho_w A \frac{dh}{dt} \qquad (7.3.3a)$$

or

$$\dot{m}_v = \rho_w A E \qquad (7.3.3b)$$

where $E = -dh/dt$ is the evaporation rate.

Considering the vapor phase, the extensive property is $B = m_v$ (mass of water vapor), the intensive property is $\beta = q_v$ (specific humidity), and $dB/dt = \dot{m}_v$. Continuity for the vapor phase is

$$\dot{m}_v = \frac{d}{dt} \int_{CV} q_v \rho_a d\forall + \sum_{CS} q_v \rho_a \mathbf{V} \cdot \mathbf{A} \qquad (7.3.4)$$

Considering steady flow over the pan, the time derivative of water vapor in the control volume is
zero, $\dfrac{d}{dt} \int_{CV} q_v \rho_a d\forall = 0$, and $\dot{m}_v$ from equation (7.3.3) is substituted into equation (7.3.4) to obtain

$$\rho_w AE = \sum_{CS} q_v \rho_a \mathbf{V} \cdot \mathbf{A} \tag{7.3.5}$$

which is the continuity equation for the pan. Equation (7.3.5) can be rearranged to yield

$$E = \left(\frac{1}{\rho_w A}\right) \sum_{CS} q_v \rho_a \mathbf{V} \cdot \mathbf{A} \tag{7.3.6}$$

Next, consider the heat energy equation (3.3.19):

$$\frac{dH}{dt} - \frac{dW_s}{dt} = \frac{d}{dt} \int_{CV} \left(e_u + \frac{1}{2}V^2 + gz\right)\rho_w \, d\forall + \sum_{CS} \left(\frac{p}{\rho_w} + e_u + \frac{1}{2}V^2 + gz\right)\rho_w \, \mathbf{V} \cdot \mathbf{A} \tag{3.3.19}$$

using ρ_w to represent density of water; dH/dt is the rate of heat input from an external source, $dW_s/dt = 0$ because there is no work done by the system, e_u is the specific internal heat energy of the water, $V = 0$ for the water in the pan, and the rate of change of the water surface elevation is small (dz/dt (0). Then the above equation is reduced to

$$\frac{dH}{dt} = \frac{d}{dt} \int_{CV} e_u \rho_w \, d\forall \tag{7.3.7}$$

The rate of heat input from external sources can be expressed as

$$\frac{dH}{dt} = R_n - H_s - G \tag{7.3.8}$$

where R_n is the *net radiation flux* (watts per meter squared), H_s is the *sensible heat* to the air stream supplied by the water, and G is the *ground heat flux* to the ground surface. Net radiation flux is the net input of radiation at the surface at any instant. The net radiation flux at the earth's surface is the major energy input for evaporation of water, defined as the difference between the radiation absorbed, $R_i(1 - \alpha)$ (where R_i is the *incident radiation* and α is the fraction of radiation reflected, called the *albedo*), and that emitted, R_e:

$$R_n = R_i(1 - \alpha) - R_e \tag{7.3.9}$$

The amount of radiation emitted is defined by the *Stefan–Boltzmann law*

$$R_e = e\sigma T_p^4 \tag{7.3.10}$$

where e is the *emissivity* of the surface, σ is the *Stefan–Boltzmann constant* ($5.67 \times 10^{-8}\,\text{W/m}^2 \cdot \text{K}^4$) and T_p is the absolute temperature of the surface in degrees Kelvin.

Assuming that the temperature of the water in the control volume is constant in time, the only change in the heat stored within the control volume is the change in the internal energy of the water evaporated $l_v \dot{m}_v$, where l_v is the latent heat of vaporization, so that

$$\frac{d}{dt} \int_{CV} e_u \rho_w \, d\forall = l_v \dot{m}_v \tag{7.3.11}$$

Substituting equations (7.3.8) and (7.3.11) into (7.3.7) results in

$$R_n - H_s - G = l_v \dot{m}_v \tag{7.3.12}$$

From equation (7.3.3b), $\dot{m}_v = \rho_w AE$. Substituting in (7.3.12) with $A = 1\,\text{m}^2$ and solving for E (to denote energy balance) gives the *energy balance equation for evaporation*,

$$E = \frac{1}{l_v \rho_w}(R_n - H_s - G) \tag{7.3.13}$$

Assuming the sensible heat flux H_s and the ground heat flux G are both zero, then an evaporation rate E_r, which is the rate at which all incoming net radiation is absorbed by evaporation, can be calculated as;

$$E_r = \frac{R_n}{l_v \rho_w} \qquad (7.3.14)$$

EXAMPLE 7.3.1

For a particular location the average net radiation is 185 W/m², air temperature is 28.5°C, relative humidity is 55 percent, and wind speed is 2.7 m/s at a height of 2 m. Determine the open water evaporation rate in mm/d using the energy method.

SOLUTION

Latent heat of vaporization in joules J per kg varies with T (°C), or

$l_v = 2.501 \times 10^6 - 2370\,T$, so $l_v = 2501 - 2.36 \times 28.5 = 2433$ kJ/kg. $\rho_w = 996.3$ kg/m³. The evaporation rate by the energy balance method is determined using equation (7.3.14) with $R_n = 185$ W/m²:
$E_r = R_n/(l_v \rho_w) = 185/(2433 \times 10^3 \times 996.3) = 7.63 \times 10^{-8}$ m/s $= 6.6$ mm/d.

7.3.2 Aerodynamic Method

As mentioned previously, the aerodynamic method considers the ability to transport water vapor away from the water surface, that is generated by the humidity gradient in the air near the surface and the wind speed across the surface. These processes can be analyzed by coupling the equation for mass and momentum transport in the air. Considering the control volume in Figure 7.3.4, the vapor flux $\dot{m}_v$ passing upward by convection can be defined along with the momentum flux (as a function of the humidity gradient and the wind velocity gradient, respectively (see Chow et al., 1988). The ratio of the vapor flux and the momentum flux can be used to define the *Thornthwaite–Holzman equation* for vapor transport (Thornthwaite and Holzman, 1939). Chow et al. (1988) present details of this derivation. The final form of the evaporation equation for the aerodynamic method expresses the evaporation rate E_a as a function of the difference of the vapor pressure at the surface e_{as}, which is the *saturation vapor pressure* at ambient air temperature (when the rate of evaporation and condensation are equal) and the vapor pressure at a height z_2 above the water surface, which is taken as the ambient vapor pressure in air e_a.

$$E_a = B(e_{as} - e_a) \qquad (7.3.15)$$

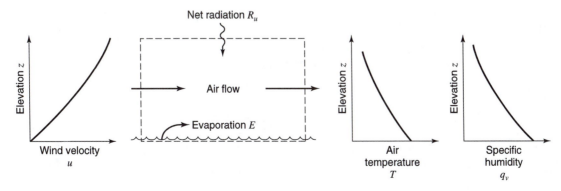

Figure 7.3.4 Evaporation from an open water surface (from Chow et al. (1988)).

where E_a has units of mm/day, and B is the vapor transfer coefficient with units of (mm/day · Pa), given as

$$B = \frac{0.102u_2}{\left[\ln(z_2/z_0)\right]^2} \qquad (7.3.16)$$

where u_2 is the wind velocity (m/s) measured at height z_2 (cm) and z_0 is the roughness height (0.01–0.06 cm) of the water surface. The vapor pressures have units of Pa (N/m²). The saturation vapor pressure is approximated as

$$e_{as} = 611 \exp\left(\frac{17.27T}{237.3+T}\right) \qquad (7.3.17)$$

and the vapor pressure is

$$e_a = R_h\, e_{as} \qquad (7.3.18)$$

where T is the air temperature in °C and R_h is the relative humidity ($R = e_a/e_{as}$) and ($0 \le R_h \le 1$).

EXAMPLE 7.3.2

Solve example 7.3.1 using the aerodynamic method, by using a roughness height $z_0 = 0.03$ cm.

SOLUTION

From equation (7.3.17), the saturated vapor pressure is

$$e_{as} = 611 \exp\left[17.27T/(237.3+T)\right] = 611 \exp\left[17.27 \times 28.5/(237.3+28.5)\right]$$
$$= 3893 \text{ Pa}$$

The ambient vapor pressure e_a is determined from equation (7.3.18); for a relative humidity $R_h = 0.55$, $e_a = e_{as}\,R_h = 3893 \times 0.55 = 2141$ Pa. The vapor transfer coefficient B is given by equation (7.3.16) in which $u_2 = 2.7$ m/s, $z_2 = 2$ m and $z_0 = 0.03$ cm for an open water surface, so that $B = 0.102u_2/\left[\ln(z_2/z_0)\right]^2 = 0.102 \times 2.7/\left[\ln(200/0.03)\right]^2 = 0.0036$ mm/d · Pa then, the evaporation rate by the aerodynamic method is given by equation (7.3.15):

$$E_a = B(e_{as} - e_a) = 0.0036(3893 - 2141) = 6.31 \text{ mm/d}.$$

7.3.3 Combined Method

When the energy supply is not limiting, the aerodynamic method can be used and when the vapor transport is not limiting, the energy balance method can be used. However, both of these factors are not normally limiting, so a combination of these methods is required. The combined method equation is

$$E = \left(\frac{\Delta}{\Delta+\gamma}\right)E_r + \left(\frac{\gamma}{\Delta+\gamma}\right)E_a \qquad (7.3.19)$$

in which ( )E_r is the vapor transport term and ( )E_a is the aerodynamic term. γ is the *psychrometric constant* (approximately 66.8 Pa/°C) and Δ is the *gradient of the saturated vapor pressure curve* $\Delta = de_{as}/dT$ at air temperature T_α given as

$$\Delta = \frac{4098e_{as}}{(237.3+T_\alpha)^2} \qquad (7.3.20)$$

in which e_{as} is the *saturated vapor pressure* (the maximum moisture content the air can hold for a given temperature).

The combination method is best for application to small areas with detailed climatological data including net radiation, air temperature, humidity, wind speed, and air pressure. For very large areas, energy (vapor transport) largely governs evaporation. Priestley and Taylor (1972) discovered that the aerodynamic term in equation (7.3.19) is approximately 30 percent of the energy term, so that equation (7.3.19) can be simplified to

$$E = 1.3\left(\frac{\Delta}{\Delta + \gamma}\right)E_r \tag{7.3.21}$$

which is known as the *Priestley–Taylor evaporation equation*.

EXAMPLE 7.3.3   Solve example 7.3.1 using the combined method.

SOLUTION   The gradient of the saturated vapor pressure curve is, from equation (7.3.20),

$$\Delta = 4098e_{as}/(237.3 + T)^2 = 4098 \times 3893/(237.3 + 28.5)^2 = 225.8 \text{ Pa}/°\text{C}$$

The psychrometric constant γ is approximately 66.8 Pa/°C; then E_r and E_a may be combined according to equation (7.3.19) to give

$$E = \Delta/(\Delta + \gamma)E_r + \gamma/(\Delta + \gamma)E_a$$

$$= (225.8/(225.8 + 66.8))6.6 + (66.8/(225.8 + 66.8))6.31 = 0.772(6.6) + 0.228(6.31)$$

$$= 5.10 + 1.44 = 6.54 \text{ mm/d}$$

EXAMPLE 7.3.4   Solve example 7.3.1 using the Priestley–Taylor method.

SOLUTION   The evaporation is computed using equation (7.3.21):

$$E = 1.3\left(\frac{\Delta}{\Delta + \gamma}\right)E_r = 1.3\left(\frac{225.8}{225.8 + 66.8}\right)6.6$$

$$= 6.62 \text{ mm/d}$$

7.4 INFILTRATION

The process of water penetrating into the soil is *infiltration*. The rate of infiltration is influenced by the condition of the soil surface, vegetative cover, and soil properties including porosity, hydraulic conductivity, and moisture content. In order to discuss infiltration, we must first consider the division of subsurface water (see Figure 6.1.2) and the various subsurface flow processes shown in Figure 7.4.1. These processes are infiltration of water to become *soil moisture, subsurface flow* (unsaturated flow) through the soil, and *groundwater flow* (saturated flow). *Unsaturated flow* refers to flow through a porous medium when some of the voids are occupied by air. *Saturated flow* occurs when the voids are filled with water. The *water table* is the interface between the saturated and unsaturated flow where atmospheric pressure prevails. Saturated flow occurs below the water table and unsaturated flow occurs above the water table.

7.4.1 Unsaturated Flow

The cross-section through an unsaturated porous medium (Figure 7.4.2a) is now used to define *porosity*, η:

$$\eta = \frac{\text{volume of voids}}{\text{total volume}} \tag{7.4.1}$$

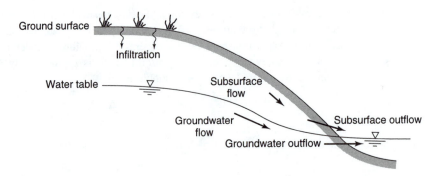

Figure 7.4.1 Subsurface water zones and processes (from Chow et al. (1988)).

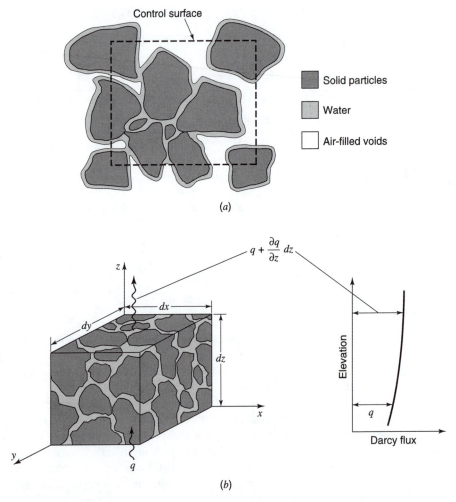

Figure 7.4.2 (*a*) Cross-section through an unsaturated porous medium; (*b*) Control volume for development of the continuity equation in an unsaturated porous medium (from Chow et al. (1988)).

in which η is $0.25 < \eta < 0.75$, and the *soil moisture content*, θ,

$$\theta = \frac{\text{volume of water}}{\text{total volume}} \tag{7.4.2}$$

in which θ is $0 \le \theta \le \eta$. For saturated conditions, $\theta = \eta$.

Consider the control volume in Figure 7.4.2b for an unsaturated soil with sides of lengths dx, dy, and dz with a volume of $dxdydz$. The volume of water contained in the control volume is $\theta dxdydz$. Flow through the control volume is defined by the *Darcy flux*, $q = Q/A$, which is the volumetric flow rate per unit of soil area. For this derivation, the horizontal fluxes are ignored and only the vertical (z) direction is considered, with z positive upwards.

With the control volume approach, the extensive property B is the mass of soil water, so the intensive property $\beta = dB/dm = 1$ and $dB/dt = 0$, because no phase changes are occurring in water. The general control volume equation for continuity, equation (3.2.1), is applicable:

$$0 = \frac{d}{dt} \int_{CV} \rho \, d\forall + \int_{CS} \rho \mathbf{V} \cdot d\mathbf{A} \tag{7.4.3}$$

The time rate of change of mass stored in the control volume is

$$\frac{d}{dt} \int_{CV} \rho \, d\forall = \frac{d}{dt}(\rho\theta dxdydz) = \rho dxdydz \frac{\partial \theta}{\partial t} \tag{7.4.4}$$

where the density is assumed constant. The net outflow of water is the difference between the volumetric inflow at the bottom ($qdxdy$) and the volumetric outflow at the top $[q + (\partial q/\partial z)dz]dxdy$, so

$$\int_{CS} \rho \mathbf{V} \cdot d\mathbf{A} = \rho \left(q + \frac{\partial q}{\partial z} dz \right) dxdy - \rho q dxdy = \rho dxdydz \frac{\partial q}{\partial z} \tag{7.4.5}$$

Substituting equations (7.4.4.) and (7.4.5) into (7.4.3) and dividing by $\rho dxdydz$ results in the following *continuity equation for one-dimensional unsteady unsaturated flow in a porous medium*:

$$\frac{\partial \theta}{\partial t} + \frac{\partial q}{\partial z} = 0 \tag{7.4.6}$$

Darcy's law relates the *Darcy flux* q to the rate of headloss per unit length of medium. For flow in the vertical direction the headloss per unit length is the change in total head ∂h over a distance, ∂z, i.e., $-\partial h/\partial z$, where the negative sign indicates that total head decreases (as a result of friction) in the direction of flow. Darcy's law can now be expressed as

$$q = -K \frac{\partial h}{\partial z} \tag{7.4.7}$$

where K is the *hydraulic conductivity*. This law applies to areas that are large compared with the cross-section of individual pores and grains of the medium. Darcy's law describes a steady uniform flow of constant velocity with a net force of zero in a fluid element. In unconfined saturated flow the forces are gravity and friction. For unsaturated flow the forces are gravity, friction, and the *suction force* that binds water to soil particles through surface tension.

In unsaturated flow the void spaces are only partially filled with water, so that water is attracted to the particle surfaces through electrostatic forces between the water molecule polar bonds and the particle surfaces. This in turns draws water up around the particle surfaces, leaving air in the center of the voids. The energy due to the soil suction forces is referred to as the *suction head* ψ in unsaturated flow, which varies with moisture content. Total head is then the sum of the suction and gravity heads:

$$h = \psi + z \tag{7.4.8}$$

Note that the velocities are so small that there is no term for velocity head in this expression for total head.

Darcy's law can now be expressed as

$$q = -K \frac{\partial(\psi + z)}{\partial z} \tag{7.4.9}$$

Darcy's law was originally conceived for saturated flow and was extended by Richards (1931) to unsaturated flow with the provision that the hydraulic conductivity is a function of the suction head, i.e., $K = K(\psi)$. Also, the hydraulic conductivity can be related more easily to the degree of saturation, so that $K = K(\theta)$. Because the soil suction head varies with moisture content and moisture content varies with elevation, the *suction gradient* can be expanded by using the chain rule to obtain

$$\frac{\partial \psi}{\partial z} = \frac{d\psi}{d\theta} \frac{\partial \theta}{\partial z} \tag{7.4.10}$$

in which $\partial\theta/\partial z$ is the *wetness gradient* and the reciprocal of $d\psi/d\theta$, i.e., $d\theta/d\psi$, is the *specific water capacity*. Now equation (7.4.9) can be modified to

$$q = -K\left(\frac{\partial \psi}{\partial z} + \frac{\partial z}{\partial z}\right) = -K\left(\frac{d\psi}{d\theta}\frac{\partial \theta}{\partial z} + 1\right) = -\left(K\frac{d\psi}{d\theta}\frac{\partial \theta}{\partial z} + K\right) \tag{7.4.11}$$

The *soil water diffusivity* $D(L^2/T)$ is defined as

$$D = K\frac{d\psi}{d\theta} \tag{7.4.12}$$

so substituting this expression for D into equation (7.4.11) results in

$$q = -\left(D\frac{\partial \theta}{\partial z} + K\right) \tag{7.4.13}$$

Using the continuity equation (7.4.6) for one–dimensional, unsteady, unsaturated flow in a porous medium yields

$$\frac{\partial \theta}{\partial t} = \frac{\partial q}{\partial z} = \frac{\partial}{\partial z}\left(D\frac{\partial \theta}{\partial z} + K\right) \tag{7.4.14}$$

which is a one-dimensional form of *Richards' equation*. This equation is the governing equation for unsteady unsaturated flow in a porous medium (Richards, 1931). For a homogeneous soil, $\partial K/\partial z = 0$, so that $\partial\theta/\partial t = \partial(D\partial\theta/\partial z)/\partial z$.

EXAMPLE 7.4.1   Determine the flux for a soil in which the hydraulic conductivity is expressed as a function of the suction head as $K = 250(-\psi)^{-2.11}$ in cm/d at depth $z_1 = 80$ cm, $h_1 = -145$ cm, and $\psi_1 = -65$ cm at depth $z_2 = 100$ cm, $h_2 = -160$ cm, and $\psi_2 = -60$ cm.

SOLUTION   The flux is determined using equation (7.4.7). First the hydraulic conductivity is computed using an average value of $\psi = [-65 + -(60)]/2 = -62.5$ cm. Then $K = 250(-\psi)^{-2.11} = 250 (62.5)^{-2.11} = 0.041$ cm/d. The flux is then

$$q = -K\left(\frac{h_1 - h_2}{z_1 - z_2}\right) = -0.041\left[\frac{-145 - (-160)}{-80 - (-100)}\right] = -0.03 \text{cm/d}$$

The flux is negative because the moisture is flowing downwards in the soil.

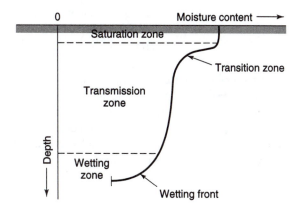

Figure 7.4.3 Moisture zones during infiltration (from Chow et al. (1988)).

EXAMPLE 7.4.2 Determine the soil water diffusivity for a soil in which $\theta = 0.1$ and $K = 3 \times 10^{-11}$ mm/s from a relationship of $\psi(\theta)$ at $\theta = 0.1$, $\Delta\psi = 10^7$ mm, and $\Delta\theta = 0.35$.

SOLUTION Using equation (7.4.12), the soil water diffusivity is $D = Kd\psi/d\theta = (3 \times 10^{-11} \text{ mm/s})(10^7 \text{ mm}/0.35) = 8.57 \times 10^{-4} \text{ mm}^2/\text{s}$.

7.4.2 Green–Ampt Method

Figure 7.4.3 illustrates the distribution of soil moisture within a soil profile during downward movement. These moisture zones are the *saturated zone*, the *transmission zone*, a *wetting zone*, and a *wetting front*. This profile changes as a function of time as shown in Figure 7.4.4.

The *infiltration rate f* is the rate at which water enters the soil surface, expressed in in/hr or cm/hr. The *potential infiltrate rate* is the rate when water is ponded on the soil surface, so if no ponding occurs the actual rate is less than the potential rate. Most infiltration equations describe a potential infiltration rate. *Cumulative infiltration F* is the accumulated depth of water infiltrated, defined mathematically as

$$F(t) = \int_0^t f(\tau)d\tau \tag{7.4.15}$$

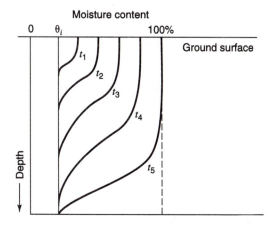

Figure 7.4.4 Moisture profile as a function of time for water added to the soil surface.

and the infiltration rate is the time derivative of the cumulative infiltration given as

$$f(t) = \frac{dF(t)}{dt} \tag{7.4.16}$$

Figure 7.4.5 illustrates a rainfall hyetograph with the infiltration rate and cumulative infiltration curves. (See Section 8.2 for further details on the rainfall hyetograph.)

Green and Ampt (1911) proposed the simplified picture of infiltration shown in Figure 7.4.6. The *wetting front* is a sharp boundary dividing soil with moisture content θ_i below from saturated soil with moisture content η above. The wetting front has penetrated to a depth L in time t since infiltration began. Water is ponded to a small depth h_0 on the soil surface.

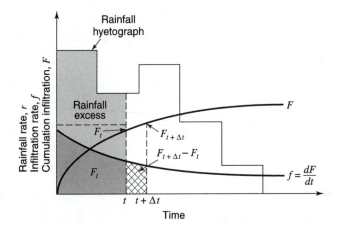

Figure 7.4.5 Rainfall infiltration rate and cumulative infiltration. The rainfall hyetograph illustrates the rainfall pattern as a function of time. The cumulative infiltration at time t is F_t or $F(t)$ and at time $t + \Delta t$ is $F_{t + \Delta t}$ or $F(t + \Delta t)$ is computed using equation 7.4.15. The increase in cumulative infiltration from time t to $t + \Delta t$ is $F_{t + \Delta t} - F_t$ or $F(t + \Delta t) - F(t)$ as shown in the figure. Rainfall excess is defined in Chapter 8 as that rainfall that is neither retained on the land surface nor infiltrated into the soil.

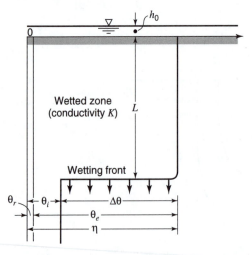

Figure 7.4.6 Variables in the Green–Ampt infiltration model. The vertical axis is the distance from the soil surface, the horizontal axis is the moisture content of the soil (from Chow et al. (1988)).

Consider a vertical column of soil of unit horizontal cross-sectional area (Figure 7.4.7) with the control volume defined around the wet soil between the surface and depth L. For a soil initially of moisture content θ_i throughout its entire depth, the *moisture content* will increase from θ_i to η (the porosity) as the wetting front passes. The moisture content θ is the ratio of the volume of water to the total volume within the control surface. $L(\eta - \theta_i)$ is then the increase in the water stored within the control volume as a result of infiltration, through a unit cross-section. By definition this quantity is equal to F, the cumulative depth of water infiltrated into the soil, so that

$$F(t) \;=\; L(\eta - \theta_i) = L\Delta\theta \tag{7.4.17}$$

where $\Delta\theta = (\eta - \theta_i)$.

Darcy's law may be expressed (using equation 7.4.7) as

$$q = K\frac{\partial h}{\partial z} = -K\frac{\Delta h}{\Delta z} \tag{7.4.18}$$

The Darcy flux q is constant throughout the depth and is equal to $-f$, because q is positive upward while f is positive downward. If points 1 and 2 are located respectively at the ground surface and just on the dry side of the wetting front, then equation (7.4.18) can be approximated by

$$f = K\left[\frac{h_1 - h_2}{z_1 - z_2}\right] \tag{7.4.19}$$

The head h_1 at the surface is equal to the ponded depth h_0. The head h_2, in the dry soil below the wetting front, equals $-\psi - L$. Darcy's law for this system is written as

$$f = K\left[\frac{h_0 - (-\psi - L)}{L}\right] \tag{7.4.20a}$$

and if the ponded depth h_0 is negligible compared to ψ and L,

$$f \approx K\left[\frac{\psi + L}{L}\right] \tag{7.4.20b}$$

This assumption ($h_0 = 0$) is usually appropriate for surface water hydrology problems because it is assumed that ponded water becomes surface runoff.

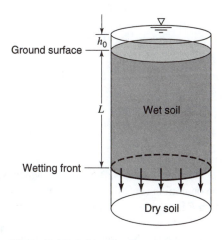

Figure 7.4.7 Infiltration into a column of soil of unit cross-sectional area for the Green–Ampt model (from Chow et al. (1988)).

From equation (7.4.17), the wetting front depth is $L = F/\Delta\theta$, and assuming $h_0 = 0$, substitution into equation (7.4.20) gives

$$f = K\left[\frac{\psi\Delta\theta + F}{F}\right] \tag{7.4.21}$$

Since $f = dF/dt$, equation (7.4.21) can be expressed as a differential equation in the one unknown F as

$$\frac{dF}{dt} = K\left[\frac{\psi\Delta\theta + F}{F}\right]$$

To solve for F, cross-multiply to obtain

$$\left[\frac{F}{F + \psi\Delta\theta}\right]dF = Kdt$$

Then divide the left-hand side into two parts

$$\left[\frac{F + \psi\Delta\theta}{F + \psi\Delta\theta} - \frac{\psi\Delta\theta}{F + \psi\Delta\theta}\right]dF = Kdt$$

and integrate

$$\int_0^{F(t)}\left[1 - \frac{\psi\Delta\theta}{F + \psi\Delta\theta}\right]dF = \int_0^t Kdt$$

to obtain

$$F(t) - \psi\Delta\theta\{\ln[F(t) + \psi\Delta\theta] - \ln(\psi\Delta\theta)\} = Kt \tag{7.4.22a}$$

or

$$F(t) - \psi\Delta\theta\ln\left(1 + \frac{F(t)}{\psi\Delta\theta}\right) = Kt \tag{7.4.22b}$$

Equation (7.4.22) is the *Green–Ampt equation* for cumulative infiltration. Once F is computed using equation (7.4.22), the infiltration rate f can be obtained from equation (7.4.21) or

$$f(t) = K\left[\frac{\psi\Delta\theta}{F(t)} + 1\right] \tag{7.4.23}$$

When the ponded depth h_0 is not negligible, the value of $\psi - h_0$ is substituted for ψ in equations (7.4.22) and (7.4.23).

Equation (7.4.22) is a nonlinear equation in F that can be solved by the method of successive substitution by rearranging (7.4.22)

$$F(t) = Kt + \psi\Delta\theta\ln\left(1 + \frac{F(t)}{\psi\Delta\theta}\right) \tag{7.4.24}$$

Given K, t, ψ, and $\Delta\theta$, a trial value F is substituted on the right-hand side (a good trial value is $F = Kt$) and a new value of F calculated on the left-hand side, which is substituted as a trial value on the right-hand side, and so on until the calculated values of F converge to a constant. The final value of cumulative infiltration F is substituted into (7.4.23) to determine the corresponding infiltration rate f.

Equation (7.4.22) can also be solved by Newton's method, which is more complicated than the method of successive substitution but converges in fewer iterations. Newton's iteration method is explained in Appendix A. Referring to equation (7.4.24), application of the Green–Ampt model requires estimates of the hydraulic conductivity K, the wetting front soil suction head ψ (see Table 7.4.1), and $\Delta\theta$.

Table 7.4.1   Green–Ampt Infiltration Parameters for Various Soil Classes*

Soil class	Porosity η	Effective Porosity θ_e	Wetting Front Soil Suction Head ψ (cm)	Hydraulic Conductivity K (cm/h)
Sand	0.437	0.417	4.95	11.78
	(0.374–0.500)	(0.354–0.480)	(0.97–25.36)	
Loamy sand	0.437	0.401	6.13	2.99
	(0.363–0.506)	(0.329–0.473)	(1.35–27.94)	
Sandy loam	0.453	0.412	11.01	1.09
	(0.351–0.555)	(0.283–0.541)	(2.67–45.47)	
Loam	0.463	0.434	8.89	0.34
	(0.375–0.551)	(0.334–0.534)	(1.33–59.38)	
Silt loam	0.501	0.486	16.68	0.65
	(0.420–0.582)	(0.394–0.578)	(2.92–95.39)	
Sandy clay loam	0.398	0.330	21.85	0.15
	(0.332–0.464)	(0.235–0.425)	(4.42–108.0)	
Clay loam	0.464	0.309	20.88	0.10
	(0.409–0.519)	(0.279–0.501)	(4.79–91.10)	
Silty clay loam	0.471	0.432	27.30	0.10
	(0.418–0.524)	(0.347–0.517)	(5.67–131.50)	
Sandy clay	0.430	0.321	23.90	0.06
	(0.370–0.490)	(0.207–0.435)	(4.08–140.2)	
Silty clay	0.479	0.423	29.22	0.05
	(0.425–0.533)	(0.334–0.512)	(6.13–139.4)	
Clay	0.475	0.385	31.63	0.03
	(0.427–0.523)	(0.269–0.501)	(6.39–156.5)	

*The numbers in parentheses below each parameter are one standard deviation around the parameter value given.

Source: Rawls, Brakensiek, and Miller (1983).

The *residual moisture content* of the soil, denoted by θ_r, is the moisture content after it has been thoroughly drained. The *effective saturation* is the ratio of the available moisture $(\theta - \theta_r)$ to the maximum possible available moisture content $(\eta - \theta_r)$, given as

$$s_e = \frac{\theta - \theta_r}{\eta - \theta_r} \tag{7.4.25}$$

where $\eta - \theta_r$ is called the *effective porosity* θ_e.

The effective saturation has the range $0 \le s_e \le 1.0$, provided $\theta_r \le \theta \le \eta$. For the initial condition, when $\theta = \theta_i$, cross-multiplying equation (7.4.25) gives $\theta_i - \theta_r = s_e\theta_e$, and the change in the moisture content when the wetting front passes is

$$\Delta\theta = \eta - \theta_i = \eta - (s_e\theta_e + \theta_r)$$

$$\Delta\theta = (1 - s_e)\theta_e \tag{7.4.26}$$

A logarithmic relationship between the effective saturation s_e and the soil suction head ψ can be expressed by the *Brooks–Corey equation* (Brooks and Corey, 1964):

$$s_e = \left[\frac{\psi_b}{\psi}\right]^\lambda \tag{7.4.27}$$

in which ψ_b and λ are constants obtained by draining a soil in stages, measuring the values of s_e and ψ at each stage, and fitting equation (7.4.27) to the resulting data.

Brakensiek et al. (1981) presented a method for determining the Green–Ampt parameters using the Brooks–Corey equation. Rawls et al. (1983) used this method to analyze approximately 5000 soil horizons across the United States and determined average values of the Green–Ampt parameters η, θ_e, ψ, and K for different soil classes, as listed in Table 7.4.1. As the soil becomes finer, moving from sand to clay, the wetting front soil suction head increases while the hydraulic conductivity decreases. Table 7.4.1 also lists typical ranges for η, θ_e, and ψ. The ranges are not large for η and θ_e, but ψ can vary over a wide range for a given soil. K varies along with ψ, so the values given in Table 7.4.1 for both ψ and K should be considered typical values that may show a considerable degree of variability in application (American Society of Agricultural Engineers, 1983; Devaurs and Gifford, 1986).

EXAMPLE 7.4.3

Use the Green–Ampt method to evaluate the infiltration rate and cumulative infiltration depth for a silty clay soil at 0.1-hr increments up to 6 hrs from the beginning of infiltration. Assume an initial effective saturation of 20 percent and continuous ponding.

SOLUTION

From Table 7.4.1, for a silty clay soil, $\theta_e = 0.423$, $\psi = 29.22$ cm, and $K = 0.05$ cm/hr. The initial effective saturation is $s_e = 0.2$, so $\Delta\theta = (1 - s_e)\theta_e = (1 - 0.20)0.423 = 0.338$, and $\psi\Delta\theta = 29.22 \times 0.338 = 9.89$ cm. Assuming continuous ponding, the cumulative infiltration F is found by successive substitution in equation (7.4.24):

$$F = Kt + \psi\Delta\theta \ln[1 + F/(\psi\Delta\theta)] = 0.05t + 9.89 \ln[1 + F/9.89]$$

For example, at time $t = 0.1$ hr, the cumulative infiltration converges to a final value $F = 0.29$ cm. The infiltration rate f is then computed using equation (7.4.23):

$$f = K (1 + \psi\Delta\theta/F) = 0.05 (1 + 9.89/F)$$

As an example, at time $t = 0.1$ hr, $f = 0.05(1 + 9.89/0.29) = 1.78$ cm/hr. The infiltration rate and the cumulative infiltration are computed in the same manner between 0 and 6 hrs at 0.1-hr intervals; the results are listed in Table 7.4.2.

Table 7.4.2 Infiltration Computations Using the Green–Ampt Method

Time t (hr)	0.0	0.1	0.2	0.3	0.4	0.5	1.0	1.5	2.0	2.5	3.0	3.5	4.0	4.5	5.0	5.5	6.0
Infiltration rate f (cm/hr)	∞	1.78	1.20	0.97	0.84	0.75	0.54	0.44	0.39	0.35	0.32	0.30	0.28	0.27	0.26	0.25	0.24
Infiltration depth F (cm)	0.00	0.29	0.43	0.54	0.63	0.71	1.02	1.26	1.47	1.65	1.82	1.97	2.12	2.26	2.39	2.51	2.64

EXAMPLE 7.4.4

Ponding time t_p is the elapsed time between the time rainfall begins and the time water begins to pond on the soil surface. Develop an equation for ponding time under a constant rainfall intensity i, using the Green–Ampt infiltration equation (see Figure 7.4.8).

SOLUTION

The infiltration rate f and the cumulative infiltration F are related in equation (7.4.22). The cumulative infiltration at ponding time t_p is $F_p = it_p$, in which i is the constant rainfall intensity (see Figure 7.4.8). Substituting $F_p = it_p$ and the infiltration rate $f = i$ into equation (7.4.23) yields

$$i = K\left(\frac{\psi\Delta\theta}{it_p} + 1\right)$$

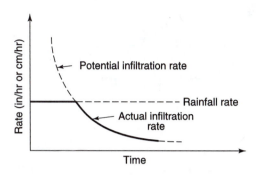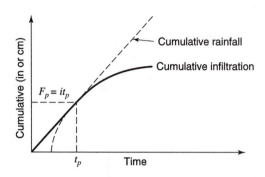

Figure 7.4.8 Ponding time. This figure illustrates the concept of ponding time for a constant intensity rainfall. Ponding time is the elapsed time between the time rainfall begins and the time water begins to pond on the soil surface.

and solving, we get

$$t_p = \frac{K\psi\Delta\theta}{i(i-K)}$$

which is the ponding time for a constant rainfall intensity.

7.4.3   Other Infiltration Methods

The simplest accounting of abstraction is the Φ-*index* (refer to Figure 7.4.9), which is a constant rate of abstraction (in/h or cm/h). Other cumulative infiltration and infiltration rate equations include Horton's and the SCS method. Horton's equation (Horton, 1933) is an empirical relation that assumes infiltration begins at some rate f_o and exponentially decreases until it reaches a constant rate f_c (refer to Figure 7.4.9). The infiltration capacity is expressed as

$$f_t = f_c + (f_o - f_c)e^{-kt} \tag{7.4.28}$$

and the cumulative infiltration capacity is expressed as

$$F_t = f_c t + \frac{(f_0 - f_c)}{k}(1 - e^{-kt}) \tag{7.4.29}$$

where k is a decay constant. Many other empirical infiltration equations have been developed that can be found in the various hydrology texts. There are many other infiltration methods, including the SCS method described in Sections 8.6–8.8. In summary, the Green–Ampt and the SCS are both used in the U.S. Army Corps of Engineers HEC-1 model, and both are used widely in the United States.

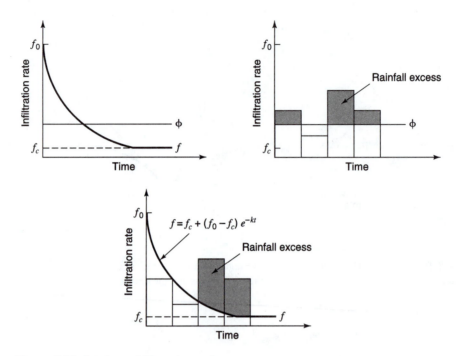

Figure 7.4.9 Φ-index and Horton's equation.

PROBLEMS

7.2.1 Determine the 25-year return period rainfall depth for 30 min duration in Chicago, Illinois.

7.2.2 Determine the 2-, 10-, 25-, and 100-year precipitation depths for a 15-min duration storm in Memphis, Tennessee.

7.2.3 Determine the 2- and 25-year intensity-duration-frequency curves for Memphis, Tennessee.

7.2.4 Determine the 10- and 50-year intensity-duration-frequency curves for Chicago, Illinois.

7.2.5 Determine the 2-, 5-, 10-, 25-, 50-, and 100-year depths for a 1-hr duration storm in Phoenix, Arizona. Repeat for a 6-hr duration storm.

7.2.6 Develop the Types I, IA, II, and III 24-hour storms for a 24-hour rainfall of 20 in. Plot and compare these storms.

7.2.7 Develop the SCS 6-hr storm for a 6-hr rainfall of 12 in.

7.2.8 Determine the design rainfall intensities (mm/hr) for a 25-year return period, 60 minute duration storm using equation (7.2.5) with $c = 12.1$, $m = 0.25$, $e = 0.75$, and $f = 0.125$.

7.2.9 What is the all-season PMP (in) for 200 mi near Chicago, Illinois?

7.3.1 Solve example 7.3.1 for an average net radiation of 92.5 W/m^2. Compare the resulting evaporation rate with that in example 7.3.1.

7.3.2 Solve example 7.3.2 for a roughness height $z_0 = 0.04$ cm. Compare the resulting evaporation rate with that in example 7.3.2.

7.3.3 Solve example 7.3.3 for an average net radiation of 92.5 W/m^2. Compare the resulting evaporation rate with that in example 7.3.3.

7.3.4 Solve example 7.3.4 for an average net radiation of 92.5 W/m^2. Compare the resulting evaporation rate with that in example 7.3.4.

7.3.5 At a certain location during the winter, the average air temperature is 10°C and the net radiation is 40 W/m^2, and during the summer the net radiation is 200 W/m^2 and the temperature is 25°C. Compute the evaporation rates using the Priestley–Taylor method.

7.4.1 Determine the infiltration rate and cumulative infiltration curves (0 to 5 h) at 1-hr increments for a clay loam soil. Assume an initial effective saturation of 40 percent and continuous ponding.

7.4.2 Rework problem 7.4.1 using an initial effective saturation of 20 percent.

7.4.3 Rework example 7.4.3 for a sandy loam soil.

7.4.4 Compute the ponding time and cumulative infiltration at ponding for a sandy clay loam soil with a 30-percent initial effective saturation, subject to a rainfall intensity of 2 cm/h.

7.4.5 Rework problem 7.4.4 for a silty clay soil.

7.4.6 Determine the cumulative infiltration and the infiltration rate on a sandy clay loam after 1 hr of rainfall at 2 cm/hr if the initial effective saturation is 25 percent. Assume the ponding depth is negligible in the calculations.

7.4.7 Rework problem 7.4.6 assuming that any ponded water remains stationary over the soil so that the ponded depth must be accounted for in the calculations.

7.4.8 Derive the equation for cumulative infiltration using Horton's equation.

7.4.9 Use the Green–Ampt method to compute the infiltration rate and cumulative infiltration for a silty clay soil ($\eta = 0.479$, $\psi = 29.22$ cm, $K = 0.05$ cm/hr) at 0.25-hr increments up to 4 hr from the beginning of infiltration. Assume an initial effective saturation of 30 percent and continuous ponding.

7.4.10 The parameters for Horton's equation are $f_0 = 3.0$ in/h, $f_c = 0.5$ in/h, and $K = 4.0$ h^{-1}. Determine the infiltration rate and cumulative infiltration at 0.25-hr increments up to 4 hr from the beginning of infiltration. Assume continuous ponding.

7.4.11 Derive an equation for ponding time using Horton's equation.

7.4.12 Compute the ponding time and cumulative infiltration at ponding for a sandy clay loam soil of 25 percent initial effective saturation for a rainfall intensity of (a) 2 cm/h (b) 3 cm/h and (c) 5 cm/h.

7.4.13 Rework problem 7.4.12 considering a silt loam soil.

REFERENCES

American Society of Agricultural Engineers, "Advances in Infiltration," in *Proc. National Conf. on Advances in Infiltration*, Chicago, IL, ASAE Publication 11–83, St. Joseph, MI. 1983.

Bedient, P. B., and W. C. Huber, *Hydrology and Floodplain Analysis*, second edition, Addison-Wesley, Reading, MA, 1992.

Bhowmik, N., "River Basin Management: An Integrated Approach," *Water International*, International Water Resources Association, vol. 23, no. 2, pp. 84–90, June 1998.

Brakensiek, D. L., R. L. Engleman, and W. J. Rawls, "Variation Within Texture Classes of Soil Water Parameters," *Trans. Am Soc. Agric. Eng.*, vol. 24, no. 2, pp. 335–339, 1981.

Bras, R. L., *Hydrology: An Introduction to Hydrologic Science*, Addison-Wesley, Reading, MA, 1990.

Brooks, R. H., and A. T. Corey, "Hydraulic Properties of Porous Media," *Hydrology Papers*, no. 3, Colorado State University, Fort Collins, CO, 1964.

Chow, V. T., *Handbook of Applied Hydrology*, McGraw Hill, New York, 1964.

Chow, V. T., D. R. Maidment, and L. W. Mays, *Applied Hydrology*, McGraw-Hill, New York, 1988.

Demissee, M., and L. Keefer, "Watershed Approach for the Protection of Drinking Water Supplies in Central Illinois," *Water International*, IWRA, vol. 23, no. 4, pp. 272–277, 1998.

Devaurs, M., and G. F. Gifford, "Applicability of the Green and Ampt Infiltration Equation to Rangelands," *Water Resource Bulletin*, vol. 22, no.1, pp. 19–27, 1986.

Feynman, R. P., *Six Easy Pieces*, Perseus Books, Reading, MA, 1995.

Feynman, R. P., R. B. Leighton, and M. Sands, *The Feynman Lecture Notes on Physics*, vol. I, Addison-Wesley, Reading, MA, 1963.

Flood Control District of Maricopa County, *Drainage Design Manual for Maricopa County, Arizona*, vol. I, Hydrology, Phoenix, AZ, 1992.

Frederick, R. H., V. A. Meyers, and E. P. Auciello, " Five to 60-Minute Precipitation Frequency for the Eastern and Central United States," NOAA Technical Memo NWS HYDRO-35, National Weather Service, Silver Spring, MD, June 1977.

Green, W. H. and G. A. Ampt, "Studies on Soil Physics. Part I: The Flow of Air and Water Through Soils," *J. Agric. Sci.*, vol. 4, no. 1, pp. 1–24, 1911.

Gupta, R. S., *Hydrology and Hydraulic Systems*, Prentice-Hall, Englewood Cliffs, NJ, 1989.

Hansen, E. M., L. C. Schreiner, and J. F. Miller, "Application of Probable Maximum Precipitation Estimates—United States East of 105th Meridian," NOAA Hydrometeorological Report No. 52, U.S. National Weather Service, Washington, DC, 1982.

Hershfield, D. M., "Rainfall Frequency Atlas of the United States for Durations from 30 Minutes to 24 Hours and Return Periods from 1 to 100 Years," Tech. Paper 40, U.S. Dept. of Comm., Weather Bureau, Washington, DC, 1961.

Horton, R. E., "The Role of Infiltration in the Hydrologic Cycle," *Trans. Am. Geophysical Union*, vol. 14, pp. 446–460, 1933.

Huff, F. A., Time Distribution of Rainfall in Heavy Storms, *Water Resources Research*, vol. 3, no. 4, pp. 1007–1019, 1967.

Maidment, D. R. (editor-in-chief), *Handbook of Hydrology*, McGraw-Hill, New York, 1993.

Marsh, W. M., *Earthscape: A Physical Geography*, John Wiley and Sons, New York, 1987.

Masch, F. D., *Hydrology*, Hydraulic Engineering Circular No. 19, FHWA-10-84-15, Federal Highway Administration, U.S. Department of the Interior, McLean, VA, 1984.

Mays, L. W., and Y. K. Tung, *Hydrosystems Engineering and Management*, McGraw-Hill, New York, 1992.

McCuen, R. H., *Hydrologic Analysis and Design*, second editon Prentice-Hall, Englewood Cliffs, NJ, 1998.

Miller, J. F., R. H. Frederick, and R. J. Tracey, *Precipitation Frequency Atlas of the Coterminous Western United States (by States)*, NOAA Atlas 2, 11 vols., National Weather Service, Silver Spring, MD, 1973.

Moore, W. L., E. Cook, R. S. Gooch, and C. F. Nordin, Jr., "The Austin, Texas, Flood of May, 24–25, 1981," National Research Council, Committee on Natural Disasters, Commission on Engineering and Technical Systems, National Academy Press, Washington, DC, 1982.

Parrett, C., N. B. Melcher, and R. W. James, Jr., "Flood Discharges in the Upper Mississippi River Basin, in *Floods in the Upper Mississippi River Basin*," U.S. Geological Survey Circular, 1120-A, U.S. Government Printing Office, Washington, DC, 1993.

Ponce, V. M., *Engineering Hydrology: Principles and Practices*, Prentice-Hall, Englewood Cliffs, NJ, 1989.

Priestley, C. H. B., and R. J. Taylor, "On the Assessment of Surface Heat Flux and Evaporation Using Large-Scale Parameter," *Monthly Weather Review*, vol. 100, pp. 81–92, 1972.

Rawls, W. J., D. L. Brakensiek, and N. Miller, "Green–Ampt Infiltration Parameters from Soils Data," *J. Hydraulic Div., ASCE*, vol. 109, no.1, pp. 62–70, 1983.

Richards, L. A., "Capillary Conduction of Liquids Through Porous Mediums," *Physics*, vol. 1, pp. 318–333, 1931.

Roberson, J. A., J. J. Cassidy, and M. H. Chaudhry, *Hydraulic Engineering*, second edition, John Wiley & Son, 1998.

Schreiner, L. C., and J. T. Reidel, Probable Maximum Precipitation Estimates, United States East of the 105th Meridian, NOAA Hydrometeorological Report no. 51, National Weather Service, Washington, DC, June 1978.

Singh, V. P., *Elementary Hydrology*, Prentice-Hall, Englewood Cliffs, NJ, 1992.

Thornthwaite, C. W., and B. Holzman, "The Determination of Evaporation from Land and Water Surface," *Monthly Weather Review*, vol. 67, pp. 4–11, 1939.

U.S. Army Corps of Engineers, Hydrologic Engineering Center, HEC-1, *Flood Hydrograph Package*, User's Manual, Davis, CA, 1990.

U.S. Department of Agriculture Soil Conservation Service, *A Method for Estimating Volume and Rate of Runoff in Small Watersheds*, Tech. Paper 149, Washington, DC, 1973.

U.S. Department of Agriculture Soil Conservation Service, *Urban Hydrology for Small Watersheds*, Tech. Release no. 55, Washington, DC, 1986.

U.S. Department of Commerce, *Probable Maximum Precipitation Estimates, Colorado River and Great Basin Drainages*, Hydrometeorological Report no. 49, NOAA, National Weather Service, Silver Spring, MD, 1977.

U.S. Department of Commerce, *Seasonal Variation of 10-Square-Mile Probable Maximum Precipitation Estimates, United States East of the 105th Meridian*, Hydrometeorological Report no. 53, NOAA, National Weather Service, Silver Spring, MD, April 1980.

U.S. Environmental Data Services, *Climate Atlas of the U.S.*, U.S. Government Printing Office, Washington, DC, pp. 43–44, 1968.

U.S. National Academy of Sciences, *Understanding Climate Change*, National Academy Press, Washington, DC, 1975.

U.S. National Academy of Sciences, *Safety of Existing Dams: Evaluation and Improvement*, National Academy Press, Washington, DC, 1983.

U.S. National Research Council, Committee on Opportunities in the Hydrologic Sciences, Water Science and Technology Board, *Opportunities in the Hydrologic Sciences*, National Academy Press, Washington, DC, 1991.

U.S. National Research Council, Global Change in the Geosphere-Biosphere, National Academy Press, Washington, DC, 1986.

U.S. Weather Bureau, *Generalized Estimates of Probable Maximum Precipitation West of the 105th Meridian*, Tech. Paper no. 38, Washington, DC, 1960.

U.S. Weather Bureau, *Generalized Estimates of Probable Maximum Precipitation and Rainfall-Frequency Data for Puerto Rico and Virgin Islands*, Tech. Paper no. 42, Washington, DC, 1961.

U.S. Weather Bureau, *Probable Maximum Precipitation in the Hawaiian Islands*, Hydrometeorological Report no. 39, Washington, DC, 1963a.

U.S. Weather Bureau, *Probable Maximum Precipitation Rainfall-Frequency Data for Alaska*, Tech. Report no. 47, Washington, DC, 1963b.

U.S. Weather Bureau, *Two- to Ten-Day Precipitation for Return Periods of 2 to 100 Years in the Contiguous United States*, Tech. Paper 49, Washington, DC, 1964.

U.S. Weather Bureau, *Probable Maximum Precipitation, Northwest States*, Hydrometeorological Report no. 43, Washington, DC, 1966.

U.S. Weather Bureau, *Interim Report-Probable Maximum Precipitation in California*, Hydrometeorological Report no. 36, Washington, DC, 1961, with revisions in 1969.

Viessman, W. Jr., and G. L. Lewis, *Introduction to Hydrology*, fourth edition, Harper and Row, New York, 1996.

Wanielista, M., R. Kersten, and R. Eaglin, *Hydrology: Water Quantity and Quality Control*, John Wiley & Sons, New York, 1997.

Chapter **8**

Surface Runoff

8.1 DRAINAGE BASINS

As defined in Chapter 7, *drainage basins*, *catchments*, and *watersheds* are three synonymous terms that refer to the topographic area that collects and discharges surface streamflow through one outlet or mouth. The study of topographic maps from various physiographic regions reveals that there are several different types of drainage patterns (Figure 8.1.1). *Dendritic patterns* occur where rock and weathered mantle offer uniform resistance to erosion. Tributaries branch and erode headward in a random fashion, which results in slopes with no predominant direction or orientation. *Rectangular patterns* occur in faulted areas where streams follow a more easily eroded fractured rock in fault lines. *Trellis patterns* occur where rocks being dissected are of unequal resistance so that the extension and daunting of tributaries is most rapid on least resistant areas.

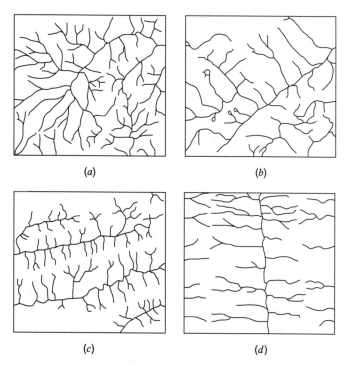

(a) (b)

(c) (d)

Figure 8.1.1 Common drainage patterns (*a*) Dendritic; (*b*) Rectangular. (*c*) Trellis on folded terrain. (*d*) Trellis on mature, dissected coastal plain (from Hewlett and Nutter (1969)).

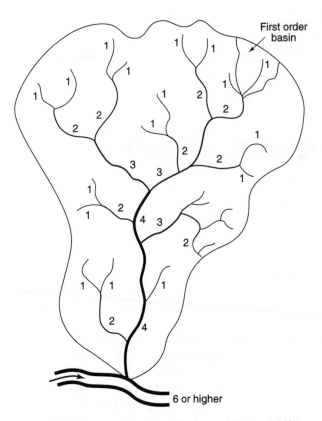

Figure 8.1.2 Stream orders (from Hewlett and Nutter (1969)).

Streams can also be classified within a basin by systematically ordering the network of branches. Horton (1945) originated the quantitative study of stream networks by developing an ordering system and laws relating to the number and length of streams of different order (see Chow et al., 1988). Strahler (1964) slightly modified Horton's stream ordering to that shown in Figure 8.1.2. Essentially, each non-branching channel segment is a *first-order stream*. Streams, which receives only first-order segments, are *second order*, and so on. When a channel of lower order joins a channel of higher order, the channel downstream retains the higher of the two orders. The order of a drainage basin is the order of the stream draining its outlet.

Streams can also be classified by the period of time during which flow occurs. *Perennial streams* have a continuous flow regime typical of a well-defined channel in a humid climate. *Intermittent streams* generally have flow occurring only during the wet season (50 percent of the time or less). *Ephemeral streams* generally have flow occurring during, and for short periods after storms. These streams are typical of climates without very well-defined streams.

The *stream flow hydrograph* or *discharge hydrograph* is the relationship of flow rate (discharge) and time at a particular location on a stream (see Figure 8.1.3*a*). The *hydrograph* is "an integral expression of the physiographic and climatic characteristics that govern the relation between rainfall and runoff of a particular drainage basin" (Chow, 1964). Figures 8.1.3*b* and *c* illustrate the rising and falling of the water table in response to rainfall. Figures 8.1.3*d* and *e* illustrate that the flowing stream channel network expands and contracts in response to rainfall.

The spatial and temporal variations of rainfall and the concurrent variation of the abstraction processes define the runoff characteristics from a given storm. When the local abstractions have been accomplished for a small area of a watershed, water begins to flow overland as *overland flow* and eventually into a drainage channel (in a gulley or stream valley). When this occurs, the hydraulics of the natural drainage channels have a large influence on the runoff characteristics

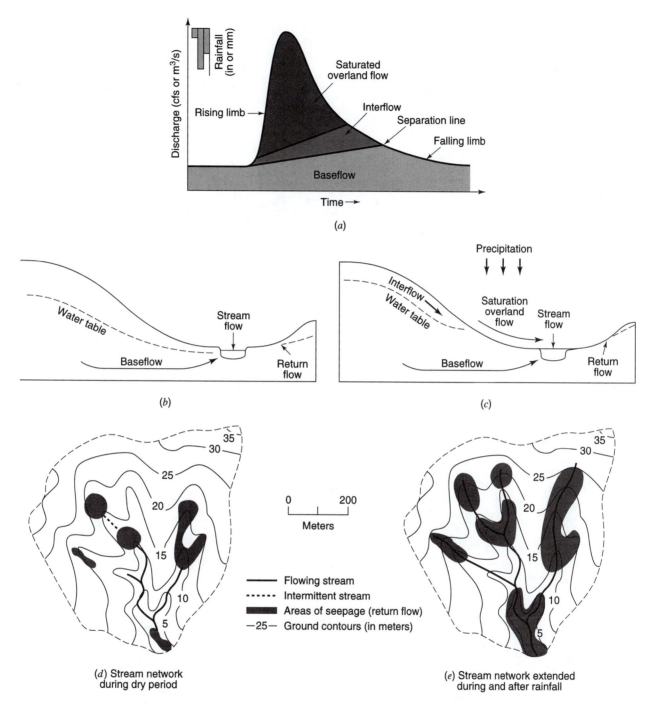

Figure 8.1.3 (*a*) Separation of sources of streamflow on an idealized hydrograph; (*b*) Sources of streamflow on a hillslope profile during a dry period. (*c*) During a rainfall event. (*d*) Stream network during dry period. (*e*) Stream network extended during and after rainfall (from Mosley and McKerchan (1993)).

from the watershed. Some of the factors that determine the hydraulic character of the natural drainage system include: (a) drainage area, (b) slope, (c) hydraulic roughness, (d) natural and channel storage, (e) stream length, (f) channel density, (g) antecedent moisture condition, and (h) other factors such as vegetation, channel modifications, etc. The individual effects of each of these

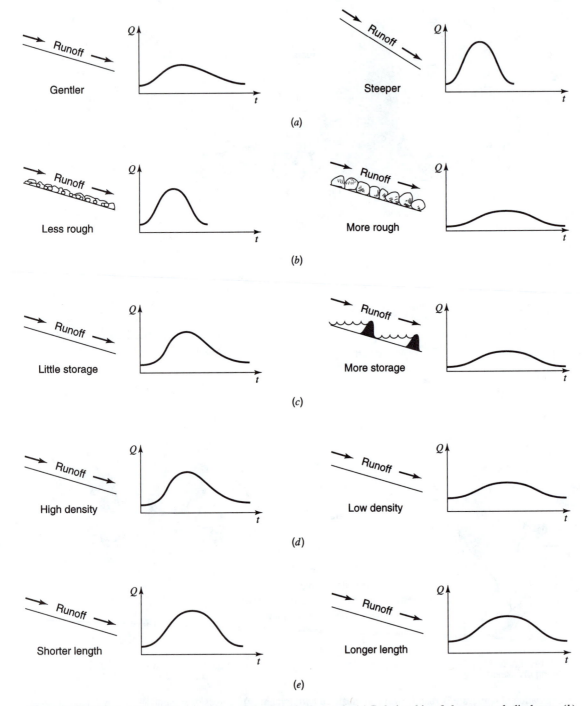

Figure 8.1.4 Effects of basin characteristics on the flood hydrograph. (*a*) Relationship of slope to peak discharge; (*b*) Relationship of hydraulic roughness to runoff. (*c*) Relationship of storage to runoff. (*d*) Relationship of drainage density to runoff. (*e*) Relationship of channel length to runoff (from Masch (1984)).

factors are difficult, and in many cases impossible, to quantify. Figure 8.1.4 illustrates the effects of some of the drainage basin characteristics on the surface runoff (discharge hydrographs) and Figure 8.1.5 illustrates the effects of storm shape, size, and movement on surface runoff.

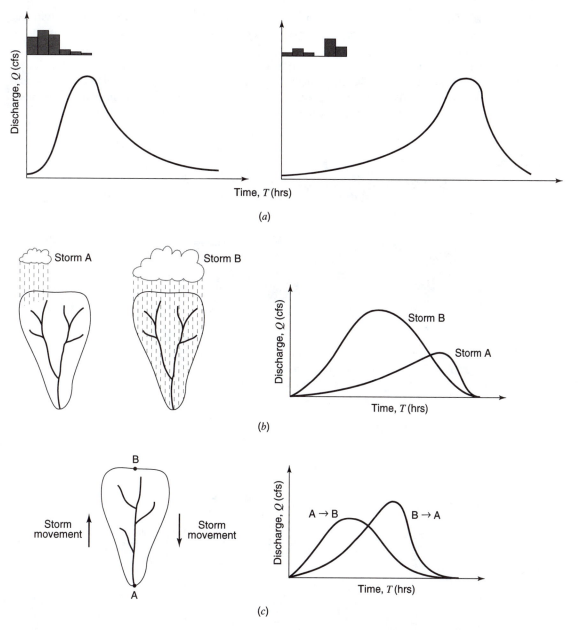

Figure 8.1.5 Effects of storm shape, size, and movement on surface runoff. (*a*) Effect of time variation of rainfall intensity on the surface runoff; (*b*) Effect of storm size on surface runoff. (*c*) Effect of storm movement on surface runoff (from Masch 1984)).

8.2 HYDROLOGIC LOSSES AND RAINFALL EXCESS

Rainfall excess, or *effective rainfall*, is that rainfall that is neither retained on the land surface nor infiltrated into the soil. After flowing across the watershed surface, rainfall excess becomes direct runoff at the watershed outlet. The graph of rainfall excess versus time is the rainfall excess hyetograph. As shown in Figure 8.2.1, the difference between the observed total rainfall hyetograph and the rainfall excess hyetograph is the *abstractions*, or *losses*. Losses are primarily water absorbed by infiltration with some allowance for interception and surface storage. The relationships of rainfall, infiltration rate, and cumulative infiltration are shown in Figure 8.2.2. Figure 8.2.2 illustrates the relationships for rainfall and runoff data of an actual storm that can be obtained from data recorded by the U.S. Geological Survey. Using the rainfall data, rainfall hyetographs can be computed.

The objective of many hydrologic design and analysis problems is to determine the surface runoff from a watershed due to a particular storm. This process is commonly referred to as *rainfall-runoff analysis*. The processes (steps) are illustrated in Figure 8.2.3 to determine the *storm runoff hydrographs* (or streamflow or discharge hydrograph) using the unit hydrograph approach.

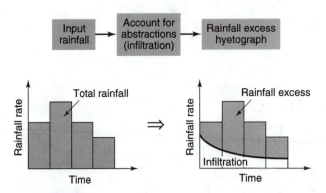

Figure 8.2.1 Concept of rainfall excess. The difference between the total rainfall hyetograph on the left and the total rainfall excess hyetograph on the right is the abstraction (infiltration).

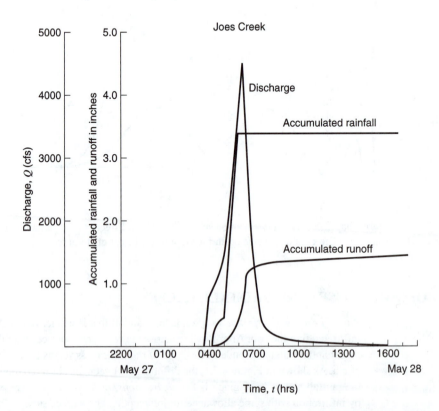

Figure 8.2.2 Precipitation and runoff data for Joes Creek, storm of May 27–28, 1978 (from Masch (1984)).

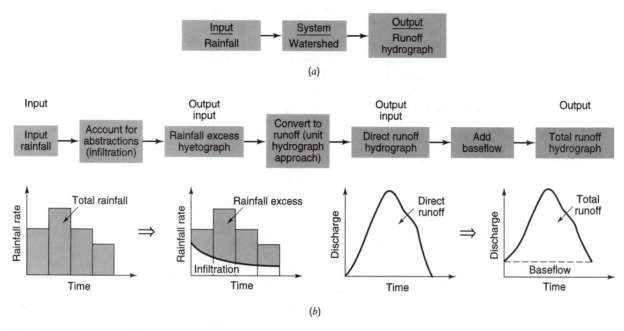

Figure 8.2.3 Storm runoff hydrographs. (*a*) Rainfall-runoff modeling; (*b*) Steps to define storm runoff.

8.3   RAINFALL-RUNOFF ANALYSIS USING UNIT HYDROGRAPH APPROACH

The objective of *rainfall-runoff analysis* is to develop the runoff hydrograph as illustrated in Figure 8.2.3*a*, where the system is a watershed or river catchment, the input is the rainfall hyetograph, and the output is the runoff or discharge hydrograph. Figure 8.2.3*b* defines the processes (steps) to determine the runoff hydrograph from the rainfall input using the *unit hydrograph approach*.

A *unit hydrograph* is the direct runoff hydrograph resulting from 1 in (or 1 cm in SI units) of excess rainfall generated uniformly over a drainage area at a constant rate for an effective duration. The unit hydrograph is a simple linear model that can be used to derive the hydrograph resulting from any amount of excess rainfall. The following basic assumptions are inherent in the unit hydrograph approach:

1. The excess rainfall has a constant intensity within the effective duration.
2. The excess rainfall is uniformly distributed throughout the entire drainage area.
3. The base time of the direct runoff hydrograph (i.e., the duration of direct runoff) resulting from an excess rainfall of given duration is constant.
4. The ordinates of all direct runoff hydrographs of a common base time are directly proportional to the total amount of direct runoff represented by each hydrograph.
5. For a given watershed, the hydrograph resulting from a given excess rainfall reflects the unchanging characteristics of the watershed.

The following *discrete convolution equation* is used to compute direct runoff hydrograph ordinates Q_n, given the rainfall excess values P_m and given the unit hydrograph ordinates U_{n-m+1} (Chow et al., 1988):

$$Q_n = \sum_{m=1}^{n \le M} P_m U_{n-m+1} \qquad \text{for } n = 1, 2, \ldots, N \qquad (8.3.1)$$

where n represents the direct runoff hydrograph time interval and m represents the precipitation time interval ($m = 1, ..., n$).

The reverse process, called *deconvolution*, is used to derive a unit hydrograph given data on P_m and Q_n. Suppose that there are M pulses of excess rainfall and N pulses of direct runoff in the storm considered; then N equations can be written for Q_n, $n = 1, 2 ..., N$, in terms of $N - M + 1$ unknown values of the unit hydrograph, as shown in Table 8.3.1. Figure 8.3.1 diagramatically illustrates the calculation and the runoff contribution by each rainfall input pulse.

Table 8.3.1   The Set of Equations for Discrete Time Convolution

$$
\begin{aligned}
Q_1 &= P_1 U_1 \\
Q_2 &= P_2 U_1 + P_1 U_2 \\
Q_3 &= P_3 U_1 + P_2 U_2 + P_1 U_3 \\
&\cdots \\
Q_M &= P_M U_1 + P_{M-1} U_2 + \ldots + P_1 U_M \\
Q_{M+1} &= \quad 0 + P_M U_2 \; + \ldots + P_2 U_M + P_1 U_{M+1} \\
&\cdots \\
Q_{N-1} &= \quad 0 + \quad 0 + \ldots + \quad 0 + \quad 0 + \ldots + P_M U_{N-M} + P_{M-1} U_{N-M+1} \\
Q_N &= \quad 0 + \quad 0 + \ldots + \quad 0 + \quad 0 + \ldots + \quad 0 \quad + U_{N-M+1}
\end{aligned}
$$

Once the unit hydrograph has been determined, it may be applied to find the direct runoff and streamflow hydrographs for given storm inputs. When a rainfall hyetograph is selected, the abstractions are subtracted to define the excess rainfall hyetograph. The time interval used in defining the excess rainfall hyetograph ordinates must be the same as that for which the unit hydrograph is specified.

EXAMPLE 8.3.1

The 1-hr unit hydrograph for a watershed is given below. Determine the runoff from this watershed for the storm pattern given. The abstractions have a constant rate of 0.3 in/h.

Time (h)	1	2	3	4	5	6
Precipitation (in)	0.5	1.0	1.5	0.5		
Unit hydrograph (cfs)	10	100	200	150	100	50

SOLUTION

The calculations are shown in Table 8.3.2. The 1-hr unit hydrograph ordinates are listed in column 2 of the table; there are $L = 6$ unit hydrograph ordinates, where $L = N - M + 1$. The number of excess rainfall intervals is $M = 4$. The excess precipitation 1-hr pulses are $P_1 = 0.2$ in, $P_2 = 0.7$ in, $P_3 = 1.2$ in, and $P_4 = 0.2$ in, as shown at the top of the table. For the first time interval $n = 1$, the discharge is computed using equation (8.3.1):

$$Q_1 = P_1 U_1 = 0.2 \times 10 = 2 \text{ cfs}$$

For the second time interval, n = 2,

$$Q_2 = P_1 U_2 + P_2 U_1 = 0.2 \times 100 + 0.7 \times 10 = 27 \text{ cfs}$$

and similarly for the remaining direct runoff hydrograph ordinates. The number of direct runoff ordinates is $N = L + M - 1 = 6 + 4 - 1 = 9$; i.e., there are nine nonzero ordinates, as shown in Table 8.3.2. Column 3 of Table 8.3.2 contains the direct runoff corresponding to the first rainfall pulse, $P_1 = 0.2$ in, and column 4 contains the direct runoff from the second rainfall pulse, $P_2 = 0.2$ in, etc. The direct runoff hydrograph, shown in column 7 of the table, is obtained, from the principle of superposition, by adding the values in columns 3–6.

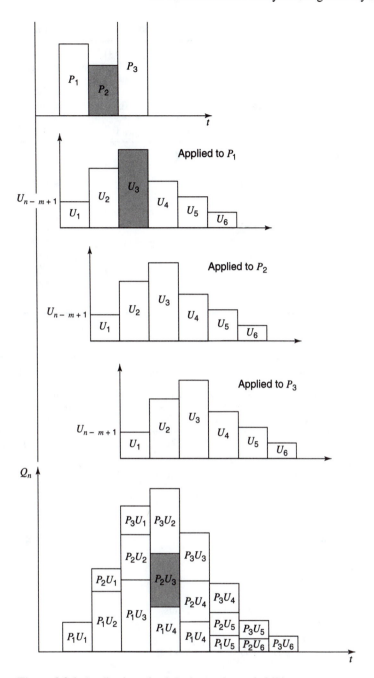

Figure 8.3.1 Application of unit hydrograph to rainfall input.

Table 8.3.2   Calculation of the Direct Runoff Hydrograph

(1)	(2)	(3)	(4)	(5)	(6)	(7)
			Total Precipitation (in)			
	Unit	0.5	1	1.5	0.5	Direct
Time	Hydrograph		Excess Precipitation (in)			Runoff
(hr)	(cfs/m)	0.2	0.7	1.2	0.2	(cfs)
0	0	0	0			0
1	10	2	0	0		2
2	100	20	7	0	0	27
3	200	40	70	12	0	122
4	150	30	140	120	2	292
5	100	20	105	240	20	385
6	50	10	70	180	40	300
7	0	0	35	120	30	185
8			0	60	20	80
9				0	10	10
10					0	0

EXAMPLE 8.3.2

Determine the 1-hr unit hydrograph for a watershed using the precipitation pattern and runoff hydrograph below. The abstractions have a constant rate of 0.3 in/hr, and the baseflow of the stream is 0 cfs.

Time (h)	1	2	3	4	5	6	7	8	9	10
Precipitation (in)	0.5	1.0	1.5	0.5						
Runoff (cfs)	2	27	122	292	385	300	185	80	10	0

SOLUTION

Using the deconvolution process, we get $Q_1 = P_1 U_1$

so that for $P_1 = 0.5 - 0.3 = 0.2$ in and $Q_1 = 2$ cfs,

$U_1 = Q_1/P_1 = 2/0.2 = 10$ cfs.

$Q_2 = P_1 U_2 + P_2 U_1$, so that

$U_2 = (Q_2 - P_2 U_1)/P_1$

where

$P_2 = 1.0 - 0.3 = 0.7$ in and $Q_2 = 27$ cfs.

$U_2 = (27 - 0.7(10))/0.2 = 100$ cfs and

$Q_3 = P_1 U_3 + P_2 U_2 + P_3 U_1$

then

$U_3 = (Q_3 - P_2 U_2 - P_3 U_1)/P_1$, so that

$U_3 = (122 - 0.7(100) - 1.2(10))/0.2 = 200$ cfs.

The rest of the unit hydrograph ordinates can be calculated in a similar manner.

8.4   SYNTHETIC UNIT HYDROGRAPHS

When observed rainfall-runoff data are not available for unit hydrograph determination, a *synthetic unit hydrograph*, can be developed. A unit hydrograph developed from rainfall and streamflow data in a watershed applies only to that watershed and to the point on the storm where the streamflow data were measured. Synthetic unit hydrograph procedures are used to develop unit hydrographs for other locations on the stream in the same watershed or other watersheds that are of similar character.

One of the most commonly used synthetic unit hydrograph procedures is Snyder's synthetic unit hydrograph. This method relates the time from the centroid of the rainfall to the peak of the unit hydrograph to geometrical characteristics of the watershed. To determine the regional parameters C_t and C_p, one can use values of these parameters determined from similar watersheds. C_t can be determined from the relationship for the *basin lag:*

$$t_p = C_1 C_t (L \cdot L_c)^{0.3} \tag{8.4.1}$$

where C_1, L, and L_c are defined in Table 8.4.1. Solving equation (8.4.1) for C_t gives

$$C_t = \frac{t_p}{C_1 (L \cdot L_c)^{0.3}} \tag{8.4.2}$$

Table 8.4.1   Steps to Compute Snyder's Synthetic Unit Hydrograph

Step 0	Measured information from topography map of watershed: • L = main channel length in mi (km) • L_c = length of the main stream channel from outflow point of watershed to a point opposite the centroid of the watershed in mi (km) • A = watershed area in mi^2 (km^2) Regional parameters C_t and C_p determined from similar watersheds.
Step 1	Determine time to peak (t_p) and duration (t_r) of the standard unit hydrograph: $t_p = C_1 C_t (L \cdot L_c)^{0.3}$ (hours) $t_r = t_p/5.5$ (hours) where $C_1 = 1.0 (0.75$ for SI units)
Step 2	Determine the time to peak t_{PR} for the desired duration t_R: $t_{PR} = t_p + 0.25(t_R - t_r)$ (hours)
Step 3	Determine the peak discharge, Q_{PR} in cfs/in ((m^3/s)/cm in SI units) $Q_{PR} = \dfrac{C_2 C_P A}{t_{PR}}$ where $C_2 = 640$ (2.75 for SI units)
Step 4	Determine the width of the unit hydrograph at $0.5 Q_{PR}$ and $0.75 Q_{PR}$. W_{50} is the width at 50% of the peak given as $W_{50} = \dfrac{C_{50}}{(Q_{PR}/A)^{1.08}}$ where $C_{50} = 770$ (2.14 for SI units). W_{75} is the width at 75% of the peak given as $W_{75} = \dfrac{C_{75}}{(Q_{PR}/A)^{1.08}}$ where $C_{75} = 440$ (1.22 for SI units)

Table 8.4.1 Steps to Compute Snyder's Synthetic Unit Hydrograph (*continued*)

Step 5 Determine the base, T_B, such that the unit hydrograph represents 1 in (1 cm in SI units) of direct runoff volume:

$$1 \text{ in} = \left[\left(\frac{W_{50}+T_B}{2}\right)(0.5Q_{PR}) + \left(\frac{W_{75}+W_{50}}{2}\right)(0.25Q_{PR}) + \frac{1}{2}W_{75}(0.25Q_{PR})\right]\left(\text{hr} \times \frac{\text{ft}^3}{\text{sec}}\right)$$

$$\left(\frac{1}{A(\text{mi})^2} \times \frac{1 \text{ mi}^2}{(5280)^2 \text{ ft}^2} \times \frac{12 \text{ in}}{\text{ft}} \times \frac{3600 \text{ sec}}{\text{hr}}\right)$$

Solving for t_b, we get

$$t_b = 2581\frac{A}{Q_{PR}} - 1.5W_{50} - W_{75}$$

for A in mi^2, Q_{PR} in cfs, W_{50} and W_{75} in hours.

Step 6 Define known points of the unit hydrograph. $(T_P = t_{PR} + \frac{t_R}{2})$

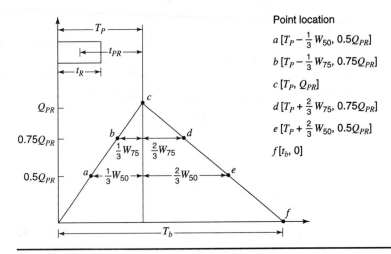

Point location

$a\ [T_P - \frac{1}{3}W_{50}, 0.5Q_{PR}]$

$b\ [T_P - \frac{1}{3}W_{75}, 0.75Q_{PR}]$

$c\ [T_P, Q_{PR}]$

$d\ [T_P + \frac{2}{3}W_{75}, 0.75Q_{PR}]$

$e\ [T_P + \frac{2}{3}W_{50}, 0.5Q_{PR}]$

$f\ [t_b, 0]$

To compute C_t for a gauged basin, L and L_c are determined for the gauged watershed and t_p from the derived unit hydrograph for the gauged basin.

To compute the other required parameter C_p, the expression for peak discharge of the standard unit hydrograph can be used:

$$Q_p = \frac{C_2 C_p A}{t_p} \tag{8.4.3}$$

or for a unit discharge (discharge per unit area)

$$q_p = \frac{C_2 C_p}{t_p} \tag{8.4.4}$$

Solving equation (8.4.4) for C_p gives

$$C_p = \frac{q_p t_p}{C_2} \tag{8.4.5}$$

This relationship can be used to solve for C_p for the ungauged watershed, knowing the terms in the right-hand side.

Section 8.8 discusses the SCS-unit hydrograph procedure.

EXAMPLE 8.4.1

A watershed has a drainage area of 5.42 mi^2; the length of the main stream is 4.45 mi and the main channel length from the watershed outlet to the point opposite the center of gravity of the watershed is 2.0 mi. Using $C_t = 2.0$ and $C_p = 0.625$, determine the standard synthetic unit hydrograph for this basin. What is the standard duration? Use Snyder's method to determine the 30-min unit hydrograph parameter.

SOLUTION

For the standard unit hydrograph, equation (8.4.1) gives

$$t_p = C_1 C_t (LL_c)^{0.3} = 1 \times 2 \times (4.45 \times 2)^{0.3} = 3.85 \text{ hr}$$

The standard rainfall duration $t_r = 3.85/5.5 = 0.7$ hr. For a 30-min unit hydrograph, $t_R = 30$ min $= 0.5$ hr. The basin lag $t_{PR} = t_p - (t_r - t_R)/4 = 3.85 - (0.7 - 0.5)/4 = 3.80$ hr. The peak flow for the required unit hydrograph is $= q_p t_p/t_{PR}$ and, substituting equation (8.4.4) in the previous equation, $q_{PR} = q_p t_p/t_{PR} = (C_2 C_p/t_p) t_p/t_{PR} = C_2 C_p/t_{PR}$, so that $q_{PR} = 640 \times 0.625/3.80 = 105.26$ cfs/(in · mi^2) and the peak discharge is $Q_{PR} = q_{PR} A = 105.26 \times 5.42 = 570$ cfs/in.

The widths of the unit hydrograph are computed next. At 75 percent of the peak discharge, $W_{75} = C_{W_{75}} q_{PR}^{-1.08} = 440 \times 105.26^{-1.08} = 2.88$ hr. At 50 percent of the peak discharge, $W_{50} = C_{W_{50}} q_{PR}^{-1.08} = 770 \times 105.26^{-1.08} = 5.04$ hr.

The base time t_b may be computed assuming a triangular shape. This, however, does not guarantee that the volume under the unit hydrograph corresponds to one inch (or one cm, for SI units) of excess rainfall. To overcome this, the value of t_b may be exactly computed taking into account the values of W_{50} and W_{75} by solving the equation in step 5 of Table 8.4.1 for t_b:

$$t_b = 2581 \, A/Q_{PR} - 1.5 \, W_{50} - W_{75}$$

so that, with $A = 5.42$ mi^2, $W_{50} = 5.04$ hr, $W_{75} = 2.88$ hr, and $Q_{PR} = 570$ cfs/in, $t_b = 2581(5.42)/570 - 1.5 \times 5.04 - 2.88 = 14.1$ hr.

8.5   S-HYDROGRAPHS

In order to change a unit hydrograph from one duration to another, the *S-hydrograph method*, which is based on the principle of superposition, can be used. An S-hydrograph results theoretically from a continuous rainfall excess at a constant rate for an indefinite period. This curve (see Figure 8.5.1) has an S-shape with the ordinates approaching the rate of rainfall excess at the time of equilibrium.

Basically the S-curve (hydrograph) is the summation of an infinite number of t_R duration unit hydrographs, each lagged from the preceding one by the duration of the rainfall excess, as illustrated in Figure 8.5.2.

A unit hydrograph for a new duration, t_R', is obtained by: (1) lagging the S-hydrograph (derived with the t_R duration unit hydrographs) by the new (desired) duration t_R', (2) *subtracting* the two S-hydrographs from one another, and (3) and *multiplying* the resulting hydrograph ordinates by the ratio t_R/t_R'. Theoretically the S-hydrograph is a smooth curve because the input rainfall excess is assumed to be a constant, continuous rate. However, the numerical processes of the procedures may result in an undulatory form that may require smoothing or adjustment of the S-hydrograph.

EXAMPLE 8.5.1

(Adapted from Sanders, 1980)

Using the 2-hr unit hydrograph in Table 8.5.1, construct a 4-hr unit hydrograph.

SOLUTION

See computations in Table 8.5.1.

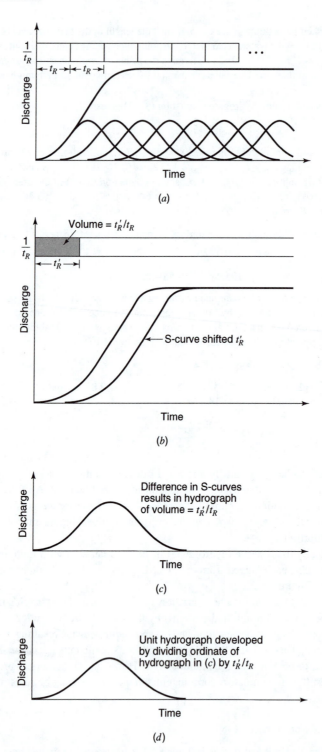

Figure 8.5.1 Development of a unit hydrograph for duration t_R' from a unit hydrograph for duration t_R.

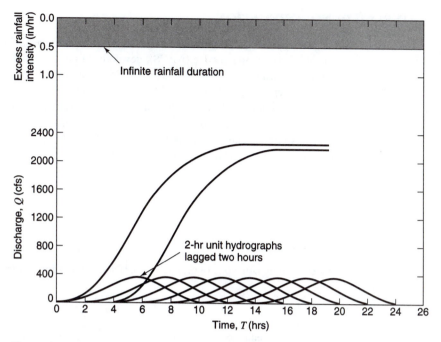

Figure 8.5.2 Graphical illustration of the S-curve construction (from Masch (1984)).

Table 8.5.1   S-Curve Determined from a 2-hr Unit Hydrograph to Estimate a 4-hr Unit Hydrograph

Time Hrs	2-hr Unit Hydrograph (cfs/in)	Lagged 2-hr Unit Hydrograph (cfs/in)			S-Curve	Lagged S-curve	4-hr Hydrograph	4-hr Unit Hydrograph (cfs/in)
0	0				0	—	0	0
2	69	0			69	—	69	34
4	143	69	0	...	212	0	212	106
6	328	143	69	...	540	69	471	235
8	389	328	143	...	929	212	717	358
10	352	389	328		1281	540	741	375
12	266	352	389		1547	929	618	309
14	192	266	352		1739	1281	458	229
16	123	192	·		1862	1547	315	158
18	84	123	·		1946	1739	207	103
20	49	84	·		1995	1862	133	66
22	20	49	·		2015	1946	69	34
24	0	20	·		*2015	1995	20	10
26	0	0	·	...	*2015	2015	0	0

*Adjusted values

Source: Sanders (1980).

8.6 SCS RAINFALL-RUNOFF RELATION

The U.S. Department of Agriculture Soil Conservation Service (SCS)(1972), now the National Resources Conservation Service (NRCS), developed a rainfall-runoff relation for watershed. For the storm as a whole, the depth of excess precipitation or direct runoff P_e is always less than or equal to the depth of precipitation P; likewise, after runoff begins, the additional depth of water retained in the watershed F_a, is less than or equal to some *potential maximum retention S* (see Fig. 8.6.1). There is some amount of rainfall I_a (initial abstraction before ponding) for which no runoff will occur, so the potential runoff is $P - I_a$. The SCS method assumes that the ratios of the two actual to the two potential quantities are equal, that is,

$$\frac{F_a}{S} = \frac{P_e}{P - I_a} \tag{8.6.1}$$

From continuity,

$$P = P_e + I_a + F_a \tag{8.6.2}$$

so that combining equations (8.6.1) and (8.6.2) and solving for P_e gives

$$P_e = \frac{(P - I_a)^2}{P - I_a + S} \tag{8.6.3}$$

which is the basic equation for computing the depth of excess rainfall or direct runoff from a storm by the SCS method.

From the study of many small experimental watersheds, an empirical relation was developed for I_a:

$$I_a = 0.2S \tag{8.6.4}$$

so that equation (8.6.3) is now expressed as

$$P_e = \frac{(P - 0.2S)^2}{P + 0.8S} \tag{8.6.5}$$

Empirical studies by the SCS indicate that the potential maximum retention can be estimated as

$$S = \frac{1000}{CN} - 10 \tag{8.6.6}$$

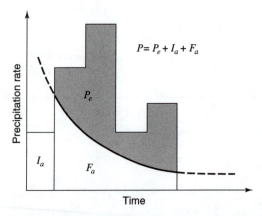

Figure 8.6.1 Variables in the SCS method of rainfall abstractions: I_a = initial abstraction, P_e = rainfall excess, F_a = continuing abstraction, and P = total rainfall.

where CN is a runoff curve number that is a function of land use, antecedent soil moisture, and other factors affecting runoff and retention in a watershed. The curve number is a dimensionless number defined such that $0 \le CN \le 100$. For impervious and water surfaces $CN = 100$; for natural surfaces $CN < 100$.

The SCS rainfall-runoff relation (equation (8.6.5)) can be expressed in graphical form using the curve numbers as illustrated in Figure 8.6.2. Equation (8.6.5) or Figure 8.6.2 can be used to estimate the volume of runoff when the precipitation volume P and the curve number CN are known.

8.7   CURVE NUMBER ESTIMATION AND ABSTRACTIONS

8.7.1   Antecedent Moisture Conditions

The curve numbers shown in Figure 8.6.2 apply for normal *antecedent moisture conditions* (AMC II). Antecedent moisture conditions are grouped into three categories:

AMC I—Low moisture

AMC II—Average moisture condition, normally used for annual flood estimates

AMC III—High moisture, heavy rainfall over preceding few days

For dry conditions (AMC I) or wet conditions (AMC III), equivalent curve numbers can be computed using

$$CN(\text{I}) = \frac{4.2CN(\text{II})}{10 - 0.058CN(\text{II})} \tag{8.7.1}$$

and

$$CN(\text{III}) = \frac{23CN(\text{II})}{10 + 0.13CN(\text{II})} \tag{8.7.2}$$

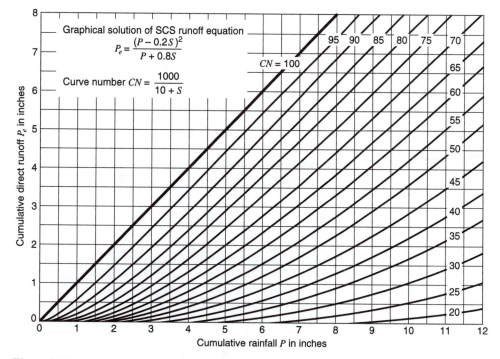

Figure 8.6.2 Solution of the SCS runoff equations (from U.S. Department of Agriculture Soil Conservation Service (1972)).

The range of antecedent moisture conditions for each class is shown in Table 8.7.1. Table 8.7.2 lists the adjustment of curve numbers to conditions I and III for known II conditions.

Table 8.7.1 Classification of Antecedent Moisture Classes (AMC) for the SCS Method of Rainfall Abstractions

| | Total 5-Day Antecedent Rainfall (in) | |
AMC Group	Dormant Season	Growing Season
I	Less than 0.5	Less than 1.4
II	0.5 to 1.1	1.4 to 2.1
III	Over 1.1	over 2.1

Source: U.S. Department of Agriculture Soil Conservation Service (1972).

Table 8.7.2 Adjustment of Curve Numbers for Dry (Condition I) and Wet (Condition III) Antecedent Moisture Conditions

| | Corresponding CN for Condition | |
CN for Condition II	I	III
100	100	100
95	87	99
90	78	98
85	70	97
80	63	94
75	57	91
70	51	87
65	45	83
60	40	79
55	35	75
50	31	70
45	27	65
40	23	60
35	19	55
30	15	50
25	12	45
20	9	39
15	7	33
10	4	26
5	2	17
0	0	0

Source: U.S. Department of Agriculture Soil Conservation Service (1972).

8.7.2 Soil Group Classification

Curve numbers have been tabulated by the Soil Conservation Service on the basis of soil type and land use in Table 8.7.3. The four soil groups in Table 8.7.3 are described as:

Group A: Deep sand, deep loess, aggregated silts

Group B: Shallow loess, sandy loam

Group C: Clay loams, shallow sandy loam, soils low in organic content, and soils usually high in clay

Group D: Soils that swell significantly when wet, heavy plastic clays, and certain saline soils

The values of *CN* for various land uses on these soil types are given in Table 8.7.3. For a watershed made up of several soil types and land uses, a composite *CN* can be calculated.

Minimum infiltration rates for the various soil groups are:

Group	Minimum Infiltration Rate (in/hr)
A	0.30 − 0.45
B	0.15 − 0.30
C	0 − 0.05

Table 8.7.3 Runoff Curve Numbers (Average Watershed Condition, $I_a = 0.2S$)

Land Use Description		Curve Numbers for Hydrologic Soil Group			
		A	B	C	D
Fully developed urban areas[a] (vegetation established)					
Lawns, open spaces, parks, golf courses, cemeteries, etc.					
Good condition; grass cover on 75% or more of the area		39	61	74	80
Fair condition; grass cover on 50% to 75% of the area		49	69	79	84
Poor condition; grass cover on 50% or less of the area		68	79	86	89
Paved parking lots, roofs, driveways, etc.		98	98	98	98
Streets and roads					
Paved with curbs and storm sewers		98	98	98	98
Gravel		76	85	89	91
Dirt		72	82	87	89
Paved with open ditches		83	89	92	93
	Average % impervious[b]				
Commercial and business areas	85	89	92	94	95
Industrial districts	72	81	88	91	93
Row houses, town houses, and residential with lot sizes 1/8 acre or less	65	77	85	90	92
Residential: average lot size					
1/4 acre	38	61	75	83	87
1/3 acre	30	57	72	81	86
1/2 acre	25	54	70	80	85
1 acre	20	51	68	79	84
2 acre	12	46	65	77	82
Developing urban areas[c] (no vegetation established)					
Newly graded area		77	86	91	94

Land Use	Cover Treatment of Practice	Hydrologic Condition[d]				
Cultivated agricultural land						
Fallow	Straight row		77	86	91	94
	Conservation tillage	Poor	76	85	90	93
	Conservation tillage	Good	74	83	88	90
Row crops	Straight row	Poor	72	81	88	91
	Straight row	Good	67	78	85	89
	Conservation tillage	Poor	71	80	87	90
	Conservation tillage	Good	64	75	82	85

Table 8.7.3 Runoff Curve Numbers (*continued*)

Cover			Curve Numbers for Hydrologic Soil Group			
Land Use	Treatment of Practice	Hydrologic Condition[d]	A	B	C	D
	Contoured	Poor	70	79	84	88
	Contoured	Good	65	75	82	86
	Contoured and conservation tillage	Poor	69	78	83	87
		Good	64	74	81	85
	Contoured and terraces	Poor	66	74	80	82
	Contoured and terraces	Good	62	71	78	81
	Contoured and terraces and conservation tillage	Poor	65	73	79	81
		Good	61	70	77	80
Small grain	Straight row	Poor	65	76	84	88
	Straight row	Good	63	75	83	87
	Conservation tillage	Poor	64	75	83	86
	Conservation tillage	Good	60	72	80	84
	Contoured	Poor	63	74	82	85
	Contoured	Good	61	73	81	84
	Contoured and conservation tillage	Poor	62	73	81	84
		Good	60	72	80	83
	Contoured and terraces	Poor	61	72	79	82
	Contoured and terraces	Good	59	70	78	81
	Contoured and terraces and conservation tillage	Poor	60	71	78	81
		Good	58	69	77	80
Close-seeded legumes or rotation meadow[e]	Straight row	Poor	66	77	85	89
	Straight row	Good	58	72	81	85
	Contoured	Poor	64	75	83	85
	Contoured	Good	55	69	78	83
	Contoured and terraces	Poor	63	73	80	83
	Contoured and terraces	Good	51	67	76	80
Noncultivated agricultural land Pasture or range	No mechanical treatment	Poor	68	79	86	89
	No mechanical treatment	Fair	49	69	79	84
	No mechanical treatment	Good	39	61	74	80
	Contoured	Poor	47	67	81	88
	Contoured	Fair	25	59	75	83
	Contoured	Good	6	35	70	79
Meadow		—	30	58	71	78
Forestland—grass or orchards—evergreen or deciduous		Poor	55	73	82	86
		Fair	44	65	76	82
		Good	32	58	72	79
Brush		Poor	48	67	77	83
		Good	20	48	65	73
Woods		Poor	45	66	77	83
		Fair	36	60	73	79
		Good	25	55	70	77
Farmsteads		—	59	74	82	86
Forest–range Herbaceous		Poor		79	86	92
		Fair		71	80	89
		Good		61	74	84

Table 8.7.3 Runoff Curve Numbers (*continued*)

Cover		Hydrologic Condition[d]	Curve Numbers for Hydrologic Soil Group			
Land Use	Treatment of Practice		A	B	C	D
Oak–aspen		Poor		65	74	
		Fair		47	57	
		Good		30	41	
Juniper–grass		Poor		72	83	
		Fair		58	73	
		Good		41	61	
Sage–grass		Poor		67	80	
		Fair		50	63	
		Good		35	48	

[a] For land uses with impervious areas, curve numbers are computed assuming that 100% of runoff from impervious areas is directly connected to the drainage system. Pervious areas (lawn) are considered to be equivalent to lawns in good condition and the impervious areas have a *CN* of 98.

[b] Includes paved streets.

[c] Use for the design of temporary measures during grading and construction. Impervious area percent for urban areas under development vary considerably. The user will determine the percent impervious. Then using the newly graded area *CN* and Figure 8.7.1*a* or *b*, the composite *CN* can be computed for any degree of development.

[d] For conversation tillage poor hydrologic condition, 5% to 20% of the surface is covered with residue (less than 750-lb/acre row crops or 300-lb/acre small grain).

For conservation tillage good hydrologic condition, more than 20% of the surface is covered with residue (greater than 750-lb/acre row crops or 300-lb/acre small grain).

[e] Close-drilled or broadcast.

For noncultivated agricultural land:
Poor hydrologic condition has less than 25% ground cover density.
Fair hydrologic condition has between 25% and 50% ground cover density.
Good hydrologic condition has more than 50% ground cover density.

For forest–range:
Poor hydrologic condition has less than 30% ground cover density.
Fair hydrologic condition has between 30% and 70% ground cover density.
Good hydrologic condition has more than 70% ground cover density.

Source: U.S. Department of Agriculture Soil Conservation Service (1986).

8.7.3 Curve Numbers

Table 8.7.3 gives the curve numbers for average watershed conditions, $I_a = 0.2S$, and antecedent moisture condition II. For watersheds consisting of several subcatchments with different *CN*s, the area-averaged composite *CN* can be computed for the entire watershed. This analysis assumes that the impervious areas are directly connected to the watershed drainage system (Figure 8.7.1*a*). If the percent imperviousness is different from the value listed in Table 8.7.3 or if the impervious areas are not directly connected, then Figures 8.7.1*a* or *b*, respectively can be used. The pervious *CN* used in these figures is equivalent to the open-space *CN* in Table 8.7.3. If the total imperious area is less than 30 percent, Figure 8.7.1*b* is used to obtain a composite *CN*. For natural desert landscaping and newly graded areas, Table 8.7.3 gives only the *CN*s for pervious areas.

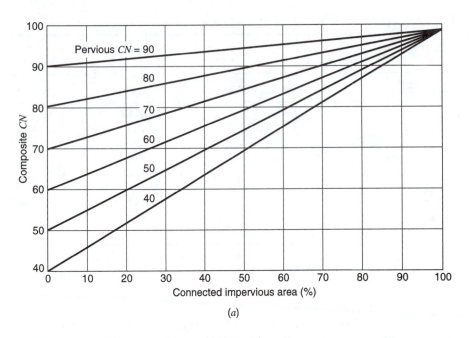

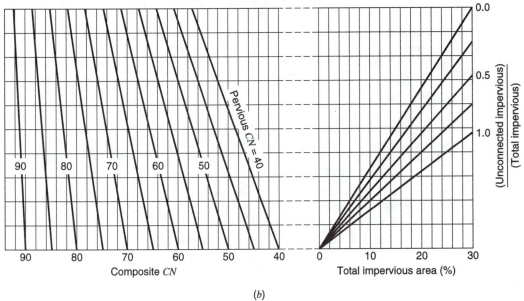

Figure 8.7.1 Relationships for determining composite CN. (*a*) Connected impervious area; (*b*) Unconnected impervious area (from U.S. Department of Agriculture Soil Conservation Service (1986)).

EXAMPLE 8.7.1 Determine the weighted curve numbers for a watershed with 40% residential (1/4-acre lots), 25% open space, good condition, 20% commercial and business (85% impervious), and 15% industrial (72% impervious), with corresponding soil groups of C, D, C, and D.

SOLUTION

The corresponding curve numbers are obtained from Table 8.7.3:

Land Use (%)	Soil Group	Curve Number
40%	C	83
25%	D	80
20%	C	94
15%	D	93

The weighted curve number is

$$CN = 0.40(83) + 0.25(80) + 0.20(94) + 0.15(93)$$

$$= 33.2 + 20 + 18.8 + 13.95$$

$$= 85.95 \text{ (use 86)}$$

EXAMPLE 8.7.2

The watershed in example 8.7.1 experienced a rainfall of 6 in, what is the runoff volume?

SOLUTION

Using equation (8.6.5), P_e = runoff volume is

$$P_e = \frac{(P - 0.2S)^2}{P + 0.8S}$$

where S is computed with the weighted curve number of 86 from example 8.7.1:

$$S = \frac{1000}{86} - 10 = 1.63$$

So

$$P_e = \frac{[6 - 0.2(1.63)]^2}{6 + 0.8(1.63)} = \frac{32.19}{7.3}$$

$$= 4.41 \text{ inches of runoff}$$

EXAMPLE 8.7.3

For the watershed in examples 8.7.1 and 8.7.2, the 6-inch rainfall pattern was 2 in the first hour, 3 in the second hour and 1 in the third hour. Determine the cumulative rainfall and cumulative rainfall excess as functions of time.

SOLUTION

The initial abstractions are computed as $I_a = 0.2S$ with $S = 1.63$ from example 8.7.2, so $I_a = 0.2(1.63) = 0.33$ in. The remaining losses for time period (the first hour) are computed using the following equation, derived by combining equations (8.6.1) and (8.6.2):

$$F_{a,t} = \frac{S(P_t - I_a)}{P_t - I_a + S} = \frac{1.63(P_t - 0.33)}{P_t - 0.33 + 1.63} = \frac{1.63(P_t - 0.33)}{P_t + 1.3}$$

$$F_{a,1} = \frac{1.63(2 - 0.33)}{2 + 1.3} = 0.82 \text{ in}$$

The total loss for the first hour is $0.33 + 0.82 = 1.15$ in and the excess is

$$P_{e1} = P_1 - I_a - F_{a,1} = 2 - 0.33 - 0.82 = 0.85 \text{ in}$$

For the second hour, $P_t = 2 + 3 = 5$ in, so

$$F_{a,2} = \frac{1.63(5 - 0.33)}{5 + 1.3} = 1.21 \text{ in}$$

and the cumulative rainfall excess is $P_{e_2} = 5 - 0.33 - 1.21 = 3.46$ in.

For the third hour, $P_3 = 2 + 3 + 1 = 6$ in, so

$$F_{a,3} = \frac{1.63(6 - 0.33)}{6 + 1.3} = 1.27 \text{ in}$$

and $P_{e_3} = 6 - 0.33 - 1.27 = 4.40$ in (which compares well with the results of example 8.7.2).

The results are summarized below, along with the rainfall excess hyetograph.

Time (h)	Cumulative Rainfall P_t (in)	Cumulative Abstractions I_a (in)	$F_{a,t}$ (in)	Cumulative Rainfall Excess P_e (in)	Rainfall Excess Hyetograph (in)
1	2	0.33	0.82	0.85	0.85
2	5	0.33	1.21	3.46	2.61
3	6	0.33	1.27	4.40	0.94

8.8 SCS UNIT HYDROGRAPH PROCEDURE

The SCS dimensionless unit hydrograph and mass curve are shown in Figure 8.8.1 and tabulated in Table 8.8.1. The SCS dimensionless equivalent triangular unit hydrograph is also shown in Figure 8.8.1. The following section discusses how to develop a unit hydrograph from these dimensionless unit hydrographs.

Table 8.8.1   Ratios for Dimensionless Unit Hydrograph and Mass Curve

Time Ratios, t/T_p	Discharge Ratios, q/q_p	Mass Curve Ratios, Q_d/Q
0	0.000	0.000
0.1	0.030	0.001
0.2	0.100	0.006
0.3	0.190	0.012
0.4	0.310	0.035
0.5	0.470	0.065
0.6	0.660	0.107
0.7	0.820	0.163
0.8	0.930	0.228
0.9	0.990	0.300
1.0	1.000	0.375
1.1	0.990	0.450
1.2	0.930	0.522
1.3	0.860	0.589
1.4	0.780	0.650
1.5	0.680	0.700
1.6	0.560	0.751
1.7	0.460	0.790
1.8	0.390	0.822
1.9	0.330	0.849
2.0	0.280	0.871
2.2	0.207	0.908
2.4	0.147	0.934
2.6	0.107	0.953
2.8	0.077	0.967

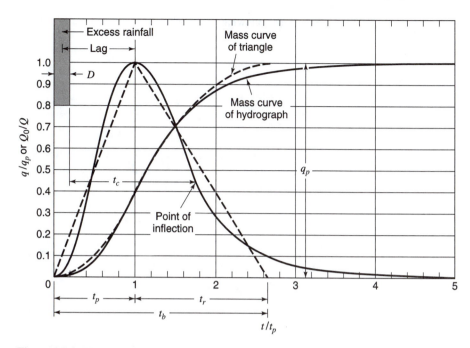

Figure 8.8.1 Dimensionless curvilinear unit hydrograph and equivalent triangular hydrograph (from U.S. Department of Agriculture Soil Conservation Service (1986)).

Table 8.8.1 Ratios for Dimensionless Unit Hydrograph and Mass Curve (*continued*)

Time Ratios, t/T_p	Discharge Ratios, q/q_p	Mass Curve Ratios, Q_d/Q
3.0	0.055	0.977
3.2	0.040	0.984
3.4	0.029	0.989
3.6	0.021	0.993
3.8	0.015	0.995
4.0	0.011	0.997
4.5	0.005	0.999
5.0	0.000	1.000

Source: U.S. Department of Agriculture Soil Conservation Service (1972).

8.8.1 Time of Concentration

The *time of concentration* for a watershed is the time for a particle of water to travel from the hydrologically most distant point in the watershed to a point of interest, such as the outlet of the watershed. SCS has recommended two methods for time of concentration, the *lag method* and the *upland*, or *velocity method.*

The lag method relates the *time lag* (t_L), defined as the time in hours from the center of mass of the rainfall excess to the peak discharge, to the slope (Y) in percent, the hydraulic length (L) in feet, and the potential maximum retention (S), expressed as

$$t_L = \frac{L^{0.8}(S+1)^{0.7}}{1900 Y^{0.5}} \qquad (8.8.1)$$

The SCS uses the following relationship between the time of concentration (t_c) and the lag (t_L):

$$t_c = \frac{5}{3}t_L \tag{8.8.2}$$

or

$$t_c = \frac{L^{0.8}(S+1)^{0.7}}{1140Y^{0.5}} \tag{8.8.3}$$

where t_c is in hours. Refer to Figure 8.8.1 to see the SCS definition of t_c and t_L.

The velocity (upland) method is based upon defining the time of concentration as the ratio of the hydraulic flow length (L) to the velocity (V):

$$t_c = \frac{L}{3600V} \tag{8.8.4}$$

where t_c is in hours, L is in feet and V is in ft/s. The velocity can be estimated knowing the land use and the slope in Figure 8.8.2. Alternatively, we can think of the concentration as being the sum of travel times for different segments

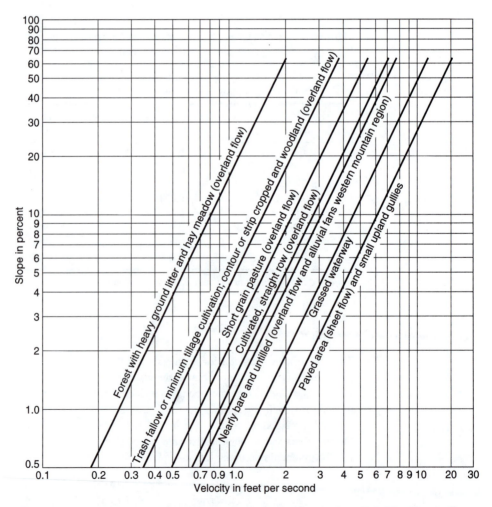

Figure 8.8.2 Velocities for velocity upland method of estimating t_c (from U.S. Department of Agriculture Soil Conservation Service (1986)).

$$t_c = \frac{1}{3600} \sum_{i=1}^{k} \frac{L_i}{V_i} \qquad (8.8.5)$$

for k segments, each with different land uses.

8.8.2   Time to Peak

Time to peak (t_p) is the time from the beginning of rainfall to the time of the peak discharge (Figure 8.8.1)

$$t_p = \frac{t_R}{2} + t_L \qquad (8.8.6)$$

where t_p is in hours, t_R is the duration of the rainfall excess in hours, and t_L is the lag time in hours. The SCS recommends that t_R be 0.133 of the time of concentration of the watershed, t_c:

$$t_R = 0.133 t_c \qquad (8.8.7)$$

and because $t_L = 0.6\ t_c$ by equation (8.8.2), then by equation (8.8.6) we get

$$t_p = \frac{0.133 t_c}{2} + 0.6 t_c$$

$$t_p = 0.67 t_c \qquad (8.8.8)$$

8.8.3   Peak Discharge

The area of the unit hydrograph equals the volume of direct runoff Q, which was estimated by equation (8.6.5). With the equivalent triangular dimensionless unit hydrograph of the curvilinear dimensionless unit hydrograph in Figure 8.8.1, the time base of the dimensionless triangular unit hydrograph is 8/3 of the time to peak t_p, as compared to $5t_p$ for the curvilinear. The areas under the rising limb of the two dimensionless unit hydrographs are the same (37 percent).

Based upon geometry (Figure 8.8.1), we see that

$$Q = \frac{1}{2} q_p (t_p + t_r) \qquad (8.8.9)$$

for the direct runoff Q, which is 1 inch where t_r is the recession time of the dimensionless triangular unit hydrograph and q_p is the peak discharge. Solving equation (8.8.9) for q_p gives

$$q_p = \frac{Q}{t_p} \left[\frac{2}{1 + t_r / t_p} \right] \qquad (8.8.10)$$

Letting $K = \left[\dfrac{2}{1 + t_r / t_p} \right]$, then

$$q_p = \frac{KQ}{t_p} \qquad (8.8.11)$$

where Q is the volume, equals to one inch for a unit hydrograph.

The above equation can be modified to express q_p in ft^3/sec, t_p in hours, and Q in inches:

$$q_p = 645.33 K \frac{AQ}{t_p} \qquad (8.8.12)$$

The factor 645.33 is the rate necessary to discharge 1 in of runoff from 1 mi^2 in 1 hr. Using $t_r = 1.67t_p$ gives $K = [2/(1+1.67)] = 0.75$; then equation (8.8.12) becomes

$$q_p = \frac{484AQ}{t_p} \qquad (8.8.13)$$

For SI units,

$$q_p = \frac{2.08AQ}{t_p} \qquad (8.8.14)$$

where A is in square kilometers.

The steps in developing a unit hydrograph are:

Step 1 Compute the time of concentration using the lag method (equation 8.8.3) or the velocity method (equation (8.8.4) or (8.8.5)).

Step 2 Compute the time to peak $t_p = 0.67t_c$ (equation (8.8.8)) and then the peak discharge q_p using equation (8.8.13) or (8.8.14).

Step 3 Compute time base t_b and the recession time t_r:

Triangular hydrograph: $t_b = 2.67t_p$
Curvilinear hydrograph: $t_b = 5t_p$
$t_r = t_b - t_p$

Step 4 Compute the duration $t_R = 0.133\,t_c$ and the lag $t_L = 0.6\,t_c$ by using equations (8.8.7) and (8.8.2), respectively.

Step 5 Compute the unit hydrograph ordinates and plot. For the triangular only t_p, q_p, and t_r are needed. For the curvilinear, use the dimensionless ratios in Table 8.8.1.

EXAMPLE 8.8.1

For the watershed in example 8.7.1, determine the triangular SCS unit hydrograph. The average slope of the watershed is 3 percent and the area is 3.0 mi^2. The hydraulic length is 1.2 mi.

SOLUTION

Step 1 The time of concentration is computed using equation (8.8.1), with S = 1.63 from example 8.7.2:

$$t_L = \frac{(6336)^{0.8}(1.63+1)^{0.7}}{1900\sqrt{3}} = 0.66 \text{ hr}$$

and $t_c = \dfrac{5}{3}t_L = 1.1 \text{ hr}$

Step 2 The time to peak $t_p = 0.67t_c = 0.67(1.1) = 0.74$ hr.

Step 3 The time base is $t_b = 2.67t_p = 1.97$ hr.

Step 4 The duration is $t_R = 0.133t_c = 0.133(1.1) = 0.15$ hr, and t_L is 0.66 hr.

Step 5 The peak is (for $Q = 1$ in)

$$q_p = \frac{484AQ}{t_p} = \frac{484(3)(1)}{0.74} = 1962 \text{ cfs.}$$

In summary, the triangular unit hydrograph has a peak of 1962 cfs at the time to peak of 0.74 hr with a time base of 1.97 hr. This is a 0.15-hr duration unit hydrograph.

8.9   KINEMATIC-WAVE OVERLAND FLOW RUNOFF MODEL

Hortonian overland flow occurs when the rainfall rate exceeds the infiltration capacity and sufficient water ponds on the surface to overcome surface tension effects and fill small depressions. Overland flow is surface runoff that occurs in the form of sheet flow on the land surface without concentrating in clearly defined channels (Ponce, 1989). For the purposes of rainfall-runoff analysis, this flow can be viewed as a one-dimensional flow process (Figure 8.9.1) in which the flux is proportional to some power of the storage per unit area, expressed as (Woolhiser et al., 1990):

$$Q = \alpha h^m \tag{8.9.1}$$

where Q is the discharge per unit width, h is the storage of water per unit area (or depth if the surface is a plane), and α and m are parameters related to slope, surface roughness, and whether the flow is laminar or turbulent.

The mathematical description of overland flow can be accomplished through the continuity equation in one-dimensional form and a simplified form of the momentum equation. This model is referred to as the *kinematic wave model*. *Kinematics* refers to the study of motion exclusive of the influence of mass and force. A *wave* is a variation in flow, such as a change in flow rate or water surface elevation. *Wave celerity* is the velocity with which this variation travels. *Kinematic waves* govern flow when inertial and pressure forces are negligible.

The *kinematic wave equations* (also see Chapter 9) for a one-dimensional flow are expressed as Continuity:

$$\frac{\partial A}{\partial t} + \frac{\partial Q}{\partial x} = q(x,t) \tag{8.9.2}$$

Momentum:

$$S_0 - S_f = 0 \tag{8.9.3}$$

where A is the cross-sectional area of flow, Q is the discharge, t is time, x is the spatial coordinate, $q(x, t)$ is the lateral inflow rate, S_0 is the overland flow slope, and S_f is the friction slope.

Equation (8.9.3) indicates that the gravity and friction forces are balanced, so that flow does not accelerate appreciably. The inertial (local and convective acceleration) term and pressure term are neglected in the kinematic wave model (refer to Section 9.4). Eliminating these terms eliminates the mechanism to describe backwater effects and flood wave peak attenuation.

Considering that h is the storage per unit area or depth, then $A = h$, so that equation (8.9.2) becomes

$$\frac{\partial h}{\partial t} + \frac{\partial Q}{\partial x} = q(x,t) \tag{8.9.4}$$

Substituting equation (8.9.1) into (8.9.4) gives

$$\frac{\partial h}{\partial t} + \frac{\partial(\alpha h^m)}{\partial x} = q(x,t) \tag{8.9.5}$$

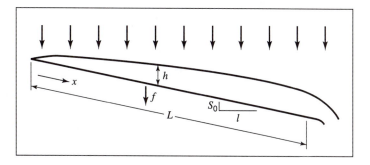

Figure 8.9.1 Definition sketch of overland flow on a plane as a one-dimensional flow (from Woolhiser et al. (1990)).

The one-dimensional overland flow on a plane surface (illustrated in Figure 8.9.1) is not the type of flow found in most watershed situations (Woolhiser et al., 1990). The kinematic assumption does not require sheet flow as shown; it requires only that the discharge be some unique function of the amount of water stored per unit of area.

Woolhiser and Liggett (1976) and Morris and Woolhiser (1980) showed that the kinematic-wave formulation is an excellent approximation for most overland flow conditions. Keep in mind that these equations are a simplification of the Saint-Venant equations (see Chapter 9).

The kinematic-wave equation (8.9.5) for overland flow can be solved numerically using a four-point implicit method where the finite-difference approximations for the spatial and temporal derivatives are respectively

$$\frac{\partial h}{\partial x} = \theta \frac{h_{i+1}^{j+1} - h_i^{j+1}}{\Delta x} + (1-\theta)\frac{h_{i+1}^j - h_i^j}{\Delta x} \tag{8.9.6}$$

and

$$\frac{\partial h}{\partial t} = \frac{1}{2}\left[\frac{h_i^{j+1} - h_i^j}{\Delta t} + \frac{h_{i+1}^{j+1} - h_{i+1}^j}{\Delta t}\right]$$

or

$$\frac{\partial h}{\partial t} = \frac{h_i^{j+1} + h_{i+1}^{j+1} - h_i^j - h_{i+1}^j}{2\Delta t} \tag{8.9.7}$$

and

$$q = \frac{1}{2}\left(\bar{q}_{i+1} + \bar{q}_i\right) \tag{8.9.8}$$

where θ is a weighting parameter for spatial derivative, $\theta = \Delta t'/\Delta t$ (see Figure 8.9.2). The derivative $\partial h/\partial t$ is the average of the temporal derivatives at locations i and $i + 1$ or for the midway locations between i and $i + 1$, and $\bar{q}_i$ and $\bar{q}_{i+1}$ are the average lateral inflows at i and $i + 1$, respectively. Notation for the finite-difference grid is shown in Figure 8.9.2. Substituting these finite difference expressions (8.9.6), (8.9.7), and (8.9.8) into (8.9.5) and simplifying results in the following finite-difference equation:

$$h_{i+1}^{j+1} - h_{i+1}^j + h_i^{j+1} - h_i^j$$

$$+ \frac{2\Delta t}{\Delta x}\left\{\theta\left[\alpha_{i+1}^{j+1}(h_{i+1}^{j+1})^m - \alpha_i^{j+1}(h_i^{j+1})^m\right] + (1-\theta)\left[\alpha_{i+1}^j(h_{i+1}^j)^m - \alpha_i^j(h_i^j)^m\right]\right\}$$

$$- \Delta t(\bar{q}_{i+1} + \bar{q}_i) = 0 \tag{8.9.9}$$

The only unknown in the above equation is h_{i+1}^{j+1}, which must be solved by using Newton's method (see Appendix A). Using Manning's equation to express equation (8.9.1), $Q = \alpha h^m$, we find

$$Q = \left[\frac{1.49S^{1/2}}{n}\right]h^{5/3} \tag{8.9.10}$$

where

$$\alpha = \frac{1.49S_0^{1/2}}{n}, \, m = 5/3 \tag{8.9.11}$$

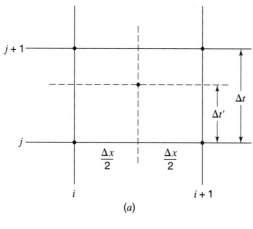

(a)

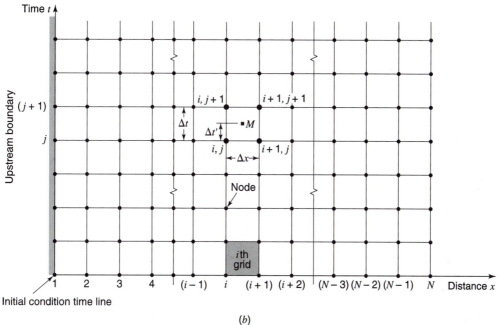

(b)

Figure 8.9.2 The x-t solution plane: The finite-difference forms of the Saint-Venant equations are solved at a discrete number of points (values of the independent variables x and t) arranged to the time axis represent locations along the plane, and those parallel to the distance axis represent times. (a) Four points of finite-difference grid; (b) Finite difference grid.

where n is Manning's roughness coefficient and S_0 is the slope of the overland flow plane. Recommended values of Manning's roughness coefficients for overland flow are given in Table 8.9.1. The *time to equilibrium* of a plane of length L and slope S_0 can be derived using Manning's equation as

$$t_c = \frac{nL}{1.49 \, S_0^{1/2} \, h^{2/3}} \tag{8.9.12}$$

Table 8.9.1 Recommended Manning's Roughness Coefficients for Overland Flow

Cover or treatment	Residue rate, Tons/Acre	Value Recommended	Range
Concrete or asphalt		0.011	0.010–0.013
Bare sand		.01	.010–.016
Graveled surface		.02	.012–.03
Bare clay loam (eroded)		.02	.012–.033
Fallow - no residue		.05	006–.16
Chisel plow	<1/4	.07	.006–.17
	<1/4–1	.18	.07–.34
	1–3	.30	.19–.47
	>3	.40	.34–.46
Disk/harrow	<1/4	.08	.008–.41
	1/4–1	.16	.10–.25
	1–3	.25	.14–.53
	>3	.30	—
No till	<1/4	.04	.03–.07
	1/4–1	.07	.01–.13
	1–3	.30	.16–.47
Moldboard plow (fall)		.06	.02–.10
Colter		.10	.05–.13
Range (natural)		.13	.01–.32
Range (clipped)		.10	.02–.24
Grass (bluegrass sod)		.45	.39–.63
Short grass prairie		.15	.10–.20
Dense grass[1]		.24	.17–.30
Bermuda grass[1]		.41	.30–.48

[1]Weeping lovegrass, bluegrass, buffalo grass, blue gamma grass, native grass mix (OK), alfalfa, lespedeza (from Palmer, 1946).

Sources: Woolhiser (1975), Engman (1986), Woolhiser et al. (1990).

The U.S. Department of Agriculture Agricultural Research Service (Woolhiser et al., 1990) has developed a KINematic runoff and EROSion model referred to as KINEROS. This model is event-oriented; i.e., it is a physically based model describing the processes of interception, infiltration, surface runoff, and erosion from small agricultural and urban watersheds. The model is distributed because flows are modeled for both the watershed and the channel elements, as illustrated in Figures 8.9.3 and 8.9.4. The model is *event-oriented* because it does not have components describing evapo-transpiration and soil water movement between storms. In other words, there is no hydrologic balance between storms.

Figures 8.9.3 and 8.9.4 illustrate that the approach to describing a watershed is to divide it into a branching system of channels with plane elements contributing lateral flow to channels. The KINEROS model takes into account interception, infiltration, overland flow routing, channel

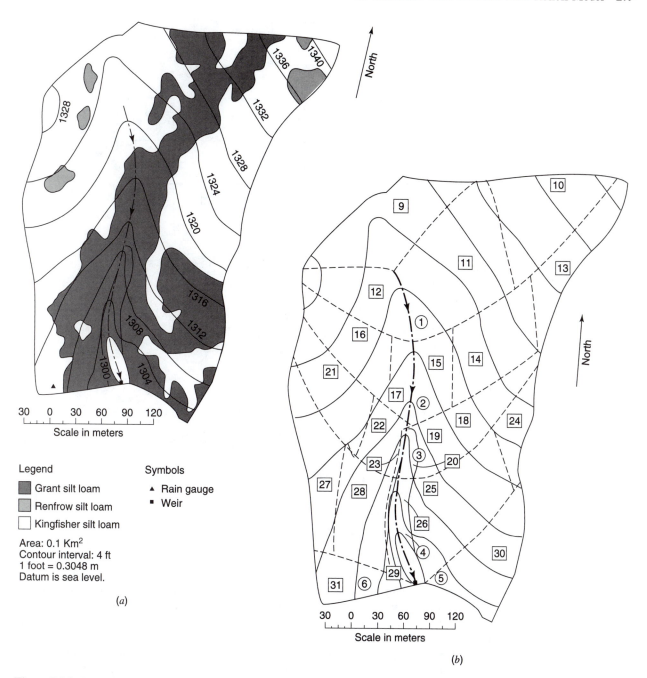

Figure 8.9.3 R-5 catchment, Chickasha, OK. (*a*) Contour map; (*b*) Division into plane and channel elements (from Woolhiser et al. (1990)).

routing, reservoir routing, erosion, and sediment transport. Overland flow routing has been described in this section. Channel routing is performed using the kinematic-wave approximation described in Chapter 9. The reservoir routing in KINEROS is basically a level-pool routing procedure, as described in Chapter 9.

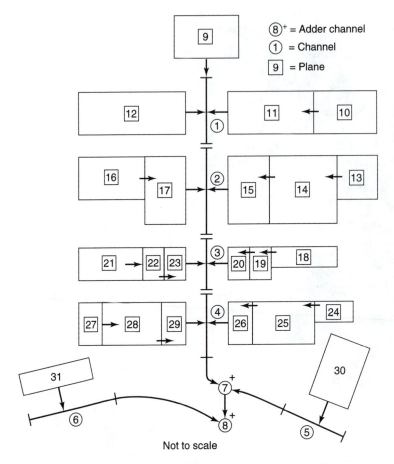

Figure 8.9.4 Schematic of R-5 plane and channel configuration (from Woolhiser et al. (1990)).

8.10   COMPUTER MODELS FOR RAINFALL-RUNOFF ANALYSIS

Computer models for runoff analysis can be classified as event-based models or continuous simulation models. *Event-based models* are used to simulate the discrete rainfall-runoff events using, for example, the unit hydrograph approach. These models emphasize infiltration and surface-runoff components with the objective of determining direct runoff, and are applicable to excess-water flow calculations in cases where direct runoff is the major contributor to streamflow. In Table 8.10.1 the HEC-1 and the TR-20 and TR-55 models are single-event models. The KINEROS model (Woolhiser et al., 1990) described in the previous section is an overland flow model based on the kinematic-wave routing. This model is a *distributed event-based model*. Other examples include the kinematic-wave model for overland flow routing in the HEC-1 model (see Maidment, 1993, and Mays, 1996, for other models).

Continuous-simulation models account for the overall moisture balance of the basin, including moisture accounting between storm events. These models explicitly account for all runoff components including surface flow and indirect runoff such as inter flow and baseflow, and are well suited for *long-term runoff forecasting*. The SSARR and HSPF models listed in Table 8.10.1 are examples of continuous-simulation models.

Table 8.10.1 Computer Models Hydrologic Analysis

Computer Model	What Is Modeled (Primary Use)	Whom to Contact for Information	What Is Available
HEC-1	Catchment runoff	U.S. Army Corps of Engineers (USACE) Hydrologic Engineering Center (HEC) 609 Second St. Davis, CA 95616 www.hec.usace.mil *or* National Technical Information Service U.S. Dept. of Commerce 5285 Port Royal Rd. Springfield, VA 22161 www.ntis.gov/ordering.htm	Program in FORTRAN; source is available. PC-version available. Detailed user's manual is available. HEC technical reports, training documents, and project reports available. HEC distributes to U.S. government users. Vendors provide to others for a fee, but the program is public domain. Program and documentation are also available from U.S. National Technical Information Service (NTIS). Training is available from HEC (federal) and universities (other users).
TR-20, TR-55	Catchment runoff	National Resources Conservation Service (NRCS) P.O. Box 2890 Washington, DC 20013 www.nrcs.usda.gov	PC versions of both TR-20 and TR-55 are available from offices of the Soil Conservation Service throughout the United States, along with documentation of the software. Both programs are public-domain so are available from a variety of other sources, including vendors and academic institutions. Technical reports describing math models are available from NTIS.
SSARR	Catchment runoff	USACE North Pacific Division Attn: CENPD-EN-WM-HES P.O. Box 2870 Portland, OR 97208-2870	The program was developed for and is used by the Corps on mainframe computers. PC/workstation versions are available. Program user's manuals and project reports are available.
HSPF	Catchment runoff	Environmental Protection Agency (EPA) Environmental Research Laboratory Athens, GA 30613 www.epa.gov www.usgs.gov	A PC version of HSPF is available, but as the program is written in FORTRAN, it may be compiled and executed on any computer for which a compiler is available. Documentation is available from EPA.

Source: Adapted from Ford and Hamilton (1996).

PROBLEMS

8.3.1 Determine the 4-hr unit hydrograph using the following data for a watershed having a drainage area of 200 km², assuming a constant rainfall abstraction rate and a constant baseflow of 20 m³/s.

Four-hour period	1	2	3	4	5	6	7	8	9	10	11
Rainfall (cm)	1.0	2.5	4.0	2.0							
Storm flow (m³/s)	25	75	175	225	180	100	80	60	40	25	20

8.3.2 Using the unit hydrograph developed in problem 8.3.1, determine the direct runoff from the 200 km² watershed using the following rainfall excess pattern.

Four-hour period	1	2	3	4
Rainfall excess (cm)	2.0	3.0	0.0	1.5

8.3.3 The ordinates at 1-hr intervals of a 1-hr unit hydrograph are 100, 300, 500, 700, 400, 200, and 100 cfs. Determine the direct runoff hydrograph from a 3-hr storm in which 1 in of excess rainfall occurred in the first hour, 2 in the second hour, and 0.5 in the third hour. What is the area of the watershed in mi²?

8.4.1 A watershed has a drainage area of 14 km²; the length of the main stream is 7.16 km, and the main channel length from the watershed outlet to the point opposite the center of gravity of the watershed is 3.22 km. Use $C_t = 2.0$ and $C_p = 0.625$ to determine the standard synthetic unit hydrograph for the watershed. What is the standard duration? Use Snyder's method to determine the 30-min unit hydrograph for the watershed.

8.4.2 Watershed A has a 2-hr unit hydrograph with $Q_p = 276$ m³/s, $t_{PR} = 6$ hr, $W_{50} = 4.0$ hr, and $W_{75} = 2$ hr. The watershed area $= 259$ km², $L_c = 16.1$ km, and $L = 38.6$ km. Watershed B is assumed to be hydrologically similar with an area of 181 km², $L = 25.1$ km, and $L_c = 15.1$ km. Determine the 1-hr synthetic unit hydrograph for watershed B. Determine the direct runoff hydrograph for a 2-hr storm that has 1.5 cm of excess rainfall the first hour and 2.5 cm of excess rainfall the second hour.

8.5.1 Using the 4-hr unit hydrograph developed in problem 8.3.1, use the S-curve method to develop the 8-hr unit hydrograph for this 200 km² watershed.

8.5.2 Using the one-hour unit hydrograph given in problem 8.3.3, develop the 3-hr unit hydrograph using the S-curve method.

8.7.1 Determine the weighted curve numbers for a watershed with 60 percent residential (1/4 acre lots), 20 percent open space, good condition, and 20 percent commercial and business (85 percent impervious) with corresponding soil groups of C, D, and C.

8.7.2 Rework example 8.7.1 with corresponding soil groups of B, C, and D.

8.7.3 The watershed in problem 8.7.1 experienced a rainfall of 5 in; what is the runoff volume?

8.7.4 Rework example 8.7.2 with a 7-in rainfall.

8.7.5 Calculate the cumulative abstractions and the excess rainfall hyetograph for the situation in problems 8.7.1 and 8.7.3.

8.7.6 Calculate the cumulative abstraction and the excess rainfall hyetograph for the situation in problems 8.7.2 and 8.7.4.

8.8.1 Develop the SCS triangular unit hydrograph for a 400-acre watershed that has been commercially developed. The flow length is 1500 ft, the slope is 3 percent and the soil group is group B.

8.8.2 Prior to development of the 400-acre watershed in problem 8.8.1, the land use was controlled pasture land with fair condition. Compute the SCS triangular unit hydrograph and compare with the one for commercially developed conditions.

8.8.3 Using the watershed defined in problems 8.8.1 and 8.8.2, determine the SCS triangular unit hydrograph assuming residential lot size. Compare with the results in problem 8.8.2.

8.8.4 A 20.7 km² watershed has a time of concentration of 1.0 hr. Calculate the 10-min unit hydrograph for the watershed using the SCS triangular unit hydrograph method. Determine the direct runoff hydrograph for a 30-min storm having 1.5 cm of excess rainfall the first 10 min, 0.5 cm the second 10 min, and 1.0 cm the third 10 min.

8.9.1 Develop a flowchart of the kinematic overland flow runoff model described in Section 8.9.

8.9.2 Develop the appropriate equations to solve equation 8.9.9 by Newton's method.

8.9.3 Derive equation (8.9.12).

REFERENCES

Chow, V. T. (editor), *Handbook of Applied Hydrology*, McGraw-Hill, New York, 1964.

Chow, V. T., D. R. Maidment, and L. W. Mays, *Applied Hydrology*, McGraw-Hill, New York, 1988.

Engman, E. T., Roughness Coefficients for Routing Surface Runoff, *Journal of Irrigation and Drainage Engineering*; American Society of Civil Engineers, 112(1), pp. 39–53, 1986.

Ford, D., and D. Hamilton, "Computer Models for Water-Excess Management," Chapter 28 in *Water Resources Handbook*, (edited by L. W. Mays), McGraw-Hill, New York, 1996.

Hewlett, J. D., and W. L. Nutter, *An Outline of Forest Hydrology*, University of Georgia Press, Athens, GA, 1969.

Horton, R. E., "Erosional Development of Streams and Their Drainage Basins; Hydrological Approach to Quantitative Morphology," *Bull. Geol. Soc. Am.*, vol. 56, pp. 275–370, 1945.

Maidment, D. R., *Handbook of Hydrology*, McGraw-Hill, New York, 1993.

Masch, F. D., *Hydrology*, Hydraulic Engineering Circular No. 19, FHWA-10-84-15, Federal Highway Administration, U.S. Department of the Interior, McClean, VA, 1984.

Mays, L. W., *Water Resources Handbook*, McGraw-Hill, New York, 1996.

Morris, E. M., and D. A. Woolhiser, "Unsteady One-Dimensional Flow over a Plane: Partial Equilibrium and Recession Hydrographs," *Water Resources Research*, vol. 16, no. 2, pp. 355–360, 1980.

Mosley, M. P., and A. I. McKerchar, "Streamflow," in *Handbook of Hydrology* (edited by D. R. Maidment), McGraw-Hill, New York, 1993.

Palmer, V. J., "Retardance Coefficients for Low Flow in Channels Lined with Vegetation," *Transactions of the American Geophysical Union*, 27(11), pp. 187–197, 1946.

Ponce, V. M., *Engineering Hydrology: Principles and Practices*, Prentice-Hall, Englewood Cliffs, NJ, 1989.

Sanders, T. G. (editor), *Hydrology for Transportation Engineers*, U.S. Dept. of Transportation, Federal Highway Administration, 1980.

Strahler, A. N., "Quantitative Geomorphology of Drainage Basins and Channel Networks," section 4-II in *Handbook of Applied Hydrology*, (edited by V. T. Chow), McGraw-Hill, New York, 1964.

U.S. Department of Agriculture Soil Conservation Service, National Engineering Handbook, Section 4, *Hydrology*, available from U.S. Government Printing Office, Washington, DC, 1972.

U.S. Department of Agriculture Soil Conservation Service, "A Method for Estimating Volume and Rate of Runoff in Small Watersheds," Tech. Paper 149, Washington, DC, April, 1973.

U.S. Department of Agriculture Soil Conservation Service, "Urban Hydrology for Small Watersheds," Tech. Release No. 55, Washington, DC, June, 1986.

Woolhiser, D. A., and J. A. Liggett, "Unsteady, One-Dimensional Flow over a Plane—the Rising Hydrograph," *Water Resources Research*, vol. 3(3), pp. 753–771, 1967.

Woolhiser, D. A., R. E. Smith, and D. C. Goodrich, "*KINEROS, A Kinematic Runoff and Erosion Model: Documentation and User Manual*, U.S. Department of Agricultural Research Service, ARS-77, Tucson, AZ, 1990.

Chapter 9

Reservoir and Stream Flow Routing

9.1 ROUTING

Figure 9.1.1 illustrates how stream flow increases as the *variable source area* extends into the drainage basin. The variable source area is the area of the watershed that is actually contributing flow to the stream at any point. The variable source area expands during rainfall and contracts thereafter.

Flow routing is the procedure to determine the time and magnitude of flow (i.e., the flow hydrograph) at a point on a watercourse from known or assumed hydrographs at one or more points upstream. If the flow is a flood, the procedure is specifically known as flood routing. Routing by lumped system methods is called *hydrologic (lumped) routing*, and routing by distributed systems methods is called *hydraulic (distributed) routing*.

For hydrologic routing, input $I(t)$, output $Q(t)$, and storage $S(t)$ as functions of time are related by the continuity equation (3.2.10)

$$\frac{dS}{dt} = I(t) - Q(t) \tag{9.1.1}$$

Even if an inflow hydrograph $I(t)$ is known, equation (9.1.1) cannot be solved directly to obtain the outflow hydrograph $Q(t)$, because both Q and S are unknown. A second relationship, or storage function, is required to relate S, I, and Q; coupling the storage function with the continuity equations provides a solvable combination of two equations and two unknowns.

The specific form of the storage function depends on the nature of the system being analyzed. In reservoir routing by the level pool method (Section 9.2), storage is a nonlinear function of Q, $S = f(Q)$ and the function $f(Q)$ is determined by relating reservoir storage and outflow to reservoir water level. In the Muskingum method (Section 9.3) for flow routing in channels, storage is linearly related to I and Q.

The effect of storage is to redistribute the hydrograph by shifting the centroid of the inflow hydrograph to the position of that of the outflow hydrograph in a *time of redistribution*. In very long channels, the entire flood wave also travels a considerable distance and the centroid of its hydrograph may then be shifted by a time period longer than the time of redistribution. This additional time may be considered the *time of translation*. The total time of flood movement between the centroids of the inflow and outflow hydrographs is equal to the sum of the time of redistribution and the time of translation. The process of redistribution modifies the shape of the hydrograph, while translation changes its position.

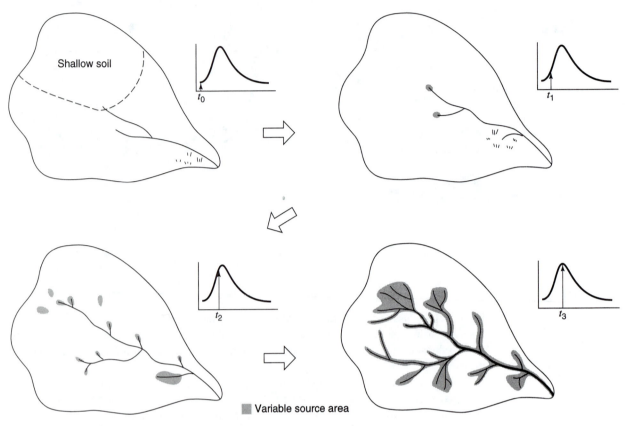

Shallow soil

Variable source area

Figure 9.1.1 The small arrows in the hydrographs show how streamflow increases as the variable source extends into swamps, shallow soils, and ephemeral channels. The process reverses as streamflow declines (from Hewlett (1982)).

9.2 HYDROLOGIC RESERVOIR ROUTING

Level pool routing is a procedure for calculating the outflow hydrograph from a reservoir assuming a horizontal water surface, given its inflow hydrograph and storage-outflow characteristics. Equation (9.1.1) can be expressed in the in finite-difference form to express the change in storage over a time interval (see Figure 9.2.1) as

$$S_{j+1} - S_j = \frac{I_j + I_{j+1}}{2}\Delta t - \frac{Q_j + Q_{j+1}}{2}\Delta t \tag{9.2.1}$$

The inflow values at the beginning and end of the jth time interval are I_j and I_{j+1}, respectively, and the corresponding values of the outflow are Q_j and Q_{j+1}. The values of I_j and I_{j+1} are pre-specified. The values of Q_j and S_j are known at the jth time interval from calculations for the previous time interval. Hence, equation (9.2.1) contains two unknowns, Q_{j+1} and S_{j+1}, which are isolated by multiplying (9.1.1) through by $2/\Delta t$, and rearranging the result to produce:

$$\left[\frac{2S_{j+1}}{\Delta t} + Q_{j+1}\right] = (I_j + I_{j+1}) + \left[\frac{2S_j}{\Delta t} - Q_j\right] \tag{9.2.2}$$

In order to calculate the outflow Q_{j+1}, a storage-outflow function relating $2S/\Delta t + Q$ and Q is needed. The method for developing this function using elevation-storage and elevation-outflow relationships is shown in Figure 9.2.2. The relationship between water surface elevation and reservoir storage can be derived by planimetering topographic maps or from field surveys. The elevation-discharge relation is derived from hydraulic equations relating head and discharge for

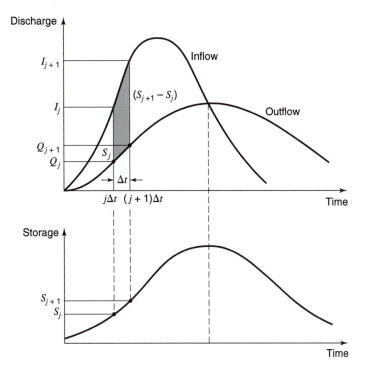

Figure 9.2.1 Change of storage during a routing period Δt.

various types of spillways and outlet works. (See Chapter 17.) The value of Δt is taken as the time interval of the inflow hydrograph. For a given value of water surface elevation, the values of storage S and discharge Q are determined (parts (a) and (b) of Figure 9.2.2), and then the value of $2S/\Delta t + Q$ is calculated and plotted on the horizontal axis of a graph with the value of the outflow Q on the vertical axis (part (c) of Figure 9.2.2).

In routing the flow through time interval j, all terms on the right side of equation (9.2.2) are known, and so the value of $2S_{j+1}/\Delta t + Q_{j+1}$ can be computed. The corresponding value of Q_{j+1} can be determined from the storage-outflow function $2S/\Delta t + Q$ versus Q, either graphically or by linear interpolation of tabular values. To set up the data required for the next time interval, the value of $(2S_{j+1}/\Delta t - Q_{j+1})$ is calculated using

$$\left[\frac{2S_{j+1}}{\Delta t} - Q_{j+1}\right] = \left[\frac{2S_{j+1}}{\Delta t} + Q_{j+1}\right] - 2Q_{j+1} \tag{9.2.3}$$

The computation is then repeated for subsequent routing periods.

EXAMPLE 9.2.1

Consider a 2-acre stormwater detention basin with vertical walls. The triangular inflow hydrograph increases linearly from zero to a peak of 60 cfs at 60 min and then decreases linearly to a zero discharge at 180 min. Route the inflow hydrograph through the detention basin using the head-discharge relationship for the 5-ft diameter pipe spillway in columns (1) and (2) of Table 9.2.1. The pipe is located at the bottom of the basin. Assuming the basin is initially empty, use the level pool routing procedure with a 10-min time interval to determine the maximum depth in the detention basin.

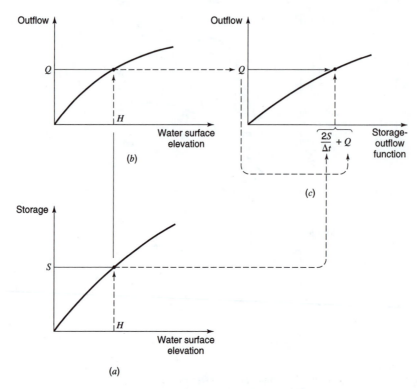

Figure 9.2.2 Development of the storage-outflow function for level pool routing on the basis of storage-elevation-outflow curves (from Chow et al. (1988)).

Table 9.2.1 Elevation-Discharge-Storage Data for Example 9.2.1

1	2	3	4
Head H (ft)	Discharge Q (cfs)	Storage S (ft^3)	$\dfrac{2S}{\Delta t} + Q$ (cfs)
0.0	0	0	.00
0.5	3	43,500	148.20
1.0	8	87,120	298.40
1.5	17	130,680	452.60
2.0	30	174,240	610.80
2.5	43	217,800	769.00
3.0	60	261,360	931.20
3.5	78	304,920	1094.40
4.0	97	348,480	1258.60
4.5	117	392,040	1423.80
5.0	137	435,600	1589.00

SOLUTION

The inflow hydrograph and the head-discharge (columns 1 and 3) and discharge-storage (columns 2 and 3) relationships are used to determine the routing relationship in Table 9.2.1. A routing interval of 10 min is used to determine the routing relationship $2S/\Delta t + Q$ vs. Q, which is columns 2 and 4 in Table 9.2.1. The routing computations are presented in Table 9.2.2, with the sequence of computations indicated by the arrows. These computations are carried out using equation (9.2.3). For the first time interval, $S_1 = Q_1 = 0$ because the reservoir is empty at $t = 0$; then $(2S_1/\Delta t - Q_1) = 0$. The value of the storage-outflow function at the end of the time interval is

$$\left[\frac{2S_2}{\Delta t} + Q_2\right] = (I_1 + I_2) + \left[\frac{2S_1}{\Delta t} - Q_1\right] = (0 + 10) + 0 = 10$$

The value of Q_2 is determined using linear interpolation, so that

$$Q_2 = 0 + \frac{(3-0)}{(148.2-0)}(10-0) = 0.2 \text{ cfs}$$

With $Q_2 = 0.2$, then $2S_2/\Delta t - Q_2$ for the next iteration is

$$\left[\frac{2S_2}{\Delta t} - Q_2\right] = \left[\frac{2S_2}{\Delta t} + Q_2\right] - 2Q_2 = 10 - 2(0.2) = 9.6 \text{ cfs}$$

The computation now proceeds to the next time interval. Refer to Table 9.2.1 for the remaining computations.

Table 9.2.2 Routing of Flow Through Detention Reservoir by the Level Pool Method (example 9.2.1)

Time t	Inflow I_j	$I_j + I_{j+1}$	$\frac{2S_j}{\Delta t} - Q_j$	$\frac{2S_{j+1}}{\Delta t} + Q_{j+1}$	Outflow
(min)	(cfs)	(cfs)	(cfs)	(cfs)	(cfs)
.00	.00				.00
10.00	10.00	10.00	.00	10.00	.20
20.00	20.00	30.00	9.60	39.60	.80
30.00	30.00	50.00	37.99	87.99	1.78
40.00	40.00	70.00	84.43	154.43	3.21
50.00	50.00	90.00	148.01	238.01	5.99
60.00	60.00	110.00	226.04	336.04	10.20
70.00	55.00	115.00	315.64	430.64	15.72
80.00	50.00	105.00	399.21	504.21	21.24
90.00	45.00	95.00	461.72	556.72	25.56
100.00	40.00	85.00	505.61	590.61	28.34
110.00	35.00	75.00	533.93	608.93	29.85
120.00	30.00	65.00	549.24	614.24	30.28
130.00	25.00	55.00	553.67	608.67	29.83
140.00	20.00	45.00	549.02	594.02	28.62
150.00	15.00	35.00	536.78	571.78	26.79
160.00	10.00	25.00	518.19	543.19	24.44
170.00	5.00	15.00	494.30	509.30	21.66
180.00	.00	5.00	465.98	470.98	18.51
190.00	.00	.00	433.96	433.96	15.91
200.00	.00	.00	402.14	402.14	14.05
210.00	.00	.00	374.03	374.03	12.41
220.00	.00	.00	349.20	349.20	10.97
230.00	.00	.00	327.27	327.27	9.69
240.00	.00	.00	307.90	307.90	8.55

9.3 HYDROLOGIC RIVER ROUTING

The *Muskingum method* is a commonly used hydrologic routing method that is based upon a variable discharge-storage relationship. This method models the storage volume of flooding in a river channel by a combination of wedge and prism storage (Figure 9.3.1). During the advance of a flood wave, inflow exceeds outflow, producing a wedge of storage. During the recession, outflow exceeds inflow, resulting in a negative wedge. In addition, there is a prism of storage that is formed by a volume of constant cross-section along the length of prismatic channel.

Assuming that the cross-sectional area of the flood flow is directly proportional to the discharge at the section, the *volume of prism storage* is equal to KQ, where K is a proportionality coefficient (approximate as the travel time through the reach), and the *volume of wedge storage* is equal to $KX(I - Q)$, where X is a weighting factor having the range $0 \le X \le 0.5$. The total storage is defined as the sum of two components,

$$S = KQ + KX(I - Q) \tag{9.3.1}$$

which can be rearranged to give the storage function for the Muskingum method

$$S = K[XI + (1 - X)Q)] \tag{9.3.2}$$

and represents a linear model for routing flow in streams.

The value of X depends on the shape of the modeled wedge storage. The value of X ranges from 0 for reservoir-type storage to 0.5 for a full wedge. When $X = 0$, there is no wedge and hence no backwater; this is the case for a level-pool reservoir. In natural streams, X is between 0 and 0.3, with a mean value near 0.2. Great accuracy in determining X may not be necessary because the results of the method are relatively insensitive to the value of this parameter. The parameter K is the time of travel of the flood wave through the channel reach. For hydrologic routing, the values of K and X are assumed to be specified and constant throughout the range of flow.

The values of storage at time j and $j + 1$ can be written, respectively, as

$$S_j = K[XI_j + (1 - X)Q_j] \tag{9.3.3}$$

$$S_{j+1} = K[XI_{j+1} + (1 - X)Q_{j+1}] \tag{9.3.4}$$

Using equations (9.3.3) and (9.3.4), the change in storage over time interval Δt is

$$S_{j+1} - S_j = K\{[XI_{j+1} + (1 - X)Q_{j+1}] - [XI_j + (1 - X)Q_j]\} \tag{9.3.5}$$

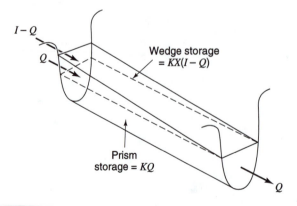

Figure 9.3.1 Prism and wedge storages in a channel reach.

The change in storage can also be expressed using equation (9.2.1). Combining equations (9.3.5) and (9.2.1) and simplifying gives

$$Q_{j+1} = C_1 I_{j+1} + C_2 I_j + C_3 Q_j \qquad (9.3.6)$$

which is the routing equation for the Muskingum method, where

$$C_1 = \frac{\Delta t - 2KX}{2K(1-X) + \Delta t} \qquad (9.3.7)$$

$$C_2 = \frac{\Delta t + 2KX}{2K(1-X) + \Delta t} \qquad (9.3.8)$$

$$C_3 = \frac{2K(1-X) - \Delta t}{2K(1-X) + \Delta t} \qquad (9.3.9)$$

Note that $C_1 + C_2 + C_3 = 1$.

The routing procedure can be repeated for several sub-reaches (N_{steps}) so that the total travel time through the reach is K. To insure that the method is computationally stable and accurate, the U.S. Army Corps of Engineers (1990) uses the following criterion to determine the number of routing reaches:

$$\frac{1}{2(1-X)} \le \frac{K}{N_{steps}\Delta t} \le \frac{1}{2X} \qquad (9.3.10)$$

If observed inflow and outflow hydrographs are available for a river reach, the values of K and X can be determined. Assuming various values of X and using known values of the inflow and outflow, successive values of the numerator and denominator of the following expression for K, derived from equations (9.3.5) and (9.3.8), can be computed using

$$K = \frac{0.5\Delta t\left[(I_{j+1} + I_j) - (Q_{j+1} + Q_j)\right]}{X(I_{j+1} - I_j) + (1-X)(Q_{j+1} - Q_j)} \qquad (9.3.11)$$

The computed values of the numerator (storage) and denominator (weighted discharges) are plotted for each time interval, with the numerator on the vertical axis and the denominator on the horizontal axis. This usually produces a graph in the form of a loop, as shown in Figure 9.3.2. The value of X that produces a loop closest to a single line is taken to be the correct value for the reach, and K, according to equation (9.3.11), is equal to the slope of the line. Since K is the time required for the incremental flood wave to traverse the reach, its value may also be estimated as the observed time of travel of peak flow through the reach.

EXAMPLE 9.3.1

The objective of this example is to determine K and X for the Muskingum routing method using the February 26 to March 4, 1929 data on the Tuscasawas River from Dover to Newcomerstown. This example is taken from the U.S. Army Corps of Engineers (1960) as used in Cudworth (1989). Columns 2 and 3 in Table 9.3.1 are the inflow and outflow hydrographs for the reach. The numerator and denominator of equation (9.3.11) were computed (for each time period) using four values of $X = 0, 0.1, 0.2$, and 0.3. The accumulated numerators are in column 9 and the accumulated denominators (weighted discharges) are in columns 11, 13, 15, and 17. In Figure 9.3.2, the accumulated numerator (storages) from column (9) are plotted against the corresponding accumulated denominator (weighted discharges) for each of the four X values. According to Figure 9.3.2, the best fit (linear relationship) appears to be for $X = 0.2$, which has a resulting $K = 1.0$. To perform a routing, K should equal Δt, so that if $\Delta t = 0.5$ day, as in this case, the reach should be subdivided into two equal reaches ($N_{steps} = 2$) and the value of K should be 0.5 day for each reach.

Table 9.3.1 Determination of Coefficients K and X for the Muskingum Routing Method. Tuscarawas River, Muskingum Basin, Ohio Reach from Dover to Newcomerstown, February 26 to March 4, 1929

(1) Date Δt = 0.5 day	(2) In-flow[1], ft³/s	(3) Out-flow[2], ft³/s	(4) I_2+I_1, ft³/s	(5) O_2+O_1, ft³/s	(6) I_2-I_1, ft³/s	(7) O_2-O_1, ft³/s	(8) N[3]	(9) ΣN	(10) D[4] X=0	(11) ΣD X=0	(12) D X=0.1	(13) ΣD X=0.1	(14) D X=0.2	(15) ΣD X=0.2	(16) D X=0.3	(17) ΣD X=0.3
2-26-29 a.m.	2,200	2,000	16,700	9,000	12,300	5,000	1,900	—	5,000	—	5,700	—	6,500	—	7,200	—
p.m.	14,500	7,000	42,900	18,700	13,900	4,700	6,100	1,900	4,700	5,000	5,600	5,700	6,500	6,500	7,500	7,200
2-27-29 a.m.	28,400	11,700	60,200	28,200	3,400	4,800	8,000	8,000	4,800	9,700	4,600	11,300	4,500	13,000	4,300	14,700
p.m.	31,800	16,500	61,500	40,500	-2,100	7,500	5,200	16,000	7,500	14,500	6,700	15,900	5,600	17,500	4,600	19,000
2-28-29 a.m.	29,700	24,00	55,000	53,100	-4,400	5,100	500	21,200	5,100	22,000	4,100	22,600	3,200	23,100	2,300	23,600
p.m.	25,300	29,100	45,700	57,500	-4,900	-700	-2,900	21,700	-700	27,100	-1,100	26,700	-1,500	26,300	-2,000	25,900
3-01-29 a.m.	20,400	28,400	36,700	52,200	-4,100	-4,600	-3,900	18,800	-4,600	26,400	-4,600	25,600	-4,500	24,800	-4,400	23,900
p.m.	16,300	23,800	28,900	43,200	-3,700	-4,400	-3,600	14,900	-4,400	21,800	-4,300	21,000	-4,300	20,300	-4,200	19,500
3-02-29 a.m.	12,600	19,300	21,900	34,700	-3,300	-4,100	-3,200	11,300	-4,100	17,400	-4,000	16,700	-3,900	16,000	-3,900	15,300
p.m.	9,300	15,300	16,000	26,500	-2,600	-4,100	-2,500	8,100	-4,100	13,300	-4,000	12,700	-3,800	12,100	-3,600	11,400
3-03-29 a.m.	6,700	11,200	11,700	19,400	-1,700	-3,000	-1,900	5,500	-3,000	9,200	-2,800	8,700	-2,800	8,300	-2,600	7,800
p.m.	5,000	8,200	9,100	14,600	-900	-1,800	-1,400	3,600	-1,800	6,200	-1,700	5,900	-1,600	5,500	-1,600	5,200
3-04-29 a.m.	4,100	6,400	7,700	11,600	-500	-1,200	-1,000	2,200	-1,200	4,400	-1,200	4,200	-1,100	3,900	-900	3,600
p.m.	3,600	5,200	6,000	9,800	-1,200	-600	-1,000	1,200	-600	3,200	-600	3,000	-700	2,800	-800	2,700
3-05-29 a.m.	2,400	4,600	—	—	—	—	—	200	—	2,600	—	2,400	—	2,100	—	1,900

[1]Inflow to reach was adjusted to equal volume of outflow.
[2]Outflow is the hydrograph at Newcomerstown.
[3]Numerator, N, is $\Delta t/2$, column (4) – column (5).
[4]Denominator, D, is column (7) + X[column (6) – column (7)].

Note: From plottings of column (9) versus columns (11), (13), (15), and (17), the plot giving the best fit is considered to define K and X.

$$K = \frac{\text{Numerator, } N}{\text{Denominator, } D} = \frac{0.5\Delta t[(I_2+I_1)-(O_2+O_1)]}{X(I_2-I_1)+(1-X)(O_2-O_1)}$$

Source: Cudworth (1989).

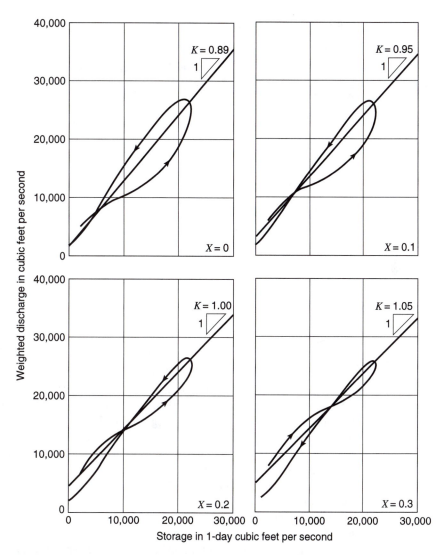

Figure 9.3.2 Typical valley storage curves.

EXAMPLE 9.3.2

Route the inflow hydrograph below using the Muskingum method; $\Delta t = 1$ hr, $X = 0.2$, $K = 0.7$ hrs.

Time (hrs)	0	1	2	3	4	5	6	7
Inflow (cfs)	0	800	2000	4200	5200	4400	3200	2500

Time (hrs)	8	9	10	11	12	13
Inflow (cfs)	2000	1500	1000	700	400	0

$$C_1 = \frac{1.0 - 2(0.7)(0.2)}{2(0.7)(1 - 0.2) + 1.0} = 0.3396$$

$$C_2 = \frac{1.0 + 2(0.7)(0.2)}{2(0.7)(1 - 0.2) + 1.0} = 0.6038$$

$$C_3 = \frac{2(0.7)(1 - 0.2) - 1.0}{2(0.7)(1 - 0.2) + 1.0} = 0.0566$$

(Adapted from Masch (1984).)

Check to see if $C_1 + C_2 + C_3 = 1$:

$0.3396 + 0.6038 + 0.0566 = 1$

Using equation (9.3.6) with $I_1 = 0$ cfs, $I_2 = 800$ cfs, and $Q_1 = 0$ cfs, compute Q_2 at $t = 1$ hr:

$$Q_2 = C_1 I_2 + C_2 I_1 + C_3 Q_1$$

$$= (0.3396)(800) + 0.6038(0) + 0.0566(0)$$

$$= 272 \text{ cfs } (7.7 \text{ m}^3/\text{s})$$

Next compute Q_3 at $t = 2$ hr:

$$Q_3 = C_1 I_3 + C_2 I_2 + C_3 Q_2$$

$$= (0.3396)(2000) + 0.6038(800) + 0.0566(272)$$

$$= 1178 \text{ cfs } (33 \text{ m}^3/\text{s})$$

The remaining computations result in

Time (hrs)	0	1	2	3	4	5	6	7
Q (cfs)	0	272	1178	2701	4455	4886	4020	3009

Time (hrs)	8	9	10	11	12	13	14	15
Q (cfs)	2359	1851	1350	918	610	276	16	1

9.4 HYDRAULIC (DISTRIBUTED) ROUTING

Distributed routing or *hydraulic routing*, also referred to as *unsteady flow routing,* is based upon the one-dimensional unsteady flow equations referred to as the *Saint–Venant equations*. The hydrologic river routing and the hydrologic reservoir routing procedures presented previously are lumped procedures and compute flow rate as a function of time alone at a downstream location. Hydraulic (distributed) flow routings allow computation of the flow rate and water surface elevation (or depth) as function of both space (location) and time. The Saint–Venant equations are presented in Table 9.4.1 in both the *velocity-depth (nonconservation) form* and the *discharge-area (conservation) form*.

The momentum equation contains terms for the physical processes that govern the flow momentum. These terms are: the *local acceleration term*, which describes the change in momentum due to the change in velocity over time, the *convective acceleration term*, which describes the change in momentum due to change in velocity along the channel, the *pressure force term*, proportional to the change in the water depth along the channel, the gravity force term, proportional to the bed slope S_0, and the friction force term, proportional to the friction slope S_f. The local and convective acceleration terms represent the effect of inertial forces on the flow.

Alternative distributed flow routing models are produced by using the full continuity equation while eliminating some terms of the momentum equation (refer to Table 9.4.1). The simplest distributed model is the *kinematic wave model*, which neglects the local acceleration, convective acceleration, and pressure terms in the momentum equation; that is, it assumes that $S_0 = S_f$ and the friction and gravity forces balance each other. The *diffusion wave model* neglects the local and convective acceleration terms but incorporates the pressure term. The *dynamic wave model* considers all the acceleration and pressure terms in the momentum equation.

The momentum equation can also be written in forms that take into account whether the flow is steady or unsteady, and uniform or nonuniform, as illustrated in Table 9.4.1. In the continuity equation, $\partial A/\partial t = 0$ for a steady flow, and the lateral inflow q is zero for a uniform flow.

Table 9.4.1   Summary of the Saint–Venant Equations*

Continuity equation

Conservation form

$$\frac{\partial Q}{\partial x} + \frac{\partial A}{\partial t} = 0$$

Nonconservation form

$$V\frac{\partial y}{\partial x} + \frac{\partial V}{\partial x} + \frac{\partial y}{\partial t} = 0$$

Momentum equation

Conservation form

$$\frac{1}{A}\frac{\partial Q}{\partial t} \quad + \quad \frac{1}{A}\frac{\partial}{\partial x}\left(\frac{Q^2}{A}\right) \quad + \quad g\frac{\partial y}{\partial x} \quad - \quad g(S_0 \quad - \quad S_f) \quad = 0$$

Local acceleration term	Convective acceleration term	Pressure force term	Gravity force term	Friction force term

Nonconservation form (unit with element)

$$\frac{\partial V}{\partial t} \quad + \quad V\frac{\partial V}{\partial x} \quad + \quad g\frac{\partial y}{\partial x} \quad - \quad g(S_0 \quad - \quad S_f) \quad = 0$$

|———————————— Kinematic wave
|—————————————————— Diffusion wave
|————————————————————————————— Dynamic wave

*Neglecting lateral inflow, wind shear, and eddy losses, and assuming $\beta = 1$.

x = longitudinal distance along the channel or river, t = time, A = cross-sectional area of flow, h = water surface elevation, S_f = friction slope, S_0 = channel bottom slope, g = acceleration due to gravity, V = velocity of flow, and y = depth of flow.

9.4.1   Unsteady Flow Equations: Continuity Equation

The *continuity equation* for an unsteady variable-density flow through a control volume can be written as in equation (3.2.1):

$$0 = \frac{d}{dt}\int_{CV}\rho \, d\forall + \int_{CS}\rho \mathbf{V}\cdot \mathbf{dA} \tag{9.4.1}$$

Consider an elemental control volume of length dx in a channel. Figure 9.4.1 shows three views of the control volume: (*a*) an elevation view from the side, (*b*) a plan view from above, and (*c*) a channel cross-section. The inflow to the control volume is the sum of the flow Q entering the control volume at the upstream end of the channel and the lateral inflow q entering the control volume as a distributed flow along the side of the channel. The dimensions of q are those of flow per unit length of channel, so the rate of lateral inflow is qdx and the mass inflow rate is

$$\int_{inlet}\rho \mathbf{V}\cdot \mathbf{dA} = -\rho(Q + qdx) \tag{9.4.2}$$

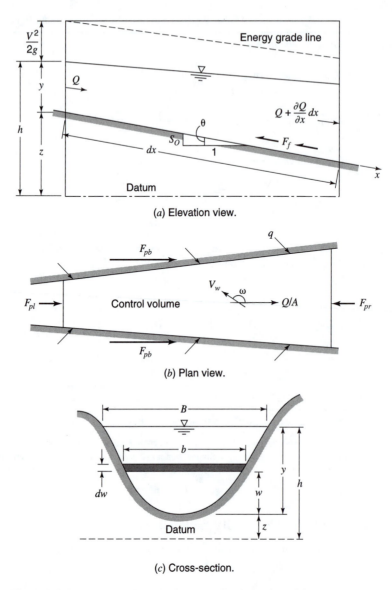

Figure 9.4.1 An elemental reach of channel for derivation of Saint–Venant equations.

This is negative because inflows are considered negative in the control volume approach (Reynolds transport theorem). The mass outflow from the control volume is

$$\int_{outlet} \rho \mathbf{V} \, \mathbf{dA} = \rho \left(Q + \frac{\partial Q}{\partial x} dx \right) \qquad (9.4.3)$$

where $\partial Q/\partial x$ is the rate of change of channel flow with distance. The volume of the channel element is $A dx$, where A is the average cross-sectional area, so the rate of change of mass stored within the control volume is

$$\frac{d}{dt} \int_{CV} \rho \, d\forall = \frac{\partial(\rho A dx)}{\partial t} \qquad (9.4.4)$$

where the partial derivative is used because the control volume is defined to be fixed in size (though the water level may vary within it). The net outflow of mass from the control volume is found by substituting equations (9.4.2)–(9.4.4) into (9.4.1):

$$\frac{\partial(\rho A dx)}{dt} - \rho(Q + q dx) + \rho\left(Q + \frac{\partial Q}{\partial x} dx\right) = 0 \tag{9.4.5}$$

Assuming the fluid density ρ is constant, equation (9.4.5) is simplified by dividing through by ρdx and rearranging to produce the *conservation form* of the continuity equation,

$$\frac{\partial Q}{\partial x} + \frac{\partial A}{\partial t} - q = 0 \tag{9.4.6}$$

which is applicable at a channel cross-section. This equation is valid for a *prismatic* or a *nonprismatic* channel; a prismatic channel is one in which the cross-sectional shape does not vary along the channel and the bed slope is constant.

For some methods of solving the Saint–Venant equations, the *nonconservation form* of the continuity equation is used, in which the average flow velocity V is a dependent variable, instead of Q. This form of the continuity equation can be derived for a unit width of flow within the channel, neglecting lateral inflow, as follows. For a unit width of flow, $A = y \times 1 = y$ and $Q = VA = Vy$. Substituting into equation (9.4.6) yields

$$\frac{\partial(Vy)}{\partial x} + \frac{\partial y}{\partial t} = 0 \tag{9.4.7}$$

or

$$V\frac{\partial y}{\partial x} + y\frac{\partial V}{\partial x} + \frac{\partial y}{\partial t} = 0 \tag{9.4.8}$$

9.4.2 Momentum Equation

Newton's second law is written in the form of Reynolds transport theorem as in equation (3.4.5):

$$\sum \mathbf{F} = \frac{d}{dt}\int_{CV} \mathbf{V}\rho\, d\forall + \sum_{CS} \mathbf{V}\rho\mathbf{V}\cdot d\mathbf{A} \tag{9.4.9}$$

This states that the sum of the forces applied is equal to the rate of change of momentum stored within the control volume plus the net outflow of momentum across the control surface. This equation, in the form $\Sigma F = 0$, was applied to steady uniform flow in an open channel in Chapter 5. Here, unsteady nonuniform flow is considered.

Forces. There are five forces acting on the control volume:

$$\sum F = F_g + F_f + F_e + F_p \tag{9.4.10}$$

where F_g is the *gravity force* along the channel due to the weight of the water in the control volume, F_f is the *friction force* along the bottom and sides of the control volume, F_e is the *contraction/expansion force* produced by abrupt changes in the channel cross-section, and F_p is the *unbalanced pressure force* (see Figure 9.4.1). Each of these four forces is evaluated in the following paragraphs.

Gravity. The volume of fluid in the control volume is Adx and its weight is $\rho gAdx$. For a small angle of channel inclination θ, $S_0 \approx \sin\theta$ and the gravity force is given by

$$F_g = \rho gAdx \sin\theta \approx \rho gAS_0 dx \tag{9.4.11}$$

where the channel bottom slope S_0 equals $-\partial z/\partial x$.

Friction. Frictional forces created by the shear stress along the bottom and sides of the control volume are given by $-\tau_0 P dx$, where $\tau_0 = \gamma R S_f = \rho g(A/P)S_f$ is the bed shear stress and P is the wetted perimeter. Hence the friction force is written as

$$F_f = -\rho g A S_f dx \tag{9.4.12}$$

where the friction slope S_f is derived from resistance equations such as Manning's equation.

Contraction/expansion. Abrupt contractions or expansions of the channel cause energy losses through eddy motion. Such losses are similar to minor losses in a pipe system. The magnitude of eddy losses is related to the change in velocity head $V^2/2g = (Q/A)^2/2g$ through the length of channel causing the losses. The drag forces creating these eddy losses are given by

$$F_e = -\rho g A S_e dx \tag{9.4.13}$$

where S_e is the eddy loss slope

$$S_e = \frac{K_e}{2g} \frac{\partial (Q/A)^2}{\partial x} \tag{9.4.14}$$

in which K_e is the nondimensional expansion or contraction coefficient, negative for channel expansion (where $\partial (Q/A)^2/\partial x$ is negative) and positive for channel contractions.

Pressure. Referring to Figure 9.4.1, the unbalanced pressure force is the resultant of the hydrostatic force on the each side of the control volume. Chow et al. (1988) provide a detailed derivation of the pressure force F_p as simply

$$F_p = \rho g A \frac{\partial y}{\partial x} dx \tag{9.4.15}$$

The sum of the forces in equation (9.4.10) can be expressed, after substituting equations (9.4.11), (9.4.12), (9.4.13), and (9.4.15), as

$$\Sigma F = \rho A S_0 dx - \rho g A S_f dx - \rho g A S_e dx - \rho g A \frac{\partial y}{\partial x} dx \tag{9.4.16}$$

Momentum. The two momentum terms on the right-hand side of equation (9.4.9) represent the rate of change of storage of momentum in the control volume, and the net outflow of momentum across the control surface, respectively.

Net momentum outflow. The mass inflow rate to the control volume (equation (9.4.2)) is $-\rho(Q + qdx)$, representing both stream inflow and lateral inflow. The corresponding momentum is computed by multiplying the two mass inflow rates by their respective velocity and a *momentum correction factor* β:

$$\int_{inlet} V \rho V\, dA = -\rho(\beta VQ + \beta v_x q dx) \tag{9.4.17}$$

where $-\rho \beta VQ$ is the momentum entering from the upstream end of the channel, and $-\rho \beta v_x q dx$ is the momentum entering the main channel with the lateral inflow, which has a velocity v_x in the x direction. The term β is known as the *momentum coefficient* or *Boussinesq coefficient*; it accounts for the nonuniform distribution of velocity at a channel cross-section in computing the momentum. The value of β is given by

$$\beta = \frac{1}{V^2 A} \int v^2 dA \tag{9.4.18}$$

where v is the velocity through a small element of area dA in the channel cross-section. The value of β ranges from 1.01 for straight prismatic channels to 1.33 for river valleys with floodplains (Chow, 1959; Henderson, 1966).

The momentum leaving the control volume is

$$\int_{\text{outlet}} \mathbf{V}\rho\mathbf{V} \ \mathbf{dA} = \rho\left[\beta VQ + \frac{\partial(\beta VQ)}{\partial x}dx\right] \qquad (9.4.19)$$

The net outflow of momentum across the control surface is the sum of equations (9.4.17) and (9.4.19):

$$\int_{\text{CS}} \mathbf{V}\rho\mathbf{V}\mathbf{dA} = -\rho(\beta VQ + \beta v_x qdx) + \rho\left[\beta VQ + \frac{\partial(\beta VQ)}{\partial x}dx\right]$$

$$= -\rho\left[\beta v_x q - \frac{\partial(\beta VQ)}{\partial x}\right]dx \qquad (9.4.20)$$

Momentum storage. The time rate of change of momentum stored in the control volume is found by using the fact that the volume of the elemental channel is $A dx$, so its momentum is $\rho A dx V$, or $\rho Q dx$, and then

$$\frac{d}{dt}\int_{\text{CV}} \mathbf{V}\rho \, d\forall = \rho\frac{\partial Q}{\partial x}dx \qquad (9.4.21)$$

After substituting the force terms from equation (9.4.16) and the momentum terms from equations (9.4.20) and (9.4.21) into the momentum equation (9.4.9), it reads

$$\rho gAS_0 dx - \rho gAS_f dx - \rho gAS_e dx - \rho gA\frac{\partial y}{\partial x}dx = -\rho\left[\beta \, v_x q - \frac{\partial(\beta VQ)}{\partial x}\right]dx + \rho\frac{\partial Q}{\partial t}dx \quad (9.4.22)$$

Dividing through by ρdx, replacing V with Q/A, and rearranging produces the conservation form of the momentum equation:

$$\frac{\partial Q}{\partial t} + \frac{\partial(\beta Q^2/A)}{\partial t} + gA\left(\frac{\partial y}{\partial x} - S_0 + S_f + S_e\right) - \beta q v_x = 0 \qquad (9.4.23)$$

The depth y in equation (9.4.23) can be replaced by the water surface elevation h, using

$$h = y + z \qquad (9.4.24)$$

where z is the elevation of the channel bottom above a datum such as mean sea level. The derivative of equation (9.4.24) with respect to the longitudinal distance x along the channel is

$$\frac{\partial h}{\partial x} = \frac{\partial y}{\partial x} + \frac{\partial z}{\partial x} \qquad (9.4.25)$$

but $\partial z/\partial x = -S_0$, so

$$\frac{\partial h}{\partial x} = \frac{\partial y}{\partial x} - S_0 \qquad (9.4.26)$$

The momentum equation can now be expressed in terms of h by using equation (9.4.26) in (9.4.23):

$$\frac{\partial Q}{\partial t} + \frac{\partial(\beta \, Q^2/A)}{\partial x} + gA\left(\frac{\partial h}{\partial x} + S_f + S_e\right) - \beta q v_x = 0 \qquad (9.4.27)$$

The Saint–Venant equations, (9.4.6) for continuity and (9.4.27) for momentum, are the governing equations for one-dimensional, unsteady flow in an open channel. The use of the terms S_f and S_e in equation (9.4.27), which represent the rate of energy loss as the flow passes through the channel, illustrates the close relationship between energy and momentum considerations in describing the flow. Strelkoff (1969) showed that the momentum equation for the Saint–Venant equations can also be derived from energy principles, rather than by using Newton's second law as presented here.

The nonconservation form of the momentum equation can be derived in a similar manner to the nonconservation form of the continuity equation. Neglecting eddy losses, wind shear effect, and lateral inflow, the nonconservation form of the momentum equation for a unit width in the flow is

$$\frac{\partial V}{\partial t} + V\frac{\partial V}{\partial x} + g\left(\frac{\partial y}{\partial x} - S_0 + S_f\right) = 0 \tag{9.4.28}$$

9.5 KINEMATIC WAVE MODEL FOR CHANNELS

In Section 8.9, a kinematic wave overland flow runoff model was presented. This is an implicit nonlinear kinematic model that is used in the KINEROS model. This section presents a general discussion of the kinematic wave followed by brief description of the very simplest linear models, such as those found in the U.S. Army Corps of Engineers HEC-1, and the more complicated models such as the KINEROS model (Woolhiser et al., 1990).

Kinematic waves govern flow when inertial and pressure forces are not important. Dynamic waves govern flow when these forces are important, as in the movement of a large flood wave in a wide river. In a kinematic wave, the gravity and friction forces are balanced, so the flow does not accelerate appreciably.

For a kinematic wave, the energy grade line is parallel to the channel bottom and the flow is steady and uniform ($S_0 = S_f$) within the differential length, while for a dynamic wave the energy grade line and water surface elevation are not parallel to the bed, even within a differential element.

9.5.1 Kinematic Wave Equations

A *wave* is a variation in a flow, such as a change in flow rate or water surface elevation, and the *wave celerity* is the velocity with which this variation travels along the channel. The celerity depends on the type of wave being considered and may be quite different from the water velocity. For a kinematic wave the acceleration and pressure terms in the momentum equation are negligible, so the wave motion is described principally by the equation of continuity. The name kinematic is thus applicable, as *kinematics* refers to the study of motion exclusive of the influence of mass and force; in *dynamics* these quantities are included.

The kinematic wave model is defined by the following equations.

Continuity:

$$\frac{\partial Q}{\partial x} + \frac{\partial A}{\partial t} = q(x, t) \tag{9.5.1}$$

Momentum:

$$S_0 = S_f \tag{9.5.2}$$

where $q(x, t)$ is the net lateral inflow per unit length of channel.

The momentum equation can also be expressed in the form

$$A = \alpha Q^\beta \tag{9.5.3}$$

For example, Manning's equation written with $S_0 = S_f$ and $R = A/P$ is

$$Q = \frac{1.49 S_0^{1/2}}{n P^{2/3}} A^{5/3} \tag{9.5.4}$$

which can be solved for A as

$$A = \left(\frac{n P^{2/3}}{1.49 \sqrt{S_0}} \right)^{3/5} Q^{3/5} \tag{9.5.5}$$

so $\alpha = \left[n P^{2/3} / \left(1.49 \sqrt{S_0} \right) \right]^{0.6}$ and $\beta = 0.6$ in this case.

Equation (9.5.1) contains two dependent variables, A and Q, but A can be eliminated by differentiating equation (9.5.3):

$$\frac{\partial A}{\partial t} = \alpha \beta Q^{\beta - 1} \left(\frac{\partial Q}{\partial t} \right) \tag{9.5.6}$$

and substituting for $\partial A / \partial t$ in equation (9.5.1) to give

$$\frac{\partial Q}{\partial x} + \alpha \beta Q^{\beta - 1} \left(\frac{\partial Q}{\partial t} \right) = q \tag{9.5.7}$$

Alternatively, the momentum equation could be expressed as

$$Q = a A^B \tag{9.5.8}$$

where a and B are defined using Manning's equation. Using

$$\frac{\partial Q}{\partial x} = \frac{dQ}{dA} \frac{\partial A}{\partial x} \tag{9.5.9}$$

the governing equation is

$$\frac{\partial A}{\partial t} + \frac{dQ}{dA} \frac{\partial A}{\partial x} = q \tag{9.5.10}$$

where dQ/dA is determined by differentiating equation (9.5.8):

$$\frac{dQ}{dA} = a B A^{B-1} \tag{9.5.11}$$

and substituting in equation (9.5.10):

$$\frac{\partial A}{\partial t} = a B A^{B-1} \frac{\partial A}{\partial t} = q \tag{9.5.12}$$

The kinematic wave equation (9.5.7) has Q as the dependent variable and the kinematic wave equation (9.5.12) has A as the dependent variable. First consider equation (9.5.7), by taking the logarithm of (9.5.3):

$$\ln A = \ln \alpha + \beta \ln Q \tag{9.5.13}$$

and differentiating

$$\frac{dQ}{Q} = \frac{1}{\beta} \left(\frac{dA}{A} \right) \tag{9.5.14}$$

This defines the relationship between relative errors dA/A and dQ/Q. For Manning's equation $\beta < 1$, so that the discharge estimation error would be magnified by the ratio $1/\beta$ if A were the dependent variable instead of Q.

Next consider equation (9.5.12); by taking the logarithm of (9.5.8):

$$\ln Q = \ln a + B \ln A \tag{9.5.15}$$

$$\frac{dA}{A} = \frac{1}{B}\left(\frac{dQ}{Q}\right)$$

or

$$\frac{dQ}{Q} = B\left(\frac{dA}{A}\right) \tag{9.5.16}$$

In this case $B > 1$, so that the discharge estimation error would be decreased by B if A were the dependent variable instead of Q. In summary, if we use equation (9.5.3) as the form of the momentum equation, then Q is the dependent variable with equation (9.5.7) being the governing equation; if we use equation (9.5.8) as the form of the momentum equation, then A is the dependent variable with equation (9.5.12) being the governing equation.

9.5.2 U.S. Army Corps of Engineers HEC-1 Kinematic Wave Model for Overland Flow and Channel Routing

The HEC-1 computer program actually has two forms of the kinematic wave. The first is based upon equation (9.5.12) where an explicit finite difference form is used (refer to Figures 9.5.1 and 8.9.2):

$$\frac{\partial A}{\partial t} = \frac{A_{i+1}^{j+1} - A_{i+1}^{j}}{\Delta t} \tag{9.5.17}$$

$$\frac{\partial A}{\partial x} = \frac{A_{i+1}^{j} - A_{i}^{j}}{\Delta x} \tag{9.5.18}$$

and

$$A = \frac{A_{i+1}^{j} + A_{i}^{j}}{2} \tag{9.5.19}$$

$$q = \frac{q_{i+1}^{j+1} + q_{i+1}^{j}}{2} \tag{9.5.20}$$

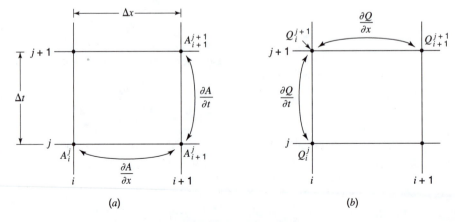

Figure 9.5.1 Finite difference forms. (*a*) HEC-1 "standard form;" (*b*) HEC-1 "conservation form."

Substituting these finite-difference approximations into equation (9.5.12) gives

$$\frac{1}{\Delta t}\left(A_{i+1}^{j+1} - A_{i+1}^{j}\right) + aB\left[\frac{A_{i+1}^{j} + A_{i}^{j}}{2}\right]^{B-1}\left[\frac{A_{i+1}^{j} - A_{i}^{j}}{\Delta x}\right] = \frac{q_{i+1}^{j+1} + q_{i+1}^{j}}{2} \tag{9.5.21}$$

The only unknown in equation (9.5.21) is A_{i+1}^{j+1}, so

$$A_{i+1}^{j+1} = A_{i+1}^{j} - aB\left(\frac{\Delta t}{\Delta x}\right)\left[\frac{A_{i+1}^{j} + A_{i}^{j}}{2}\right]^{B-1}\left(A_{i+1}^{j} - A_{i,j}\right) + \left(q_{i+1}^{j+1} + q_{i+1}^{j}\right)\frac{\Delta t}{2} \tag{9.5.22}$$

After computing A_{i+1}^{j+1} at each grid along a time line going from upstream to downstream (see Figure 8.9.2), compute the flow using equation (9.5.8):

$$Q_{i+1}^{j+1} = a\left(A_{i+1}^{j+1}\right)^{B} \tag{9.5.23}$$

The HEC-1 model uses the above kinematic wave model as long as a stability factor $R < 1$ (Alley and Smith, 1987), defined by

$$R = \frac{a}{q\Delta x}\left[\left(q\Delta t + A_{i}^{j}\right)^{B} - \left(A_{i}^{j}\right)^{B}\right] \text{ for } q > 0 \tag{9.5.24a}$$

$$R = aB\left(A_{i}^{j}\right)^{B-1}\frac{\Delta t}{\Delta x} \text{ for } q = 0 \tag{9.5.24b}$$

Otherwise HEC-1 uses the form of equation (9.5.1), where (see Figure 9.5.1)

$$\frac{\partial Q}{\partial x} = \frac{Q_{i+1}^{j+1} - Q_{i}^{j+1}}{\Delta x} \tag{9.5.25}$$

$$\frac{\partial A}{\partial t} = \frac{A_{i}^{j+1} - A_{i}^{j}}{\Delta t} \tag{9.5.26}$$

so

$$\frac{Q_{i+1}^{j+1} - Q_{i}^{j+1}}{\Delta x} + \frac{A_{i}^{j+1} - A_{i}^{j}}{\Delta t} = q \tag{9.5.27}$$

Solving for the only unknown Q_{i+1}^{j+1} yields

$$Q_{i+1}^{j+1} = Q_{i}^{j+1} + q\Delta x - \frac{\Delta x}{\Delta t}\left(A_{i}^{j+1} - A_{i}^{j}\right) \tag{9.5.28}$$

Then solve for A_{i+1}^{j+1} using equation (9.5.23):

$$A_{i+1}^{j+1} = \left(\frac{1}{a}Q_{i+1}^{j+1}\right)^{1/B} \tag{9.5.29}$$

The *initial condition* (values of A and Q at time 0 along the grid, referring to Figure 8.9.2) are computed assuming uniform flow or nonuniform flow for an initial discharge. The *upstream boundary* is the inflow hydrograph from which Q is obtained.

The kinematic wave schemes used in the HEC-1 model are very simplified. Chow et al. (1988) presented both linear and nonlinear kinematic wave schemes based upon the equation (9.5.7) formulation. An example of a more desirable kinematic wave formulation is that by Woolhiser et al. (1990) presented in the next subsection.

9.5.3 KINEROS Channel Flow Routing Model

The KINEROS channel routing model uses the equation (9.5.10) form of the kinematic wave equation (Woolhiser et al., 1990):

$$\frac{\partial A}{\partial t} + \frac{dQ}{dA}\frac{\partial A}{\partial x} = q(x,\ t) \tag{9.5.10}$$

where $q(x, t)$ is the net lateral inflow per unit length of channel. The derivatives are approximated using an implicit scheme in which the spatial and temporal derivatives are, respectively,

$$\frac{\partial A}{\partial x} = \theta\frac{A_{i+1}^{j+1} - A_i^{j+1}}{\Delta x} + (1-\theta)\frac{A_{i+1}^j - A_i^j}{\Delta x} \tag{9.5.30}$$

$$\frac{dQ}{dA}\frac{\partial A}{\partial x} = \theta\left(\frac{dQ}{dA}\right)^{j+1}\left(\frac{A_{i+1}^{j+1} - A_i^{j+1}}{\Delta x}\right) + (1-\theta)\left(\frac{dQ}{dA}\right)^{j+1}\left(\frac{A_{i+1}^j - A_i^j}{\Delta x}\right) \tag{9.5.31}$$

and

$$\frac{\partial A}{\partial t} = \frac{1}{2}\left[\frac{A_i^{j+1} - A_i^j}{\Delta t} + \frac{A_{i+1}^{j+1} - A_{i+1}^j}{\Delta t}\right] \tag{9.5.32}$$

or

$$\frac{\partial A}{\partial t} = \frac{A_i^{j+1} + A_{i+1}^{j+1} - A_i^j - A_{i+1}^j}{2\Delta t} \tag{9.5.33}$$

Substituting equations (9.5.31) and (9.5.33) into (9.5.10), we have

$$\frac{A_{i+1}^{j+1} - A_{i+1}^j + A_i^{j+1} - A_i^j}{2\Delta t} + \left\{\theta\left[\left(\frac{dQ}{dA}\right)^{j+1}\left(\frac{A_{i+1}^{j+1} - A_i^{j+1}}{\Delta x}\right)\right] + (1-\theta)\left[\left(\frac{dQ}{dA}\right)^{j+1}\left(\frac{A_{i+1}^j - A_i^j}{\Delta x}\right)\right]\right\}$$

$$= \frac{1}{2}\left(q_{i+1}^{j+1} + q_i^{j+1} + q_{i+1}^j + q_i^j\right) \tag{9.5.34}$$

The only unknown in this equation is A_{i+1}^{j+1}, which must be solved for numerically by use of an iterative scheme such as the Newton–Rhapson method (see Appendix A).

Woolhiser et al. (1990) use the following relationship between channel discharge and cross-sectional area, which embodies the kinematic wave assumption:

$$Q = \alpha R^{m-1} A \tag{9.5.35}$$

where R is the hydraulic radius and $\alpha = 1.49 S^{1/2}/n$ and $m = 5/3$ for Manning's equation.

9.5.4 Kinematic Wave Celerity

Kinematic waves result from changes in Q. An increment in flow dQ can be written as

$$dQ = \frac{\partial Q}{\partial x}dx + \frac{\partial Q}{\partial t}dt \tag{9.5.36}$$

Dividing through by dx and rearranging produces:

$$\frac{\partial Q}{\partial x} + \frac{dt}{dx}\frac{\partial Q}{\partial t} = \frac{dQ}{dx} \tag{9.5.37}$$

Equations (9.5.7) and (9.5.37) are identical if

$$\frac{dQ}{dt} = q \qquad (9.5.38)$$

and

$$\frac{dx}{dt} = \frac{1}{\alpha\beta Q^{\beta-1}} \qquad (9.5.39)$$

Differentiating equation (9.5.3) and rearranging gives

$$\frac{dQ}{dA} = \frac{1}{\alpha\beta Q^{\beta-1}} \qquad (9.5.40)$$

and by comparing equations (9.5.38) and (9.5.40), it can be seen that

$$\frac{dx}{dt} = \frac{dQ}{dA} \qquad (9.5.41)$$

or

$$c_k = \frac{dx}{dt} = \frac{dQ}{dA} \qquad (9.5.42)$$

where c_k is the kinematic wave celerity. This implies that an observer moving at a velocity $dx/dt = c_k$ with the flow would see the flow rate increasing at a rate of $dQ/dx = q$. If $q = 0$ the observer would see a constant discharge. Equations (9.5.38) and (9.5.42) are the *characteristic equations* for a kinematic wave, two ordinary differential equations that are mathematically equivalent to the governing continuity and momentum equations.

The kinematic wave celerity can also be expressed in terms of the depth y as

$$c_k = \frac{1}{B}\frac{dQ}{dy} \qquad (9.5.43)$$

where $dA = Bdy$.

Both kinematic and dynamic wave motion are present in natural flood waves. In many cases the channel slope dominates in the momentum equation; therefore, most of a flood wave moves as a kinematic wave. Lighthill and Whitham (1955) proved that the velocity of the main part of a natural flood wave approximates that of a kinematic wave. If the other momentum terms ($\partial V/\partial t$, $V(\partial V/\partial x)$ and $(1/g)\partial y/\partial x$) are not negligible, then a dynamic wave front exists that can propagate both upstream and downstream from the main body of the flood wave.

9.6 MUSKINGUM-CUNGE MODEL

Cunge (1969) proposed a variation of the kinematic wave method based upon the Muskingum method (see Chapter 8). With the grid shown in Figure 9.6.1, the unknown discharge Q_{i+1}^{j+1} can be expressed using the Muskingum equation ($Q_{j+1} = C_1 I_{j+1} + C_2 I_j + C_3 Q_j$):

$$Q_{i+1}^{j+1} = C_1 Q_i^{j+1} + C_2 Q_i^j + C_3 Q_{i+1}^j \qquad (9.6.1)$$

where $Q_{i+1}^{j+1} = Q_{j+1}$; $Q_i^{j+1} = I_{j+1}$; $Q_i^j = I_j$; and $Q_{i+1}^j = Q_j$. The Muskingum coefficients are

$$C_1 = \frac{\Delta t - 2KX}{2K(1-X) + \Delta t} \qquad (9.6.2)$$

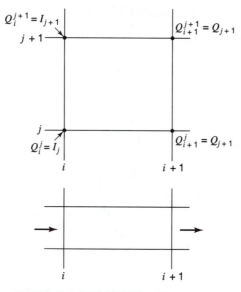

$Q_i^{j+1} = I_{j+1}$

$Q_{i+1}^{j+1} = Q_{j+1}$

$j+1$

j

$Q_i^j = I_j$

$Q_{i+1}^j = Q_{j+1}$

i

$i+1$

i

$i+1$

Figure 9.6.1 Finite-difference grid for Muskingum–Cunge method.

$$C_2 = \frac{\Delta t + 2KX}{2K(1-X) + \Delta t} \tag{9.6.3}$$

$$C_3 = \frac{2K(1-X) - \Delta t}{2K(1-X) + \Delta t} \tag{9.6.4}$$

Cunge (1969) showed that when K and Δt are considered constant, equation (9.6.10) is an approximate solution of the kinematic wave. He further demonstrated that (9.6.1) can be considered an approximation of a modified diffusion equation if

$$K = \frac{\Delta x}{c_k} = \frac{\Delta x}{dQ/dA} \tag{9.6.5}$$

and

$$X = \frac{1}{2}\left(1 - \frac{Q}{Bc_k S_0 \Delta x}\right) \tag{9.6.6}$$

where c_k is the celerity corresponding to Q and B, and B is the width of the water surface. The value of $\Delta x/(dQ/dA)$ in equation (9.6.5) represents the time propagation of a given discharge along a channel reach of length Δx. Numerical stability requires $0 \le x \le 1/2$. The solution procedure is basically the same as the kinematic wave.

9.7 IMPLICIT DYNAMIC WAVE MODEL

The conservation form of the Saint–Venant equations is used because this form provides the versatility required to simulate a wide range of flows from gradual long-duration flood waves in rivers to abrupt waves similar to those caused by a dam failure. The equations are developed from equations (9.4.6) and (9.4.25) as follows.

Weighted four-point finite-difference approximations given by equations (9.7.1)–(9.7.3) are used for dynamic routing with the Saint–Venant equations. The spatial derivatives $\partial Q/\partial x$ and $\partial h/\partial x$ are estimated between adjacent time lines:

$$\frac{\partial Q}{\partial x} = \theta \frac{Q_{i+1}^{j+1} - Q_i^{j+1}}{\Delta x_i} + (1-\theta)\frac{Q_{i+1}^j - Q_i^j}{\Delta x_i} \tag{9.7.1}$$

$$\frac{\partial h}{\partial x} = \theta \frac{h_{i+1}^{j+1} - h_i^{j+1}}{\Delta x_i} + (1-\theta)\frac{h_{i+1}^j - h_i^j}{\Delta x_i} \tag{9.7.2}$$

and the time derivatives are:

$$\frac{\partial(A + A_0)}{\partial t} = \frac{(A + A_0)_i^{j+1} + (A + A_0)_{i+1}^{j+1} - (A + A_0)_i^j - (A + A_0)_{i+1}^j}{2\Delta t_j} \tag{9.7.3}$$

$$\frac{\partial Q}{\partial t} = \frac{Q_i^{j+1} + Q_{i+1}^{j+1} - Q_i^j - Q_{i+1}^j}{2\Delta t_j} \tag{9.7.4}$$

The nonderivative terms, such as q and A, are estimated between adjacent time lines, using:

$$q = \theta \frac{q_i^{j+1} + q_{i+1}^{j+1}}{2} + (1-\theta)\frac{q_i^j + q_{i+1}^j}{2} = \theta \bar{q}_i^{j+1} + (1-\theta)\bar{q}_i^{-j} \tag{9.7.5}$$

$$A = \theta \left[\frac{A_i^{j+1} + A_{i+1}^{j+1}}{2}\right] + (1-\theta)\left[\frac{A_i^j + A_{i+1}^j}{2}\right] = \theta \bar{A}_i^{j+1} + (1-\theta)\bar{A}_i^j \tag{9.7.6}$$

where $\bar{q}_i$ and $\bar{A}_i$ indicate the lateral flow and cross-sectional area averaged over the reach Δx_i.

The finite-difference form of the continuity equation is produced by substituting equations. (9.7.1), (9.7.3), and (9.7.5) into (9.4.6):

$$\theta\left(\frac{Q_{i+1}^{j+1} - Q_i^{j+1}}{\Delta x_i} - \bar{q}_i^{j+1}\right) + (1-\theta)\left(\frac{Q_{i+1}^j - Q_i^j}{\Delta x_i} - \bar{q}_i^j\right)$$

$$+ \frac{(A + A_0)_i^{j+1} + (A + A_0)_{i+1}^{j+1} - (A + A_0)_i^j - (A + A_0)_{i+1}^j}{2\Delta t_j} = 0 \tag{9.7.7}$$

Similarly, the momentum equation (9.4.27) is written in finite-difference form as:

$$\frac{Q_i^{j+1} + Q_{i+1}^{j+1} - Q_i^j - Q_{i+1}^j}{2\Delta t_j}$$

$$+ \theta\left[\frac{\left(\beta Q^2/A\right)_i^{j+1} - \left(\beta Q^2/A\right)_{i+1}^{j+1}}{\Delta x_i} + g\bar{A}_i^{j+1}\left(\frac{h_{i+1}^{j+1} - h_i^{j+1}}{\Delta x_i} + \left(\bar{S}_f\right)_i^{j+1} + \left(\bar{S}_e\right)_i^{j+1}\right) - \left(\overline{\beta q v_x}\right)_i^{j+1}\right]$$

$$+ (1-\theta)\left[\frac{\left(\beta Q^2/A\right)_i^j - \left(\beta Q^2/A\right)_{i+1}^j}{\Delta x_i} + g\bar{A}_i^j\left(\frac{h_{i+1}^j - h_i^j}{\Delta x_i} + \left(\bar{S}_f\right)_i^j + \left(\bar{S}_e\right)_i^j\right) - \left(\overline{\beta q v_x}\right)_i^j\right] = 0 \tag{9.7.8}$$

The four-point finite-difference form of the continuity equation can be further modified by multiplying equation (9.7.7) by Δx_i to obtain

$$\theta(Q_{i+1}^{j+1} - Q_i^{j+1} - \bar{q}_i^{j+1}\Delta x_i) + (1-\theta)(Q_{i+1}^j - Q_i^j - \bar{q}_i^j\Delta x_i)$$

$$+ \frac{\Delta x_i}{2\Delta t_i}\left[(A + A_0)_i^{j+1} + (A + A_0)_{i+1}^{j+1} - (A + A_0)_i^j - (A + A_0)_{i+1}^j\right] = 0 \tag{9.7.9}$$

Similarly, the momentum equation can be modified by multiplying by Δx_i to obtain

$$\frac{\Delta x_i}{2\Delta t_j}\left(Q_i^{j+1} + Q_{i+1}^{j+1} - Q_i^j - Q_{i+1}^j\right)$$

$$+\theta\left\{\left(\frac{\beta Q^2}{A}\right)_{i+1}^{j+1} - \left(\frac{\beta Q^2}{A}\right)_i^{j+1} + g\overline{A}_i^{j+1}\left[h_{i+1}^{j+1} - h_i^{j+1} + \left(\overline{S}_f\right)_i^{j+1}\Delta x_i + \left(\overline{S}_e\right)_i^{j+1}\Delta x_i\right] - \left(\overline{\beta q v_x}\right)_i^{j+1}\Delta x_i\right\}$$

$$+(1-\theta)\left\{\left(\frac{\beta Q^2}{A}\right)_{i+1}^j - \left(\frac{\beta Q^2}{A}\right)_i^j + g\overline{A}_i^j\left[h_{i+1}^j - h_i^j + \left(\overline{S}_f\right)_i^j\Delta x_i + \left(\overline{S}_e\right)_i^j\Delta x_i\right] - \left(\overline{\beta q v_x}\right)_i^j\Delta x_i\right\} = 0$$

$$(9.7.10)$$

where the average values (marked with an overbar) over a reach are defined as

$$\overline{\beta}_i = \frac{\beta_i + \beta_{i+1}}{2} \tag{9.7.11}$$

$$\overline{A}_i = \frac{A_i + A_{i+1}}{2} \tag{9.7.12}$$

$$\overline{B}_i = \frac{B_i + B_{i+1}}{2} \tag{9.7.13}$$

$$\overline{Q}_i = \frac{Q_i + Q_{i+1}}{2} \tag{9.7.14}$$

Also,

$$\overline{R}_i = \overline{A}_i/\overline{B}_i \tag{9.7.15}$$

for use in Manning's equation. Manning's equation may be solved for S_f and written in the form shown below, where the term $|Q|Q$ has magnitude Q^2 and sign positive or negative depending on whether the flow is downstream or upstream, respectively:

$$\left(\overline{S}_f\right)_i = \frac{\overline{n}_i^2|\overline{Q}_i|\overline{Q}_i}{2.208\overline{A}_i^2\overline{R}_i^{4/3}} \tag{9.7.16}$$

The minor headlosses arising from contraction and expansion of the channel are proportional to the difference between the squares of the downstream and upstream velocities, with a contraction/expansion loss coefficient K_e:

$$\left(\overline{S}_e\right)_i = \frac{(K_e)_i}{2g\Delta x_i}\left[\left(\frac{Q}{A}\right)_{i+1}^2 - \left(\frac{Q}{A}\right)_i^2\right] \tag{9.7.17}$$

The terms having superscript j in equations (9.7.9) and (9.7.10) are known either from initial conditions or from a solution of the Saint–Venant equations for a previous time line. The terms g, Δx_i, β_i, K_e, C_w, and V_w are known and must be specified independently of the solution. The unknown terms are Q_i^{j+1}, Q_{i+1}^{j+1}, h_{i+1}^{j+1}, A_i^{j+1}, $A_{i+1}^{j+1}B_i^{j+1}$, and B_{i+1}^{j+1}. However, all the terms can be expressed as functions of the unknowns Q_i^{j+1}, Q_{i+1}^{j+1}, h_i^{j+1}, and h_{i+1}^{j+1}, so there are actually four unknowns. The unknowns are raised to powers other than unity, so equations (9.7.9) and (9.7.10) are nonlinear equations.

The continuity and momentum equations are considered at each of the N-1 rectangular grids shown in Figure 9.7.1, between the upstream boundary at $i = 1$ and the downstream boundary at

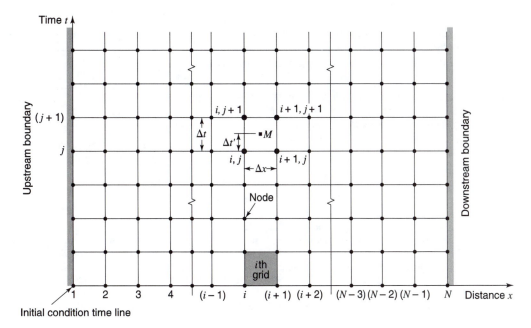

Figure 9.7.1 The *x-t* solution plane. The finite-difference forms of the Saint–Venant equations are solved at a discrete number of points (values of the independent variables *x* and *t*) arranged to form the rectangular grid shown. Lines parallel to the time axis represent locations along the channel, and those parallel to the distance axis represent times (from Fread (1974)).

$i = N$. This yields 2N-2 equations. There are two unknowns at each of the N grid points (Q and h), so there are 2N unknowns in all. The two additional equations required to complete the solution are supplied by the upstream and downstream boundary conditions. The upstream boundary condition is usually specified as a known inflow hydrograph, while the downstream boundary condition can be specified as a known stage hydrograph, a known discharge hydrograph, or a known relationship between stage and discharge, such as a rating curve. The U.S. National Weather Service FLDWAV model (hsp.nws.noaa.gov/oh/hrl/rvmech) uses the above to describe implicit dynamic wave model formulation.

PROBLEMS

9.1.1 The storage-outflow characteristics for a reservoir are given below. Determine the storage-outflow function $2S/\Delta t + Q$ versus Q for each of the tabulated values using $\Delta t = 1.0$ hr. Plot a graph of the storage-outflow function.

Storage (10^6 m³)	70	80	85	100	115
Outflow (m³)	0	50	150	350	700

9.2.1 Route the inflow hydrograph given below through the reservoir with the storage-outflow characteristics given in problem 3.6.1 using the level pool method. Assume the reservoir has an initial storage of 70×10^6 m³.

Time (h)	0	1	2	3	4	5	6	7	8
Inflow (m³/s)	0	40	60	150	200	300	250	200	180

Time (h)	9	10	11	12	13	14	15	16
Inflow(m³/s)	220	320	400	280	190	150	50	0

9.2.2 Rework problem 9.2.1 assuming the reservoir storage is initially 80×10^3 m³.

9.2.3 Write a computer program to solve problems 9.2.1 and 9.2.2.

9.2.4 Rework example 9.1.1 using a 1.5-acre detention basin.

9.2.5 Rework example 9.1.1 using a triangular inflow hydrograph that increases linearly from zero to a peak of 90 cfs at 120 min and then decreases linearly to a zero discharge at 240 min. Use a 30-min routing interval.

9.2.6 Rework example 9.2.2 using $\Delta t = 2$ hrs.

9.2.7 Rework example 9.2.2 assuming $X = 0.3$ hrs.

9.3.1 Rework example 9.2.2 assuming $K = 1.4$ hr.

9.3.2 Calculate the Muskingum routing K and number of routing steps for a 1.25-mi long channel. The average cross-section dimensions for the channel are a base width of 25 ft and an average depth of 2.0 ft. Assume the channel is rectangular and has Manning's n 0.04 and a slope of 0.009 ft/ft.

9.3.3 Route the following upstream inflow hydrograph through a downstream flood control channel reach using the Muskingum method. The channel reach has a $K = 2.5$ hr and $X = 0.2$. Use a routing interval of 1 hr.

Time (h)	1	2	3	4	5	6	7
Inflow (cfs)	90	140	208	320	440	550	640

Time (h)	8	9	10	11	12	13	14
Inflow (cfs)	680	690	630	570	470	390	

Time (h)	15	16	17	18	19	20	
Inflow (cfs)	330	250	180	130	100	90	

9.3.4 Use the U.S. Army Corps of Engineers HEC-1 computer program to solve Problem 9.3.3.

REFERENCES

Alley, W. M., and P. E. Smith, Distributed Routing Rainfall-Runoff Model, Open File Report 82–344, U.S. Geological Survey, Reston, VA, 1987.

Bradley, J., *Hydraulics of Bridge Water Way*, Hydraulic Design Sevies No. 1, Federal Highway Administration. U.S. Department of Transportation, Washington, DC, 1978.

Chow, V. T., *Open Channel Hydraulics*, McGraw-Hill, New York, 1959.

Chow, V. T. (editor-in-chief), *Handbook of Applied Hydrology*, McGraw-Hill, New York, 1964.

Chow, V. T., D. R. Maidment, and L. W. Mays, *Applied Hydrology*, McGraw-Hill, New York, 1988.

Cudworth, A. G., Jr., *Flood Hydrology Manual*, U. S. Department of the Interior, Bureau of Reclamation, Denver, CO, 1989.

Cunge, J. A., "On the Subject of a Flood Propagation Method (Muskingum Method)," *Journal of Hydraulics Research*, International Association of Hydraulic Research, vol. 7, no, 2, pp. 205–230, 1969.

Fread, D. L., "Discussion of 'Implicit Flood Routing in Natural Channels,' by M. Amein and C. S. Fang," *Journal of the Hydraulics Division*, ASCE, vol. 97, no HY.7, pp. 1156–1159, 1971.

Fread, D. L., *Numerical Properties of Implicit Form-Point Finite Difference Equation of Unsteady Flow*, NOAA Technical Memorandum NWS HYDRO 18, National Weather Service, NOAA, U.S. Dept. of Commerce, Silver Spring, MD, 1974.

Fread, D. L., "Theoretical Development of Implicit Dynamic Routing Model," Dynamic Routing Service at Lower Mississippi River Forecast Center, Slidell, Louisiana, National Weather Service, NOAA, Silver Spring, MD, 1976.

Henderson, F. M., *Open Channel Flow*, Macmillan, New York, 1966.

Hewlett, J. D., *Principles of Forest Hydrology*, University of Georgia Press, Athens, GA, 1982.

Lighthill, M. J., and G. B. Whitham, "On Kinematic Waves, I: Flood Movement in Long Rivers," *Proc. Roy. Soc. London* A, vol. 229, no. 1178, pp. 281–316, 1955.

Maidment, D. R. (editor-in-chief), *Handbook of Hydrology*, McGraw-Hill, New York, 1993.

Masch, F. D., *Hydrology*, Hydraulic Engineering Circular No. 19, FHWA-10-84-15, Federal Highway Administration, U.S. Department of the Interior, McLean, VA, 1984.

Mays, L. W., and Y. K. Tung, *Hydrosystems Engineering and Management*, McGraw-Hill, New York, 1992.

McCuen, R. H., *Hydrologic Analysis and Design*, Prentice-Hall, Englewood Cliffs, NJ, 1989.

Morris, E. M., and D. A. Woolhiser, 1980, "Unsteady One-Dimensional Flow over a Plane: Partial Equilibrium and Recession Hydrographs," *Water Resources Research* 16(2): 355–360.

Mosley, M. P., and A. I. McKerchar, "Streamflow," in *Handbook of Hydrology* (edited by D. R. Maidment), McGraw-Hill, New York, 1993.

Ponce, V. M., *Engineering Hydrology: Principles and Practices*, Prentice-Hall, Englewood Cliffs, NJ, 1989.

Strahler, A. N., "Quantitative Geomorphology of Drainage Basins and Channel Networks," section 4–II in *Handbook of Applied Hydrology*, edited by V. T. Chow, pp. 4–39, 4–76, McGraw-Hill, New York, 1964.

Strelkoff, T., "Numerical Solution of Saint–Venant Equation," *Journal of the Hydraulics Division*, ASCE, vol. 96, no. HY1, pp. 223–252, 1970.

Strelkoff, T., The One-Dimensional Equations of Open-Channel Flow, *Journal of the Hydraulics Division*, American Society of Civil Engineers, vol. 95, no. Hy3, pp. 861–874, 1969.

U.S. Army Corps of Engineers, "Routing of Floods Through River Channels," *Engineer Manual*, 1110–2–1408, Washington, DC, 1960.

U.S. Army Corps of Engineers, Hydrologic Engineering Center, *HEC-1, Flood Hydrograph Package, User's Manual*, Davis, CA, 1990.

U.S. Department of Agriculture, Soil Conservation Service, "A Method for Estimating Volume and Rate of Runoff in Small Watersheds," Tech. Paper 149, Washington, DC, 1973.

U.S. Department of Agriculture, Soil Conservation Service, "Urban Hydrology for Small Watersheds," Tech. Release no. 55, Washington, DC, 1986.

U.S. Environmental Data Services, *Climate Atlas of the U.S.*, U.S. Government Printing Office, Washington, DC, pp. 43–44, 1968.

U.S. National Research Council, Committee on Opportunities in the Hydrologic Sciences, Water Science and Technology Board, *Opportunities in the Hydrologic Sciences*, National Academy Press, Washington, DC, 1991.

Viessman, W., Jr., and G. L. Lewis, *Introduction to Hydrology*, fourth edition, Harper and Row, New York, 1996.

Woolhiser, D. A., and J. A. Liggett, Unsteady, One-Dimensional Flow Over a Plane—the Rising Hydrograph, *Water Resources Research*, vol. 3(3), pp. 753–771, 1967.

Woolhiser, D. A., R. E. Smith, and D. C. Goodrich, *KINEROS, A Kinematic Runoff and Erosion Model: Documentation and User Manual*, U. S. Department of Agricultural Research Service, ARS-77, Tucson, AZ, 1990.

Chapter 10

Probability, Risk, and Uncertainty Analysis for Hydrologic and Hydraulic Design

10.1 PROBABILITY CONCEPTS

This section very briefly covers probability concepts that are important in the probabilistic, risk and uncertainty analysis for hydrologic and hydraulic design and analysis. Table 10.1.1 provides definitions of the various probability concepts needed for analysis. Many hydraulic and hydrologic variables must be treated as random variables because of the uncertainties involved in the respective hydraulic and hydrologic processes. As an example, the extremes that occur are random hydrologic events and can therefore be treated as such.

A *random variable X* is a variable described by a *probability distribution*. The distribution specifies the chance that an observation x of the variable will fall in a specified range of X. A set of observations $x_1, x_2,..., x_n$, of the random variable X is called a *sample*. It is assumed that samples are drawn from a population (generally unknown) possessing constant statistical properties, while the properties of a sample may vary from one sample to another. The possible range of variation of all of the samples that could be drawn from the population is called the *sample space*, and an *event* is a subset of the sample space.

A *probability distribution* is a function representing the frequency of occurrence of the value of a random variable. By fitting a distribution to a set of data, a great deal of the probabilistic information in the sample can be compactly summarized in the function and its associated parameters. Fitting distributions can be accomplished by the method of moments or the method of maximum likelihood (see Chow et al., 1988). Between the two methods, the method of moments is more widely used, primarily for its computational simplicity. The method relates the parameters in a probability distribution model to the statistical moments to which the parameter-moment relationships for commonly used distributions in frequency analysis and reliability analysis are immediately available (see Table 10.1.2). In practice, the true mechanism that generates the observed random process is not entirely known. Therefore, to estimate the parameter values in a probability distribution model by the method of moments, sample moments are used.

Table 10.1.1 Various Parameter and Statistics Used to Describe Populations and Samples

Concept	Population Value, Discrete Case	Population Value, Continuous Case	Sample Value
Cumulative distribution function (cdf)	Describes the probability that a random variable is less than or equal to a specified value x	Describes the probability that a random variable is less than or equal to a specified value x	Empirical distribution function (edf): describes the observed frequency of a random variable being less than or equal to a specified value x
Probability mass function (pmf) and probability density function (pdf)	pmf: the probability that X is equal to k	pdf: first derivative of the cumulative distribution function	Histogram: observed frequency with which random variable X falls into the assigned ranges
Mean, average, or expected value	$\mu \equiv \sum_{i=1}^{\infty} P(X = x_i) x_i$	$f(x) \equiv \dfrac{dF(x)}{dx}$ $\mu \equiv \int_{-\infty}^{\infty} x f(x) dx$	$\overline{X} \equiv \sum_{i=1}^{n} \dfrac{X_i}{n}$
Variance	$\sigma^2 \equiv \sum_{i=1}^{\infty} P(X = x_i)(x_i - \mu)^2$	$\sigma^2 \equiv \int_{-\infty}^{\infty} (x - \mu)^2 f(x) dx$	$S^2 = \sum_{i=1}^{n} \dfrac{(X_i - \overline{X})^2}{n-1}$
kth central moment	$M_k = \sum_{i=1}^{\infty} P(X = x_i)(x_i - \mu)^k$	$M_k \equiv \int_{-\infty}^{\infty} (x - \mu)^k f(x) dx$	$\tilde{M}_k \equiv \sum_{i=1}^{n} \dfrac{(X_i - \overline{X})^k}{n}$
Standard deviation		$\sigma \equiv \sqrt{\sigma^2}$	$S \equiv \sqrt{S^2}$
Coefficient of variation or relative standard deviation (if $\mu \neq 0$)		$CV \equiv \dfrac{\sigma}{\mu}$	$CV \equiv \dfrac{S}{\overline{X}}$
Coefficient of skew (a measure of asymmetry)		$\gamma \equiv \dfrac{M_3}{\sigma^3}$	$G \equiv \dfrac{\tilde{M}_3}{S^3}$
Quantiles	x_p is any value of X that has the properties that $P[X < x_p] \leq p$ $P[X > x_p] \leq 1 - p$		$\hat{X}_p$ is the pth quantile of edf
Median (useful for describing central tendency regardless of skewness)	$x_{0.5}$ Any value of X that has the property that $P[X < x_p] \leq 0.5$ $P[X > x_p] \leq 0.5$		$\hat{X}_{0.5}$ The middle observation in a sorted sample, or the average of the two middle observations if the sample size is even.

Table 10.1.1 Various Parameter and Statistics Used to Describe Populations and Samples (*continued*)

Concept	Population Value, Discrete Case	Population Value, Continuous Case	Sample Value
Upper quartile, lower quartile, and hinges	Upper quartile = $x_{0.75}$	Lower quartile = $x_{0.25}$	Upper hinge = $\hat{X}_{0.75}$ This is an approximation to the sample upper quartile; it is defined as the median of all sample values of $X \le x_{0.50}$. The lower hinge, $\hat{X}_{0.25}$, is defined analogously.
Interquartile range (useful for describing spread of data regardless of symmetry)	$x_{0.75} - x_{0.25}$ Width of central region of population containing probability of 0.5		$\hat{X}_{0.75} - \hat{X}_{0.25}$ Width of central region of data set encompassing approximately half the data

Source: Hirsh et al (1993).

Table 10.1.2 Probability Distributions Commonly Used

Distribution	Probability Density Function	Range	Parameter-Moment Relations
Normal	$f(x) = \dfrac{1}{\sqrt{2\pi}\,\sigma} e^{-(x-\mu)^2/2\sigma^2}$	$-\infty < x < \infty$	
Lognormal	$f(x) = \dfrac{1}{\sqrt{2\pi x}\,\sigma_{\ln x}} e^{-(\ln x - \mu_{\ln x})^2/(2\sigma_{\ln x}^2)}$	$x > 0$	$\mu_{\ln x} = \dfrac{1}{2}\ln\left[\dfrac{\mu_x^2}{1+\Omega_x^2}\right]$ $\sigma_{\ln x}^2 = \ln(1+\Omega_x^2)$ $\Omega_x = \sigma_x/\mu_x$
Exponential	$f(x) = \lambda e^{-\lambda x}$	$x \ge 0$	$\lambda = \dfrac{1}{\mu_x}$
Gamma	$f(x) = \dfrac{\lambda^\beta x^{\beta-1} e^{-\lambda x}}{\Gamma(\beta)}$ where Γ = gamma function	$x \ge 0$	$\lambda = \dfrac{\mu_x}{\sigma_x^2}, \beta = \dfrac{\mu_x^2}{\sigma_x^2} = \dfrac{1}{C_v^2}$
Extreme Value Type I	$f(x) = \dfrac{1}{\alpha} e^{-(x-\beta)/\alpha - e^{-(x-\beta)/\alpha}}$	$-\infty < x < \infty$	$\alpha = \sqrt{6}\,\sigma_x/\pi$ $\beta = \mu_x - 0.5772\alpha$
Log Person Type 1	$f(x) = \dfrac{\lambda^\beta (y-\epsilon)^{\beta-1} e^{-\lambda(v-\epsilon)}}{x\Gamma(\beta)}$ where $y = \log x$	$\log x \ge \epsilon$	$\lambda = \dfrac{s_y}{\sqrt{\beta}},$ $\beta = \left[\dfrac{2}{G_s(y)}\right]^2$ $\epsilon = \bar{y} - s_y\sqrt{\beta}$ (assuming $G_s(y)$ is positive)

10.2   COMMONLY USED PROBABILITY DISTRIBUTIONS

Of the distributions presented in Table 10.2.1, only the normal and log-normal distributions are discussed in this subsection. Section 10.4 discusses the Pearson Type III distribution. Section 10.6 discusses the exponential distribution.

10.2.1   Normal Distribution

The normal distribution is a well-known probability distribution, also called the *Gaussian distribution*. Two parameters are involved in a normal distribution: the mean and the variance. A normal random variable having a mean μ and a variance σ^2 is herein denoted as $X \sim N(\mu, \sigma^2)$ with a PDF of

$$f(x) = \frac{1}{\sqrt{2\pi}\sigma} \exp\left[-\frac{1}{2}(\frac{x-\mu}{\sigma})^2\right] \quad \text{for} -\infty < x < \infty \tag{10.2.1}$$

A normal distribution is bell-shaped and symmetric with respect to $x = \mu$. Therefore, the skew coefficient for a normal random variable is zero. A random variable Y that is a linear function of a normal random variable X is also normal. That is, if $X \sim N(\mu, \sigma^2)$ and $Y = aX + b$ then $Y \sim N(a\mu + b, a^2\sigma^2)$. An extension of this theorem is that the sum of normal random variables (independent or dependent) is also a normal random variable.

Probability computations for normal random variables are made by first transforming to the standardized variate as

$$Z = (X - \mu)/\sigma \tag{10.2.2}$$

in which Z has a zero mean and unit variance. Since Z is a linear function of the random variable X, Z is also normally distributed. The PDF of Z, called the *standard normal distribution*, can be expressed as

$$\phi(z) = \frac{1}{\sqrt{2\pi}} \exp\left[-\frac{z^2}{2}\right] \quad \text{for} -\infty < z < \infty \tag{10.2.3}$$

A table of the CDF of Z is given in Table 10.2.1. Computations of probability for $X \sim N(\mu, \sigma^2)$ can be performed using

$$P(X \le x) = P\left[\frac{X-\mu}{\sigma} \le \frac{x-\mu}{\sigma}\right] = P[Z \le z] = \Phi(z) \tag{10.2.4}$$

where $\Phi(z)$ is the CDF of the standard normal random variable Z defined as

$$\Phi(z) = \int_{-\infty}^{z} \phi(z)dz \tag{10.2.5}$$

10.2.2   Log-Normal Distribution

The log-normal distribution is a commonly used continuous distribution in hydrologic event analysis when random variables cannot be negative. A random variable X is said to be log-normally distributed if its logarithmic transform $Y = \ln(X)$ is normally distributed with mean $\mu_{\ln x}$ and variance $\sigma_{\ln x}^2$. The PDF of the log-normal random variable is

$$f(X) = \frac{1}{\sqrt{2\pi}X\sigma_{\ln X}} \exp\left[-\frac{1}{2}\left(\frac{\ln X - \mu_{\ln X}}{\sigma_{\ln X}}\right)^2\right] \quad \text{for} \ 0 < X < \infty \tag{10.2.6}$$

which can be derived from the normal PDF, that is, equation (10.2.1). Statistical properties of a log-normal random variable of the original scale can be computed from those of the log-transformed variable. To compute the statistical moments of X from those of $\ln X$, the following formulas are useful:

$$\mu_X = \exp(\mu_{\ln X} + \sigma_{\ln X}^2/2) \tag{10.2.7}$$

$$\sigma_X^2 = \mu_X^2 [\exp(\sigma_{\ln X}^2) - 1] \tag{10.2.8}$$

$$\Omega_X^2 = \exp(\sigma_{\ln X}^2) - 1 \tag{10.2.9}$$

$$\lambda_X = \Omega_X^2 + 3\Omega_X \tag{10.2.10}$$

From equation (10.2.10) it is obvious that log-normal distributions are always positively skewed because $\Omega_X > 0$. Conversely, the statistical moments of $\ln X$ can be computed from those of X by

$$\mu_{\ln X} = \frac{1}{2}\ln\left[\frac{\mu_X^2}{1+\Omega_X^2}\right] \tag{10.2.11}$$

$$\sigma_{\ln X}^2 = \ln\left(\Omega_X^2 + 1\right) \tag{10.2.12}$$

Since the sum of normal random variables is normally distributed, the multiplication of log-normal random variables is also log-normally distributed. Several properties of log-normal random variables are useful:

1. If X is a log-normal random variable and $Y = aX^b$, then Y has a log-normal distribution with mean $\mu_{\ln Y} = \ln a + b\mu_{\ln X}$ and variance $\sigma_{\ln Y}^2 = b^2\sigma_{\ln X}^2$.
2. If X and Y are independently log-normally distributed, $W = XY$ has a log-normal distribution with mean $\mu_{\ln W} = \mu_{\ln X} + \mu_{\ln Y}$ and variance $\sigma_{\ln W}^2 = \sigma_{\ln X}^2 + \sigma_{\ln Y}^2$.
3. If X and Y are independent and log-normally distributed then $R = X/Y$ is log-normal with $\mu_{\ln R} = \mu_{\ln X} - \mu_{\ln Y}$ and variance $\sigma_{\ln R}^2 = \sigma_{\ln X}^2 + \sigma_{\ln Y}^2$.

EXAMPLE 10.2.1

The annual maximum series of flood magnitudes in a river is assumed to follow a log-normal distribution with a mean of 6000 m³/s and a standard deviation of 4000 m³/s. (a) What is the probability in each year that a flood magnitude would exceed 7000 m³/s? (b) Determine the flood magnitude with a return period of 100 years.

SOLUTION

(a) Let Q be a random variable representing the annual maximum flood magnitude. Since Q is assumed to follow a log-normal distribution, $\ln(Q)$ is normally distributed with mean and variance that can be computed using $\mu_Q = 6000$ and $\Omega_Q = \sigma_Q/\mu_Q = 4000/6000 = 0.667$. By equations (10.2.11) and (10.2.12), respectively, we find

$$\mu_{\ln Q} = \frac{1}{2}\ln\left[\frac{\mu_Q^2}{1+V_Q^2}\right] = \frac{1}{2}\ln\left[\frac{6000^2}{1+0.667^2}\right] = 8.515$$

$$\sigma_{\ln Q}^2 = \ln(\Omega_Q^2 + 1) = \ln(0.667^2 + 1) = 0.368$$

The probability that the flood magnitude exceeds 7000 m³/s is

$$P(Q > 7000) = P(\ln Q > \ln 7000) = 1 - P(\ln Q \le \ln 7000) =$$

$$= 1 - P\{[(\ln Q - \mu_{\ln Q})/\sigma_{\ln Q}] \le [(\ln 7000 - \mu_{\ln Q})/\sigma_{\ln Q}]\}$$

$$= 1 - P[Z \le (\ln 7000 - 8.515)/\sqrt{0.368}]$$

$$= 1 - P[Z \le 0.558]$$

$$= 1 - F(0.558) = 1 - 0.712 = 0.288$$

(b) A 100-year event in hydrology represents the event that occurs, on the average, once every 100 years. Therefore, the probability in every single year that a 100-year event is equaled or exceeded is 0.01, i.e., $P(Q \ge q_{100}) = 0.01$ in which q_{100} is the magnitude of the 100-year flood. This part of the problem is to determine q_{100}, which is the reverse of part (a).

$$P(Q \leq q_{100}) = 1 - P(Q = q_{100}) = 1 - 0.01 = 0.99$$

Since $P(Q \leq q_{100}) = P[\ln Q \leq \ln q_{100}] = P[Z \leq (\ln q_{100} - \mu_{\ln Q})/\sigma_{\ln Q}] = 0.99$,

$$0.99 = P[Z \leq (\ln q_{100}) - 8.515)/\sqrt{0.368}\,]$$

$$0.99 = \Phi\{(\ln q_{100} - 8.515)/\sqrt{0.368}\,\}$$

$$0.99 = \Phi(z)$$

From the standard normal probability table (Table 10.2.1), $z = 2.33$ for $\Phi(2.33) = 0.99$. Solving $z = (\ln q_{100} - 8.515)/\sqrt{0.368}$ for q_{100} first yields $\ln q_{100} = 9.928$, then $q_{100} = 20,500 \text{ m}^3/\text{s}$.

Table 10.2.1 Cumulative Probability of the Standard Normal Distribution*

z	.00	.01	.02	.03	.04	.05	.06	.07	.08	.09
0	0.5000	0.5040	0.5080	0.5120	0.5160	0.5199	0.5239	0.5279	0.5319	0.5359
0.1	0.5398	0.5438	0.5478	0.5517	0.5557	0.5596	0.5636	0.5675	0.5714	0.5753
0.2	0.5793	0.5832	0.5871	0.5910	0.5948	0.5987	0.6026	0.6064	0.6103	0.6141
0.3	0.6179	0.6217	0.6255	0.6293	0.6331	0.6368	0.6406	0.6443	0.6480	0.6517
0.4	0.6554	0.6591	0.6628	0.6664	0.6700	0.6736	0.6772	0.6808	0.6844	0.6879
0.5	0.6915	0.6950	0.6985	0.7019	0.7054	0.7088	0.7123	0.7157	0.7190	0.7224
0.6	0.7257	0.7291	0.7324	0.7357	0.7389	0.7422	0.7454	0.7486	0.7517	0.7549
0.7	0.7580	0.7611	0.7642	0.7673	0.7704	0.7734	0.7764	0.7794	0.7823	0.7852
0.8	0.7881	0.7910	0.7939	0.7967	0.7995	0.8023	0.8051	0.8078	0.8106	0.8133
0.9	0.8159	0.8186	0.8212	0.8238	0.8164	0.8289	0.8315	0.8340	0.8365	0.8389
1.0	0.8413	0.8438	0.8461	0.8485	0.8508	0.8531	0.8554	0.8577	0.8599	0.8621
1.1	0.8643	0.8665	0.8686	0.8708	0.8729	0.8749	0.8770	0.8790	0.8810	0.8830
1.2	0.8849	0.8869	0.8888	0.8907	0.8925	0.8944	0.8962	0.8980	0.8997	0.9015
1.3	0.9032	0.9049	0.9066	0.9082	0.9099	0.9115	0.9131	0.9147	0.9162	0.9177
1.4	0.9192	0.9207	0.9222	0.9236	0.9251	0.9265	0.9279	0.9292	0.9306	0.9319
1.5	0.9332	0.9345	0.9357	0.9370	0.9382	0.9394	0.9406	0.9418	0.9429	0.9441
1.6	0.9452	0.9463	0.9474	0.9484	0.9495	0.9505	0.9515	0.9525	0.9535	0.9545
1.7	0.9554	0.9564	0.9573	0.9582	0.9591	0.9599	0.9608	0.9616	0.9625	0.9633
1.8	0.9641	0.9649	0.9656	0.9664	0.9671	0.9678	0.9686	0.9693	0.9699	0.9706
1.9	0.9713	0.9719	0.9726	0.9732	0.9738	0.9744	0.9750	0.9756	0.9761	0.9767
2.0	0.9772	0.9778	0.9783	0.9788	0.9793	0.9798	0.9803	0.9808	0.9812	0.9817
2.1	0.9821	0.9826	0.9830	0.9834	0.9838	0.9842	0.9846	0.9850	0.9854	0.9857
2.2	0.9861	0.9864	0.9868	0.9871	0.9875	0.9878	0.9881	0.9884	0.9887	0.9890
2.3	0.9893	0.9896	0.9898	0.9901	0.9904	0.9906	0.9909	0.9911	0.9913	0.9916
2.4	0.9918	0.9920	0.9922	0.9925	0.9927	0.9929	0.9931	0.9932	0.9934	0.9936
2.5	0.9938	0.9940	0.9941	0.9943	0.9945	0.9946	0.9948	0.9949	0.9951	0.9952
2.6	0.9953	0.9955	0.9956	0.9957	0.9959	0.9960	0.9961	0.9962	0.9963	0.9964
2.7	0.9965	0.9966	0.9967	0.9968	0.9969	0.9970	0.9971	0.9972	0.9973	0.9974
2.8	0.9971	0.9975	0.9976	0.9977	0.9977	0.9978	0.9979	0.9979	0.9980	0.9981
2.9	0.9981	0.9982	0.9982	0.9983	0.9984	0.9984	0.9985	0.9985	0.9986	0.9986
3.0	0.9987	0.9987	0.9987	0.9988	0.9988	0.9989	0.9989	0.9989	0.9990	0.9990
3.1	0.9990	0.9991	0.9991	0.9991	0.9992	0.9992	0.9992	0.9992	0.9993	0.9993
3.2	0.9993	0.9993	0.9994	0.9994	0.9994	0.9994	0.9994	0.9995	0.9995	0.9995
3.3	0.9995	0.9995	0.9995	0.9996	0.9996	0.9996	0.9996	0.9996	0.9996	0.9997
3.4	0.9997	0.9997	0.9997	0.9997	0.9997	0.9997	0.9997	0.9997	0.9997	0.9998

*To employ the table for $z < 0$, use where $F_z(z) = 1 - $ where $F_z(|z|)$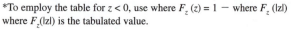
where $F_z(|z|)$ is the tabulated value.

Source: Grant and Leavenworth (1972).

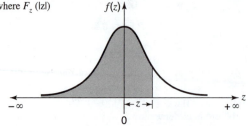

10.3   HYDROLOGIC DESIGN FOR WATER EXCESS MANAGEMENT

Hydrologic design is the process of assessing the impact of hydrologic events of a water resource system and choosing values for the key variables of the system so that it will perform adequately (Chow et al., 1988). This section focuses on water excess management; however, many of the concepts are applicable to water supply (use) management.

10.3.1   Hydrologic Design Scale

The *hydrologic design scale* is the range in magnitude of the design variable (such as the design discharge) within which a value must be selected to determine the inflow to the system (see Figure 10.3.1). The most important factors in selecting the design value are cost and safety. The optimal magnitude for design is one that balances the conflicting considerations of cost and safety. The practical upper limit of the hydrologic design scale is not infinite, since the global hydrologic cycle is a closed system; that is, the total quantity of water on earth is essentially constant. Although the true upper limit is unknown, for practical purposes an estimated upper limit may be determined. This *estimated limiting value* (ELV) is defined as the largest magnitude possible for a hydrologic event at a given location, based on the best available hydrologic information.

The concept of an estimated limiting value is implicit in the *probable maximum precipitation* (PMP) and the corresponding *probable maximum flood* (PMF). The probable maximum precipitation is defined by the World Meteorological Organization (1983) as a "quantity of precipitation that is close to the physical upper limit for a given duration over a particular basin." Based on

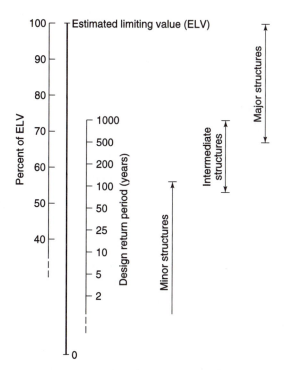

Figure 10.3.1 Hydrologic design scale. Approximate ranges of the design level for different types of structures are shown. Design may be based on a percentage of the ELV or on a design return period. The values for the two scales shown in the diagram are illustrative only and do not correspond directly with one another (from Chow et al. (1988)).

worldwide records, the PMP can have a return period of as long as 500,000,000 years. However, the return period varies geographically. Some arbitrarily assign a return period, say 10,000 years, to the PMP or PMF, but this has no physical basis.

Generalized design criteria for water-control structures have been developed, as summarized in Table 10.3.1. According to the potential consequence of failure, structures are classified as *major, intermediate,* and *minor*; the corresponding approximate ranges on the design scale are shown in Figure 10.3.1. The criteria for dams in Table 10.3.1 pertain to the design of spillway capacities, and are taken from the National Academy of Sciences (1983). The Academy defines a *small dam* as having 50–1000 acre-feet of storage or being 25–40 ft high, an *intermediate dam* as having 1000–50,000 acre-ft of storage or being 40–100 ft high, and a *large dam* as having more than 50,000 acre-ft of storage or being more than 100 ft high. In general, there would be considerable loss of life and extensive damage if a major structure failed. In the case of an intermediate structure, a small loss of life would be possible and the damage would be within the financial capability of the owner. For minor structures, there generally would be no loss of life, and the damage would be of the same magnitude as the cost of replacing or repairing the structure.

Table 10.3.1 Generalized Design Criteria for Water-Control Structures

Type of Structure	Return Period (Years)	ELV
Highway culverts		
Low traffic	5–10	—
Intermediate traffic	10–25	—
High traffic	50–100	—
Highway bridges		
Secondary system	10–50	—
Primary system	50–100	—
Farm drainage		
Culverts	5–50	—
Ditches	5–50	—
Urban drainage		
Storm sewers in small cities	2–25	—
Storm sewers in large cities	25–50	—
Airfields		
Low traffic	5–10	—
Intermediate traffic	10–25	—
High traffic	50–100	—
Levees		
On farms	2–50	—
Around cities	50–200	—
Dams with no likelihood of loss of life (low hazard)		
Small dams	50–100	—
Intermediate dams	100 +	—
Large dams	—	50–100%
Dams with probable loss of life (significant hazard)		
Small dams	100 +	50%
Intermediate dams	—	50–100%
Large dams	—	100%
Dams with high likelihood of considerable loss of life (high hazard)		
Small dams	—	50–100%
Intermediate dams	—	100%
Large dams	—	100%

Source: Chow et al. (1988).

10.3.2 Hydrologic Design Level (Return Period)

A *hydrologic design level* on the design scale is the magnitude of the hydrologic event to be considered for the design of a structure or project. As it is not always economical to design structures and projects for the estimated limiting values, the ELV is often modified for specific design purposes. The final design value may be further modified according to engineering judgment and the experience of the designer or planner. Table 10.3.1 presents generalized criteria for water-control structures. A large number of the structures are designed using return periods.

An extreme hydrologic event is defined to have occurred if the magnitude of the event X is greater than or equal to some level x_T, i.e., $X \geq x_T$. The *return period T* of the event $X = x_T$ is the expected value of the *recurrence interval* (time between occurrences). The expected value $E( )$ is the average value measured over a very large number of occurrences. Consequently, the return period of a hydrologic event of a given magnitude is defined as the *average recurrence interval* between events that equal or exceed a specified magnitude.

10.3.3 Hydrologic Risk

The *probability of occurrence* $P(X \geq x_T)$ of the hydrologic event $(X \geq x_T)$ for any observation is the inverse of the return period, i.e.,

$$P(X \geq x_T) = \frac{1}{T} \qquad (10.3.1)$$

For a 100-year peak discharge, the probability of occurrence in any given year is $P(X \geq x_{100}) = 1/100 = 0.01$.

The probability of nonexceedance is

$$P(X < x_T) = 1 - \frac{1}{T} \qquad (10.3.2)$$

Because each hydrologic event is considered independent, the *probability of nonexceedance* for n years is

$$P(X < x_T \text{ each year for } n \text{ years}) = \left(1 - \frac{1}{T}\right)^n$$

The complement, the *probability of exceedance* at least once in n years, is

$$P(X \geq x_T \text{ at least once in } n \text{ years}) = 1 - \left(1 - \frac{1}{T}\right)^n$$

which is the probability that a T-year return period event will occur at least once in n years. This is also referred to as the *natural, inherent,* or *hydrologic risk of failure* $\bar{R}$:

$$\bar{R} = 1 - \left(1 - \frac{1}{T}\right)^n = 1 - \left[1 - P(X \geq x_T)\right]^n \qquad (10.3.3)$$

where n is referred to as the expected life of the structure. The hydrologic risk relationship is plotted in Figure 10.3.2.

EXAMPLE 10.3.1   Determine the hydrologic risk of a 100-year flood occurring during the 30-year service life of a project.

SOLUTION   Use equation (10.3.3) to determine the risk:

$$\bar{R} = 1 - \left(1 - \frac{1}{T}\right)^n = 1 - \left(1 - 1/100\right)^{30} = 0.26$$

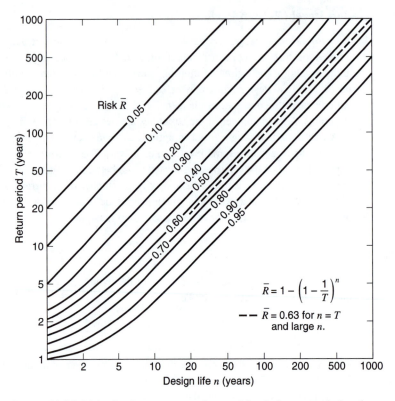

Figure 10.3.2 Risk of at least one exceedance of the design event during the design life (from Chow et al. (1988)).

10.3.4 Hydrologic Data Series

Figure 10.3.3a shows all the data available (that have been collected) for a hydrologic event. This represents a *complete-duration series*. A *partial-duration series* includes data that are selected so that their values are greater than some base value. An *annual-exceedance series* has a base value so that the number of values in the series is equal to the number of years of record. Figure 10.3.3b illustrates the annual exceedance series. An *extreme-value series* consists of the largest or smallest values occurring in each of the equally long time intervals of the record. If the time interval length is one year, the series is an *annual series*. An *annual maximum* series over the largest values in each respective year (Figure 10.3.3c) consists of the largest annual values and an *annual minimum series* consists of the smallest annual values in each of the respective years. Figure 10.3.4 illustrates the annual-exceedence series and the annual maximum series of the hypothetical data in Figure 10.3.3.

The return periods for annual exceedance series T_E are related to the corresponding annual maximum series return period T by (Chow, 1964)

$$T_E = \left[\ln\left(\frac{T}{T-1} \right) \right]^{-1} \tag{10.3.4}$$

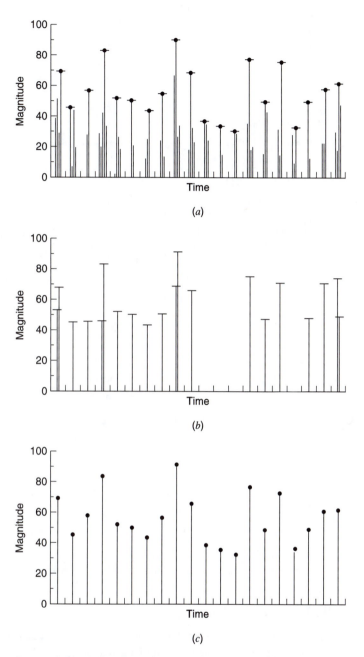

Figure 10.3.3 Hydrologic data arranged by time of occurrence. (*a*)
Original data: *N* = 20 years; (*b*) Annual exceedances. (*c*) Annual
maxima (from Chow (1964)).

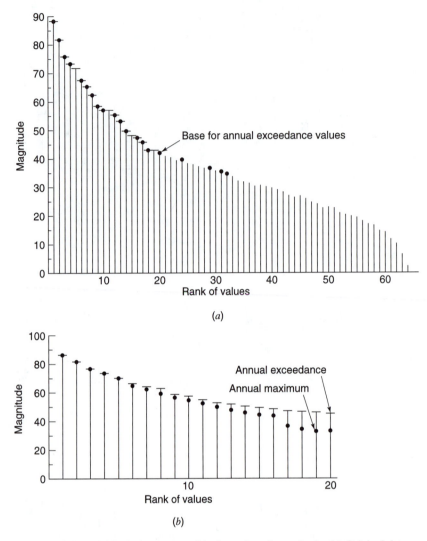

Figure 10.3.4 Hydrologic data arranged in the order of magnitude. (*a*) Original data; (*b*) Annual exceedance and maximum values (from Chow (1964)).

10.4 HYDROLOGIC FREQUENCY ANALYSIS

One of the primary objectives of the frequency analysis of hydrologic data is to determine the recurrence interval of a hydrologic event of a given magnitude. The *recurrence interval*, which is the same as the return period, may also be defined as the average interval of time within which the magnitude of a hydrologic event will be equaled or exceeded once, on the average. The term "frequency" is often used interchangeably with "recurrence interval"; however, it should not be construed to indicate a regular or stated interval of occurrence or recurrence. *Hydrologic frequency analysis* is the approach of using probability and statistical analysis to estimate future frequencies (probabilities of hydrologic events occurring) based upon information contained in hydrologic records. Through the use of statistical methods, observed data is analyzed so as to provide not only a more accurate estimate of future frequencies than is indicated by the observed data, but also criteria for determining the reliability of frequency estimates. The emphasis of hydrologic frequency analysis in this section is for the determination of flood frequency curves for streams and rivers.

The results of flood flow frequency analysis can be used for many engineering purposes: (1) for the design of dams, bridges, culverts, water supply systems, and flood control structures; (2) to determine the economic value of flood control projects; (3) to determine the effect of encroachments in the floodplain; (4) to determine a reservoir stage for real estate acquisition and reservoir-use purposes; (5) to the select of runoff magnitudes for interior drainage, pumping plant, and local protection project design; and (6) for flood-plain zoning, etc.

In the application of statistical methods to hydrologic frequency analysis, theoretical probability distributions are utilized. The hydrologic events that have occurred are assumed to constitute a random sample (observed set of events) and then are used to make inferences about the true population (all possible events) for the theoretical distribution considered. These inferences are subject to considerable uncertainty because a set of observed hydrologic events represents only a sample or small subset of the many sets of physical conditions that could represent the population described by the theoretical probability distribution.

The existing methods of frequency analysis are numerous, with many diverse and confusing viewpoints and theories. Several types of probability distributions have been used in the past for hydrologic frequency determination. The most popular of these for flood flow frequency determination have been the log-normal, Gumbel (extreme value type I in Table 10.2.1), and log-Pearson Type III Table 10.2.1 distributions. Because of the range of uncertainty and diversity of methods in determining flood flow estimates and the varying results that can be obtained using the various methods, the U.S. Water Resources Council (1981) attempted to promote a uniform or consistent approach to flood-flow frequency studies.

Computation of the magnitudes of extreme events, such as flood flows, requires that the probability distribution function be invertible; that is, given a value for T or $[f(x_T) = T/(T - 1)]$, the corresponding value of x_T can be determined. Some probability distribution functions are not readily invertible, including the normal and Pearson Type III distributions, and an alternative method of calculating the magnitudes of extreme events is required for these distributions.

The magnitude x_T of a hydrologic event may be represented as the mean $\bar{x}$ plus a departure of the variate from the mean. This departure is equal to the product of the standard deviation S_x and a frequency factor K_T. The departure Δx_T and the frequency factor K_T are functions of the return period and the type of probability distribution to be used in the analysis. Chow (1951) proposed the following frequency factor equation:

$$x_T = \bar{x} + K_T S_x \tag{10.4.1}$$

When the variable analyzed is $y = \log x$, then the same method is applied to the statistics for the logarithms of the data, using

$$y_T = \bar{y} + K_T S_y \tag{10.4.2}$$

and the required value of x_T is found by taking the antilog of y_T. For a given distribution, a K–T relationship can be determined between the frequency factor and the corresponding return period.

For the log-Pearson Type III distribution, the first step is to compute the logarithms of the hydrologic data, $y = \log x$. Usually, logarithms to base 10 are used. The mean $\bar{y}$, standard deviation S_y, and coefficient of skewness G_s are calculated for the logarithms of the data. Note that while γ is generally used for skew coefficient, G_s is used for a sample space (see Table 10.1.1). The frequency factor depends on the return period T and the coefficient of skewness G_s. When $G_s = 0$, the frequency factor is equal to the standard normal variable z. When $G_s \neq 0$, K_T is approximated by Kite (1977) as

$$K(T, G_s) = z + (z^2 - 1)k + \frac{1}{3}(z^3 - 6z)k^2 - (z^2 - 1)k^3 + zk^4 + \frac{1}{3}k^5 \tag{10.4.3}$$

where $k = G_s/6$. Table 10.4.1 lists values of the frequency factor for the Pearson Type III (and log-Pearson Type III) distribution for various values of the return period and coefficient of skewness.

The U.S. Water Resources Council (WRC) recommended that the log-Pearson Type III be used as a base method for flood flow frequency studies (U.S. Water Resources Council, 1981). This was an attempt to promote a consistent, uniform approach to flood flow frequency determination for use in all federal planning involving water and related land resources. This choice of the log-Pearson Type III is, however, subjective to some extent, in that no rigorous statistical criteria exist on which a comparison of distributions can be made.

The frequency factor equation for the log-Pearson Type III distribution is written in terms of discharge as

$$\log Q_T = \bar{y} + K(T, G_s) \cdot S_y \tag{10.4.4}$$

where Q_T is the discharge for the T-year return period.

The steps in the procedure to compute the discharge Q_T of return period T are as follows:

Step 1 Transform all discharge values to log Q_1, log Q_2 ..., log Q_n.

Step 2 Determine the mean ($\bar{y}$), standard deviation (S_y) and skew (G_s) of the log-transformed values.

Step 3 Use Table 10.4.1 to determine the frequency factors for the return periods of interest.

Step 4 Apply the frequency factor equation (10.4.4) and compute Q_T = antilog (log Q_T).

EXAMPLE 10.4.1

The mean, standard deviation, and skew of the log transformed discharges for the Medina River, Texas, are 3.639, 0.394, and 0.200, respectively, where the discharges are in ft³/s. Compute the 10-year and 100-year peak discharges.

SOLUTION

Assume that the peak discharges follow a log-Pearson Type III distribution and the procedures recommended by WRC will be used. First, we need to determine the frequency factors for 10-year and 100-year storm events from Table 10.4.1. To determine the frequency factors, we need to know the exceedance probability and the skew coefficient. Since the exceedance probability for a T-year storm event is $1/T$, the exceedance probabilities for 10-year and 100-year storm events are 0.1 and 0.01, respectively. Since the skew coefficient is given as $G_s = 0.200$, the frequency factors for the 10-year and 100-year storm events can be found from Table 10.4.1 as $K_{10} = 1.301$ and $K_{100} = 2.472$. Next, compute the 10-year and 100-year peak discharges:

$$\log Q_T = \bar{y} + K(T, G_s)S_y$$

$$\log Q_{10} = 3.639 + 1.301(0.394)$$

$$Q_{10} = 14,180 \text{ cfs}$$

$$\log Q_{100} = 3.639 + 2.472(0.394)$$

$$Q_{10} = 41,020 \text{ cfs}$$

Table 10.4.1 K_T Values for Pearson Type III Distribution

				Recurrence Interval in Years						
1.0101	1.0526	1.1111	1.2500	2	5	10	25	50	100	200
				Exceedance Probability						
Skew coeff.	.99	.95	.90	.80	.50	.20	.10	.04	.02	.01	.005
3.0	−0.667	−0.665	−0.660	−0.636	−0.396	0.420	1.180	2.278	3.152	4.051	4.970
2.9	−0.690	−0.688	−0.681	−0.651	−0.390	0.440	1.195	2.277	3.134	4.013	4.909
2.8	−0.714	−0.711	−0.702	−0.666	−0.384	0.460	1.210	2.275	3.114	3.973	4.847
2.7	−0.740	−0.736	−0.724	−0.681	−0.376	0.479	1.224	2.272	3.093	3.932	4.783
2.6	−0.769	−0.762	−0.747	−0.696	−0.368	0.499	1.238	2.267	3.071	3.889	4.718
2.5	−0.799	−0.790	−0.771	−0.711	−0.360	0.518	1.250	2.262	3.048	3.845	4.652
2.4	−0.832	−0.819	−0.795	−0.725	−0.351	0.537	1.262	2.256	3.023	3.800	4.484
2.3	−0.867	−0.850	−0.819	−0.739	−0.341	0.555	1.274	2.248	2.997	3.753	4.515

Table 10.4.1 K_T Values for Pearson Type III Distribution (*continued*)

	Recurrence Interval in Years										
	1.0101	1.0526	1.1111	1.2500	2	5	10	25	50	100	200
	Exceedance Probability										
Skew coeff.	.99	.95	.90	.80	.50	.20	.10	.04	.02	.01	.005
2.2	−0.905	−0.882	−0.844	−0.752	−0.330	0.574	1.284	2.240	2.970	3.705	4.444
2.1	−0.946	−0.914	−0.869	−0.765	−0.319	0.592	1.294	2.230	2.942	3.656	4.372
2.0	−0.990	−0.949	−0.895	−0.777	−0.307	0.609	1.302	2.219	2.912	3.605	4.298
1.9	−1.037	−0.984	−0.920	−0.788	−0.294	0.627	1.310	2.207	2.881	3.553	4.223
1.8	−1.087	−1.020	−0.945	−0.799	−0.282	0.643	1.318	2.193	2.848	3.499	4.147
1.7	−1.140	−1.056	−0.970	−0.808	−0.268	0.660	1.324	2.179	2.815	3.444	4.069
1.6	−1.197	−1.093	−0.994	−0.817	−0.254	0.675	1.329	2.136	2.780	3.388	3.990
1.5	−1.256	−1.131	−1.018	−0.825	−0.240	0.690	1.333	2.146	2.743	3.330	3.910
1.4	−1.318	−1.168	−1.041	−0.832	−0.225	0.705	1.337	2.128	2.706	3.271	3.838
1.3	−1.383	−1.260	−1.064	−0.838	−0.210	0.719	1.339	2.108	2.666	3.211	3.745
1.2	−1.449	−1.243	−1.086	−0.844	−0.195	0.732	1.340	2.087	2.626	3.149	3.661
1.1	−1.518	−1.280	−1.107	−0.848	−0.180	0.745	1.341	2.066	2.585	3.087	3.575
1.0	−1.588	−1.317	−1.128	−0.852	−0.164	0.758	1.340	2.043	2.542	3.022	3.489
.9	−1.660	−1.353	−1.147	−0.854	−0.148	0.769	1.339	2.018	2.498	2.957	3.401
.8	−1.733	−1.388	−1.166	−0.856	−0.132	0.780	1.336	1.993	2.453	2.891	3.312
.7	−1.806	−1.423	−1.183	−0.857	−0.116	0.790	1.333	1.967	2.407	2.824	3.223
.6	−1.880	−1.458	−1.200	−0.857	−0.099	0.800	1.328	1.939	2.359	2.755	3.132
.5	−1.955	−1.491	−1.216	−0.856	−0.083	0.808	1.323	1.910	2.311	2.686	3.041
.4	−2.029	−1.524	−1.231	−0.855	−0.066	0.816	1.317	1.880	2.261	2.615	2.949
.3	−2.104	−1.555	−1.245	−0.853	−0.050	0.824	1.309	1.849	2.211	2.544	2.856
.2	−2.178	−1.586	−1.258	−0.850	−0.033	0.830	1.301	1.818	2.159	2.472	2.763
.1	−2.252	−1.616	−1.270	−0.846	−0.017	0.836	1.292	1.785	2.107	2.400	2.670
.0	−2.326	−1.645	−1.282	−0.842	0	0.842	1.282	1.751	2.054	2.326	2.576
−.1	−2.400	−1.673	−1.292	−0.836	0.017	0.846	1.270	1.716	1.000	2.252	2.482
−.2	−2.472	−1.700	−1.301	−0.830	0.033	0.850	1.258	1.680	1.945	2.178	2.388
−.3	−2.544	−1.726	−1.309	−0.824	0.050	0.853	1.245	1.643	1.890	2.104	2.294
−.4	−2.615	−1.750	−1.317	−0.816	0.066	0.855	1.231	1.606	1.834	2.029	2.201
−.5	−2.686	−1.774	−1.323	−0.808	0.083	0.856	1.216	1.567	1.777	1.955	2.108
−.6	−2.755	−1.797	−1.328	−0.800	0.099	0.857	1.200	1.528	1.720	1.880	2.016
−.7	−2.824	−1.819	−1.333	−0.790	0.116	0.857	1.183	1.488	1.663	1.806	1.929
−.8	−2.891	−1.839	−1.336	−0.780	0.132	0.856	1.166	1.448	1.606	1.733	1.837
−.9	−2.957	−1.858	−1.339	−0.769	0.148	0.854	1.147	1.407	1.549	1.660	1.749
−1.0	−3.022	−1.877	−1.340	−0.758	0.164	0.852	1.128	1.366	1.492	1.588	1.664
−1.1	−3.087	−1.894	−1.341	−0.745	0.180	0.848	1.107	1.324	1.435	1.518	1.581
−1.2	−3.149	−1.910	−1.340	−0.732	0.195	0.844	1.086	1.282	1.379	1.449	1.501
−1.3	−3.211	−1.925	−1.339	−0.719	0.210	0.838	1.064	1.240	1.324	1.383	1.424
−1.4	−3.271	−1.938	−1.337	−0.705	0.225	0.832	1.041	1.198	1.270	1.318	1.351
−1.5	−3.330	−1.951	−1.333	−0.690	0.240	0.825	1.018	1.157	1.217	1.256	1.282
−1.6	−3.388	−1.962	−1.329	−0.675	0.254	0.817	0.994	1.116	1.166	1.197	1.216
−1.7	−3.444	−1.972	−1.324	−0.660	0.268	0.808	0.970	1.075	1.116	1.140	1.155
−1.8	−3.499	−1.981	−1.318	−0.643	0.282	0.799	0.945	1.035	1.069	1.087	1.097
−1.9	−3.553	−1.989	−1.310	−0.627	0.294	0.788	0.920	0.996	1.023	1.037	1.044
−2.0	−3.605	−1.996	−1.302	−0.609	0.307	0.777	0.895	0.959	0.980	0.990	0.995
−2.1	−3.656	−2.001	−1.294	−0.592	0.319	0.765	0.869	0.923	0.939	0.946	0.949
−2.2	−3.705	−2.006	−1.284	−0.574	0.330	0.752	0.844	0.888	0.900	0.905	0.907
−2.3	−3.753	−2.009	−1.274	−0.555	0.341	0.739	0.819	0.855	0.864	0.867	0.869
−2.4	−3.800	−2.011	−1.262	−0.537	0.351	0.725	0.795	0.823	0.830	0.832	0.833

Table 10.4.1 K_T Values for Pearson Type III Distribution (*continued*)

				Recurrence Interval in Years							
1.0101	1.0526	1.1111	1.2500	2	5	10	25	50	100	200
				Exceedance Probability							
Skew coeff.	.99	.95	.90	.80	.50	.20	.10	.04	.02	.01	.005
−2.5	−3.845	−2.012	−1.250	−0.518	0.360	0.711	0.771	0.793	0.798	0.799	0.800
−2.6	−3.889	−2.013	−1.238	−0.499	0.368	0.696	0.747	0.764	0.768	0.769	0.769
−2.7	−3.932	−2.012	−1.224	−0.479	0.376	0.681	0.724	0.738	0.740	0.740	0.741
−2.8	−3.973	−2.010	−1.210	−0.460	0.384	0.666	0.702	0.712	0.714	0.714	0.714
−2.9	−4.013	−2.007	−1.195	−0.440	0.390	0.651	0.681	0.683	0.689	0.690	0.690
−3.0	−4.051	−2.003	−1.180	−0.420	0.396	0.636	0.660	0.666	0.666	0.667	0.667

EXAMPLE 10.4.2

The annual maximum series for the U.S.G.S. gauge on the Wichita River near Cheyenne, Oklahoma, is listed in Table 10.4.2. The objective is to compute the 25-year and 100-year peak discharges using the log-Pearson Type III distribution. Also compute plotting position using the Weibull plotting position formula, $P(X > x_T) = 1/T = m/(n + 1)$, where m is the rank of descending values and n is the number of peaks in the annual maximum series. (Adapted from Cudworth, 1989.)

Table 10.4.2 Annual Peak Discharges for Each Year of Record for the Drainage Area Above Foss Dam

(1) Year	(2) Annual Peak Discharge, ft³/s	(3) Ranked Annual Peak Discharge, ft³/s	(4) Weibull Plotting Position	(5) Logarithm of Discharge y	(6) y^2	(7) y^3
1938	14,600	69,800	0.021	4.84386	23.46298	113.65139
1939	3,070	40,000	.042	4.60206	21.17896	97.46683
1940	1,080	14,600	.063	4.16435	17.34181	72.217.37
1941	40,000	14,000	.083	4.14613	17.19039	71.27359
1942	14,000	11,900	.104	4.07555	16.61011	67.69533
1943	2,190	9,900	.125	3.99563	15.96506	63.79047
1944	1,240	8,900	.146	3.94939	15.59768	61.60133
1945	9,900	8,900	.167	3.94939	15.59768	61.60133
1946	8,900	8,450	.189	3.92686	15.42023	60.55308
1947	7,100	7,310	.208	3.86392	14.92988	57.68785
1948	8,900	7,100	.229	3.85126	14.83220	57.12267
1949	11,900	6,420	.250	3.80754	14.49736	55.19928
1950	8,450	5,830	.271	3.76567	14.18027	53.39822
1951	5,040	5,040	.292	3.70243	13.70799	50.75287
1952	465	4,710	.313	3.67302	13.49108	49.55299
1953	3,550	4,650	.333	3.66839	13.45709	49.36584
1954	69,800	4,470	.354	3.65031	13.32476	48.63952
1955	5,830	4,210	.375	3.62428	13.13541	47.60639
1956	3,890	3,890	.396	3.58995	12.88774	46.26635
1957	4,210	3,550	.417	3.55023	12.60413	44.74757
1958	1,750	3,070	.438	3.48714	12.16015	42.40413
1959	6,420	2,990	.458	3.47567	12.08028	41.98707
1960	1,510	2,930	.479	3.46687	12.01919	41.66896
1961	7,310	2,280	.500	3.35793	11.27569	37.86299

Table 10.4.2   Annual Peak Discharges for Each Year of Record for the
Drainage Area Above Foss Dam (*continued*)

(1)	(2)	(3)	(4)	(5)	(6)	(7)
Year	Annual Peak Discharge ft³/s	Ranked Annual Peak Discharge, ft³/s	Weibull Plotting Position	Logarithm of Discharge y	y^2	y^3
1962	2.930	2,190	.521	3.34044	11.15854	37.27443
1963	574	1,960	.542	3.29226	10.83898	35.68473
1964	159	1,800	.563	3.25527	10.59678	34.49539
1965	1,400	1,750	.583	3.24304	10.51731	34.10805
1966	1,800	1,510	.604	3.17898	10.10591	32.12650
1967	2,990	1,420	.625	3.15229	9.93693	31.32409
1968	4,470	1,400	.646	3.14613	9.89813	31.14082
1969	2,280	1,360	.667	3.13354	9.81907	30.76846
1970	734	1,240	.688	3.09342	9.56925	29.60170
1971	4,710	1,080	.708	3.03342	9.20164	27.91243
1972	1,360	1,050	.729	3.02119	9.12759	27.57618
1973	265	734	.750	2.86570	8.21224	23.53381
1974	592	592	.771	2.77232	7.68576	21.30738
1975	1,050	574	.792	2.75891	7.61158	20.99968
1976	1,960	560	.813	2.74819	7.55255	20.75584
1977	4,660	465	.833	2.66745	7.11529	18.97968
1978	297	427	.854	2.63043	6.91916	18.20037
1979	400	400	.875	2.60206	6.77072	17.61781
1980	560	297	.896	2.47276	6.11454	15.11980
1981	38	265	.917	2.42325	5.87214	14.22967
1982	1,420	159	.938	2.20140	4.84616	10.66834
1983	427	119	.958	2.07555	4.30791	8.94128
1984	119	38	.979	<u>1.57978</u>	<u>2.49570</u>	<u>3.94266</u>
			Totals	156.87558	543.2220	1940.4225

Source: Cudworth (1989).

SOLUTION

Step 1 Transform data using logarithms, computed as given in Table 10.4.2.

Step 2 Determine the mean, standard deviation and skew of the log-transformed values:

$$\bar{y} = \frac{\Sigma y}{n} = \frac{156.876}{47} = 3.338$$

$$S_y = \sqrt{\frac{\Sigma y^2 - \frac{\left(\Sigma y\right)^2}{n}}{n-1}} = \sqrt{\frac{543.222 - \frac{(156.876)^2}{47}}{47-1}} = 0.653$$

$$G_s = \frac{n^2\left(\Sigma y^3\right) - 3n\left(\Sigma y\right)\left(\Sigma y^2\right) + 2\left(\Sigma y\right)^3}{n(n-1)(n-2)(S_y^3)}$$

$$= \frac{(47)^2(1940.423) - (3)(47)(156.876)(543.222) + (2)(156.876)^3}{(47)(47-1)(47-2)(0.653)^3}$$

$$= -0.294$$

$$G_s \approx -0.3$$

Step 3 Determine frequency factors for the 25-year and the 100-year events using Table 10.4.1:

$K(25, -0.3) = 1.643$

$K(100, -0.3) = 2.104$

Step 4 Apply frequency factor equations to determine Q_{25} and Q_{100}

$$\log Q_{25} = \bar{y} + K(25, -0.3) S_y$$

$$= 3.338 + 1.643(0.653)$$

$$= 4.411$$

$Q_{25} = $ antilog $(4.411) = 25,765$ cfs

$$\log Q_{100} = \bar{y} + K(100, -0.3) S_y$$

$$= 3.338 + 2.104(0.653)$$

$$= 4.712$$

$Q_{100} = $ antilog $(4.712) = 51,525$ cfs

The annual peak discharges are ranked in descending order in column 3 of Table 10.4.2 and the Weibull plotting positions are listed in column 4 for the corresponding discharge in column 3. As an example, the largest discharge has a plotting position of $m = 1$, so $1/T = 1/(47 + 1) = 0.021$. The second largest, $m = 2$, has a Weiball plotting position of $2/(47 + 1) = 0.042$.

10.5   U.S. WATER RESOURCES COUNCIL GUIDELINES FOR FLOOD FLOW FREQUENCY ANALYSIS

Flood flow frequency analysis is another method of discharge determination using statistical methods when gauged data are available to develop annual maximum series. Section 10.4 introduced hydrologic frequency analysis for flood flow. This section extends the concepts to the so-called *U.S. Water Resources Council (WRC) method.*

10.5.1   Procedure

The skew coefficient is very sensitive to the size of the sample; thus, it is difficult to obtain an accurate estimate from small samples. Because of this, the U.S. Water Resources Council (1981) recommended using a generalized estimate of the skew coefficient when estimating the skew for short records. As the length of record increases, the skew is usually more reliable. The guidelines recommend the use of a *weighted skew* G_w, based upon the equation

$$G_w = WG_s + (1 - W)G_m \tag{10.5.1}$$

where W is a weight, G_s is the skew coefficient computed using the sample data, and G_m is a map skew, values of which for the United States are found in Figure 10.5.1. The generalized skew is derived as a weighted average between skew coefficients computed from sample data (sample skew) and regional or map skew coefficients (referred to as a generalized skew in U.S. Water Resources Council, 1981).

A weighting procedure was derived that is a function of the variance of the sample skew and the variance of the map skew. Such a procedure considers the uncertainty of deriving skew coefficients from both sample data and regional or map values to obtain a generalized skew that minimizes uncertainty based upon information known.

The estimates of the sample skew coefficient and the map skew coefficient in equation (10.5.1) are assumed to be independent with the same mean and respective variances. Assuming independence of G_s and G_m, the variance (mean square error) of the weighted skew, $V(G_w)$, can be expressed as

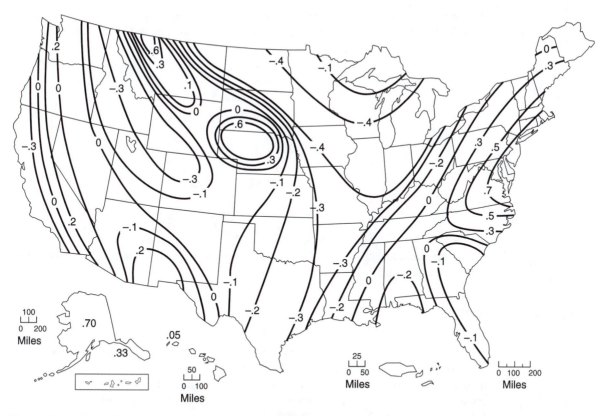

Figure 10.5.1 Generalized skew coefficients of annual maximum streamflow (from U.S. Water Resources Council (1981)).

$$V(G_w) = W^2 \cdot V(G_s) + (1 - W)^2 \cdot V(G_m) \qquad (10.5.2)$$

where $V(G_s)$ is the variance of the sample skew and $V(G_m)$ is the variance of the map skew. The skew weight that minimizes the variance of the weighted skew can be determined by differentiating equation (10.5.2) with respect to W and solving $d[V(G_w)]/dW = 0$ for W to obtain

$$W = \frac{V(G_m)}{V(G_s) + V(G_m)} \qquad (10.5.3)$$

Equation (10.5.3) is a convex function and the second derivative of equation (10.5.2) is greater than 0, proving that equation (10.5.3) gives the value of W that minimizes the weighted skew.

Determination of W using equation (10.5.3) requires the values of $V(G_m)$ and $V(G_s)$. $V(G_m)$ can be estimated from the map of the skew coefficients as the squared value of the standard deviation of station values of skew coefficients about the isolines of the skew map. The value of $V(G_m)$, estimated for the skew map in U.S. Water Resources Council (1981) is 0.3025. Alternatively, $V(G_m)$ could be derived from a regression study relating the skew to physiographical and meteorological characteristics of the basins and determining $V(G_m)$ as the square of the standard error of the regression equation (Tung and Mays, 1981).

The weighted skew G_w can be determined by substituting equation (10.5.3) into equation (10.5.1), resulting in

$$G_w = \frac{V(G_m) \cdot G_s + V(G_s) \cdot G_m}{V(G_m) + V(G_s)} \qquad (10.5.4)$$

The variance (mean square error) of the station skew for log-Pearson Type III random variables can be obtained from the results of Monte Carlo experiments by Wallis et al. (1974). Their results showed that $V(G_s)$ of the logarithmic station skew is a function of record length and population skew. For use in calculating G_w, this function $V(G_s)$ can be approximated with sufficient accuracy using

$$V(G_s) = 10^{A - B[\log_{10}(n/10)]} \qquad (10.5.5)$$

where

$$A = -0.33 + 0.08 \, |G_s| \quad \text{if} \quad |G_s| \leq 0.90 \qquad (10.5.6a)$$

$$A = -0.52 + 0.30 \, |G_s| \quad \text{if} \quad |G_s| > 0.90 \qquad (10.5.6b)$$

$$B = 0.94 - 0.26 \, |G_s| \quad \text{if} \quad |G_s| \leq 1.50 \qquad (10.5.7a)$$

$$B = 0.55 \quad \text{if} \quad |G_s| > 1.50 \qquad (10.5.7b)$$

in which $|G_s|$ is the absolute value of the station skew (used as an estimate of population skew) and n is the record length in years. The same steps used in section 10.4 can be used to compute the discharge Q_T of return period T.

10.5.2 Testing for Outliers

The WRC method recommends that outliers be identified and adjusted according to their recommended methods. *Outliers* are data points that depart significantly from the trend of the remaining data. The retention or deletion of these outliers can significantly affect the magnitude of statistical parameters computed from the data, especially for small samples. Procedures for treating outliers require judgment involving both mathematical and hydrologic considerations. According to the U.S. Water Resources Council (1981), if the station skew is greater than +0.4, tests for high outliers are considered first. If the station skew is less than −0.4, tests for low outliers are considered first. Where the station skew is between ±0.4, tests for both high and low outliers should be applied before eliminating any outliers from the data set.

The following frequency equation can be used to detect high outliers:

$$\overline{y}_H = y + K_N S_y \qquad (10.5.8)$$

where y_H is the high outlier threshold in log units and K_N is the K value from Table 10.5.1 for sample size N. If the logarithms of peaks in the sample are greater than y_H in the above equation, then they are considered high outliers. Flood peaks considered high outliers should be compared with historic flood data and flood information at nearby sites. According to the U.S. Water Resources Council (1981), if information is available indicaing that a high outlier(s) is the maximum in an extended period of time, the outlier(s) is treated as historic flood data. If useful historic information is not available to adjust for high outliers, then the outliers should be retained as part of the systematic record.

The following frequency equation can be used to detect low outliers:

$$\overline{y}_L = y - K_N S_y \qquad (10.5.9)$$

where y_L is the low outlier threshold in log units. Flood peaks considered low outliers are deleted from the record and a conditional probability adjustment described in the U.S. Water Resources Council (1981) is applied. Use of the K values in Table 10.5.1 is equivalent to a one-sided test that detects outliers at the 10 percent level of significance. The K values are based on a normal distribution for detection of single outliers.

Table 10.5.1 Outlier Test K Values: 10 Percent Significance Level K Values

Sample size	K value	Sample size	K value	Sample size	K value	Sample size	K value
10	2.036	45	2.727	80	2.940	115	3.064
11	2.088	46	3.736	81	2.945	116	3.067
12	2.134	47	2.744	82	2.949	117	3.070
13	2.175	48	2.753	83	2.953	118	3.073
14	2.213	49	2.760	84	2.957	119	3.075
15	2.247	50	2.768	85	2.961	120	3.078
16	2.279	51	2.775	86	2.966	121	3.081
17	2.309	52	2.783	87	2.970	122	3.083
18	2.335	53	2.790	88	2.973	123	3,086
19	2.361	54	2.798	89	2.977	124	3.089
20	2.385	55	2.804	90	2.981	125	3.092
21	2.408	56	2.811	91	2.984	126	3.095
22	2.429	57	2.818	92	2.989	127	3.097
23	2.448	58	2.824	93	2.993	128	3.100
24	2.467	59	2.831	94	2.996	129	3.102
25	2.486	60	2.837	95	3.000	130	3.104
26	2.502	61	2.842	96	3.003	131	3.107
27	2.519	62	2.849	97	3.006	132	3.109
28	2.534	63	2.854	98	3.011	133	3.112
29	2.549	64	2.860	99	3.014	134	3.114
30	2.563	65	2.866	100	3.017	135	3.116
31	2.577	66	2.871	101	3.021	136	3.119
32	2.591	67	2.877	102	3.024	137	3.122
33	2.604	68	2.883	103	3.027	138	3.124
34	2.616	69	2.888	104	3.030	139	3.126
35	2.628	70	2.893	105	3.033	140	3.129
36	2.639	71	2.897	106	3.037	141	3.131
37	2.650	72	2.903	107	3.040	142	3.133
38	2.661	73	2.908	108	3.043	143	3.135
39	2.671	74	2.912	109	3.046	144	3.138
40	2.682	75	2.917	110	3.049	145	3.140
41	2.692	76	2.922	111	3.052	146	3.142
42	2.700	77	2.927	112	3.055	147	3.144
43	2.710	78	2.931	113	3.058	148	3.146
44	2.719	79	2.935	114	3.061	149	3.148

This table contains one-sided 10 percent significance level K values for a normal distribution.

Source: U.S. Water Resources Council (1981).

EXAMPLE 10.5.1

Using the data for example 10.4.2 (annual maximum series for the gauge on the Wichita River near Cheyenne, Oklahoma), compute the 25-year and the 100-year peak discharges using the U.S. Water Resource Council (1981) guidelines. The map skew for this location is -0.015.

SOLUTION

Step 1–2 Transform data using logarithms and compute statistics as in example 10.4.2, where $\bar{y} = 3.338$, $S_y = 0.653$, and $G_s = -0.294 \approx -0.3$.

Step 3 Compute the weighted skew coefficient, G_w.

Step 3a Compute A and B using equations (10.5.6) and (10.5.7):

$$A = -0.33 + 0.08 \,|{-0.294}| = -0.306$$

$$B = 0.94 - 0.26 \, |-0.294| = 0.864$$

Step 3b Compute $V(G_s)$.

$$V(G_s) = 10^{A-B\log(n/10)} = 10^{-0.306-0.864\log(47/10)} = 0.130$$

Step 3c Use equation (10.5.4) to compute the weighted skew coefficient using $V(G_m) = 0.302$ (as estimated in U.S. Water Resouces Council (1981)):

$$G_w = \frac{0.302(-0.294)+0.130(-0.015)}{0.302+0.130} = -0.210$$

Step 4 Use Table 10.4.1 to obtain the frequency factors using the weighted skew:

$$K(25, -0.210) = 1.676$$

$$K(100, -0.210) = 2.171$$

Step 5 Apply the frequency factor equation to determine Q_{25} and Q_{100}:

$$\log Q_{25} = \bar{y} + K(25, -0.251) \, S_y$$

$$= 3.338 + 1.676 \, (0.653) = 4.432$$

$$Q_{25} = 27{,}070 \text{ cfs}$$

$$\log Q_{100} = 3.338 + 2.171 \, (0.653) = 4.756$$

$$Q_{100} = 56{,}975 \text{ cfs}$$

Note that in example 10.4.2 we computed $Q_{25} = 25{,}765$ cfs and $Q_{100} = 51{,}525$ cfs when the sample skew was used.

EXAMPLE 10.5.2

Using the data in examples 10.4.2 and 10.5.1, determine if any outliers exist in the annual maximum series (Table 10.4.2).

SOLUTION

Determine whether the data includes any high or low outliers. First, consider high outliers. From Table 10.5.1, $K_n = 2.744$ for $n = 47$, and using equation (10.5.8) to determine the threshold value y_H, we find

$$y_H = \bar{y} + K_n S_y = 3.338 + 2.744 \, (0.653) = 5.130$$

so

$$Q_H = 10^{y_H} = (10)^{5.130} = 134{,}900 \text{ cfs}$$

The largest recorded value of 69,800 cfs for the year 1954 in Table 10.4.2 does not exceed the threshold value of 134,900 cfs.

Next consider low outliers and use equation (10.5.9) to determine the threshold:

$$y_L = \bar{y} - K_N S_y = 3.338 - 2.744 \, (0.653) = 1.546$$

so

$$Q_L = (10)^{y_L} = 10^{1.546} = 35 \text{ cfs}$$

The smallest recorded peak, 38 cfs for 1963, is larger than 35 cfs, so there are no low outliers by this methodology.

10.6 ANALYSIS OF UNCERTAINTIES

In the design and analysis of hydrosystems, many quantities of interest are functionally related to a number of variables, some of which are subject to uncertainty. For example, hydraulic engineers frequently apply weir flow equations such as $Q = CLH^{1.5}$ to estimate spillway capacity in which the coefficient C and head H are subject to uncertainty. As a result, discharge over the spillway is not certain. A rather straightforward and useful technique for the approximation of such uncertainties is the *first-order analysis of uncertainties*, sometimes called the *delta method*.

The use of the first-order analysis of uncertainties is quite popular in many fields of engineering because of its relative ease in application to a wide array of problems. First-order analysis is used to estimate the uncertainty in a deterministic model formulation involving parameters that are uncertain (not known with certainty). More specifically, first-order analysis enables one to estimate the mean and variance of a random variable that is functionally related to several other variables, some of which are random. By using first-order analysis, the combined effect of uncertainty in a model formulation, as well as the use of uncertain parameters, can be assessed.

Consider a random variable y that is a function of k random variables (multivariate case):

$$y = g(x_1, x_2, ..., x_k) \tag{10.6.1}$$

This can be a deterministic equation such as the weir equation mentioned above, or the rational formula or Manning's equation; or this function can be a complex model that must be solved on a computer. The objective is to treat a deterministic model that has uncertain inputs in order to determine the effect of the uncertain parameters $x_1, ..., x_k$ on the model output y.

Equation (10.6.1) can be expressed as $y = g(\mathbf{x})$, where $\mathbf{x} = x_1, x_2, ..., x_k$. Through a Taylor series expansion about k random variables, ignoring the second- and higher-order terms, we get

$$y \approx g(\overline{\mathbf{x}}) + \sum_{i=1}^{k}\left[\frac{\partial g}{\partial x_i}\right]_{\overline{\mathbf{x}}}\left(X_i - \overline{x}_i\right) \tag{10.6.2}$$

The derivation $[\partial g/\partial x_i]_{\overline{\mathbf{x}}}$ are the *sensitivity coefficients* that represent the rate of change of the function value $g(\mathbf{x})$ at $\mathbf{x} = \overline{\mathbf{x}}$

Assuming that the k random variables are independent, then the variance of y is approximated as

$$\sigma_y^2 = \text{Var}[y] = \sum a_i^2 \sigma_{x_i}^2 \tag{10.6.3}$$

and the coefficient of variation is Ω_y:

$$\Omega_y = \left[\sum_{\tau=1}^{k} a_i^2 \left(\frac{\overline{x}_i}{\mu_y}\right)^2 \Omega_{x_i}^2\right]^{1/2} \tag{10.6.4}$$

where $a_i = (\partial g/\partial x)_{\overline{x}}$. Refer to Mays and Tung (1992) for a detailed derivation of equations (10.6.3) and (10.6.4).

EXAMPLE 10.6.1

Apply the first-order analysis to the rational equation $Q = CiA$, in which C is the runoff coefficient, i is the rainfall intensity in in/m, and A is the drainage area in acres, to determine formulas for σ_Q and Ω_Q. Consider C, i, and A to be uncertain.

SOLUTION

The first-order approximation of Q is determined using equation (10.6.2), so that

$$Q \approx \overline{Q} + \left[\frac{\partial Q}{\partial C}\right]_{\overline{C},\overline{i},\overline{A}}(C - \overline{C}) + \left[\frac{\partial Q}{\partial i}\right]_{\overline{C},\overline{i},\overline{A}}(i - \overline{i}) + \left[\frac{\partial Q}{\partial A}\right]_{\overline{C},\overline{i},\overline{A}}(A - \overline{A})$$

For $C = \overline{C}$, $i = \overline{i}$, and $A = \overline{A}$, $\overline{Q} = \overline{C}\,\overline{i}\,\overline{A}$. The variance is computed using equation (10.6.3):

$$\sigma_Q^2 = \left[\frac{\partial Q}{\partial C}\right]_{(\overline{C},\overline{i},\overline{A})}^2 \sigma_C^2 + \left[\frac{\partial Q}{\partial i}\right]_{(\overline{C},\overline{i},\overline{A})}^2 \sigma_i^2 + \left[\frac{\partial Q}{\partial A}\right]_{(\overline{C},\overline{i},\overline{A})}^2 \sigma_A^2$$

$$\sigma_Q = \left\{ \left[\frac{\partial Q}{\partial C} \right]^2_{(\bar{C}, \bar{i}, \bar{A})} \sigma_C^2 + \left[\frac{\partial Q}{\partial i} \right]^2_{(\bar{C}, \bar{i}, \bar{A})} \sigma_i^2 + \left[\frac{\partial Q}{\partial A} \right]^2_{(\bar{C}, \bar{i}, \bar{A})} \sigma_A^2 \right\}^{1/2}$$

The coefficient of variation is computed using equation (10.6.4):

$$\Omega_Q^2 = \left[\frac{\partial Q}{\partial C} \right]^2 \left(\frac{\bar{C}}{\bar{Q}} \right)^2 \Omega_C^2 + \left[\frac{\partial Q}{\partial i} \right]^2 \left(\frac{\bar{i}}{\bar{Q}} \right)^2 \Omega_i^2 + \left[\frac{\partial Q}{\partial A} \right]^2 \left(\frac{\bar{A}}{\bar{Q}} \right)^2 \Omega_A^2$$

$$= (\bar{i}\bar{A})^2 \left(\frac{\bar{C}}{\bar{Q}} \right)^2 \Omega_C^2 + (\bar{C}\bar{A})^2 \left(\frac{\bar{i}}{\bar{Q}} \right)^2 \Omega_i^2 + (\bar{C}\bar{i})^2 \left(\frac{\bar{A}}{\bar{Q}} \right)^2 \Omega_A^2$$

$$= \Omega_C^2 + \Omega_i^2 + \Omega_A^2$$

$$\Omega_Q = \left[\Omega_C^2 + \Omega_i^2 + \Omega_A^2 \right]^{1/2}$$

EXAMPLE 10.6.2

Determine the mean coefficient of variation, and standard deviation of the runoff using the rational equation with the following parameter values:

Parameter	Mean	Coefficient of Variation
C	0.8	0.09
i	100 mm/h	0.5
A	0.1 km^2	0.005

$Q = KCiA$, where $K = 0.28$ for SI units for i in mm/hr and A in km^2.

SOLUTION

Using $\bar{Q} = 0.28\,\bar{C}\,\bar{i}\,\bar{A}$ from example (10.6.1), we find

$$\bar{Q} = 0.28(0.8)(100)(0.1) = 2.24 \text{ m}^3/\text{s}$$

Using $\Omega_Q = \left[\Omega_C^2 + \Omega_i^2 + \Omega_A^2 \right]^{1/2}$ from example 10.6.1, we find

$$\Omega_Q = \left[0.09^2 + 0.5^2 + 0.005^2 \right]^{1/2} = 0.508$$

The standard deviation of Q can be determined using

$$\sigma_Q = \bar{Q}\Omega_Q = 2.24(0.508) = 1.138 \text{ m}^3/\text{s}$$

EXAMPLE 10.6.3

Apply the first-order analysis to Manning's equation for full pipe flow, given in U.S. customary units as

$$Q = \frac{0.463}{n} S^{1/2} D^{8/3}$$

to determine for computing equations for σ_Q and Ω_Q. Consider the diameter D to be deterministic without any uncertainty, and consider n and S to be uncertain.

SOLUTION

Since n and S are uncertain, Manning's equation can be rewritten as

$$Q = Kn^{-1}S^{1/2}$$

where $K = 0.463D^{8/3}$. The first-order approximation of Q is determined using equation (10.6.2), so that

$$Q \approx \overline{Q} + \left[\frac{\partial Q}{\partial n}\right]_{\overline{n},\overline{S}} (n - \overline{n}) + \left[\frac{\partial Q}{\partial S}\right]_{\overline{n},\overline{S}} (S - \overline{S})$$

$$= \overline{Q} + \left[-K\overline{n}^{-2}\overline{S}^{1/2}\right](n - \overline{n}) + \left[0.5K\overline{n}^{-1}\overline{S}^{-1/2}\right](S - \overline{S})$$

where $\overline{Q} = K\overline{n}^{-1}\overline{S}^{1/2}$.

Next compute the variance of the pipe capacity using equation (10.6.3):

$$\sigma_Q^2 = \left[\frac{\partial Q}{\partial n}\right]_{(\overline{n},\overline{S})}^2 \sigma_n^2 + \left[\frac{\partial Q}{\partial S}\right]_{(\overline{n},\overline{S})}^2 \sigma_S^2$$

$$\sigma_Q = \left\{\left[\frac{\partial Q}{\partial n}\right]_{(\overline{n},\overline{S})}^2 \sigma_n^2 + \left[\frac{\partial Q}{\partial S}\right]_{(\overline{n},\overline{S})}^2 \sigma_S^2\right\}^{1/2}$$

Determine the coefficient of variation of Q using equation (10.6.4):

$$\Omega_Q^2 = \sum_{i=1}^{2} \left[\frac{\partial Q}{\partial x_i}\right]^2 \left[\frac{\overline{x}_i}{\overline{Q}}\right]^2 \Omega_{x_i}^2$$

$$= \left[\frac{\partial Q}{\partial n}\right]^2 \left[\frac{\overline{n}}{\overline{Q}}\right]^2 \Omega_n^2 + \left[\frac{\partial Q}{\partial S}\right]^2 \left[\frac{\overline{S}}{\overline{Q}}\right]^2 \Omega_S^2$$

$$= \left[\frac{-K\overline{S}^{1/2}}{\overline{n}^2}\right]^2 \left[\frac{\overline{n}}{\overline{Q}}\right]^2 \Omega_n^2 + \left[\frac{0.5K}{\overline{n}\overline{S}^{1/2}}\right]^2 \left[\frac{\overline{S}}{\overline{Q}}\right]^2 \Omega_S^2$$

$$= \left[\frac{-K\overline{S}^{1/2}}{\overline{Q}}\right]^2 \left[\frac{1}{\overline{n}^2}\right] \Omega_n^2 + [0.5]^2 \left[\frac{K}{\overline{n}\overline{S}^{1/2}}\right]^2 \left[\frac{\overline{S}}{\overline{Q}}\right]^2 \Omega_S^2$$

$$= \left[\overline{n}^2\right] \left[\frac{1}{\overline{n}^2}\right] \Omega_n^2 + 0.25 \left[\frac{1}{\overline{S}}\right] \left[\overline{S}\right] \Omega_S^2$$

$$= \Omega_n^2 + 0.25 \Omega_S^2$$

$$\Omega_Q = \left[\Omega_n^2 + 0.25 \Omega_S^2\right]^{1/2}$$

EXAMPLE 10.6.4

Determine the mean capacity of a storm sewer pipe, the coefficient of variation of the pipe capacity, and the standard deviation of the pipe capacity using Manning's equation for full pipe flow. Refer to example 10.6.3. The following parameter values are to be considered:

Parameter	Mean	Coefficient of Variation
n	0.015	0.01
D	1.5 m	0
S	0.001	0.05

SOLUTION

Manning's equation in SI units for full pipe flow is

$$Q = \frac{0.311}{n} S^{1/2} D^{8/3}$$

so for first-order analysis we have

$$\overline{Q} = \frac{0.311}{\overline{n}} \overline{S}^{1/2} D^{8/3} = \frac{0.311}{0.015} (0.001)^{1/2} (1.5)^{8/3}$$

$$= 1.93 \text{ m}^3/\text{s}$$

Using example 10.6.3, we find

$$\Omega_Q = \left[(0.01)^2 + 0.25(0.05)^2\right]^{1/2} = 0.027$$

$$\sigma_Q = \overline{Q}\,\Omega_Q = 1.93(0.027)$$
$$= 0.052 \text{ m}^3/\text{s}$$

10.7 RISK ANALYSIS: COMPOSITE HYDROLOGIC AND HYDRAULIC RISK

The *resistance* or strength of a component is defined as the ability of the component to fulfill its required purpose satisfactorily without a failure when subjected to an external stress. *Stress* is the loading of the component, which may be a mechanical load, an environmental exposure, a flow rate, temperature fluctuation, etc. The stress or loading tends to cause failure of the component. When the strength of the component is less than the stress imposed on it, failure occurs. This type of analysis can be applied to the reliability analysis of components of various hydraulic systems. The *reliability of a hydraulic system* is defined as the probability of the resistance to exceed the loading, i.e., the probability of survival. The terms "stress" and "strength" are more meaningful to structural engineers, whereas the terms "loading" and "resistance" are more descriptive to water resources engineers. The *risk of a hydraulic component*, subsystem, or system is defined as the probability of the loading exceeding the resistance, i.e., the *probability of failure*. The mathematical representation of the reliability R can be expressed as

$$R = P(r > \ell) = P(r - \ell > 0) \tag{10.7.1}$$

where $P(\ )$ refers to probability, r is the resistance, and ℓ is the loading. The relationship between reliability R and risk $\overline{R}$ is

$$R = 1 - \overline{R} \tag{10.7.2}$$

The resistance of a hydraulic system is essentially the flow-carrying capacity of the system, and the loading is essentially the magnitude of flows through or pressure imposed on the system by demands. Since the loading and resistance are random variables due to the various hydraulic and demand uncertainties, a knowledge of the probability distributions of r and ℓ is required to develop reliability models. The computation of risk and reliability can be referred to as *loading-resistance interference*. Probability distributions for loading and resistance are illustrated in Figure 10.7.1. The *reliability* is the probability that the resistance is greater than the loading for all possible values of the loading.

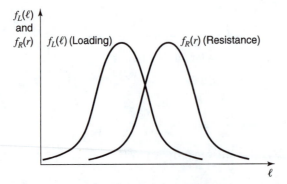

Figure 10.7.1 Load-resistance interference reliability analysis.

The word "static," from the reliability computation point of view, represents the worst single stress, or load, applied. Actually, the loading applied to many hydraulic systems is a random variable. Also, the number of times a loading is imposed is random.

10.7.1   Reliability Computation by Direct Integration

Following the reliability definition given in equation (10.7.1), the reliability can be expressed as

$$R = \int_0^\infty f_R(r) \left[\int_0^r f_L(\ell) \, d\ell \right] dr = \int_0^\infty f_R(r) F_L(r) \, dr \tag{10.7.3}$$

in which $f_R(\ )$ and $f_L(\ )$ represent the probability density functions of resistance and loading, respectively. The reliability computations require the knowledge of the probability distributions of loading and resistance. A schematic diagram of the reliability computation by equation (10.7.3) is shown in Figure 10.7.2.

To illustrate the computation procedure involved, we consider that the loading ℓ and the resistance r are exponentially distributed, i.e.,

$$f_L(\ell) = \lambda_\ell e^{-\lambda_\ell \ell}, \quad \ell \geq 0 \tag{10.7.4}$$

$$f_R(r) = \lambda_r e^{\lambda_r r}, \quad r \geq 0 \tag{10.7.5}$$

Then the static reliability can be derived by applying equation (10.7.3) in a straightforward manner as

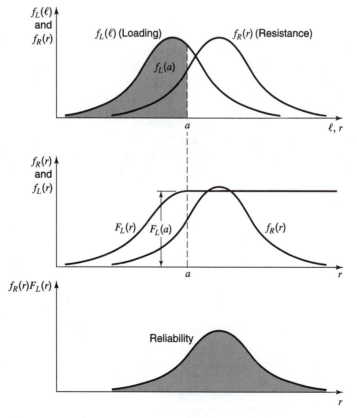

Figure 10.7.2 Graphical illustration of the steps involved in reliability computation, by equation (10.6.3).

$$R = \int_0^\infty \lambda_r e^{-\lambda_r r} \left[\int_0^r \lambda_\ell e^{-\lambda_\ell \ell} \, d\ell \right] dr = \int_0^\infty \lambda_r e^{-\lambda_r r} \left[1 - e^{-\lambda_\ell r} \right] dr = \frac{\lambda_\ell}{\lambda_r + \lambda_\ell} \qquad (10.7.6)$$

For some special combinations of load and resistance distributions, the static reliability can be derived analytically in the closed form. In cases in which both the loading ℓ and resistance r are log-normally distributed, the reliability can be computed as (Kapur and Lamberson, 1977)

$$R = \int_{-\infty}^\infty \phi(z) \, dz = \Phi(z) \qquad (10.7.7)$$

where $\phi(z)$ and $\Phi(z)$ are the probability density function and the cumulative distribution function, respectively, for the standard normal variate z given as

$$z = \frac{\mu_{\ln r'} - \mu_{\ln \ell}}{\sqrt{\sigma_{\ln r'}^2 + \sigma_{\ln \ell}^2}} \qquad (10.7.8)$$

The values of the cumulative distribution function $\Phi(z)$ for the standard normal variate are given in Table 10.2.1.

In cases in which the loading ℓ is exponentially distributed and the resistance is normally distributed, the reliability can be expressed as (Kapur and Lamberson, 1977)

$$R = 1 - \Phi\left(\frac{\mu_r}{\sigma_r}\right) - \exp\left[-\frac{1}{2}\left(2\mu_r \lambda_\ell - \lambda_\ell^2 \sigma_r^2\right)\right] \times$$

$$\left[1 - \Phi\left(-\frac{\mu_r - \lambda_\ell \sigma_r^2}{\sigma_r} \right) \right] \qquad (10.7.9)$$

EXAMPLE 10.7.1   Consider a water distribution system (see Figure 10.7.3) consisting of a storage tank serving as the source and a 2-ft diameter cast-iron pipe 1 mile long, leading to a user. The head elevation at the source is maintained at a constant height of 100 ft above the elevation at the user end. It is also known that, at the user end, the required pressure head is fixed at 20 psi with variable demand on flow rate. Assume that the demand in flow rate is random, having a log-normal distribution with mean 1 cfs and standard deviation 0.3 cfs. Because of the uncertainty in pipe roughness, the supply to the user is not certain. We know that the pipe has been installed for about three years. Therefore, our estimation of the pipe roughness in the Hazen–Williams equation is about 130, with some errors of ± 20. Again, we further assume that the Hazen–Williams C coefficient has a log-normal distribution with a mean of 130 and a standard

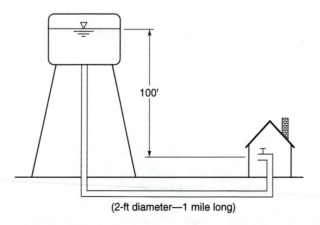

100'

(2-ft diameter—1 mile long)

Figure 10.7.3 Example system.

deviation of 20. It is required to estimate the reliability with which the water demand by the user will be satisfied.

SOLUTION

In this example, the resistance of the system is the water supply from the source, while the loading is the water demand by the user. Both supply and demand are random variables. By the Hazen–Williams equation, the supply is calculated as

$$r = Q_s = \frac{C}{149.2}\left(\frac{\Delta h}{L}\right)^{0.54} D^{2.63}$$

where Δh is the head difference (in ft) between the source and the user, D is the pipe diameter in feet, and L is the pipe length in feet. Because the roughness coefficient C is a random variable, so is the supply. Due to the multiplicative form of the Hazen–Williams equation, the logarithmic transformation leads to a linear relation among variables, i.e.,

$$\ln Y = \ln C - \ln(149.2) + 0.54\ln\left[\frac{100 - \dfrac{(20)(144)}{62.4}}{5280}\right] + 2.63\ln(2)$$

$$= \ln C - 5.659$$

Assume that the roughness coefficient C is log-normally distributed; then $\ln C$ is normally distributed, as is the log-transformed water supply (resistance). From the moment relations given in Table 10.1.2 for log-normal distribution, the mean and the standard deviation of $\ln C$ are determined as follows. From Table 10.1.2,

$$\mu_{\ln r} = \mu_{\ln C} = \ln C - \sigma_{\ln r}^2/2$$

where $\sigma_{\ln r} = \sqrt{\ln\left(1+\Omega_C^2\right)} = \sqrt{\ln\left[1+(20/130)^2\right]} = 0.153$. Thus, $\mu_{\ln r} = \ln 130 - (0.153^2/2) = 4.856$.

From these results, the mean and the standard deviation of $\ln Y$ are -0.803 and 0.153, respectively.

Because the water demand (loading) has a log-normal distribution, the mean and standard deviation of its log-transformed scale can be calculated in the same manner as for roughness coefficient C. That is,

$$\sigma_{\ln \ell} = \sqrt{\ln\left[1+(0.3/1)^2\right]} = 0.294 \quad \text{and} \quad \mu_{\ln \ell} = 1 - \left(0.294^2/2\right) = -0.043.$$

Knowing the distributions and statistical properties of the load (water demand) and resistance (water supply), both log-normal in this example, we can calculate the reliability of the system by equation (10.7.7) as

$$R = \Phi\left[z = \frac{-0.803 - (-0.043)}{\sqrt{0.153^2 + 0.294^2}}\right] = \Phi(z = -2.293) = 1 - \Phi(2.293) = 0.013$$

This means that the water demanded by the user will be met with a probability of 1.3 percent.

10.7.2   Reliability Computation Using Safety Margin/Safety Factor

Safety Margin

The *safety margin* (*SM*) is defined as the difference between the project capacity (resistance) and the value calculated for the design loading, $SM = r - \ell$. The reliability is equal to the probability that $r > \ell$, or equivalently,

$$R = p(r - \ell > 0) = P(SM > 0) \tag{10.7.10}$$

If r and ℓ are independent random variables, then the mean value of SM is given by

$$\mu_{SM} = \mu_r - \mu_\ell \tag{10.7.11}$$

and its variance by

$$\sigma_{SM}^2 = \sigma_r^2 + \sigma_\ell^2 \tag{10.7.12}$$

If the safety margin is normally distributed, then

$$z = (SM - \mu_{SM})/\sigma_{SM} \tag{10.7.13}$$

is a standard normal variate z. By subtracting μ_{SM} from both sides of the inequality in equation (10.7.10) and dividing both sides by σ_{SM}, it can be seen that

$$R = P\left(z > \frac{\mu_{SM}}{\sigma_{SM}}\right) = \Phi\left(\frac{\mu_{SM}}{\sigma_{SM}}\right) \tag{10.7.14}$$

The key assumption of this analysis is that it considers that the safety margin is normally distributed but does not specify the distributions of loading and capacity. Ang (1973) indicated that, provided $R > 0.001$, R is not greatly influenced by the choice of distribution for r and ℓ and the assumption of a normal distribution for SM is satisfactory. For lower risk than this (e.g., $R = 0.00001$), the shape of the tails of the distributions for r and ℓ becomes critical, in which case accurate assessment of the distribution of SM of direct integration procedure should be used to evaluate the risk or probability of failure.

EXAMPLE 10.7.2

Apply the safety margin approach to evaluate the reliability of the simple water distribution system described in Example 10.7.1.

SOLUTION

Calculate the mean and standard deviation of the resistance (i.e., water supply) as

$$\mu_r = \exp\left(\mu_{\ln\ell} + \frac{1}{2}\sigma_{\ln\ell}^2\right) = \exp\left[-0.803 + \frac{1}{2}(0.153)^2\right] = 0.453 \text{ cfs}$$

and

$$\sigma_r = \sqrt{\mu_r^2\left[\exp\left(\sigma_{\ln\ell}^2\right) - 1\right]} = \sqrt{0.453^2\left[\exp(0.153)^2 - 1\right]} = 0.070 \text{ cfs}$$

From the problem statement, the mean and standard deviation of the load (water demand) are $\mu\ell = 1$ cfs and $\sigma\ell = 0.3$ cfs, respectively. Therefore, the mean and variance of the safety margin can be calculated as

$$\mu_{SM} = \mu_r - \mu_\ell = 0.453 - 1.0 = -0.547 \text{ cfs}$$
$$\sigma_{SM}^2 = \sigma_r^2 + \sigma_\ell^2 = (0.070)^2 + (0.3)^2 = 0.095 \text{ (cfs)}^2$$

Now, the reliability of the system can be assessed, by the safety margin approach, as

$$R = \Phi\left[\frac{-0.453}{\sqrt{0.095}}\right] = \Phi[-1.470] = 1 - \Phi[1.470] = 1 - 0.929 = 0.071$$

The reliability computed by the safety margin method is not identical to that of direct integration, although the difference is practically negligible. It should, however, be pointed out that the distribution of the safety margin in this example is not exactly normal, as assumed. Thus, the reliability obtained should be regarded as an approximation to the true reliability.

EXAMPLE 10.7.3

Determine the risk (probability) that the surface runoff (loading) exceeds the capacity of the storm sewer pipe for the problems in examples 10.6.2 and 10.6.4. Use the safety margin approach.

SOLUTION

From example 10.6.2, $\bar{Q}_\ell = 2.24$ m³/s and $\sigma_\ell = 1.138$ m³/s. From example 10.6.4, $\bar{Q}_r = 1.93$ m³/s and $\sigma_r = 0.052$ m³/s. Compute μ_{SM} using equation (10.7.11):

$$\mu_{SM} = \mu_r - \mu_\ell = \bar{Q}_r - \bar{Q}_\ell = 1.93 - 2.24 = -0.31 \text{ m}^3/\text{s}$$

Compute σ_{SM} using equation (10.7.12):

$$\sigma_{SM} = \left[\sigma_r^2 + \sigma_\ell^2\right]^{1/2} = \left[(0.052)^2 + (1.138)^2\right]^{1/2} = 1.139 \text{ m}^3/\text{s}$$

Compute the reliability using equation (10.7.14):

$$R = \Phi\left(\frac{\mu_{SM}}{\sigma_{SM}}\right) = \Phi\left(\frac{-0.310}{1.139}\right) = \Phi(-0.272) = 0.607$$

The risk (the probability of $Q_\ell > Q_r$) is thus

$$\bar{R} = 1 - R = 1 - 0.607$$

$$= 0.393 \text{ or } 39.3\%$$

Safety Factor

The *safety factor* (SF) is given by the ratio r/ℓ and the reliability can be specified by $P(SF > 1)$. Several safety factor measures and their usefulness in hydraulic engineering are discussed by Mays and Tung (1992) and Yen (1978). By taking logarithms of both sides of this inequality, we find

$$R = P(SF > 1) = P[\ln(SF) > 0] = P[\ln(r/\ell) > 0]$$

$$= P\left(z \le \frac{\mu_{\ln SF}}{\sigma_{\ln SF}}\right) = \Phi\left(\frac{\mu_{\ln SF}}{\sigma_{\ln SF}}\right) \tag{10.7.15}$$

If the resistance and loading are independent and log-normally distributed, then the risk can be expressed as

$$\bar{R} = \Phi\left\{\frac{\ln\left[\frac{\mu_r}{\mu_\ell}\sqrt{\frac{1+\Omega_\ell^2}{1+\Omega_r^2}}\right]}{\ln\left[\left(1+\Omega_\ell^2\right)\left(1+\Omega_r^2\right)\right]^{1/2}}\right\} \tag{10.7.16}$$

EXAMPLE 10.7.4

Determine the risk (probability) that the surface runoff (loading) exceeds the capacity of the storm sewer pipe for the problems in examples 10.6.2 and 10.6.4. Use the safety factor approach.

SOLUTION

$\bar{Q}_\ell = 2.24$ m³/s, $\Omega_\ell = 0.508$, $\bar{Q}_r = 1.93$ m³/s, $\Omega_r = 0.027$. Use equation (10.7.16) to compute the risk:

$$\bar{R} = \Phi\left\{\frac{\ln\left[\frac{\bar{Q}_r}{\bar{Q}_\ell}\sqrt{\frac{1+\Omega_\ell^2}{1+\Omega_r^2}}\right]}{\ln\sqrt{\left(1+\Omega_\ell^2\right)\left(1+\Omega_r^2\right)}}\right\}$$

$$= \Phi \left\{ \frac{\ln \left[\frac{1.93}{2.24} \sqrt{\frac{1+0.508^2}{1+0.027^2}} \right]}{\ln \sqrt{(1+0.508^2)(1+0.027^2)}} \right\}$$

$$= \Phi(-3.00)$$

$$= 0.382 = 38.2\%$$

Note that the risk values (magnitudes) calculated for the same problem by the safety margin approach in example 10.7.3 and by the safety factor approach in this example are very close.

10.8   COMPUTER MODELS FOR FLOODFLOW FREQUENCY ANALYSIS

Table 10.8.1 describes the features of the HEC-FFA (U.S. Army Corps of Engineers, 1992) and PEAKFQ (U.S. Geological Survey) models that are used for flood flow frequency analysis, based upon fitting the log-Pearson Type III distribution to observed annual maximum flood series. The World Wide Web (www) site for the U.S. Army Corps of Engineers is www.usace.army.mil and for the U.S. Geological Survey is www.usgs.gov.

Table 10.8.1   HEC-FFA and PEAKFQ Features

Feature	Analysis Procedure
Parameter estimation	Estimate parameters with method of movements; this assumes sample mean, standard deviation, skew coefficient = parent population mean, standard deviation, and skew coefficient. To account for variability in skew computed from small samples, use weighted sum of station skew and regional skew.
Outliers	These are observations that "... depart significantly from the trend of the remaining data." Models identify high and low outliers. If information available indicates that a high outlier is maximum in the extended time period, it is treated as historical flow. Otherwise, they are treated as part of a systematic sample. Low outliers are deleted from the sample, and conditional probability adjustment is supplied.
Zero flows	If the annual maximum flow is zero (or below a specified threshold), the observations are deleted from the sample. The model parameters are estimated with the remainder of the sample. The resulting probability estimates are adjusted to account for the conditional probability of exceeding a specified discharge, given that a nonzero flow occurs.
Historical flood information	If information is available indicating that an observation represents the greatest flow in a period longer than that represented by the sample, model parameters are computed with "historically" weighted moments.
Broken record	If observations are missed due to "... conditions not related to flood magnitude," different sample segments are analyzed as a single sample with the size equal to the sum of the sample sizes.

Table 10.8.1 HEC-FFA and PEAKFQ Features (*continued*)

Feature	Analysis Procedure
Expected probability adjustment	This adjustment is made to the model results "... to incorporate the effects of uncertainty in application of the [frequency] curve."

Source: Ford and Hamilton (1996).

PROBLEMS

10.2.1 Solve example 10.2.1 to determine the flood magnitude having a return period of 50 years.

10.2.2 The annual maximum series of flood magnitude in a river has a log-normal distribution with a mean of 8000 m^3/s and a standard deviation of 3000 m^3/s. (a) What is the probability in each year that a flood magnitude would exceed 12,000 m^3/s? (b) Determine the flood magnitude for return periods of 25 and 100 years.

10.2.3 Solve problem 10.2.2 for a mean of 5000 m^3/s.

10.3.1 Determine the hydrologic risk of a 100-year flood occurring during the 20-year service life of a project. What is the chance that the 25-year flood will not occur?

10.3.2 Determine the hydrologic risk of a 25-year flood occurring during the 20-year service life of a project. What is the chance that the 25-year flood will not occur?

10.3.3 Determine the corresponding return periods of an annual exceedance series for corresponding annual maximum series return periods of 2, 5, 10, 25, and 100 years.

10.4.1 The mean, standard deviation, and skew coefficient of the log-transformed discharges (in ft^3/s) for a river are 4.5, 0.6, and 0.2 respectively. Compute the 10-year and 100-year peak discharges.

10.4.2 Solve example 10.4.1 to compute the 25-year and 200-year peak discharges using the log-Pearson Type III distribution.

10.4.3 Solve example 10.4.2 using the annual peak discharges for 1938–1960 using the log-Pearson Type III distribution.

10.4.4 Solve example 10.4.2 to compute the 10-year and 200-year peak discharges using the log-Pearson Type III distribution.

10.5.1 Use the annual flood data for the Floyd River, James, Iowa, in the table below to perform a flood frequency analysis using the U.S. Water Resources Council Guidelines. Take the map skew for this location as −0.4.

Year	Discharge (cfs)	Year	Discharge (cfs)
1935	1460	1955	2260
1936	4050	1956	318
4937	3570	1957	1330
1938	2060	1958	970
1939	1300	1959	1920
1940	1390	1960	15100
1941	1720	1961	2870

Year	Discharge (cfs)	Year	Discharge (cfs)
1942	6280	1962	20600
1943	1360	1963	3810
1944	7440	1964	726
1945	5320	1965	7500
1946	1400	1966	7170
1947	3240	1967	2000
1948	2710	1968	829
1949	4520	1969	17300
1950	4840	1970	4740
1951	8320	1971	13400
1952	13900	1972	2940
1953	71500	1973	5660
1954	6250		

10.5.2 Solve problem 10.5.1 using only the maximum series for 1935–1950.

10.5.3 Solve problem 10.5.1 using only the maximum series for 1935–1970.

10.5.4 Consider the annual maximum series for the U.S. Geological Survey gauging station on Clear Creek near Pearland, Texas. The statistics of the log-transformed flows are $\bar{y} = 2.98$ cfs, $S_y = 0.27$, $G_s = -1.1$ and $n = 46$. Using the log-Pearson Type III distribution, compare the 100-year discharges with and without the U.S. Water Resources Council (1981) recommendation on skew coefficients. Take the map skew for this location as −0.3.

10.5.5 In problem 10.5.4, what is the 100-year flood using the normal distribution?

10.6.1 Apply the first-order analysis of uncertainty to the weir flow equation for spillways: $Q = CLH^{1.5}$. Consider C and H to be uncertain.

10.6.2 Solve example 10.6.1 considering only C and i to be uncertain.

10.6.3 Solve example 10.6.2 assuming no uncertainty in A.

10.6.4 Apply first-order analysis to Manning's equation for full pipe flow assuming n, S, and D to be uncertain.

10.7.1 Calculate the risk of failure (walls are overtopped) of an open channel, using the safety margin approach (SM is normally distributed). The capacity is computed using Manning's equation and a first-order analysis is used to determine the coefficient of variation of the capacity. The mean loading is 2500 cfs and the coefficient of variation of loading is 0.20. The slope of the channel is 0.002 with a coefficient of variation of 0.10. Manning's roughness factor is 0.04 and has a coefficient of variation of 0.10. The channel cross-section is rectangular with a width of 50 ft and a wall height of 10 ft.

10.7.2 Solve problem 10.7.1 using the safety factor approach.

10.7.3 Solve problem 10.7.1 assuming the capacity and loading to be log-normally distributed.

10.7.4 Solve example 10.7.1 assuming that the demand in flow rate is log-normal with a mean of 0.5 cfs and a standard deviation of 0.3 cfs. The Hazen–Williams C has a log-normal distribution with a mean of 120 and a standard deviation of 20.

10.7.5 What would be the reliability in example 10.7.1 if the safety margin approach were used, assuming the SM is distributed normally?

10.7.6 Compare the risks of a discharge of 2000 cfs being equaled or exceeded at least once over the next 10 years with and without the U.S. Water Resources Council recommendation on the skew coefficient. Assume $G_w = -0.7$.

10.8.1 Use the HEC-FFA model to solve problem 10.5.1.

10.8.2 Use the HEC-FFA model to solve example 10.4.2.

REFERENCES

Ang, A. H. S., "Structural Risk Analysis and Reliability-Based Design," *Journal of the Structural Engineering Division*, American Society of Civil Engineers, vol. 99, no. ST9, pp. 1891–1910, 1973.

Chow, V. T., "A General Formula for Hydrologic Frequency Analysis," *Trans. Am. Geophysical Union*, vol. 32, no. 2, pp. 231–237, 1951.

Chow, V. T., "Statistical and Probability Analysis of Hydrologic Data," Sec. 8-I in *Handbook of Applied Hydrology*, (edited by V. T. Chow), McGraw-Hill, New York, 1964.

Chow, V. T., D. R. Maidment, and L. W. Mays, *Applied Hydrology*, McGraw-Hill, New York, 1988.

Cudworth, A. D., Jr., *Flood Hydrology Manual*, U.S. Dept. of the Interior, Bureau of Reclamation, U.S. Government Printing Office, Denver, CO, 1989.

Ford, D., and D. Hamilton, "Computer Models for Water-Excess Management," in *Water Resources Handbook*, edited by L.W. Mays, McGraw-Hill, New York, 1996.

Grant, E. L., and R. S. Leavenworth, *Statistical Quality and Control*, McGraw-Hill, New York, 1972.

Henley, E. J., and H. Kumamoto, *Reliability Engineering and Risk Assessment*, Prentice Hall, Englewood Cliffs, N.J., 1981.

Kapur, K. C., and L. R. Lamberson, *Reliability in Engineering Design*, John Wiley and Sons, New York, 1977.

Kite, G. W., *Frequency and Risk Analysis in Hydrology*, Water Resources Publications, Fort Collins, CO, 1977.

Hirsch, R. M., D. R. Helsel, T. A. Cohn, and E. J. Gilroy, "Statistical Analysis of Hydrologic Data," Chapter 17 in *Handbook of Hydrology*, edited by D. R. Maidment, McGraw-Hill, New York, 1993.

Mays, L. W., (editor), *Reliability Analysis of Water Distribution Systems*, American Society of Civil Engineers, New York, 1989.

Mays, L. W., and Y. K. Tung, *Hydrosystems Engineering and Management*, McGraw-Hill, New York, 1992.

Mays, L. W., et al., "Methodologies for the Assessment of Aging Water Distribution Systems," Report No. CRWR 227, Center for Research in Water Resources, The University of Texas at Austin, (July) 1989.

National Academy of Sciences, *Safety of Existing Dams: Evaluation and Improvement*, National Academy Press, Washington, DC, 1983.

Tung, Y. K., and L. W. Mays, "Risk Analysis for Hydraulic Design," *Journal of the Hydraulics Division*, American Society of Civil Engineers, vol. 106, no. HY5, pp. 893–913, 1980.

Tung, Y. K., and L. W. Mays, "Reducing Hydrologic Parameter Uncertainty," *Journal of Water Resources Planning and Management Division*, American Society of Civil Engineers, vol. 107, no. WR1, pp. 245–262, 1981.

U.S. Army Corps of Engineers, "HEC-FFA: Flood Frequency Analysis: User's Manual," Hydrologic Engineering Center, Davis, CA, 1992.

U.S. Water Resources Council (now called Interagency Advisory Committee on Water Data), *Guidelines for Determining Flood Flow Frequency*, Bulletin 17B, available from Office of Water Data Coordination, U.S. Geological Survey, Reston, VA, 1981.

Wallis, J. R., N. C. Matalas, and J. R. Slack, "Just a Moment," *Water Resources Research*, vol. 10, no. 2, pp. 211–219, 1974.

World Meteorological Organization, *Guide to Hydrological Practices*, Vol. II, *Analysis, Forecasting, and Other Applications*, WMO no. 168, 4th edition, Geneva, Switzerland, 1983.

Yen, B. C., "Safety Factor in Hydrologic and Hydraulic Engineering Design," *Proceedings, International Symposium on Risk Reliability in Water Resources*, University of Waterloo, Waterloo, Ontario, Canada, (June) 1978.

Yen, B. C., and Y. K. Tung, (editors), *Reliability and Uncertainty Analysis in Hydraulic Design*, compiled by ASCE Subcommittee on Uncertainty and Reliability Analysis in Design of Hydraulic Structures of the Technical Committee on Probabilistic Approaches to Hydraulics of the Hydraulics Division of ASCE, New York, 1993.

Chapter 11

Water Withdrawals and Uses

11.1 WATER-USE DATA—CLASSIFICATION OF USES

Care should be taken in using or reading the terms that describe water uses, as these terms are often used inconsistently and misleadingly in the water literature. Gleick (1998) points out that the term "water use" has encompassed many different ideas to mean the withdrawal of water, gross water use, and the consumptive use of water. According to Gleick, *withdrawal* should refer to the act of taking water from a source for storage or use. *Gross water use* is distinguished from water withdrawal by the inclusion of recirculated or reused water. *Water consumption* or *consumptive use* should refer to the use of water in a manner that prevents its immediate reuse, such as through evaporation, plant transpiration, contamination, or incorporation into a finished product. As an example, water for cooling power plants may be withdrawn from a river or lake, used and then returned to the river or lake. According to Gleick, this should not be considered consumptive use. Agricultural water may have both consumptive and nonconsumptive uses, as some water is transpired or incorporated into plant material, with the remainder returning to the groundwater or surface water sources.

The U.S. Geological Survey (USGS) conducts the National Water Use Information (NWUI) program, which established the national system of water-use accounting. Water-use circulars are prepared by the USGS at five-year intervals. *Water use* is defined from a hydrologic perspective as all water flows that are a result of human intervention within the hydrologic cycle. The USGS uses the following seven water-use flows: (1) water withdrawals for offstream purposes, (2) water deliveries at point of use or quantities released after use, (3) consumptive use, (4) conveyance loss, (5) reclaimed waste water, (6) return flow, and (7) instream flow (Solley et al., 1993). A schematic of water-use flows and losses is presented in Figure 1.3.1.

Table 11.1.1 provides the USGS definition of water-use terms. Figure 1.3.2 is a diagram that tracks the sources, uses, and disposition of freshwater using the hydrologic accounting system (see Figure 1.3.1). Table 1.3.1 provides definitions of the major water-use purposes. Water use defined by purpose of actual use is a more restrictive definition of water use.

Table 1.2.3 illustrates the dynamics of water use in the world. For the world as whole, water requirements are growing and will continue to grow in all types of economic activity. This trend continues despite the progressive trends towards stabilization of water needs that have clearly taken shape in a number of countries (Shiklomanov, 1996). Table 1.2.4 compares annual runoff and water consumption by continents and by physiographic and economic regions of the world. This table can be used to compare the amounts of water withdrawal and consumption throughout the world with stream flow resources.

Table 11.1.1   Definitions of Water-Use Terms

Term	Definition
Consumptive use	That part of water withdrawn that is evaporated, transpired, incorporated into products or crops, consumed by humans or livestock, or otherwise removed from the immediate water environment.
Conveyance loss	The quantity of water that is lost in transit from a pipe, canal, conduit, or ditch by leakage or evaporation.
Delivery and release	The amount of water delivered to the point of use and the amount released after use.
Instream use	Water that is used, but not withdrawn, from a surface- or ground-water source for such purposes as hydroelectric-power generation, navigation, water-quality improvement, fish propagation, and recreation.
Offstream use	Water withdrawn or diverted from a surface- or ground-water source for public water supply, industry, irrigation, livestock, thermoelectric-power generation, and other uses.
Public supply	Water withdrawn by public or private water suppliers and delivered to users.
Return flow	The water that reaches a surface- or ground-water source after release from the point of use and thus becomes available for further use.
Reclaimed wastewater	Wastewater-treatment-plant effluent that has been diverted for beneficial use before it reaches a natural waterway or aquifer.
Self-supplied water	Water withdrawn from a surface- or groundwater source by a user rather than being obtained from a public supply.
Withdrawal	Water removed from the ground or diverted from a surface-water source for offstream use.

Source: Adapted from Solley et al. (1993) as presented in Dziegielewski et al. (1996).

Trends in freshwater withdrawals and consumption in the United States for 1960–1990 are illustrated in Figure 11.1.1. The per-capita consumption of water in 1990 was less than in 1965, despite an almost 50 percent increase in real per-capita income over that period (Rogers, 1993). Total water demand continued to decrease from the high in 1980, although the 1990 demand was slightly higher than the 1985 demand. Large reductions in water use by agriculture and industry appear to have occurred between 1980 and 1990 (Rogers, 1993). Tables 11.1.2–11.1.5 provide estimates of water use for industrial water use, municipal water use, and residential interior water use, respectively.

Table 11.1.2   Typical Rates of Water Use for Various Industries

Industry	Range of Flow, (gal/ton product)
Cannery	
Green beans	12,000–17,000
Peaches and pears	3,600–4,800
Other fruits and vegetables	960–8,400
Chemical	
Ammonia	24,000–72,000
Carbon dioxide	14,400–21,600
Lacrose	144,000–192,000
Sulfur	1,920–2,400
Food and beverage	
Beer	2,400–3,840

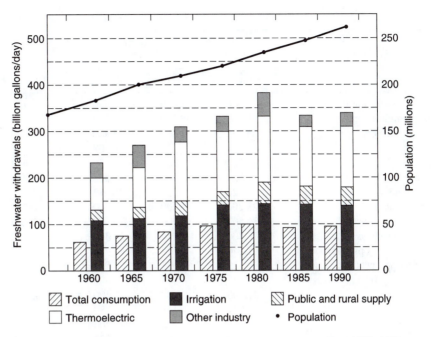

Figure 11.1.1 Trends in U.S. freshwater withdrawals and consumption, 1960–1990 (from Solley et al. (1993), as presented in Rogers (1993)).

Table 11.1.2 Typical Rates of Water Use for Various Industries (*continued*)

Industry	Range of Flow, (gal/ton product)
Bread	480–960
Meat packing	3,600–4,800[1]
Milk products	2,400–4,800
Whisky	14,000–19,200
Pulp and paper	
Pulp	60,000–190,000
Paper	29,000–38,000
Textile	
Bleaching	48,000–72,000[2]
Dyeing	7,200–14,400[2]

[1]Live weight.
[2]Cotton.
Source: Metcalf and Eddy, Inc. (1991).

Table 11.1.3 Water Used by Industries in the United States, 1983[a]

Industry Group	Gross Water Used (10⁶ m³/yr)			Water Discharged		
	Total	Intake	Reuse	Total (10⁶ m³/yr)	Untreated (%)	Treated (%)
All mineral industries	12,600	4,540	8,080	3,930	31.9	68.1
Metal mining	2,780	645	2,140	504	39.7	60.3
Anthracite mining	20	8	12	28	12	87
Bituminous coal and lignite mining	449	172	278	440	26.2	73.8
Oil and gas extraction	5,500	2,280	3,220	1,800	31.0	69.0

Table 11.1.3   Water Used by Industries in the United States, 1983[a] (*continued*)

| Industry Group | Gross Water Used (10^6 m³/yr) | | | Water Discharged | | |
	Total	Intake	Reuse	Total (10^6 m³/yr)	Untreated (%)	Treated (%)
Nonmetallic minerals, except fuels	3,860	1,430	2,430	1,150	32.6	67.4
All manufacturing industries	128,000	38,500	90,200	33,800	54.9	45.1
Food and kindred products	5,330	2,560	2,880	2,090	64.5	35.5
Tobacco products	128	20	108	15	D	D
Textile mill products	1,260	503	759	438	52.9	47.1
Lumber and wood products	827	326	501	269	63.2	36.8
Furniture and fixtures	26	13	13	13	88	12
Paper and allied products	28,200	7,200	21,000	6,700	27.1	72.9
Chemicals and allied products	36,500	12,900	23,600	11,300	67.0	33.0
Petroleum and coal products	23,400	3,100	20,300	2,650	46.2	53.8
Rubber and miscellaneous plastic products	1,240	288	954	237	63.6	36.4
Leather and leather products	25	23	2	22	D	D
Stone, clay, and glass products	1,280	586	689	503	74.9	25.1
Primary metal industries	22,300	8,950	13,400	8,000	58.1	41.9
Fabricated metal products	977	248	730	233	48.4	51.6
Machinery, except electrical	1,170	455	706	398	67.9	32.2
Electrical and electronic equipment	1,270	281	988	266	60.5	39.5
Transportation equipment	3,830	579	3,250	528	67.5	32.5
Instruments and related products	424	113	312	105	49.3	50.4
Miscellaneous manufacturing industries	58.4	16	42.1	15	D[b]	D

[a] Percentages may not add to 100.0% due to rounding.
[b] D, Withheld to avoid disclosing data for individual companies.

Source: Gleick (1993a).

Table 11.1.4   Water Requirements for Municipal Establishments

Type	Unit	Average Use	Peak Use
Hotels	Liter/day/square meter	10.4	17.6
Motels	Liter/day/square meter	9.1	63.1
Barber shops	Liter/day/barber chair	207	1,470
Beauty shops	Liter/day/station	1,020	4,050
Restaurants	Liter/day/seat	91.6	632.0
Night clubs	Liter/day/person served	5	5
Hospitals	Liter/day/bed	1,310	3,450
Nursing homes	Liter/day/bed	503	1,600
Medical offices	Liter/day/square meter	25.2	202
Laundy	Liter/day/square meter	10.3	63.9
Laundromats	Liter/day/square meter	88.4	265.0
Retail space	Liter/day/sales square meter	4.3	11
Elementary schools	Liter/day/student	20.4	186
High schools	Liter/day/student	25.1	458
Bus-rail depot	Liter/day/square meter	136	1,020
Car washes	Liter/day/inside square meter	194.7	1,280
Churches	Liter/day/member	0.5	17.8
Golf-swim clubs	Liter/day/member	117	84
Bowling alleys	Liter/day/alley	503	503
Residential colleges	Liter/day/student	401	946
New office buildings	Liter/day/square meter	3.8	21.2

Table 11.1.4 Water Requirements for Municipal Establishments (*continued*)

Type	Unit	Average Use	Peak Use
Old office buildings	Liter/day/square meter	5.8	14.4
Theaters	Liter/day/seat	12.6	12.6
Service stations	Liter/day/inside square meter	10.2	1,280
Apartments	Liter/day/occupied unit	821	1,640
Fast food restaurants	Liter/day/establishment	6,780	20,300

Source: Gleick (1993b).

Table 11.1.5 Typical Household Water Use in the United States

Use	Unit	Range
Washing machine	Liters per load	130–270
Standard toilet	Liters per flush	10–30
Ultra volume toilet	Liters per flush	6 or less
Silent leak	Liters per day	150 or more
Nonstop running toilet	Liters per minute	20 or less
Dishwasher	Liters per load	50–120
Water-saver dishwasher	Liters per load	40–100
Washing dishes with tap running	Liters per minute	20 or less
Washing dishes in a filled sink	Liters	20–40
Running the garbage disposal	Liters per minute	10–20
Bathroom faucet	Liters per minute	20 or less
Brushing teeth	Liters	8
Shower head	Liters per minute	20–30
Low-flow shower head	Liters per minute	6–11
Filling a bathtub	Liters	100–300
Watering a 750-square meter lawn	Liters per month[a]	7,600–16,000
Standard sprinkler	Liters per hour	110–910
One drip-irrigation emitter	Liters per hour	1–10
1/2 inch diameter hose	Liters per hour	1,100
5/8 inch diameter hose	Liters per hour	1,900
3/4 inch diameter hose	Liters per hour	2,300
Slowly dripping faucet	Liters per month	1,300–2,300
Fast-leaking faucet	Liters per month	7,600 or more
Washing a car with running water	Liters in 20 minutes	400–800
Washing a car with pistol-grip faucet	Liters in 20 minutes	60 or more
Uncovered pool (60 square meters)	Liters lost per month[a]	3,000–11,000+
Covered pool	Liters lost per month[a]	300–1,200

[a] Depending on climate.

Source: Gleick (1993b).

Sustainable water use has been defined by Gleick et al. (1995) as "the use of water that supports the ability of human society to endure and flourish into the indefinite future without undermining the integrity of the hydrological cycle or the ecological systems that depend on it." The following seven sustainability requirements were presented:

1. A basic water requirement will be guaranteed to all humans to maintain human health.
2. A basic water requirement will be guaranteed to restore and maintain the health of ecosystems.
3. Water quality will be maintained to meet certain minimum standards. These standards will vary depending on location and how the water is to be used.

4. Human actions will not impair the long-term renewability of freshwater stocks and flows.

5. Data on water-resources availability, use, and quality will be collected and made accessible to all parties.

6. Institutional mechanisms will be set up to prevent and resolve conflicts over water.

7. Water planning and decision making will be democratic, ensuring representation of all affected parties and fostering direct participation of affected interests.

11.2 WATER FOR ENERGY PRODUCTION

Freshwater and energy are two resources that are intricately connected. Energy is used to help clean and transport water and water is used to help produce energy. This section looks at the connection between demand for and use of energy and water. As a comparison, Table 11.2.1 summarizes water use for energy production and Table 11.2.2 summarizes consumptive water use for electricity production.

Table 11.2.1 Consumptive Water Use for Energy Production

Energy Technology	Consumptive Use $(m^3/10^{12}J(th))$
Nuclear fuel cycle	
Open pit uranium mining	20
Underground uranium mining	0.2
Uranium milling	8–10
Uranium hexafluoride conversion	4
Uranium enrichment: gasoline diffusion	11–13[a]
Uranium enrichment: gas centrifuge	2
Fuel fabrication	1
Nuclear fuel reprocessing	50
Coal fuel cycle	
Surface mining: no revegetation	2
Surface mining: revegetation	5
Underground mining	3–20[b]
Beneficiation	4
Slurry pipeline	40–85
Other plant operations	90[d]
Oil fuel cycle	
Onshore oil exploration	0.01
Onshore oil extraction and production	3–8
Enhanced oil recovery	120
Water flooding	600
Thermal steam injection	100–180
Forward combustion/air injection	50
Micellar polymer	8,900[c]
Caustic injection	100
Carbon dioxide	640[c]
Oil refining (traditional)	25–65
Oil refining (reforming and hydrogeneration)	60–120
Other plant operations	70[d]
Nuclear fuel cycle	
Natural gas fuel cycle	
Onshore gas exploration	Negligible
Onshore gas extraction	Negligible
Natural gas processing	6

Table 11.2.1   Consumptive Water Use for Energy Production (*continued*)

Energy Technology	Consumptive Use $(m^3/10^{12}J(th))$
Gas pipeline operation	3
Other plant operations	100[d]
Synthetic fuels	
Solvent refined and H–coal	175
Lurgi with subbituminous	175
Lurgi with lignite	225
In situ gasification	90–130
Coal gasification	40–95
Coal liquefaction	35–70
TOSCO II shale oil retorting	100
In situ retorting of oil shale	30–60
Tar sands (Athabasca)	70–180
Other technologies	
Solar active space heat	265
Solar passive space heat	Negligible

[a] Excluding water use by additional power plants required for the energy-intensive uranium enrichment process.

[b] Top end of range reflects once-through system with no recycle.

[c] Median of a wide range.

[d] Other plant operations includes plant service, potable water requirements, and boiler makeup water. For coal facilities, this also includes ash handling and flue gas desulfurization process makeup water.

Source: Gleick (1993a).

Table 11.2.2   Consumptive Water Use for Electricity Production

Energy Technology	System Efficiency[a] (%)	Consumptive Use $(m^3/10^3 kWh)$
Conventional coal combustion		
Once-through cooling	35	1.2
Cooling towers	35	2.6
Fluidized bed coal combustion		
Once-through cooling	36	0.8
Oil and natural gas combustion		
Once-through cooling	36	1.1
Cooling towers	36	2.6
Nuclear generation (LWR)		
Cooling towers	31	3.2
Nuclear generation (HTGR)		
Cooling towers	40	2.2
Geothermal generation (vapor-dominated)		
Cooling towers (Geysers, US)	15	6.8
Once-through cooling (Wairakei, NZ)	7.5	13
Geothermal generation (water-dominated)		
Cooling towers (Heber, US)	10	15
Wood-fired generation		
Cooling towers	32	2.3
Renewable energy systems		
Photovoltaics: residential		Negligible
Photovoltaics: central utility		0.13[b]

Table 11.2.2 Consumptive Water Use for Electricity Production (*continued*)

Energy Technology	System Efficiency[a] (%)	Consumptive Use (m³/10³ kWh)
Solar thermal: Luz system		4
Wind generation		Negligible
Ocean thermal		No freshwater
Hydroelectric systems[c]		
United States (average)		17
California (median of range)		5.4
California (average of range)		26

[a] Efficiency of conversion of thermal energy to electrical energy.
[b] Maximum water use for array washing and portable water needs.
[c] Assumes all evaporative losses are attributable to the hydroelectrical facilities. For reservoirs with significant non-hydroelectric uses, such as recreation and flood control.

Source: Gleick (1993a).

The largest consumptive use of water for electricity production is, by far, for thermal power production. Thermal power plants (nuclear and fossil steam plant) need boiler makeup water, condenser-cooling water, potable water, and plant service water, with the most important use being for cooling purposes.

Figure 11.2.1 presents an overview of the processes and materials flows with emphasis on activities outside the basic electricity generation processes. Figure 11.2.2 displays in greater detail the processes contained in the box for the electric power plant (in Figure 11.2.1). Figure 11.2.2 identifies the major flows of water, steam, and fuel-related materials. Three basic types of cooling water systems have been developed for thermal power plants, *once-through cooling*, *pond cooling*, and *evaporative cooling towers*. These are illustrated in Figure 11.2.3.

Once-through cooling (Figure 11.2.3*a*) withdraws water from the source, passes the water through condensers, and then immediately returns the water to the source at an elevated temperature. The advantages of once-through cooling include: least costly to construct, imposes the smallest capacity penalty (i.e., power loss due to fan and pump requirements and turbine back pressure), requires less water treatment, and evaporates less water than cooling ponds or evaporation cooling towers. The major drawback of once-through cooling is thermal pollution. As a result of the 1972 Federal Water Pollution Control Act Amendments, all new power plants must adapt the "best available control technology" (BACT) for thermal pollution. BACT was subsequently defined by the U.S. Environmental Protection Agency as closed cycle evaporative cooling towers.

Cooling ponds (Figure 11.2.3*b*) are basically large ponds where warm water is stored or detained to allow cooling before returning to the boilers to once again produce steam. Most evaporatively cooled facilities use cooling towers rather than ponds, because ponds require more land (1 to 5 acres/MW) and conserve more water. If land and water are cheap, or if water storage is desired for other purposes, these cooling ponds may be considered because of the low capital and maintenance costs.

Evaporative-cooling tower systems pass water in a secondary cooling loop through an air stream, where it is cooled and then recycled to the condensers (Figure 11.2.3*c*).

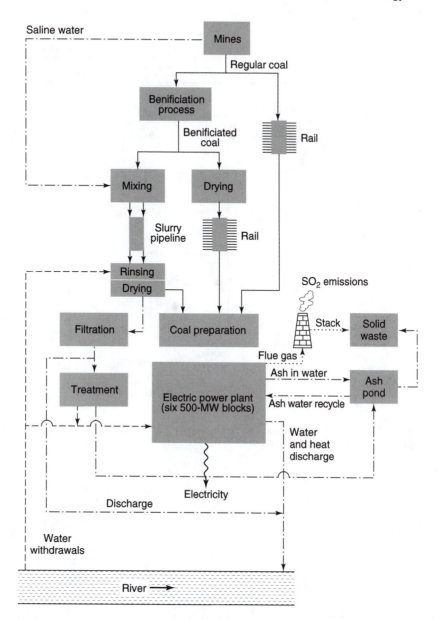

Figure 11.2.1 Flows of processes and materials in the generation of electricity with emphasis on coal handling and combustion, where - - - denotes river water, –·– denotes polluted water, ——— denotes coal, and ···· denotes flue gas or solid waste (from Kindler and Russell (1984)).

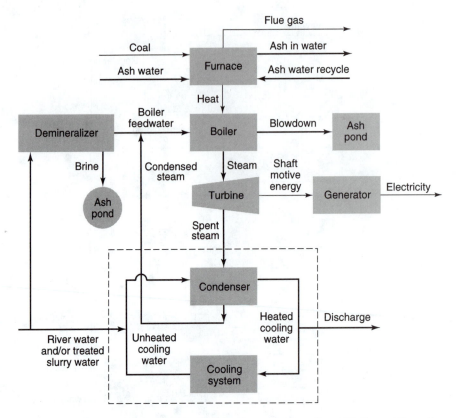

Figure 11.2.2 Basic unit processes for the electric power plant (from Kindler and Russell (1984)).

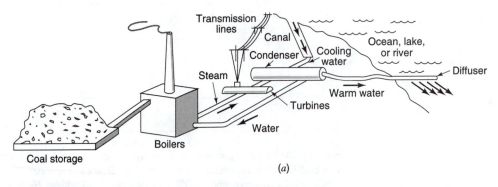

(a)

Figure 11.2.3 Three methods of providing cooling water to condenser for a thermal power plant. (a) Once-through cooling.

Figures 11.2.4–11.2.7 illustrate the basic types of cooling towers. In the *conventional mechanical-draft wet-cooling tower system* in Figure 11.2.4, air is induced by fans. Air can also be induced through the chimney effect by building a 200-ft or taller hyperbolic shell. Figure 11.2.5 shows two types of *mechanical cooling towers* and two types of *natural-draft cooling towers*. Natural-draft cooling towers require more capital cost but less energy cost to operate. Natural-draft towers also require favorable climatic conditions. The actual costs of either type of system depend strongly on climate (Hobbs et al., 1996).

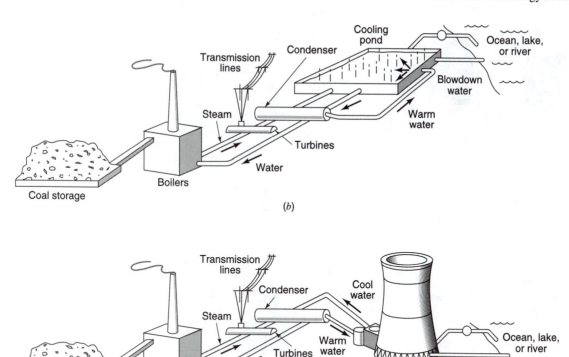

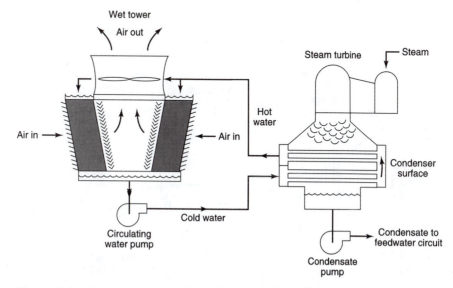

Figure 11.2.3 (*continued*) Three methods of providing cooling water to condenser for a thermal power plant. (*b*) Pond cooling; (*c*) Wet-tower cooling (from Roberson et al. (1998)).

Figure 11.2.4 Conventional mechanical-draft wet-cooling-tower system (from Mitchell (1989)).

Figure 11.2.6 shows two alternative *indirect dry-cooling tower systems*, one using mechanical draft and the other a natural-draft tower. Indirect systems contain a secondary loop that picks up heat at the condensers and rejects it at the towers. Direct dry-cooling systems pass the condensing system from the turbine directly to the heat exchangers.

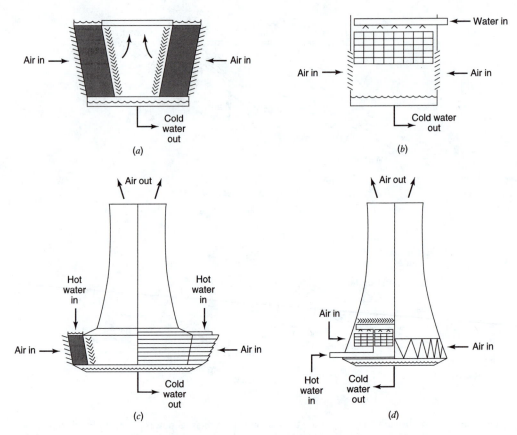

Figure 11.2.5 Different types of wet-cooling towers. (*a*) Mechanical-draft crossflow; (*b*) Mechanical-draft counterflow; (*c*) Natural-draft crossflow; (*d*) Natural-draft counterflow (from Mitchell (1989)).

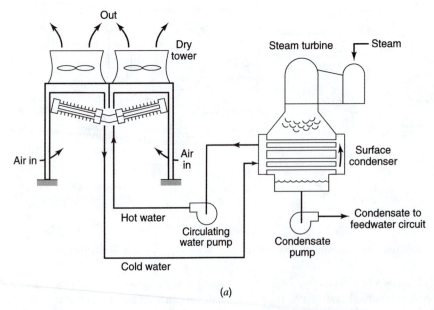

Figure 11.2.6 Indirect dry cooling-tower system. (*a*) With surface condenser and mechanical-draft tower.

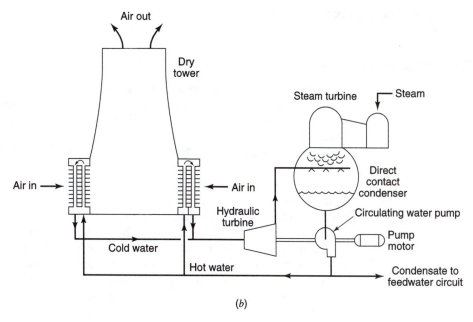

Figure 11.2.6 (*continued*) (*b*) With direct-contact condenser and natural-draft tower (from Mitchell (1989)).

Figure 11.2.7 shows a *combined wet/dry-cooling tower system* with separate wet and dry towers. The dry section handles the heat load for most of the year and during warm periods, some of the thermal load is directed to the wet tower.

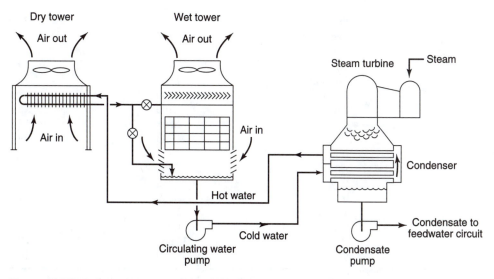

Figure 11.2.7 Series-connected wet/dry cooling-tower system with separate wet and dry towers (from Mitchell (1989)).

11.3 WATER FOR AGRICULTURE

11.3.1 Irrigation Trends and Needs

During the 1970s, irrigation expanded worldwide, followed in the 1980s by a drastically slowed rate of expansion, with a general worldwide emphasis on rehabilitation and management of existing projects during the 1990s (Replogle et al., 1996). Worldwide irrigation development has historically paralleled world population growth. According to Postel (1993), today one-third of the global harvest comes from the 16 percent of the world's croplands that is irrigated (see Table 11.3.1). Many counties—such as China, Egypt, India, Indonesia, Israel, Japan, North Korea and South Korea, Pakistan, and Peru—rely on such land for more than half their domestic food production (Postel, 1993). Gleick (1993c) and the Food and Agriculture Organization (FAO) (1993) provide detailed data and statistics on various aspects throughout the world.

According to the FAO (1993c) and the U.S. Department of Agriculture (1992), approximately 47 million acres (19 million ha) are irrigated in the United States. Throughout the world each year some areas are withdrawn from irrigation and others are added. In the United States there has been a decrease of around 9 percent of irrigation area during the last decade. On a worldwide basis, salinity problems have reduced the area irrigated in some areas, while irrigation is constantly added in other areas.

Table 11.3.1   Net Irrigated Area, Top 20 Countries and World, 1989

Country	Net Irrigated Area[a] (Thousand Hectares)	Share of Cropland that Is Irrigated (%)
China	43,379	47
India	45,039	25
Soviet Union	21,064	9
United States	20,162	11
Pakistan	16,220	78
Indonesia	7,550	36
Iran	5,750	39
Mexico	5,150	21
Thailand	4,230	19
Romania	3,450	33
Spain	3,360	17
Italy	3,100	26
Japan	2,868	62
Bangladesh	2,738	29
Brazil	2,700	3
Afghanistan	2,660	33
Egypt	2,585	100
Iraq	2,550	47
Turkey	2,220	8
Sudan	1,890	15
Other	36,664	7
World	235,299	16

[a] Area actually irrigated; does not take into account double cropping.

Source: Postel (1992).

11.3.2   Irrigation Infrastructure

The three basic types of irrigation methods are: *surface irrigation*, *sprinkler irrigation*, and *micro-irrigation*. Some methods actually cross these three traditional boundaries and are therefore difficult to categorize. Surface irrigation is the distribution of water across fields by flowing over the field surfaces. Sprinkler and micro-irrigation in contrast use closed pipes (conduits) to distribute water. Micro-irrigation is a general category that includes various types of low-emission rate devices, including drip irrigation, trickle irrigation, subsurface irrigation, bubbler irrigation, and others. Table 11.3.2 illustrates the use of micro-irrigation throughout the world.

Table 11.3.2   Use of Microirrigation, Leading Countries and World 1991[a]

Country	Area Under Microirrigation (Hectares)	Share of Total Irrigated Area Under Microirrigation[b] (%)
United States	606,000	3.0
Spain	160,000	4.8
Australia	147,000	7.8
Israel[c]	104,302	48.7
South Afica	102,250	9.0
Egypt	68,450	2.6
Mexico	60,600	1.2
France	50,953	4.8
Thailand	41,150	1.0
Columbia	29,500	5.7
Cyprus	25,000	71.4
Portugal	23,565	3.7
Italy	21,700	0.7
Brazil	20,150	0.7
China	19,000	<0.1
India	17,000	<0.1
Jordan	12,000	21.1
Taiwan	10,005	2.4
Morocco	9,766	0.8
Chile	8,830	0.7
Other	39,397	—
World[d]	1,576,618	0.7

[a] Micro-irrigation includes primarily drop (surface and subsurface) methods and micro-sprinklers.
[b] Irrigated areas are for 1989, the latest available.
[c] Israel's drip and total irrigated area are down 18 and 15 percent, respectively, from 1986, reflecting water allocation cutbacks due to drought.
[d] 13,280 hectares (11,200 of them in the Soviet Union) were reported in 1981 by countries that did not report at all in 1991; world total does not include this area.

Source: Postel (1992).

11.3.2.1   Water Supply Infrastructure

Water supplies for irrigation depend on: (1) managing precipitation runoff, either diverted from streams or from reservoir storage, (2) existing groundwater sources, or (3) reclaimed water from municipalities, either directly or as recharge to groundwater (Repogle et al., 1996). Most irrigation water supplies require the management of reservoirs and/or groundwater aquifers, while some water supplies are directly from rivers. Figure 11.3.1 illustrates the changes in world-irrigated area per capita from 1960 to 1990.

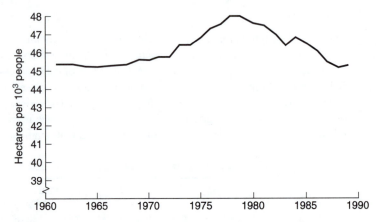

Figure 11.3.1 World irrigated area per capita, 1960–1990. Increases in irrigated area are now failing to keep up with increases in population. After increasing through the 1970s, per-capita irrigated area worldwide is now falling. If this trend continues, feeding the world's growing population will require even greater improvements in yields per hectare or greater food production from nonirrigated lands. Such improvements are likely to be increasingly difficult (from Postel (1992)).

Figures 11.3.2 and 11.3.3 illustrate the Chief Joseph Dam Project located in the north central part of the state of Washington in the United States This project extends approximately 140 miles along the Columbia and Okanogan Rivers from the city of Wenatchee to the Canadian border. This is an all-pressure system, which means that water is pumped from a river or reservoir through pipelines and delivered to the individual farm tracts under sufficient pressure for irrigation to use sprinklers without a booster pump. Figures 11.3.2 and 11.3.3 illustrate the East and Brays Landing Units, respectively, showing the vertical schematic layouts of the units. Each lateral serves a considerable range of elevations. Difference in terrain of these two units dictated a somewhat different system layout. The land for the Brays Landing Unit is located on benches (level areas) so that this system was built using a series of lifts, as compared to the design of the East landing unit.

Water supply infrastructure for irrigation systems at the project level requires an institutional framework such as irrigation districts, mutual companies, or commercial companies (Replogle et al., 1996). The ASCE Manuals and Reports on Engineering Practice No. 57 (1991) provides information on the management, operation, and maintenance of irrigation and drainage systems, providing additional details on the institutional framework.

On a project level, the infrastructure includes structures such as reservoirs, well fields, river diversion, canals, high-pressure pipelines, low-head pipe lines, semi-closed pipelines, and various hydraulic structures such as inlet structures, drop structures, check structures, diversion boxes, and measurement structures.

11.3.2.2   *Farm Infrastructure*

The three basic types of irrigation, methods have been previously defined as surface irrigation, sprinkler irrigation, and micro-irrigation.

Surface Irrigation
Surface irrigation includes the following types (Replogle et al., 1996):

(a) *continuous-flood* or *paddy irrigation*, in which small basins are flooded during essentially the entire growing season;

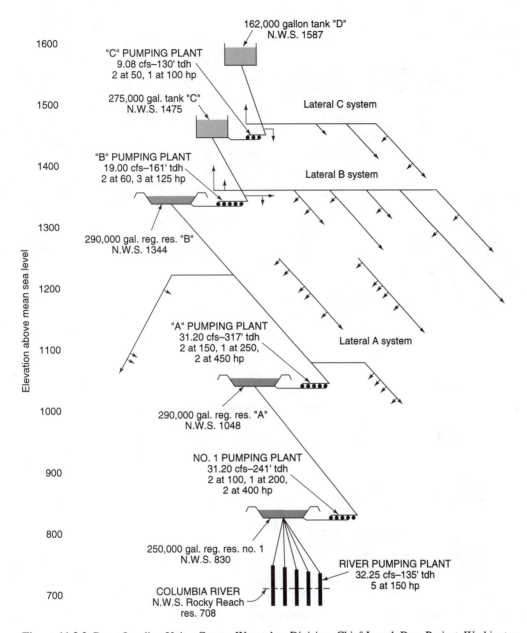

Figure 11.3.2 Brays Landing Unit—Greater Wenatchee Division, Chief Joseph Dam Project, Washington (from Walter (1971)).

(b) *basin irrigation* confines water to a given area by ponding over the area but does not remain ponded for a long time such as one day (Figure 11.3.4);

(c) *furrow irrigation* uses furrows that are dug between crops planted in rows to control and guide water for either steep land or very level land (Figure 11.3.5);

(d) *level basin irrigation* has no soil-surface slope in any direction;

(e) *border-strip irrigation* applies water to one end of a rectangular strip of sloping land so that water advances down slope and either runs off the end or ponds behind a dike;

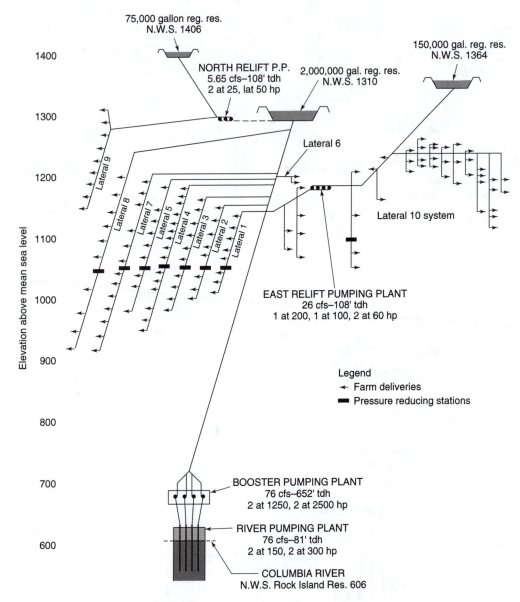

Figure 11.3.3 East Unit—Greater Wenatchee Division, Chief Joseph Dam Project, Washington (from Walter (1971)).

(f) *surge-flow irrigation* applies water to the head of a furrow for a time period, then stops until the flow has infiltrated, then a second surge is applied, and so on, until a satisfactory amount of irrigation has been completed; and

(g) *reuse irrigation* collects irrigation-runoff water (tail water) from a field and uses the water.

Control structures are required in open canals (*ditch delivery systems*) to regulate velocity, head, and the quantity of water released into distribution laterals, basins, borders, and furrows. *Division boxes*, as shown in Figure 11.3.6, direct or divide flow from a supply pipe or channel between two or more distribution laterals. Figure 11.3.6*a* shows a fixed proportional flow divider-type division box. Some division boxes have movable splitters to change the flow proportions. Figure 11.3.6*b* shows a division box with a weir-type overflow outlet. Figure 11.3.7 illustrates commonly used drop structures in open-channel delivery systems.

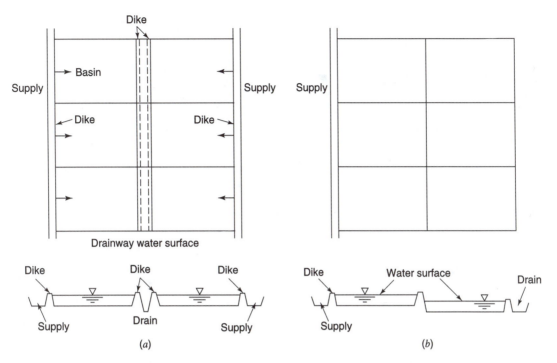

Figure 11.3.4 Basin irrigation system layouts. (*a*) A drainway midway between supply laterals; (*b*) A tier arrangement (from James (1988)).

Figure 11.3.5 Furrow irrigation system. (Courtesy of U.S. Bureau of Reclamation.)

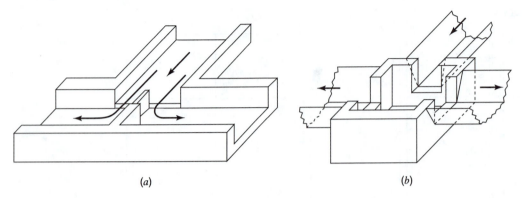

(a)   (b)

Figure 11.3.6 Division boxes. (*a*) A fixed proportional flow divider; (*b*) Weir-type overflow outlets (from James (1988)).

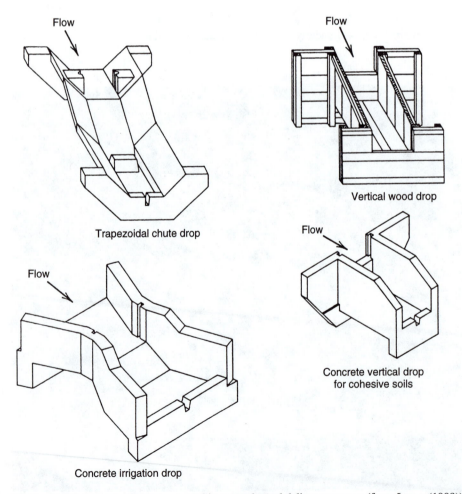

Flow

Flow

Trapezoidal chute drop

Vertical wood drop

Flow

Flow

Concrete vertical drop
for cohesive soils

Concrete irrigation drop

Figure 11.3.7 Some drop structures used in open-channel delivery systems (from James (1988)).

Sprinkler Irrigation

Sprinkler-irrigation systems are distinguished by whether or not the structures move in the field during the irrigation event, so there are two basic categories, moving (Figure 11.3.8) and fixed-in-place systems (Figure 11.3.9). Sprinkler irrigation systems include the following types:

(a) *permanent, solid-set sprinklers* are fixed on risers from buried lines or lines suspended above a crop or over trees (Figure 11.3.9);

(b) *hand-move sprinklers* are fixed sprinklers that are dissembled, moved, and reassembled between irrigations;

(c) *continuous-move sprinkler systems* move continuously during irrigation; and

(d) *center-pivot irrigation systems* (the most common moving irrigation systems) supply water at a central point and a lateral line rotates around this center.

Micro-irrigation

Micro-irrigation is a general category of various types of low-emission-rate devices, including drip irrigation, trickle irrigation, subsurface irrigation, bubbler irrigation, and the moving low-energy precision-application (LEPA) systems (Replogle et al., 1996). Micro-irrigation methods do not wet the entire soil surface or volume, but only that portion that needs to be watered for the particular crop.

Figure 11.3.8 Moving sprinkler-irrigation system (center-pivot irrigation system). (Courtesy of the U.S. Bureau of Reclamation.)

Trickle irrigation is the frequent, slow application of water either directly to the land surface or into the root zone of the crop. In other words, trickle irrigation irrigates only the root zone of the crop and maintains the water content of the root zone at near-optimal level. Figure 11.3.10 illustrates the components of a trickle irrigation system.

James (1988) categorizes trickle irrigation as including drip irrigation, bubbler irrigation, and spray irrigation, defined as follows:

(a) *drip irrigation* applies water as discrete drops on a slow, nearly continuous basis;

(b) *subsurface irrigation* applies water below the soil surface using point and line source emitters;

(c) *bubbler irrigation* applies water to the land surface on small streams from tubes that are attached to buried laterals; and

(d) *spray irrigation* sprays water as a mist over the land surface using small sprinkler-like devices often called *micro sprinklers*.

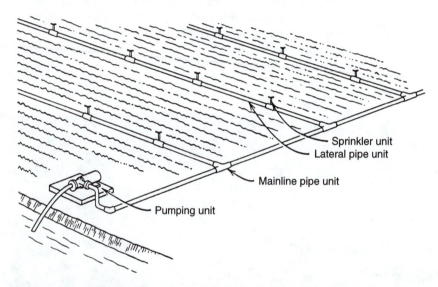

Figure 11.3.9 A typical sprinkle irrigation system consists of four basic units: a pumping unit, mainline pipes, lateral pipes, and one or more sprinklers (from James (1988)).

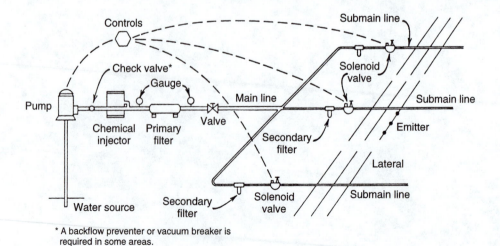

Figure 11.3.10 The components of a trickle irrigation system (from James (1988)).

11.3.3   Irrigation System Selection and Performance

Characteristics of Systems

Table 11.3.3 provides a qualitative overview of the characteristics of several alternative irrigation types. The advantages and disadvantages of each may also depend on site-specific properties and economic considerations.

Table 11.3.3   Characteristics of Alternative Irrigation Systems

| Site and Situation Factor | Surface Systems | | Sprinkler System | | | Drip Systems | |
	Redesigned Surface System	Level Basins	Intermittent Mechanical Move	Continuous Mechanical Move	Solid Set	Emitters and Porous Tubes	Bubblers and Splitters
Average efficiency rating	60–70%	80%	70–80%	80%	70–80%	80–90%	80–90%
Soil	Uniform soils with moderate to low infiltration	Uniform soils with moderate to low infiltration	All	Sandy or high infiltration rate soils	All	All	All, basin required for medium and low-intake soils
Topography	Moderate slopes	Small slopes	Level to rolling	Level to rolling	Level to rolling	All	All
Crops	All	All	Generally shorter crops	All but trees and vineyards	All	High value required	High value required
Water supply	Large streams	Very large streams	Small streams nearly continuous	Small streams nearly continuous	Small streams	Small streams continuous and clean	Small streams continuous
Water quality	All but very high salts	All	Salty water may harm plants	Salty water may harm plants	Salty water may harm plants	All, can potentially use high salt waters	All, can potentially use high salt waters
Labor requirement	High, training required	Low, some training	Moderate, some training	Low, some training	Low to high, little training	Low to high, some training	Low, little training
Energy requirement	Low	Low	Moderate to high	Moderate to high	Moderate	Low to moderate	Low
Management skill	Moderate	Moderate	Moderate	Moderate to high	Moderate	High	High
Machinery operations	Medium to long fields	Short fields	Medium field length, small interference	Circular fields, some interference	Some interference	May have considerable interference	Some interference
Duration of use	Short to long term	Long term	Short to medium term	Short to medium term	Long term	Long term, durability unknown	Long term
Weather	All	All	Poor in windy conditions	Better in windy conditions than other sprinklers	Windy conditions reduce performance; good for cooling	All	All

Source: Wade (1986), as presented in Gleick (1993c).

Selection of Irrigation Systems

Table 11.3.4 provides a list of factors affecting the selection of the appropriate irrigation method. Table 11.3.5 provides a guide for selection of a method of irrigation. This table, however, provides some information on the cost of water provided by the different irrigation methods. Cost is often a limiting factor in any new development. Also, this table does not address water and soil qualities, which are also important factors.

Table 11.3.4 Factors Affecting the Selection of an Appropriate Irrigation method

Irrigation method	Factors Affecting Selection					
	Land	Soil	Crop	Climate	Plusses	Minuses
Surface	Level or graded to central slope and surface smoothness	Suited for medium to fine textures but not for infiltrability	For most crops, except those sensitive to standing water or poor aeration	For most climates only slightly affected by wind	Low cost, simple low pressure required	Prone to overirrigate and rising water table
Sprinklers	For all lands	For most soils	For most crops, except sensitive to fungus disease and leaf scorch by salts	Affected by wind (drift, evaporation, and poor distribution)	Control of rate and frequency allows irrigation of sloping and sandy soils	Initial costs and water pressure requirements
Drip	For all regular and irregular slopes	For all soils and intake rates	For row crops and orchards, but not close-growing crops	Not affected by wind, adapted to all climates	High-frequency and precise irrigation, can use saline water and rough land, reduced evaporation	Initial and annual costs, requires expert management, prone to clogging, requires filtration
Microsprayer	For all lands	For all intake rates	For row crops and orchards	May be affected by wind	High-frequency and precise irrigation, less prone to clog	High costs and maintenance
Bubbler	Flat lands and gentle slopes	For all intake rate	For tree crops	Not affected by wind	High-frequency irrigation, no clogging, simple	Not a commercial product

Source: Hillel (1987), as presented in Gleick (1993c).

Table 11.3.5 Guide for Selecting a Method of Irrigation

Irrigation Method	Topography	Crops	Remarks
Widely spaced borders	Land slopes capable of being graded to less than 1% slope and preferably 0.2%	Alfalfa and other deep rooted close-growing crops and orchards	The most desirable surface method for irrigating close-growing crops where topographical conditions are favorable. Even grade in the direction of irrigation is required on flat land and is desirable but not essential on slopes of more than 0.5%. Grade changes should be slight and reverse grades must be avoided. Cross slope is permissible when confined to differences in elevation between border strips of 6–9 cm.
Closely spaced borders	Land slopes capable of being graded to 4% slope of less than preferably less than 1%	Pastures	Especially adapted to shallow soils underlain by claypan or soils that have a lower water intake rate. Even grade in the direction of irrigation is desirable but not essential. Sharp grade changes and reverse grades should be smoothed out. Cross slope is permissible when confined to differences in elevation between borders of 6–9 cm. Since the border strips may have less width, a greater total cross slope is permissible than for border-irrigated alfalfa.

Table 11.3.5   Guide for Selecting a Method of Irrigation (*continued*)

Irrigation Method	Topography	Crops	Remarks
Check back and cross furrows	Land slopes capable of being graded to 0.2% slope or less	Fruit	This method is especially designed to obtain adequate distribution and penetration of moisture in soils with low-water intake rates.
Corrugations	Land slopes capable of being graded to slopes between 0.5% and 12%	Alfalfa, pasture, and grain	This method is especially adapted to steep land and small irrigation streams. An even grade in the direction of changes and reverse grades should at least be smoothed out. Due to the tendency of corrugations to clog and overflow and cause serious erosion, cross slopes should be avoided as much as possible.
Graded contour furrows	Variable land slopes of 2–25% but preferably less	Row crops and fruit	Especially adapted to row crops on steep land, though hazardous due to possible erosion from heavy rainfall. Unsuitable for rodent-infested fields or soils that crack excessively. Actual grade in the direction of irrigation 0.5–1.5%. No grading required beyond filling gullies and removal of abrupt ridges.
Contour ditches	Irregular slopes up to 12%	Hay, pasture, and grain	Especially adapted to foothill condition. Requires little or no surface grading.
Rectangular checks (levees)	Land slopes capable of being graded so single or multiple tree basins will be leveled within 6 cm	Orchards	Especially adapted to soil that have either a relatively high- or low-water intake rate. May require considerable grading.
Contour levee	Slightly irregular land slopes of less than 1%	Fruits, rice, grain, and forage crops	Reduces the need to grade land. Frequently employed to avoid altogether the necessity of grading. Adapted best to soils that have either a high- or low-intake rate.
Portable pipes	Irregular slopes up to 12%	Hay, pasture, and grain	Especially adapted to foothill conditions. Requires little or no surface grading.
Subirrigation	Smooth-flat	Shallow-rooted crops such as potatoes or grass	Requires a water table, very permeable subsoil conditions, and precise leveling. Very few areas adapted to this method.
Sprinkler	Undulating 1–>35% slope	All crops	High operation and maintenance costs. Good for rough or very sandy lands in areas of high production and good markets. Good method where power costs are low. May be the only practical method in areas of steep or rough topography. Good for high rainfall areas where only a small supplemental water supply is needed.
Contour bench terraces	Sloping land—best for slopes under 3% but useful to 6%	Any crop but particularly suited to cultivated crops	Considerable loss of productive land due to berms. Requires expensive drop structures for water erosion control.
Subirrigation (installed pipes)	Flat to uniform slopes up to 1% surface should be smooth	Any crop; row crops of high-value crops usually used	Requires installation of perforated plastic pipe in root zone at narrow spacings. Some difficulties in roots plugging the perforations. Also a problem as to correct spacing. Field trials on different soils are needed. This is still in the development stage.
Localized (drip, trickle, etc.)	Any topographic condition suitable for row crop farming	Row crops or fruit	Perforated pipe on the soil surface drips water at base of individual vegetable plans or around fruit trees. Has been successfully used in Israel with saline irrigation water. Still in development stage.

Source: Doneen and Westcot (1984), as presented in Gleick (1993c).

Performance

From a crop-production perspective, the irrigation system performance is often described by the *irrigation efficiency* (Replogle et al., 1996):

$$E_l = \frac{\text{volume of irrigation water beneficially used}}{\text{volume of irrigation water supplied}} \times 100\% \qquad (11.3.1)$$

which describes water use in terms of benefit for agricultural production. Beneficial uses include crop evapotranspiration (ET), water needed for removing excess salt from the soil, climate control, soil preparation, and weed control. Some of these uses actually consume water, while others do not.

From a hydrologic balance perspective, the use of water by agriculture can be described by the *consumptive use coefficient* (Replogle et al., 1996):

$$C_{cu} = \frac{\text{volume of water consumptively used}}{\text{volume of irrigation water supplied}} \times 100\% \qquad (11.3.2)$$

which includes uses that are not directly beneficial for agricultural production, such as evapotranspiration and sprinkler and reservoir evaporation. These two definitions of water use are not the same, but are related.

Different terms are often used for defining the performance of the irrigation system (i.e., separated from crop management). Common terms include *the application efficiency E_a* and *low-quarter distribution uniformity, DU_{lq}*, defined in terms of that 25 percent of the field receiving the least depth of infiltrated water (American Society of Agricultural Engineers, 1993):

$$E_a = \frac{\text{average depth of water stored in the root zone}}{\text{average depth applied}} \qquad (11.3.3)$$

$$DU_{lq} = \frac{\text{average low-quarter depth of water infiltrated}}{\text{average depth of water infiltrated}} \qquad (11.3.4)$$

11.3.4   Water Requirements for Irrigation

The *total water requirement* or *diversion requirement* for an irrigation system includes the water needed by the crop in addition to the losses associated with the application and delivery of water. The *crop water requirement* or the *consumptive use* can be determined experimentally by planting a crop in a lysimeter or tank of soil and keeping account of water added and soil moisture changes. The overall consumptive use for large areas can be defined by a moisture accounting or balance, as illustrated in Figure 11.3.11. Table 11.3.6 provides a list of crop water requirements and crop evapotranspiration for various crops.

Table 11.3.6   Crop Water Requirements and Crop Evapotranspiration

Crop	Southwest United States Range (mm)	Missouri and Arkansas Basins Range (mm)	FAO Guidelines Range (mm)	United States and Canada[a] Range (mm)
Farm crops				
Alfalfa/forage	1,060–1,550	591–799	600–1,600	594–1,890
Barley	378–558	405–555		384–643
Broomcorn	296–351			
Buckwheat		320–396		
Cocoa			800–1,200	
Coffee			800–1,200	
Corn (maize)	439–607	375–558	400–750	373–617
Cotton	716–1,070		550–950	912–1,050

Table 11.3.6   Crop Water Requirements and Crop Evapotranspiration (*continued*)

Crop	Southwest United States Range (mm)	Missouri and Arkansas Basins Range (mm)	FAO Guidelines Range (mm)	United States and Canada[a] Range (mm)
Emmer	363–570			
Feterita	296–335			
Flax	375–485	448–564	450–900	381–795
Grains (small)			300–450	
Grass				579–1,320
Groundnut			500–700	
Kafir	402–469	436–479		
Millet	277–332	247–287		
Milo	293–509	332–518		
Oats	579–637	411–552		
Oil seeds			300–600	
Potatoes	485–622	421–518	350–625	455–617
Rhodes grass	1,060–1,350			
Rice			500–950	920
Safflower				635–1,150
Sisal			550–800	
Sorghum	515–634	323–448	300–650	549–645
Soybeans	506–856		450–825	399–564
Sudan grass	878–963			
Sugar beets	539–829	488–762	450–850	546–1,050
Sugar cane	1,060–1,390		1,000–1,500	
Sunflowers		366–427		
Tobacco			300–500	
Wheat	445–683	415–549	450–650	414–719
Vegetable crops				
Beans		396–488	250–500	396–417
Beans, snap	253–439			
Beets, Table	265–418			
Broccoli				500
Cabbage	287–454			437–622
Carrots	387–488			422
Cauliflower	436–539			472
Corn, sweet				386–498
Cucumbers		528–1,140		
Lettuce	219–411			216
Melons	756–1,040			
Onions	223–463		350–600	592
Onions, Greens				445
Peas	369–475	415–591		340
Spinach	244–326			
Sweet potatoes	539–686		400–675	
Tomatoes	290–433	640–853	300–600	366–681
Fruit				
Apples		640–792		531–1,060
Avocados			650–1,000	
Bananas			700–1,700	
Cantaloupes		457–701		485
Dates			900–1,300	
Deciduous trees			700–1,050	
Grapefruit			650–1,000	1,220
Oranges			600–950	933
Plums				1,070
Vineyards			450–900	
Walnuts			700–1,000	

[a]from Kammerer (1982).
Source: Gleick (1993c).

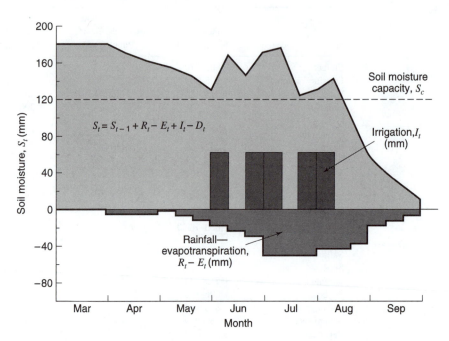

Figure 11.3.11 Soil moisture balance with irrigation (from Gouevsky and Maidment (1984)).

EXAMPLE 11.3.1

Determine the water requirements for irrigation of corn, per acre land per day, during the crop's active growing season. Assume a consumptive use coefficient (C_{cu}, equation (11.3.2)) of 50 percent. Use the Blaney–Criddle formula for determining the reference crop evapotranspiration ET_0 (mm/day), given as (FAO, 1977)

$$ET_0 = c\left[p(0.46T + 8)\right]$$

where T is the mean daily temperature in °C over the month considered, p is the mean daily percentage of total annual daytime hours, given as a function of the month used and the latitude of the location, and c is an adjustment factor that depends on minimum relative humidity, sunshine hours, and daytime wind estimates. Use $T = 33$°C, $p = 0.33$, $c = 1.0$. Note that the evapotranspiration for a given crop must be corrected from the reference evapotranspiration using the formula $ET_c = K_c ET_0$, where K_c is the crop coefficient. K_c for corn during its active growing season can be taken as 0.8.

SOLUTION

From the given values of T, p, and c, $ET_0 = 1.0\{0.33[0.46(33) + 8]\} = 7.65$ mm/day. For corn, $ET_c = K_c ET_0 = 0.8(7.65) = 6.12$ mm/day. In terms of the volume of water consumptively used per acre, we get

volume of water consumptively used = 6.12 mm/day (1 acre) = 6.12 mm/day (10^{-3}m/mm) (3.281ft/m) (1 acre) = 0.02 acre-ft/day

Using equation (11.3.2), we find

$$\text{volume of irrigation water required} = \frac{\text{volume of water consumptively used}}{C_{cu}} = 0.02 / 0.5$$

$$= 0.04 \text{ acre - ft/day}$$

11.3.5   Impacts of Irrigation

Table 11.3.7 illustrates some of the impacts of irrigation development, as presented in Gleick (1993c). Table 11.3.8 illustrates some of the environmental effects of agriculture on water quality, as presented in Gleick (1993c).

Table 11.3.7   Impacts of Irrigation Development

Casual activity	Possible Impact	Possible Remedies
Surface irrigation	1. Waterlogging 2. Soil salinization 3. Increase of diseases 4. Degradation of water quality	1. Increased irrigation efficiency 2. Construction of drainage systems 3. Disease control measures 4. Control of irrigation water quality
Sewage irrigation	1. Contamination of food crops 2. Direct contamination of humans 3. Dispersion in air 4. Contamination of grazing animals	1. Regulatory control 2. Tertiary treatment and sterilization of sewage
Use of fertilizers	1. Pollution of groundwater, especially with nitrates 2. Pollution of surface flow	1. Controlled use of fertilizers 2. Increased irrigation efficiency
Use of pesticides	1. Pollution of surface flow 2. Destruction of fish	1. Limited use of pesticides 2. Coordination with schedule of irrigation
Irrigation with high silt load	1. Clogging of canals 2. Raising of level of fields 3. Harmful sediment deposits on fields and crops	1. Avoiding use of flow with high silt load 2. Soil conservation measures on upstream watershed
High-velocity surface flow	1. Erosion of earth canals 2. Furrow erosion 3. Surface erosion	1. Proper design of canals 2. Proper design of furrows 3. Land leveling 4. Correctly built and maintained terraces
Intensive sprinkling of sloping land	1. Soil erosion	1. Correctly designed and operated system

Source: Gleick (1993c).

Table 11.3.8   Selected Environmental Effects of Agriculture on Water Quality

Agricultural Practices	Soil	Groundwater	Surface Water
Land development, land consolidation programs	Inadequate management leading to soil degradation	Other water management influencing groundwater table	Soil degradation, siltation, water pollution with soil particles
Irrigation, drainage	Excess salts, waterlogging	Loss of quality (more salts), drinking water supply affected	Soil degradation, siltation, water pollution with soil particles
Tillage	Wind erosion, water erosion		Soil degradation, siltation, water pollution with soil particles
Mechanization: large or heavy equipment	Soil compaction, soil erosion		Soil degradation, siltation, water pollution with soil particles
Nitrogen fertilizer use		Nitrate leaching	
Phosphate fertilizer use	Accumulation of heavy metals (such as cadmium)		Runoff leaching or direct discharge leading to eutrophication
Manure, slurry use	Excess accumulation of phosphates, copper (pig slurry)	Nitrate, phosphate combination (by use of excess slurry)	Runoff leaching or direct discharge leading to eutrophication
Sewage sludge compost	Accumulation of heavy metals contaminates		Runoff leaching or direct discharge leading to eutrophication

Table 11.3.8   Selected Environmental Effects of Agriculture on Water Quality (*continued*)

Agricultural Practices	Soil	Groundwater	Surface Water
Applying pesticides	Accumulation of pesticides and degradation products	Leaching of mobile pesticide residues and degradation products	
Input of feeder additives, medicines	Adverse effects depend on input		
Modern building (e.g., silos) and intensive livestock farming	Excess accumulation of phosphates, copper (pig slurry)	Nitrate, phosphate contamination (by use of excess slurry)	Runoff leaching or direct discharge leading to eutrophication

Source: Gleick (1993c).

11.4   WATER SUPPLY/WITHDRAWALS

11.4.1   Withdrawals

Most fresh water is supplied by pumping groundwater and withdrawing water from lakes and rivers. The demand for fresh water supplies increases because of population growth and increasing development. Figure 11.4.1 illustrates water demand that can be supplied at different levels of mobilization of the potentially available water. Table 11.4.1 provides six estimates of the global stocks of fresh and saline water, as presented by Gleick (1993d). Table 11.4.2 presents the total water withdrawals by major users in selected OCED (Organization for Economic Cooperation and Development) countries.

Table 11.4.1   Global Stocks of Water (10^3 km^3)

Stock	Nace (1967)	UNESCO (1974)	L'vovich (1974)	Baumgartner and Reichel (1975)	Berner and Berner (1987)	WRI (1988)
Freshwater lakes	125	91	280[e]	225[f]	125[e]	100
Saline lakes and inland seas	104	854	[e]	[f]	[e]	105
Rivers[a]	1.25	2.12	1.2	[f]	1.7	1.7
Soil moisture	67[b]	16.5	85	[g]	65	70
Groundwater	8,350[c]	23,400[k]	60,000[h]	8,062[g]	9,500[c]	8,200
Ice caps and glaciers	29,200	24,064	24,000	27,820	29,000	27,500
Underground permafrost ice	[d]	300	[d]	[d]	[d]	[d]
Swamp water	[d]	11.47	[d]	[d]	[d]	[d]
Biota	[d]	1.12	[d]	[d]	0.6	1.1
Total inland water	37,800					
Atmospheric water	13	12.9	14	13	13	13
Ocean water	1,320,000	1,338,000	1,370,323	1,348,000	1,370,000	1,350,000
Total stocks[j]	1,360,000	1,385,985	1,454,193	1,384,120	1,408,700	[i]

[a]Average instantaneous volume.
[b]Includes vadose water (subsurface water above the water table level).
[c]To depth of 4 km.
[d]These values included in other categories or not included.
[e]Fresh and saline water.
[f]All lake and river water included in the value for freshwater lakes.

[g]Soil water included in groundwater value.
[h]Refers to volume of water in upper 5 km of earth's crust, excluding chemically bound water.
[i]No total given in original source.
[j]Total given in original source. May not add up to sum of individual stocks.
[k]Of this total, 10,530 are freshwater.

Sources: Baumgartner and Reichel (1975), Berner and Berner (1987), Gleick (1993d), L'vovich (1974), Nace (1967), UNESCO (1974), WRI (1988).

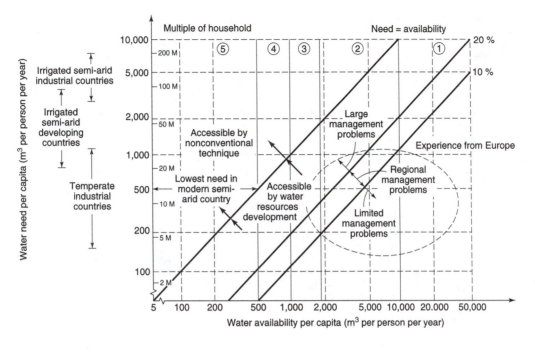

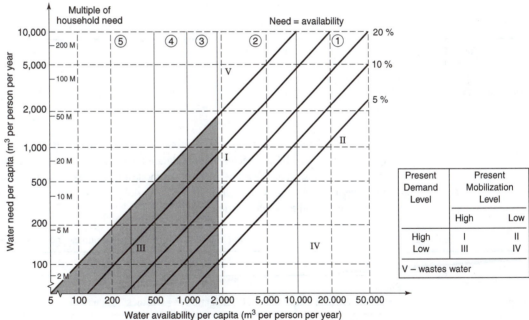

Figure 11.4.1 Logarithmic diagram showing water demand possible to supply at different levels of mobilization of potentially available water. The horizontal axis shows per capita availability (m³ per person and year). Circled code numbers at the top of the diagrams refer to the water competition levels. The vertical axis shows water demand expressed both as m³ per person and year and as multiples of a household demand H, assumed to be 100l per person and day. Crossing lines show different mobilization levels of water availability, achieved through water storage, flow control, and other measures of water resources development. Roman numbers refer to water predicament positions (from Falkenmark et al. (1989) and Falkenmark and Lindh (1993)).

Table 11.4.2 U.S. Water Withdrawals, by Region, 1960 to 1990 (km³/yr).

Region	1960	1965	1970	1975	1980	1985	1990
New England	8.9	10.0	13.4	19.4	18.0	22.5	19.2
Mid-Atlantic	37.5	46.5	62.3	71.9	71.9	60.5	65.8
South Atlantic-Gulf	26.1	39.3	48.4	59.5	67.8	60.2	61.1
Great Lakes	40.1	45.7	54.0	49.8	52.6	44.1	44.8
Ohio	33.2	41.5	49.8	49.8	52.6	43.2	42.0
Tennessee	10.4	11.3	10.9	15.2	16.6	12.7	12.7
Upper Mississippi	15.2	22.1	22.1	26.3	31.8	23.4	28.7
Lower Mississippi	7.3	7.2	18.0	22.1	29.1	24.3	26.4
Souris-Red-Rainy	0.2	0.4	0.4	0.5	0.3	0.4	0.4
Missouri	29.9	28.8	33.2	48.4	54.0	47.7	51.9
Arkansas-White-Red	14.4	14.4	16.6	20.8	33.2	21.2	21.7
Texas-Gulf	30.4	22.1	29.1	30.4	23.5	26.0	25.6
Rio Grande	[a]	10.1	8.7	7.5	6.5	7.8	8.3
Upper Colorado	19.4	9.3	11.2	5.7	11.8	10.5	9.8
Lower Colorado	[b]	9.1	10.0	11.8	12.0	10.2	10.7
Great Basin	9.7	9.5	9.3	9.5	10.4	11.4	10.1
Pacific Northwest	40.1	40.1	41.5	45.7	47.0	49.1	50.2
California	45.7	52.6	66.4	70.6	74.7	69.0	65.1
Alaska	0.3	0.1	0.3	0.3	0.3	0.6	0.9
Hawaii	2.2	2.8	3.7	3.5	3.5	3.0	3.8
Caribbdan	1.7	2.4	4.2	5.7	4.6	3.8	4.4
Total[c]	374	429	512	581	625[d]	552	564

[a]Included in Texas-Gulf.
[b]Included in Upper Colorado.
[c]Figures may not add to total due to independent rounding.
[d]This total revised in 1988 to 600 km³/yr.

Source: Gleick (1993b).

Table 11.4.3 provides a summary of groundwater withdrawals in the United States by water resource region and by use sector. All the data are for withdrawals, not consumption. This table presents only freshwater, as a few regions use modest quantities of saline groundwater, primarily for power plant cooling (Gleick, 1993b). Table 11.4.4 presents groundwater resources and use in selected Asian and Pacific countries.

Table 11.4.3 Groundwater Withdrawals in the United States, by Water Resources and Sector, 1990[a]

Region	Freshwater Withdrawals (10^6 m³/yr)								
	Public Supply	Domestic	Commercial	Irrigation	Livestock	Industrial	Mining	Thermo-electric	Total
New England	453	233	113	12	7.5	133	1.8	4.0	958
Mid-Atlantic	1,934	547	130	141	98	497	290	6.4	3,640
South Atlantic-Gulf	3,468	910	166	3,178	274	1,238	531	57	9,820
Great Lakes	636	390	37	182	70	325	30	4.4	1,680
Ohio	1,069	486	80	39	75	735	1,083	90	3,660
Tennessee	151	77	77	5.2	44	32	35	0.0	421
Upper Mississippi	1,603	513	185	489	300	482	15	29	3,620
Lower Mississippi	968	124	28	8,607	981	692	11	104	11,500
Souris-Red-Rainy	47	30	0.1	77	22	1.8	0.3	0.0	179
Missouri Basin	844	191	46	9,947	334	157	133	69	11,700
Arkansas-White-Red	503	163	37	9,118	218	93	68	43	10,200

Table 11.4.3   Groundwater Withdrawals in the United States, by Water Resources and Sector, 1990[a] (*continued*)

| Region | Freshwater Withdrawals (10^6 m³/yr) | | | | | | | | |
	Public Supply	Domestic	Commercial	Irrigation	Livestock	Industrial	Mining	Thermo-electric	Total
Texas-Gulf	1,451	109	65	5,485	75	195	116	64	7,560
Rio Grande	511	32	26	2,238	28	15	90	22	2,960
Upper Colorado	44	14	7.7	44	7.3	4.0	54	0.0	175
Lower Colorado	709	51	32	3,095	41	68	192	65	4,250
Great Basin	489	18	10	1,948	37	106	106	10	2,730
Pacific Northwest	1,004	293	68	10,873	816	464	6.9	14	13,500
California	4,504	290	80	14,644	278	174	21	6.4	20,000
Alaska	47	8.6	12	0.1	0.1	72	5.9	6.5	87
Hawaii	305	12	54	276	4.7	28	1.9	131	813
Caribbean	112	5.1	0.6	75	7.2	15	2.6	3.6	221
Total	20,900	4,500	1,250	70,500	3,720	5,460	2,790	728	110,000

[a]Figures may not add to totals due to independent rounding.

Source: Gleick (1993b).

Table 11.4.4   Groundwater Resources and Use in Selected Asian and Pacific Countries

| | Groundwater Utilization (10^6 m³/year) by Various Sectors of the Economy | | | | Exploitable Potential (10^6 m³/yr) | Utilization as Percent of Exploitable Potential | Groundwater Use as Percent of Total Fresh-water Use | Groundwater Extraction Cost at Wellhead (U.S. \$/m³) | Year |
	Domestic and municipal	Irrigation	Industry	Total					
Australia	1,424	1,297	111	2,832	37,377	8			
Bangladesh	1,710	5,997	24	7,731				0.01	1979
Guam				35	70	50	75		
India	4,600	143,500	1,900	150,000				0.02	1973
Indonesia	3,640	11	33	3,684	455,520	1			
Malaysia	60%	5%	35%	100%	23,200			0.06-0.09	1981
Pakistan		40,000		40,000				0.01	1981
Philippines	2,000		1,000	3,000	33,000	9	35	0.03	1983
Thailand	541	202	181	924					
Iran							41		
Japan							16	0.01-0.04	1980
Kiribata							100		
Northern Mariana Islands								0.03	1983
Samoa								0.04	1983
Mongolia				6,000					
Korea Rep				15,000					
USSR (Asian)				279,000					

Source: Gleick (1993b).

Table 11.4.5 presents some relatively new water supply techniques and compares their costs, stage of development, physical requirements, advantages and disadvantages, reliability, and common application. These water systems include desalination, transport—tankers, transport—icebergs, water reuse, and cloud seeding. The advantages and disadvantages of each method obviously depend on the alternatives available in a region, level of technical knowledge of water managers, location, and climate characteristics, among many other factors. The characteristics in the table are applicable only generally. Figure 11.4.2 shows the Yuma Desalting Plant near Yuma, Arizona.

Table 11.4.5 Characteristics of Unconventional Water-Supply Systems

	Desalination	Transport-Tankers	Transport-Icebergs	Water Reuse	Cloud Seeding
Cost of water ($1985/m^3)	Brackish Seawater $0.25–1.00 $1.30–8.00	$1.25–7.50	Saudi Arabian project $0.02–0.85	$0.07–1.80	$0.01
Stage of development	Moderate to high	Low to moderate	Very low	Moderate to high	Low to moderate
Special physical requirements	1. Source of clean brackish or seawater 2. Method or place to dispose of a brine solution 3. Major equipment and facility construction	1. Port facilities 2. Storage facilities	1. Large tugboats 2. A need for water adjacent to a coastal area 3. Deep draught channels 4. Storage facilities 5. Equipment to melt ice and collect water	1. Source of wastewater 2. Major equipment and facility construction	1. Suitable clouds 2. Scientific staff 3. Structures and land use to take advantage of increased rainfall
Advantages	1. Proven systems available 2. Wide range of suppliers 3. Provides independence from external sources of supply	1. Low level of technology required 2. Can be implemented quickly for emergency use		1. Proven techniques available 2. Wide range of suppliers 3. Nonpotable applications 4. Can reduce problems associated with present methods of wastewater disposal	1. Only moderate to small capital investment needed
Disadvantages	1. Requires skilled techniques for operation and repair 2. Generally requires use of foreign exchange to purchase equipment 3. Energy-intensive process	1. Deep channels and facilities needed for large vessels 2. Must depend on sources outside country. The potential for a supply interruption because of storms, conflicts boycotts, or strikes is relatively high 3. Can only meet low-volume demand	1. Essentially untried method 2. Not suitable for small-scale experimentation	1. Requires some operation and maintenance 2. Improper operation could create the potential for advers public health effects 3. May not be aesthetically or culturally acceptable	1. Still in the limited development stage with many uncertainties 2. Cannot control the results
Certainty of operation	High	High	Low	High	Low to moderate
Typical applications for water	Potable Industial	Potable Agriculture Industrial	Potable Agriculture Industrial	Agriculture Industrial	Potable Agricultural Industrial

Source: Gleick (1993b), modified and updated from United Nations (1988).

Figure 11.4.2 (*a*) Yuma Desalting Plant in the Colorado River near Yuma, Arizona. This is a reverse osmosis desalting plant that was constructed to reduce salinity levels of water delivered to Mexico in the Colorado River. (Courtesy of U.S. Bureau of Reclamation.)

Figure 11.4.2 (*b*) Yuma Desalting Plant process area which is located in large building in Figure 11.4.2*a*. Operators are shown loading the membrane elements into the reverse osmosis vessels. (Courtesy of U.S. Bureau of Reclamation.)

11.4.2   Examples of Regional Water Supply Systems

Central Arizona Project

An example of a regional water supply system is the Central Arizona Project (CAP), shown on the map in Figure 11.4.3. CAP is a 336-mile long system of aqueducts, tunnels, pumping plants, and pipelines. Constructed by the U.S. Bureau of Reclamation the CAP delivers Colorado River water from Lake Havasu on Arizona's western border to Maricopa, Pinal, and Pima counties in Central and Southern Arizona. Operated and maintained by the Central Arizona Water Conservation District (CAWCD), the CAP delivers water to users in the three-county service area. The CAP can deliver an average of 1.5 maf of water each year to cities, industries, Indian communities, and farmers. During shortages, cities, industries, and Indian communities have priority for the water. Once these uses are met, non-Indian agricultural users will receive water.

Figure 11.4.4 shows the Havasu, the project's largest pumping plant, which lifts the water intake from Lake Havasu to the Buckskin Mountains Tunnel's inlet 824 feet above the lake's surface. The water then flows seven miles through the tunnel to the first section of canal.

The first 17 miles of the Hayden–Rhodes Aqueduct play a specific role in the project's operation. This oversized section of canal between Havasu and Bouse Hills pumping plants acts as an

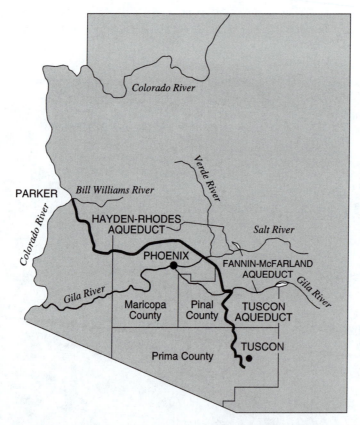

Figure 11.4.3 Central Arizona Project map. Colorado River water is delivered to users by pumping it from Lake Havasu into the conveyance system, then relifting it through a series of pumping plants across the state. From Lake Havasu to the end of the aqueduct, the water is lifted nearly 2,900 feet in elevation by 14 pumping plants. (Courtesy of Central Arizona Project.)

in-line storage reservoir. Because the Havasu plant will use about half the energy needed for project pumping, it will be operated primarily when energy costs are lowest to reduce pumping costs. The water will then be pumped through the rest of the system as demand requires. Figure 11.4.5 shows the CAP through a residential area in Scottsdale, Arizona. A part of this overall project is the recharge of water in facilities as shown in Figure 11.4.7. Figure 11.4.6 shows the control room at CAP headquarters in Phoenix, Arizona. The CAP SCADA (Supervisory Control and Data Acquisition) is used for real-time monitoring and control of the CAP.

Southeast Anatolian Project (GAP)

The Southeast Anatolian Project (GAP) is a regional development scheme in the southeastern region of Turkey that involves the construction of 22 dams, 19 hydroelectric generation stations, and the irrigation of 1.7 million hectares of land. This project shown in Figure 11.4.8a is known by the Turkish acronym GAP (for Guneydogu Anadolu Projesi). The Atatürk Dam (shown in Figure 11.4.8b), one of the largest in the world, is the centerpiece of the GAP project. Figure 13.2.2 shows closer views of the hydroelectric generation facility at the Atatürk Dam.

Israel's National Water Carrier System

The main water resources of Israel (Figure 11.4.9) are concentrated in Lake Kinneret (also known as Lake Galilee), the Coastal Aquifer, and the Western Mountain Aquifer (or Limestone Aquifer). About 80 percent of all the water used in Israel is derived from these three sources (Bruins, 1999). Lake Kinneret is the principal natural storage reservoir of fresh surface water, having an area of 165 km^2 and a maximum capacity of almost 4,000 mcm (million cubic meter) of water. Water is supplied to the lake by the Jordan River and its tributaries, and to a lesser extent by wadis and springs. Lake Kinneret remains the main water source of the National Water Carrier.

Figure 11.4.4 Havasu intake and pumping plant with Lake Havasu shown in background. (Courtesy of Central Arizona Project.)

Figure 11.4.5 CAP through a residential area in Scottsdale, Arizona. (Courtesy of Central Arizona Project.)

Figure 11.4.6 CAP control room. The CAP's operating features—pumping plants, check structures and turnouts—are remotely operated from the project's headquarters in north Phoenix by a computer assisted control and communications system. Cables buried along the aqueduct carry operating commands from the Control Center's computer to check structures and turnouts. The commands raise or lower gates in these structures so water can be moved through the system or delivered to users. A microwave system is used for communicating with the pumping plants. Commands are sent from the Control Center through the microwave system to pump units at the plants, starting and stopping them as needed. The communications systems also carry information from the operating features back to operators at the Control Center. The operators, who plan system operations, also monitor the system's performance and adjust it as necessary, correcting problems that may occur. The microwave and cable systems can be used inter-changeably to operate the project features. If this remote operations capability is temporarily lost, each individual operating feature can be operated locally. (Courtesy of Central Arizona Project.)

Figure 11.4.7 Avra Valley Recharge Facility near Tucson, Arizona. Recharge facilities consist of five major components: (1) the pipeline that carries the water (or treated effluent) from the source (or wastewater treatment plant); (2) percolation (infiltration) basins where the treated effluent infiltrates into the ground; (3) the soil immediately below the infiltration basins (vadose zone); (4) the aquifer where water is stored for a long duration; and (5) the recovery well where water is pumped from the aquifer for potable or non-potable reuse. (Courtesy of Central Arizona Project, photograph by M. Early)

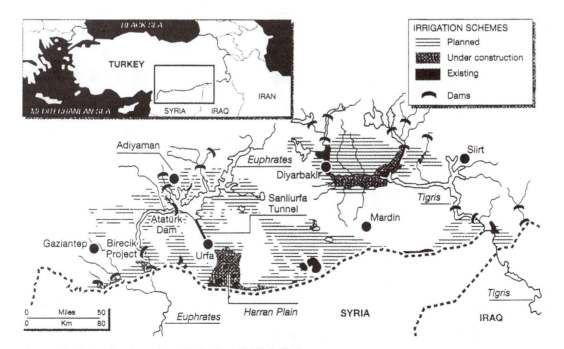

Figure 11.4.8 (*a*) Southeast Anatolia Project (GAP) in Turkey.

Figure 11.4.8 (*continued*) (*b*) Atatürk Dam. The Atatürk Dam is the largest structure ever built in Turkey for irrigation and hydropower generation. Located on the Euphrates, it is the key structure for the development of the Lower Euphrates Project as well as for the Southeastern Anatolian Project. Also refer to Figure 13.2.2 and 17.2.15. (Courtesy of the Southwest Anatolia Project.)

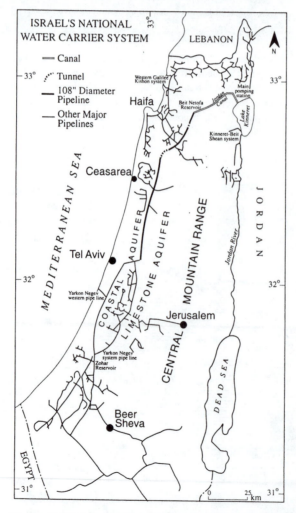

Figure 11.4.9 Israel's National Water Carrier System (Bruin, 1999)

11.5 WATER DEMAND AND PRICE ELASTICITY

11.5.1 Price Elasticity of Water Demand

From an economic viewpoint, *demand* is a general concept denoting the willingness of consumers or users to purchase goods, services, or inputs to production processes. Demand is often used interchangeably with requirement in discussing water use. A common-sense rule concerning demand is that for a single consumer or group of consumers, the quantity demanded increases as the price (cost) per unit decreases. A *requirement*, however, is a purchase at the same quantity no matter what the price. Water demand is a function of several factors, including price.

The *elasticity of demand* is the responsiveness of consumers' purchases to varying price. The most frequently used elasticity concept is *price elasticity*, which is defined as the percentage change in quantity taken if price is changed 1 percent. Young (1996) states that "the price elasticity of demand for water measures the willingness of consumers to give up water use in the face of rising prices, or conversely, the tendency to use more as price falls."

The price elasticity of water is defined as

$$\eta_P = \frac{\Delta d}{\overline{d}} \div \frac{\Delta P}{\overline{P}} \tag{11.5.1}$$

where η_P is the price elasticity, $\overline{d}$ is the average quantity of water demanded, $\overline{P}$ is the average price, Δd is the change in the demand, and ΔP is the change in the price. For a continuous-demand function, the following more general formula is applicable:

$$\eta_P = \frac{dd}{d} \div \frac{dP}{P} \tag{11.5.2}$$

Table 11.5.1 summarizes some of the values of price elasticity of water demand reported in the literature. Two different ways have been used to formulate the price elasticity of demand for water, one based upon average price and the other based upon marginal price. Agthe and Billings (1980) state that the elasticity determined based upon average price overestimates the result; therefore, they (like several others) recommend that the marginal price be used.

Table 11.5.1 Summary of Some of the Price Elasticity Values

No.	Researchers	Research Area	Estimated Price Elasticity	Estimated Income Elasticity	Remarks
1	Howe & Linaweaver (1967)	Eastern U.S.	−0.860		
2	Howe & Linaweaver (1967)	Western U.S.	−0.52		
3	Wong (1972)	Chicago	−0.02	0.20	
4	Wong (1972)	Chicago suburb	−0.28	0.26	
5	Young (1973)	Tucson	−0.60 – −0.65		Exponential and linear models used
6	Gibbs (1978)	Metropolitan Miami	−0.51	0.51	Elasticity measured with the mean marginal price
7	Gibbs (1978)	Metroplitan Miami	−0.62	082	Elasticity measured with the average price
8	Agthe & Billings (1980)	Tucson	−0.27 – −0.71		Long-run model
9	Agthe & Billings (1980)	Tucson	−0.18 – −0.36		Short-run model
10	Howe (1982)		−0.06		
11	Howe (1982)	Eastern U.S.	−0.57		
12	Howe (1982)	Western U.S.	−0.43		
13	Hanke & de Maré (1982)	Malmo, Sweden	−0.15		

Table 11.5.1   Summary of Some of the Price Elasticity Values (*continued*)

No.	Researchers	Research Area	Estimated Price Elasticity	Estimated Income Elasticity	Remarks
14	Jones & Morris (1984)	Metropolitan Denver	−0.14 – −0.44	040 – 0.55	Linear and log-log models used
15	Moncur (1989)	Honolulu	−0.27		Short-run model
16	Moncur (1989)	Honolulu	−0.35		Long-run model
17	Jordan (1994)	Spalding County Georgia	−0.33		A price elasticity of −0.07 was also reported for no rate structure, but increased price level

The use of the price elasticity of water has been applied to some cities, and some important achievements have been made. The following schematic may depict the general trend of this principle, as derived from the conclusion reached by Jordan (1994):

$$\Uparrow \text{(price)} \Rightarrow \Downarrow \text{(water demand)} \& \Uparrow \text{(revenue)} \qquad (11.5.3)$$

An increase by less than 40 percent of the price resulted in a 10 percent decrease in the demand in Honolulu, Hawaii—the announced goal of the restrictions imposed in the drought episodes of 1976 to 1978 and in 1984 (Moncur, 1987). This was achieved using a price elasticity of only −0.265. In Tucson, Arizona, an inverted rate structure was credited with reducing public demand from about 200 gallons per capita per day (gpcd) to 140–160 gpcd (Maddock and Hines, 1995).

The manner in which water utilities are structured is probably the most important factor, which complicates the study of price elasticity. For instance, some customers who own homes or who pay water bills react more or less to the price change, whereas those who rent apartments or who do not pay water bills are indifferent. Furthermore, the water necessities for residential, commercial, and industrial purposes are not equally important. Because of these factors, different researchers had to study demand elasticity by categorizing water distribution systems for industrial, commercial, and residential uses. The demand patterns under these categories are not uniform.

One of the most comprehensive studies on price elasticity of water demand, done by Schneider and Whitlatch (1991) for six user categories (residential, commercial, industrial, government, school, and total metered), showed different results for these categories. Residential water use is further complicated by different factors: many residents who rent housing do not pay for water and thus are indifferent to demand regulations; the patterns for indoor and outdoor water demand differ quite significantly and hence necessitate different approaches to demand analysis. The climatic conditions of a given area and the time of the year are basic reasons why apparently different elasticity values are reported for the eastern and the western United States and for winter and summer uses.

From the studies enumerated so far, however, a general conclusion can be reached: that demand is elastic to price increase. Almost all research has reinforced this hypothesis. However, differences exist between the elasticity values calculated for different geographic locations. For instance, Howe (1982) obtained values of −0.57 and −0.43 for the eastern and the western United States, respectively. On the other hand, no clear consistency exists in how elasticity is calculated: some use average price, some use marginal price, and some include an intramarginal rate structure.

EXAMPLE 11.5.1   If the price elasticity for water is −0.5, interpret the meaning of this.

SOLUTION   A 1.0 percent increase in the price of water results in a 0.5 percent decrease in water demand, with all other factors remaining constant. This pertains only to the situation where price-quantities data pairs are close to each other, or where a smooth demand function (section 11.5.2) can be fitted statistically to the known data (see Kindler and Russell, 1984).

11.5.2   Demand Models

A *demand function* (Figure 11.5.1) relates the quantity of a commodity that a consumer is willing to purchase to price, income, and other variables. Demand curves are generally negatively sloped; that is, the lower the price, the greater the quantity demanded. Typically, demand is expressed as a function of different variables. A general form of *demand models* is

$$d = f(x_1, x_2, ..., x_k) + \epsilon \tag{11.5.4}$$

where f is the function of variables x_1, x_2, ..., x_k, and ϵ is a random error (random variable) describing the joint effect on q of all the factors not explicitly considered by the variables.

Several explicit linear, semilogarithmic, and logarithmic models have been developed. Agthe and Billings (1980), for example, gave the following water demand function for Tucson, Arizona:

$$\ln(d) = -7.36 - 0.267\ln(P) + 1.61\ln(I) - 0.123\ln(DIF) + 0.0897\ln(W) \tag{11.5.5}$$

In the above equation, d is the monthly water consumption of the average household in 100 ft³; P is the marginal price facing the average household in cents per 100 ft³; DIF is the difference between the actual water and sewer use bill minus what would have been paid if all water were sold at the marginal rate ($); I is the personal income per household ($/month); and W is the evapotranspiration minus rainfall (inches). The above equation implicitly relates demand to the hydrologic index W. The positive coefficient of W shows that demand increases exponentially with W, which indirectly indicates increases in demand with the dryness of weather conditions.

Equation (11.5.5) may be rearranged as

$$d = 0.00006362 P^{-0.267} I^{1.61} (DIF)^{-0.123} W^{0.0897} \tag{11.5.6}$$

or, in more general terms,

$$d = a' P^{b'} I^{c'} (DIF)^{d'} W^{e'} \tag{11.5.7}$$

where a', b', c', d', and e' are constants. The price elasticity of demand for equation (11.5.6) is -0.267. Therefore, changing the price while keeping the other variables constant results in different average demand values $\bar{d}_{P_i}$. Again, varying W while keeping the other variables constant gives a general relation of the average demand associated with the return period T.

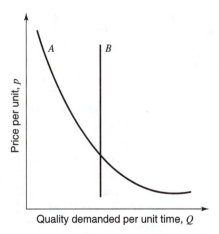

Figure 11.5.1 Demand functions. Curve A is a typical demand curve and curve B is the demand curve for a requirement for which the same quantity is demanded no matter what the price.

EXAMPLE 11.5.2

Foster and Beattie (1979) developed a demand function for urban residential water demand for application in the United States:

$$\ln Q = -1.3895 - 0.1278 P_{av} + 0.4619 \ln(I) - 0.1699 \ln(F) + 0.4345 \ln(H)$$

where Q is the quantity of water demanded at a meter (1000 cubic feet per year); P_{av} is the average water price (dollars per 1000 ft^3); I is the median household income (dollars per year); F is the precipitation in inches during the growing season; and H is the average number of residents per meter. Determine the price elasticity of demand calculated at the mean price, $3.67/1,000 gallons.

SOLUTION

The demand function can be expressed as

$$Q = e^{-1.3895 - 0.1278 P_{av}} I^{0.4619} F^{-0.1699} H^{0.4345}$$

and the derivative of Q with respect to price is

$$dQ/dP = I^{0.4619} F^{-0.1699} H^{0.4345} e^{-1.3895 - 0.1278 P_{av}} (-0.1278) = -0.1278 Q$$

Price elasticity of demand is then

$$\epsilon = \frac{P}{Q} \frac{dQ}{dP} = \frac{P}{Q}(-0.1278 Q) = -0.1278(3.67) = -0.469$$

The price elasticity of -0.469 indicates that for a 1.0 percent increase in price, a 0.469 percent decrease in quantity demanded would be expected; conversely, a 1.0 percent decrease in price would produce a 0.469 percent increase in quantity demanded.

11.6   DROUGHT MANAGEMENT

11.6.1   Drought Management Options

Droughts are generally associated with sustained periods of significantly lower soil moisture levels and water supply than the normal levels around which the local environment and society have stabilized (Rasmusson et al., 1993). Droughts have been defined from several viewpoints, including: (1) meteorological definition; (2) agricultural definition; (3) hydrologic definition; and (4) economic definition. *Hydrologic drought* typically refers to periods of below-normal streamflow and/or depleted reservoir storage. *Economic drought* concerns the economic areas of human activity affected by drought as a result of physical processes. *Agricultural drought* typically refers to periods where the soil moisture is inadequate to initiate and sustain crop growth. *Meteorological drought* is the time when the actual cumulative moisture supply falls short.

Droughts continue to rate as one of the most severe weather-induced problems around the world. Global attention to natural hazard reduction includes drought as one of the major hazards.

Shortage of water supply during drought periods is such a significant factor for the general welfare that its effect cannot easily be overstated. Domestic water supply shortages during these periods in particular have been crucial in some cases and as a result various measures have been initiated by different water supply agencies to reduce water demand during such periods. These measures, which may be considered as semi-empirical to empirical, include water metering, leak detection and repair, rate structures, regulations on use, educational programs, drought contingency planning, water recycling and reuse, pressure reduction, and so on. Such efforts are collectively termed *water conservation*, although there has not been a uniform definition among authors.

Experience from past droughts have shown that the action of water managers can greatly influence the magnitude of the monetary and nonmonetary losses from drought. A variety of drought management options have been undertaken in response to anticipated shortages of water, which can be categorized as (Dziegielewski et al., 1986): (1) demand reduction measures; (2) efficiency improvements in water supply and distribution system; and (3) emergency water supplies. A topology of drought management options is given in Table 11.6.1.

Table 11.6.1 A Topology of Drought Management Options

I.   Demand Reduction Measures
    1. Public education campaign coupled with appeals for voluntary conservation
    2. Free distribution and/or installation of particular water-saving devices:
        2.1 Low-flow showerheads
        2.2 Shower flow restrictors
        2.3 Toilet dams
        2.4 Displacement devices
        2.5 Pressure-reducing valves
    3. Restrictions on nonessential uses:
        3.1 Filling of swimming pools
        3.2 Car washing
        3.3 Lawn sprinkling
        3.4 Pavement hosing
        3.5 Water-cooled air conditioning without recirculation
        3.6 Street flushing
        3.7 Public fountains
        3.8 Park irrigation
        3.9 Irrigation of golf courses
    4. Prohibition of selected commercial and institutional uses:
        4.1 Car washes
        4.2 School showers
    5. Drought emergency pricing:
        5.1 Drought surcharge on total water bills
        5.2 Summer use charge
        5.3 Excess use charge
        5.4 Drought rate (special design)
    6. Rationing programs:
        6.1 Per capita allocation of residential use
        6.2 Per household allocation of residential use
        6.3 Prior use allocation of residential use
        6.4 Percent reduction of commercial and institutional use
        6.5 Percent reduction of industrial use
        6.6 Complete closedown of industries and commercial establishments with heavy uses of water
II. System Improvements
    1. Raw water sources
    2. Water treatment plant
    3. Distribution system:
        3.1 Reduction of system pressure to minimum possible levels
        3.2 Implementation of a leak detection and repair program
        3.3 Discontinuing hydrant and main flushing
III. Emergency Water Supplies
    1. Inter-district transfers:
        1.1 Emergency interconnections
        1.2 Importation of water by trucks
        1.3 Importation of water by railroad cars
    2. Cross-purpose diversions:
        2.1 Reduction of reservoir releases for hydropower production
        2.2 Reduction of reservoir releases for flood control
        2.3 Diversion of water from recreation water bodies
        2.4 Relaxation of minimum steamflow requirements
    3. Auxiliary emergency sources:
        3.1 Utilization of untapped creeks, ponds, and quarries

Table 11.6.1 A Topology of Drought Management Options (*continued*)

3.2 Utilization of dead reservoir storage
3.3 Construction of a temporary pipeline to an abundant source of
     water (major river)
3.4 Reactivation of abandoned wells
3.5 Drilling of new wells
3.6 Cloud seeding

Source: Dziegielewski (1986).

Not only is water conservation necessary during drought periods, but its economic merits are also important to consider. In the United States, federal mandates urge that opportunities for water conservation be included as a part of the economic evaluation of proposed water supply projects (Griffin and Stoll, 1983). Water conservation during drought periods, however, requires important attention because our demand for water may exceed the available resource in the demand environment. Conservation may be achieved through different activities. According to the U.S. Water Resources Council (1979), these activities include but are not limited to:

1. reducing the level and/or altering the time pattern of demand by metering, leak detection and repair, rate structure changes, regulations on use (e.g., plumbing codes), education programs, drought contingency planning;

2. modifying management of existing water development and supplies by recycling, reuse, and pressure reduction; and

3. increasing upstream watershed management and conjunctive use of ground and surface water (Griffin and Stoll, 1983).

The effort to conserve water started out with metering. Both domestic and sprinkling demands reduced significantly as a result of the introduction of water meters (Hanke, 1970). Grunewald et al. (1976) stated: "Traditionally, water utility managers have adjusted water quantity (rather) than prices as changes in demand occurred."

Increasing the price of domestic water supply has been a focus of many studies. These studies were conducted to analyze the effect of urban water pricing and how it contributes to water conservation during a drought period (Agthe and Billings, 1980; Moncur, 1989). However, variations have been observed in the approaches followed. According to Jordan (1994), *water pricing* is an effective way of conserving water, compared to the other measures mentioned above. An increase in the price of water contributes to water conservation because of the fact that customers have limited money. For every percent increase in the price, there is some decrease in the demand, as explained by the price elasticity (defined in Section 1.5.2). A significant number of studies have been undertaken in different regions to determine price elasticity associated with pricing. The following section discusses price elasticity.

Figure 11.6.1 shows three of the municipal pricing options commonly used: (1) the *uniform block rate* (average cost pricing) with a service charge; (2) an *increasing block* (tier) *rate* with minimum allowance; and (3) an *increasing seasonal block rate* with minimum allowance. The increasing block pricing schemes are increasingly common. They are typically used by water agencies facing water shortages or limited supply capacity. Increasing block rates are used to encourage water conservation and more efficient use of water.

11.6.2 Drought Severity

Every natural phenomenon with which detrimental effects to human beings and their environment are associated need our keen attention on how and when it occurs. Unfortunately, the degree of some such phenomena, including drought, is difficult to determine as accurately as desirable before they occur. A study by the National Research Council (1986) indicated that there is not a

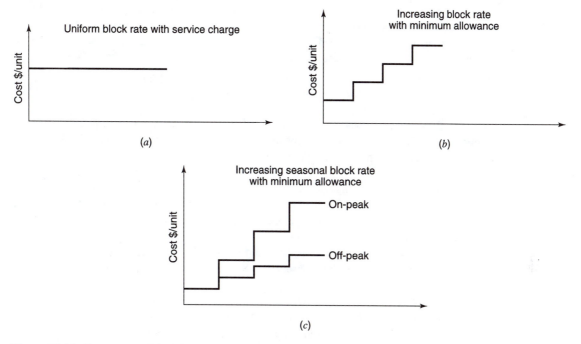

Figure 11.6.1 Common municipal pricing options. (*a*) Quantity used plus monthly service charge; (*b*) Quantity used plus monthly service charge for minimum amount. (*c*) Quantity used plus monthly service charge for minimum amount.

firm rationale or explanation of the drought mechanism. It adds that, though empirical relations have been documented so far, why and when these relations trigger the occurrence of significant drought is not understood.

No single definite method is used as a drought severity indicator. Nonetheless, there are some which are being used in different fields. According to Wilhite (1993), the simplest drought index in widespread use is the percent of normal precipitation. This, indeed, is a good approach to infer the status of the available supply. However, it does not imply an obvious forecast to enable a risk-management body to be prepared for a forthcoming drought period.

Several drought-severity indices have been used so far. Some of them are used to assess past drought event's severity, and a few others are used for forecasting. The *Palmer Drought Severity Index (PDSI)* is an example of the former category, while the *Surface Water Supply Index (SWSI)* and the *Southern Oscillation Index (SOI)* are examples of the latter.

Palmer (1965) expressed the severity of a drought event by the following equation (Steila, 1972; Puckett, 1981):

$$PDSI_i = 0.897 PDSI_{i-1} + \tfrac{1}{3} Z_i \tag{11.6.1}$$

where *PDSI* is the Palmer Drought Severity Index and *Z* is an adjustment to soil moisture for carry-over from one month to the next, expressed as

$$Z_i = k_j \left[PPT_i - \left(\alpha_j PE_i + \beta_j G_i + \gamma_j R_i - \delta_j L_i \right) \right] \tag{11.6.2}$$

in which the subscript *j* represents one of the calendar months and *i* is a particular month in a series of months. PPT_i is the precipitation, PE_i is the potential evapotranspiration, G_i is the soil moisture recharge, R_i is the surface runoff (excess precipitation), and L_i is the soil moisture loss for month *i*. The coefficients α_j, β_j, γ_j, and δ_j are the ratios for long-term averages of actual to potential magnitudes for *E*, *G*, *R*, and *L* based on a standard 30-year climatic period.

The Surface Water Supply Index (SWSI) gives a forecast of a drought event. It was introduced as a better indicator of water availability in the western United States than the Palmer Drought

Index (Garen, 1993). It is a weighted index that generally expresses the potential availability of the forthcoming season's water supply (U.S. Soil Conservation Service, 1988). It is formulated as a re-scaled weighted of nonexceedance probabilities of four hydrologic components: snowpack, precipitation, streamflow, and reservoir storage (Garen, 1993):

$$SWSI = \frac{\alpha p_{snow} + \beta p_{prec} + \gamma p_{strm} + \omega p_{resv} - 50}{12} \qquad (11.6.3)$$

where α, β, γ, and ω are weights for each hydrologic component and add up to unity; P_i is the probability of nonexceedance (in percent) for component i; and the subscripts *snow*, *prec*, *strm*, and *resv* stand for the snowpack, precipitation, streamflow, and reservoir storage hydrologic components, respectively. This index has a numerical value for a given basin that varies between -4.17 to $+4.17$. The following are the ranges for the index for practical purposes: $+2$ or above, $-2 - +2$, $-3 - -2$, $-4 - -3$, and -4 or below. These ranges are associated with the qualitative expressions of abundant water supply, near-normal, moderate drought, severe drought, and extreme drought conditions, respectively.

The *SWSI* has been in use to forecast different basins' monthly surface water supply availabilities (see, for example, the U.S. Soil Conservation Service, 1988). In fact, it gives a forecast of both wet and dry (drought) months.

The Southern Oscillation Index (SOI) is also used to forecast drought based on the Southern Oscillation, a phenomenon that affects large-scale atmospheric and oceanographic features of the tropical Pacific Ocean (Kawamura et al., 1998). The oscillation is characterized by either sea-surface temperatures or differences in barometric pressures. The SOI's best-known extremes are El Niño events (Kawamura et al., 1998). Wilhite (1993) reports that several scientists agree that it has been possible to forecast drought for up to six months in Australia by using the SOI.

11.6.3 Economic Aspects of Water Shortage

The general trend of the average demand with the return period may be shown by the demand curve in Figure 11.6.2. Demand increases with the return period of the drought severity because the more severe the drought, the more the customers are prompted to use more water. Different demand curves are illustrated in Figure 11.6.3 for different price levels. As shown in this figure, the higher the price, the lower the demand for given hydrologic conditions.

In equation (11.5.7), the demand d is related to the hydrologic index W, which is related to the return period. The available supply (flow) q is also related to the return period (Hudson and Hazen, 1964). Thus the general relationships between demand and return period and supply and return period shown in Figure 11.6.2 are based on these trends.

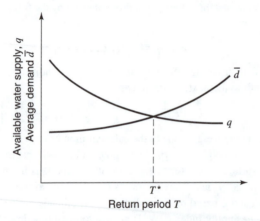

Figure 11.6.2 Water supply availability and average demand as related to the return period T.

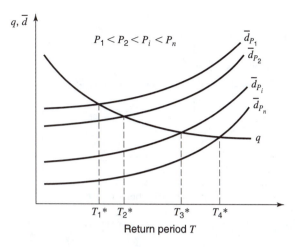

Figure 11.6.3 Water supply and average demands for different price values as related to T.

Shortage of water supply during drought periods results in different types of losses in the economy, including, but not limited to, agricultural, commercial, and industrial. In agriculture, lack of water supply results in crop failures, in commerce it may result in a recession of business and in industry it may result in underproduction of commodities. The loss in each production or service sector depends on the purpose of the sector. For instance, the economic impact of drought on agriculture depends on the crop type, etc. (Easterling, 1993). There is no single common way of assessing the economic impact of drought on any one of the sectors. Evaluating and comparing what actually happens during a drought period with what would have happened had there been no drought may be one way of assessing the effects of drought. Dixon et al. (1996) adopted the concept of *willingness-to-pay* to value changes in well-being. They define willingness-to-pay as the maximum individuals would have been willing to pay to avoid the drought management strategies imposed by water agencies.

On the other hand, since water is supplied during a drought period at a greater price, it can be viewed as a revenue generator. Therefore, when the demand exceeds the available supply, the revenue collected by the water supply agency is less than what could have been collected had there been more supply than that actually available. In other words, if the demand exceeds the supply, the problem is not limited to lack of water; there will also be economic loss, since the customers would pay for more supply if there were enough. Depending on the risk level, it is possible to decide whether supply augmentation is necessary or the pressure for more demand could be tolerated with the available supply.

Some water shortage relief efforts can be undertaken so that emergency water supplies may be made available to users. This can be implemented by well drilling, trucking in potable supplies, or transporting water through small-diameter emergency water lines. In such cases, it may be required that the emergency supply construction costs be paid by the users (Dziegielewski et al., 1991). The estimation of the expected financial loss can be used to determine and inform the users of its extent and advise them of the necessity, if any, of paying for the emergency supply construction costs.

If the option for emergency supply construction is justified, then the design needs to take into consideration the possibilities of optimization. The construction can be designed such that the financial risk and the cost of construction are optimal. Figure 11.6.4 illustrates this optimization process.

The economic loss (damage) can be calculated as shown below and the cost of emergency construction must be determined from the physical conditions at the disposal of the water supply agency. The damage that would result if a certain drought event occurred is one of the main

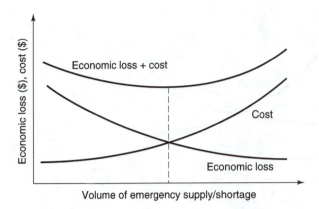

Figure 11.6.4 Optimization for emergency water supply construction.

decision factors. Since the time of occurrence of the drought event that causes the damage is difficult to determine, only the expected value is assessed by associating its magnitude with its probability of occurrence.

The *expected annual damage cost D_T* for the event $x > x_T$ is (Ejeta and Mays, 1998)

$$D_T = \int_{x_T}^{\infty} D(x) f(x)\, dx \qquad (11.6.4)$$

where $f(x)\, dx$ is the probability that an event of magnitude x will occur in any given year and $D(x)$ is the damage cost that would result from that event. The event x in this case can be taken as the demand and x_T can be the available supply during a drought event of return period T.

Breaking down the expected damage cost into intervals, we get

$$\Delta D_i = \int_{x_{i-1}}^{x_i} D(x) f(x)\, dx \qquad (11.6.5)$$

from which the finite-difference approximation is obtained

$$\begin{aligned} \Delta D_i &= \left[\frac{D(x_{i-1}) + D(x_i)}{2} \right] \int_{x_{i-1}}^{x_i} f(x)\, dx \\ &= \left[\frac{D(x_{i-1}) + D(x_i)}{2} \right] \left[p(x \geq x_{i-1}) - p(x \geq x_i) \right] \end{aligned} \qquad (11.6.6)$$

Thus the annual damage cost for a return period T is given as

$$D_T = \sum_{i=1}^{\infty} \left[\frac{D(x_{i-1}) + D(x_i)}{2} \right] \left[p(x \geq x_{i-1}) - p(x \geq x_i) \right] \qquad (11.6.7)$$

To determine the annual expected damage in the above equation, the damage that results from drought events of different severity levels must be quantified.

The magnitude of the drought (in monetary units) may be obtained by estimating the volume of water shortage that would result from that drought. In other words, not having the water results in some financial loss to the water supply customer. The resulting financial loss to the customer from a certain drought event is thus considered as the damage from that drought event.

As shown in Figure 11.6.5, after the critical return period T^* the divergence between the demand and the supply increases with the return period. Expressing the demand and the supply as a function of return period T of drought events enables one to estimate the annual expected water supply shortage volume, as given by the following:

$$S_V = \int_{T^*}^{T} \left[d(T) - q(T) \right] dt \qquad (11.6.8)$$

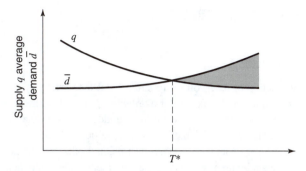

Figure 11.6.5 Demand and supply, showing water shortage volume when demand exceeds supply.

The *shortage volume* S_V is illustrated by the shaded area in Figure 11.6.5. The shortage volume for a drought event of a higher return period above the critical one results in higher shortage volume and consequently a higher associated damage. The relationship between the shortage volume and the associated damage generally depends on several factors, including water use category—residential, industrial, commercial, agricultural, and so on. To use the procedure presented herein for assessing the damage that results from certain water-shortage volume, the damage given by equation (11.6.8) must be developed for a specific user category.

EXAMPLE 11.6.1

The hypothetical annual damage cost that would occur if a flood of return period T were to occur is given, in \$1000, as

$$D(T) = T^2 - T = \frac{1}{p^2} - \frac{1}{p}$$

where $p = 1/T$, that is, annual exceedance probability. Determine

(a) $D(T)$ for $T = 1, 2, 5, 10, 20, 50, 100, 200, 500$, and 1000;

(b) The damages between the return period intervals given in part (*a*); and

(c) The annual expected damage cost for a structure designed for each of the return periods given in (*a*).

SOLUTION

The solutions to this problem are given in the table below.

Column 3 gives $D(T)$ for T = 1, 2, 5, 10, 20, 50, 100, 200, 500, and 1000. Column 6 gives the damages between the return periods indicated and column 7 gives the total expected damage cost for a structure designed for these return periods.

T	p	$D(T)$	$p(T_i) - p(T_{i-1})$	$[(D(T_{i-1}) + D(T_i))/2]$	(4) * (5)	Total expected damage
(1)	(2)	(3)	(4)	(5)	(6)	(7)
1	1.000	0				0.000
2	0.500	2	0.500	1	0.5	0.500
5	0.200	20	0.300	11	3.3	3.800
10	0.100	90	0.100	55	5.5	9.300
20	0.050	380	0.050	235	11.75	21.050
40	0.025	1560	0.025	970	24.25	45.300
100	0.010	9900	0.015	5730	85.95	131.250
200	0.005	39800	0.005	24850	124.25	255.500
500	0.002	249500	0.003	144650	433.95	689.450
1000	0.001	999000	0.001	624250	624.25	1313.700

11.7 ANALYSIS OF SURFACE WATER SUPPLY

11.7.1 Surface-Water Reservoir Systems

The primary function of reservoirs is to smooth out the variability of surface-water flow through control and regulation and make water available when and where it is needed. The use of reservoirs for temporary storage would result in an undesirable increase in water loss through seepage and evaporation. However, the benefits that can be derived through regulating the flow for water supplies, hydropower generation, irrigation uses, and other activities can offset such losses. The net benefit associated with any reservoir development project is dependent on the size and operation of the reservoir as well as the various purposes of the project.

Reservoir system operations may be grouped into two general operation purposes: conservation and flood control. Conservation purposes include water supply, low-flow augmentation for water quality, recreation, navigation, irrigation and hydroelectric power, and any other purpose for which water is saved for later release. Flood control is simply the retention or detention of water during flood events for the purpose of reducing downstream flooding. This chapter focuses only on surface-water reservoir system for conservation. The flood control aspect of reservoir system operation is discussed in section 14.6.

Generally, the total reservoir storage space in a multipurpose reservoir has three major parts (see Figure 11.7.1): (1) the *dead storage zone*, mainly required for sediment collection, recreation, or hydropower generation; (2) the *active storage*, used for conservation purposes, including water supplies, irrigation, navigation, etc.; (3) the *flood control storage* reserved for storage of excessive flood volume to reduce potential downstream flood damage. In general, these storage spaces can be determined separately and combined later to arrive at a total storage volume for the reservoir.

11.7.2 Storage—Firm Yield Analysis for Water Supply

The determination of storage-yield relationships for a reservoir project is one of the basic hydrologic analyses associated with the design of reservoirs. Two basic problems in *storage-yield studies* (U.S. Army Corps of Engineers, 1977) are: (1) determination of storage required to supply a specified yield; and (2) determination of yield for a given amount of storage. The former is usually encountered in the planning and early design phases of a water resources development study, while the latter often occurs in the final design phases or in the reevaluation of an existing project for a more comprehensive analysis. Other objectives of storage-yield analysis include: (1) the determination of complementary or competitive aspects of multiple-project development; (2) determination of complementary or competitive aspects of multiple-purpose development in a single project; and (3) analysis of alternative operation rules for a project or group of projects.

The procedures used to develop a storage-yield relationship include (U.S. Army Corps of Engineers, 1977): (1) simplified analysis; and (2) detailed sequential analysis. The simplified techniques are satisfactory when the study objectives are limited to preliminary or feasibility studies. Detailed methods that include both simulation and optimization analysis are usually required when the study objectives advance to the design phase. The objective of simplified methods is to obtain a reasonably good estimate of the results, which can be further improved by a detailed sequential analysis. Factors affecting the selection of method for analysis are: (1) study requirements; (2) degrees of accuracy required; and (3) the basic data required and available.

Firm yield is defined as the largest quantity of flow or flow rate that is dependable at the given site along the stream at all times. More specifically, Chow et al. (1988) define the firm yield of a reservoir as the mean annual withdrawal rate that would lower the reservoir to its minimum allowable level just once during the critical drought of record. The most commonly used method to determine the firm yield of an unregulated river is to construct a *flow-duration curve*, which is a graph of the discharge as a function of the percent of time that flow is equaled or exceeded.

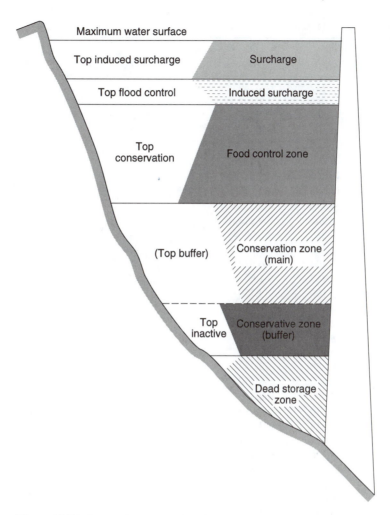

Figure 11.7.1 Reservoir storage allocation zones (from U.S. Army Corps of Engineers (1977)).

The flow-duration curve can be developed for a given location on the stream by arranging the observed flow rates in order of descending magnitude. From this, the percentage of time for each flow magnitude to be equaled can be computed. Then, this percentage of time of exceedance is plotted against the flow magnitude to define the flow-duration relationship. The firm yield is the flow magnitude that is equaled or exceeded 100 percent of the time for a historical sequence of flows. Flow-duration curves are used in determining of the water supply potential for the planning and design of water resource projects, in particular hydropower plants.

EXAMPLE 11.7.1   Use the monthly flow data for the Little Weiser River given in Table 11.7.1 to develop the flow-duration curve and determine the firm yield for this site.

Table 11.7.1 Monthly Flows in the Little Weiser River Near Indian Valley, Idaho, for Water Years 1966–1970

t	Year	Month	Flow (AF)	ΣQF_t (AF)	t	Year	Month	Flow (AF)	ΣQF_t (AF)
1	1965	10	742	742	31		4	6,720	129,105
2		11	1,060	1,802	32		5	13,290	142,395
3		12	1,000	2,802	33		6	9,290	151,685
4	1966	1	1,500	3,302	34		7	1,540	153,225
5		2	1,080	4,382	35		8	915	154,140
6		3	6,460	10,842	36		9	506	154,646
7		4	10,000	20,842	37		10	886	155,532
8		5	13,080	33,922	38		11	3,040	158,572
9		6	4,910	38,832	39		12	2,990	161,562
10		7	981	39,813	40	1969	1	8,170	169,732
11		8	283	40,096	41		2	2,800	172,532
12		9	322	40,398	42		3	4,590	177,122
13		10	404	40,822	43		4	21,960	199,082
14		11	787	41,609	44		5	30,790	229,872
15		12	2,100	43,709	45		6	14,320	244,192
16	1967	1	4,410	48,119	46		7	2,370	246,562
17		2	2,750	50,869	47		8	709	247,271
18		3	3,370	54,239	48		9	528	247,799
19		4	5,170	59,409	49		10	859	248,658
20		5	19,680	79,089	50		11	779	249,437
21		6	19,630	98,719	51		12	1,250	250,687
22		7	3,590	102,309	52	1970	1	11,750	262,437
23		8	710	103,019	53		2	5,410	267,849
24		9	518	103,537	54		3	5,560	273,407
25		10	924	104,461	55		4	5,610	279,017
26		11	1,020	105,481	56		5	24,330	303,347
27		12	874	106,355	57		6	32,870	336,217
28	1968	1	1,020	107,375	58		7	7,280	343,497
29		2	8,640	116,015	59		8	1,150	344,647
30		3	6,370	122,385	60		9	916	345,563

SOLUTION

To develop a flow-duration curve, first arrange the flows in descending order with the largest flow ordered as 1 and the smallest as N, where N is the total number of months in the record (Table 11.7.2). The percentage of time flow magnitude is equaled or exceeded is computed by 100 m/N where m is the rank of flow magnitude shown in column 1. Using columns 2 and 3, one can construct the flow-duration curve as shown in Figure 11.7.2. The firm yield of the river at this particular site is 283 acre-ft per month, which is equivalent to an average flow rate of 4.76 cfs. In general, a firm yield based upon average daily flows is less than a firm yield based upon average monthly flows: the shorter the time period, the more erratic the flow-duration curve appears.

Table 11.7.2 Ranking of Flows for the Little Weiser River

Rank (m)	Flow (AF)	% Time Equaled and Exceeded	Rank (m)	Flow (AF)	% Time Equaled and Exceeded
1	32,870	1.7	31	2,750	51.7
2	30,790	3.3	32	2,370	53.3
3	24,330	5.0	33	2,100	55.0
4	21,960	6.7	34	1,540	56.7
5	19,680	8.3	35	1,500	58.3
6	19,630	10.0	36	1,250	60.0
7	14,320	11.7	37	1,150	61.7
8	13,290	13.3	38	1,080	63.3

Table 11.7.2   Ranking of Flows for the Little Weiser River (*continued*)

Rank (m)	Flow (AF)	% Time Equaled and Exceeded	Rank (m)	Flow (AF)	% Time Equaled and Exceeded
9	13,080	15.0	39	1,060	65.0
10	11,750	16.7	40	1,020	66.7
11	10,000	18.3	41	1,020	68.3
12	9,290	20.0	42	1,000	70.0
13	8,640	21.7	43	981	71.7
14	8,170	23.3	44	924	73.3
15	7,280	25.0	45	916	75.0
16	6,720	26.7	46	915	76.7
17	6,460	28.3	47	886	78.3
18	6,370	30.0	48	874	80.0
19	5,610	31.7	49	859	81.7
20	5,560	33.3	50	787	83.3
21	5,410	35.0	51	779	85.0
22	5,170	36.7	52	742	86.7
23	4,910	38.3	53	710	88.3
24	4,590	40.0	54	709	90.0
25	4,410	41.7	55	528	91.7
26	3,590	43.3	56	518	93.3
27	3,370	45.0	57	506	95.0
28	3,040	46.7	58	404	96.7
29	2,990	48.3	59	322	98.3
30	2,800	50.0	60	283	100.0

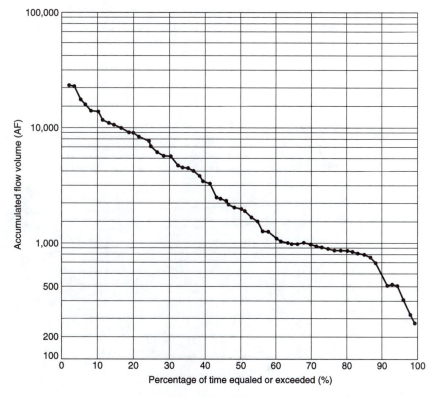

Figure 11.7.2 Flow-duration curve for the Little Weiser River near Indian Valley, Idaho (1966–1970).

Mass-Curve Analysis.   To increase the firm yield of an unregulated river, surface impoundment facilities are constructed to regulate the river. Two methods, *mass-curve analysis* and *sequent-peak analysis*, can be used to develop storage-yield relationships for specific locations along a river. A *mass curve* is a plot of the cumulative flow volumes as a function of time. The mass-curve was first developed by Ripple in 1883. The method uses historical or synthetic stream flow sequences over a time interval [0,*T*]. Implicitly the analysis assumes that the time interval includes the *critical period*, which is the time period over which the flows have reached a minimum, causing the greatest drawdown of a reservoir. Mass-curve analysis can be implemented using graphical procedures to determine the critical period and firm yield.

A critical period (Figure 11.7.3) always begins at the end of a preceding high-flow period that leaves the reservoir full. A critical period ends when the reservoir has refilled after the drought period. The *critical drawdown period* (Figure 11.7.3) begins when the reservoir is full and ends when the reservoir is empty. The mass curve for the critical period is presented in Figure 11.7.4.

The *Ripple method* is applicable when the release is constant. In cases where the releases vary, however, it is easier to compute the difference between the cumulative reservoir in flows and cumulative reservoir releases. The required active storage volume is the maximum difference. Of course, this alternative approach can be applied to a constant release case and can be implemented graphically. The procedures can be applied repeatedly by varying releases to derive the storage-yield curve at a given reservoir site.

The assumption implicitly built into the mass-curve method is that the total release over the time interval of analysis does not exceed the total reservoir inflows. In mass-curve analysis, the critical sequence of flows might occur at the end of the streamflow record. When this occurs, the period of analysis is doubled from [0,*T*] to [0, 2*T*] with the inflow sequence repeating itself in the second period, and the analysis proceeds. If the required total release exceeds the total historical inflow over the recorded period, the mass-curve analysis does not yield a finite reservoir capacity.

EXAMPLE 11.7.2

Using the monthly data in Table 11.7.1 for the Little Weiser River, construct the cumulative mass curve over the five-year period and determine the required active storage capacity to produce a firm yield of 2000 acre-ft/month.

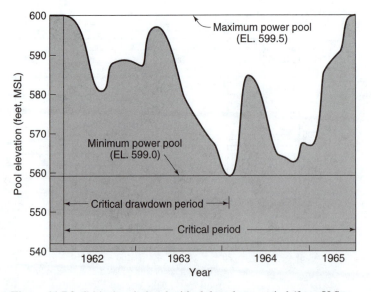

Figure 11.7.3 Critical period and critical drawdown period (from U.S. Army Corps of Engineers (1985)).

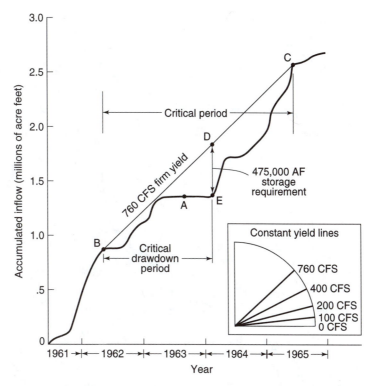

Figure 11.7.4 Mass curve and constant yield lines (from U.S. Army Corps of Engineers (1985)).

SOLUTION

The cumulative monthly flows in column 5 of Table 11.7.1 are obtained by successively adding monthly flow in column 4. Cumulative monthly flow is plotted as a function of time to develop the mass curve shown in Figure 11.7.5.

Points $A_1, A_2, A_3,$ and A_4 define the beginning of low-flow periods and points $B_1, B_2, B_3,$ and B_4 define the end of the low-flow periods. A line representing the constant reservoir release of 2000 acre-ft per month is placed tangent to the beginning of each low-flow period, i.e., $A_1, A_2, A_3,$ and A_4. Then the locations in each low-flow period with the largest vertical distance between the mass curve at points $B_1, B_2, B_3,$ and B_4 and the tangent lines are identified. The largest of the four vertical distances at B_1 is 7200 acre-ft, which is the required active storage.

EXAMPLE 11.7.3

For the critical period shown by the mass curve in Figure 11.7.4, determine if the unregulated inflow can be increased to a firm flow of 760 cfs. (Adapted from U.S. Army Corps of Engineers, 1985.)

SOLUTION

The 760-cfs firm yield curve is applied to a positive point of tangency on the mass curve (point B) and is extended to the point where it again intersects the mass curve (point C). Period B–C is a complete storage draft-refill cycle (which corresponds to the critical period on Figure 11.7.3). The length of the vertical coordinate between the 760-cfs yield curve and the mass diagram represents the amount of storage drafted from the reservoir at any point in time, and the point where this ordinate is at its maximum length (point D) represents the total amount of reservoir storage required to maintain a firm flow of 760 cfs during this particular flow period.

Period B–C is the most adverse sequence of flows in the period of record; a volume of 475,000 acre-ft is required to assure a firm yield of 760 cfs at the project.

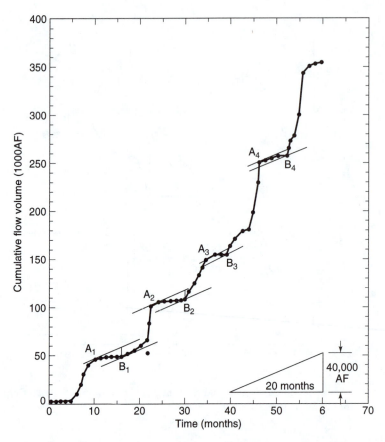

Figure 11.7.5 Cumulative mass curve for Little Weiser River, and application of mass curve analysis to compute firm yield (Example 11.7.2).

Sequent-Peak Analysis. The *sequent-peak* method computes the cumulative sum of inflows QF_t minus the reservoir releases R_t, that is, $\Sigma_t u_t = \Sigma_t (QF_t - R_t)$, for all time periods t over the time interval of analysis $[0,T]$. To solve this problem graphically, the cumulative sum of u_t is plotted against t. The required storage for the interval is the vertical difference between the first peak and the low point before the sequent peak. The method has the same two assumptions as the mass-curve analysis.

Algebraically, the sequent-peak method can be implemented using the following equation recursively:

$$K_t = \begin{cases} R_t - QF_t + K_{t-1} & \text{if positive} \\ 0 & \text{otherwise} \end{cases} \qquad (11.7.1)$$

where K_t is the required storage capacity at the beginning of period t. The initial value of K_t at $t = 0$ is set to zero. In general, the method using equation (11.7.1) is applied repeatedly, up to twice the length of the recorded time span, to take into account the possibility that the critical flow sequence occurs at the end of the streamflow record. The maximum value of the calculated K_t, is the required active reservoir storage capacity for the flow sequence and the considered releases.

In reality, the hydrological components, precipitation, evaporation, and seepage, in addition to the streamflow inflows, determine the storage volume in a reservoir.

Precipitation that falls directly on the reservoir surface contributes to the storage volume. Evaporation and seepage result in losses to the available water in active reservoir storage. Depending on the location and the geological conditions of the reservoir site, the total losses from evaporation and seepage are an important influence on the mass balance of the reservoir system. Neglecting such factors would result in serious overestimation of the water availability and, consequently, underestimation of the requited reservoir storage capacity to support the desired releases. In arid and semiarid areas, such as the southwestern United States, the quantity of water loss through evaporation may be large enough significantly to lessen the positive effects of impounding the water.

The amount of water loss through evaporation and seepage is a function of storage, impounding surface area, and geological and meteorological factors. The net inflows to a reservoir must be adjusted and used in the mass-curve and the sequent-peak methods. The adjusted reservoir inflow QF_{ta} in period t can be estimated as

$$QF_{ta} = QF_t + PP_t - EV_t - SP_t \qquad (11.7.2)$$

in which PP_t is the precipitation amount on the reservoir surface, EV_t is the evaporation, and SP_t is the seepage loss during period t. The elements on the right hand side of equation (11.7.2) depend on the storage and reservoir surface area during time period t, which are, in turn, a function of those hydrological components.

Mass-curve analysis and the sequent-peak method are used in the planning stages to determine the capacity of a single-surface reservoir for a specified release pattern. It enables engineers to develop storage-yield curves for the reservoirs under consideration. However, the ability of the two methods to analyze a reservoir system involving several reservoirs is severely restricted. Furthermore, active storage capacity of a reservoir depends on various hydrologic elements whose contributions to the mass balance, in turn, are a function of unknown reservoir storage. Such an implicit relationship cannot be accounted for directly by the mass-curve and sequent-peak analysis.

Optimization models, on the other hand, can explicitly consider such implicit relationships and can be solved directly by appropriate methods. In addition, systems consisting of several multiple-purpose reservoirs can be modeled and their interrelationships accounted for in an optimization model. Refer to Mays and Tung (1992) for further details on this topic.

EXAMPLE 11.7.4

Use the following monthly evaporation loss and precipitation and use the sequent-peak method to determine the required active storage for producing 2000 acre-ft/month firm yield. Seepage losses are negligible.

Month	10	11	12	1	2	3	4	5	6	7	8	9
EV(AF)	270	275	280	350	470	450	400	350	370	330	300	290
PP(AF)	3	5	5	10	30	50	100	150	70	10	2	3

SOLUTION

Computations by the sequent-peak method considering other hydrologic components are shown in Table 11.7.3. Columns 2–5 contain data for monthly required release, surface inflow, precipitation, and evaporation, respectively. Columns 3–5 are used to compute the adjusted inflow according to equation (11.7.2). The adjusted inflow for each month is used in equation (11.7.1) to compute K_t. The active storage required is 8840 acre-ft, as indicated in Table 11.7.3. The presence of evaporation loss results in an increase in required active storage. It should be pointed out that, in this example, the monthly precipitation and evaporation amounts are constants and are assumed independent of storage. In actuality, monthly values for PP_t and EV_t are functions of storage, which is an unknown quantity in the exercise. To account accurately for the values of PP_t and EV_t as storage changes, a trial-and-error procedure is needed to determine the required K_a for a given firm yield.

Table 11.7.3 Computations of Sequent-Peak Method Considering Other Hydrologic Components

t (month) (1)	R_t (AF/mon) (2)	QF_t (AF/mon) (3)	PP_t (AF/mon) (4)	EV_t (AF/mon) (5)	K_{t-1} (AF/mon) (6)	K_t (AF/mon) (7)
1	2000	742	3	270	0	1525
2	2000	1060	5	275	1525	2735
3	2000	1000	5	280	2735	4010
4	2000	1500	10	350	4010	4850
5	2000	1080	30	470	4850	6210
6	2000	6460	50	450	6210	2150
7	2000	10000	100	400	2150	0
8	2000	13080	150	350	0	0
9	2000	4910	70	370	0	0
10	2000	981	10	330	0	1339
11	2000	283	2	300	1339	3354
12	2000	322	3	290	3354	5319
13	2000	404	3	270	5319	7182
14	2000	787	5	275	7182	8665
15	2000	2100	5	280	8665	8840
16	2000	4410	10	350	8840	6770
17	2000	2750	30	470	6770	6460
18	2000	3370	50	450	6460	5490
19	2000	5170	100	400	5490	2620
20	2000	19680	150	350	2620	0
21	2000	19630	70	370	0	0
22	2000	3590	10	330	0	0
23	2000	710	2	300	0	1588
24	2000	518	3	290	1588	3357
25	2000	924	3	270	3357	4700
26	2000	1020	5	275	4700	5950
27	2000	874	5	280	5950	7351
28	2000	1020	10	350	7351	8671
29	2000	8640	30	470	8671	2471
30	2000	6370	50	450	2471	0
31	2000	6720	100	400	0	0
32	2000	13290	150	350	0	0
33	2000	9290	70	370	0	0
34	2000	1540	10	330	0	780
35	2000	915	2	300	780	2163
36	2000	506	3	290	2163	3944
37	2000	886	3	270	3944	5325
38	2000	3040	5	275	5325	4555
39	2000	2990	5	280	4555	3840
40	2000	8170	10	350	3840	0
41	2000	2800	30	470	0	0
42	2000	4590	50	450	0	0
43	2000	21960	100	400	0	0
44	2000	30790	150	350	0	0
45	2000	14320	70	370	0	0
46	2000	2370	10	330	0	0
47	2000	709	2	300	0	1589
48	2000	528	3	290	1589	3348
49	2000	859	3	270	3348	4756
50	2000	779	5	275	4756	6247
51	2000	1250	5	280	6247	7272

Table 11.7.3   Computations of Sequent-Peak Method Considering Other Hydrologic Components (*continued*)

t (month) (1)	R_t (AF/mon) (2)	QF_t (AF/mon) (3)	PP_t (AF/mon) (4)	EV_t (AF/mon) (5)	K_{t-1} (AF/mon) (6)	K_t (AF/mon) (7)
52	2000	11750	10	350	7272	0
53	2000	5410	30	470	0	0
54	2000	5560	50	450	0	0
55	2000	5610	100	400	0	0
56	2000	24330	150	350	0	0
57	2000	32870	70	370	0	0
58	2000	7280	10	330	0	0
59	2000	1150	2	300	0	1148
60	2000	916	3	290	1148	2519

Source: Mays and Tung (1992).

11.7.3   Reservoir Simulation

11.7.3.1   *Operation Rules*

Operating rules (policies) are used to specify how water is managed in a reservoir and throughout a reservoir system. These rules are specified to achieve system stream-flow requirements and system demands in a manner that maximizes objectives, which may be expressed in the form of benefits. *System demands* may be expressed as minimum desired and minimum required flows to be met at selected locations in the system. Operation rules may be designed to vary seasonally in response to the seasonal demands for water and the stochastic nature of supplies. Operating rules, often established on a monthly basis, prescribe how water is to be regulated during the subsequent month (or months) based on the current state of the system.

In reservoir operation, the *benefit function* used should indicate that shortages cause severe adverse consequences while surpluses may enhance benefits only moderately. It is common practice to define operating rules in terms of a minimum yield or target value. If water supply to all demand points is rigidly constrained when droughts occur, it may be impossible to satisfy all demands.

Reservoir storage is commonly divided into different zones, as shown in Figure 11.7.1. *Rule curves* indicate the boundary of various zones (see Figure 11.7.6) throughout the year. In developing rule curves for a multipurpose reservoir, consideration must be given to whether or not conflicts in serving various purposes occur. When a number of reservoirs serve the same purpose, system rule curves should be developed.

It is essential that operating rules be formulated with information that will be available at the time when operation decisions are made. If forecasts are used in operation, their degree of reliability should be taken into account in deriving operating rules. Likewise, all physical, legal, and other constraints should be considered in formulating and evaluating operation rules. Further, uncertainties associated with the rule curves and changes in physical and legal conditions should be incorporated in developing the rule curves, if possible.

Rule curves are developed to provide guidance on what operational policy is to be employed at a reservoir or dam site. The *operational decision* is based on the current state of the system and the time of year in order to take into account the seasonal variation of reservoir inflows. A simple rule curve may specify the next period's release based solely on the storage level in the current month. A more complicated rule curve might consider storage at other reservoirs, specifically at downstream control points, and perhaps a forecast of future expected inflows to the reservoir.

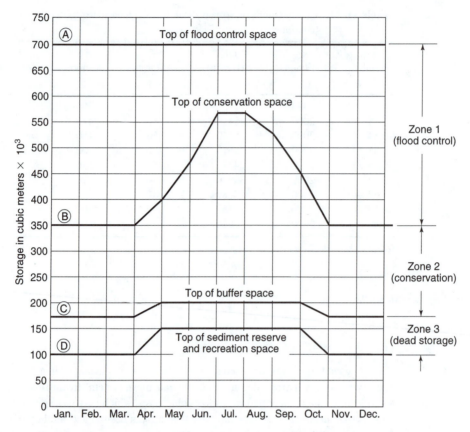

Figure 11.7.6 Example of seasonally varying storage boundaries for multipurpose reservoir (from U.S. Army Corps of Engineers (1977)).

Three basic methods have been used in planning, design, and operation of reservoir systems: (a) simplified methods such as nonsequential analysis; (b) simulation; and (c) optimization. Simple methods are generally used for analyzing systems involving one reservoir with one purpose, using data for only a critical flow period. Simulation models can handle much more complex system configurations and can preserve much more fully the stochastic, dynamic characteristics of reservoir systems. The search for an optimal alternative is dependent on the engineer's ability to manipulate design variables and operating policies in an efficient manner. There may be no guarantee that a globally optimal alternative is found. Optimization models may have a greater number of assumptions and approximations than the simulation models; these are generally needed to make the model mathematically tractable; the combined use of simulation and optimization models can overcome this difficulty. Refer to Mays and Tung (1992) and Mays (1997) for more detail.

11.7.3.2 Conservation Simulation

Reservoir simulation refers to the mathematical simulation of a river network with reservoir(s). The planning and operation of a reservoir system requires the simulation of these systems to determine if demands can be met for water supply (municipal, industrial, and/or agricultural users, hydropower, instream flow maintenance for water quality, and flood control). For purposes of discussion here, water uses are considered in two categories: flood control and conservation, where conservation use refers to all nonflood control uses.

The purpose of *reservoir simulation* for a given multiple-purpose, multiple-reservoir system is to determine the reservoir operation (reservoir releases) over a given time period with known streamflows at input points to these reservoirs and other control points throughout the system. The objective is to operate the reservoirs so as best to meet flow demands for water uses. Reservoir simulation can be used to determine if a reservoir operation policy for a particular system can be used to meet demands. Reservoir simulation can also be used in a trial-and-error fashion to develop reservoir operation strategies (policies), and for determining reservoir storage requirements.

The general procedure for conducting a reservoir system analysis for conservation purposes using simulation involves (U.S. Army Corps of Engineers, 1977):

1. identifying the system;

2. determining the study objectives and specifying the criteria used to measure the objectives;

3. examining the availability of the system data;

4. formulating a model that is mathematically and quantitatively representative of the system's components hydrology, and operating criteria;

5. validating the model;

6. organizing and solving the model; and

7. analyzing and evaluating the results according to how well they achieve the objectives of the study.

PROBLEMS

11.3.1 Rework example 11.3.1 for areas that have T values of 20°C, 25°C, and 30°C and compare the results. Assume that all the other parameters remain the same.

11.5.1 Determine the price elasticity of demand for the demand function (equation (11.5.6)) developed by Billings and Agthe (1980).

11.5.2 The linear demand model derived by Hanke and de Maré (1984) for Malmö, Sweden, is

$Q = 64.7 + 0.00017(\text{Inc}) + 4.76(\text{Ad}) + 3.92(\text{Ch}) - 0.406(R) + 29.03(\text{Age}) - 6.42(P)$

where

Q = quantity of metered water used per house, per semiannual period (in m³).

Inc = real gross income per house per annum (in Swedish crowns; actual values reported per annum and interpolated values used for mid-year periods).

Ad = number of adults per house, per semiannual period.

Ch = number of children per house, per semiannual period.

R = rainfall per semiannual period (in mm).

Age = age of house, with the exception that it is a dummy variable with a value of 1 for houses built in 1968 and 1969, and a value of 0 for houses built between 1936 and 1946.

P = real price in Swedish crowns per m³ of water, per semiannual period (includes all water and sewer commodity charges that are a function of water use).

Using the average values of $P = 1.7241$ and $Q = 75.2106$ for the Malmö data, determine the elasticity of demand. Explain the meaning of the result.

11.5.3 Determine the elasticities of demand for the water-demand model in problem 11.5.2 using $P = 1.5$ and $Q = 75.2106$; $P = 2.0$ and $Q = 75.2106$; $P = 1.7241$ and $Q = 50$; and $P = 1.7241$ and $Q = 100$.

11.6.1 In example 11.6.1, assume that the capital cost as a function of the return period can be given as

$\text{Cost} = 10.5 + 0.0166T - 8 * 10^{-6} T^2$

for $T \le 1000$. Determine the optimum design return period.

11.7.1 Table 11.7.1 contains monthly runoff volumes for water years 1966 to 1970 for Little Weiser River near Indian Valley, Idaho. Construct the flow-duration curve using the years 1967 to 1970 and determine the firm-yield for this site.

11.7.2 Using the monthly flow data in Table 11.7.1 for the Little Weiser River, construct the cumulative mass curve over a five-year period and determine the required active storage capacity to produce a firm yield of 2500 acre-ft/month.

11.7.3. Use problem 11.7.2 to determine the active storage capacity required to produce 2000 acre-ft/month firm yield by computing the cumulative difference between supply and demand.

REFERENCES

Agthe, D. E., and R. B. Billings, "Dynamic Models of Residential Water Demand," *Water Resources Research* 3, pp. 476–480, 1980.

American Society of Agricultural Engineers (ASAE), *Soil and Water Engineering Terminology*, Standard ASAE 5526, St. Joseph, MI, 1993.

American Society of Civil Engineers (ASCE), *Management, Operation, and Maintenance of Irrigation and Drainage Systems*, ASCE Manuals and Reports on Engineering Practice No. 57, ASCE, New York, 1991.

Baumgartner, A., and E. Reichel, *The World Water Balance: Mean Annual Global, Continental and Maritime Precipitation*, Evaporation, and Runoff, Elsevier, Amsterdam, 1975.

Billings, R. B., and D. E. Agthe, "Price Elasticities for Water: A Case of Increasing Block Rates," *Land Economics*, vol. 56, no. 1, pp. 73–84, 1980.

Bruins, H. J., "Drought Management and Water Supply Systems in Israel," *Drought Management Planning in Water Supply Systems*, (edited by E. Cabrera and J. Garcia-Serra), pp. 299–321, Kluwer Academic Publishers, Dordrecht, 1999.

Chow, V. T., D. R. Maidment, and L. W. Mays, *Applied Hydrology*, McGraw-Hill, New York, 1988.

Dixon, L. S., et al., *Drought Management Policies and Economic Effects in Urban Areas of California, 1987–1992*, RAND, 1996.

Doneen, L. D., and D. W. Westcot, *Irrigation Practice and Water Management*, FAO Irrigation and Drainage Paper 1, Rev. 1, Food and Agriculture Organization of the United Nations, Rome, 1984.

Dziegielewski, B., et al., "Drought Management Option," *Drought Management and Its Impact on Public Water Systems*, Report on a Colloquium Sponsored by the Water Science and Technology Board, Sept. 5, 1985, National Academy Press, Washington, DC, 1986.

Dziegielewski, B., et al., *National Study of Water Management During Drought: A Research Assessment for the U.S. Army Corps of Engineers Water Resources Support Center*, Institute for Water Resources, IWR Report 91-NDS-3, 1991.

Dziegielewski, B., E. M. Optiz, and D. R. Maidment, "Water Demand Analysis," *Water Resources Handbook*, (edited by L. W. Mays), McGraw-Hill, New York, 1996.

Easterling, W. E., "Assessing the Regional Consequences of Drought: Putting the MINK Methodology to Work on Today's Problems," *Drought Assessment, Management and Planning: Theory and Case Studies* (edited by Donald A. Wilhite), Kluwer Academic Publishers, Boston, MA, 1993.

Ejeta, M. Z., and L. W. Mays, "Urban Water Pricing and Drought Management: A Risk-Based Approach," *Drought Management Planning in Water Supply Systems* (edited by E. Cabrera), Kluwer Academic Publishers, 1998.

E. K. Berner and R. A. Berner, *The Global Water Cycle: Geochemistry and Environment*, pp. 13–14, reprinted/adapted by permission of Prentice-Hall, Englewood Cliffs, New Jersey, 1987.

Falkenmark, M., J. Lundquist, and C. Widstrand, "Marco-Scale Water Scarcity Requires Micro-Scale Approaches: Aspects of Vulnerability in Semiarid Development," *Natural Resources Forum*, vol. 13, no. 4, pp. 258–267, 1989.

Falkenmark, M., and G. Lindh, "Water and Economic Development," *Water in Crisis* (edited by P. H. Gleick), Oxford University Press, Oxford, 1993.

Food and Agriculture Organization, *Crop Water Requirements*, Irrigation and Drainage Paper 24, U.N., Rome, 1977.

Food and Agriculture Organization, *FAO Production Yearbook 1990*, FAO Statistical Series, vol. 44, no. 99, Statistics Division of the Economic and Social Policy Department, Rome, 1990.

Food and Agriculture Organization, *FAO Production Yearbook 1992*, FAO Statistical Series, vol. 46, no. 112, Rome, 1993.

Foster, H. S., Jr., and B. R. Beattie, "Price Elasticities for Water: A Case of Increasing Block Rates," *Land Economics*, vol. 56, no. 1, pp. 73–84, 1980.

Garen, D. C., "Revised Surface-Water Supply Index for Western United States," *Journal of Water Resources Planning and Management*, ASCE, vol. 119, no. 4, 437–454, 1993.

Gibbs, K., "Price Variables in Residential Water Demand Models," *Water Resources Research*, vol. 14, no. 1, pp. 15–18, 1978.

Gleick, P. H., "Water and Energy Appendix G," *Water in Crisis* (edited by P. H. Gleick), Oxford University Press, Oxford, 1993a.

Gleick, P. H., "Water and Human Use Appendix H," *Water in Crisis* (edited by P. H. Gleick), Oxford University Press, Washington, DC, 1993b.

Gleick, P. H., "Water and Agriculture Appendix E," *Water in Crisis* (edited by P. H. Gleick), Oxford University Press, Washington, DC, 1993c.

Gleick, P. H., "Global and Regional Fresh Water Resources," *Water in Crisis* (edited by P. H. Gleick), Oxford University Press, Oxford, 1993d.

Gleick, P. H., *The World Water: The Biennial Report on Freshwater Resources*, Island Press, Washington, DC, 1998.

Gleick, P. H., P. Loh, S. Gomes, and J. Morrison, "California Water 2020: A Sustainable Vision," Pacific Institute for Studies in Development, Environment, and Security, Oakland, CA, 1995.

Gouevsky, I. U., and D. R., Maidment, "Agricultural Water Demands," *Modeling Water Demands*, (edited by J. Kindler and C. S. Russell), Academic Press, London, 1984.

Griffin, R. C., and J. Stoll, "The Enhanced Role of Water Conservation in the Cost-Benefit Analysis of Water Projects," *Water Resources Bulletin*, American Water Resources Association, vol. 19, no. 3, pp. 447–457, 1983.

Grunewald, O. C., et al., "Rural Residential Water Demand: An Econometric and Simulation Analysis," *Water Resources Bulletin*, vol. 12, no. 5, pp. 951–961, 1976.

Hanke, S. H., and L. de Maré, "Residential Water Demand: A Pooled Time Series, Cross Section Study of Malmö," *Water Resources Bulletin*, vol. 18, no. 4, pp. 621–625, 1982.

Hanke, S. H., "Demand for Water Under Dynamic Conditions," *Water Resources Research*, vol. 6, no. 5, pp. 1253–1261, 1970.

Hillel, D., *The Efficient Use of Water in Irrigation*, World Bank Technical Paper No. 64, The International Bank for Reconstruction and Development (World Bank), Washington, DC, 1987.

Hobbs, B. F., R. L. Mittelstadt, and J. R. Lund, "Energy and Water," *Water Resources Handbook* (edited by L. W. Mays), McGraw-Hill, New York, 1996.

Howe, C. W., "The Impact of Price on Residential Water Demand: Some New Insights," *Water Resources Research*, vol. 18, no. 4, pp. 713–716, 1982.

Howe, C. W., and F. P. Linaweaver, Jr., "The Impact of Price on Residential Water Demand and Its Relation to System Design and Price Structure," *Water Resources Research*, vol. 3, no. 1, pp. 13–32, 1967.

Hudson, H. E., Jr., and R. Hazen, "Drought and Low Stream Flows," *Handbook of Applied Hydrology* (edited by V. T. Chow), McGraw-Hill, New York, 1964.

James, L. G., *Principles of Farm Irrigation System Design*, John Wiley & Sons, New York, 1988.

Jones, C. V., and J. R. Morris, "Instrumental Price Estimates and Residential Water Demand," *Water Resources Research*, vol. 20, no. 2, pp. 197–202, 1984.

Jordan, J. L., "The Effectiveness of Pricing as a Stand-Alone Water Conservation Program," *Journal of the American Water Resources Association*, vol. 30, no. 5, pp. 871–877, 1994.

Kammerer, "Estimated Demand of Water for Different Purposes," *Water for Human Consumption*, Tycuoly International, Dublin, 1982.

Kawamura, A., et al., "Chaotic Characteristics of the Southern Oscillation Index Time Series," *Journal of Hydrology*, vol. 204, pp. 168–181, 1998.

Kindler, J., and C. S. Russell, *Modeling Water Demands*, Academic Press, London, 1984.

L'vovich, M. I., *World Water Resources and Their Future*, Copyright by American Geophysical Union, Washington, DC (English translation of a 1974 USSR publication edited by R. Nace), 1979.

Lindh, G., *Water and the City*, UNESCO, Paris, France, 1985.

Maddock, T. S., and W. G. Hines, "Meeting Future Public Water Supply Needs: A Southwest Perspective," *Journal of the American Water Resources Association*, vol. 31, no. 2, pp. 317–329, 1995.

Mays, L. W. (editor), *Water Resources Handbook*, McGraw-Hill, New York, 1996.

Mays, L. W., *Optimal Control of Hydrosystems*, Marcel Dekker, New York, 1997.

Mays, L. W., and Y. K. Tung, *Hydrosystems Engineering and Management*, McGraw-Hill, New York, 1992.

Metcalf and Eddy, Inc., *Wastewater Engineering Treatment, Disposal, and Reuse*, 3rd ed., McGraw-Hill, New York, 1991.

Mitchell, R. D., *Survey of Water-Conserving Heat Rejection Systems*, Report EPRI GS-6252, Electric Power Research Institute, Palo Alto, CA, 1989.

Moncur, J. E. T., "Drought Episodes Management: The Role of Price," *Water Resources Bulletin, The American Water Resources Association*, vol. 25, no. 3, pp. 499–505, 1989.

Moncur, J. E. T., "Urban Water Pricing and Drought Management," *Water Resources Research*, vol. 23, no. 3, pp. 393–398, 1987.

Moncur, J. E. T., "Drought Episodes Management: The Role of Price," *Water Resources Bulletin*, vol. 25, no. 3, pp. 499–505, 1989.

National Research Council, *Drought Management and Its Impact on Public Water Systems*, Colloquium 1, National Academy Press, Washington, DC, 1986.

Palmer, W. C., *Meteorological Drought*, U.S. Weather Bureau Research Paper No. 45, 1965.

Postel, S., *Last Oasis: Facing Water Scarcity*, W. W. Norton, New York, 1992.

Postel, S., "Water and Agriculture," *Water in Crisis* (edited by P. H. Gleick), Oxford University Press, Oxford, 1993.

Puckett, L. J., *Dendroclimatic Estimates of a Drought Index for Northern Virginia*, Geological Survey Water Supply Paper 2080, United States Government Printing Office, Washington, DC, 1981.

Rasmusson, E. M, R. E. Dickinson, J. E. Kutzback, and M. K. Cleveland, "Climatology," *Handbook of Hydrology* (edited by D. R. Maidment), McGraw-Hill, New York, 1993.

Replogle, J. A., A. J. Clemmens, and M. E. Jensen, "Irrigation Systems," *Water Resources Handbook* (edited by L. W. Mays), McGraw-Hill, New York, 1996.

R. Nace, "Are we running out of water?" U.S. Geological Surey Circular no. 536, Washington, DC, 1967.

Roberson, J. A., J. J. Cassidy, and M. H. Chaudhry, *Hydraulic Engineering*, 2nd Ed., John Wiley & Sons, New York, 1998.

Rogers, P., *America's Water: Federal Roles and Responsibilities*, MIT Press, Cambridge, MA, 1993.

Schneider, M. L., and E. E. Whitlatch, "User-Specific Water Demand Elasticities," *Journal of Water Resources Planning and Management*, vol. 117, no. 1, pp. 52–73, 1991.

Sheer, D. P., "Analyzing the Risk of Drought: The Occoquan Experience," *Journal of American Water Works Association*, vol. 72, no. 12, pp. 246–253, 1980.

Shiklomanov, I. A., "World Fresh Water Resources," *Water in Crisis*, (edited by P. H. Gleick), Oxford University Press, Oxford, 1993.

Solley, W., R. R. Pierce, and H. A. Perlman, "Estimated Use of Water in the United States, 1990," U.S. Geological Survey Circular 1081, U.S. Geological Survey, Washington, DC, 1993.

Solley, W. B., *Estimates of Water Use in the Western United States*, Report to the Western Water Policy Review Advisory Commission, U.S. Geological Survey, Reston, VA, 1997.

Steila, D., *Drought in Arizona: A Drought Identification Methodology and Analysis*, University of Arizona, Tucson, AZ, 1972.

Tchobanozlous, G., and E. D. Schroeder, *Water Quality*, Addison-Wesley, Reading, MA, 1985.

Tung, Y. K., "Uncertainty and Reliability Analysis," *Water Resources Handbook* (edited by L. W. Mays), McGraw-Hill, New York, 1996.

United Nations, "The Demand for Water: Procedure and Methodologies for Projecting Water Demand in the Context of Regional and National Planning," Natural Resources of Water Series No. 3, United Nations Publication, New York, 1976.

United Nations, "Use of Nonconventional Water Resources in Developing Countries," National Resource/Water Series No. 14, New York, 1988.

U.S. Army Corps of Engineers, *Hydrologic Engineering Methods for Water Resources Development: Reservoir System Analysis for Conservation*, vol. 9, Davis, CA, 1977.

U.S. Department of Agriculture, *Agricultural Statistics—1992*, U.S. Government Printing Office, Washington, DC, 1992.

U.S. Department of Agriculture, *Irrigation Water Requirement*, Technical Release No. 21, Soil Conservation Service, Engineering Division, P.83, 1967.

U.S. Department of Housing and Urban Development, *Residential Water Conservation Projects, Summary Report*, 1984.

U.S. Soil Conservation Service, *Colorado Water Supply Outlook*, United States Department of Agriculture, U.S. Government Documents, 1988.

USSR Committee for the International Hydrological Decade, *World Water Balance and Water Resources of the Earth*, Studies and Reports in Hydrology, no. 25. UNESCO, Paris and Gidrometeoizdat, Leningrad (English translation of a 1974 USSR publication), 1978.

U.S. Water Resources Council, "Procedures for Evaluation of National Economic Development (NED) Benefits and Costs in Water Resources Planning (Level C): Final Rule," *Federal Register*, vol. 44, no. 242, pp. 72891–72976, 1979.

Wade, J. C., "Efficiency and Optimization in Irrigation Analysis," *Energy and Water Management in Western Irrigation Agriculture* (edited by N. K. Whitlesey), Studies in Water Policy and Management No. 7, West View Press, Boulder, CO, pp. 73–100, 1986.

Walter, B. H., "Construction and Operation of Pump Irrigation Facilities Along the Columbia and Okanugan Rivers," *Optimization of Irrigation and Drainage Systems*, Proceedings, ASCE Irrigation and Drainage Specialty Conference, Lincoln, Nebraska, ASCE, New York, 1971.

Wilhite, D. A., "Planning for Drought: A Methodology," *Drought Assessment, Management and Planning: Theory and Case Studies* (edited by D. A. Wilhite), Kluwer Academic Publishers, Norwell, MA, pp. 87–108, 1993.

Wong, S. T., "A Model on Municipal Water Demand: A Case Study of Northern Illinois," *Land Economics*, pp. 34–44, 1972.

World Resources Institutes, *World Resources 1988–89*, World Resources Institute and the International Institute for Environment and Development in collaboration with the United Nations Environment Programme, Basic Books, New York, 1988.

Young, R. A., "Price Elasticity of Demand for Municipal Water: A Case Study of Tucson, Arizona," *Water Resources Research*, vol. 9, no. 4, pp. 1068–1072, 1973.

Young, R. A., "Water Economics," *Water Resources Handbook* (edited by L. W. Mays), McGraw-Hill, New York, 1996.

Chapter **12**

Water Distribution

The purpose of this chapter is to provide a detailed introduction to water distribution systems (networks). The distribution systems (networks) considered in this chapter are pressurized conduit systems; we do not consider water distribution systems consisting of canals with open-channel flow. A much more extensive treatment of water distribution systems is provided in the *Water Distribution Systems Handbook* (Mays, 2000).

12.1 INTRODUCTION

12.1.1 Description, Purpose, and Components of Water Distribution Systems

The purpose of a water distribution network is to supply the system's users with the amount of water demanded and to supply this water with adequate pressure under various loading conditions. A *loading condition* is defined as a time pattern of demands. A municipal water supply system may be subject to a number of different loading conditions: fire demands at different nodes; peak daily demands; a series of patterns varying throughout a day; or a critical load when one or more pipes are broken. In order to insure that a design is adequate, a number of loading conditions, including critical conditions, must be considered. The ability to operate under a variety of load patterns is required of a reliable network.

Water distribution systems have three major components: *pumping stations*, *distribution storage*, and *distribution piping*. These components may be farther divided into subcomponents, which in turn can be divided into sub-subcomponents. For example, the pumping station component consists of structural, electrical, piping, and pumping unit subcomponents. The pumping unit can be divided farther into sub-subcomponents: pump, driver, controls, power transmission. The exact definition of components, subcomponents and sub-subcomponents depends on the level of detail of the required analysis and to a somewhat greater extent on the level of detail of available data. In fact, the concept of component–subcomponent–sub-subcomponent merely defines a hierarchy of building blocks used to construct the water distribution system. Figure 12.1.1 shows the hierarchical relationship of system, components, subcomponents, and sub-subcomponents for a water distribution system.

A water distribution system operates as a system of independent components. The hydraulics of each component is relatively straightforward; however, these components depend directly upon each other and as a result affect one another's performance. The purpose of design and analysis is

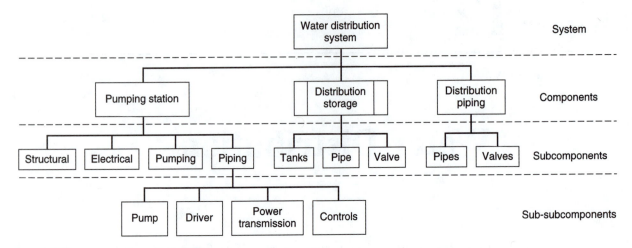

Figure 12.1.1 Hierarchical relationship of system, components, subcomponents, and sub-subcomponents for a water distribution system (from Cullinane (1989)).

to determine how the systems perform hydraulically under various demands and operation conditions. Such analyses are used for the following situations:

- Design of a new distribution system
- Modification and expansion of an existing system
- Analysis of system malfunction such as pipe breaks, leakage, valve failure, pump failure
- Evaluation of system reliability
- Preparation for maintenance
- System performance and operation optimization

Figure 12.1.2 illustrates a typical municipal water utility showing the water distribution system as a part of this overall water utility. In some locations where excellent quality groundwater is available, water treatment may involve only chlorination.

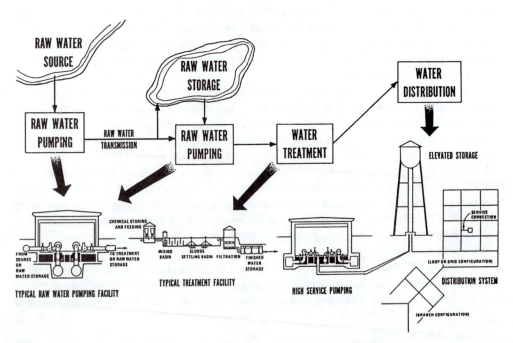

Figure 12.1.2 Functional components of a water utility (from Cullinane (1989)).

Figure 12.1.3 Cement-lined and -coated steel pipe (42-inch diameter) being transported to trench. (Courtesy of Northwest Pipe Company.)

Pipe sections or *links* are the most abundant elements in the network. These sections are constant in diameter and may contain fittings and other appurtenances, such as valves, storage facilities, and pumps. Pipes are manufactured in different sizes and are composed of different materials, such as steel, cast or ductile iron, reinforced or prestressed concrete, asbestos cement, polyvinyl chloride, polyethylene, and fiberglass. The American Water Works Association publishes standards for pipe construction, installation, and performance in the C-series standards (continually updated). Pipes are the largest captial investment in a distribution system. Figure 12.1.3 shows a steel pipeline that is cement coated and lined. Figure 12.1.4 shows steel pipelines being installed that are polyethylene-coated. Various types of joints and couplings to connect pipes are shown in Figure 12.1.5.

A *node* refers to either end of a pipe. Two categories of nodes are junction nodes and fixed-grade nodes. Nodes where the inflow or the outflow is known are referred to as junction nodes. These nodes have lumped demand, which may vary with time. Nodes to which a reservoir is attached are referred to as fixed-grade nodes. These nodes can take the form of tanks or large constant-pressure mains.

Control valves regulate the flow or pressure in water distribution systems. If conditions exist for flow reversal, the valve will close and no flow will pass. The most common type of control valve is the *pressure-reducing (pressure-regulating) valve (PRV)*, which is placed at pressure zone boundaries to reduce pressure. The PRV maintains a constant pressure at the downstream side of the valve for all flows with a pressure lower than the upstream head. When connecting high-pressure and low-pressure water distribution systems, the PRV permits flow from the high-pressure system if the pressure on the low side is not excessive.

Figure 12.1.4 Parallel 60-in and 42-in diameter steel water transmission lines being installed in Aurora, Colorado. Lightweight polyethylene tape coating and O-ring bell and spigot joints helped speed installation. (Courtesy of Northwest Pipe Company.)

The headloss through the valve varies, depending upon the downstream pressure and not on the flow in the pipe. If the downstream pressure is greater than the PRV setting, then the pressure in chamber A will close the valve. Another type of check valve, a *horizontal swing valve*, operates under similar principle. *Pressure-sustaining valves* operated similarly to PRVs monitoring pressure at the upstream side of the valve. Figure 12.1.6 illustrates a pressure-reducing and check valve. Figure 12.1.6*c* illustrates the typical application of these valves.

There are many other types of valves, including isolation valves to shut down a segment of a distribution system; direction-control (check) valves to allow the flow of water in only one direction, such as swing check valves, rubber-flapper check valves, slanting check disk check valves, and double-door check valves; and air-release/vacuum-breaker valves to control flow in the main. Figure 12.1.7 show the typical application of a two-way altitude valve.

Distribution-system storage is needed to equalize pump discharge near an efficient operation point in spite of varying demands, to provide supply during outages of individual components, to provide water for fire fighting, and to dampen out hydraulic transients (Walski, 1996). Distribution storage in a water distribution network is closely associated with the water tank. An elevated storage tank installation is illustrated in Figure 12.1.8. Tanks are usually made of steel and can be built at ground level or be elevated at a certain height above the ground. The water tank is used to supply water to meet the requirements during high system demands or during emergency conditions when pumps cannot adequately satisfy the pressure requirements at the demand nodes. If a minimum volume of water is kept in the tank at all times, then unexpected high demands cannot be met during critical conditions. The higher the pump discharge, the lower the pump head becomes. Thus, during a period of peak demands, the amount of available pump head is low.

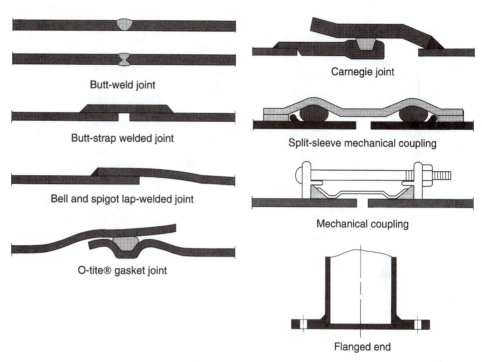

Figure 12.1.5 Variety of joints and couplings used for water pipe systems. Gasketed joints can be used for pressures up to 400 psi and welded joints are recommended for higher pressure applications. Couplings and flanges are used for valve connections or where the diameter changes. (Courtesy of Northwest Pipe Company.)

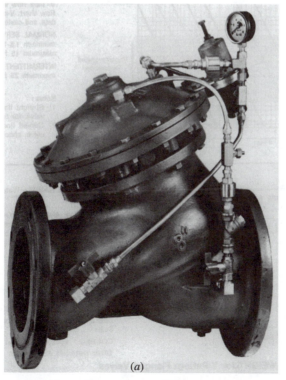

(a)

Figure 12.1.6 Pressure-reducing valve and check valve.
(a) Pressure-reducing valve;

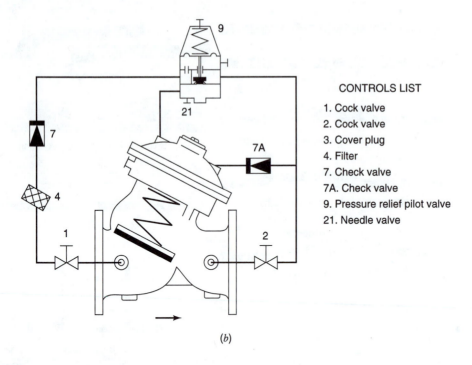

CONTROLS LIST

1. Cock valve
2. Cock valve
3. Cover plug
4. Filter
7. Check valve
7A. Check valve
9. Pressure relief pilot valve
21. Needle valve

(b)

1) Maintains a constant reduced pressure
2) Prevents return flow

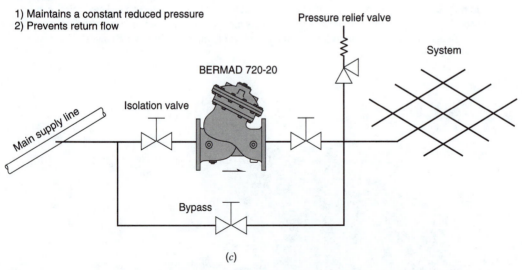

(c)

Figure 12.1.6 (*continued*) Pressure-reducing valve and check valve. (*b*) control diagram of pressure-reducing valve; (*c*) Typical application of pressure-reducing and check valve. (Courtesy of Bermad.)

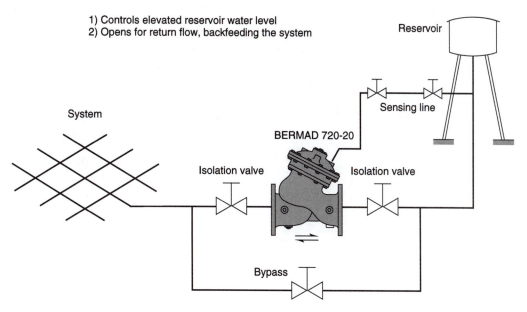

1) Controls elevated reservoir water level
2) Opens for return flow, backfeeding the system

Reservoir

System

BERMAD 720-20

Sensing line

Isolation valve

Isolation valve

Bypass

Figure 12.1.7 Two-way altitude valve. Controls water level in reservoirs by sensing built-up reservoir head without using external devices such as floats. Valve opens to fill reservoir and closes at a predetermined reservoir level. Valve reopens to discharge reservoir water back into system when system pressure drops below reservoir static head. (Courtesy of Bermad.)

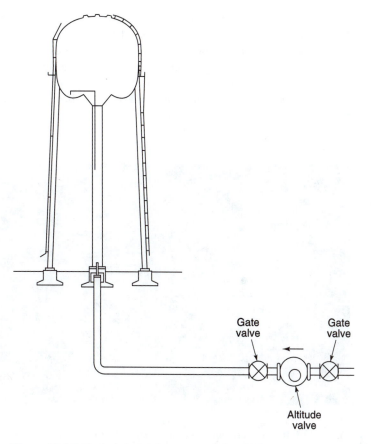

Gate valve

Gate valve

Altitude valve

Figure 12.1.8 Typical elevated storage tank installations.

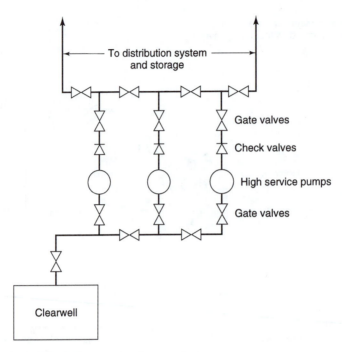

Figure 12.1.9 Schematic of a typical water distribution system
pumping station.

Pumps are used to increase the energy in a water distribution system. A pumping station in
Mesa, Arizona, is shown in Figure 12.1.9. There are many different types of pumps (positive-
displacement pumps, kinetic pumps, turbine pumps, horizontal centrifugal pumps, vertical pumps,
and horizontal pumps). The most commonly used type of pump used in water distribution systems
is the *centrifugal pump*. Figure 12.1.10 illustrates a pumping station with centrifugal pumps.
Pumping stations house the pump, motors, and the auxiliary equipment. Figure 12.1.11 shows a
pump at a well site in Mesa, Arizona. Figure 12.1.12 shows the Elmwood pumping station and
storage reservoir in Mesa, Arizona.

(a)

Figure 12.1.10 Pumping station in Mesa, Arizona. (a) Pump house;

(*b*)

(*c*)

Figure 12.1.10 (*continued*) Pumping station in Mesa, Arizona. (*b*) Centrifugal pumps; (*c*) Manifold at pump station. (Courtesy of City of Mesa, Arizona.)

Figure 12.1.11 Pump at a well site in Mesa, Arizona. (Courtesy of City of Mesa, Arizona)

Figure 12.1.12 Elmwood pumping station and reservoir in Mesa, Arizona. (Courtesy of City of Mesa, Arizona.)

The *metering* (flow measurement) of water mains involves a wide array of metering devices. These include electromagnetic meters, ultrasonic meters, propeller or turbine meters, displacement meters, multijet meters, proportional meters, and compound meters. *Electromagnetic meters* measure flow by means of a magnetic field generated around an insulated section of pipe. *Ultrasonic meters* utilize sound-generating and sound-receiving sensors (transducers) attached to the sides of the pipe. *Turbine meters* (Figure 12.1.13) have a measuring chamber that is turned by the flow of water. *Multijet meters* have a mulitblade rotor mounted on a vertical spindle within a cylindrical measuring chamber. *Proportional meters* utilize restriction in the water line to divert a portion of water into a loop that holds a turbine or displacement meter, with the diverted flow being proportional to the flow in the main line. *Compound meters* connect different sized meters in parallel.

Refer to Mays (1999, 2000), Ysusi (1999, 2000), and Walski (1996) for further descriptions of water distribution system components.

Figure 12.1.13 Turbine meter. (Courtesy of Master Meter.)

12.1.2   Pipe Flow Equations

Typically for water distribution systems, the relationship used to describe flow or velocity in pipes is the *Hazen–Williams equation*:

$$V = K_H C R^{0.63} S_f^{0.54} \tag{12.1.1}$$

where $K_H = 1.318$ for U.S. customary units and $K_H = 0.849$ for SI units, V is the average flow velocity in ft/s (m/s), C is the *Hazen–Williams roughness coefficient* as listed in Table 12.1.1 for pipes of different materials and ages, R is the hydraulic radius in ft (m), and S_f is the friction slope in ft/ft (m/m). The friction headloss as a gradient in terms of feet per 1000 ft (or meters per 1000 m) can be expressed as

$$h_L = A \left(\frac{Q}{C} \right)^{1.85} D^{-4.87} = \left(\frac{BQ}{CD^{2.63}} \right)^{1.85} \tag{12.1.2}$$

where $A = 10{,}500$ for U.S. customary units and $A = 10{,}700$ for SI units, $B = 149$ for U.S. customary units and $B = 151$ for SI units, h_L is ft/1000 ft (m/1000 m), and D is the pipe diameter in ft (m). Later in this chapter the headloss is expressed as a function of discharge:

$$h_L = KQ^n \tag{12.1.3}$$

where K is referred to as a pipe coefficient and n is an exponential flow coefficient. For the Hazen–Williams equation

$$h_L = KQ^{1.85} \tag{12.1.4}$$

where

$$K = \left[\frac{\varphi L}{C^{1.85} D^{4.87}} \right] \tag{12.1.5}$$

where $\phi = 4.73$ for U.S. customary units and $\phi = 10.66$ for SI units. The headloss h_L is in ft (m); L is the pipe length in ft (m), Q is the flow rate in cfs (m³/s), and D is the pipe diameter in ft (m).

Another pipe flow equation for headloss is the *Darcy–Weisbach equation* (4.3.13):

$$h_L = f \frac{L}{D} \frac{V^2}{2g} \tag{12.1.6}$$

where f is the Darcy–Weisbach friction factor. The friction factor is a function of the Reynold's number and the *relative roughness*, which is the absolute roughness of the interior pipe surface divided by the pipe diameter. The friction factor can be determined from a Moody diagram (Figure 4.3.5), or other friction formulas presented in Chapter 4. Equation (12.1.6) can be expressed in terms of the discharge as

$$h_L = KQ^2 \tag{12.1.7}$$

where

$$K = \left[\frac{8 f L}{\pi^2 g D^5} \right] \tag{12.1.8}$$

and h_L is the headloss in ft (m), L is the pipe length in ft (m), Q is the flow rate in cfs (m³/s), D is the pipe diameter in ft (m), and g is the acceleration due to gravity, 32.2 ft/s² (9.81 m/s²).

Another equation that can be used for headloss, which is valid for fully turbulent flow, is Manning's equation (5.1.23 and 5.1.25):

$$h_L = \frac{n^2 V^2 L}{\beta R^{4/3}} = \left[\frac{n^2 L}{\beta R^{4/3} A^2} \right] Q^2 = KQ^2 \tag{12.1.9}$$

where $\beta = 2.21$ for U.S. customary units and $\beta = 1$ for SI units, h_L is the headloss in ft (m), n is Manning's roughness factor, R is the hydraulic radius in ft (m), L is the pipe length in ft (m), and A is the cross-sectional area of the pipe in ft^2 (m^2). Also,

$$K = \left[\frac{n^2 L}{\phi D^{5.33}} \right]$$

(12.1.10)

where $\phi = 0.216$ in U.S. customary units and $\phi = 0.0972$ in SI units.

Table 12.1.1　Typical Coefficients of Pipe Friction for Design[a]

Material	Hazen–Williams C	Manning n[b]	Moody Diagram ϵ[b] mm	in
New pipe or lining				
Smooth glass or plastic[c]	150	0.009	0.919	0.00075
Centrifugally spun cement-mortar lining[d]	145	0.009	0.028	0.0015
Cement-mortar lining troweled in place	140	0.009	0.076	0.003
Commercial steel or wrought iron	140	0.009	0.076	0.003
Galvanized iron	135	0.010	0.13	0.005
Ductile or cast iron, uncoated	130	0.010	0.19	0.0075
Asbestos–cement, coated	145	0.009	0.038	0.0015
Asbestos–cement, uncoated	140	0.009	0.076	0.003
Centrifugally cast concrete pressure pipe	135	0.010	0.13	0.005
Ten-State Standards (1978)				
Cement mortar or plastic lining	120	0.011	0.41	0.016
Unlined steel or ductile iron	100	0.011	1.5	0.060
Old pipe or lining [in moderate service (20 yr. or more), nonagressive water][e]				
Smooth glass or plastic	135	0.010	0.13	0.005
Centrifugally spun cement-mortar lining[f]	130	0.010	0.19	0.0075
Cement mortar troweled in place	125	0.010	0.28	0.011
Asbestos cement, coated	130	0.010	0.19	0.0075
Asbestos cement, uncoated	125	0.010	0.28	0.011
Ductile iron or carbon steel, uncoated	100	0.013	1.5	0.060
Centrifugally cast concrete pressure pipe	130	0.010	0.19	0.0075
Wood stave	110	0.012	0.89	0.035
Riveted steel	80	0.016	5.6	0.22
Concrete, formed	80	0.016	5.6	0.22
Clay (not pressurized)	100	0.013	1.5	0.060
Wrought iron	100	0.013	1.5	0.060
Galvanized iron	90	0.014	0.30	0.012

[a]For critical problems, consult the other sources

[b]Values are calculated from C coefficients for 300-mm (12-in) pipe, a velocity of 1.1–2.1 m/s (3.7–6.9 ft/s), and a temperature of 20°C (68°F).

[c]PVC, polyethylene, polypropylene, polybutylene, reinforced thermosetting resin pipe, and polyvinyl chloride.

[d]Average value for pipes 150 to 900 mm (6 to 36 in) diameter.

[e]For conservative design, reduce old pipe C values (and increase n values) by 0.02%/mm (0.5%/in) for pipe less than 450 mm (18 in).

　Note that the Hazen–Williams and Manning equations predict headloss on the unsafe side for small pipes and/or low velocities.

[f]Conservative values for water pipe 150–500 mm (6–20 in).

Source: Sanks (1998).

EXAMPLE 12.1.1

Determine the headloss due to friction in a 1-m diameter, 300-m long pipe using Manning's equation with $n = 0.013$ and the Hazen–Williams equation with $C = 100$. Solve for $Q = 1$ m³/s and $Q = 2$ m³/s.

SOLUTION

Using Manning's equation, we find

$$h_L = \left[\frac{n^2 L}{\left[AR^{2/3}\right]^2}\right]Q^2 = \left[\frac{n^2 L}{\left[\left(\pi\left(D^2/4\right)\right)\left(D/4\right)^{2/3}\right]^2}\right]Q^2 = \left[\frac{n^2 L}{0.0972 D^{16/3}}\right]Q^2$$

$$= \left[\frac{(0.013)^2(300)}{0.0972(1)^{16/3}}\right]Q^2 = 0.522 Q^2$$

For $Q = 1$ m³/s, $h_L = 0.522$ m. For $Q = 2$ m³/s, $h_L = 2.09$ m.

Using the Hazen–Williams equation yields

$$\frac{Q}{A} = 0.849 CR^{0.63}\left(\frac{h_L}{L}\right)^{0.54}$$

$$h_L = L\left[\frac{1}{0.849 CAR^{0.63}}\right]^{1.85}Q^{1.85} = L\left[\frac{1}{0.849 C\left(\frac{\pi D^2}{4}\right)\left(\frac{D}{4}\right)^{0.63}}\right]^{1.85}Q^{1.85}$$

$$= \left[\frac{10.66 L}{C^{1.85}D^{4.87}}\right]Q^{1.85} = \left[\frac{10.66(300)}{(100)^{1.85}(1)^{4.87}}\right]Q^{1.85} = 0.638 Q^{1.85}$$

For $Q = 1$ m³/s, $h_L = 0.638$ m. For $Q = 2$ m³/s, $h_L = 2.3$ m.

12.2 SYSTEM COMPONENTS

12.2.1 Pumps

12.2.1.1 Classification

Centrifugal pumps are most commonly used in water distribution applications because of their low cost, simplicity, and reliability in the range of flows and heads encountered (see Walski, 1996, for more details). As a result, the discussion on pumps in this chapter is restricted to centrifugal pumps. A *centrifugal pump* is any pump in which fluid is energized by a rotating impeller, whether the flow is radial, axial, or a combination of both (mixed), using colloquial usage in the United States (Tchobanoglous, 1989). In Europe, centrifugal pumps are strictly defined as radial flow pumps only. Here we use the U.S. usage. Centrifugal pumps are classified into three groups according to the manner in which the fluid moves through the pump (refer to Figure 12.2.1):

- *Radial flow pumps* displace the fluid radially in the pump.
- *Axial flow pumps or propeller pumps* displace the fluid axially in the pump.
- *Mixed-flow pumps* displace the flow both radially and axially in the pump.

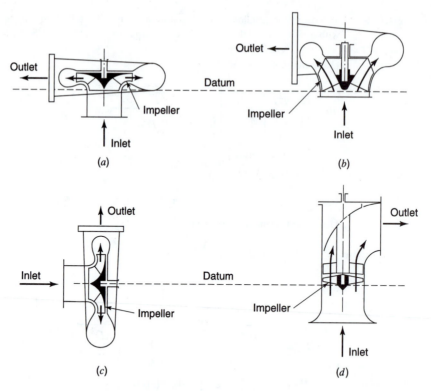

Figure 12.2.1 Typical flow paths in centrifugal pump. (*a*) Radial flow, vertical; (*b*) Mixed flow; (*c*) Radial flow, horizontal; (*d*) Axial flow (from Tchobanoglous (1998)).

The *pump capacity* (Q) is the flow rate or discharge of a pump expressed in SI units in cubic meters per second for large pumps or liters per second and cubic meters per hour for small pumps. Using U.S. customary units, pump capacity is expressed in gallons per minute, million gallons per day, or cubic feet per second. The *head* is the difference in elevation between a free surface of water above (or below) the reference datum, which varies with the type of pump, as shown in Figure 12.2.2.

Multistage pumps are pumps with more than one impeller (stage). The stages are in series so that the discharge of the first stage (first impeller) discharges directly into the suction side of the second stage (second impeller), etc. The impellers are on a single shaft and are enclosed in a single pump housing. Figure 12.2.3*a* shows a single-stage pump and Figure 12.2.3*b* shows a multistage pump.

Figure 12.2.4*a* and b show screw pumps and centrifugal pumps at a water treatment plant. Also refer to Figure 12.1.2. *Screw pumps* are classified as *positive-displacement pumps*. These pumps are used for pumping irrigation water, drainage water, and wastewater. They are based on the Archimedes screw principle in which a revolving shaft fitted with helical blades rotates in an inclined trough pushing water up the trough (as shown in Figure 12.2.4*a*). Two advantages of the screw pump over a centrifugal pump are: (1) they can pump large solids without clogging and (2) they operate at a constant speed over a wide range of flows at relatively good efficiencies.

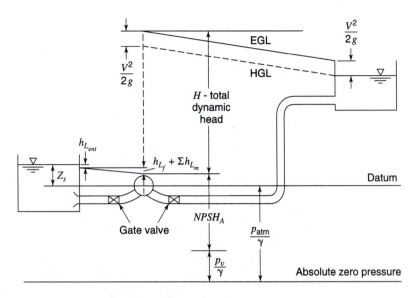

Figure 12.2.2 Illustration of total dynamic head and net positive suction head available for a simple pipe and pump station.

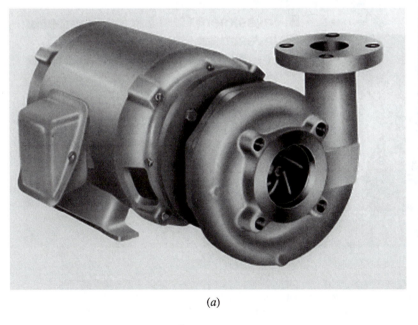

(a)

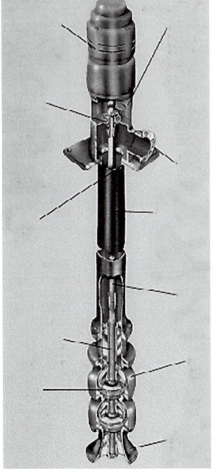

(b)

Figure 12.2.3 Single- and multistage pump. (a) Single-stage pump; (b) Multistage pump; (Courtesy of Paco Pumps.)

(a)

(b)

Figure 12.2.4 Pumps used at water treatment plants. (a) Screw pump used in water treatment plant to lift water (raw water) from pretreatment section to balance of treatment plant in Phoenix, Arizona; (b) High-service pumps at a water treatment plant in Phoenix, Arizona. (Photograph by L. W. Mays.)

12.2.1.2   Operating Characteristics

Operating characteristics of pumps are dependent upon their size, speed, and design. In centrifugal pumps, similar flow patterns occur in geometrically similar pumps. Through dimensional analysis the following three independent coefficients can be derived to describe the operation of pumps:

$$\textit{Discharge coefficient: } C_Q = \frac{Q}{nD^3} \qquad (12.2.1)$$

$$\text{Head coefficient: } C_H = \frac{H}{n^2 D^2} \tag{12.2.2a}$$

$$\text{or } C_H = \frac{gH}{n^2 D^2} \text{ (for dimensional correctness)} \tag{12.2.2b}$$

$$\text{Power coefficient: } C_P = \frac{P}{\rho n^3 D^5} \tag{12.2.3}$$

Even though equation (12.2.2a) is not dimensionally correct, it is commonly used. These equations are the same either in SI or U.S. customary units, but with different values. Q is pump capacity in m^3/sec (ft^3/min), n is the speed in radians per second (revolutions per minute), D is the impeller diameter in meter (feet), H is the head in meter (feet), g is the acceleration due to gravity (9.81 m/s^2)(32.2 ft/s^2), P is the power input in kilowatts (horsepower), and ρ is the density in kilograms per cubic meter (slugs per cubic foot).

The above coefficients can be used to define the *affinity laws* for a pump operating at two different speeds and the same diameter. Consider the ratios $(C_Q)_1 = (C_Q)_2$ for the same diameter and different speeds n_1 and n_2, then

$$\frac{Q_1}{Q_2} = \frac{n_1}{n_2} \tag{12.2.4}$$

Similarly for $(C_H)_1 = (C_H)_2$:

$$\frac{H_1}{H_2} = \left(\frac{n_1}{n_2}\right)^2 \tag{12.2.5}$$

and similarly for $(C_P)_1 = (C_P)_2$:

$$\frac{P_1}{P_2} = \left(\frac{n_1}{n_2}\right)^3 \tag{12.2.6}$$

where 1 and 2 represent corresponding points. These relationships (*affinity laws*) are used to determine the effect of changes in speed on the pump capacity, head, and power. These relationships assume that the efficiency remains the same from one point on a pump curve to a homologous point on another pump curve. Affinity laws for discharge (equation 12.2.4) and for head (equation 12.2.5) are accurate; however, the affinity law for power may not be accurate.

EXAMPLE 12.2.1

A pump operating at 1800 rpm delivers 180 gal/min at 80 ft head. If the pump is operated at 2160 rpm, what are the corresponding head and discharge?

SOLUTION

Use the affinity law, equation (12.2.4), to compute the corresponding discharge:

$$\frac{Q_{2160}}{Q_{1800}} = \frac{2160}{1800}$$

$$Q_{2160} = 180\left(\frac{2160}{1800}\right) = 216 \text{ gal/min}$$

Use the affinity law equation (12.2.5) to compute the corresponding head:

$$\frac{H_{2160}}{H_{1800}} = \left(\frac{2160}{1800}\right)^2$$

$$H_{2160} = 80\left(\frac{2160}{1800}\right)^2 = 115.2 \text{ ft}$$

12.2.1.3 Specific Speed

The *specific speed* n_S is a parameter used to select the type of centrifugal pump that is best suited to a particular application. The diameter term in equations (12.2.1) and (12.2.2a) can be eliminated by dividing $C_Q^{1/2}$ by $C_H^{3/4}$, that is

$$n_S = \frac{C_Q^{1/2}}{C_H^{3/4}} = \frac{\left(Q/nD^3\right)^{1/2}}{\left(H/n^2D^2\right)^{3/4}} = \frac{nQ^{1/2}}{H^{3/4}} \tag{12.2.7}$$

The *total dynamic head* is the total head against which a pump must operate. For a given speed of pump operation, the Q and H must be at the point of maximum efficiency. Equation (12.2.7) is dimensionally incorrect; however, it is the customary definition used in the United States Using equations (12.2.1) and (12.2.2b) results in a dimensionally correct definition of the specific speed, with a conversion factor of 17,200, resulting in n_S ranging from about 0.03 to about 0.91 in U.S. customary units. Figure 12.2.5 relates pump efficiency to specific speed and discharge showing the various impeller shapes. In determining the specific speed for multistage pumps, the head is the head per stage. Because $n_S \sim 1/H^{3/4}$, n_S decreases with an increase in H and the efficiency is small for the smaller n_S, one impeller used for large heads results in low efficiency. As a result multistage pumps can increase efficiency.

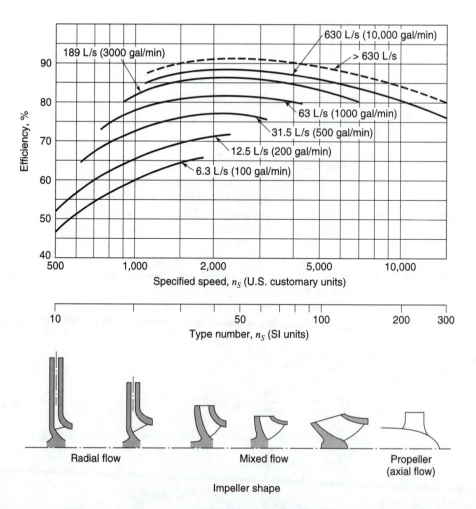

Figure 12.2.5 Pump efficiency as related to specific speed and discharge (from Tchobanoglous (1998)).

EXAMPLE 12.2.2

A flow of 0.02 m³/s must be pumped against a head of 25 m. The pump will be driven by an electric motor with a speed of 1450 rev/min. What type of pump should be used and what is the corresponding efficiency?

SOLUTION

Compute the specific speed using equation (12.2.7) and Figure 12.2.5 to select the type of pump:

$$n_S = \frac{nQ^{1/2}}{H^{3/4}} = \frac{1450(0.02)^{1/2}}{(25)^{3/4}} = 18.34$$

From Figure 12.2.5, a radial flow centrifugal pump would be used, with an efficiency around 68 percent.

EXAMPLE 12.2.3

Solve example 12.2.2 using U.S. customary units.

SOLUTION

Using the flow rate of 0.02 m³/s yields 317 gal/min and the head of 25 m is 82 ft; the specific speed is

$$n_S = \frac{1450(317)^{1/2}}{(82)^{3/4}} = 947$$

From Figure 12.2.5, a radial flow centrifugal pump is selected with an efficiency around 68 percent.

12.2.1.4   Cavitation and Net Positive Suction Head

The objective of this section is to illustrate how to determine whether or not cavitation will be a problem for a particular pump operation. *Cavitation* occurs in pumps when the absolute pressure at the pump inlet decreases below the vapor pressure of the fluid, at which time vapor bubbles form at the impeller inlet (suction side). As shown in Figure 12.2.6, the bubbles are transported through the impeller, where they reach a higher pressure and abruptly collapse. When the collapse occurs on the surface of the impeller blade, the liquid rapidly moves on to fill the space left by the bubble, impacting very small areas with very large localized pressures that pit and erode the plane surface. Collapse of the vapor bubbles can produce noise and vibration.

In order to determine if cavitation will be a problem, the concept of net positive suction head is used. The *available net positive suction head* ($NPSH_A$) at the eye of an impeller is computed and compared to the *required net positive suction head* ($NPSH_R$) of the pump, which is specified by the manufacturer to minimize cavitation. $NPSH_R$ is the absolute dynamic head in the impeller eye, defined by (refer to Figure 12.2.2)

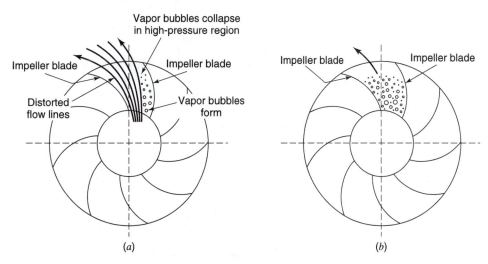

Figure 12.2.6 Formation of vapor bubbles in a pump impeller. (*a*) Partial cavitation; (*b*) Full cavitation.

$$NPSH_A = \frac{p_{atm}}{\gamma} + Z_s - \frac{p_v}{\gamma} - h_{L_{ent}} - h_{L_f} - \sum h_{L_m} \tag{12.2.8}$$

where p_{atm}/γ is the atmospheric pressure head in meters (feet), Z_s is the static suction head in meters (feet) at the impeller eye (negative if there is a suction lift), p_v/γ is the vapor pressure head in meters (feet), $h_{L_{ent}}$ is the entrance headloss in meters (feet), h_{L_p} is the suction pipe friction headloss in meters (feet), and $\sum h_{L_m}$ is the sum of minor losses of valves and fittings in meters (feet). The $NPSH_A$ should always be greater than the $NPSH_R$.

A cavitation constant known as *Thoma's cavitation constant* is defined as the ratio of the *net positive suction head at the point of cavitation inception* ($NPSH_i$) to the total dynamic head H, expressed mathematically as

$$\sigma = \frac{NPSH_i}{H} \tag{12.2.9}$$

When applied to multistage pumps, H is the total dynamic head per stage. The value of $NPSH_R$ cannot be used in the above equation.

Thoma's cavitation constant can be related approximately to specific speed and pump efficiency (Rutschi, 1960, and Tchobanoglous, 1989) as

$$\sigma = \frac{Kn_S^{4/3}}{10^6} \tag{12.2.10}$$

where values of K are given in Table 12.2.1. Equations (12.2.9) and (12.2.10) should not be used for design decisions; instead, recommended values of $NPSH_R$ from pump manufacturers should be used. Combining equations (12.2.9) and (12.2.10) we can estimate $NPSH_i$ can be made using

$$NPSH_i = \frac{HKn_S^{4/3}}{10^6} \tag{12.2.11}$$

Table 12.2.1 Values of K for Equation (12.2.10)[a]

Pump efficiency (%)	K	
	SI units[b]	U.S. customary units[c]
70	1726	9.4
80	1210	6.3
90	796	4.3

[a]For double-suction pumps, use the same formula with Q equal to half of the actual value.
[b]Use with n, values in cubic meters per second, meters, and revolutions per minute.
[c]Use with n, values in gallons per minute, feet, and revolutions per minute.

Source: Tchobanoglous (1998).

EXAMPLE 12.2.4

Estimate the available net positive suction head ($NPSH_A$) for a new system with the configuration shown in Figure 12.2.2 for a discharge of 0.12 m³/s. The suction piping and the discharge piping are both cement mortar-lined ductile iron pipe with a Hazen–Williams coefficient of 140. The suction piping has an inside diameter (ID) of 300 mm and a length of 5 m. The discharge piping has an ID of 250 mm and a length of 100 m. The static suction head is +3 m. The system has a bell mouth entrance, two 90° bends, a gate valve on the suction side, and a gate valve on the discharge line. The elevation is at 1000 m above mean sea level, the temperature is 20°C, and $p_{atm}/\gamma = 9.19$ m.

SOLUTION

The $NPSH_A$ is computed using equation (12.2.8):

$$NPSH_A = \frac{p_{atm}}{\gamma} + Z_s - \frac{p_v}{\gamma} - h_{L_{ent}} - h_{L_f} - \sum h_{L_m}$$

Referring to Table 2.1.2, $p_v/\gamma = 0.24$ m. The losses due to friction are computed using the Hazen–Williams equation (12.1.1):

$$V = 0.849CR^{0.63}S_f^{0.54}$$

The friction loss in meter per 1000 m of pipe can be expressed more conveniently as (equation 12.1.2)

$$h_{L_f} = 10,700\left(\frac{Q}{C}\right)^{1.85}D^{-4.87} = \left(\frac{151Q}{CD^{2.63}}\right)^{1.85}$$

where h_{L_f} is meters per 1000 m, Q is m³/s and D is in m.

Suction piping losses:

$$h_{L_f} = 10,700\left(\frac{0.12}{140}\right)^{1.85}(0.300)^{-4.87}$$

$$= 7.988 \text{ m per 1000 m of pipe}$$

$$= (7.988)(5\text{m}/1000 \text{ m}) = 0.0399 \text{ m total loss}$$

Discharge piping losses:

$$h_{L_f} = 10,700\left(\frac{0.12}{140}\right)^{1.85}(0.250)^{-4.87}$$

$$= 19.41 \text{ m per 1000 m of pipe}$$

$$= (19.41)(100\,\text{m}/1000 \text{ m}) = 1.941 \text{ m total loss}$$

Bend losses (discharge piping):

$$h_{L_b} = K\frac{V^2}{2g}$$

$$V = \frac{Q}{A} = \frac{0.12}{0.049} = 2.45 \text{ m/s}$$

$$K = 0.25 \text{ (Table 12.2.4)}$$

$$h_{L_b} = (0.25)\frac{(2.45)^2}{(2)(9.81)} = 0.076 \text{ m}$$

Gate valve losses (discharge piping):

$$K = 0.2 \text{ (Table 12.2.5)}$$

$$h_{L_G} = K\frac{V^2}{2g} = 0.2(0.306) = 0.061 \text{ m}$$

Gate valve loss (suction piping):

$$V = \frac{Q}{A} = \frac{0.12}{0.071} = 1.69 \text{ m/s}$$

$$h_{L_G} = (0.2)\frac{(1.69)^2}{(2)(9.81)} = (0.2)(0.145) = 0.029 \text{ m}$$

Bell mouth entrance loss:

$K = 0.05$ (Table 12.2.4)

$$h_{L_{ent}} = K\frac{V^2}{2g} = 0.05(0.145) = 0.007$$

Total minor losses:

$$\sum h_{L_m} = 0.076 + 0.076 + 0.061 + 0.029 + 0.007 = 0.249 \text{ m}$$

Applying equation (12.2.8) yields

$$NPSH_A = 9.19 + 3 - 0.24 - (0.0399 + 1.941) - 0.249 = +9.72$$

EXAMPLE 12.2.5

Estimate the net positive suction head at the point of cavitation inception ($NPSH_i$) for a system in which the total dynamic head is 40 m. The pump delivers 400 m³/h at a rotational speed of 1200 rpm at around 80 percent efficiency. The pump is to operate at an elevation of 1000 m above sea level and the temperature is 20°C.

SOLUTION

$NPSH_i$ is computed using equation (12.2.11)

$$NPSH_i = \frac{HKn_S^{4/3}}{10^6}$$

Use a K value of 1210 from Table 12.2.1. The specific speed is

$$n_S = \frac{nQ^{1/2}}{H^{3/4}} = \frac{1200\left(\dfrac{400}{3600}\right)^{1/2}}{40^{3/4}} = 25.15$$

$$NPSH_i = \frac{40(1210)(25.15)^{4/3}}{10^6} = 3.566$$

EXAMPLE 12.2.6

For the situation in example 12.2.5, the manufacturer's $NPSH_R$ is 3.3. What is the allowable suction head? The entrance loss is 0.001 m, the bend losses add up to 0.15 m, and the sum of headlosses due to friction in the suction and discharge piping is 1.5 m.

SOLUTION

To compute the allowable suction head, use equation (12.2.8) and solve for Z_s with $NPSH_R$ replacing $NPSH_A$:

$$NPSH_R = \frac{p_{atm}}{\gamma} + Z_s - \frac{p_V}{\gamma} - h_{L_{ent}} - h_{L_f} - \sum h_{L_m}$$

$p_{atm}/\gamma = 9.19$ m, $p_{V\gamma} = 0.24$ m.

Solving the above equation for Z_s yields

$$Z_s = NPSH_R - \frac{p_{atm}}{\gamma} + \frac{p_V}{\gamma} + h_{L_{ent}} + h_{L_f} + \sum h_{L_m}$$

$$Z_s = 3.3 - 9.19 + 0.24 + 0.001 + 1.5 + 0.15$$

$$Z_s = -4 \text{ m}$$

The minus sign means a suction lift. In practice, this lift should be reduced by about 0.6 m. The $NPSH_A$ should always be greater than the $NPSH_R$.

12.2.1.5 *Pump Characteristics*

A *pump head-characteristic curve* is a graphical representation of the total dynamic head versus the discharge that a pump can supply. These curves, which are determined from pump tests, are supplied by the pump manufacturer. Figure 12.2.7 shows the head-characteristic curve for various impeller diameters along with the efficiency and brake horsepower curves. When two or more pumps are operated, the *pump station losses*, which are the headlosses associated with the piping into and out of the pump, should be subtracted from the manufacturer's pump curve to derive the *modified head-characteristic curve*, as shown in Figure 12.2.8.

Two points of interest on the pump curve are the shutoff head and the normal discharge or rated capacity. The *shutoff head* is the head output by the pump at zero discharge, while the *normal discharge* (or *head*) or *rated capacity* is the discharge (or head) where the pump is operating at its most efficient level. Variable-speed motors can drive pumps at a series of rotative speeds, which would result in a set of pump curves for the single pump, as illustrated in Figure 12.2.9. Typically, to supply a given flow and head, a set of pumps is provided to operate in series or parallel and the number of pumps working depends on the flow requirements. This makes it possible to operate the pumps near their peak efficiency.

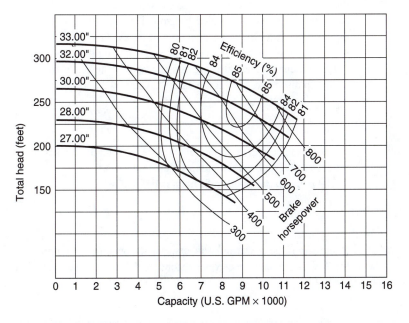

Figure 12.2.7 Manufacturer's pump performance curves.

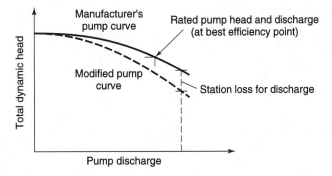

Figure 12.2.8 Modified pump curve.

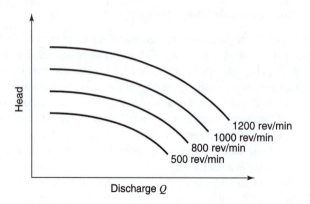

Figure 12.2.9 Pump performance curves for variable-speed pumps.

Multiple-pump operation for one or more pumps in parallel or in series requires the addition of the modified head-characteristic curves. For pumps operating in parallel, the modified head-characteristic curves are added horizontally with the respective heads remaining the same (see Figure 12.2.10a). For pumps operating in series, the modified head-characteristic curves are added vertically, with the respective discharges remaining the same (Figure 12.2.10b).

Pump manufacturers also provide curves relating the *brake horsepower* (required by pump) to the pump discharge (see Figure 12.2.7). The brake horsepower (*bhp*) is calculated using

$$bhp = \frac{\gamma Q H}{550e} \tag{12.2.12}$$

where Q is the pump discharge in cfs, H is the total dynamic head in ft, γ is the specific weight of water in lb/ft^3, and e is the pump efficiency.

The *power input* (*bhp*) in SI units is defined as

$$bhp = \frac{\gamma Q H}{e} \tag{12.2.13}$$

where *bhp* is the power in kilowatts, γ is the specific weight in kilonewtons per cubic meter, Q is the flow rate in cubic meter per second, and H is the total dynamic head in meters.

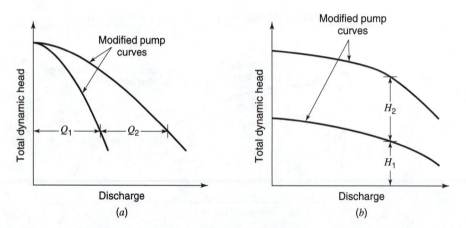

Figure 12.2.10 Pumps operating (*a*) in parallel and (*b*) series.

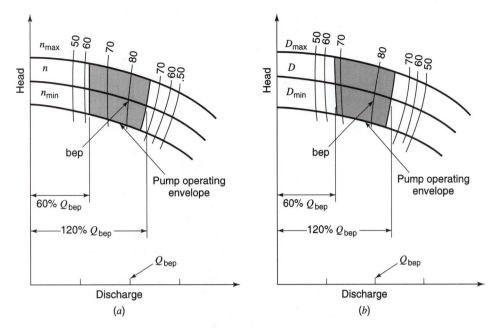

Figure 12.2.11 Pump operating envelopes based on the percentage of capacity at the best efficiency capacity. (*a*) Rotational speed; (*b*) Impeller diameter (from Tchobanoglous (1998)).

The *pump efficiency* is the power delivered by the pump to the water (water horsepower) divided by the power delivered to the pump by the motor (brake horsepower). *Efficiency curves*, as shown in Figure 12.2.7, define how well the pump is transmitting energy to water. Pumps operate best at their *best efficiency point* (bep) because of minimum radial loads on the impeller and minimum cavitation problems.

Operating ranges of a pump can be developed by establishing (1) a minimum acceptable efficiency and (2) setting upper and lower limits on the allowable impeller diameters. Figure 12.2.11 illustrates the operating range of a pump based on these criteria.

12.2.2 Pipes and Fittings

Water distribution piping can be of several types, including: ductile iron pipe (DIP), steel, polyvinyl chloride (PVC) pipe, asbestos cement (AC) pipe (ACP), reinforced concrete pressure pipe (RCPP), and others. The American Water Works Association (AWWA) publishes C-series standards that provide standards for pipe construction, installation, and performance. The following are several factors that must be considered in the selection of both exposed and buried pipe and fittings (Bosserman et al., 1998):

- Properties of the fluid
  - Corrosive or scale-forming properties
  - Unusual characteristics, e.g., viscosity of sludge
- Service conditions
  - Pressure (including surges and transients)
  - Corrosive atmosphere for exposed piping
  - Soil loads, bearing capacity and settlement, external loads, and corrosion potential for buried piping
- Availability
  - Sizes
  - Thickness

- Fittings
- Properties of the pipe
  - Strength (static and fatigue, especially for water hammer)
  - Ductility
  - Corrosion resistance
  - Fluid friction resistance of pipe or lining
- Economics
  - Required life
  - Maintenance
  - Cost (FOB plus freight to job site)
  - Repairs
  - Salvage value

Tables 12.2.2 and 12.2.3 present comparisons of pipe for exposed service and buried service, respectively. Bosserman et al. (1998) present a more detailed discussion of the factors that affect pipe selection. Table 12.2.4 presents energy loss coefficients for flanged pipe fittings.

Table 12.2.2 Comparison of Pipe for Exposed Service

Pipe	Advantages	Disadvantages/Limitations
Ductile iron pipe (DIP)	Yield strength: 290,000 kPa (42,000 lb/in^2) Ultimate strength: 414,000 kPa (60,000 lb/ in^2) $E = 166 \times 10^6$ kPa (24×10^6 lb/in^2) Ductile, elongation = 10% Good corrosion resistance Wide variety of available fittings and joints Available sizes: 100–1350 mm (4–54 in) ID Wide range of available thicknesses Good resistance to water hammer	Maximum pressure = 2400 kPa (350 lb/in^2) High cost, especially for long freight hauls No diameters above 1350 mm (54 in) Difficult to weld Class 53 is the thinnest allowed for American flanged pipe (with screwed flanges in the U.S.)

Steel	Yield strengths: 207,000–414,000 kPa
	  (30,000 to 60,000 lb/in^2)
	 Ultimate strengths: 338,000–518,000 kPa
	  (49,000–75,000 lb/in^2)
	 $E = 207 \times 10^6$ kPa (30×10^6 lb/in^2)
	 Ductile, elongation varies from 17 to 35%
	 Pressure rating to 17,000 kPa (2500 lb/in^2)
	 Diameters to 3.66 m (12 ft)
	 Widest variety of available fittings and joints
	 Custom fittings can be mitered and welded
	 Excellent resistance to water hammer
	 Low cost
		Corrosion resistance low unless lined

Source: Bosserman et al. (1998*b*).

Table 12.2.3 Comparison of Pipe for Buried Service

Pipe	Advantages	Disadvantages/Limitations
Ductile iron pipe (DIP)	See Table 12.2.2; high strength for supporting earth loads, long life	See Table 12.2.2; may require wrapping or cathodic protection in corrosive soils
Steel pipe	See Table 12.2.2; high strength for supporting earth loads	See Table 12.2.2; poor corrosion resistance unless both lined and coated or wrapped, may require cathodic protection in corrosive soils

Table 12.2.3   Comparison of Pipe for Buried Service (*continued*)

Pipe	Advantages	Disadvantages/Limitations
Polyvinyl chloride (PVC) pipe	Tensile strength (hydrostatic design basis = 26,400 kPa (4000 lb/in^2); E = 2,600,000 kPa (400,000 lb/in^2); light in weight, very durable, very smooth, liners and wrapping not required, good variety of fittings available or can use ductile or cast-iron fittings with adapters, can be solvent-welded; diameters from 100 to 375 mm (4 to 36 in)	Maximum pressure = 2400 kPa (350 lb/in^2); little reserve of strength for water hammer if ASTM D 2241 is followed, AWWA C900 includes allowances for water hammer; limited resistance to cyclic loading, unsuited for outdoor use above ground
Asbestos cement (AC) pipe (ACP)	Yield strength: not applicable; design based on crushing strength, see ASTM C 296 and C 500; E = 23,500,000 kPa (3,400,000 lb/in^2); rigid, light weight in long lengths, low cost; diameters from 100 to 1050 mm (4 to 42 in) compatible with cast-iron fittings, pressure ratings from 1600 to 3100 kPa (225 to 450 lb/in^2) for large pipe 450 mm (18 in) or more	Attacked by soft water, acids, sulfates; requires trust blocks at elbows, tees, and dead ends; maximum pressure = 1380 kPa (200 lb/in^2) for pipe up to 400 mm (16 in); health hazards of asbestos in potable water service are controversial, U.S. EPA proposed ban on most asbestos products in 1988, being reconsidered for AC pipe
Reinforced concrete pressure pipe (RCPP)	Several types available to suit different conditions; high strength for supporting earth loads, wide variety of sizes and pressure ratings, low cost, sizes from 600 to 3600 mm (24 to 144 in)	Attacked by soft water, acids, sulfides, sulfates, and chlorides, often requires protective coatings; water hammer can crack outer shell, exposing reinforcement to corrosion and destroying its strength with time; maximum pressure = 1380 kPa (200 lb/in^2)

Source: Bosserman et al. (1989*b*).

Table 12.2.4   Recommended Energy Loss Coefficients, K, for Flanged Pipe Fittings[a]

Fitting		K	Fitting		K
Entrance			Forged or cast fittings		
Bell mouth		0.05	Return bent, $r = 1.4 D$		0.40
Rounded		0.25	Tee, line flow		0.30
Sharp edged		0.5	Tee, branch flow		0.75
Projecting		0.8			

Table 12.2.4 Recommended Energy Loss Coefficients, K, for Flanged Pipe Fittings[a] (*continued*)

Fitting	K	Fitting		K
Exits		Cross, line flow		0.50
All of the above	1.0	Cross, branch flow		0.75
Bends, mitered		Wye, 45°		0.50
$\theta = 15°$	0.05			
$\theta = 22.5°$	0.075	Increasers		
$\theta = 30°$	0.10	Conical		
$\theta = 45°$	0.20			
$\theta = 60°$	0.35			
$\theta = 90°$	0.80	Conical (approximate)		
90° bend 3 × 30° = 90° 4 × 22.5° = 90°	0.30	Sudden		
		Reducers		
Forged or cast fittings		Conical		
90° elbow, standard	0.25			
90° elbow long radius	0.18	Sudden		
45° elbow	0.18			

Increasers, Conical:
$$h = K\left[1 - \left(\frac{D_1}{D_2}\right)^2\right]V_2^2/2g$$
$$K = 3.5(\tan \theta)^{1.22}$$

Conical (approximate):
$$h = 0.25(V_1^2 - V_2^2)/2g$$

Sudden:
$$h = \frac{V_1^2 - V_2^2}{2g} = \left[\left(\frac{A_1}{A_2}\right)^2 - 1\right]\frac{V_2^2}{2g}$$

Reducers, Conical:
$$h = KV_2^2/2g$$
$$K = 0.03 \pm 0.01$$

Sudden:
$$h = \frac{1}{2}\left[1 - \left(\frac{D_1}{D_2}\right)^2\right]V_2^2/2g$$

[a]$h = KV^2/2g$, where v is the maximum velocity in nonprismatic fittings. Increase K by 5% for each 25 mm (1 in) decrement in pipe smaller than 300 mm (12 in). Expect K values to vary from −20 to +30% or more.

Source: Sanks (1998).

12.2.3 Valves

Valves are very important for the proper functioning of water distribution systems. The types of valves include:

- *Isolation valves*, used to shut down portions of a system, include:
  - Ball
  - Butterfly
  - Cone
  - Eccentric plug
  - Gate (Table 4.3.3)
  - Plug
- *Check valves* (Table 4.3.3) are directional control valves that allow flow of water to only one direction
  - Swing check valves
  - Counterpost-guided (or silent) check valves

- Double-door (or double-disc or double-leaf) check valves
- Foot valves
- Ball lift valves
- Tilting (or slanting) disc check valves

Control valves are used to regulate flow or pressure by operating partly open, creating high headlosses and pressure differentials. These include *pressure-reducing valves* (PRV) and *pressure-sustaining valves* (PSV). PRVs are used to monitor (reduce) downstream pressures and PSVs are used to monitor pressures upstream of the valve. *Flow control valves* are used to maintain flow at a preset rate through throttling. Table 12.2.5 lists energy loss coefficients for fully open valves and Tables 4.3.1 and 4.3.3 provide loss coefficients for valves.

Table 12.2.5   Recommended Energy Loss Coefficients, *K*, for Valves Fully Open[a,b,c]

Valve Type	K
Angle	1.8–2.9
Ball	0.04
Butterfly	
25-lb Class	0.16
75-lb Class	0.27
150-lb Class	0.35
Check valves	
Center-guided globe style	2.6
Double door	
8 in or smaller	2.5
10 to 16 in	1.2
Foot	
Hinged disc	1–1.4
Poppet	5–14
Rubber flapper	
V < 6 ft/s	2.0
V > 6 ft/s	1.1
Slanting disc[d]	0.25–2.0
Swing[d]	0.6–2.2
Cone	0.04
Diaphragm or pinch	0.2–0.75
Gate	
Double disc	0.1–0.2
Resilient seat	0.3
Globe	4.0–6.0
Knife gate	
Metal seat	0.2
Resilient seat	0.3
Plug	
Lubricated	0.5–1.0
Eccentric	
Rectangular (80%) opening	1.0
Full bore opening	0.5

[a]$h = KV^2/2g$, where V is the velocity in the approach piping.

[b]For 300-mm (12-in) valves and velocities of about 2 m/s (6 ft/s).

   Note that K may increase significantly for smaller valves. Consult the manufacturer.

[c]Expect K to vary from –20 to + 50% or more.

[d]Depending on adjustment of closure mechanism, velocity may have to exceed 4 m/s (12 ft/s) to open the valve fully. Adjustment is crucial to prevent valve slam, so be very conservative in estimating K (which can be several times the tabulated value). Consult the manufacturer.

Source: Sanks (1998).

12.3 SYSTEM CONFIGURATION AND OPERATION

Water distribution systems are made up of networks of discrete components: pipes, fittings, pumps, valves, and storage tanks. The configurations of these systems vary significantly. These systems are typically very large and complex, especially for a large number of consumers spread over a wide service area. Hundreds to thousands of pipes may be required to distribute water to users throughout the system. Storage tanks and pumping stations are required to provide flexibility in demands as a function of time and location. The time variation in demand is illustrated in Figure 12.3.1.

Large systems may include several pressure zones where pumps, valves, and tanks maintain required service pressures. Figures 12.3.2 and 12.3.3 illustrate an example system that includes two pressure zones served from a single treatment plant. In this system the one pressure zone is served by pumps at the treatment plant and an elevated storage tank. The other pressure zone is served by a booster pumping station and an elevated storage tank.

The *piezometric* surface (surface of hydraulic grade line, HGL) is one way to visualize water-distribution-system hydraulics for various water-distribution-system configurations. Figure 12.3.4 illustrates the hydraulic grade line for a simple system under two operating conditions: (1) a low-demand condition and (2) a high-demand condition. During the low-demand conditions water is pumped from the pumping station to satisfy demand and to fill the elevated tank. During the high-demand condition the pumping station cannot supply the required demand, so water is supplied to the network by both the pumping station and the elevated storage tank. During low-demand conditions the HGL is highest at the pumping station and slopes downward to meet the free surface at the elevated storage tank. The HGL slope indicates the energy required to pump water to the elevated storage tanks. During high demand the HGL drops to a minimum in the highest-demand area.

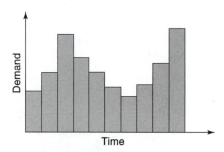

Figure 12.3.1 A demand curve.

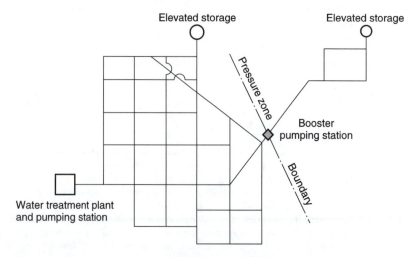

Figure 12.3.2 Schematic of example system (from Velon and Johnson (1993)).

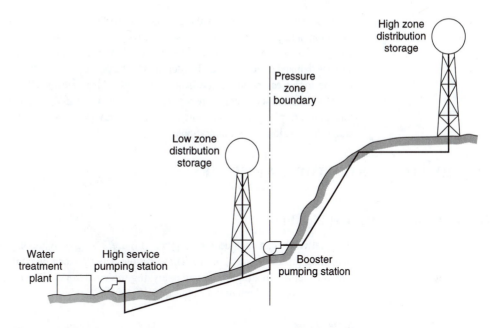

Figure 12.3.3 Typical two-pressure-zone system (from Velon and Johnson (1993)).

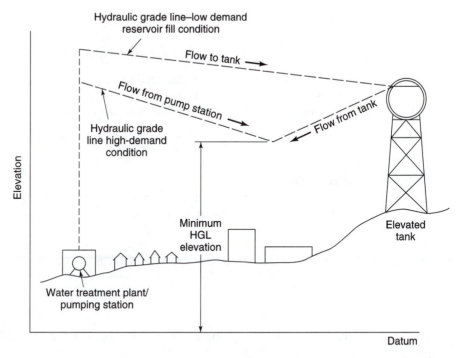

Figure 12.3.4 Hydraulic grade line under two demand conditions (from Velon and Johnson (1993)).

For more complex systems the piezometric surface, represented by contour plots of the HGL elevation, and contour plots of the pressure can be very helpful in analyzing the configuration and design of a system. Areas of a water distribution system that are subject to low pressures under various demand conditions can be identified. The portions of a water distribution system with high friction losses can be identified from these plots. Also, facilities that limit the ability of a system to meet demand and pressure requirements can be identified. Piezometric surface plots and contour pressure plots are a convenient means for reviewing and analyzing either new or existing water distribution system operations.

12.4   HYDRAULICS OF SIMPLE NETWORKS

12.4.1   Series and Parallel Pipe Flow

Pipe networks can include series pipes (Figure 12.4.1*a*), parallel pipes (Figure 12.4.1*b*), and branching pipes (Figure 12.4.1*c*). These simple networks can be converted to an *equivalent pipe*, which is helpful in simplifying and analyzing these networks. Two pipes are equivalent when, for the same headloss, both deliver the same rate of flow.

Series Pipes
For a system with two or more pipes in series, the total headloss is

$$h_{L_e} = \sum_i h_{L_i} \tag{12.4.1}$$

or

$$K_e Q^n = \sum K_i Q^{n_i} = K_1 Q^{n_1} + K_2 Q^{n_2} \dots \tag{12.4.2}$$

where K_e is the pipe coefficient for the equivalent pipe. When using the Hazen–Williams equation, all $n = 1.85$, and when using the Darcy–Weisbach equation, $n = 2$ for rough pipes and $n = 1.75$ for smooth pipes for fully turbulent flow; therefore K_e simplifies to

$$K_e = K_1 + K_2 + \dots = \sum K_i \tag{12.4.3}$$

Parallel Pipes
The concept of an equivalent pipe can be applied to two or more pipes in parallel. The headloss in each parallel pipe between junctions must be equal:

$$h_L = h_{L_1} = h_{L_2} = h_{L_3} = \dots \tag{12.4.4}$$

The total flow rate is the sum of individual flows in each pipe is given as

$$Q = Q_1 + Q_2 + \dots = \sum Q_i \tag{12.4.5}$$

Substituting $Q = \left(\dfrac{h_L}{K_e}\right)^{\frac{1}{n_e}}$, from $h_L = KQ^n$, into equation (12.4.5) yields

$$\left(\frac{h_L}{K_e}\right)^{\frac{1}{n_e}} = \left(\frac{h_L}{K_1}\right)^{\frac{1}{n_1}} + \left(\frac{h_L}{K_2}\right)^{\frac{1}{n_2}} + \dots = \sum \left(\frac{h_L}{K_i}\right)^{\frac{1}{n_i}} \tag{12.4.6}$$

When using the Hazen–Williams equation, all the exponents n_i are equal; when using the Darcy–Weisbach equation, it is customary to assume n_i equal for all pipes (i.e., fully turbulent, $n = 2$) for rough pipes. Then equation (12.4.6) can be reduced to

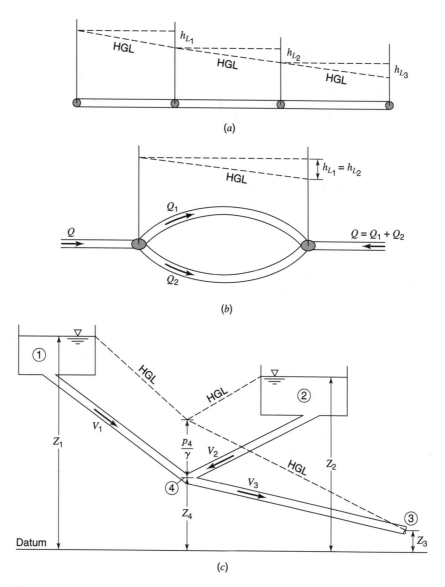

Figure 12.4.1 Simple pipe systems. (*a*) Series pipe system; (*b*) parallel pipe system; (*c*) branching pipe system.

$$\left(\frac{1}{K_e}\right)^{\frac{1}{n}} = \left(\frac{1}{K_1}\right)^{\frac{1}{n}} + \left(\frac{1}{K_2}\right)^{\frac{1}{n}} + \ldots = \sum_i \left(\frac{1}{K_i}\right)^{\frac{1}{n}} \qquad (12.4.7)$$

EXAMPLE 12.4.1

The system shown in Figure 12.4.2 consists of two reservoirs, the pump, pipe AB, and parallel pipes BC and BD. Headlosses between the lower reservoir and the pump are to be ignored. The Manning roughness factor for each pipe is $n = 0.01$. For a flow rate of 6 cfs, determine the total head to pump if $\Delta Z = 34$ ft. What is the brake horsepower required for an efficiency of 0.80?

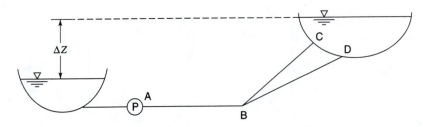

Figure 12.4.2 Example 12.4.1 pipe system.

Pipe	Diameter	Length
AB	18"	7000 ft
BC	9"	3000 ft
BD	12"	4000 ft

SOLUTION

The headloss for Manning's equation can be expressed in terms of the diameter using equation (12.1.10), so that

$$K_{AB} = \frac{(0.01)^2(7000)}{0.216(1.5)^{5.33}} = 0.373$$

Similarly $K_{BC} = 6.436$ and $K_{BD} = 1.852$. The headlosses in pipes BC and BD are equal, so that

$$h_{L_{BC}} = K_{BC}Q_{BC}^2 = h_{L_{BD}} = K_{BD}Q_{BD}^2$$

or

$$Q_{BC} = \sqrt{\frac{K_{BD}}{K_{BC}}}Q_{BD}$$

The total flow in pipes BC and BD ($Q_{BC} + Q_{BD}$) must equal the flow in pipe AB, i.e.,

$$Q_{AB} = Q_{BC} + Q_{BD} = \sqrt{\frac{K_{BD}}{K_{BC}}}Q_{BD} + Q_{BD}$$

$$= \left(\sqrt{\frac{K_{BD}}{K_{KBC}}} + 1\right)Q_{BD} = (0.536 + 1)\,Q_{BD} = 1.536\,Q_{BD}$$

$$Q_{BD} = 0.651Q_{AB}$$

$$Q_{BC} = 0.349Q_{AB}$$

The total headloss from the lower reservoir to the upper reservoir is

$$h_L = h_{L_{AB}} + h_{L_{BD}} = h_{L_{AB}} + h_{L_{BC}} = K_{AB}Q_{AB}^2 + K_{BD}Q_{BD}^2$$

$$= K_{AB}Q_{AB}^2 + K_{BD}(0.651Q_{AB})^2 = 0.373Q_{AB}^2 + 1.852(0.651Q_{AB})^2$$

$$h_L = 1.158Q_{AB}^2$$

The total head to pump is then

$$H_T = \Delta Z + 1.158Q_{AB}^2$$

(a) For $Q_{AB} = 6$ cfs and $\Delta Z = 34$ ft:

$$H_T = 34 + 1.158(6)^2 = 75.69 \text{ ft.}$$

(b) $bhp = \dfrac{QH_T\gamma}{550e} = \dfrac{6 \times 75.69 \times 62.4}{550 \times 0.80} = 64.4 \text{ hp}$

12.4.2   Branching Pipe Flow

The flow distribution in branching systems can be determined by applying the continuity and energy equations. Figure 12.4.1c illustrates a branching system with the hydraulic grade line shown. The energy equation for the pipe from reservoir surface 1 to the junction at 4 is

$$Z_1 = Z_4 + \frac{p_4}{\gamma} + h_{L_1} \tag{12.4.8}$$

From reservoir 2 to 4, the energy equation is

$$Z_2 = Z_4 + \frac{p_4}{\gamma} + h_{L_2} \tag{12.4.9}$$

and from junction 4 to the outlet at 3, the energy equation is

$$Z_3 = Z_4 + \frac{p_4}{\gamma} - h_{L_3} \tag{12.4.10}$$

The continuity equation is

$$Q_1 + Q_2 = Q_3 \tag{12.4.11}$$

Knowing the pipe sizes, lengths, and controlling elevations, we can solve these equations for the pressure head at 4, p_4/γ and the velocities, V_1, V_2, and V_3. Knowing the velocities the flow in each pipe can be determined.

EXAMPLE 12.4.2

The three reservoirs shown in Figure 12.4.3 are connected by riveted steel pipes. Determine the flow rate in each pipe, assuming fully turbulent flow.

SOLUTION

First, develop an expression for the headloss $h_L = KQ^2$ for riveted steel pipe. The relative roughness values are determined from Figure 4.3.3 and the resulting friction factors are based on fully turbulent flow from the Moody diagram (Figure 4.3.5):

Pipe 1: $\epsilon/D = 0.0015, f_1 = 0.0215$: $K_1 = \dfrac{fL}{39.7D^5} = \dfrac{0.0215(3000)}{39.7(24/12)^5} = 0.0508$

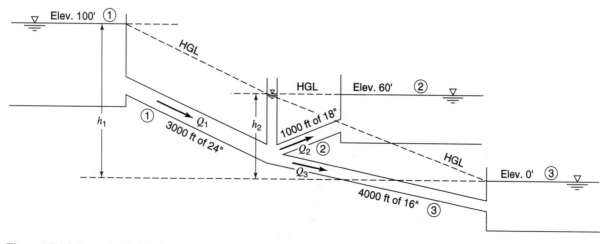

Figure 12.4.3 Example 12.4.2 pipe system.

Pipe 2: $\epsilon/D = 0.0020, f_2 = 0.0234$: $K_2 = \dfrac{fL}{39.7D^5} = \dfrac{0.0234(1000)}{39.7(18/12)^5} = 0.0776$

Pipe 3: $\epsilon/D = 0.0023, f_3 = 0.0242$: $K_3 = \dfrac{fL}{39.7D^5} = \dfrac{0.0242(4000)}{39.7(16/12)^5} = 0.579$

The expressions for headloss are

$$h_{L_1} = K_1 Q_1^2 = h_1 - h_j$$

$$h_{L_2} = K_2 Q_2^2 = h_j - h_2$$

$$h_{L_3} = K_3 Q_3^2 = h_j - h_3$$

where h_j is the HGL elevation at the pipe junction. The continuity at the pipe junction is $Q_1 - Q_2 - Q_3 = 0$. Substituting values of K_1, K_2, and K_3 and h_1, h_2, and h_3 into the above headloss equations results in

$$0.0508Q_1^2 = 100 - h_j$$

$$0.0776Q_2^2 = 60 - h_j$$

$$0.579Q_3^2 = h_j - 0$$

These three equations and the continuity relationship $Q_1 - Q_2 - Q_3 = 0$ result in four equations with four unknowns Q_1, Q_2, Q_3, and h_j. They can be solved simultaneously to obtain $Q_1 = 23.5$, $Q_2 = 12.4$, $Q_3 = 11.2$ cfs, and $h_j = 72$ ft. The Reynolds number should be checked to assume the f values are correct.

A trial-and-error procedure can be used by finding an expression of the discharge in term of h_j and substituting these into the continuity equation:

$$Q_1 = \sqrt{\frac{h_1 - h_j}{K_1}} = \sqrt{\frac{100 - h_j}{0.0508}}$$

$$Q_2 = \sqrt{\frac{h_2 - h_j}{K_2}} = \sqrt{\frac{h_j - 60}{0.0776}}$$

$$Q_3 = \sqrt{\frac{h_j - h_3}{K_3}} = \sqrt{\frac{h_j - 0}{0.579}}$$

Substituting these into the continuity equation at the pipe junction yields

$$\sqrt{\frac{100 - h_j}{0.0508}} - \sqrt{\frac{h_j - 60}{0.0776}} - \sqrt{\frac{h_j - 0}{0.579}} = 0$$

Then solve for $h_j = 72$ ft.

12.5 PUMP SYSTEMS ANALYSIS

12.5.1 System Head Curves

First we consider a pump-force main system. The head developed by the pump must equal the total dynamic head in the system. To determine the operating point, the pump head-characteristic curve (H-Q curve of the pump) and a *system-head-capacity curve* (H-Q curve of the piping system) are required. The *pump operating point* is the single point on these two H-Q curves where they intersect. Figure 12.5.1 illustrates a system-head-capacity curve in which the head is the sum of friction, fitting, and valve losses. These curves are plotted over a range of discharges, starting at zero discharge and varying up to the maximum expected. The system-head-capacity curves are approximate functions of the velocity head $V^2/2g$. These curves are typically approximated as parabolas, keeping in mind that headlosses are a function of 1.85 for the Hazen–Williams equation.

Figure 12.5.2 illustrates a typical system-head-capacity curve with a negative lift. Figure 12.5.3 illustrates a system with two different discharge heads. The combined system-head curve is obtained by adding the flow rates for the two pipes at the same head. Figure 12.5.4 illustrates a system with a varying static lift. Note that there are separate system-head curves, one for the maximum lift and one for the minimum lift. Figure 12.5.5 illustrates a system-head curve for a throttled system in which various valve openings are used to vary the head-discharge relationship.

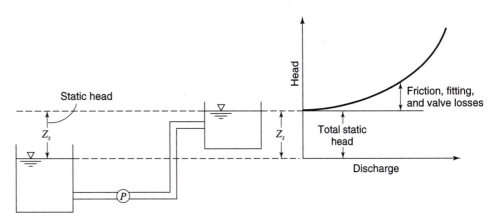

Figure 12.5.1 Typical system-head-capacity curve with positive lift.

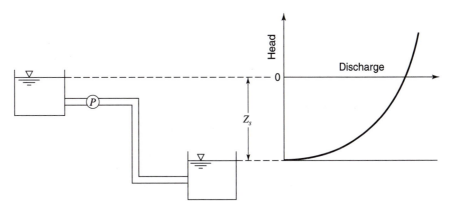

Figure 12.5.2 Typical system-head-capacity curve with negative lift.

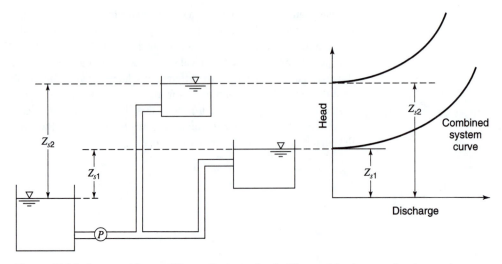

Figure 12.5.3 System with two different discharge heads. The combined system-head curve is obtained by adding the flow rates for the two pipes at the same head.

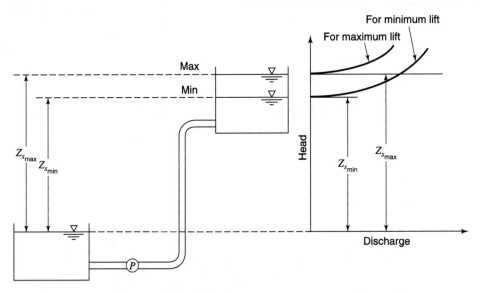

Figure 12.5.4 System with varying static lift.

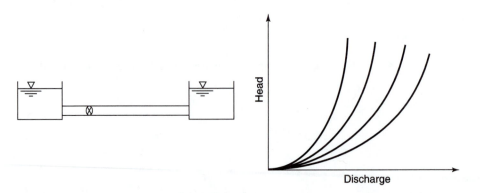

Figure 12.5.5 System throttled by valve operation.

12.5.2 Pump Operating Point

In order to determine the operating point for a pump or pumps in a piping system, the modified pump head-capacity curve is superimposed on the system-head-capacity curve. Figure 12.5.6 illustrates a single-speed pump head-capacity curve superimposed on a system-head-capacity curve for a piping system with a fixed static lift Z_s. The point of intersection of the two head-capacity curves is the pump operating point. The pump operating point should be at or near the maximum efficiency of the pump, as shown in Figure 12.5.6.

Figure 12.5.7 illustrates the operating points for two single-speed pumps operating in parallel for a piping system that has a range of static lift. Note that system-head curves are shown for the maximum lift and for the minimum lift. Also shown is the range of operating points (between minimum and maximum static lift) for a single pump.

Figure 12.5.8 shows a system with three single-speed pumps of different sizes operating in parallel. Figure 12.5.9 illustrates the operating points for a variable-speed pump with a system-head curve. The affinity laws (described in section 12.2.1.2) can be used to determine the rotational speed at any desired operating point.

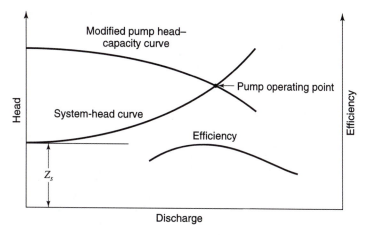

Figure 12.5.6 Determining operating points for a single-speed pump with a fixed static lift Z_s.

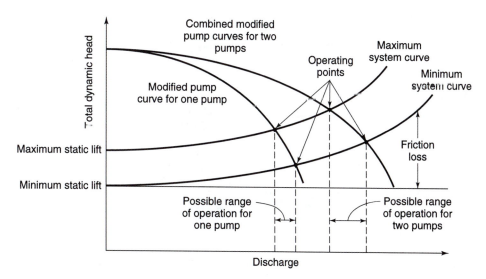

Figure 12.5.7 Determining operating points for two single-speed pumps in parallel and a variable static lift.

As pipes age, the system-head curve rises due to changes in frictional resistance, resulting from scaling (deposition) on the pipe walls. Figure 12.5.10 illustrates the changing operating point for the decrease in the Hazen–Williams coefficient that can result from aging. Figure 12.5.11 illustrates the operating point for two pumps operating in series.

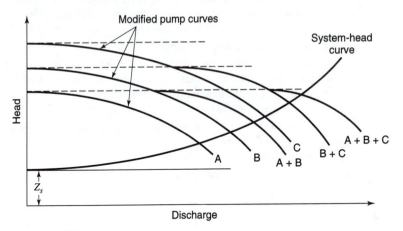

Figure 12.5.8 System with three single-speed pumps of different sizes operating in parallel.

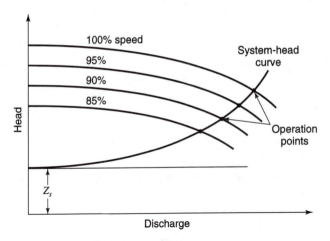

Figure 12.5.9 System with a single pump operated at variable speed.

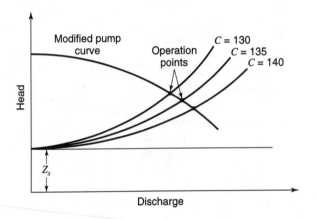

Figure 12.5.10 Change in operating points for decreasing Hazen–Williams coefficient due to pipe aging.

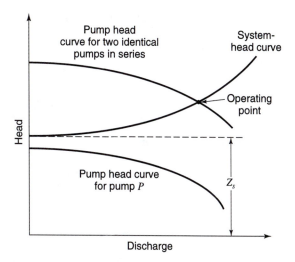

Figure 12.5.11 Operating point for two pumps operating in series.

12.5.3   System Design for Water Pumping

Water pumping stations can be considered to fall into the following general types of systems (Bosserman et al., 1998):

- Pumping into an elevated storage tank from a source such as a well
- Pumping raw water from a river or lake
- Pumping (high-service pumping) of finished water at high pressure
- Booster pumping
  - In-line booster pumping into an elevated tank or a reservoir (Figure 12.5.12)
  - Distribution system booster pumping without a storage tank

Pumping stations (except for well pumping stations) require redundant pumps, so that when each pump is taken out of service for maintenance, the other pumps can meet all demands. Table

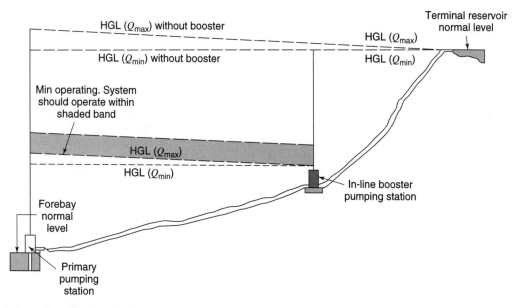

Figure 12.5.12 Transmission main with a primary pumping station and a booster pumping station (from Bosserman et al. (1998)).

12.5.1 compares water pumping station types, describing the advantages and disadvantages of each type. Vertical mixed-flow or radial flow pumps are almost always used for pumping of wells. Table 12.5.2 compares well pumps. Bosserman (1999, 2000) provides additional detail on pump system hydraulic design.

Table 12.5.1 Comparison of Water Pumping Station Types

Advantages	Disadvantages
Vertical wet well pumps	
Wide selection is available: axial flow (single stage), mixed flow (single and multistage). Francis turbine (single and multistage), and radial flow (turbine is single or multistage)	High superstructure to pull pumps or, at greater expense, pumps can be pulled through a hatch in the roof of a low superstructure by a truck crane parked outside, which may be cheaper than inside crane
Small floor area and small superstructure	Requires disconnecting the motor and pulling the pump for inspection or repair
Ground floor pump motors above flooding	Requires more shaft bearings
Pump suction always flooded; no priming	Priming may be required if air in the pump column causes problems
NPSH easily met	If idle pumps collect air in pump column, either (1) an automatic air vent valve at the discharge elbow or (2) priming must be installed
Ideal for deep installation and large variations in water level	
Vertical turbines are especially flexible in meeting head requirements by adding stages	
Wet well–dry well, horizontal pumps below water level (flooded suction)	
Pump types available include split-case centrifugal, end-suction overhung-shaft, and horizontally mounted axial-flow (propeller and mixed-flow) types	Large pump room floor area
Eliminates priming and air problems	Greater excavation
Low headroom requirement	Motors are subject to flooding
Easy maintenance and accessibility	Longer electric conduits
Service can be accomplished without disconnecting motor	Pumping stations are more costly due to additional floor area, access, lighting, ventilation, etc.
	Greater forced ventilation needed to cool motors
Horizontal pumps on floor above suction well	
Easy maintenance and accessibility	Requires priming equipment
Short electric conduits	Dependability (due to priming equipment) is reduced
Motors above floor level	Limits the choice of pumps to those for negative suction
Ventilation minimized	
Reduced excavation and below-ground construction; similar to wet pit	

Source: Klein and Sanks (1989).

Table 12.5.2 Comparison of Well Pumps

Advantages	Disadvantages
Self-priming centrifugals	
Capacity to 0.25 m^3/s (4000 gal/min)	Limited to caisson, gallery, or wells with a water table less than 4.6 m (15 ft) below the pump
Vertical turbines (V/T)	
Most common type of well pump	Efficiency as low as 50% depending on the size and application
Motor or engine driven	
Diameter: 50 to > 1200 mm (2 to > 48 in)	
Capacities exceeding 0.6 m3/s (10,000 gal/min)	
Use for finished water and booster pumping is increasing	
Can be tailored for specific head by adding bowls	

Table 12.5.2 Comparison of Well Pumps (*continued*)

Advantages	Disadvantages
Vertical turbines, lineshaft	
Minimum maintenance	Unsuited to crooked wells
Driver accessible	Somewhat noisy
Excellent for wells less than 90 m (300 ft) deep	Suitability is marginal at depths over 300 m (1000 ft). Even at depths less than 180 m (600 ft), shaft stretch wears impellers and bowls unless the thrust bearing is kept carefully adjusted
Vertical turbines, submersible	
Quiet, so they are suitable near hospitals, schools, and residences	Maintenance requires pulling the entire unit
Only solution for crooked wells	Regular maintenance is required
Practical at depths over 210 m (700 ft)	Seal problems may be severe
Water cooling is very effective	Long electric cables
No long-shaft problems	

Source: Klein and Sanks (1998).

EXAMPLE 12.5.1

Water at 60°F is pumped from reservoir A (surface elevation 20.0 ft) to reservoir B (surface elevation 50.0 ft), as shown in Figure 12.5.16. A 12-in diameter suction pipe (from reservoir A to pump) is 100 ft long and the discharge pipe (from pump to reservoir B) is 10.0 in in diameter and 1500 ft long. Assume that the friction factor for each pipe is 0.015. The suction pipe has an entrance loss coefficient of $K_e = 1.0$ and a check valve loss coefficient of $K_V = 2.5$. The discharge pipe has two gate valves $K_{GV} = 0.2$ fully open. The pump has the head-discharge curve and efficiency curves shown in Figure 12.5.13. The pump is operated at 1800 rpm.

(a) What is the flow rate?
(b) Find the power required to drive the pump.

SOLUTION

(a) Begin by developing an expression for the system curve. Write the energy equation from reservoir surface A to reservoir surface B:

$$0 + 0 + 20 + h_p = 0 + 0 + 50 + \Sigma\, h_{L_{A-B}}$$

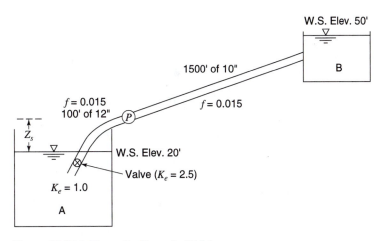

Figure 12.5.16 Figure for Example 12.5.1.

$$h_p = 30 + \frac{V_{12"}^2}{2g}\left[K_{entr} + K_V + f_{12"}\frac{L_{12"}}{D_{12"}}\right] + \frac{V_{10"}^2}{2g}\left[2K_{GV} + f_{10"}\frac{L_{10"}}{D_{10"}} + K_{exit}\right]$$

$$h_p = 30 + \frac{V_{12"}^2}{2g}\left[1.0 + 2.5 + 0.015\frac{100}{1.0}\right] + \frac{V_{10"}^2}{2g}\left[2(0.2) + 0.015\frac{1500}{10/12} + 1\right]$$

Next, obtain an expression for the system-head curve in terms of Q:

$$h_p = 30 + \frac{Q^2}{2g\left(\pi(1)^2/4\right)^2}(5) + \frac{Q^2}{2g\left(\pi(10/12)^2/4\right)^2}(28.4) = 30 + (0.126 + 1.482)Q^2$$

$$h_p = 30 + 1.61\,Q^2 \text{ (system-head curve)}$$

Plot a few points of the system-head curve on the pump-characteristic curve:

Q (cfs)	h_p (ft)
4	55.8
5	70.2
6	88.0
5.5	78.7

From the plot, the operating point is at $Q = 5.7$ cfs at a head of 82.3 ft, and $e = 74\%$.

(b) The power required to drive the pump is computed using equation (12.2.12)

with $Q = 5.7$ cfs, $h_p = 82.3$ ft, and $e = 74\%$

$$bhp = \frac{\gamma Q h_p}{550e} = \frac{(62.4 \text{ lb/ft}^3)(5.7\,cfs)(82.3 \text{ ft})}{550(0.74)} = 71.9 \text{ HP}$$

EXAMPLE 12.5.2

The pump whose characteristic curves are shown in Figure 12.5.13 is used to pump water from reservoir A to reservoir B through 1000 ft of 8.00-in diameter pipe whose friction factor is 0.017 (as shown in Figure 12.5.14). Use the entrance loss coefficient of $K_e = 0.5$ and gate valve loss of $K_{GV} = 0.2$. (a) What flow rate does it pump? (b) What input power is required to drive the pump? (c) If the critical value of σ for this pump is 0.20, what is the maximum distance L that the pump may be placed along the line from reservoir A without cavitation occurring?

SOLUTION

(a) Write the energy equation:

$$\frac{P_A}{\gamma} + Z_A + \frac{V_A^2}{2g} + h_p = \frac{P_B}{\gamma} + Z_B + \frac{V_B^2}{2g} + \sum h_L$$

and solving for h_p:

$$h_p = 30 + [0.5 + 2(0.2) + 0.017\frac{1000}{8/12} + 1.0]\frac{V_{8"}^2}{2g}$$

Since h_p is needed as a function of Q, replace $V_{8"}$ with Q/A:

$$V_{8"} = \frac{Q}{A} = \frac{Q}{\frac{\pi}{4}\left(\frac{8}{12}\right)^2} = \frac{Q}{0.349}$$

$$h_p = 30 + [1.9 + 25.5]\frac{Q^2}{2g(0.349)^2} = 30 + 3.49Q^2$$

Calculate h_p required for a few values of Q and plot on the characteristic curve:

Q (cfs)	h_p (ft)
3	61.4
4	85.5
5	117.3
4.5	100.7
4.4	97.6

From the plot, $Q = 4.4$ cfs at a head of 98 ft and $e = 83\%$.

(b) The power required to drive the pump is

$$bhp = \frac{\gamma Q h_p}{550(0.83)} = \frac{62.4(4.4)(98)}{550(0.83)} = 59 \text{ HP}$$

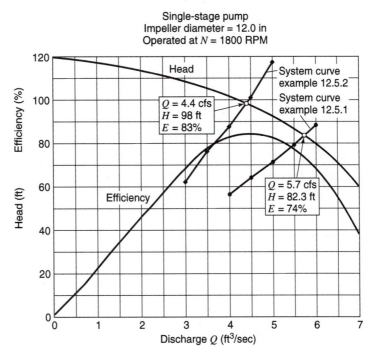

Figure 12.5.13 Characteristic curves for Examples 12.5.1 and 12.5.2.

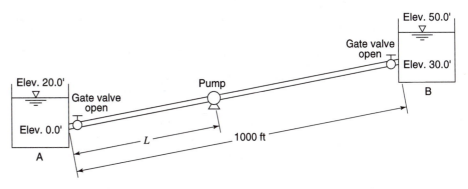

Figure 12.5.14 Example 12.5.2 pipe system.

EXAMPLE 12.5.3

Two reservoirs A, and B, are connected by a long pipe that has characteristics such that the headloss through the pipe is expressible as $h_L = 20(Q/100)^2$, where h_L is in feet and Q is the flow rate in gpm. A single pump with the head-capacity curve for 1800 rpm given below is used to pump the water from A to B. (a) For the single pump, plot a curve showing delivery rate versus difference in elevation, Δz, for ΔZ ranging from -20 to $+80$ ft. (b) Repeat for a rotative speed of 2160 rpm.

Head (ft)	Flow Rate (gpm)
100	0
90	110
80	180
60	250
40	300
20	340

SOLUTION

(a) Compute the system-head curve for $\Delta z = -20, 0, 40,$ and 80 ft.

Head (ft)	Flow Rate (gpm)	h_L (ft)	System head ($\Delta Z + h_L$)			
			$\Delta Z = -20$	$\Delta Z = 0$	$\Delta Z = 40$	$\Delta Z = 80$
100	0	0	-20	0	40	80
90	110	24.2	4.2	24.2	64.2	104.2
80	180	64.8	44.8	64.8	104.8	144.8
60	250	125	105.0	125.0	165.0	205.0
40	300	180	160.0	180.0	220.0	260.0
20	340	231.2	211.2	231.2	271.2	311.2

Next plot the pump head-characteristic curve and the system-head curve for each ΔZ shown in Figure 12.5.15a. From this figure obtain the ΔZ-discharge relationship:

Δz (ft)	-20	0	40	80
Q_{1800} (gpm)	216	192	150	80

This relationship is plotted in Figure 12.5.15b.

(b) Repeat the above process for a rotative speed of 2160 for the same pump, using the affinity laws, equations (12.2.4) and (12.2.5):

$$Q_{2160} = Q_{1800}\left(\frac{2160}{1800}\right) = 1.2 \times Q_{1800}$$

$$H_{2160} = Q_{1800}\left(\frac{2160}{1800}\right)^2 = 1.44 \times H_{1800}$$

The resulting pump-head characteristic curve is

H_{2160} (ft)	144	129.6	115.2	86.4	57.6	28.8
Q_{2160} (gpm)	0	132	216	300	360	408

This is plotted in Figure 12.5.15a. Using this figure, the ΔZ-discharge relationship is

ΔZ (ft)	-20	0	40	80
Q_{2160} (gpm)	250	232	195	158

which is plotted in Figure 12.5.15b.

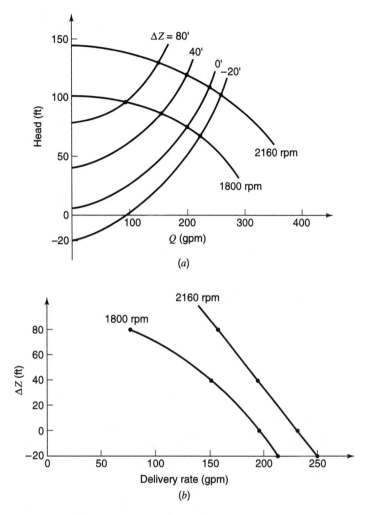

Figure 12.5.15 Example 12.5.3 results. (*a*) Head-characteristic and system-head curves; (*b*) Delivery-rate curve.

EXAMPLE 12.5.4

A water supply facility consists of two reservoirs, two pumps, and connecting pipe lines. A schematic of the installation is shown in Figure 12.5.16. Water is supplied to reservoir 2 from reservoir 1.

The elevation of the water surface of reservoir 1 is 100 ft and the elevation of the water surface of reservoir 2 is 250 ft. The suction and discharge lines are all located 10 ft below the water surfaces. The Darcy–Weisbach friction factor is 0.021 for all pipes. The pump characteristics are shown in Figure 12.5.17. Determine:

(a) When both pumps are operating, what is the total discharge to reservoir 2?
(b) What is the percentage of total discharge from line A?
(c) What is the discharge delivered through pump A? (cfs or gpm)
(d) What is the efficiency of pump A if its input horsepower is 80?

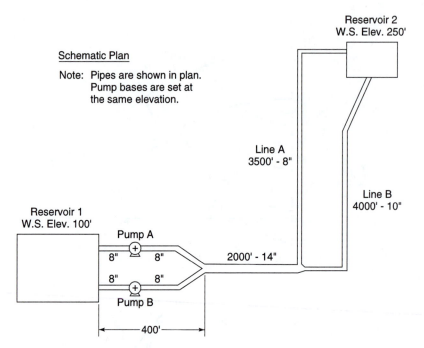

Figure 12.5.16 Example 12.5.4 system.

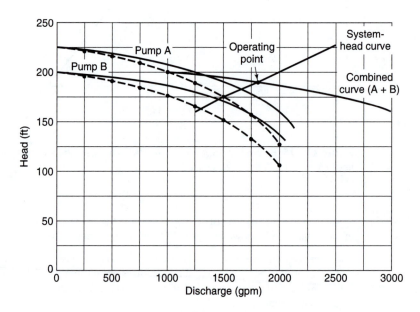

Figure 12.5.17 Pump characteristics for Example 12.5.4.

SOLUTION

The first objective is to develop an expression for the system head losses. Develop an expression for headloss in terms of the discharges. Pipe 1 is the 400 ft pipe in which pump A is placed. Using the Darcy–Weisbach equation (12.1.6), the headloss due to friction is

$$h_{L_{f1}} = f\frac{L}{D}\frac{V^2}{2g} = \left[\frac{8fL}{\pi^2 gD^5}\right]Q^2 = K_1Q^2$$

$$K_1 = \frac{8(0.021)(400)}{\pi^2 g(8/12)^5} = 1.606$$

so $h_{L_{f1}} = 1.606Q_1^2$

Pipe 2 is the 400 ft pipe in which pump B is placed. This pipe is the same size as pipe A, so $h_{L_{f2}} = 1.606Q_2^2$.

Develop modified pump curves for pumps A and B to account for headlosses in the 400 ft 8 in pipes. (1 gpm = 2.23 × 10^{-3} ft^3/s):

Q_1 (gpm)	Q_1 (cfs)	$h_{L_{f1}} = 1.606Q_1^2$
0	0	0
500	1.115	1.997
1000	2.230	7.986
1500	3.345	17.970
2000	4.460	31.946

These values are the same for pump B because its length and diameter are the same as pipe 1. The modified pump characteristic curves are shown in Figure 12.5.17. The combined pump curve for pumps A and B in parallel can now be developed using the modified pump curves, as shown in Figure 12.5.17.

Next determine the system-head curve for the remaining part of the system: $H_T = 250 - 100 + h_{L_f}$. The headloss due to

$$h_{L_f} = h_{L_{f3}} + h_{f_A} = h_{L_{f3}} + h_{f_B}$$

$$= K_3Q_3^2 + K_AQ_A^2 = K_3Q_3^2 + K_BQ_B^2$$

Pipe 3 is the 2000 ft, 14 inch diameter pipe:

$$K_3 = \frac{8(0.021)(2000)}{\pi^2 g(14/12)^5} = 0.489$$

$$h_{L_3} = 0.489Q_3^2$$

For pipe (line) A:

$$K_A = \frac{8(0.021)(3500)}{\pi^2 g(8/12)^5} = 14.050$$

$$h_{L_A} = 14.050Q_A^2$$

For pipe (line) B:

$$K_B = \frac{8(0.021)(4000)}{\pi^2 g(10/12)^5} = 5.262$$

$$h_{L_B} = 5.262Q_B^2$$

The total head is then

$$H_T = 150 + h_{L_f}$$

$$= 150 + 0.489Q_3^2 + 14.050Q_A^2$$

$$H_T = 150 + 0.489Q_3^2 + 5.262Q_B^2$$

Pipes A and B are parallel so that $h_{L_{f_A}} = h_{L_{f_B}}$:

$$14.050Q_A^2 = 5.262Q_B^2$$

$$Q_A = 0.612Q_B$$

$$Q_B = 1.634Q_A$$

From continuity,

$$Q_3 = Q_A + Q_B = Q_A + 1.634Q_A$$

$$Q_3 = 2.634Q_A$$

$$Q_A = 0.380Q_3$$

Now H_T can be expressed in terms of one flow Q_3, which is also the total flow:

$$H_T = 150 + 0.489Q_3^2 + 14.050(0.380Q_3)^2$$

$$= 150 + 0.489Q_3^2 + 2.025Q_3^2$$

$$= 150 + 2.514Q_3^2$$

$$= 150 + 2.514Q^2$$

Next plot the system-head curve on the combined pump curve to find the point of operation, as given in Figure 12.5.17.

Q (gpm)	Q (cfs)	H_T (ft)
1500	3.345	178.1
2000	4.460	200.0
2500	5.575	228.1
3000	6.690	262.5

(a) The operating point is a flow of approximately 1850 gpm, so that the total discharge to the reservoir is 1850 gpm.
Flow in line A, $Q_A = 0.380 \; Q_3 = 0.380 \, (1850) = 703$ gpm
Flow in line B, $Q_B = 1850 - 703 = 1147$ gpm

(b) The percentage of discharge in line A is 38%, or 703 gpm.

(c) From Figure 12.5.17, the discharge delivered through pump A is approximately 1200 gpm and through pump B is approximately 650 gpm.

(d) Efficiency of pump A is computed using

$$e_A = \frac{\gamma H_{\text{pump A}} Q_{\text{pump A}}}{bhp(550)}$$

$$Q_{\text{pump A}} = 1200 \text{ gpm} = 2.676 \text{ ft}^3/\text{s}$$

$$H_{\text{pump A}} = H_T = 150 + 2.514Q^2 = 150 + 2.514(4.126)^2$$

$$= 193 \text{ ft}$$

$$e_A = \frac{62.4(193)(2.676)}{80(550)} = 0.73$$

$$= 73\% \text{ (efficiency)}$$

12.6   NETWORK SIMULATION

12.6.1   Conservation Laws

The distribution of flows through a network under a certain *loading pattern* (demands as a function of time) must satisfy the conservation of mass and the conservation of energy. Figure 12.6.1 shows a simple example network consisting of 19 pipes. Assuming water is an incompressible fluid, by the conservation of mass, flow at each of the junction nodes must be conserved, that is,

$$\sum Q_{in} - \sum Q_{out} = \sum Q_{ext} \tag{12.6.1}$$

where Q_{in} and Q_{out} are the pipe flows into and out of the node, respectively, and Q_{ext} is the external demand or supply at the node.

For each *primary loop*, which is an independent closed path, the conservation of energy must hold; that is, the sum of energy or headlosses, h_L, minus the energy gains due to pumps, H_{pump}, around the loop must be equal to zero:

$$\sum_{i,j \in I_p} h_{L_{i,j}} - \sum_{k \in J_p} H_{pump,k} = 0 \tag{12.6.2}$$

where $h_{L_{i,j}}$ refers to the headloss in the pipe connecting nodes i and j, I_p is the set of pipes in the loop p, k refers to pumps, J_p is the set of pumps in loop p, and $H_{pump,\,k}$ is the energy added by pump k contained in the loop and summed over the number of pumps. Equation (12.6.2) must be written for all independent loops.

Energy must be conserved between *fixed-grade nodes* (FGN), which are points of known constant grade (elevation plus pressure head). If there are N_F such nodes, then there are $N_F - 1$ independent equations of the form

$$\Delta E_{FGN} = \sum_{i,j \in I_p} h_{L_{i,j}} - \sum_{k \in J_p} H_{pump,k} \tag{12.6.3}$$

where ΔE_{FGN} is the difference in total grade between the two FGNs. The total number of equations, $N_J + N_L + (N_F - 1)$, also defines the number of pipes in the network, in which N_J is the number of junction nodes and N_L is the total number of independent loops. The headloss due to pipe friction can be defined by equations (12.1.4), (12.1.7), or (12.1.9).

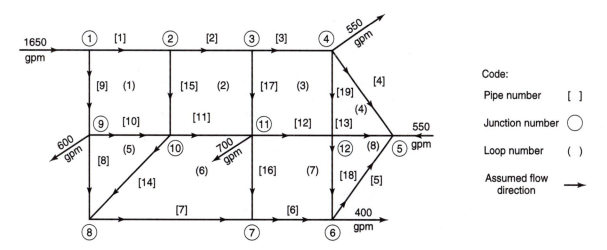

Figure 12.6.1 Example network (from Wood and Charles (1972)).

The change in head that occurs across each component is related to the flow through the component. By substituting the appropriate relationships for each component into the continuity and energy equations, it is possible to set up a system of nonlinear equations with the same number of unknowns. This set of equations can be solved by iterative techniques for the unknowns. Several computer programs have been written to automate these procedures. These models, called *network solvers* or *simulation models*, are now widely accepted and applied. This section presents the equations used to describe the relationships between headloss and flow and then discusses how each component is represented in a network simulation model.

The relationship between the added head H_{pump} and discharge Q is typically a concave curve with H_{pump} increasing as Q decreases, as shown in Figure 12.2.7. For the normal operating range, this curve is usually well approximated by a quadratic or exponential equation, that is:

$$H_{pump} = AQ^2 + BQ + H_c \tag{12.6.4}$$

or

$$H_{pump} = H_c - CQ^n \tag{12.6.5}$$

with A, B, and n being coefficients and H_c the *cutoff head* or maximum head. Also associated with a pump is an efficiency curve that defines the relationship of energy consumption and pump output (Figure 12.2.7). Efficiency e is a function of Q and appears as a function of power *bhp* from equation (12.2.12) as

$$e = \frac{QH_{pump}\gamma}{550(bhp)} \tag{12.6.6}$$

A pump achieves maximum efficiency at the *design* or *rated discharge*. Depending upon the simulation model a pump may also be described by a curve of constant power, *bhp*.

As noted in the previous section, the limiting constraints in the design problem are usually the pressure restrictions at the nodes. Since the headlosses in the system increase almost quadratically with the flow rates, as seen in the Hazen–Williams equation, less head is required for patterns with lower total demand and as the demand level increases the head needed increases but faster than linearly. This relationship is a system curve from which the least-cost operation of pumps can be determined.

12.6.2 Network Equations

The governing conservation equations can be written in terms of the unknown nodal heads or the pipe flows using loop equations, head or nodal equations, or ΔQ equations. The loop or flow equations consist of the junction relationships written with respect to the N_p unknown flow rates. The component equations with pipe flows are substituted for h_L in the energy equations to form an additional $N_L + (N_F - 1)$ equations. This results in N_p equations written with respect to the N_p unknown flow rates.

The head or nodal equations use only flow continuity and consider the nodal heads as unknown rather than the pipe flows. In this case, additional equations are required for each pump and valve, increasing the total number of equations. For a link the difference in head between the connected nodes i and j is equal to $h_{L_{i,j}}$:

$$h_{L_{i,j}} = H_i - H_j \tag{12.6.7}$$

This relationship can be substituted into the Hazen–Williams equation, which in turn is rewritten and substituted for Q in the continuity equations.

The following *nodal equation* results for the node shown in Figure 12.6.2 with assumed flow directions defined by the arrows (flow from a junction is assumed negative):

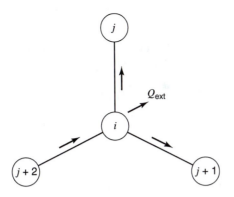

Figure 12.6.2 Node connected with three links.

$$-\left(\frac{H_i - H_j}{K_{p,i,j}}\right)^{0.54} - \left(\frac{H_i - H_{j+1}}{K_{p,i,j+1}}\right)^{0.54} + \left(\frac{H_{j+2} - H_i}{K_{p,j+2,i}}\right)^{0.54} - Q_{\text{ext},i} = 0 \qquad (12.6.8)$$

where $K_{p,i,j}$ is the coefficient defined in equations (12.1.5), (12.1.8), or (12.1.10) for the pipe connecting nodes i and j. These nodal equations can be written for each junction and component node, resulting in a system of nonlinear equations with the same number of unknowns, which is the total number of nodal heads. Similarly, the equations for the other components can be rewritten with respect to the nodal heads. The nodal equations can be linearized in an iterative solution technique.

The ΔQ equations directly use the *loop equations* and implicitly ensure that the node equations are satisfied. In this formulation the energy equation for each loop is written in terms of the flows:

$$\sum_{(i,j) \in I_p} K_{p,i,j} \left(Q_{i,j} + \Delta Q_{i,j}\right)^n = 0 \qquad (12.6.9)$$

12.6.3 Network Simulation: Hardy–Cross Method

Several iterative solution approaches have been applied to solve the sets of equations described in the previous section, including the *linear theory method*, the *Newton–Raphson method* (Appendix A), and the *Hardy–Cross technique*. Due to the nature of the equations, the linear theory method for solving the flow equations and the Newton–Raphson method for solving the node equations are considered most efficient.

The energy equation for each loop in a water distribution system expressed in equation (12.6.9) must be rewritten to take into account the direction of flow as

$$\sum h_{L,i,j} = \sum_{(i,j) \in I_P} K_{P,i,j} (Q_{i,j} + \Delta Q_{i,j}) \left|(Q_{i,j} + \Delta Q_{i,j})\right|^{n-1} = 0 \qquad (12.6.10)$$

or

$$\sum h_{L,i,j} = \sum_{(i,i) \in I_p} K_{P,i,j} \, \text{sgn}(Q_{i,j} + \Delta Q_{i,j})(Q_{i,j} + \Delta Q_{i,j})^n = 0 \qquad (12.6.11)$$

This summation can be expanded as follows:

$$\sum K_{p,i,j} Q_{i,j}^n + \sum n K_{p,i,j} (\Delta Q_{i,j})(Q_{i,j})^{n-1} + \left(\frac{n-1}{2}\right) \sum n K_{p,i,j} (\Delta Q_{i,j})^2 (Q_{i,j})^{n-2} + \ldots = 0 \quad (12.6.12)$$

Assuming the $\Delta Q_{i,j}$ are small, the $(\Delta Q_{i,j})^2$ and succeeding terms can be neglected, leaving

$$\sum K_{p,i,j} Q_{i,j}^n + \sum n K_{p,i,j} \Delta Q_{i,j} (Q_{i,j})^{n-1} = 0 \qquad (12.6.13)$$

The procedure, presented by Hardy and Cross (1936), first assumes a flow distribution in a network. These assumed flows, however, may not satisfy the energy requirement (equation 12.6.10). Therefore, the same correction $\Delta Q_{i,j}$ is made to all pipes in a particular loop p. To compute ΔQ, use the equation (12.6.13) rearranged as follows:

$$\sum K_{p,i,j} Q_{i,j}^n + \Delta Q_p \sum n K_{p,i,j} (Q_{i,j})^{n-1} = 0 \tag{12.6.14}$$

and solve for ΔQ_p:

$$\Delta Q_p = -\frac{\sum K_{p,i,j} Q_{i,j}^n}{\sum \left| n K_{p,i,j} Q_{i,j}^{n-1} \right|} \tag{12.6.15}$$

The numerator is the algebraic sum of the headloss in loop p with regard to the sign of the flow. For example, if positive is taken as clockwise, then pipes with flows in the clockwise direction have a positive headloss term $+K_{p,i,j} Q_{i,j}^n$ and pipes with flow in the counterclockwise direction have a negative headloss term, $-K_{p,i,j} Q_{i,j}^n$. The denominator is the absolute sum because in equation (12.6.15), ΔQ_p was given the same sign in all pipes in the loop.

The correction ΔQ_p for loop p is applied in the same sense to each pipe in the loop. In other words, if clockwise is positive, then ΔQ is added to flows in the clockwise direction and subtracted from flows in the counterclockwise direction. This method basically determines the ΔQ_p for each loop separately. Then the flows for each pipe are corrected using

$$Q_{i,j}(\text{new}) = Q_{i,j}(\text{old}) + \sum_{p \in M_p} \Delta Q_p \tag{12.6.16}$$

where M_p is the set of loops to which pipe i, j belongs. M_p has at most two loops because each pipe can only be common to one or two loops. The new flows are then used to repeat the process until the values of ΔQ_p are within a desired accuracy.

Basically the Hardy–Cross method is an adaptation of the Newton iterative method. One equation (loop) is solved at a time before proceeding to the next equation (loop) during each iteration. This requires the assumption that all other ΔQ except ΔQ_p are known, i.e., $\Delta Q_1, \Delta Q_2, \ldots, \Delta Q_p-1$, $\Delta Q_{p+1} \ldots$ are temporarily known, as described previously.

Dropping subscripts except for p to represent the loop yields

$$\sum h_L = \sum K \operatorname{sgn}(Q + \Delta Q_p)(Q + \Delta Q_p)^n = 0 \tag{12.6.17}$$

Applying Newton's method (Appendix A) to the above equation, we find that the function $f(\Delta Q_p)$ is

$$f(\Delta Q_p) = \sum K \operatorname{sgn}(Q + \Delta Q)(Q + \Delta Q_p)^n - \sum h_L \tag{12.6.18}$$

and the derivative is

$$f'(\Delta Q_p) = n \sum K (Q + \Delta Q_p)^{n-1} \tag{12.6.19}$$

Note that $d[\operatorname{sgn} f(x) f(x)^n]/dx = nf(x)^{n-1} df(x)/dx$.

Applying Newton's method yields the iterative equation

$$x(k+1) = x(k) - \frac{f(x(k))}{f'(x(k))} \tag{12.6.20}$$

for the kth iteration; then

$$\Delta Q_p(k+1) = \Delta Q_p(k) - \frac{f\big(\Delta Q_p(k)\big)}{f'\big(\Delta Q_p(k)\big)} \tag{12.6.21}$$

The Hardy–Cross method commonly applies one iterative correction to each equation (loop) before proceeding to the next equation (loop). Several variations of this exist in the literature. After applying one iterative correction to all equations (loops), the process is repeated until convergence is achieved. It is also common to adjust the flow rates in all pipes in the loop immediately upon computing each ΔQ. Therefore, each equation $f(\Delta Q_p(k + 1)) = 0$ is evaluated with all previous ΔQ, i.e., $\Delta Q_p(k) = 0$, so that equation (12.6.21) reduces to

$$\Delta Q_p(k+1) = -\frac{f\big(\Delta Q_p(k)\big)}{f'\big(\Delta Q_p(k)\big)} \tag{12.6.22}$$

Upon substituting equations (12.6.18) and (12.6.19), the new $\Delta Q_p(k + 1)$ or ΔQ_p (new) is computed using

$$\Delta Q_p(\text{new}) = -\frac{\sum K \, \text{sgn}\big(Q + \Delta Q_p\big)\big(Q + \Delta Q_p\big)^n}{n \sum K \big(Q + \Delta Q_p\big)^{n-1}} \tag{12.6.23}$$

Steps of the Hardy–Cross method are summarized below (modified from Jeppson, 1976)

1. Determine the values of K using equations (12.1.5), (12.1.8), or (12.1.10).
2. Start with initial guesses of the flow rate in each pipe such that continuity at all junctions are satisfied; assume initial $\Delta Q_p = 0$ for all loops.
3. Compute sum of headlosses around a loop, applying the appropriate sign, $\text{sgn}(Q + \Delta Q_p)$. This sum is the numerator in equation (12.6.23). If the direction of movement around the loop is opposite to the direction of flow, then $\text{sgn}(Q + \Delta Q_p)$ is negative.
4. Compute the denominator of equation (12.6.23) for the same loop.
5. Solve equation (12.6.23) for ΔQ_p (new).
6. Repeat steps 3 through 5 for each loop in the network
7. Define new flows using equation (12.6.16).
8. Repeat steps 3 through 7 iteratively until all the ΔQ_p computed during an iteration are small enough to be insignificant.

EXAMPLE 12.6.1

Determine the flow rate in each pipe for the simple network in Figure 12.6.3. Assume that fully turbulent flow exists for all pipes and the Darcy–Weisbach friction factor is $f = 0.02$. Use the Hardy–Cross method.

SOLUTION

Step 1

Using equation (12.1.7), $h_L = KQ^2 = \left[\dfrac{8fL}{\pi^2 g D^5}\right]Q^2 = \left[\dfrac{8fL}{39.68 D^5}\right]Q^2$

so $K = \left(\dfrac{L}{1984 D^5}\right)$

Determining the K-values yields

$K_{AB} = \left[\dfrac{5000}{1984(2)^5}\right] = 0.079$

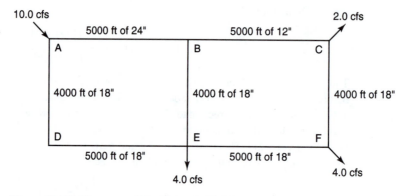

Figure 12.6.3 Pipe network for Example 12.6.1.

$K_{BC} = 2.520$

$K_{AD} = K_{BE} = K_{CF} = 0.265$

$K_{DE} = K_{EF} = 0.332$

Step 2 The initial guesses of flow for each pipe are shown in Figure 12.6.4a.

Step 3, 4, and 5 Consider loop 1 first and compute ΔQ_1:

$$\Delta Q_1 = -\frac{1}{2}\frac{\left[K_{AD}Q_{AD}^2 + K_{DE}Q_{DE}^2 - K_{BE}Q_{BE}^2 - K_{AB}Q_{AB}^2\right]}{\left[K_{AD}Q_{AD} + K_{DE}Q_{DE} + K_{BE}Q_{BE} + K_{AB}Q_{AB}\right]}$$

$$\Delta Q_1 = -\frac{1}{2}\frac{\left[(0.265)(3.8)^2 + (0.332)(3.8)^2 - (0.265)(4.0)^2 - (0.079)(6.2)^2\right]}{\left[(0.265)(3.8) + (0.332)(3.8) + (0.265)(4.0) + (0.079)(6.2)\right]}$$

$$\Delta Q_1 = -\frac{1.344}{7.637} = -0.176$$

Step 6 Next consider remaining loop, loop 2, to compute ΔQ_2:

$$\Delta Q_2 = -\frac{1}{2}\frac{\left[K_{BE}Q_{BE}^2 + K_{EF}Q_{EF}^2 - K_{CF}Q_{CF}^2 - K_{BC}Q_{BC}^2\right]}{\left[K_{BE}Q_{BE} + K_{EF}Q_{EF} + K_{CF}Q_{CF} + K_{BC}Q_{BC}\right]}$$

$$\Delta Q_2 = -\frac{1}{2}\frac{\left[(0.265)(4.0)^2 + (0.332)(3.8)^2 - (0.265)(0.2)^2 - (2.520)(2.2)^2\right]}{\left[(0.265)(4.0) + (0.332)(3.8) + (0.265)(0.2) + (2.520)(2.2)\right]}$$

$$\Delta Q_2 = -\frac{-3.173}{15.837} = 0.200$$

Step 7 Define new flows for next iteration:

$Q_{AD} = 3.8 + (-0.176) = 3.624$

$Q_{DE} = 3.8 + (-0.176) = 3.624$

$Q_{BE} = 4.0 - \Delta Q_1 + \Delta Q_2 = 4.0 - (-0.176) + 0.200 = 4.376$

$Q_{AB} = 6.2 - (-0.176) = 6.376$

$Q_{BC} = 2.2 - 0.200 = 2.000$

$Q_{EF} = 3.8 + 0.200 = 4.000$

$Q_{CF} = 0.2 - 0.200 = 0.000$

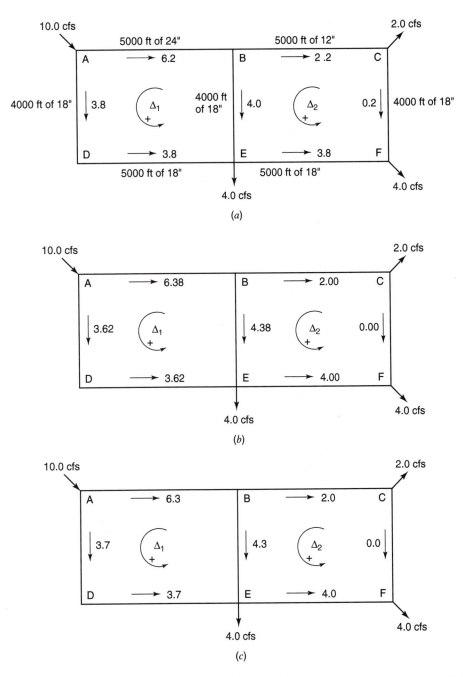

Figure 12.6.4 Solution results for (a) One; (b) Two; and (c) Three iterations for Example 12.6.1.

Steps 3–6 For the next iteration, repeat steps 3–6 with the new flows determined in Step 7 and shown in Figure 12.6.4b:

$$\Delta Q_1 = -\frac{1}{2}\frac{[-0.446]}{[3.827]} = 0.058$$

$$\Delta Q_2 = -\frac{1}{2}\frac{[0.307]}{[7.528]} = -0.020$$

Step 7 Define new flows for next iteration:

$Q_{AD} = 3.624 + 0.058 = 3.682$

$Q_{DE} = 3.624 + 0.058 = 3.682$

$Q_{BE} = 4.376 - 0.058 + (-0.02) = 4.298$

$Q_{AB} = 6.376 - 0.058 = 6.318$

$Q_{BC} = 2.000 - (-0.020) = 2.020$

$Q_{EF} = 4.000 + (-0.020) = 3.980$

$Q_{CF} = 0.000 - (-0.020) = 0.020$

For the next iteration, repeat steps 3–6 with the new flows determined above and shown in Figure 12.6.3c:

$$\Delta Q_1 = \frac{-[0.045]}{2[3.836]} = -0.006 \text{ cfs}$$

$$\Delta Q_2 = \frac{-[-0.128]}{2[7.556]} = 0.008 \text{ cfs}$$

The above corrections ΔQ_1 and ΔQ_2 are insignificant, so the final flows are those shown in Figure 12.6.4c:

EXAMPLE 12.6.2

For the network in example 12.6.1, determine the headloss from A to F.

SOLUTION

There are several sets of pipes that represent a path from node A to node F. Each of these should result in the same headloss.

(a) $h_{L_{fA-F}} = \sum h_{L_{f_i}} = h_{L_{AD}} + h_{L_{DE}} + h_{L_{EF}}$

$\qquad = \left(K_{AD} Q_{AD}^2 + K_{DE} Q_{DE}^2 + K_{EF} Q_{EF}^2 \right)$

$\qquad = \left[(0.265)(3.682)^2 + (0.332)(3.682)^2 + (0.332)(3.980)^2 \right]$

$h_{L_{fA-F}} = 13.35 \text{ ft}$

(b) or $h_{L_{fA-F}} = h_{L_{AB}} + h_{L_{BE}} + h_{L_{EF}}$

$\qquad = K_{AB} Q_{AB}^2 + K_{BE} Q_{BE}^2 + K_{EF} Q_{EF}^2$

$\qquad = \left[(0.079)(6.318)^2 + (0.265)(4.298)^2 + (0.332)(3.980)^2 \right]$

$h_{L_{fA-F}} = 13.31 \text{ ft}$

(c) or $h_{L_{fA-F}} = h_{L_{AB}} + h_{L_{BC}} + h_{L_{CF}} = K_{AB} Q_{AB}^2 + K_{BC} Q_{BC}^2 + K_{CF} Q_{CF}^2$

$\qquad = \left[(0.079)(6.318)^2 + (2.520)(2.020)^2 + (0.265)(0.020)^2 \right]$

$h_{L_{fA-F}} = 13.44 \text{ ft}$

Every path should have the same headloss ($h_{L_{fA-F}}$) from A–F; the differences are due to roundoffs.

12.6.4 Network Simulation: Linear Theory Method

The linear theory method was presented by Wood and Charles (1972) for simple networks and later extended to include pumps and other appurtenances (Wood, 1980). Comparing the other approaches, the Newton–Raphson method may converge more quickly than the linear method for small systems, whereas it may converge very slowly for large networks compared to the linear method (Wood and Charles, 1972). The linear theory method, however, has the capability to analyze all components, with more flexibility in the representation of pumps and better convergence properties. The University of Kentucky model, KYPIPE, by Wood (1980) and Wood et al. (1995) is a widely used and accepted program based on the linear theory method.

The linear theory method solves for the discharge Q using the *path (energy) equations*,

$$\Delta E = \sum h_L + \sum h_{Lm} - \sum H_{\text{pump}} \tag{12.6.24}$$

and using equation (12.1.3) for friction losses and equation (12.6.4) or (12.6.5) for the pump curve:

$$\Delta E = \sum \left(K_p Q^n + K_m Q^2 \right) - \sum \left(AQ^2 + BQ + H_c \right) \tag{12.6.25}$$

where $n = 1.85$ for the Hazen–Williams equation, $n = 2$ for the Darcy–Weisbach equation, and $n = 2$ for Manning's equation (which is valid for fully turbulent flow only).

The pressure head (grade) difference in a pipe section with a pump for $Q = Q_r$ can be expressed as

$$f(Q_r) = K_p Q_r^n + K_m Q_r^2 - \left(AQ_r^2 + BQ_r + H_c \right) \tag{12.6.26}$$

where r represents the rth iteration. The gradient $\partial f/\partial Q$ evaluated at Q_r is

$$G_r = \left[\frac{\partial f}{\partial Q_r} \right] = nK_p Q_r^{n-1} + 2K_m Q_r - \left(2AQ_r + B \right) \tag{12.6.27}$$

The nonlinear energy equations are linearized in terms of the unknown flow rate Q_{r+1} in each pipe using

$$f(Q_{r+1}) = f(Q_r) + \left[\frac{\partial f}{\partial Q} \right]_{Q_r} (Q_{r+1} - Q_r)$$

$$= f(Q_r) + G_r(Q_{r+1} - Q_r) \tag{12.6.28}$$

The *path equations* (either from one fixed grade to another one or around a loop) can be written as

$$\Delta E = \sum f(Q_{r+1}) = \sum f(Q_r) + \sum G_r(Q_{r+1} - Q_r) \tag{12.6.29}$$

where the Σ refers to summing over each pipe and $\Delta E = 0$, so that

$$\sum G_r Q_{r+1} = \sum \left(G_r Q_r - f(Q_r) \right) \tag{12.6.30}$$

For a path between two fixed-grade nodes, ΔE is a constant; then by equation (12.6.29) we find

$$\sum G_r Q_{r+1} = \sum \left(G_r Q_r - f(Q_r) \right) + \Delta E \tag{12.6.31}$$

Equations (12.6.30) and/or (12.6.31) are used to formulate $N_L + (N_F - 1)$ equations and are combined with the N_J continuity equations (12.6.1) to form a set of $N_P = N_L + (N_F - 1) + N_J$ linear equations (number of pipes) in terms of the unknown flow rate Q_{r+1} in each pipe. Using a set of initial flow rates Q_r in each pipe, the system of linear equations is solved for Q_{r+1} using a matrix procedure. This new set of flow rates Q_{r+1} is used as the known values to obtain a second solution of the linear equations. This procedure continues until the change in flow rates $|Q_{r+1} - Q_r|$ is insignificant and meets some convergence criterion.

EXAMPLE 12.6.3

Develop the system of equations to solve for the pipe flows of the 19-pipe water distribution network shown in Figure 12.6.1 (example adapted from Wood and Charles, 1972). The equations are to be based upon the linear theory method using loop equations.

SOLUTION

Let $Q_1, Q_2 \ldots$ represent the flow in pipe 1, pipe 2 ....

Conservation of flow at each node:

Node 1: $Q_1 + Q_9 = 1,650$ Node 7: $Q_7 + Q_{16} - Q_6 = 0$

Node 2: $Q_1 - Q_2 - Q_{15} = 0$ Node 8: $Q_8 + Q_{14} - Q_7 = 0$

Node 3: $Q_2 - Q_3 - Q_{17} = 0$ Node 9: $Q_9 - Q_{10} - Q_8 = 600$

Node 4: $Q_3 - Q_4 - Q_{19} = 500$ Node 10: $Q_{10} + Q_{15} - Q_{11} - Q_{14} = 0$

Node 5: $Q_4 + Q_5 - Q_{13} = -550$ Node 11: $Q_{11} + Q_{17} - Q_{12} - Q_{16} = 700$

Node 6: $Q_{18} + Q_6 - Q_5 = 400$ Node 12: $Q_{14} + Q_{19} - Q_{13} - Q_{18} = 0$

All 12 of the conservation of flow constraints are needed.

Conservation of energy (loop equations):

Loop 1: $K_{p,1}Q_1^n + K_{p,15}Q_{15}^n - K_{p,10}Q_{10}^n - K_{p,9}Q_9^n = 0$

Loop 2: $K_{p,2}Q_2^n + K_{p,17}Q_{17}^n - K_{p,11}Q_{11}^n - K_{p,15}Q_{15}^n = 0$

Loop 3: $K_{p,3}Q_3^n + K_{p,19}Q_{19}^n - K_{p,12}Q_{12}^n - K_{p,17}Q_{17}^n = 0$

Loop 4: $K_{p,4}Q_4^n - K_{p,13}Q_{13}^n - K_{p,19}Q_{19}^n = 0$

Loop 5: $K_{p,10}Q_{10}^n + K_{p,14}Q_{14}^n - K_{p,8}Q_8^n = 0$

Loop 6: $K_{p,11}Q_{11}^n + K_{p,16}Q_{16}^n - K_{p,7}Q_7^n - K_{p,14}Q_{14}^n = 0$

Loop 7: $K_{p,12}Q_{12}^n + K_{p,18}Q_{18}^n - K_{p,6}Q_6^n - K_{p,16}Q_{16}^n = 0$

Loop 8: $K_{p,13}Q_{13}^n - K_{p,5}Q_5^n - K_{p,18}Q_{18}^n = 0$

The above eight conservation of energy equations are linearized using $k = K_p Q^{n-1}$:

Loop 1: $k_1 Q_1 + k_{15}Q_{15} - k_{10}Q_{10} - k_9 Q_9 = 0$

Loop 2: $k_2 Q_2 + k_{17}Q_{17} - k_{11}Q_{11} - k_{15}Q_{15} = 0$

Loop 3: $k_3 Q_3 + k_{19}Q_{19} - k_{12}Q_{12} - k_{17}Q_{17} = 0$

Loop 4: $k_4 Q_4 + k_{13}Q_{13} - k_{19}Q_{19} = 0$

Loop 5: $k_{10}Q_{10} + k_{14}Q_{14} - k_8 Q_8 = 0$

Loop 6: $k_{11}Q_{11} + k_{16}Q_{16} - k_7 Q_7 - k_{14}Q_{14} = 0$

Loop 7: $k_{12}Q_{12} + k_{18}Q_{18} - k_6 Q_6 - k_{16}Q_{16} = 0$

Loop 8: $k_{13}Q_{13} + k_5 Q_5 - k_{18}Q_{18} = 0$

This system of 19 equations (11 conservation of flow equations and 8 energy equations) can be solved for the 19 unknown discharges.

12.6.5   Extended-Period Simulation

The equations described above are the relationships between flow and head for the main components in the network and can be solved for a single demand pattern operating in a steady state. An *extended-period simulation* (EPS) analyzes a series of demand patterns in sequence. The purpose of an EPS is to determine the variation in tank levels and their effect on the pressures in the system.

The water surface elevation in a tank varies depending upon the pressure distribution at the node at which the tank level is connected to the system.

Unlike a single-period analysis where a tank level is considered fixed, in an EPS the tank levels change with progressive simulations to take into account inflow and outflow. In an extended period simulator, flows are assumed constant throughout a subperiod. Tank levels, which are modeled as FGNs, are adjusted using simple continuity at the end of the subperiod and these new levels are then used as the fixed grades for the next subperiod. The accuracy of the simulation is dependent upon the length of the subperiods and the magnitude of flows to and from the tank.

12.7 MODELING WATER DISTRIBUTION SYSTEMS

12.7.1 Computer Models

Numerous computer models (computer software) have been developed to solve the network simulation equations. The types of models can be classified into four basic types:

- Steady-state hydraulic models
- Extended-period hydraulic simulation models
- Water quality simulation models
- Optimization models

The previous sections in this chapter have explained the mathematics and algorithms for the steady-state and extended-period simulation models. Section 12.7.4 briefly explains the water quality fundamentals in more detail.

Two of the more widely used models in the United States are the EPANET model and the KYPIPE2 (KYQUAL) model. Each of these models can simulate steady-state conditions, extended period simulation, and water quality.

EPANET (Rossman et al., 1993; Rossman, 1994; and Rossman, 2000) is a model developed by the U.S. Environmental Protection Agency that can be used for steady-state and extended-period hydraulic simulation and for water quality simulation. This model is an explicit time-driven water-quality modeling algorithm for tracking transient concentrations of substances in pipe networks. "Explicit" refers to the fact that the calculation of concentrations at a given time are directly obtained from the previously known concentration front. Substance transport is simulated directly in the modeling process. Substance mass is allocated to discrete volume elements in each pipe during each time step. Reaction occurs in each element and mass is transported between elements. Mass and flow volumes are mixed together at downstream nodes. The algorithm used for the water-quality modeling is referred to as a *discrete volume element method* (DVEM). The algorithm used to solve the hydraulic equations is a gradient algorithm.

KYPIPE2/KYQUAL (Wood et al., 1995) is a model developed at the University of Kentucky. The original model KYPIPE was developed for steady-state and extended-period hydraulic simulation. The KYQUAL model performs the water-quality simulation. The network-simulation algorithm is based upon the linear theory method presented in section 12.6.4.

12.7.2 Calibration

Calibration is the process of adjusting model input data so that simulated hydraulics and water quality results adequately reflect observed field data. Calibration is an extremely important part of the modeling process of distribution systems. The process of calibration can be difficult, time-consuming, and costly. An accurate representation of the system and components is a must adequately to perform the calibration process. Two of the major sources of error in simulation analysis for hydraulics are demands (loading distribution) and pipe-carrying capacity. The importance of each of these error sources will depend on the network application.

To simulate the hydraulics, fire-flow pressure measurements are traditionally used. This involves measuring pressures and flow in isolated pipes or pipe sections. Use the measured values in a simulation model and adjust pipe friction factors followed by a simulation. Continue this process until friction factors are selected that provide simulated pressures and flows that reflect the measured data. Other adjustments that may be necessary are adjusting pressure-regulating valves (PRVs and PSVs), redistributing demands, adjusting pipe diameters, and adjusting pump lifts. Calibration can also be performed by comparing measured water surface elevations in tanks with simulated elevations in tanks. A hydraulic model should first be calibrated with field data before being used in any design or evaluation process.

Water-quality tracers such as fluoride can be used for the calibration process. These tracers are added to the distribution system followed by observation of the concentration in various parts of the system. Adjustments of the model parameters are made followed by simulation, and continued until simulated and observed concentrations match. Refer to Ormsbee and Lingireddy (2000) for a detailed discussion of the calibration of water distribution system models.

12.7.3 Application of Models

The following steps are used in applying simulation models (Clark et al., 1988):

1. *Model selection*—Definition of model requirements and selection of a model (hydraulic and/or water quality) that fits desired requirements.
2. *Network representation*—Representation of the distribution system components in the model.
3. *Calibration*—Adjustment of model parameters so that predicted results adequately reflect observed field data.
4. *Verification*—Independent comparison of model and field results to verify the adequacy of the model representation.
5. *Problem definition*—Definition of the specific design or operational problem to be studied and incorporation of the situation (i.e., demands, system operation) into the model.
6. *Model application*—Use of the model to study the specific problem/situation.
7. *Analysis/display of results*—Following the application of the model, the results should be displayed and analyzed to determine the reasonableness of the results and to translate the results into a solution to the problem.

12.7.4 Water Quality Modeling

This section provides a very brief description of the fundamentals of dynamic water quality simulation for water distribution systems. Models such as EPANET and KYQUAL are designed to track the fate of dissolved substances flowing through a pipe network over time. The interest in water quality modeling in water distribution systems in the United States is a result of the Safe Drinking Water Act of 1974 (SDWA) and its amendment of 1986 (SDWAA). These require the U.S. Environmental Protection Agency to establish *maximum contaminant level* (MCL) goals for each contaminant that may have adverse health effects on humans. As a result of the SDWAA, there has been increased interest in understanding the variation of water quality in municipal water distribution systems. The following discussion is based upon the procedure used in EPANET.

Water quality models use flows determined from the hydraulic simulation for each time step to solve the conservation of mass equation for the substance within each link connecting nodes i and j, given as

$$\frac{\partial C_{ij}}{\partial t} = -\frac{q_{ij}}{A_{ij}}\left(\frac{\partial C_{ij}}{\partial x_{ij}}\right) + \theta(C_{ij}) \tag{12.7.1}$$

where $C_{i,j}$ is the concentration of the substance in link i, j as a function of distance $x_{i,j}$ and time t (i.e., $C_{i,j} = C_{i,j}(x_{i,j}, t)$) (mass/ft^3), $q_{i,j}$ is the flow rate in link i, j at time t (ft^3/sec), $A_{i,j}$ is the cross-sectional area of link i, j (ft^2), and $\theta(C_{i,j})$ is the reaction rate of the substance in link i, j (mass/ft^3/day).

Equation (12.7.1) is solved with a known initial condition (at time $t = 0$) and the following boundary condition at the beginning of link i, j (where $x_{i,j} = 0$):

$$C_{i,j}(0,t) = \frac{\sum\limits_{k} q_{ki} C_{ki}(L_{ki},t) + M_i}{\sum\limits_{k} q_{ki} + Q_{si}} \tag{12.7.2}$$

where M_i is the mass of substance introduced by an external source at node i, Q_{si} is the flow rate of the source, and $L_{k,i}$ is the length of link k, i. The summations are over all links k, i that have flow into node i of link i, j. The above boundary condition for link i, j depends on the end node concentration of all links k, i that deliver flow into link i, j. EPANET uses the discrete volume element method (DVEM) to numerically solve equations (12.7.1) and (12.7.2) (Rossman et al., 1993).

The rate of reaction of a constituent can be described through first-order kinetics. Reactions can occur in the bulk flow and with material along the pipe wall. The equation for first-order kinetics for bulk flow in a pipe is expressed as

$$\theta(C) = -k_b C - \left(\frac{k_f}{R_H}\right)(C - C_w) \tag{12.7.3}$$

where k_b is the first-order bulk reaction rate constant (1/sec), k_f is the mass transfer coefficient between the bulk flow and pipe wall (ft/sec), R_H is the hydraulic radius of the pipe (ft), and C_w is the concentration of the substance at the wall (mass/ft^3). The term $k_b C$ in equation (12.7.3) is the bulk flow reaction and the term $(k_f/R_H)(C - C_w)$ is the rate at which the substance is transported between the bulk flow and reaction on the pipe wall.

Assuming a first-order reaction rate at the wall and assuming no accumulation occurs, the mass balance for the wall reaction is

$$k_f(C - C_w) = k_w C_w \tag{12.7.4}$$

where k_w is the wall reaction rate (ft/sec). Solving for C_w in equation (12.7.4) and substituting into equation (12.7.3) results in

$$\theta(C) = -\left[k_b + \frac{k_w k_f}{R_H(k_w + k_f)}\right]C \tag{12.7.5}$$

The term in brackets [ ] is the overall first-order rate constant. The negative sign refers to substance decay (mass transfer from the bulk flow to the pipe wall) and a positive sign in (12.7.5) models substance growth (mass transfer from the pipe wall to the bulk flow). Determining the reaction rates k_b, k_w, and k_f is discussed in textbooks on transfer processes such as Edwards et al. (1976). Refer to Clark (2000), Clark et al. (1988), Grayman and Kirmeyer (2000), Grayman et al. (2000), and Rossman (2000) for more detailed discussions on water quality modeling.

12.8 HYDRAULIC TRANSIENTS

12.8.1 Hydraulic Transients in Distribution Systems

Hydraulic transient flow occurs when flow changes suddenly from one steady-state flow to another steady-state flow. *Transient flow* is then the intermediate-stage flow during the transition. This *hydraulic (or fluid) transient* has also been referred to as *water hammer*. The easiest explanation of transient flow or hydraulic transient is to consider the steady-state flow in the pipe, as shown in Figure 12.8.1. If the valve is instantaneously closed, a positive pressure wave is developed upstream of the valve and travels up the pipe at approximately the speed of sound. This positive pressure can be much larger than the steady-state pressure, and even large enough to cause pipe failure.

Hydraulic transients are important in water distribution systems because they can cause (1) rupture of pipe and pump casings, (2) pipe collapse, (3) vibration, (4) excessive pipe displacement, pipe line fitting and support deformation and/or failure, and (5) vapor cavity formation (cavitation, column separation)(Bosserman and Hunt, 1998). Causes of hydraulic transients in water distribution systems include (1) valve opening and closing (frequently occurs), (2) flow demand changes (rarely occurs), (3) controlled pump shutdown (rarely occurs), (4) pump failure (often occurs), (5) pump start-up (rarely occurs), (6) air-venting from lines (frequently occurs), (7) failure of flow on pressure regulation (rarely occurs), and (8) pipe rupture (rarely occurs)(Bosserman and Hunt, 1998).

This section presents the fundamentals of hydraulic transients and then discusses their control. For a more detailed discussion refer to Martin (1999, 2000).

12.8.2 Fundamentals of Hydraulic Transients

12.8.2.1 Physical Description

In order to describe the variation in pipe pressures during transient flow, consider the initial flow ($t = 0$) in Figure 12.8.2a in which the valve at the downstream end of the pipe is open and flow in the pipe has a velocity of V. For the sake of illustration headlosses are neglected, so that the initial pipe pressure is p_0 throughout the length of the pipe. The valve is instantaneously closed, at which time a pressure wave is developed that moves toward the reservoir at the speed of sound v_c. Referring to Figure 12.8.2b, water between the pressure wave and the valve is at rest, whereas water between the pressure wave and the reservoir has the initial velocity V. The pressure head in the water between the pressure wave and the valve is $\Delta p/\gamma$, where Δp refers to the increased pressure. When the pressure wave reaches the reservoir, the pressure imbalance causes water to flow from the pipe back into the reservoir at velocity V. This causes a new pressure wave that travels to the valve (see Figure 12.8.2c). As the pressure wave travels to the valve, the pressure in the pipe between the reservoir and the pressure wave returns to pressure p_0 and the velocity V is toward the

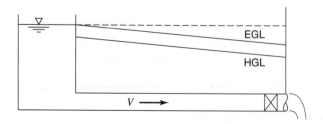

Figure 12.8.1 Steady-state flow in a pipe.

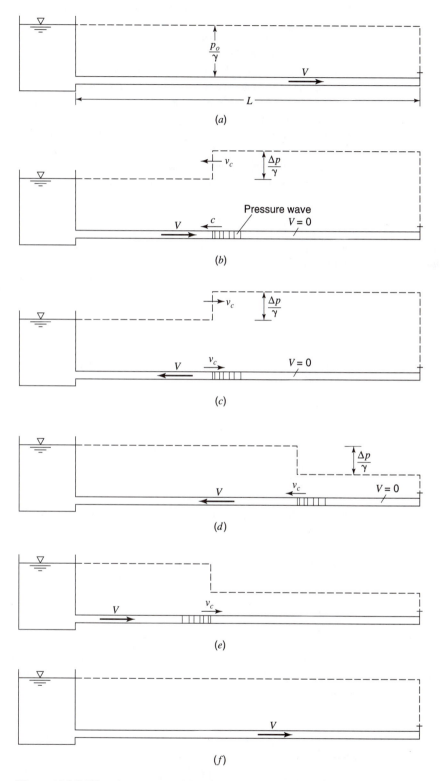

Figure 12.8.2 Water-hammer conditions. (*a*) Water-hammer process—initial condition; (*b*)
Condition during time $0 < t < L/v_c$; (*c*) Condition during time $L/v_c < t < 2L/v_c$; (*d*)
Condition during time $2L/v_c < t < 3L/v_c$; (*e*) Condition during time $3L/v_c < t < 4L/v_c$; (*f*)
Condition at time $t = 4L/v_c$ (from Roberson and Crowe (1990)).

reservoir. When this pressure wave reaches the valve, water in the pipe is flowing toward the reservoir at velocity V.

When the pressure wave reaches the valve, a rarefied pressure wave moves back toward the reservoir, as shown in Figure 12.8.2d. Note that the pressure head in the pipe between the wave and the valve is now less than that in the reservoir. The velocity in the pipe between the pressure wave and the valve is now zero. When the pressure wave reaches the reservoir, the water velocity in the pipe is zero and has a pressure less than in the reservoir at the pipe end. This pressure imbalance causes flow once again in the pipe toward the valve, as shown in Figure 12.8.2e. Now the process repeats and continues to repeat until frictional forces dampen out the pressure wave.

Figure 12.8.3a illustrates the pressure rise at the valve and the velocity at the reservoir for instantaneous closure, neglecting friction. Figure 12.8.3b illustrates the pressure rise at the valve and the velocity at the reservoir for instantaneous closure with friction included.

12.8.2.2 Wave Speed and Pressure

For purposes of deriving the wave speed and pressure, assume a rigid pipe (i.e., no expansion or contraction of the pipe). Density, pressure, and velocity of the water are ρ, p, and V on the reservoir side of the pressure wave, and are $\rho + \Delta\rho$, $p + \Delta p$, and $V = 0$ on the valve side of the wave. The momentum equation (3.4.6),

$$\sum F_x = \frac{d}{dt}\int_{CV} v_x \rho \, d\forall + \sum_{CS} v_x (\rho \cdot \mathbf{V} \cdot \mathbf{A}) \tag{3.4.6}$$

is applied to the control volume defined in Figure 12.8.4. The terms in equation (3.4.6) are

$$\sum F_x = pA - (p + \Delta p)A \tag{12.8.1}$$

$$\frac{d}{dt}\int_{cv} v_x \rho \, d\forall = \frac{d}{dt}\big[V\rho(L - v_c t)A\big] = -V\rho v_c A \tag{12.8.2}$$

$$\sum_{CS} v_x (\rho \mathbf{V} \cdot \mathbf{A}) = V\rho(-VA) \tag{12.8.3}$$

Substituting equation (12.8.1) to (12.8.3) into (3.4.6) yields

$$pA - (p + \Delta p)A = -\rho V^2 A - \rho V v_c A \tag{12.8.4}$$

which simplifies to

$$\Delta p = \rho V^2 + \rho V v_c \tag{12.8.5}$$

where ρV^2 is considered negligible as compared to the $\rho V v_c$. Then

$$\Delta p = \rho V v_c \tag{12.8.6}$$

As discussed later, v_c is considerably larger than V, allowing this assumption.

Next we apply the continuity equation (3.2.1) to determine the *pressure wave velocity* v_c. The continuity equation (3.2.1) is

$$0 = \frac{d}{dt}\int_{CV} \rho \, d\forall + \int_{CS} \rho \mathbf{V} \cdot \mathbf{dA} \tag{3.2.1}$$

where

$$\frac{d}{dt}\int_{CV} \rho \, d\forall = \frac{d}{dt}\big[\rho(L - v_c t)A + (\rho + \Delta\rho)v_c tA\big]$$

$$= -\rho v_c A + (\rho + \Delta\rho)v_c A = \Delta\rho v_c A \tag{12.8.7}$$

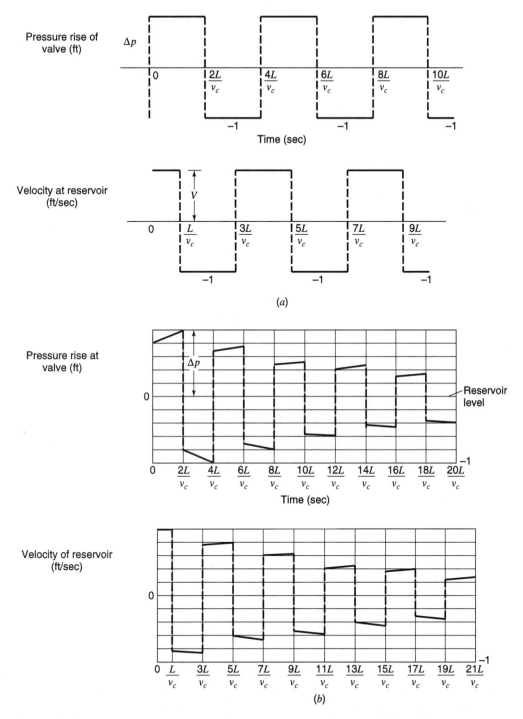

Figure 12.8.3 Pressure and velocity fluctuations for valve closure in a pipe. (*a*) Simple conduit, instantaneous closure, friction neglected; (*b*) Simple conduit, instantaneous closure, friction included.

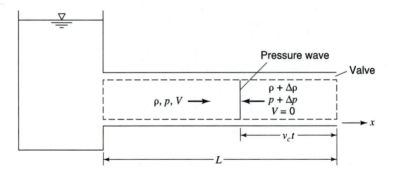

Figure 12.8.4 Control volume for pressure wave.

and

$$\int_{CS} \rho\mathbf{V}\cdot d\mathbf{A} = \rho(-VA) \tag{12.8.8}$$

Substituting equations (12.8.7) and (12.8.8) into (3.2.1) yields

$$0 = \rho(-VA) + \Delta\rho v_c A \tag{12.8.9}$$

which reduces to

$$(\Delta\rho/\rho) = (V/v_c) \tag{12.8.10}$$

The bulk modulus of elasticity E is defined in Chapter 2 as

$$E = \frac{dp}{(d\rho/\rho)} = \frac{\Delta p}{(\Delta\rho/\rho)} \tag{2.3.2}$$

so that

$$\frac{\Delta\rho}{\rho} = \frac{\Delta p}{E} \tag{12.8.11}$$

Equating (12.8.10) and (12.8.11) and solving for v_c, yields

$$v_c = \frac{VE}{\Delta p} \tag{12.8.12}$$

Substituting $\Delta p = \rho V v_c$ (equation 12.8.6) and solving for the pressure wave speed, we get

$$v_c = \sqrt{\frac{E}{\rho}} \tag{12.8.13}$$

which is for rigid pipes.

For an elastic pipe, however, this velocity of the pressure wave v_c is modified by the stretching of the pipe walls. In reality, pipes are elastic and the stretching results in the modulus of the combination of water and the pipe. A *combined modulus K* can be used to define the pressure wave speed for an elastic pipe as

$$v_c = \sqrt{\frac{K}{\rho}} \tag{12.8.14}$$

To derive K, assume that the longitudinal stress in the pipe can be ignored and let

$\forall$ = volume in pipe

$d\forall'$ = change in volume due to compression of the water

$d\forall''$ = change in volume due to stretching of the pipe wall

$d\forall$ = total change in volume

E_p = modulus of the pipe wall

The combined modulus is defined as

$$K = -\frac{\forall}{(d\forall' + d\forall'')}\,dp \tag{12.8.15}$$

which can be rearranged to produce

$$\frac{1}{K} = -\frac{d\forall'}{\forall dp} - \frac{d\forall''}{\forall dp} = \frac{1}{E} - \frac{d\forall''}{\forall dp} \tag{12.8.16}$$

By equation (2.1.5), $E = -(dp)/(d\forall/\forall)$. For a pipe wall $\forall = \pi r^2$ per unit length of pipe, where r is the radius and $d\forall'' = 2\pi r dr$.

The *modulus of elasticity of the pipe* E_p is the incremental stress divided by the increment of unit deformation. Incremental stress is rdp/t_w where t_w is the wall thickness and the increment of unit deformation is dr/r. The pipe modulus of elasticity is then

$$E_p = \frac{rdp/t_w}{dr/r} \tag{12.8.17}$$

and solving for dp yields

$$dp = \frac{E_p t_w dr}{r^2} \tag{12.8.18}$$

The term $-d\forall''/\forall dp$ in equation (12.8.16b) can now be expressed as

$$-\frac{d\forall''}{\forall dp} = \frac{2\pi r dr}{\pi r^2 (E_p t_w\, dr/r^2)} = \frac{D}{E_p t_w} \tag{12.8.19}$$

where D is the inside diameter of the pipe.

The *combined modulus K* can now be written using the above definition:

$$\frac{1}{K} = \frac{1}{E} + \frac{D}{E_p t_w}$$

or

$$K = \frac{E}{1 + (D/t_w)(E/E_p)} \tag{12.8.20}$$

Equation (12.8.14) can now be expressed as

$$v_c = \sqrt{\frac{K}{\rho}} = \sqrt{\frac{E}{\rho}\left(\frac{1}{1 + (D/t_w)(E/E_p)}\right)} \tag{12.8.21}$$

To consider various types of pipe support conditions, the above equation can be modified to include a factor k defined in Table 12.8.1:

$$V_c = \sqrt{\frac{E}{\rho}\left(\frac{1}{1 + (D/t_w)(E/E_p)k}\right)} \tag{12.8.22}$$

Table 12.8.1 Values of K from Solid Mechanics

Support Conditions	k (Thick Walls)	k (Thin Walls)
Anchored at upstream end only. Free to move longitudinally	$\dfrac{2t}{D}(1+\mu)+\left(\dfrac{D}{D+t}\right)-\left(1-\dfrac{\mu}{2}\right)$	$1-\dfrac{\mu}{2}$
Anchored against axial (longitudinally) movement so axial stress = 0	$\dfrac{2t}{D}(1+\mu)+\left(\dfrac{D}{D+t}\right)(1-\mu^2)$	$1-\mu^2$
Expansion joints throughout (axial stress = 0)	$\dfrac{2t}{D}(1+\mu)+\left(\dfrac{D}{D+t}\right)(1)$	1
Tunnels ($t=\infty$)	$\dfrac{2t}{D}(1+\mu)$	

Note: $D/D+t = 1/(1+t/D)$; μ = Poisson's ratio = 0.25 to 0.30 for many pipe materials.

The time for a pressure wave to travel from the valve to the reservoir and back to the valve is $2L/v_c$. If the time of closure is less than $2L/v_c$, then the maximum pressure developed at the valve is essentially the same as for instantaneous closure. In other words, the change in pressure is the same for a given change in velocity unless the negative wave from the reservoir mitigates the positive pressure, keeping in mind that it takes $2L/v_c$ time for the negative wave to reach the valve. The *critical time of closure* is then

$$t_c = 2L/v_c \qquad (12.8.23)$$

EXAMPLE 12.8.1

A cast iron pipe with 20-cm diameter and 15-mm wall thickness is carrying water when the outlet is suddenly closed. Use $E_{water} = 2.17 \times 10^9$ N/m^2 and $E_p = 16 \times 10^{11}$ N/m^2. If the design discharge is 40 l/s, calculate the water hammer pressure rise for:

(a) rigid pipe walls;
(b) consider stretching of pipe walls, neglecting the longitudinal stress;
(c) pipeline that is anchored at upstream end only.

SOLUTION

(a) Rigid pipe walls: The water hammer pressure rise is computed using equation (12.8.6), $\Delta p = \rho V v_c$, assuming a sudden valve closure. The pressure wave speed v_c is computed using equation (12.8.13) for a rigid pipe:

$$v_c = \sqrt{\frac{E}{\rho}} = \sqrt{\frac{2.17 \times 10^9}{1000}} = 1437 \text{ m/s}$$

The velocity of flow in the pipe is computed using continuity: $V = Q/A = (400 \times 10^{-3})/[\pi(0.2)^2/4] = 1.27$ m/s. The water hammer pressure rise is then $\Delta p = \rho V v_c = 1000\,(1.27)(1473) = 1.87 \times 10^6$ N/m^2.

(b) Consider stretching of pipe walls, neglecting longitudinal stress: The water hammer pressure is computed using equation (12.8.6), $\Delta p = \rho V v_c$, where the velocity v_c is computed using (12.8.21)

$$v_c = \sqrt{\frac{K}{\rho}} = \sqrt{\frac{E}{\rho}\left(\frac{1}{1+(D/t_w)(E/E_p)}\right)}$$

where $D = 0.2$ m, $t_w = 15 \times 10^{-3}$ m, and $E_p = 16 \times 10^{11}$ N/m:

$$v_c = \left[\frac{2.17 \times 10^9}{1000} \left(\frac{1}{1 + \left(\dfrac{0.2}{15 \times 10^{-3}} \right) \left(\dfrac{2.17 \times 10^9}{16 \times 10^{11}} \right)} \right) \right]^{1/2} = 1460 \text{ m/s}$$

The water hammer pressure increase is $\Delta p = 1000(1460)(1.27) = 1.854 \times 10^6 \text{ N/m}^2$.

(c) Pipeline that is anchored at upstream end only: The modulus is modified by $k = (1 - 0.5\mu)$ from Table 12.8.1, where μ is Poisson's ratio for the pipe. Using equation (12.8.22) yields

$$v_c = \sqrt{\frac{E}{\rho} \left(\frac{1}{1 + (D/t_w)(E/E_p)k} \right)}$$

where $\mu = 0.25$ for steel pipes:

$$v_c = \left[\frac{2.17 \times 10^9}{1000} \left(\frac{1}{1 + \left(\dfrac{0.2}{15 \times 10^{-3}} \right) \left(\dfrac{2.17 \times 10^9}{16 \times 10^{11}} \right)(1 - 0.5 \times 0.25)} \right) \right]^{1/2} = 1462 \text{ m/s}$$

$$\Delta p = 1000 \, (1462)(1.27) = 1.857 \times 10^6 \text{ N/m}^2$$

EXAMPLE 12.8.2

A steel pipe 1500 m long (Figure 12.8.5) placed on a uniform slope has a 0.5-m diameter and a 5-cm wall thickness. The pipe carries water from a reservoir and discharges it into the air at an elevation 50 m below the reservoir free surface. A valve installed at the downstream end of the pipe allows a flow rate of 0.8 m³/s. If the valve is completely closed in 1.4 s, calculate the maximum water hammer pressure at the valve and at the midlength of the pipe. Neglect longitudinal stresses. Use $E_{\text{water}} = 2.17 \times 10^9 \text{ N/m}^2$ and $E_p = 1.9 \times 10^{11} \text{ N/m}^2$.

SOLUTION

First determine the steady-state pressure. Use the energy equation between 1 and 2:

$$Z_1 = \frac{p_2}{\gamma} + \frac{V^2}{2g} + Z_2 + h_{L_{1-2}}$$

For steel pipe, use $\epsilon = 0.046$ mm, so that $\epsilon/D = 0.046/0.5 \times 10^3 = 0.000092$. The velocity of the flow in the pipe is $V = Q/A = 0.8/(0.5^2\pi/4) = 4.074$ m/s. The Reynolds number is

$$R_e = \frac{VD}{\nu} = \frac{4.074(0.5)}{1.14 \times 10^{-6}} = 1.8 \times 10^6$$

Using the Moody diagram (Figure 4.3.4), $f = 0.0129$. The pressure head at 2 is

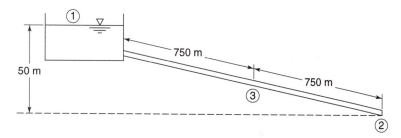

Figure 12.8.5 Example 12.8.2.

$$\frac{P_2}{\gamma} = Z_1 - Z_2 - \frac{V^2}{2g} - h_{L_{1-2}} = Z_1 - Z_2 - \left(1 + f\frac{L}{D}\right)\frac{V^2}{2g}$$

$$= 50 - 0 - \left(1 + 0.0129\frac{1500}{0.5}\right)\frac{(4.074)^2}{2(9.8)} = 16.4 \text{ m}$$

The pressure at 2 is $P_2 = 16.4\gamma = 16.4 \times 9800 = 1.6 \times 10^5$ N/m² (at valve).

At midlength (3),

$$\frac{P_3}{\gamma} = Z_1 - Z_3 - \frac{V^2}{2g} - h_{L_{1-3}} = 50 - 25 - \left(1 + 0.0129\frac{750}{0.5}\right)\frac{4.074^2}{2 \times 9.8} = 7.767 \text{ m}$$

The pressure at 3 is $p_3 = 7.6 \times 10^4$ N/m².

Determine wave speed using equation (12.8.21):

$$v_c = \sqrt{\frac{E}{\rho}\left(\frac{1}{1 + (D/t_w)(E/E_p)}\right)}$$

$$\left[\frac{2.17 \times 10^9}{1000}\left(\frac{1}{1 + \left(\frac{0.5}{0.05}\right)\left(\frac{2.17 \times 10^9}{1.9 \times 10^{11}}\right)}\right)\right]^{1/2} = 1395.6 \text{ m/s}$$

To determine maximum pressure, first determine the critical time of closure using equation (12.8.23):

$$t_c = \frac{2L}{v_c} = \frac{2(1500)}{1395.6} = 2.15 \text{ sec}$$

Since time of closure 1.4 sec < t_c = 2.15 sec, the valve closure is rapid; hence the maximum increase of pressure generated by water hammer is

$$\Delta p = \rho \, Vv_c = 1000(4.074)(1395.6) = 5.69 \times 10^6 \text{ N/m}^2$$

The maximum pressure at the valve is $p_2 + \Delta p = 1.6 \times 10^5 + 5.69 \times 10^6 = 5.85 \times 10^6$ N/m².

The maximum pressure at midlength is $p_3 + \Delta p = 7.6 \times 10^4 + 5.69 \times 10^6 = 5.77 \times 10^6$ N/m².

EXAMPLE 12.8.3

A steel pipeline (ϵ = 0.00015 ft) 2 ft in diameter and 2 miles long discharges freely at its lower end under a head of 200 ft. What water-hammer pressure would develop if a valve at the outlet were closed in 4 sec? 60 sec? Wall thickness = 0.2 in for both case of closure. Compute the stress that would develop in the walls of the pipe near the valve. If the working stress of steel is taken as 16,000 psi, what would be the minimum time of safe closure? Consider E = 4.32 × 10⁷ lb/ft² and E_p = 4.32 × 10⁹ lb/ft².

SOLUTION

We need to compute the stresses developed near the valve. First use the energy equation to determine the velocity of flow in the pipe, neglecting minor losses:

$$200 = \left(1 + f\frac{L}{D}\right)\frac{V^2}{2g}$$

Using $\epsilon = 0.00015$, we find $\epsilon/D = 0.000075$; then for fully turbulent flow, $f = 0.0115$ from the Moody diagram, so that

$$200 = \left(1 + 0.0115\frac{2 \times 5280}{2}\right)\frac{V^2}{2g}$$

and, $V = 13.8$ ft/s. Check the Reynolds number: $R_e = 13.8 \times 2/1.217 \times 10^{-5} = 2.27 \times 10^6$. Entering the Moody diagram with $R_e = 2.27 \times 10^6$ and $\epsilon/D = 0.000075$ gives $f = 0.012$ so that

$$200 = \left(1 + 0.012\frac{2 \times 5280}{2}\right)\frac{V^2}{2g}$$

and, $V = 14.1$ ft/s. Check the Reynolds number, $R_e = 14.1 \times 2/1.217 \times 10^{-5} = 2.32 \times 10^6$. Entering the Moody diagram with $R_e = 2.32 \times 10^6$ and $\epsilon/D = 0.000075$ gives $f = 0.012$. Therefore $V = 14.1$ ft/s is OK. Determine the velocity v_c using equation (12.8.21) as

$$v_c = \sqrt{\frac{E}{\rho}\left(\frac{1}{1 + (D/t_w)(E/E_p)}\right)}$$

$$= \left[\frac{4.32 \times 10^7}{1.94}\left(\frac{1}{1 + \left(\dfrac{2}{0.2/12}\right)\left(\dfrac{4.32 \times 10^7}{4.32 \times 10^9}\right)}\right)\right]^{1/2}$$

$$= \left[\frac{4.32 \times 10^7}{1.94}\left(\frac{1}{2.2}\right)\right]^{1/2} = 3181 \text{ ft/sec}$$

$$t_c = \frac{2L}{v_c} = \frac{2(2 \times 5280)}{3181} = 6.64 \text{ sec}$$

(a) For $t = 4$ sec, this is a rapid closure; then $\Delta p = \rho V v_c = 1.94\,(14.1)(3181) = 87{,}013$ lb/ft$^2 = 604$ psi. The pressure at the valve is

$$p_{\text{valve}} = p_{\text{static}} + \Delta p = \rho g h + \Delta p = 1.94\,(32.2)(200)/144 + 604$$

$$= 86.8 + 604 = 690.8 \text{ psi}$$

Stress is $\sigma = \dfrac{p_{\text{valve}}}{t_w}(r) = \dfrac{690.8}{0.2}(12) = 41{,}448$ psi.

(b) A 60-sec closure is in excess of $2L/v_c$. Let $v_c = 2L/t_c$; then

$$\Delta p = \frac{2L\rho V}{t_c} = \frac{2(2 \times 5280)(1.94)(14.1)}{60} = 9628.6 \text{ lb/ft}^2 = 66.9 \text{ psi}$$

$$p_{\text{valve}} = p_{\text{static}} + \Delta p = 86.8 + 66.9 \text{ psi} = 153.7 \text{ psi}$$

The stress is then

$$\sigma = \frac{p_{\text{valve}}}{t_w}r = \frac{153.7}{0.2}(12) = 9222 \text{ psi}$$

(c) The minimum time of closure for a working stress of 16,000 psi is now computed:

$$\Delta\sigma_a = \sigma_a - \sigma_{\text{static}} = 16{,}000 - \frac{p_{\text{static}}}{t_w}r = 16{,}000 - \frac{86.8}{0.2}(12) = 10{,}792 \text{ psi}$$

$$\Delta\sigma_a = \frac{\Delta p}{t_w}r = \frac{r}{t_w}\left(\frac{2L\rho V}{t_c}\right)$$

Solving for the time of closure gives

$$t_c = \frac{2L\rho Vr}{\Delta\sigma_a t_w} = \frac{2(2\times5280)(1.94)(14.1)(12/12)}{(10{,}792)(0.2/12)(144)} = 22.3 \text{ sec}$$

12.8.3 Control of Hydraulic Transients

12.8.3.1 Methods of Control

Various methods are used to control hydraulic transients (Martin, 1999, 2000) including:

- Pump-control valves (Figure 12.8.6)
- Air chambers (Figure 12.8.7)
- Surge tanks (Figure 12.8.8)
  - Open-end surge tank or stand pipe
  - One-way surge tank
  - Two-way surge tank
- Valves
  - Air release and vacuum relief valves
  - Swing check valves
    Cushioned swing check valves
  - Surge relief valves (Figure 12.8.9)
- High-pressure-rated piping, valves, and equipment

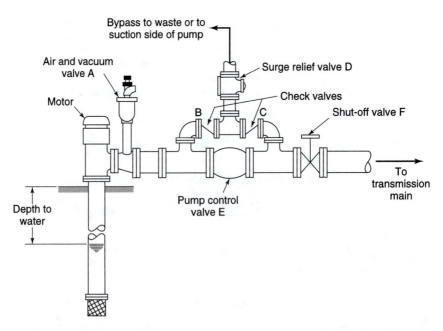

Figure 12.8.6 Upstream and downstream bypass of a pump-control valve (from Bosserman (1998)).

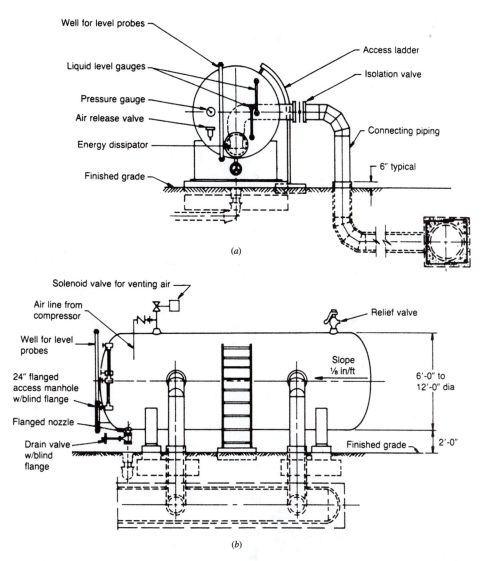

Figure 12.8.7 Horizontal air chamber for clean water service. (*a*) End elevation; (*b*) Side elevation (from Bosserman (1998)).

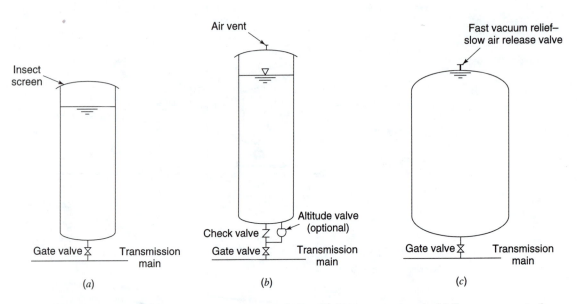

Figure 12.8.8 Surge tank. (*a*) Open-end surge tank or stand pipe; (*b*) One-way surge tank; (*c*) Two-way surge tank (from Bosserman (1998)).

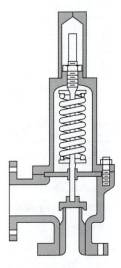

Figure 12.8.9 Pressure-actuated surge relief valve (from Bosserman (1998)).

12.8.3.2 Surge Tank Analysis

This section derives the continuity and energy equations that can be used to analyze simple surge tanks (standpipes), as shown in Figure 12.8.10. The surge tank is upstream of the valve and has the condition prior to valve closure shown in Figure 12.8.10*a*. When the valve is suddenly closed, water rises in the surge tank (or standpipe). To simplify the analysis the following derivation neglects friction and velocity head in the surge tank and losses at the pipe entrance and surge tank entrance (modified from Roberson et al., 1997).

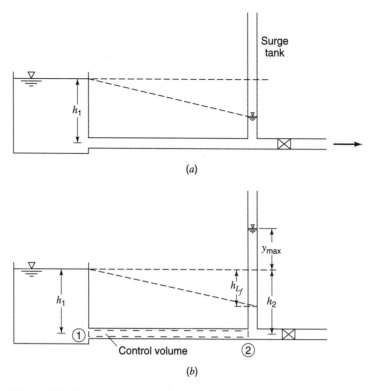

Figure 12.8.10 Surge tank analysis. (*a*) Conditions for valve open; (*b*) Condition immediately after valve closure.

Using the integral momentum equation (3.4.5), we find

$$\sum \mathbf{F} = \frac{d}{dt} \int_{CV} \mathbf{v} \rho \, d\forall + \sum_{CS} \mathbf{v} \rho \mathbf{V} \cdot \mathbf{A} \tag{3.4.5}$$

The summation of forces acting on horizontal pipe is

$$\sum \mathbf{F} = F_1 - F_2 - F_f \tag{12.8.24}$$

where F_1 is the pressure force at 1, F_2 is the pressure force acting at 2, and F_f is friction loss. F_1 is expressed as

$$F_1 = \gamma A h_1 \tag{12.8.25}$$

where A is the cross-sectional area of the pipe between 1 and 2. F_2 is expressed as

$$F_2 = \gamma A (h_1 + y) \tag{12.8.26}$$

F_f can be expressed as

$$F_f = \gamma A h_L \tag{12.8.27}$$

where h_L is the head loss due to friction. The other terms in equation (3.4.5) are

$$\frac{d}{dt} \int_{CV} \mathbf{v} \rho \, d\forall = \rho L A \frac{dV}{dt} \tag{12.8.28}$$

and

$$\sum_{CS} \mathbf{v} \rho \mathbf{V} \cdot \mathbf{A} = 0 \tag{12.8.29}$$

Substituting equations (12.8.24) though (12.8.29) into equation (3.4.5) gives

$$\gamma A h_1 - \gamma A (h_1 + y) - \gamma A h_L = \rho L A \frac{dV}{dt} + 0 \tag{12.8.30}$$

Simplifying and rearranging equation (12.8.30) produces

$$\frac{\rho}{\gamma} L \frac{dV}{dt} + y + h_L = 0 \tag{12.8.31}$$

where $h_L = f \frac{L}{D} \frac{V^2}{2g}$. A continuity equation can be derived using equation (3.2.1):

$$0 = \frac{d}{dt} \int_{CV} \rho \, d\forall + \int_{CS} \rho \mathbf{V} \cdot \mathbf{dA} \tag{3.2.1}$$

in which

$$\frac{d}{dt} \int_{CV} \rho \, d\forall = A_s \frac{dy}{dt} \tag{12.8.32}$$

and

$$\int_{CS} \rho \mathbf{V} \cdot \mathbf{dA} = -AV \tag{12.8.33}$$

where A_s is the cross-sectional area of the surge tank. Substituting equations (12.8.32) and (12.8.33) into (3.2.1) gives

$$A_s \frac{dy}{dt} = AV \tag{12.8.34}$$

EXAMPLE 12.8.4

Derive an expression for the maximum depth in a simple surge tank (standpipe) if friction is neglected.

SOLUTION

By neglecting friction losses ($h_L = 0$), equation (12.8.31) can be simplified to

$$\frac{\rho}{\gamma} L \frac{dV}{dt} + y = 0 \tag{a}$$

$$\frac{dV}{dt} = -\frac{yg}{L} \tag{b}$$

$$\frac{dV}{dt} = \frac{dV}{dy} \frac{dy}{dt} \tag{c}$$

$$\frac{dV}{dy} \frac{dy}{dt} = -\frac{yg}{L} \tag{d}$$

Using the continuity relationship, equation (12.8.34), yields

$$\frac{dy}{dt} = \frac{A}{A_s} V \tag{e}$$

and substituting this expression (e) for dy/dt into (d) gives

$$\frac{dV}{dy} \left(\frac{AV}{A_s} \right) = -\frac{yg}{L} \tag{f}$$

$$\frac{dV}{dy} = -\left(\frac{A_s}{AV}\right)\left(\frac{g}{L}\right)y \tag{g}$$

Integrating, yields

$$\int_{V_0}^{V} VdV = -\int_{0}^{y}\left[\left(\frac{A_s}{A}\right)\left(\frac{g}{L}\right)y\right]dy$$

where V_0 is the velocity in the pipe when the valve is open.

$$V^2 = V_0^2 - \left(\frac{A_s}{A}\right)\left(\frac{g}{L}\right)y^2$$

Knowing that $y \to y_{max}$ as $V \to 0$ then when the valve is completely closed $y = y_{max}$ and $V = 0$; therefore

$$y_{max} = V_0\sqrt{\left(\frac{A}{A_s}\right)\left(\frac{L}{g}\right)}$$

EXAMPLE 12.8.5

Derive an expression for the time to reach the maximum surge depth for the simple surge tank (example 12.8.4), if friction is neglected.

SOLUTION

From example 12.8.4, we have

$$V^2 = V_0^2 - \left(\frac{A_s}{A}\right)\left(\frac{g}{L}\right)y^2 \tag{a}$$

$$\frac{dy}{dt} = \frac{A}{A_s}V \quad \text{or} \quad dy = \left(\frac{A}{A_s}\right)Vdt \tag{b}$$

Substituting V from equation (a) into (b) and integrating (Morris and Wiggert, 1972) yields

$$\int_{0}^{T} dt = \int_{0}^{y}\left(\frac{A_s}{A}\right)\left[V_0^2 - \left(\frac{A_s}{A}\right)\left(\frac{g}{L}\right)y^2\right]^{-1/2}dy \tag{c}$$

$$T = \frac{A_s}{A}\left[\sin^{-1}\frac{y}{V_0\sqrt{\left(\frac{A}{A_s}\right)\left(\frac{L}{g}\right)}}\right]\sqrt{\left(\frac{A}{A_s}\right)\left(\frac{L}{g}\right)} \tag{d}$$

$$= \sqrt{\left(\frac{A_s}{A}\right)\left(\frac{L}{g}\right)}\sin^{-1}\left(\frac{y}{y_{max}}\right) \tag{e}$$

For $y = y_{max}$, $\sin^{-1}\left(\frac{y}{y_{max}}\right) = \frac{\pi}{2}$, so

$$T_{max} = \frac{\pi}{2}\sqrt{\left(\frac{A_s}{A}\right)\left(\frac{L}{g}\right)} \tag{f}$$

EXAMPLE 12.8.6

A simple surge tank of 4-ft diameter is placed near the terminus of a 6000-ft long, 24-in steel pipe. This steel pipe has 1-in thick walls with a modulus of elasticity of 30×10^6 psi. The modulus of elasticity of water is 300,000 psi. The pipe flow rate is 20 cfs. What is the maximum surge that will be experienced? How soon after the assumed instantaneous valve closure will this maximum surge be attained? Use a Darcy–Weisbach friction factor of $f = 0.012$. Also neglect the effects of friction.

SOLUTION

Neglecting the effects of friction, use the equation developed in example 12.8.4 to compute the maximum height y_{max}. First compute the velocity in the 24-in steel pipe:

$$A = \pi \times (2)^2/4 = 3.14 \text{ ft}^2$$

$$A_s = \pi \times (4)^2/4 = 12.57 \text{ ft}^2$$

$$Q = AV_0 = 20 \text{ ft}^3/\text{s}$$

$$V_0 = \frac{20}{\pi(2)^2/4} = 6.37 \text{ ft/s}$$

$$y_{max} = V_0 \sqrt{\left(\frac{A}{A_s}\right)\left(\frac{L}{g}\right)} = (6.37)\sqrt{\left(\frac{3.14}{12.57}\right)\left(\frac{6000}{32.2}\right)} = 43.46 \text{ ft}$$

The time to maximum depth in the surge tank is computed using equation (f) in example 12.8.5:

$$T_{max} = \frac{\pi}{2}\sqrt{\left(\frac{A_s}{A}\right)\left(\frac{L}{g}\right)} = \frac{\pi}{2}\sqrt{\left(\frac{12.57}{3.14}\right)\left(\frac{6000}{32.2}\right)} = 42.9 \text{ sec}$$

PROBLEMS

12.1.1 Solve example 12.1.1 using a 1.5 m diameter pipe.

12.2.1 A pump operating at 1800 rpm deliveries 180 gal/min at 80 ft head. If the pump operates at 1500 rpm, what are the corresponding head and discharge?

12.2.2 A flow of 0.03 m³/s must be pumped against a head of 30 m. The pump will be driven by an electric motor with a speed of 1800 rev/min. What type of pump should be used and what is the corresponding efficiency?

12.2.3 Solve problem 12.2.2 using U.S. customary units.

12.2.4 Solve problem 12.2.2 using a speed of 2000 rev/min.

12.2.5 Solve example 12.2.4 using a discharge of 0.18 m³/s.

12.2.6 Solve example 12.2.4 with a static suction head of −3 m.

12.2.7 Solve example 12.2.5 using a total dynamic head of 60 m.

12.2.8 Solve example 12.2.5 using a total dynamic head of 20 m.

12.2.9 Solve example 12.2.6 with the manufacturer's $NPSH_R = 5.0$.

12.2.10 Determine the brake horsepower required by a pump for a discharge of 0.2 m³/s and a total dynamic head of 20 m, with an efficiency of 80 percent.

12.4.1 Solve example 12.4.1 for a flow rate of 8 cfs and $\Delta Z = 25$ ft.

12.4.2 Solve example 12.4.1 for a flow rate of 4 cfs and $\Delta Z = 40$ ft.

12.4.3 Solve example 12.4.2 assuming the elevation of reservoir 1 is 120 ft.

12.4.4 Solve example 12.4.2 assuming the elevation of reservoir 1 is 80 ft.

12.5.1 Solve example 12.5.1 using a 12 in diameter discharge pipe from the pump to reservoir B. All other information is the same.

12.5.2 Solve example 12.5.1 assuming the reservoir elevation is increased to 60 ft.

12.5.3 Solve example 12.5.3 using a pump speed of 1500 rpm.

12.5.4 Solve example 12.5.4 assuming reservoir 2 has an elevation of 200 ft. All other information is the same.

12.5.5 Solve example 12.5.4 assuming line A is 10-in line.

12.6.1 Solve example 12.6.1 using the Hazen–Williams equation with $C = 140$.

12.6.2 Solve example 12.6.1 using Manning's equation with $n = 0.012$.

12.6.3 Solve example 12.6.2 using the Hazen–Williams equation with $C = 140$.

12.6.4 Determine the flows in each pipe of the network in Figure 12.6.3 if the demand at C changes to 4 cfs and the inflow at A changes to 12 cfs.

12.6.5 Determine the flows in each pipe of the network in Figure 12.6.3 if the demands at E and F each change to 3 cfs and the inflow at A changes to 8 cfs.

12.6.6 Determine the flows in each pipe of the network in Figure 12.6.3 if there is a demand of 4.0 cfs at D and the inflow at A changes to 14 cfs.

12.6.7 Solve example 12.6.2 using Manning's equation with $n = 0.012$.

12.6.8 Develop the system of equations to determine the discharge for the network in Figure 12.6.3 for the linear theory method.

12.7.1 Use the KYPIPE2 computer program to solve example 12.7.1.

12.7.2 Use the EPANET computer program to solve example 12.7.1.

12.7.3 Use the KYPIPE2 computer program to determine the flows in each pipe of the network in Figure 12.6.3.

12.7.4 Use the EPANET computer program to solve example 12.7.2.

12.8.1 Solve example 12.8.1 using a design discharge of 50 l/s.

12.8.2 Solve example 12.8.1 assuming the cast iron pipe has a diameter of 15 cm.

12.8.3 Solve example 12.8.2 assuming the pipe is 2000 m long.

12.8.4 Solve example 12.8.2 assuming the reservoir elevation is 100 m above the downstream end of the pipe.

12.8.5 Solve example 12.8.3 assuming the pipe diameter is 18 in.

12.8.6 Solve example 12.8.3 assuming the pipe diameter is 30 in.

12.8.7 Solve example 12.8.6 assuming the pipe diameter is 30 in.

12.8.8 Solve example 12.8.6 assuming the surge tank has a diameter of 6 ft.

12.8.9 Solve example 12.8.6 assuming the pipe is only 4000 ft long.

REFERENCES

Bhave, P. R., *Analysis of Flow in Water Distribution Systems*, Technomic, Lancaster, PA, 1991.

Bosserman, B. E. II, "Control of Hydraulic Transients," Chapter 7 in *Pumping Station Design* (edited by R. L. Sanks), Butterworth-Heinemann, Woburn, MA, 1998.

Bosserman, B. E. II, "Pump System Hydraulic Design," *Water Distribution System Handbook* (edited by L. W. Mays), McGraw-Hill, New York, 2000.

Bosserman, B. E. II, "Pump System Hydraulic Design," *Hydraulic Design Handbook* (edited by L. W. Mays), McGraw-Hill, New York, 1999.

Bosserman, B. E. II, and W. A. Hunt, "Fundamental of Hydraulic Transients," in *Pumping Station Design* (edited by R. L. Sanks), Butterworth-Heinemann, Woburn, MA, 1998.

Bosserman, B. E. II, J. C. Dowell, E. M. Huning, and R. L. Sanks, "Pipes and Fittings," *Pumping Station Design* (edited by R. L. Sanks), Butterworth-Heinemann, Woburn, MA, 1998.

Bosserman, B. E. II, R. J. Ringwood, M. D. Schmidt, and M. G. Thalhamer, "System Design for Water Pumping," in *Pumping Station Design*, (edited by R.L. Sanks), Butterworth-Heinemann, Woburn, MA, 1998.

Cersario, A. L., *Water Distribution System, Modeling, Analysis and Design*, American Water Works Association, Denver, CO 1995.

Clark, R., "Water Quality Modeling—Case Studies," *Water Distribution Systems Handbook* (edited by L. W. Mays), McGraw-Hill, New York, 2000.

Clark, R. M., W. M. Grayman, and R. M. Males, "Contaminant Propagation in Distribution Systems," *Journal of Environmental Engineering*, ASCE, vol. 114, no. 4, 1988.

Cullinane, M. J. Jr., Methodologies for the Evaluation of Water Distribution System Reliability/Availability, Ph.D. dissertation, University of Texas at Austin, 1989.

Edwards, D. K., V. E. Denny, and A. F. Mills, *Transfer Processes*, McGraw-Hill, New York, May 1976.

Grayman, W. M., and G. J. Kirmeyer, "Quality of Water in Storage," *Water Distribution Systems Handbook* (edited by L. W. Mays), McGraw-Hill, New York, 2000.

Grayman, W. M., L. A. Rossman, and E. E. Geldreich, "Water Quality," *Water Distribution Systems Handbook* (edited by L. W. Mays), McGraw-Hill, New York, 2000.

Jeppson, R. W., *Analysis of Flow in Pipe Networks*, Ann Arbor Science, Ann Arbor, MI, 1976.

Klein, F., and R. L. Sanks, "Preliminary Design Consideration," Chapter 25 in *Pumping Station Design* (edited by R. L. Sanks), Butterworth-Heinemann, Woburn, MA, 1998.

Martin, C. S., "Hydraulic Transient Design for Pipeline Systems," *Water Distribution Systems Handbook* (edited by L. W. Mays), McGraw-Hill, New York, 2000.

Martin, C. S., "Hydraulic Transient Design for Pipeline Systems," *Hydraulic Design Handbook* (edited by L. W. Mays), McGraw-Hill, New York, 1999.

Mays, L. W., (editor-in-chief), *Hydraulic Design Handbook*, McGraw-Hill, New York, 1999.

Mays, L. W., (editor-in-chief), *Water Distribution Systems Handbook*, McGraw-Hill, New York, 2000.

Mays, L. W., and Y. K. Tung, *Hydrosystems Engineering and Management*, McGraw-Hill, New York, 1992.

Morris, H. M., and J. M. Wiggert, *Applied Hydraulics in Engineering*, Ronald Press Company, New York, 1972.

Ormsbee, L. E., and S. Lingireddy, "Calibration of Hydraulic Network Models," *Water Distribution Systems Handbook* (edited by L. W. Mays) McGraw-Hill, New York, 2000.

Roberson, J. A., J. J. Cassidy, and M. H. Chaudhry, *Hydraulic Engineering*, 2nd edition, John Wiley and Sons, New York, 1997.

Roberson, J. A., and C. T. Crowe, *Engineering Fluid Mechanics*, Houghton Mifflin Company, Boston, MA, 1990.

Rossman, L. A., "Computer Models/EPANET," *Water Distribution Systems Handbook* (edited by L. W. Mays) McGraw-Hill, New York, 2000.

Rossman, L. A., "EPANET Users Manual." Risk Reduction Engineering Lab, U.S. Environmental Protection Agency, Cincinnati, OH, 1994.

Rossman, L. A., P. F. Boulos, and T. A. Altman, "Discrete Volume Element Method for Network Water Quality Models," *Journal of Water Resources Planning and Management*, ASCE, vol. 119, no. 5, pp. 505–517, 1993.

Rossman, L. A., J. G. Uber, and W. M. Grayman, "Modeling Disinfectant Residuals in Drinking Water Storage Tanks," *Journal of Environmental Engineering*, ASCE, vol. 121, no. 10, pp. 752–755, 1995.

Rutschi, K, "Die fleiderer–Sauqzahl als gutegrad der Saugfahigkeit von Kreiselpumpen," *Schweizerische Bauzeitung*, no. 12, Zurich, 1960.

Sanks, R. L., "Data for Flow in Pipes, Fittings, and Valves," Appendix B in *Pumping Station Design* (edited by R. L. Sanks), Butterworth-Heinemann, Woburn, MA, 1998.

Tchobanoglous, G., "Theory of Centrifugal Pumps," Chapter 10 in *Pumping Station Design* (edited by R. L. Sanks), Butterworth-Heinemann, Woburn, MA, 1998.

Ten-State Standards, *Recommended Standards for Sewage Works*, Great Lakes-Upper Mississippi Board of Sanitary Engineers, Health Education Service, Inc., Albany, NY, 1978.

Velon, J. P., and T. J. Johnson, "Water Distribution and Treatment," Davis' *Handbook of Applied Hydraulics*, 4th edition (edited by V. J. Zipparo and H. Hasen), McGraw-Hill, New York, 1993.

Walski, T., "Water Distribution," *Water Resources Handbook*, (edited by L. W. Mays), McGraw-Hill, New York, 1996.

Wood, D., and C. Charles, "Hydraulic Network Analysis Using Linear Theory," *Journal of Hydraulics Division, ASCE*, vol. 98, no. HY7, pp. 1157–1170, 1972.

Wood, D., "Computer Analysis of Flow in Pipe Networks Including Extended Period Simulation–User's Manual," Office of Engineering, Continuing Education and Extension, University of Kentucky, 1980.

Wood, D. J., L. E. Ormsbee, S. Reddy, and W. D. Wood, "Users' Manual, KYQUAL, Kentucky Water Quality Analysis Program," Civil Engineering Software Center, Dept. of Civil Engineering, University of Kentucky, Lexington, KY, 1995.

Ysusi, M. A., "System Design: An Overview," *Water Distribution System Design* (edited by L. W. Mays), McGraw-Hill, New York, 2000.

Ysusi, M. A., "Water Distribution System Design," *Hydraulic Design Handbook* (edited by L. W. Mays), McGraw-Hill, New York, 1999.

Chapter 13

Water for Hydroelectric Generation

13.1 ROLE OF HYDROPOWER

Water and energy are two resources that are very necessary for humankind and are intricately connected. This chapter focuses on the need for water in the production of energy, in particular water for hydroelectric generation and for thermal power production. Table 11.2.1 summarizes consumptive water use for energy production. Table 11.2.2 summarizes consumptive water use for electricity production.

Hydroelectric power production is the most obvious use of water for the production of energy. The energy in falling water is used directly to turn turbines that generate electricity. Table 13.1.1 lists the hydroelectric capacity and generation of the various continents.

Table 13.1.1 World Hydroelectric Capacity and Generation, 1990

Continent[a]	Installed Hydroelectric Capacity (10^3 MW)	Percent of Total	Hydroelectric Generation (10^6 MWh per year)	Percent of Total
North America	156.8	26	599.6	28
Central and South America	80.3	13	353.4	17
Western Europe	155.0	25	444.7	21
Eastern Europe	15.1	2	26.3	1
Soviet Union	64.4	10	217.3	10
Middle East	3.1	1	12.6	1
Africa	18.9	3	43.2	2
Far East and Oceania	121.3	20	415.8	20
Totals	614.9	100	2,112.9	100

[a]Continental sums use the country assignments of the original source. Since 1990, substantial changes in Eastern Europe and the Soviet Union have occurred.

Source: (Gleick, 1993) U.S. Department of Energy, 1992, *International Energy Annual 1990*, Energy Information Administration, DOE/EIA-0219(90), Washington, DC.

The demand for electricity, referred to as the *load*, can be expressed in terms of *energy demand* (or use power demand) or *capacity demand* (peak power demand). *The daily load shape* (power demand as a function of time of day) is illustrated in Figure 13.1.1. Load is divided into three segments (Figure 13.1.1): (1) *base load*, which is continuous for 24 hours a day; (2) *peak load*, which

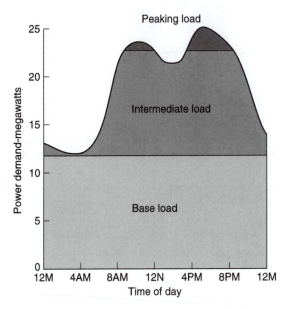

Figure 13.1.1 Daily load shape showing load types (from U.S. Army Corps of Engineers (1985)).

is the highest load occurring for a few hours a day; and (3) *intermediate load*, which is the portion between the load and the peak load. The load for the peak day of the year determines the required generating capacity, and the peak weekly or monthly load dictates the amount of energy stored in the form of fuel for thermal power plants or water for hydroelectric plants.

Load curves show the daily or weekly fluctuation in load as a function of time for an electric power system. Figure 13.1.2 shows a weekly load curve of a large electric utility. This figure illustrates how different types of generating units are used to meet daily demands. The value of a unit of energy is a function of the time of day, which explains why hydropower is used whenever possible as a peaking unit to replace high-cost energy.

A *load factor* is the ratio of the average load over a certain period of time to the peak load during that time. Load factors are expressed as daily, weekly, monthly, or yearly. If a load factor is high, the unit cost of energy is comparatively low because the system should be operating near capacity and near best efficiency. For a small load factor, much of the system's generating capacity is idle a large part of the time.

The *plant factor* is the ratio of the average load on a plant for the time period being considered to its aggregate rated capacity (installed capacity). An average annual *plant factor* is the average annual energy (KWh) divided by the installed capacity (KW) times 8760 hours.

The major types of hydroelectric developments are

- Run-of-river developments
- Pondage developments (Figure 13.1.3)
- Storage developments (Figure 13.1.4)
- Reregulating developments
- Pumped storage (Figure 13.1.5 and 13.1.6)

Run-of-river developments have a dam with a short supply pipe (penstock) that directs the water to the turbines. The natural flow of the river is used with very little alteration to the terrain stream channel at the site and there is very little impoundment of the water. Typical run-of-river projects include: (1) navigation projects, (2) irrigation diversion dams, and (3) projects lacking storage as a result of the topography. A *pure run-of-river* project has no usable storage.

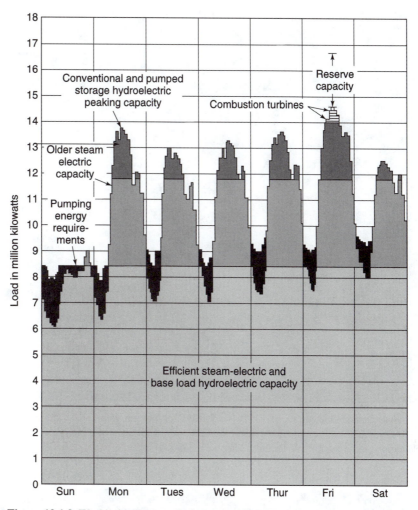

Figure 13.1.2 Weekly load curve of a large electric utility system (from U.S. Army Corps of Engineers (1979)).

Figure 13.1.3 Barkley Lock and Dam, a run-of-river project with pondage. (Courtesy of U.S. Army Corps of Engineers Nashville District.)

Figure 13.1.4 Hoover Dam and power houses on Colorado River in Arizona. Example of storage project having maximum reservoir storage of 34.852×10^6 m^3. Present rated capacity is 1,951 MW with planned rated capacity of 2,451 MW. (Courtesy of U.S. Bureau of Reclamation, photograph by A. Pernick, March 31, 1996.)

Figure 13.1.5 New Waddell Dam on the Agua Fria River in Arizona impounds Lake Pleasant. Shown downstream of dam and intake structure is the Waddell Pump/Generating Plant, which lifts Central Arizona Project (CAP) 192 feet from the canal into Lake Pleasant. The Waddell Canal connects the main CAP aqueduct with the dam. (Courtesy of Central Arizona Project.)

Figure 13.1.6 Waddell Pump/Generating Plant, which has eight units: four adjustable-speed pumps and four two-speed pump generators. Maximum power generation is 44 MW. CAP water is pumped into Lake Pleasant during colder months and released during the summer to generate electricity. (Courtesy of U.S. Bureau of Reclamation, photograph by J. Madrigal, Jr., August 22, 1995.)

Pondage developments are run-of-river projects with a small amount of storage (daily or weekly) that can be used to regulate discharges to follow daily and weekly load patterns. *Pondage* refers to the short time in ponding of water, as these projects have insufficient storage for seasonal flow regulation.

Storage developments have an extensive impoundment at the power plant or are at the reservoir upstream of the power plant. *Storage* is the long-time impounding of water to allow seasonal regulation capability.

Reregulating developments receive fluctuating discharges from large hydroelectric peaking plants and release the discharges to meet downstream flow criteria. *Reregulating reservoirs* are also called *after-bay reservoirs*.

Pumped-storage developments convert low-volume off-peak energy to high-value on-peak energy. Water is pumped from a lower reservoir to a higher reservoir using inexpensive power during periods of low energy demand. Water is then discharged through the turbine to produce to meet peak demands. Pumped storage projects can also be operated on a seasonal basis as the New Waddell Dam (Figure 13.1.5).

Hydropower can be used in power systems for peaking, for meeting intermediate loads, for base load operation, or for meeting a combination of these loads (U.S. Army Corps of Engineers, 1985). A *load-duration curve* expresses the load as a function of percent of time and is commonly used to describe system operation. Figures 13.1.7*a* and *b* show the operation of a system with the hydropower facility operating in the peaking mode and the base load, respectively. Figure 13.1.8*a* shows operation of a system with the hydropower facility operating in both the peaking and base modes. Figure 13.1.8*b* shows the operation of a run-of-river plant.

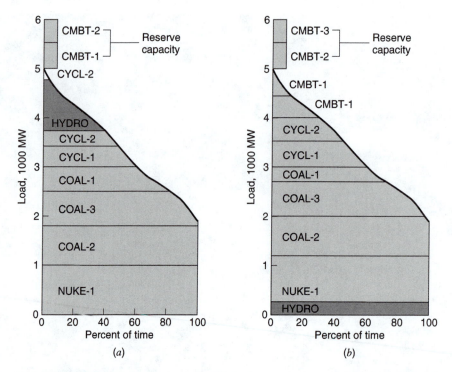

Figure 13.1.7 Duration curve showing operation of system with hydro plant. (*a*) In peaking mode; (*b*) In base load (from U.S. Army Corps of Engineers (1985)).

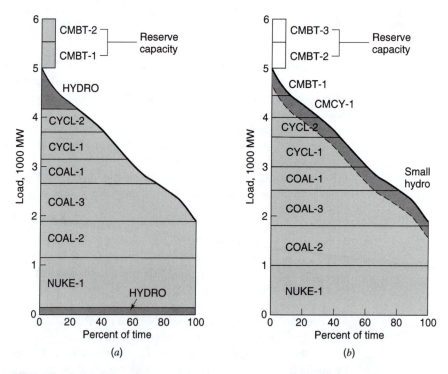

Figure 13.1.8 Duration curve showing operation of system. (*a*) With hydro plant carrying both base and peaking loads; (*b*) With pure run-of-river hydropower plant (from U.S. Army Corps of Engineers (1985)).

EXAMPLE 13.1.1   Determine the energy in MWh for the 1000-MW peak capacity hydroelectric plant placed in the base load as shown in Figure 13.1.7b. The plant has an average power output of 250 MW for the week.

SOLUTION   The energy is computed as Energy = 250 MW × 168 hours = 42,000 MWh

EXAMPLE 13.1.2   Determine the energy in MWh for the 1000-MW peak capacity hydroelectric plant placed in the peaking mode as shown in Figure 13.1.7a. The plant factor is 25 percent.

SOLUTION   The energy is computed using

Energy = (capacity, MW) × (plant factor, %) × 168 hrs/100

= 1000 MW × 25% × 168 hrs/100

= 42,000 MWh

Note that this is the same energy as for the hydropower plant placed in the base in Example 13.1.1.

13.2   COMPONENTS OF HYDROELECTRIC PLANTS

13.2.1   Elements to Generate Electricity

Three basic elements are needed for a power generation facility:

- A means to create a head
- Conduit to convey water
- Power plant

Figure 13.2.1 illustrates some of the elements in a hydroelectric plant needed to generate electricity. The dam has the two major functions of creating the head necessary to move the turbines and impounding the storage used to maintain the necessary flow release pattern. The height of the dam establishes the generating head and the available storage for power plant operation. The *reservoir* is the water impoundment behind a dam. *Storage capacity* is the volume of reservoir available to store water. *Intake structures* direct water from the reservoir into the penstock. *Penstocks* convey water from the intake structure to the powerhouse. The *powerhouse* shelters the turbines, generating units, and control and auxiliary equipment. *Turbines* convert the potential energy of water into mechanical energy, which drives the generator. *Draft tubes* convey water from the discharge side of the turbine to the tailrace. The *tailrace* maintains a minimum tailwater elevation below the power plant and keeps the draft tube submerged. Figure 13.2.2 shows the penstocks, powerhouse, and tailrace for the Atatürk Dam in Turkey.

The various energy elements are illustrated in Figure 13.2.3. For a constant discharge, the energy relation between the forebay and any other section is $z_1 + (V_1^2/2g) = z_2 + (V_2^2/2g) + h_c$, where h_c is the loss between the two sections. The various losses are illustrated in Figure 13.2.3. A *forebay* is a regulating reservoir that temporarily stores water to facilitate one or more of the following: (1) low-approach velocity to intake, (2) surge reduction, (3) sediment removal (descending), or (4) storage. They are designed to maintain the approach flow conditions as smoothly as possible (see Coleman et al. (1999) for more detail).

There are two basic types of turbines: (1) *impulse turbines* (commonly called *Pelton turbines*), in which a free jet of water impinges on a revolving element of the machine that is exposed to the atmosphere (Figure 13.2.4) and (2) *reaction turbines*, in which the flow takes place under pressure in a closed chamber (Figures 13.2.5 and 13.2.6). Reaction turbines extract power from both the kinetic energy of the water and the difference in pressure between the front end and the back of

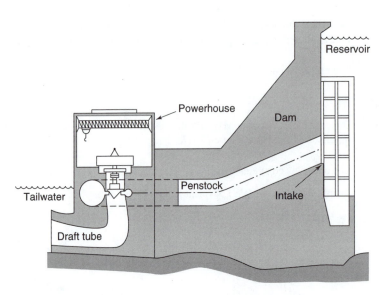

Figure 13.2.1 Components of a hydropower project (from U.S. Army Corps of Engineers (1985)).

Figure 13.2.2 View of the Atatürk Dam Hydroelectric Facility on the Euphrates River in Turkey. Shown are the powerhouse and tailrace. The rated capacity is 2400 MW. Dam height is 184 m, crest length is 1820 m, dam volume is 84.5×10^6 m^3, and maximum reservoir capacity is 48.7×10^6 m^3. Also refer to Figures 11.4.8 and 17.2.15. (Photograph by L.W. Mays.)

Figure 13.2.2 (*continued*) View of the Atatürk Dam Hydroelectric Facility on the Euphrates River in Turkey. Shown are the power-house and tailrace (Photograph by L.W. Mays).

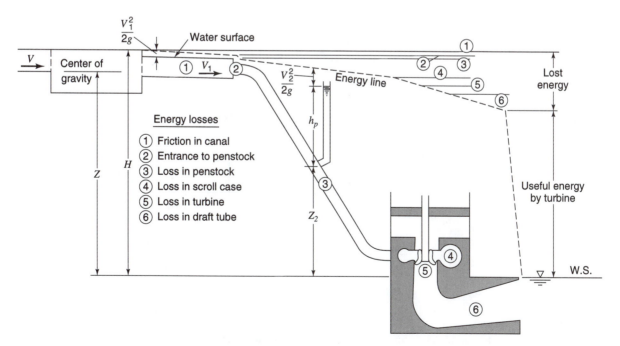

Figure 13.2.3 Energy relations in a typical hydroelectric plant (from Hasen and Antonopoulos (1993)).

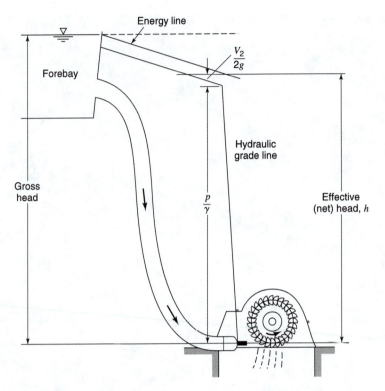

Figure 13.2.4 Definition sketch for impulse-turbine installation (from Linsley et al. (1992)).

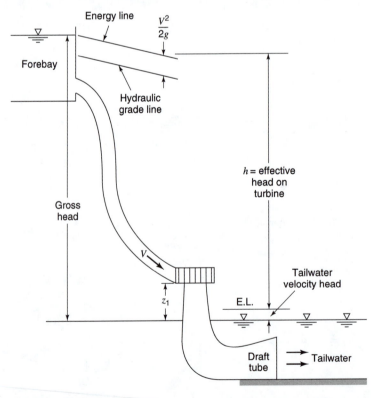

Figure 13.2.5 Definition sketch for a reaction-turbine installation (from Linsley et al. (1992)).

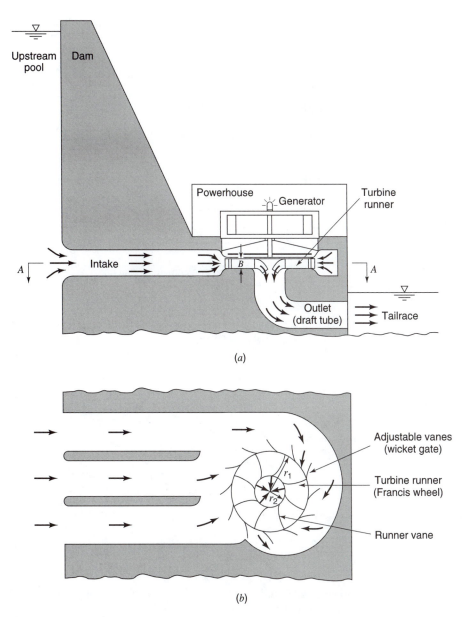

Figure 13.2.6 Schematic view of reaction-turbine installation. (*a*) Elevation view; (*b*) Plane view—section *A–A* (from Roberson et al. (1998)).

each runner blade. Figure 13.2.7 presents the application ranges for conventional turbines and Figure 13.2.8 presents turbine efficiency curves.

Impulse or Pelton turbines have a runner with numerous spoon-shaped buckets attached to the periphery and are driven by one or more jets of water issuing from fixed or adjustable nozzles, as shown in Figure 13.2.9. Large Pelton turbines can be used at heads above 1000 feet and smaller turbines can be used at heads of less than 100 feet.

Reaction turbines include *Francis turbines*, which are constructed so that water enters the runner radially and then flows towards the center and along a turbine shaft axis. These turbines have been typically used for heads ranging from 100 to 1500 feet and are the economical choice for heads ranging from 150 to 1500 feet. Figure 13.2.10 illustrates a Francis turbine.

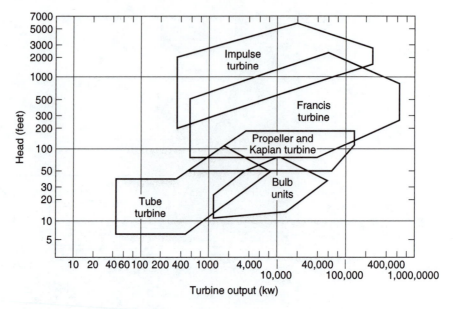

Figure 13.2.7 Application ranges for standard and custom hydraulic turbines (from U.S. Army Corps of Engineers (1985)).

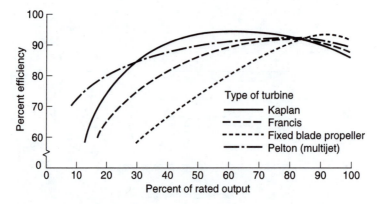

Figure 13.2.8 Turbine efficiency curves (from U.S. Army Corps of Engineers (1985)).

Fixed-blade propeller turbines are constructed so that water passes through the propeller blades in an axial direction. These turbines are typically used in the 10–200 feet head range and are the economical choice in the 50–150 feet head range. Kaplan turbines are propeller turbines with adjustable pitch blades that operate in the same general range as propeller turbines.

Tubular turbines have a guide vane assembly that is in line with the turbine and contributes to the tubular shape. These turbines may be vertical, horizontal, or slant-mounted axial flow turbines. Tubular turbines can be the economical choice for heads less than 50 feet. *Bulb turbines* are horizontal axial-flow turbines with a turbine runner connected directly to a generator or through a speed-increasing gearbox. A *rim turbine* is similar to the bulb turbine with the generator mounted on the periphery of the turbine runner blades. Rim turbines are suitable for 10- to 50-feet heads.

Figure 13.2.9 Pelton turbine and nozzle layout. (Courtesy of Sulzer-Escher Wyss Ltd.)

Figure 13.2.10 Francis turbine (turbine runner) used at Glenn Canyon Dam, on the Colorado River near Page, Arizona (Photograph by L.W. Mays).

13.2.2   Hydraulics of Turbines

This subsection briefly describes the hydraulic action of turbines; see Figure 13.2.11 for impulse turbines and Figure 13.2.12 for reaction turbines. The impulse turbine rotates at a velocity u at the center line of the buckets. When water enters the bucket, it is in position A and the bucket moves to position B, where the water leaves the bucket. The velocity changes from V_1 at the entrance to V_2 at the exit. The velocity v represents the velocity of the water relative to the bucket, so that at entry $V_1 = u + v$. Ignoring fluid friction, the magnitude of the water velocity relative to the bucket remains constant with a change in direction so that it is tangential to the bucket. The force F exerted by the discharge of water Q on the bucket is described by the impulse-momentum, neglecting friction, as

$$F = \rho Q(V_1 - V_2 \cos \alpha) \qquad (13.2.1)$$

Equation (13.2.1) can also be expressed in terms of relative velocities as

$$F = \rho Q(v - v\cos \beta) = \rho Q(V_1 - u)(1 - \cos \beta) \qquad (13.2.2)$$

where the relative velocity $v = V_1 - u$.

The power transmitted to the buckets is the product of force and velocity

$$P = Fu = \rho Q(V_1 - u)(1 - \cos \beta)u \qquad (13.2.3)$$

From equation (13.2.3) it is apparent that no power is developed when $u = 0$ or $u = V_1$. Maximum power with respect to u occurs when

$$dP/du = \rho Q(1 - \cos \beta)(V_1 - u - u) = 0 \qquad (13.2.4)$$

which can be solved for

$$u = V_1/2 \qquad (13.2.5)$$

This argument says that, when friction is negligible, the best hydraulic efficiency occurs when the peripheral speed of the wheel is half of the velocity of the jet of water.

The *brake power* delivered by a turbine to the generator is

$$Bh_p = \frac{\gamma he}{550}Q \qquad (13.2.6)$$

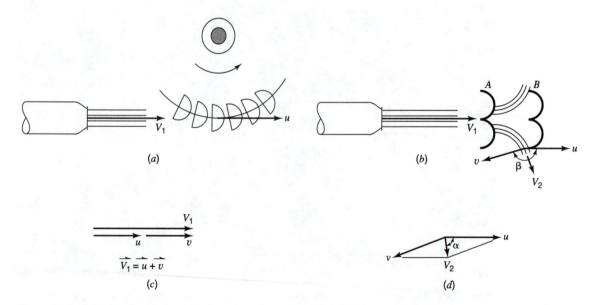

Figure 13.2.11 Hydraulic relationships for an impulse turbine (from Linsley et al. (1992)).

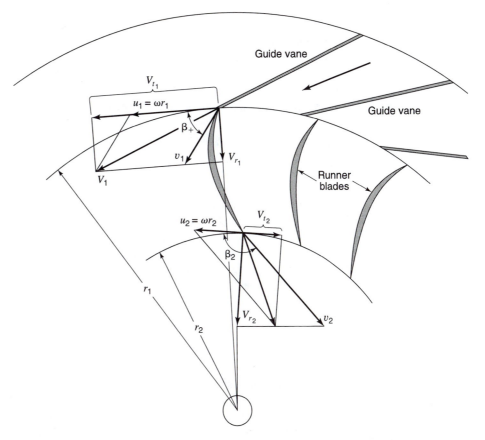

Figure 13.2.12 Vector diagrams for a reaction turbine (from Linsley et al. (1992)).

in English units where Bh_p is in horse power, h is the effective head, and γ, Q, and h are expressed in customary units of lb/ft^3, ft^3/s, and ft, respectively. In SI units the brake power is expressed as

$$Bk_w = \gamma Qhe \tag{13.2.7}$$

where Bk_w is brake kilowatts and, γ, Q, and h are expressed in kN/m^3, m^3/s, and m, respectively. Note that 1 hp = 550 ft-lb/s = 0.746 kW.

The flow through the radial-flow runner of a Francis turbine is used to explain the action of reaction turbines. In Figure 13.2.12, the velocity is $u_1 = \omega r_1$ at the entrance edge of the runner blade, where ω is the rotative speed of the runner in radians per second. This entering velocity to the blade is essentially tangential to the exit end of the guide vane. The component of V_1 that is tangent $\left(V_{t_1} \right)$ to the entrance of the runner blade is

$$V_{t_1} = \omega r_1 + V_{r_1} \cot \beta_1 \tag{13.2.8}$$

where V_{r_1} is the radial component of the velocity. The tangential component of the velocity at the exit is

$$V_{t_2} = \omega r_2 + V_{r_2} \cot \beta_2 \tag{13.2.9}$$

The torque T exerted on the runner blades is

$$T = \rho Q(V_{t_1} r_1 - V_{t_2} r_2) \tag{13.2.10}$$

where Q is the total discharge through the turbine. The power transmitted to the turbine from the water is

$$P = T\omega = \rho Q\omega(V_{t_1} r_1 - V_{t_2} r_2) \tag{13.2.11}$$

The following continuity equation can be used to determine the radial velocity components:

$$Q = 2\pi r_1 Z V_{r_1} = 2\pi r_2 Z V_{r_2} \tag{13.2.12}$$

where Z is the height of the turbine blades.

As with pumps, the concept of a *turbine-specific speed* is also used. For turbines, we are more interested in the power of the turbine than the discharge. The dimensional form of the specific speed n_s used by the hydraulic turbine industry is

$$n_s = \frac{NP^{1/2}}{H^{5/4}} \tag{13.2.13}$$

where N is the rotational speed in rpm, P is the power in horsepower, and H is the head on the turbine in feet. For a given Q and H, the actual speed is directly proportional to the specific speed.

Cavitation in turbines is very undesirable, since it causes pitting, mechanical vibration, and loss of efficiency. The concept of cavitation is discussed in more detail in Chapter 12. Cavitation can be avoided by designing, installing, and operating turbines so that the local absolute pressure never drops to the vapor pressure of the water. The susceptibility to cavitation in turbines is given by the *cavitation index*, defined as

$$\sigma = \frac{p_0/\gamma - p_v/\gamma - (z_t - z_0)}{H} \tag{13.2.14}$$

where p_0 = absolute atmospheric pressure, p_v = absolute vapor pressure of the water, z_t = elevation of the downstream side of the turbine above the water surface in the tailrace, z_0 = elevation of tailrace water surface, and H = net head across the turbine (head change from upstream of turbine to the downstream end of draft tube). As previously described the tailrace is the channel into which the flow from the draft tube discharges. See Figures 13.2.2, 13.2.5, and 13.2.6.

For a given turbine operating with a given H and speed, if z_t is increased or p_0 decreased, the pressure acting on the blades of the turbine decreases and eventually reaches a point at which cavitation would occur. Then lower values of σ indicate a greater tendency for cavitation. Cavitation susceptibility also changes with the speed of the impeller because greater speed means greater relative velocities and less pressure on the downstream side of the impeller. Critical σ_c values are obtained from experiments.

13.2.3 Power System Terms and Definitions

Many useful power system terms are defined in this subsection. *Power* is the rate of energy and the *energy* produced by a power-generating unit is equal to the power multiplied by the time period of production. The SI unit for power is kilowatt (KW), where one horsepower is 0.7457 KW. The common units for energy are kilowatt-hr (KWh), or the amount of power that can be generated and produced with little or no interruption, and *firm energy* is the corresponding energy. *Firm power* is typically thought of as being available 100 percent of the time. The power generated in excess of firm power is called *secondary* (or *surplus, interruptible power*). Secondary power cannot be relied upon, and therefore the rate of secondary power is generally well below that of firm power. Secondary power is interruptible but is available more than 50 percent of the time. The third type of power, *dump power*, is much less reliable and is available less than 50 percent of the time.

The *capacity* is the maximum amount of power that a generating unit or power plant can deliver at any given time in KW. The *rated capacity* of a generating unit is the capacity that the unit is designed to deliver. The *overflow capacity* is the capacity that a generating unit can deliver for a limited time in excess of normal rated capacity (or *nameplate capacity*).

The *installed capacity* is the nominal capacity of a power plant determined as the sum of the rated (or nameplate) capacities of all units in a power plant. The *dependable* or *firm capacity* is the capacity that a power plant can reliably contribute to meet peak power demands. *Hydraulic capacity* is the maximum discharge that a hydroelectric plant can use for power generation.

EXAMPLE 13.2.1

A reaction turbine is supplied with water through a 150-cm pipe (penstock) ($e = 1.0$ mm) that is 40 m long. The water surface in the reservoir is 20 m above the draft tube inlet, which is 4.5 m above the water level in the tailrace. If the turbine efficiency is 92 percent and the discharge is 12 m³/s, what is the power output of the turbine in kilowatts? Use $f = 0.0185$.

SOLUTION

To determine the power, use equation (13.2.7). The velocity through the 40-m penstock is computed using continuity, $V = Q/A = 12/(\pi(1.5)^2/4) = 6.8$ m/s. The headloss due to friction in the pipe is computed using the Darcy–Weisbach equation:

$$h_L = f\frac{L}{D}\frac{V^2}{2g} = 0.0185\left(\frac{40}{1.5}\right)\left(\frac{(6.8)^2}{2(9.81)}\right) = 1.16 \text{ m}$$

The effective head is

$$h = \frac{p}{\gamma} + z - h_L = 20 + 4.5 - 1.16 = 23.34 \text{ m}$$

Therefore, the power output of the turbine is computed as

$$BK_w = \gamma Qhe = 9.81(12)(24.34)(0.92) = 2,631 \text{ kW}$$

EXAMPLE 13.2.2

A turbine is to be installed at a location where the net available head is 350 ft and the available flow will average 1000 cfs. What type of turbine is recommended? If the turbine has an operating speed of 300 rpm and an efficiency of 0.90, what is the specific speed?

SOLUTION

The brake horsepower delivered by the turbine is computed using equation (13.2.6):

$$Bh_p = P = \frac{\gamma Qhe}{550} = \frac{62.4(1000)(350)(0.9)}{550} = 35,738 \text{ hp} = 26,650 \text{ kW}$$

For a head of 350 ft and 26650 KW output, according to Figure 13.2.7, this falls into the range of application of the Francis turbine. The specific speed is computed using equation (13.2.13):

$$n_s = \frac{NP^{1/2}}{H^{5/4}} = \frac{300(35,738)^{1/2}}{(350)^{5/4}} = 37.46$$

13.3   DETERMINING ENERGY POTENTIAL

13.3.1   Hydrologic Data

This section briefly discusses the types and sources of hydrologic data needed for hydropower studies. The most important data for a hydropower feasibility study is stream-flow data, which is used to develop estimates of water available for power generation. The most commonly used stream-flow data is mean daily flows, mean weekly flows, and mean monthly flows. This data is used to develop flow-duration curves, which show the percentage of time that flow equals or exceeds various values during the period of record, as shown in Figure 13.3.1. *Flow-duration curves* summarize the stream-flow characteristics and can be constructed from daily, weekly, or monthly stream-flow data. These curves, unfortunately, do not present flow in chronological sequence, do not describe the seasonal distribution of streamflows, and do not take into account variations of head independent of streamflow. Figure 13.3.2 illustrates a monthly flow distribution, which describes the season distribution. Other types of hydrologic data required include: tailwater rating curves, reservoir elevation-area-capacity relationships, sediment data, water quality data, downstream flow information, water surface fluctuation data, and evaporation seepage loss analysis.

13.3.2 Water Power Equations

The quantity of water available (discharge Q), the net hydraulic head H across the turbine, and the efficiency of the turbine e_t define the amount of power that a hydraulic turbine can provide. This relationship is expressed by the *water power equation* as

$$H_p = \frac{QHe_t}{8.815} \tag{13.3.1}$$

where H_p is the theoretical horsepower available, Q is in cfs, and H is in ft. Equation (13.3.1) can also be expressed in terms of kilowatts of electrical output as

$$\text{KW} = \frac{QHe}{11.81} \tag{13.3.2}$$

where e is the *overall efficiency* defined as

$$e = e_g e_t \tag{13.3.3}$$

where e_g is the generator efficiency. For preliminary studies a turbine and generator efficiency of 80 to 85 percent is typically used (U.S. Army Corps of Engineers, 1985).

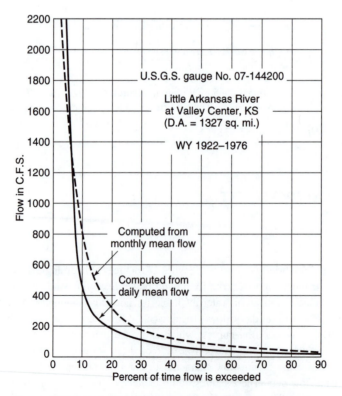

Figure 13.3.1 Flow duration curves (from U.S. Army Corps of Engineers (1979)).

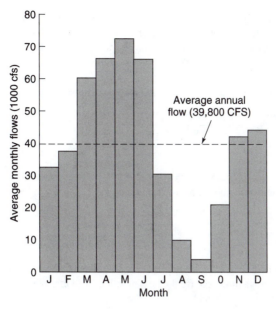

Figure 13.3.2 Monthly flow distribution (from U.S. Army Corps of Engineers (1985)).

To convert the power output (KW) to energy (KWh), equation (13.3.2) must be integrated over time:

$$\text{KWh} = \frac{1}{11.81}\int Q(t)H(t)e\,dt \qquad (13.3.4)$$

The discharges in this equation are those available for power generation. The integration process can be performed using either a *sequential streamflow routing* procedure or *flow-duration analysis*, in which the series of expected flows are represented by a flow-duration curve. For either approach the streamflow used must represent the usable flow available for power generation. The *gross* or *static head* is determined by subtracting the tailwater elevation from the water surface elevation in the forebay (see Figure 13.3.3). The *net head* is the actual head available for power generation and is used in computing the energy. The net head is the gross head minus the head losses due to intake structures, penstocks, and outlet works. The losses within the turbine are accounted for in the turbine efficiency.

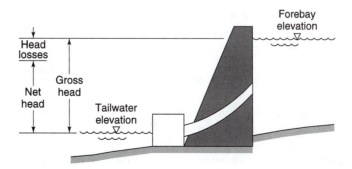

Figure 13.3.3 Gross head versus net head (from U.S. Army Corps of Engineers (1985)).

EXAMPLE 13.3.1

Determine the average discharge from a reservoir required to generate 4,700 MWh of energy at the hydropower plant for a month (30 days). The average head is 95 ft and the overall efficiency is 0.85. What is the storage draft, that is, the volume of water that must be released, assuming this is a critical drawdown period for the reservoir?

SOLUTION

The average discharge is determined using a simplified form of equation (13.3.4),

$$Q = \frac{11.81 \text{ KWh}}{Het} = \frac{11.81(47000000 \text{ KWh})}{95(0.85)(720)}$$

$$= 955 \text{ cfs}$$

The storage draft is computed considering that 955 cfs is continuous over the 30-day time period. For 30 days compute the number of acre-ft/cfs, which can be shown to be 59.5 ac-ft/cfs. Therefore,

(955 cfs) (59.5 acre-ft/cfs) = 56,800 acre-ft

EXAMPLE 13.3.2

The release for hydropower production over a 30-day month resulted in a loss of head of 10 ft in the reservoir. The average discharge for the next month is 1824 cfs. What is the loss of energy over the next 30 days due to the drop in head?

SOLUTION

Use the energy equation to determine the energy loss:

$$\text{Energy loss (KWh)} = \frac{QHe}{11.81}t = \frac{1824(0.85)(10)}{11.81}(720) = 945,000 \text{ kWh} = 945 \text{ MWh}$$

13.3.3 Turbine Characteristics and Selection

13.3.3.1 Turbine Characteristics

The efficiency, usable head range, and minimum discharge are turbine characteristics that can have an effect on energy output. For preliminary studies, a fixed efficiency and ignoring the minimum discharge and head range limitation are sufficient; however, for feasibility studies these characteristics must be taken into account as they may have a significant impact on the results.

Design head is the head at which a turbine operates at maximum efficiency. This design head is normally specified at or near the average head, where the project will operate most of the time. However, the design head should be selected so that the desired range of heads is within the permissible operating range of the turbine. The design head for run-of-river projects can be determined from a head-duration curve as the midpoint of the head range where the project is generating power (see Figure 13.3.4). Design head is normally based on an entire year of operation; however, it may be based on the peak months only, where dependable capacity is important. For pondage projects, the design head can be based on a weighted average head, where the weight is a function of the generation at each head:

$$\text{Weighted average head} = \frac{\sum (\text{head generation})}{\sum (\text{generation})}$$

Rated head is the head at which rated power is obtained with the wicket gates fully open; thus, rated head is the minimum head at which rated output can be obtained. As shown in Figure 13.3.5, the rated head capacity of a generator limits the power output, and as a consequence full-rated capacity can be obtained for all heads above the rated head. Below the rated head, the maximum achievable power output is less than the rated capacity, even with the turbines completely open. Figure 13.3.5 illustrates a turbine performance curve for a specific design. The range of head for

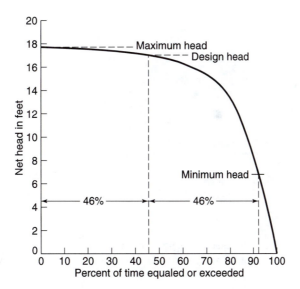

Figure 13.3.4 Head-duration curve for run-of-river project, showing how design head can be determined (from U.S. Army Corps of Engineers (1985)).

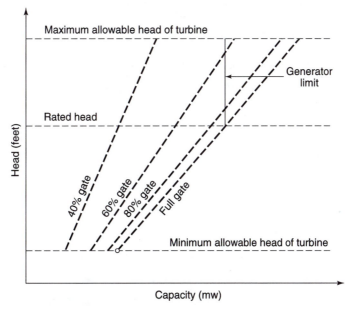

Figure 13.3.5 Turbine performance curve for a specific design (solid line represents maximum output of unit) (from U.S. Army Corps of Engineers (1985)).

the normal operation of some projects is small enough that the rated output is accomplished over the entire operating range, while for other projects, the range of head is such that the operating head falls below the rated head, so that the generator capability decreases. This occurs at storage projects with a large drawdown and at pondage projects with a large installed capacity where the tailwater elevation increases at high plant discharges. This is further illustrated in Figure 13.3.6, showing capacity versus discharge curves for different numbers of 5 MW units at a low head and run-of-river projects. Power output drops off at higher discharges as a result of tailwater encroachment.

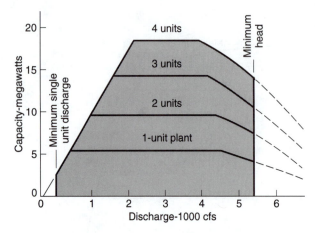

Figure 13.3.6 Capacity versus discharge for run-of-river project for alternative plant sizes (from U.S. Army Corps of Engineers (1985)).

Rated discharge is the discharge at rated head with the wicket gates fully open. Because of cavitation and vibration problems, the minimum discharge in turbines is 30 to 50 percent of the rated discharge. Table 13.3.1 lists minimum discharges. For preliminary power studies, minimum discharges normally can be ignored; however, for feasibility and more abnormal studies, the minimum discharge must be taken into account.

Table 13.3.1   Discharge and Head Ranges for Different Types of Turbines

Turbine Type	Ratio of Minimum Discharge to Rated Discharge	Ratio of Minimum Head to Maximum Head
Francis	0.40	0.50
Vertical-shaft Kaplan	0.40	0.40
Horizontal-shaft Kaplan	0.35	0.33
Fixed-blade propeller	0.65	0.40
Fixed-gate adjustable blade propeller	0.50	0.40
Fixed geometry units (pumps as turbines)	——	0.80
Pelton (adjustable nozzles)	0.20	0.80

Source: U.S. Army Corps of Engineers (1985)

The efficiency used in power studies is the combined efficiency of the turbine and the generator. Generator efficiency is usually assumed at 98 percent for large units and 95 to 96 percent for units smaller than 5 MW (U.S. Army Corps of Engineers, 1985). Turbine efficiency, on the other hand, varies with the percent of discharge head, as illustrated in Figure 13.3.7 for a typical Francis-type turbine.

13.3.3.2   Turbine Selection

The selection of a turbine is an iterative process in which preliminary power studies provide approximate plant capacity, expected head range, and an estimated design head. Preliminary turbine designs are selected and used as input for more detailed studies, making it possible to better identify the desired operating characteristics, final turbine design, and plant configuration.

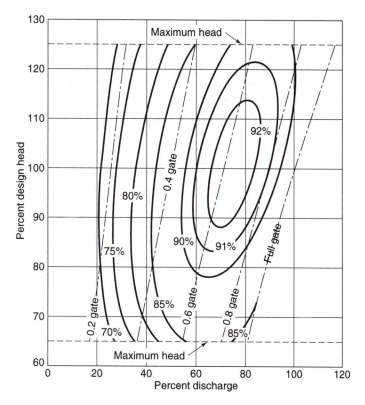

Figure 13.3.7 Typical Francis turbine performance curve (from U.S. Army Corps of Engineers (1985)).

Performance data for various types of turbines is required in the turbine selection process. This data can be obtained from manufacturers. For turbine selection, the U.S. Army Corps of Engineers (1985) recommends the following data: (a) expected head range; (b) head-duration data (not required but very useful); (c) design head (optional); (d) total plant capacity (either hydraulic capacity in cfs or generator installed capacity in megawatts); (e) minimum discharge at which generation is desired; (f) alternative combination of size and number of units to be considered (optional); (g) head range at which full-rated capacity should be provided if possible (optional); and (h) tailwater rating curve.

The *generator rated output* is selected to match the turbine output at rated head and discharge. The head at which a turbine is rated can vary with type of operation.

13.3.4 Flow Duration Method

The *flow duration method* is limited to the analysis of small hydro projects, particularly run-of-river projects, and for preliminary analysis only of other projects. The sequential streamflow-routing method (see section 13.3.5) should be used for storage projects.

The flow-duration curve previously described is the basis for this method. Flow-duration curves are typically developed from historical records to represent the percent of time different levels of streamflow are equaled or exceeded. Flow-duration curves can be converted to a *power-duration curve* by using the water power equation. The power-duration curve can then be used to estimate a site's energy potential. These types of *duration curve–energy analyses* are based on flow for an entire year for engineering purposes. For marketing purposes it is necessary to develop duration curves monthly or seasonally. Dependable capacity for small projects is usually based on the average capacity available during peak demand months.

Data requirements. The data required include (U.S. Army Corps of Engineers, 1985): routing interval (daily time interval), streamflow data, minimum length of record (representative period), streamflow losses (both consumptive and nonconsumptive), reservoir characteristics (elevation versus discharge, or assume fixed elevation), tailwater data (tailwater curve or fixed value), installed capacity (specific capacity in all but preliminary studies), turbine characteristics (specific maximum and minimum discharges and maximum and minimum heads), efficiency (fixed efficiency or efficiency versus discharge curve), headlosses (fixed value or headloss versus discharge curve), and nonpower operating criteria (use flow data that incorporate criteria).

The following is a summary of the basic steps for computing average annual energy and dependable capacity using the flow duration method (see U.S. Army Corps of Engineers, 1985 for further details).

Step 1—Develop flow-duration curve. Use available streamflow records that have been adjusted to reflect depletion and streamflow regulation. The flow duration curve for example 13.3.1 is shown in Figure 13.3.8.

Step 2—Adjust flow-duration curve. The U.S. Army Corps of Engineers (1983) recommends that if less than 30 years of flow data is available, streamflow records from nearby stations with larger periods should be analyzed to determine if substantially wetter or drier periods than the long-term average occurred. If there were substantially wetter or drier periods, then the flow-duration should be adjusted by correlating with the flow duration coming from the larger records.

Step 3—Determine flow losses. Flow losses such as consumptive losses include reservoir surface evaporation losses and diversion such as for irrigation and water supply. Nonconsumptive losses include: irrigation lock requirements, fish-passage facility requirements, leakage through or around dams and embankment structures, leakage around spillway or regulating outlet gate, and leakage through turbine wicket gates.

Step 4—Develop head data. A head versus discharge curve can be developed that reflects the variation of tailwater elevation with discharge (and forebay elevation with discharge if this relation exists). A second approach is to include the head computation directly in the water power

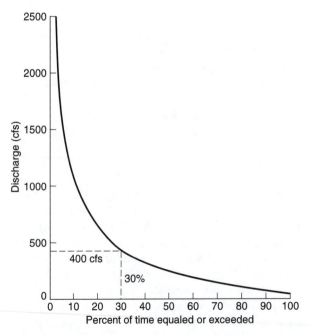

Figure 13.3.8 Flow-duration curve (from U.S. Army Corps of Engineers (1985)).

equation so that the net head is the forebay elevation minus the tailwater elevation minus the trashrack and penstock head losses.

Step 5—Select plant size. First the plant hydraulic maximum discharge that can be passed through the turbine is selected. For preliminary studies the initial plant size can be based on the average annual flow or a point between 15 and 30 percent exceedance on the flow-duration curve. Next the net head corresponding to the assumed hydraulic capacity is identified. For pure run-of-river plants the head at hydraulic capacity is the rated head. The water power equation is now used with the hydraulic capacity, rated head, and an assumed overall efficiency to compute the plant's installed capacity. Data from Table 13.3.1 can be used to establish the minimum discharge and corresponding head and the minimum head.

Step 6—Define usable flow range and derive head-duration curve. The flow-duration curve is reduced to include only the usable flow range, because the turbine characteristics limit the portion of streamflow that can be used for power generation. Using the flow-duration data and the head versus discharge data, a head-duration curve can be constructed.

Step 7—Derive the power-duration curve. Use 20 to 30 points from the flow-duration curve and determine the power for each using the water power equation. Heads are computed for each point or are taken from the head-discharge curve. Losses are subtracted from the flow in the flow-duration curve. The results are plotted as a usable generation curve with the generation values plotted at the percent exceedance points corresponding to the discharges on the flow-duration curve. The data from the usable generation-duration curve can be used to develop a true duration curve.

Step 8—Compute average annual energy. The power-duration curve, being based on all years of record, is an annual generation curve. The area under this curve represents the *average annual energy*, expressed as

$$\text{Annual energy (KWh)} = \frac{8760}{100} \int_0^{100} P \, dp \tag{13.3.5}$$

where P is the power in KW and p is the percent of time.

Step 9—Compute dependable capacity. The dependable capacity is the average power obtained from the generation-duration curve. For a run-of-river project, the generation-duration curve is based on streamflows for the peak demand months. The *dependable capacity* is

$$\text{Dependable capacity} = \text{ average generation} = \frac{1}{100} \int_0^{100} P \, dp \tag{13.3.6}$$

For pondage projects a peaking capacity-duration curve is developed to determine dependable capacity. This curve is similar to a capacity-duration curve, which shows the percent of time different levels of peaking capacity are available. The power-duration curve and the capacity-duration for run-of-river projects are identical.

EXAMPLE 13.3.3

The flow-duration curve method is used to compute the average annual energy and dependable capacity for a typical low-head run-of-river project with no pondage. The flow-duration curve is presented in Figure 13.3.1 (adapted from U.S. Army Corps of Engineers (1985)).

SOLUTION

Step 1 The area under the flow-duration curve represents an average annual flow of 390 cfs.

Step 2 No adjustments are made to the flow-duration curve.

Step 3 An average loss of 20 cfs will be used for leakage around gates, and the dam structure and evaporation losses are ignored.

Step 4 The head-discharge curve and the tailwater elevation-discharge curve are shown in Figure 13.3.9. The head curve is based on the tailwater curve with a fixed forebay elevation of 268 ft and an average headloss of 10 ft.

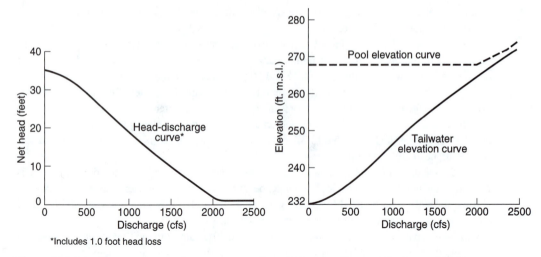

*Includes 1.0 foot head loss

Figure 13.3.9 Tailwater and head-discharge curves (from U.S. Army Corps of Engineers (1985)).

Step 5 To select a plant size, the initial size will be based on the 30-percent exceedance point (400 cfs as shown in Figure 13.3.8). The installed capacity is computed as

$$KW = \frac{QHe}{11.81} = \frac{(400 - 20)(31)(0.85)}{11.81} = 850KW$$

A single turbine with movable blades (horizontal-shaft Kaplan) is to be installed. Using Table 13.3.1, the ratio of minimum to rated discharge is 0.35 and the ratio of minimum to maximum head is 0.33. The minimum discharge is $0.35 \times (400 - 20) = 135$ cfs, which has a corresponding streamflow discharge of $135 + 20 = 155$ cfs. This corresponds to a head of 34 ft (from Figure 13.3.10). This plant is a pure run-of-river plant so that heads greater than 34 ft occur only at streamflows of less than the minimum generation streamflow of 155 cfs. Then the maximum generating head is 34 ft and the minimum head is approximately 33 percent of this, or 11 ft.

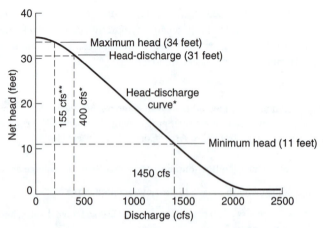

* Includes 1.0 foot head loss
** 135 cfs minimum turbine discharge plus 20 cfs flow loss

Figure 13.3.10 Net head-discharge curve showing maximum head, minimum head, and rated head (from U.S. Army Corps of Engineers (1985)).

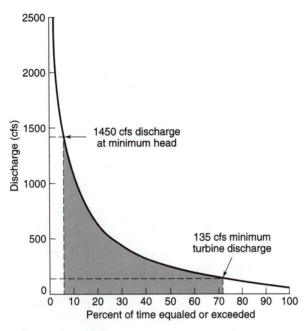

Figure 13.3.11 Total flow-duration curve showing limits imposed by minimum head and maximum discharge (from U.S. Army Corps of Engineers (1985)).

Step 6 The usable flow range is a minimum discharge of 155 cfs and a maximum head of 11 ft. (For a pure run-of-river project, the upper flow limit is defined by the minimum generating capacity.) The flow-duration data in Figure 13.3.11 and the head-discharge data in Figure 13.3.10 are used to develop the head-duration curve in Figure 13.3.12. The design head shown is the midpoint of the usable head range.

Step 7 Use points on the flow-duration curve (Figure 13.3.12) and compute the power for each discharge using the water power equation. As an example, for 270 cfs the head is 33.2 ft, which would be obtained from the head-discharge curve; then

$$\text{KW} = \frac{QHe}{11.81} = \frac{(270 - 20)(33.2)(0.85)}{11.81} = 597.38$$

After performing similar computations for all the desired points, the usable generation curve (solid line) is constructed as shown in Figure 13.3.13. This figure, however, is not a true power-duration curve because the generation values actually correspond to percent exceedances of flows. At flows greater than the rated discharge (32 percent) there is no reduction in power output due to reduced head and other factors.

Step 8 The power-duration curve (Figure 13.3.14) is integrated to determine the area. With equation (13.3.5), the average annual energy is computed as 3,390,000 KWh.

Step 9 Figure 13.3.15 is constructed for the project using the peak demand month. The dependable capacity, the average power from this curve, is computed from equation (13.3.6) as 538 KW.

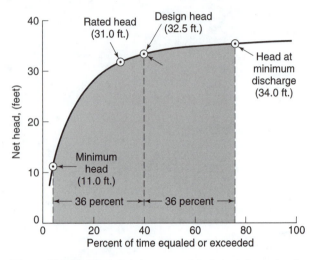

Figure 13.3.12 Head-duration curve showing minimum head, maximum head, design head, and rated head (from U.S. Army Corps of Engineers (1985)).

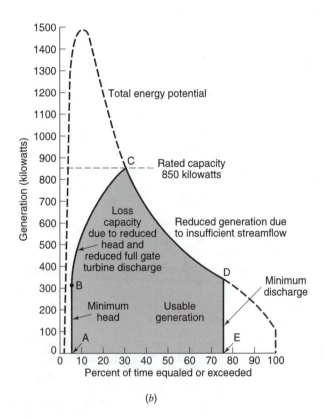

(b)

Figure 13.3.13 Usable generation-duration curve (from U.S. Army Corps of Engineers (1985)).

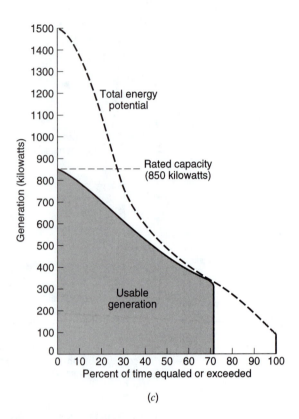

(c)

Figure 13.3.14 Usable power-duration curve (from U.S. Army Corps of Engineers (1985)).

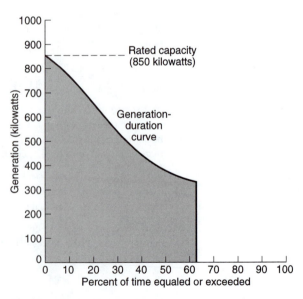

Figure 13.3.15 Generation-duration curve peak demand months (from U.S. Army Corps of Engineers (1985)).

13.3.5   Sequential Streamflow-Routing Method

The *sequential streamflow-routing method* sequentially computes the energy output for each time interval in the period of analysis. A continuity equation is used to route the streamflows through the project, taking into account the variations in reservoir elevation as a result of the reservoir regulation. Use of the sequential routing in the continuity equation allows the simulation of not only the hydropower but also flood control operation and water supply operation.

This method was developed primarily to evaluate storage projects and systems of storage projects. The method is based upon the continuity equation

$$\Delta S = I - O - L \tag{13.3.7}$$

where ΔS is the change in reservoir storage, I is the reservoir inflow, O is the reservoir outflow, and L is the sum of losses due to evaporation, diversions, and so on.

The sequential streamflow-routing method can be applied to basically any type of hydropower analysis. These include run-of-river projects; run-of-river projects with pondage; projects with flood control storage only; projects with conservation storage not regulated for power; projects with storage regulated only for power; and projects with storage regulated for multiple purposes including power, peaking hydro projects, and pumped-storage hydro projects.

The basic types of data needed are the historical streamflows and other information similar to the frequency-duration method. The basic steps for this procedure are as follows (U.S. Army Corps of Engineers, 1985):

Step 1—Select plant capacity

Step 2—Compute stream flow available for power generation

Step 3—Determine average pond elevation

Step 4—Compute net head

Step 5—Estimate efficiency

Step 6—Compute generation

Step 7—Compute average annual energy

To perform the routing, the continuity equation (13.3.7) is expanded to

$$\Delta S = I - (Q_p + Q_L + Q_S) - (E + W) \tag{13.3.8}$$

where Q_p is the power discharge, Q_L is the leakage and nonconsumptive project water requirements, Q_S is the spill, E is the net evaporation losses (evaporation minus precipitation on the reservoir), and W includes the withdrawals for water supply, irrigation, and so on. The change in storage ΔS for a given time interval can be defined as

$$\Delta S = \frac{(S_{t+\Delta t} - S_t)}{C_s} \tag{13.3.9}$$

where S_t is the beginning-of-period storage, $S_{t+\Delta t}$ is the end-of-period storage, Δt is the routing period, and C_s is a discharge to storage conversion factor. As an example, if the routing interval Δt is a 30-day month, then $C_s = 59.50$ ac-ft/ft³/s-month; if the routing interval is one week, $C_s = 13.99$ ac-ft/ft³/s-week; if the routing interval is one day, $C_s = 1.983$ ac-ft/ft³/s-day; and if the routing interval is one hour, $C_s = 0.08264$ ac-ft/ft³/s-hour.

Substituting equation (13.3.9) into equation (13.3.8) and rearranging gives

$$S_{t+\Delta t} = S_t - C_s(I - Q_P - Q_L - Q_S - E - W) \tag{13.3.10}$$

The customary U.S. unit for this equation is ac-ft. For the first iteration through the critical period, a preliminary estimate of the firm energy would be used. During the critical period the spill Q_s would normally be zero.

13.3.6 Power Rule Curve

In general *rule curves* are guidelines for reservoir operation. They are generally based on detailed sequential analysis of combinations of critical hydrologic conditions and critical demand conditions. Rule curves are developed for flood control operations and for conservation storage for irrigation, water supply, hydropower, and other purposes. Our objective here is to discuss single-purpose rule curves for power operation.

A *power rule curve* is defined as a curve, or family of curves, indicating how a reservoir is to be operated under specific conditions to obtain best or predetermined results. Figure 13.3.16 illustrates a power rule curve for the operation of a typical storage project. The general shape of a power rule curve is governed by the hydrologic and power demands. The curve defines the minimum reservoir elevation (corresponding minimum storage) that is required to generate firm power any time of the year.

Firm energy is the generation that exactly draws the reservoir level to the bottom of the power pond during the most severe drought of record. The following general steps can be taken to determine the energy output of a project in which the primary objective is to minimize firm energy (Hobbs et al., 1996):

- Identify the critical period
- Make a preliminary estimate of the firm energy potential
- Make one or more critical SSR routings to determine the actual firm energy capability and to define operating criteria that will guide year-by-year reservoir operation.
- Make an SSR routing for the total period of record to determine average annual energy
- If desired, make additional period-of-record routings using alternative operating strategies to determine which one optimizes power benefits.

With a computer model for the SSR routing such as HEC-5 (U.S. Army Corps of Engineers, 1983), these operations are done automatically.

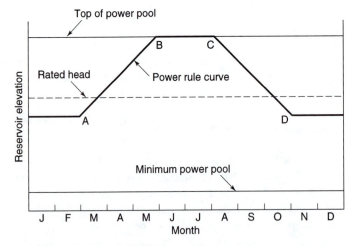

Figure 13.3.16 Rule curve for power operation of a typical storage project (from U.S. Army Corps of Engineers (1985)).

13.3.7   Multipurpose Storage Operation

Most storage projects that have power storage also have space for flood control regulation and conservation storage space for other water needs. The storage zones of a multipurpose operation are illustrated in Figure 13.3.17. A *joint-use storage zone* is one that can be used for flood regulation during part of a year and for conservation storage the remainder of the year. This is allowed in many river basins because major floods are concentrated in one season of the year. Joint-use storage allows less total reservoir storage than having separate storage zones for flood control and conservation. Figure 13.3.18 illustrates a rule curve with joint-use storage. Additional sources of information and hydroelectric and related topics can be found in Gulliver and Arndt (1991), Mays (1996, 1999), and Warnick (1989).

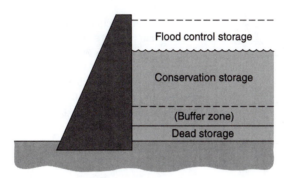

Figure 13.3.17 Storage zones (from U.S. Army Corps of Engineers (1985)).

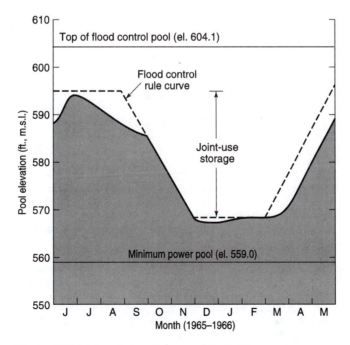

Figure 13.3.18 Regulation of a reservoir with joint-use storage through an average water year (from U.S. Army Corps of Engineers (1985)).

PROBLEMS

13.1.1 Consider the system in Figure 13.1.3*a* with the hydropower plant in the peaking load and determine the cost of weekly operation.

Plant symbol	Capacity (MW)	Plant factor (%)	Unit cost (mills/KWh)
CMBT-2	500	0	90
CMBT-1	500	0	80
HYDRO	1000	25	0
CYCL-2	500	15	30
CYCL-1	500	55	20
COAL-1	500	72	15
COAL-3	750	95	9
COAL-2	750	100	8
NUKE-1	1000	100	6

13.1.2 Consider the system in Figure 13.1.3*b* with the hydropower plant in the base load and determine the cost of weekly operation.

Plant symbol	Capacity (MW)	Plant factor (%)	Unit cost (mills/KWh)
CMBT-3	500	0	100
CMBT-2	500	0	90
CBMT-1	500	1	80
CMCY-1	500	13	60
CYCL-2	500	33	30
CYCL-1	500	48	20
COAL-1	500	62	15
COAL-3	750	85	9
HYDRO	(250)	100	0
COAL-2	750	100	8
NUKE-1	1000	100	6

13.1.3 Compare the results of problems 13.1.1 and 13.1.2. If hydropower was not used and the cost of operating an all-thermal base system for one week were $7,256,000, what would be your conclusions?

13.2.1 An impulse turbine with a single nozzle of 300 mm diameter receives water through a 1.0 m diameter riveted-steel pipe ($e = 1.0$ mm) that is 300 m long. The water level in the reservoir is 300 m above the nozzle of the turbine. If the overall turbine efficiency is 85 percent and the loss coefficient of the nozzle is 0.04, what maximum power output can be expected? How much additional power is possible if a 1200 mm riveted-steel pipe is used?

13.2.2 A reaction turbine is supplied with water through a 200 cm pipe ($e = 1.0$ mm) that is 50 m long. The water surface in the reservoir is 30 m above the draft tube inlet, which is 4.5 m above the water level in the tailrace. If the turbine efficiency is 92 percent and the discharge is 12 m³/s, what is the power output of the turbine in kilowatts? Use $f = 0.0185$.

13.3.1 Rework example 13.3.1 assuming the average head is 90 ft.

13.3.2 Rework example 13.3.2 assuming a loss of head of 5 ft.

REFERENCES

Coleman, H. W., C. Y. Wei, and J. E. Lindell, "Hydraulic Design for Energy Generation," *Hydraulic Design Handbook* (edited by L. W. Mays), McGraw-Hill, New York, 1999.

Federal Energy Regulatory Commission (FERC), *Hydroelectric Power Evaluation,* U.S. Department of Energy, draft, 1978.

Gleick, P. H., "Water and Energy," in *Water in Crisis* (edited by P. H. Gleick), Oxford University Press, New York, 1994.

Greager, W. P., and J. D. Austin, *Hydroelectric Handbook*, 2nd ed., John Wiley & Sons, New York, 1950.

Gulliver, J. S., and R. E. A. Arndt, *Hydropower Engineering Handbook*, McGraw-Hill, New York, 1991.

Hasen, H., and G. C. Antonopoulos, "Hydroelectric Plants," in *Davis' Handbook of Applied Hydraulics* (edited by V. J. Zipparo and H. Hasen), McGraw-Hill, New York, 1993.

Hobbs, B. F., R. L. Mittelstadt, and J. R. Lund, "Energy and Water," in *Water Resources Handbook* (edited by L.W. Mays), McGraw-Hill, New York, 1996.

Linsley, R. K., J. B. Franzini, D. L. Freyberg, and G. Tchobanoglous, *Water Resources Engineering*, 4th edition, McGraw-Hill, New York, 1999.

Mays, L. W. (editor-in-chief), *Water Resources Handbook*, McGraw-Hill, New York, 1996.

Mays, L. W. (editor-in-chief), *Hydraulic Design Handbook*, McGraw-Hill, New York, 1999.

Roberson, J. A., J. J. Cassidy, and M. H. Chaudhry, *Hydraulic Engineering*, Second Edition, John Wiley & Sons, New York, 1998.

U.S. Army Corps of Engineers, "Feasibility Studies for Small Scale Hydropower Additions," U.S. Army Corps of Engineers, Hydrologic Engineering Center, Davis, CA, 1979.

U.S. Army Corps of Engineers, "Application of the HEC-5 Hydropower Routines," Training Document No. 12, U.S. Army Corps of Engineers, Hydrologic Engineering Center, Davis, CA. 1983.

U.S. Army Corps of Engineers, *Hydropower*, EM 1110-2-1701, Office of the Chief of Engineers, Department of the Army, Washington, DC, 1985.

Warnick, C. C., *Hydropower Engineering*, Prentice-Hall, Englewood Cliffs, NJ, 1989.

Chapter 14

Flood Control

14.1 INTRODUCTION

Floods are natural events that have always been an integral part of the geologic history of earth. Flooding occurs (a) along rivers, streams and lakes, (b) in coastal areas, (c) on alluvial fans, (d) in ground-failure areas such as subsidence, (e) in areas influenced by structural measures, and (f) in areas that flood due to surface runoff and locally inadequate drainage. Human settlements and activities have always tended to use floodplains. Their use has frequently interfered with the natural floodplain processes, causing inconvenience and catastrophe to humans. This chapter focuses on the management of water excess (floods).

A recent example of flooding on a very large scale was the Mississippi River Basin flood of 1993 (Interagency Floodplain Management Review Committee, 1994 [commonly called the Galloway Report; Brigadier General Gerald E. Galloway was the executive director of the review committee]). Figures 7.29a and b illustrate, respectively, the general area of flooding streams and the areal distribution of total precipitation. Figures 14.1.1 a–h from the Illinois State Water Survey show various flooding as a result of the Mississippi River Basin flood of 1993.

Figure 14.1.1 Pictures of the Mississippi River Basin Flood of 1993. (*a*) Confluence of Mississippi and Illinois Rivers. (Courtesy of the Illinois State Water Survey.)

(b)

(c)

Figure 14.1.1 (b) Route 100 under water, Grafton, Illinois; (c) Confluence of Mississippi and Missouri Rivers.

(d)

(e)

Figure 14.1.1 (d) Flooding due to seepage south of St. Louis, Missouri; (e) Levee break at Kaskaskia Island, Mississippi River.

(f)

(g)

Figure 14.1.1 (f) Levee break at Miller City, Mississippi River; (g) Sand bags over a levee.

(h)

Figure 14.1.1 (h) Flooding along the Mississippi River.

14.2 FLOODPLAIN MANAGEMENT

14.2.1 Floodplain Definition

A *floodplain* is the normally dry land area adjoining rivers, streams, lakes, bays, or oceans that is inundated during flood events. The most common causes of flooding are the overflow of streams and rivers and abnormally high tides resulting from severe storms. The floodplain can include the full width of narrow stream valleys, or broad areas along streams in wide, flat valleys. As shown in Figure 14.2.1, the channel and floodplain are both integral parts of the natural conveyance of a stream. The floodplain carries flow in excess of the channel capacity and the greater the discharge, the greater the extent of flow over the flood plain. Floodplains may be defined either as natural geologic features or from a regulatory perspective. The 100-year floodplain (see Chapter 10) is the standard most commonly used in the United States for management and regulatory purposes. Flooding concerns are not limited to riverine and coastal flooding. Also of concern are floods associated with alluvial fans, unstable channels, ice jams, mudflows, and subsidence.

Alluvial fans are characterized by a cone or fan-shaped deposit of boulders, gravel, and fine sediments that have been eroded from mountain slopes and transported by flood flows, debris flows, erosion, sediment movement and deposition, and channel migration, as illustrated in Figure 14.2.2. Alluvial fans are common throughout many parts of the world. In the United States they are common in Arizona, California, Idaho, Montana, Nevada, New Mexico, Utah, Washington and Wyoming. Illustrated in Figure 14.2.3 is the flood insurance rate zone defined for alluvial fan systems.

14.2.2 Hydrologic and Hydraulic Analysis of Floods

Hydrologic and hydraulic analyses of floods are required for the planning, design, and management of many types of facilities, including hydrosystems within a floodplain or watershed. These analyses are needed for determining potential flood elevations and depths, areas of inundation, sizing of channels, levee heights, right of way limits, design of highway crossings and culverts, and many others. The typical requirements include (Hoggan, 1997):

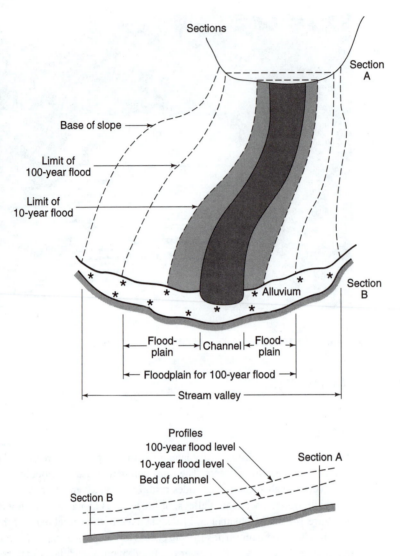

Figure 14.2.1 Typical sections and profiles in an unobstructed reach of stream valley (from Waananen et al. (1977)).

1. *Floodplain information studies.* Development of information on specific flood events such as the 10-, 100-, and 500-year frequency events.

2. *Evaluation of future land-use alternatives.* Analysis of a range of flood events (different frequencies) for existing and future land uses to determine flood-hazard potential, flood damage, and environmental impact.

3. *Evaluation of flood-loss reduction measures.* Analysis of a range of flood events (different frequencies) to determine flood damage reduction associated with specific design flows.

4. *Design studies.* Analysis of specific flood events for sizing facilities to assure their safety against failure.

5. *Operation studies.* Evaluation of a system to determine if the demands placed upon it by specific flood events can be met.

The methods used in hydrologic and hydraulic analysis are determined by the purpose and scope of the project and the data availability. Figure 14.2.4 is a schematic of the components of a

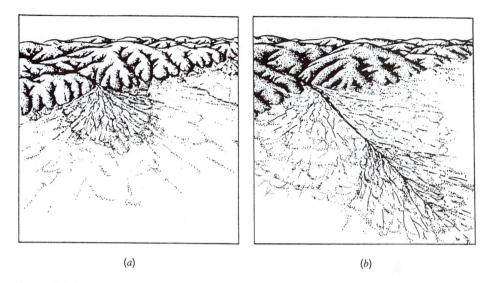

Figure 14.2.2 Two types of alluvial fans. (*a*) Unincised fan with the area of present deposition next to the mountain; (*b*) Alluvial fan with the area of present deposition downslope from the mountains due to steam-channel entrenchment at the apex of the fan (note gully extension at the toe of the fan) (from W. B. Bull (1984)).

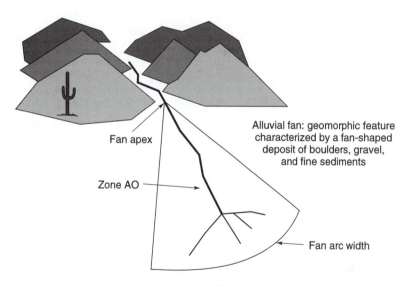

Figure 14.2.3 Alluvial fan system.

hydrologic and hydraulic analysis for floodplain studies. Hydrologic analysis for floodplains entails either a rainfall-runoff analysis or a flood-flow frequency analysis. If information from an adequate number of historical annual instantaneous peak discharges (*annual maximum series*) is available, the flood-flow frequency analysis can be performed to determine peak discharges for various return periods (as described in Chapter 10). Otherwise, a rainfall-runoff analysis must be performed using a historical storm or design storm for a particular return period to develop a storm-runoff hydrograph (as described in Chapter 8).

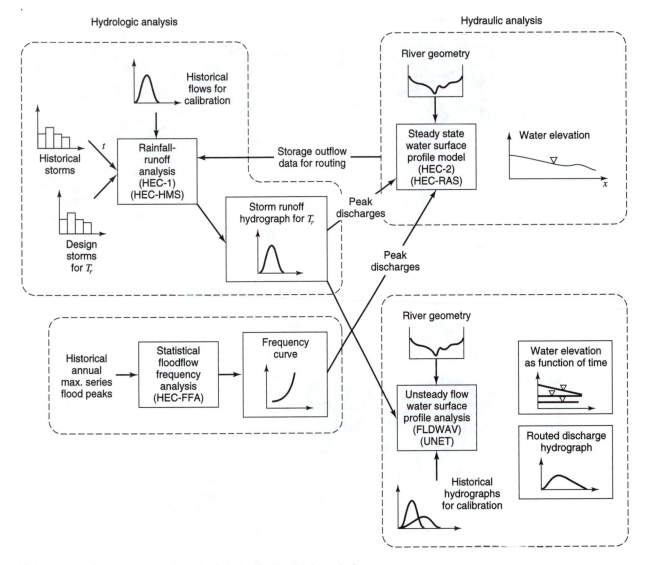

Figure 14.2.4 Components of a hydrologic-hydraulic floodplain analysis.

Determination of water-surface elevations can be performed using a steady-state water-surface profile analysis (see Chapter 5) if only peak discharges are known, or one can select the peak discharges from generated storm-runoff hydrographs. Refer to Dodson (1999) for further information on floodplain analysis. For a more detailed and comprehensive analysis, an unsteady-flow analysis based upon a hydraulic-routing model (see Chapter 9) and requiring the storm-runoff hydrograph can be used to define more accurately maximum water-surface elevations. The unsteady-flow analysis also provides more detailed information such as the routed-discharge hydrographs at various locations throughout a river reach.

14.2.3   Floodways and Floodway Fringes

Encroachment on floodplains, such as by artificial fill material, reduces flood-carrying capacity, increases the flood heights of streams, and increases flood hazards in areas beyond the encroachment. One aspect of floodplain management involves balancing the economic gain from flood-plain development against the resulting increase in flood hazard. For purposes of the Federal

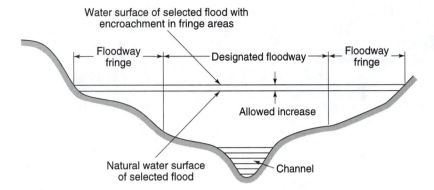

Figure 14.2.5 Definition of floodway and floodway fringe. The floodway fringe is the area between the designated floodway limit and the limit of the selected flood. The floodway limit is defined so that encroachment limited to the floodway fringe will not significantly increase flood elevation. The 100-year flood is commonly used and a 1-ft allowable increase is standard in the United States.

Emergency Management Agency (FEMA) studies, the 100-year flood area is divided into a floodway and a floodway fringe, as shown in Figure 14.2.5. The *floodway* is the channel of a stream plus any adjacent floodplain areas that must be kept free of encroachment in order for the 100-year flood to be carried without substantial increases in flood heights. FEMA's minimum standards allow an increase in flood height of 1.0 foot, provided that hazardous velocities are not produced. The *floodway fringe* is the portion of the floodplain that could be completely obstructed without increasing the water surface elevation of the 100-year flood by more than 1.0 foot at any point.

Two types of floodplain inundation maps, flood-prone area and flood-hazard maps, have been used. *Flood-prone area maps* show areas likely to be flooded by virtue of their proximity to a river, stream, bay, ocean, or other watercourse, as determined from readily available information. *Flood-hazards maps,* such as Figure 14.2.6 for Clear Creek near Houston, Texas, show the extent of inundation as determined from a study of flooding at the given location. Flood-hazard maps are commonly used in floodplain information reports and require updating when changes have occurred in the channels, on the floodplains, and in upstream areas.

14.2.4   Floodplain Management and Floodplain Regulations

According to the National Flood Insurance Program (NFIP) regulations administered by FEMA, *floodplain management* is "the operation of an overall program of corrective and preventive measures for reducing flood damage, including but not limited to emergency preparedness plans, flood control works, and floodplain management regulations." Floodplain management regulations are the most effective method for preventing future flood damage in developing communities with known flood hazards.

Floodplain management investigates problems that have arisen in developed areas and potential problems that can be forecast due to future developments. The basic approaches to floodplain management are: (1) actions to reduce susceptibility to floods, (2) actions that modify the flood, and (3) actions that assist individuals and communities in responding to floods. *Floodplain regulation* is the centerpiece of any floodplain management program, and is particularly effective in underdeveloped areas, where the ability exists to control future development.

A key component in floodplain regulation is the definition of the *flood-hazard area* (usually defined as the 100-year floodplain) and the floodway. The floodway includes the channel of the stream and the adjacent land areas that must be reserved in order to discharge the design flood without cumulatively increasing the water surface by more than a given amount. For example, the

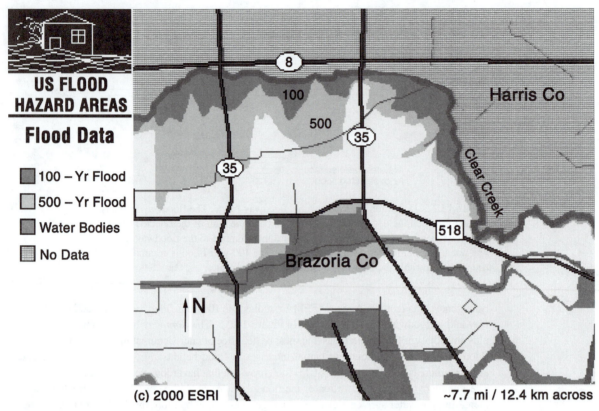

Figure 14.2.6 Floodplain map for a portion of Clear Creek in Brazoria County near Houston, Texas. Harris County is to the north and east of Clear Creek and the flood plain of Clear Creek in Harris County is not shown.

maximum rise allowed by the National Flood Insurance Program (NFIP) is one foot, but in many situations a lesser amount may be more appropriate. An adequate floodplain management plan that eliminates development from the flood-hazard areas may be a major step in the right direction. Not allowing other obstructions such as fill and detention basins to be placed in the flood-hazard area is another step forward. A floodplain provides both a conveyance mechanism and a temporary storage area for excess water. Allowing obstructions be placed in the floodplain eliminates the temporary storage areas and in turn increases the hydraulic heads to increase flood levels both downstream and upstream of the floodplain developments.

14.2.5 National Flood Insurance Program

In 1968 the U.S. Congress created the National Flood Insurance Program (NFIP) through the passage of the National Flood Insurance Act. The Flood Disaster Protection Act of 1973 and the National Flood Insurance Reform Act of 1994 further defined the NFIP. The purpose of the NFIP is to minimize future flood loss and to allow the floodplain occupants to be responsible for flood-damage costs instead of the taxpayer. The NFIP, administered by FEMA, provides federally backed flood insurance that encourages communities to enact and enforce floodplain regulations. If a state or community does not participate in the NFIP, the following consequences occur:

1. The community will not be eligible for flood disaster relief in the event of a federally declared flood disaster

2. Federal or federally related financial assistance for acquisition or construction purposes for structures in flood-prone areas will not be available

3. Flood insurance will not be available

For a state or community to be eligible for participation in the NFIP, it must agree to adopt floodplain management regulations that meet minimum standards as defined by FEMA. These minimum standards include but are not limited to:

1. Require permits for all proposed development within a flood-hazard area

2. Assure that all necessary governmental permits have been obtained

3. Ensure that proper materials and methods are used in new construction to protect new buildings from future floods (elevate the lowest finished floor of residential structures of flood-proof nonresidential structures above the base flood elevation)

4. Assure that all proposed development within a flood-hazard area is consistent with the need to minimize flood damage within the flood-prone area

5. Notify adjacent communities and the state prior to the alterations or relocation of a watercourse

6. Assure that the flood-carrying capacity within the altered or relocated portion of any watercourse is maintained

7. Prohibit encroachments, including fill, new construction, substantial improvements, and other development within the adopted regulatory floodway unless it has been demonstrated through hydrologic and hydraulic analysis performed in accordance with standard engineering practice that the proposed encroachment would not result in any increase in flood levels within the community during the occurrence of the base flood flow

A state is considered a "community" and state agencies are required to comply with minimum standards just as local communities do. A state may comply with the floodplain regulations of the local community in which state land is located, or the state may establish and enforce its own floodplain regulations for state agencies.

14.2.6   Stormwater Management and Floodplain Management

Stormwater management plans are most successful when they are implemented at the start of development in an area and should be administered as part of a land-use planning process. The implementation of a stormwater management plan, in a remedial mode, to correct stream deterioration resulting from previous uncontrolled development is a much more difficult task. Stormwater detention programs can be very effective; however in some cases may have little effect because the flood peak caused by detention diminishes as the flood passes downsteam, while the increase in total runoff caused by the development swells the total mass of the flood wave. The cumulative effect downstream of any number of detention basins would mainly be to delay the arrival of the flood crest by a few hours, and many have little or no effect on reducing the peak discharge. A partial solution to this is to provide retention over a long time period. The increases in peak flood discharges can be controlled but only through coordinated, extensive planning prior to development. Zoning to preserve undeveloped areas, particularly those in the floodplain, can be a very effective measure.

Stormwater management and floodplain management are generally separate and different programs; however, their interfaces, such as detention basins built in floodplains, are unavoidable issues. Detention basins have been placed in the floodplains of many areas of the United States. Detention basins generally should be placed out of the floodplain, particularly for small streams of relatively small drainage areas. In such cases the same storms affect the development site and the floodplain simultaneously. In other words, the time during which the detention basin is needed to store stormwater from the development is basically the same time during which the floodplain is flooded and the detention basin location would already be filled by floodwater. To the extent that the flood at the development coincides with the flooding in the floodplain, detention storage in the floodplain is ineffective. An additional factor that has decreased the effectiveness of detention basins in many areas is the large amount of fill incidental with development. The placement of effective regional detention and retention, along with improved drainage and conveyance structures and other hydraulic structures, may be required to alleviate drainage problems.

14.3   FLOOD-CONTROL ALTERNATIVES

Flooding results from conditions of hydrology and topography in floodplains such that the flows are large enough that the channel banks overflow, resulting in overbank flow that can extend over the floodplain. In large floods, the floodplain acts both as a conveyance and as a temporary storage for flood flows. The main channel is usually a defined channel that can meander through the floodplain carrying low flows. The overbank flow is usually shallow compared to the channel flow and also flows at a much lower velocity than the channel flow.

The *objective of flood control* is to reduce or to alleviate the negative consequences of flooding. Alternative measures that modify the flood runoff are usually referred to as flood-control facilities and consist of engineering structures or modifications. Construction of flood-control facilities, referred to as *structural measures*, are usually designed to consider the flood characteristics including reservoirs, diversions, levees or dikes, and channel modifications. Flood-control measures that modify the damage susceptibility of floodplains are usually referred to as *nonstructural measures* and may require minor engineering works. Nonstructural measures are designed to modify the damage potential for permanent facilities and provide for reducing potential damage during a flood event. Nonstructural measures include flood proofing, flood warning, and land-use controls. Structural measures generally require large sums of capital investment. *Floodplain management* takes an integrated view of all engineering, nonstructural, and administrative measures for managing (minimizing) losses due to flooding on a comprehensive scale.

Table 14.3.1 presents a checklist summarizing the critical requirements for *without-project condition analysis*. These represent the base condition for determining the economic value, performance, and environmental/social impacts of flood-damage-reduction measures and plans (U.S. Army Corps of Engineers, 1996a).

Table 14.3.1   Checklist for Without-Project Conditions

Study Components	√	Issues
Layout		Review/assemble available information
		Conduct field reconnaissance for historic flood data and survey specification
		Establish local contacts
		Assist in establishing study limits, damage reaches
Economic studies		Determine existing and future without-project conditions discharge-frequency and associated uncertainty
		Determine existing and future with-project conditions stage-discharge and associated uncertainty
Performance		Determine expected capacity-exceedance probability
		Determine expected life-exceedance probability
		Evaluate existing project operations/stability for range of events and key assumptions
		Describe consequences of capacity exceedances
		Perform reliability analyses
Environmental and social		Evaluate without-project riparian impacts
		Evaluate without-project social impacts

Source: U.S. Army Corps of Engineers (1996a).

14.3.1   Structural Alternatives

Table 14.3.2 summarizes several flood-damage reduction measures and the parametric relationships that are modified. The basic functional relationships required to assess the value of flood-damage reduction alternatives are shown in Figure 14.3.1. *Stage-damage relationships* define the flood severity in terms of damage cost for various stages. *Stage-discharge relationships*, also referred to as *rating curves*, are modified by various flood-control alternatives. *Flood-flow frequency relationships* (described in Chapter 10) define the recurrence of nature in terms of the flood magnitudes. *Flood-control alternatives* are designed to modify the flood characteristics by altering one or more of the above relationships. The major types of flood-control structures are reservoirs, diversions, levees or dikes, and channel modifications. Each of these are discussed below, defining the resulting changes in the basic relationships.

Table 14.3.2   Impacts of Flood-Damage-Reduction Measures

Measures	Impact of measure		
	Modifies Discharge-Frequency Function	Modifies Stage-Discharge Function	Modifies Stage-Damage Function
Reservoir	Yes	Maybe, if stream and downstream channel erosion and deposition due to change in discharge	Maybe, if increased development in floodplain
Diversion	Yes	Maybe, if channel erosion and deposition due to change in discharge	Maybe, if increased development in floodplain
Channel improvement	Maybe, if channel affects timing and storage altered significantly	Yes	Not likely
Levee or floodwall	Maybe, if floodplain storage no longer available for flood flow	Yes	Yes
Floodproofing	Not likely	Not likely	Yes
Relocation	Not likely	Maybe, if flow obstructions removed	Yes
FWP plan	Not likely	Not likely	Yes
Land-use and construction regulation	Not likely	Maybe, if flow obstructions removed	Yes
Acquisition	Not likely	Maybe, if flow obstructions removed	Yes

Source: U.S. Army Corps of Engineers (1996a).

14.3.1.1   *Flood-Control Reservoir*

Flood-control reservoirs are used to store flood waters for release after the flood event, reducing the magnitude of the peak discharge. Reservoirs modify the flood-flow frequency curve, which is lowered because of the decrease of the peak discharge of a specific event. Figure 14.3.2 illustrates the effect of flood-control reservoirs. Long-term effects of reservoir storage modify the streamflow regime and can result in channel aggradation or degradation at downstream locations, altering the rating curve.

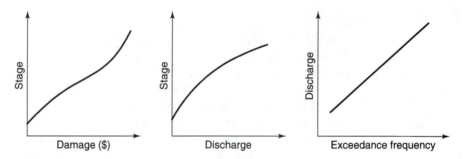

Figure 14.3.1 Flood assessment functional relationships.

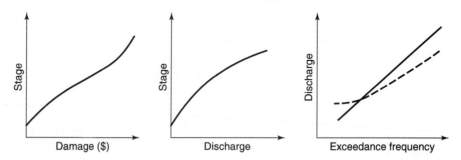

Figure 14.3.2 Effect of reservoir.

Reservoirs reduce damage by reducing discharge directly. Table 14.3.3 is a checklist that summarizes critical requirements for reservoirs. A reservoir is well suited for damage reduction in the following cases (U.S. Army Corps of Engineers, 1996a):

1. Damageable property is spread over a large geographical area downstream from the reservoir site, with several remote damage centers and relatively small local-inflow areas between them.
2. A high degree of protection, with little residual damage, is desired.
3. A variety of properties, including infrastructure, structures, contents, and agricultural property, is to be protected.
4. Water impounded may be used for other purposes, including water supply, hydropower, and recreation.
5. Sufficient real estate is available for location of the reservoir at reasonable economic, environmental, and social cost.
6. The economic value of damageable property protected will justify the cost of constructing the reservoir.

Table 14.3.3   Checklist for Reservoir

Study Components	√	Issues
Layout		Consider alternative sites based on drainage area versus capacity considerations
		Delineate environmentally sensitive aquatic and riparian habitat
		Identify damage centers, delineate developed areas, define land uses for site selection
		Determine opportunities for system synergism due to location

Table 14.3.3 Checklist for Reservoir (*continued*)

Study Components	√	Issues
Economics		Determine with-project modifications to downstream frequency function for existing and future conditions
		Quantify uncertainty in frequency function
		Formulate and evaluate range of outlet configurations for various capacities using risk-based analysis procedures
Performance		Determine expected annual-exceedance probability
		Determine expected life-exceedance probability
		Describe operation for range of events and analyze sensitivity of critical assumptions
		Describe consequences of capacity exceedances
		Determine reliability for range of events
		Conduct dam-safety evaluation
Design		Formulate and evaluate preliminary spillway and outlet configurations
		Conduct pool sedimentation analysis
		Evaluate all downstream hydrologic and hydraulic impacts
		Formulate preliminary operation plans
Environmental and social		Evaluate with-project riparian habitat
		Evaluate aquatic and riparian habitat impact and identify enhancement opportunities
		Anticipate and identify incidental recreation opportunities

Source: U.S. Army Corps of Engineers (1996a).

14.3.1.2 *Diversion*

Diversion structures are used to reroute or bypass flood flows from damage centers in order to reduce the peak flows at the damage centers. Diversion structures are designed to modify (lower) the frequency curve so that the flow magnitude for a specific event is lowered at the damage center. Figure 14.3.3 illustrates the effect of diversions on the functional relationships. The stage-damage and stage-discharge relationships remain the same if there are no other induced effects. Long-term effects of diversions can cause aggradation or degradation at downstream locations and result in sediment depositions in bypass channels.

A diversion is well suited for damage reduction in the following cases (U.S. Army Corps of Engineers, 1996a):

1. Damageable property is spread over a large geographical area with relatively minor local inflows for diversions removing water from the system.

2. A high degree of protection, with little residual damage, is desired.

3. A variety of property, including infrastructure, structures, contents, and agricultural property, is to be protected.

4. Sufficient real estate is available for location of the diversion channel or tunnel at reasonable cost.

5. The value of damageable property protected will economically justify the cost of the diversion.

Table 14.3.4 is a checklist that summarizes critical requirements for diversions.

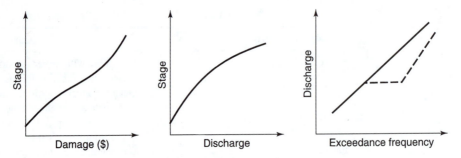

Figure 14.3.3 Effect of diversion.

Table 14.3.4 Checklist for Diversion

Study Components	√	Issues
Layout		Delineate environmentally sensitive aquatic and riparian habitat
		Identify damage centers, delineate developed areas, define land uses for site selection
		Determine right-of-way
		Identify infrastructure/utility-crossing conflicts
Economics		Determine with-project modifications to downstream frequency function for existing and future conditions
		Quantify uncertainty in frequency function
		Formulate and evaluate range of outlet configurations for various capacities using risk-based analysis procedures
Performance		Determine expected annual-exceedance probability
		Determine expected life-exceedance probability
		Describe operation for range of events and analyze sensitivity of critical assumptions
		Describe consequences of capacity exceedances
		Determine reliability for range of events
Design		Formulate and evaluate preliminary spillway and outlet configurations
		Conduct diversion channel sedimentation analysis
		Evaluate all downstream hydrologic and hydraulic impacts
		Formulate preliminary operation plans
Environmental and social		Evaluate with-project riparian habitat
		Evaluate aquatic and riparian habitat impact and identify enhancement opportunities

Source: U.S. Army Corps of Engineers (1996a).

14.3.1.3 *Levees and Floodwalls*

Levees or dikes are used to keep flood flows from floodplain areas where damage can occur. Levees essentially modify all three of the functional relationships. The effect of levees is to reduce the damage in protected areas from water surface stages within the stream or main channel. This effect essentially truncates the stage-damage relationship for all stages below the design elevation of the levee, as illustrated in Figure 14.3.4. Excluding flood flows from portions of the floodplain outside the levees constricts the flow to a smaller conveyance area, resulting in an increased stage for the various discharges. This upward shift of the stage-discharge relationship is shown in Figure

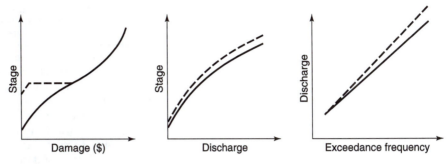

Figure 14.3.4 Effect of levee.

14.3.4. Constricting the flow to within levees reduces the amount of natural storage of a flood-wave, causing an increase in peak discharges downstream. This effect increases the discharge for various exceedance frequencies, shifting the frequency upwards, as also shown in Figure 14.3.4. Long-term effects of levees can cause aggradation or degradation of channels in downstream reaches. Even though levees have the purpose of protecting property and lives, they also bring the potential for major disasters when design discharges are exceeded and areas are inundated that had been considered safe.

Levees and floodwalls are effective damage-reduction measures in the following circumstances (U.S. Army Corps of Engineers, 1996a):

1. Damageable property is clustered geographically.
2. A high degree of protection, with little residual damage, is desired.
3. A variety of properties, including infrastructure, structures, contents, and agricultural property, is to be protected.
4. Sufficient real estate is available for location of the channel modification at reasonable economic, environmental, and social cost.
5. The economic value of damageable property protected will justify the cost of constructing the reservoir.

Table 14.3.5 is a checklist of the requirements for levees and floodwalls, and Table 14.3.6 is a checklist for interior areas.

Table 14.3.5 Checklist for Levees and Floodwalls

Study Components	√	Issues
Layout		Minimize contributing interior runoff areas (flank levees, diversion, collector system)
		Minimize area protected to reduce potential future development
		Investigate levee setback versus height tradeoffs
		Determine right-of-way available for levee or wall alignment
		Minimize openings requiring closure during flood events
Economics		Determine with-project modifications to stage-discharge function for all existing and future events
		Quantify uncertainty in stage-damage function
		Formulate and evaluate range of levee and interior area configurations for various capacities using risk-based analysis procedures
		Determine expected capacity- and stage-exceedance probability

Table 14.3.5 Checklist for Levees and Floodwalls (*continued*)

Study Components	√	Issues
Performance		Determine expected annual-exceedance probability
		Determine expected life-exceedance probability
		Describe operation for range of events and sensitivity analysis of critical assumptions
		Describe consequences of capacity exceedances
		Determine reliability (expected probability of exceeding target) for range of events
Design		Design for levee or floodwall superiority at critical features (such as pump stations, high-risk damage centers)
		Design overtopping locations and downstream end, remote from major damage centers
		Provide levee height increments to accommodate settlement, wave run-up
		Design levee exterior erosion protection
		Develop flood-warning-preparedness plan for events that exceed capacity
Environmental and social		Evaluate aquatic and riparian habitat impact and identify enhancement opportunities
		Anticipate and identify incidental recreation opportunities

Source: U.S. Army Corps of Engineers (1996a).

Table 14.3.6 Checklist for Interior Areas

Study Components	√	Issues
Layout		Define hydraulic characteristics of interior system (storm-drainage system, outlets, ponding areas, and so forth)
		Delineate environmentally sensitive aquatic and riparian habitat
		Identify damage centers, delineate developed areas, define land uses for site selection
Economics		Determine with-project modifications to interior stage-frequency function for all conditions
		Quantify uncertainty in frequency function
		Formulate and evaluate range of pond, pump, outlet configurations for various capacities using risk-based analysis procedures
Performance		Determine expected annual-exceedance probability
		Determine expected life-exceedance probability
		Describe operation for range of events and sensitivity analysis of critical assumptions
		Describe consequences of capacity exceedances
		Determine reliability (expected probability of exceeding target) for range of events
Design		Formulate and evaluate preliminary inlet and outlet configurations for facilities
		Formulate preliminary operation plans

Source: U.S. Army Corps of Engineers (1996a).

14.3.1.4   *Channel Modifications*

Channel modifications (*channel improvements*) are performed to improve the conveyance characteristics of a stream channel. The improved conveyance lowers the stages for various discharges, having the effect of lowering the stage-discharge relationship, as illustrated in Figure 14.3.5. The peak discharges for flood events are passed at lower stages, decreasing the effect of natural valley storage during passage of a flood wave. This effect results in higher peak discharges downstream than would occur without the channel modifications, causing an upward shift of the frequency curve, as illustrated in Figure 14.3.5. Long-term effects of channel modification can cause aggradation and degradation of downstream channel reaches. Channel modifications are usually for local protection but can be integrated with other flood-control alternatives to provide a more efficient flood-control system.

Channel modifications are effective flood-damage-reduction measures in the following cases (U.S. Army Corps of Engineers, 1996a):

1. Damageable property is locally concentrated.
2. A high degree of protection, with little residual damage, is desired.
3. A variety of properties, including infrastructure, structures, contents, and agricultural property, is to be protected.
4. Sufficient real estate is available for location of the channel modification at reasonable economic, environmental, and social cost.
5. The economic value of damageable property protected will justify the cost of constructing the reservoir.

Table 14.3.7 is a checklist of critical requirements for channel modifications.

Table 14.3.7   Checklist for Channel Modification

Study Components	√	Issues
Layout		Determine right-of-way restriction
		Delineate environmentally sensitive aquatic and riparian habitat
		Identify damage centers, delineate developed areas, define land uses for site selection
		Identify infrastructure/utility-crossing conflicts
Economics		Determine with-project modifications to stage-discharge function for all conditions
		Determine any downstream effects due to frequency discharge changes due to loss of channel storage
		Quantify uncertainty in stage-discharge function
		Formulate and evaluate range of channel configurations using risk-based analysis procedures
Performance		Determine expected annual-exceedance probability
		Determine expected life-exceedance probability
		Describe operation for range of events and analyze sensitivity of critical assumptions
		Describe consequences of capacity exceedances
		Determine reliability for range of events
Design		Account for ice and debris, erosion, deposition, sediment transport, and high velocities
		Evaluate straightening effects on stability
		Evaluate all impact of restrictions or obstructions
Environmental and social		Evaluate aquatic and riparian habitat impact and identify enhancement opportunities
		Anticipate and identify incidental recreation opportunities

Source: U.S. Army Corps of Engineers (1996a).

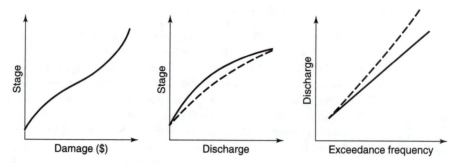

Figure 14.3.5 Effect of channel modification.

14.3.2 Nonstructural Measures

Nonstructural measures are used to modify the damage potential of permanent structures and facilities in order to decrease the susceptibility of flooding to reduce potential damages. Nonstructural measures include flood proofing, flood warning, and various types of land-use control alternatives. These measures are characterized by their value in reducing future or potential unwise floodplain use. Of the nonstructural measures mentioned above, only flood proofing has the potential to modify present damage potential.

Flood proofing consists of a range of nonstructural measures designed to modify the damage potential of individual structures susceptible to flood damage. These measures include elevating structures, waterproofing exterior walls, and rearrangement of structural working space. Flood proofing is most desirable on new facilities, and changes only the stage-damage relation, as illustrated in Figure 14.3.6, shifting the relationship upwards.

Table 14.3.8 is a checklist for nonstructural measures that include flood proofing, relocation, and flood-warning preparedness (FWP) plans. Table 14.3.9 lists the performance requirements for flood proofing.

Table 14.3.8 Checklist for Measures that Reduce Existing-Condition Damage Susceptibility

Study Components	√	Issues
Layout		Based on qualification of flood hazard, identify structures for which measures are appropriate
Economics		Determine with-project modifications to stage-damage function for all existing and future conditions
		Quantify uncertainty in stage-damage function
		Formulate and evaluate range of flood-proofing, relocation, and/or FWP plans, using risk-based analysis procedures
Performance		Determine expected annual-exceedance probability
		Determine expected life-exceedance probability
		Determine operation for range of events and sensitivity analysis of critical assumptions
		Describe consequences of capacity exceedances
		Determine reliability for range of events
Design		Develop, for all these measures, FWP plans
Environmental and social		Evaluate aquatic and riparian habitat impact and identify enhancement opportunities
		Anticipate and identify incidental recreation opportunities

Source: U.S. Army Corps of Engineers (1996a).

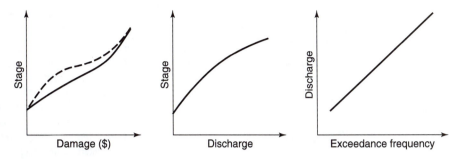

Figure 14.3.6 Effect of flood proofing.

Table 14.3.9 Performance Requirements for Flood Proofing

Flood Proofing Method	Performance Requirement
Window or door closure	Provide adequate forecasting and warning to permit installation of closures
	Identify *all* openings for closure, including fireplace cleanouts, weep holes, and so forth
	Ensure structural adequacy to prevent failure due to hydrostatic pressure or floating of structure
	Ensure watertightness to minimize and drainage to accommodate leakage
	Arrange adequate, ongoing public training to ensure proper operation
Small wall or levee	Requirements similar to major levee, but on a smaller scale, including: (1) providing for closure of openings in wall or levee, (2) ensuring structural stability of levee or wall, and (3) providing for proper interior damage
	Arrange adequate, ongoing public training to ensure proper operation
	Plan for emergency access to permit evacuation if protected area is isolated by rising floodwaters
Raising in-place	Protected beneath raised structure, as hazard is not eliminated
	Ensure structural stability of raised structure
	Plan for emergency access to permit evaluation if protected area is isolated by rising floodwaters

Source: U.S. Army Corps of Engineers (1996a).

A *flood-warning-preparedness plan (FWP plan)* reduces flood damage by giving the public an opportunity to act before flood stages increase to damaging levels. Table 14.3.10 lists the components of an FWP system. The savings due to a FWP plan may arise from reduced inundation damage, reduced cleanup costs, reduced cost of disruption of services due to opportunities to shut off utilities and make preparations, and reduced costs due to reduction of health hazards. Furthermore, FWP plans may reduce social disruption and risk to life of floodplain occupants.

Table 14.3.10   Components of a FWP System

Component	Purposes
Flood-threat-recognition subsystem	Collection of data and information; transmission of data and information; receipt of data and information; organization and display of data and information; prediction of timing and magnitude of flood events
Warning-dissemination subsystem	Determination of affected areas; identification of affected parties; preparation of warning messages; distribution of warning messages
Emergency-response subsystem	Temporary evacuation; search and rescue; mass care center operations; public-property protection; flood fight; maintenance of vital services
Post-flood-recovery subsystem	Evacuee return; debris clearance; return of services; damage assessment; provision for assistance
Continued system management	Public-awareness programs; operation, maintenance, and replacement of equipment; periodic drills; update and arrangements

Source: U.S. Army Corps of Engineers (1988).

An FWP plan is a critical component of other flood-damage-reduction measures. In addition, federal Flood Plain Management Services (FPMS) staff may provide planning services in support of local agency requests for assistance in implementation of a FWP plan; this is authorized by Section 206 of the Flood Control Act of 1960.

An FWP provides lead-time notice to floodplain occupants in order to reduce potential damage. The lead time provides the opportunity to elevate contents of structures, to perform minor proofing, and to remove property susceptible to flooding. The greatest value of flood warning is to reduce or eliminate the loss of life. Flood warning requires real-time flood forecasting and communication facilities to warn inhabitants of floodplains.

Land-use controls refer to the many administrative and other actions in order to modify floodplain land use so that the uses are compatible with the potential flood hazard. These controls consist of zoning and other building ordinances, direct acquisition of land and property, building codes, flood insurance, and information programs by local, state, and federal agencies.

14.4   FLOOD DAMAGE AND NET BENEFIT ESTIMATION

14.4.1   Damage Relationships

Flood damages are usually reported as *direct damage* to property, but this is only one of five empirical categories of damages: direct damages, indirect damages, secondary damages, intangible damages, and uncertainty damages. *Indirect damages* result from lost business and services, cost of alleviating hardship, rerouting traffic, and other related damages. *Secondary damages* result from adverse effects by those who depend on output from the damaged property or hindered services. *Intangible damages* include environmental quality, social well being, and aesthetic values. *Uncertainty damages* result from the ever-present uncertainty of flooding.

Various techniques have been used to calculate direct damages. Grigg and Helweg (1975) used three categories of techniques: aggregate formulas, historical damage curves, and empirical depth-damage curves. One of the more familiar aggregate formulas is that suggested by James (1972):

$$C_D = K_D U M_S h A \qquad (14.4.1)$$

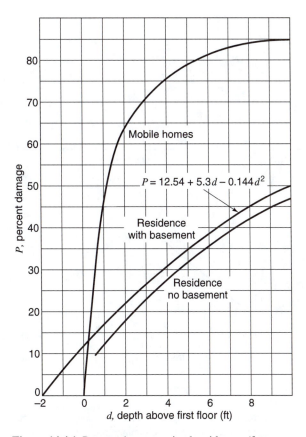

Figure 14.4.1 Percent damage, mixed residences (from Corry et al. (1980)).

where C_D is the flood damage cost for a particular flood event, K_D is the flood damage per foot of flood depth per dollar of market value of the structure, U is the fraction of floodplain in urban development, M_S is the market value of the structure inundated in dollars per developed acre, h is the average flood depth over the inundated area in feet, and A is the area flooded in acres. Eckstein (1958) presents the historical damage curve method in which the historical damages of floods are plotted against the flood stage.

The use of empirical depth-damage curves (such as shown in Figure 14.4.1) requires a property survey of the floodplain and either an individual or aggregated estimate of depth (stage) versus damage curves for structures, roads, crops, utilities, etc., that are in the floodplain. This stage damage is then related to the relationship for the stage discharge to derive the damage-discharge relationship, which is then used along with the discharge-frequency relationship to derive the damage-frequency curve, as illustrated in Figure 14.4.2.

14.4.2   Expected Damages

The *annual expected damage cost E(D)* is the area under the damage-frequency curve as shown in Figure 14.4.1, which can be expressed as

$$E(D) = \int_{q_c}^{\infty} D(q_d) f(q_d)\, dq_d = \int_{q_c}^{\infty} D(q_d)\, dF(q_d) \tag{14.4.2}$$

where q_c is the threshold discharge beyond which damage would occur, $D(q_d)$ is the flood damage for various discharges q_d, which is the damage-discharge relationship, and $f(q_d)$ and $F(q_d)$ are the

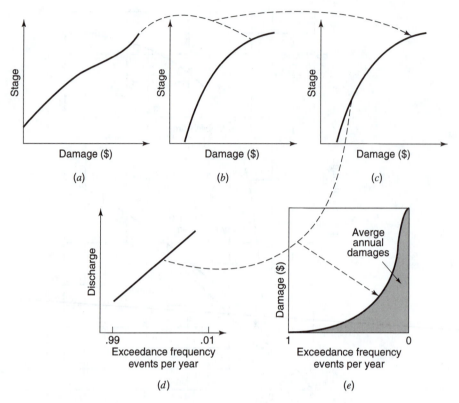

Figure 14.4.2 Computation of average annual damages. (*a*) Stage-damage relation;
(*b*) Stage-discharge relation; (*c*) Discharge-damage relation; (*d*) Discharge-frequency relation;
(*e*) Damage-frequency relation.

probability density functions (pdf) and the cumulative distribution function (cdf), respectively, of
discharge q_d. In practical applications, the evaluation of $E(D)$ by equation (14.4.2) is carried out
using numerical integration because of the complexity of damage functions and probability distribution functions. Therefore, the shaded area in Figure 14.4.2e can be approximated, numerically,
by the trapezoidal rule; as an example,

$$E(D) = \sum_{j=1}^{n} \frac{\left[D(q_j) + D(q_{j+1})\right]}{2}\left[F(q_{j+1}) - F(q_j)\right], \text{ for } q_c = q_1 \leq q_2 \leq \ldots \leq q_n < \infty \quad (14.4.3)$$

in which q_j is the discretized discharge in the interval (q_c, ∞).

EXAMPLE 14.4.1

Use the damage-frequency relationships in Table 14.4.1 for the flood-control alternatives to rank their
merits on the basis of expected flood-damage reduction.

Table 14.4.1 Example 14.4.1: Damage–Frequency Relationships

Exceed Prob. (%)	Damage[0] (10^6)	Damage[1] (10^6)	Damage[2] (10^6)	Damage[3] (10^6)	Damage[4] (10^6)
20	0	0	0	0	0
10	6	0	0	0	0
7	10	0	0	0	0
5	13	13	2	4	3
2	22	22	10	12	10
1	30	30	20	18	12

Table 14.4.1   Example 14.4.1: Damage–Frequency Relationships (*continued*)

Exceed Prob. (%)	Damage[0] (10^6)	Damage[1] (10^6)	Damage[2] (10^6)	Damage[3] (10^6)	Damage[4] (10^6)
0.5	40	40	30	27	21
0.2	50	50	43	40	35
0.1	54	54	47	43	45
0.05	57	57	55	50	56

Note: 0 — existing condition; 1 — dike system; 2 — upstream diversion; 3 — channel modification;
4 — detention basin.

SOLUTION

The economic merit of each flood-control alternative can be measured by the annual expected savings in flood damage of each alternative, which can be calculated as the difference between the annual expected damage at the existing condition (without flood-control measures) and the annual expected damage with a given flood-control measure under consideration. From Table 14.4.1, we determine Table 14.4.2, damage reduction associated with each flood-control measure at different return periods.

Table 14.4.2   Example 14.4.1: Damage Reduction as a Function of Exceedance Probability

Exceed. Prob. (%)	Damage Reduction (10^6)			
	Dike System	Upstream Diversion	Channel-ization	Detention Basin
20	0	0	0	0
10	6	6	6	6
7	10	10	10	10
5	0	9	7	10
2	0	11	9	12
1	0	10	12	18
0.5	0	10	13	19
0.2	0	7	10	15
0.1	0	5	9	9
0.05	0	2	7	1

From the data in Table 14.4.2, the average damage reduction for each flood-control alternative and incremental probability can be developed as shown in Table 14.4.3. The optimal alternative that maximizes the benefit of annual expected flood reduction is to build a detention basin upstream.

Table 14.4.3   Example 14.4.1: Annual Expected Damage Computation

Increm. Prob. (ΔF)	Damage Reduction, ΔD (10^6)			
	Dike System	Upstream Diversion	Channeli-zation	Detention Basin
0.1	3	3	3	3
0.03	8	8	8	8
0.02	5	9.5	8.5	10
0.03	0	10	8	11
0.01	0	10.5	10.5	15
0.005	0	10	12.5	18.5
0.003	0	8.5	11.5	17
0.001	0	6	9.5	12
0.0005	0	3.5	8	5
$\Sigma(\Delta D \cdot \Delta F)$	0.64	1.218	1.251	1.378

EXAMPLE 14.4.2

Columns 1–3 of Table 14.4.4 summarize the annual damage and annual costs for a dike system as a function of return period. Determine the return period for the dike system that maximizes the annual expected benefit.

Table 14.4.4 Example 14.4.4: Data

(1) T (yr)	(2) Annual Damage ($M/yr)	(3) Annual Cost ($M/yr)	(4) Total Cost ($M/yr)	(5) Benefit ($M/yr)	(6) Net Benefit ($M/yr)
5	1.94475	—	1.94475	0.0	0.0
10	1.64475	0.2	1.84475	0.3	0.1
20	1.17475	0.6	1.77475	0.77	0.17
50	0.64975	1.0	1.64975	1.295	0.295
100	0.38975	1.4	1.78975	1.555	0.155
200	0.21475	1.8	2.01475	1.730	−0.07
500	0.07975	2.1	2.17975	1.865	−0.235
1000	0.02775	2.3	2.32775	1.917	−0.383
2000	0.00000	2.5	2.50000	1.94475	−0.5555

SOLUTION

The annual total cost, shown in column 4, corresponding to different levels of production, in terms of return period, can be obtained by adding columns 2 and 3. The optimal return period associated with the least total annual expected cost is 50 years.

The same conclusion can be made by considering the annual net benefit. In this case, the annual benefit associated with different levels of protection is the saving (or reduction) in expected flood damage. For example, with a 10-year protection, the associated expected flood-damage reduction is $0.30 M/year as compared with the existing condition; with a 50-year protection, the expected benefit is $0.30 M + $0.24 M + $0.23 M + $0.525 M = $1.295 M/year. This information can be derived using the accumulated expected damage given in column 5. The annual expected net benefit then can be obtained by subtracting the annual cost in column 3 from column 5 and the results are listed in column 6. From column 6, the return period of 50-year is the optimal one associated with a net benefit of $0.295 M/year.

14.4.3 Risk-Based Analysis

Conventional risk-based design procedures for hydraulic structures consider only the inherent hydrologic uncertainty. The probability of failure of the hydraulic structure is generally evaluated by means of probability analysis as described in Chapter 10. Other aspects of hydrologic uncertainties are seldom included. Methodologies have been developed to integrate various aspects of hydrologic uncertainties into risk-based design of hydraulic structures (Mays and Tung, 1992).

Risk-based design approaches integrate the procedures of uncertainty analysis and reliability analysis in design. Such approaches consider the economic trade-offs between project costs and expected damage costs through the risk relationships. The risk-based design procedure can be incorporated into an optimization framework to determine the *optimal risk-based design*. Therefore, in an optimal risk-based design, the expected annual damage is taken into account in the objective function, in addition to the installation cost. The problem is to determine the optimal structural sizes/capacities associated with the least total expected annual cost (TEAC). Mathematically, the optimal risk-based design can be expressed as

$$\text{Min } TEAC = FC(\mathbf{x}) \cdot CRF + E(D|\mathbf{x}) \qquad (14.4.4a)$$
$$\mathbf{x}$$

$$\text{Subject to design specifications, } \mathbf{g}(\mathbf{x}) = \mathbf{0} \qquad (14.4.4b)$$

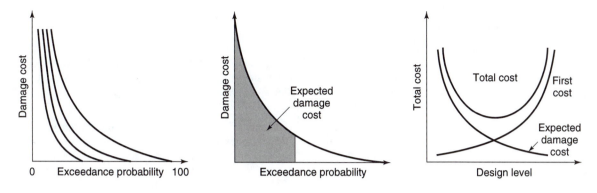

Figure 14.4.3 Computation of expected damage cost and total cost for risk-based design.

where *FC* is the total installation cost (first cost) of the structure, *CRF* is the capital recovery factor for conversion of cost to an annual basis, **x** is a vector of decision variables relating to structural sizes and/or capacities, and $E(D|\mathbf{x})$ is the expected annual damage cost due to structural failure. A diagram illustrating the cost computations for optimal risk-based design is shown in Figure 14.4.3.

In general, quantification of the first cost is straightforward. The thrust of risk-based design is the assessment of the annual expected damage cost. Depending on the types of uncertainty to be considered, assessment of $E(D|\mathbf{x})$, varies. Current practice in risk-based water-resources engineering design considers only the inherent hydrologic uncertainty due to the randomness of the hydrologic process. The mathematical representation of the annual expected damage under this condition can be computed using equation (14.4.2).

The procedure for computations of expected annual damages illustrated in Figure 14.4.2 ignores uncertainty in the various relationships. Traditional uncertainty has been accounted for by the use of factors of safety such as large return periods or freeboard for levees. The more recent advances in risk analysis has allowed the explicit accounting of uncertainty. As a result, the U.S. Army Corps of Engineers (1996b) has adapted a policy by which flood-damage reduction studies use the risk-based analysis described in the next section.

14.5 U.S. ARMY CORPS OF ENGINEERS RISK-BASED ANALYSIS FOR FLOOD-DAMAGE REDUCTION STUDIES

14.5.1 Terminology

The U.S. Army Corps of Engineers (1996b) has developed guidance and a procedure for risk-based analysis for flood-damage reduction studies that depends on

(a) Qualitative description of errors or uncertainty in selecting the proper hydrologic, hydraulic, and economic function to use when evaluating economic and engineering performance of flood-damage reduction measures

(b) Qualitative description of errors or uncertainty in selecting the parameters of these functions

(c) Conceptual techniques that determine the combined impact on plan evolution of errors in the function and their parameters

The concepts of flood risks and uncertainty made in this procedure are defined in Table 14.5.1.

Table 14.5.1 Terminology

Term	Definition
Function uncertainty (also referred to as distribution uncertainty and model uncertainty)	Lack of complete knowledge regarding the form of a hydrologic, hydraulic, or economic function to use in a particular application. This uncertainty arises from incomplete scientific or technical understanding of the hydrologic, hydraulic, or economic process.
Parameter	A quantity in a function that determines the specific form of the relationship of known input and unknown output. An example is Manning's roughness coefficient in energy loss calculations. The value of this parameter determines the relationship between a specified discharge rate and the unknown energy loss in a specific channel reach.
Parameter uncertainty	Uncertainty in a parameter due to limited understanding of the relationship or due to lack of accuracy with which parameters can be estimated for a selected hydrologic, hydraulic, or economic function.
Sensitivity analysis	Computation of the effect on the output of changes in input values or assumption.
Exceedance probability	The probability that a specified magnitude will be exceeded. Unless otherwise noted, this term is used herein to denote annual exceedance probability: the likelihood of exceedance in any year.
Median exceedance probability	In a sample of estimates of exceedance probability of a specified magnitude, this is the value that is exceeded by 50 percent of the estimates.
Capacity exceedance	Capacity exceedance implies exceedance of the capacity of a water conveyance, storage facility, or damage-reduction measure. This includes levee or reservoir capacity exceeded before over-topping, channel capacity exceedance, or rise of water above the level of raised structures.
Conditional probability	The probability of capacity exceedance, given the occurrence of a specified event.
Long-term risk	The probability of capacity exceedance during a specified period. For example, 30-year risk refers to the probability of one or more exceedances of the capacity of a measure during a 30-year period.

Source: U.S. Army Corps of Engineers (1996b).

A *flood-damage-reduction plan* includes measures that reduce damage by reducing discharge, reducing stage, or reducing damage susceptibility (U.S. Army Corps of Engineers, 1996b). For U.S. government projects, the objective of the plan is to solve the problem so that the solution will contribute to rational economic development (RED). *A flood-damage-reduction planning study* is conducted to determine (1) which measures to include in the plan, (2) where to locate the measures, (3) what size to make the measures, and (4) how to operate measures in order to satisfy the federal objective and constraints (U.S. Army Corps of Engineers, 1996b).

Figure 14.5.1 provides an overview of the risk-based analysis procedure that quantifies the uncertainty in the relationships of discharge frequency, storage discharge, and stage damage, and incorporates an analysis of the economic efficiency of alternatives.

The expected value $E[]$ of inundation damage X can be calculated using equation (14.4.2), or more simply

$$E[X] = \int_{-\infty}^{\infty} x f(x)\, dx \tag{14.5.1}$$

where $f(x)$ is the probability density function of x.

14.5.2 Benefit Evaluation

The *net benefit* of a plan is the measure of the flood-damage-reduction plan's contribution to *national economic development (NED)*. The net benefit is computed as the sum of location benefit,

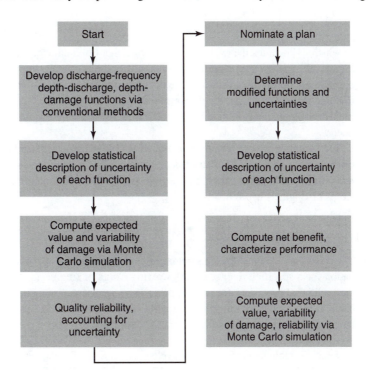

Figure 14.5.1 Risk-based analysis procedure (from U.S. Army Corps of Engineers (1996b)).

intensification benefit, and flood-inundation reduction benefit, less the total cost of implementation (operating, maintaining, repairing, replacing, and rehabilitating [OMRR&R]) of the plan. The *location benefit* is the increased net income of additional floodplain development due to a plan. The *intensification benefit* is the increased net income of existing floodplain activities. The *inundation-reduction benefit* is the plan-related reduction in physical economic damage, income loss, and emergency cost.

The *economic efficiency* of a proposed flood-damage-reduction alternative is defined as

$$NB = (B_L + B_I + B_{IR}) - C \tag{14.5.2}$$

in which NB = net benefit, B_L = location benefit, B_I = intensification benefit, B_{IR} = inundation-reduction benefit, and C = total cost of implementing, operating, maintaining, repairing, replacing, and rehabilitating the plan (the OMRR&R cost). The inundation-reduction benefit may be expressed as

$$B_{IR} = (D_{\text{without}} - D_{\text{with}}) \tag{14.5.3}$$

in which D_{without} = economic flood damage without the plan, and D_{with} = economic flood damage if the plan is implemented.

The random nature of flooding complicates determination of the inundation-reduction benefit. For example, a flood-damage-reduction plan that eliminates all inundation damage one year may be too small to eliminate all damage in an extremely wet year and much larger than required in an extremely dry year. U.S. Water Resources Council (1983) guidelines address this problem by calling for use of expected annual flood damage. Expected damage takes into account the risk of various magnitudes of flood damage each year, weighing the damage caused by each flood by the probability of occurrence. Combining equations (14.5.2) and (14.5.3) and rewriting them in terms of expected values yields:

$$NB = B_L + B_I + E[D_{\text{without}}] - E[D_{\text{with}}] - C \qquad (14.5.4)$$

in which $E[]$ denotes the expected value. For urban flood damages, this generally is computed on an annual basis because significant levels of flood damage are limited to annual recurrence. For agricultural flood damages, it may be computed as the expected damage per flood, as more than one damaging flood may occur in a given year. The NED plan is then the alternative plan that yields maximum net benefit, taking into account the full range of hydrologic conditions that might occur.

The "without project" condition in equation (14.5.4) represents existing and future system conditions in the absence of a plan. It is the base condition upon which alternative plans are formulated, from which all benefits are measured, and against which all impacts are assessed (U.S. Army Corps of Engineers, 1989).

14.5.3 Uncertainty of Stage-Damage Function

The stage-damage function is a summary statement of the direct economic cost of floodwater inundation for a specified river reach (U.S. Army Corps of Engineers, 1996). Table 14.5.2 illustrates the traditional procedure for development of a stage-damage function and Table 14.5.3 identifies some of the sources of uncertainty. From these descriptions, an aggregated description of uncertainty of the stage-damage function can be developed using the procedure in Figure 14.5.2.

Table 14.5.2   Traditional Procedure for Development of Stage-Damage Function

Step	Task
1	Identify and categorize each structure in the study area based upon its use and construction
2	Establish the first-floor elevation of each structure using topographic maps, aerial photographs, surveys, and/or hand levels
3	Estimate the value of each structure using real estate appraisals, recent sales prices, property tax assessments, replacement cost estimates, or surveys
4	Estimate the value of the contents of each structure using an estimate of the ratio of contents value to structure value for each unique structure category
5	Estimate damage to each structure due to flooding to various water depths at the structure's site using a depth-percent damage function for the structure's category along with the value from Step 3
6	Estimate damage to the contents of each structure due to flooding to various water depths using a depth-percent damage function for contents for the structure category along with the value from Step 4
7	Transform each structure's depth-damage function to a stage-damage function at an index location for the floodplain using computed water-surface profiles for reference floods
8	Aggregate the estimated damages for all structures by category for common stages

Source: U.S. Army Corps of Engineers (1996b).

Table 14.5.3   Components and Sources of Uncertainty in Stage-Damage Function

Parameter/Model	Source of Uncertainty
Number of structures in each category	Errors in identifying structures; errors in classifying structures
First-floor elevation of structure	Survey errors; inaccuracies in topographic maps; errors in interpolation of contour lines
Depreciated replacement value of structure	Errors in real estate appraisal; errors in estimation of replacement cost estimation-effective age; errors in estimation of depreciation; errors in estimation of market value

Table 14.5.3   Components and Sources of Uncertainty in Stage-Damage Function (*continued*)

Parameter/Model	Source of Uncertainty
Structure depth-damage function	Errors in post-flood damage survey; failure to account for other critical factors: floodwater velocity, duration of flood; sediment load; building material; internal construction; condition; flood warning
Depreciated replacement value of contents	Errors in content-inventory survey; errors in estimates of ratio of content to structure value
Content depth-damage function	Errors in post-flood damage survey; failure to account for other critical factors: floodwater velocity, duration of flood; sediment load; content location, flood warning

Source: U.S. Army Corps of Engineers (1996b).

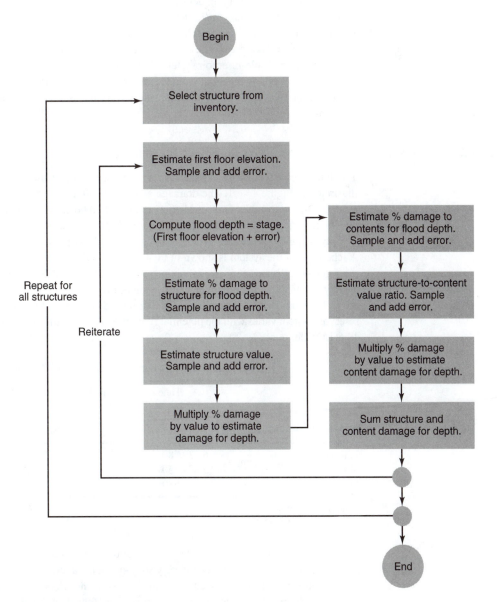

Figure 14.5.2 Development of stage-damage function with uncertainty description (from U.S. Army Corps of Engineers (1996b)).

14.6   OPERATION OF RESERVOIR SYSTEMS FOR FLOOD CONTROL

14.6.1   Flood-Control Operation Rules

Flood control is one of the general operation purposes of reservoir system operation, as discussed in Section 11.7. *Operation rules (policies)* are used to specify how water is managed throughout a system. These rules are specified in order to achieve discharge requirements and system demands that to maximize the objective expressed in the form of benefits. Operation rules may be established on seasonal, monthly, or other time periods to prescribe how water is to be regulated during subsequent time periods based on the current state of the system. Even though operating rules are usually defined for each within-year period, these rules can vary over successive years, as a result for instance of changing demand.

Flood-control operating rules define the release to be made in the current time period as a function of one or more of the following: current storage in the reservoir, forecast inflow to the reservoir, current and forecast downstream flow, and current storage in and forecast flow to other reservoirs in a multiple-reservoir system. Operating rules must be defined for controlled reservoirs as a component of any plan that includes such a reservoir. Figure 11.7.6 illustrates an example of an operation rule that shows the seasonally varying storage boundaries for a multipurpose reservoir.

14.6.2   Tennessee Valley Authority (TVA) Reservoir System Operation

The Tennessee River Basin is located in a seven-state area in the southeastern United States, as illustrated in Figure 14.6.1. The drainage area is 40,900 mi^2 (105,930 km^2). This basin is located a short distance from the Gulf of Mexico and the Caribbean Sea, which provide major sources of moisture, so that it is one of the wettest regions in the United States.

Sixty major dams are operated on in integrated multipurpose water control system by the TVA. A schematic of the system is shown in Figure 14.6.2. Of these dams, 45 are owned by the TVA in the Tennessee River Basin, another six provide flow into the TVA system but are owned and operated by others; the remaining nine dams (eight Army Corps of Engineers dams and one TVA dam) are in the Cumberland River Basin. The amount of available flood-control storage in the TVA reservoir system varies with the potential flood threat and the time of the year, as indicated in Table 14.6.1 (Miller et al., 1996).

Table 14.6.1   TVA Reservoir System Flood-Control Storage

Time of Year	Detention Storage (acre-feet)		Total System Runoff (in)
	Above Chattanooga	Total System	
January 1	6,339,780	11,330,480	5.3
March 15	5,189,750	10,030,760	4.7
June 1	1,571,580	2,137,540	1.0

Note: 1 acre-foot = 1,233 m^3
Source: Miller et al. (1996).

The typical operating rules for the TVA system reservoirs are illustrated in Figure 14.6.3. The tributary reservoir operations depend very strongly on the annual hydrologic cycle (Figure 14.6.4). The operations can be categorized into four periods: winter flood season (January 1 to March 15), fill period (March 15 to June 1), recreation season (June 1 to August 1), and drawdown season (August 1 to December 31).

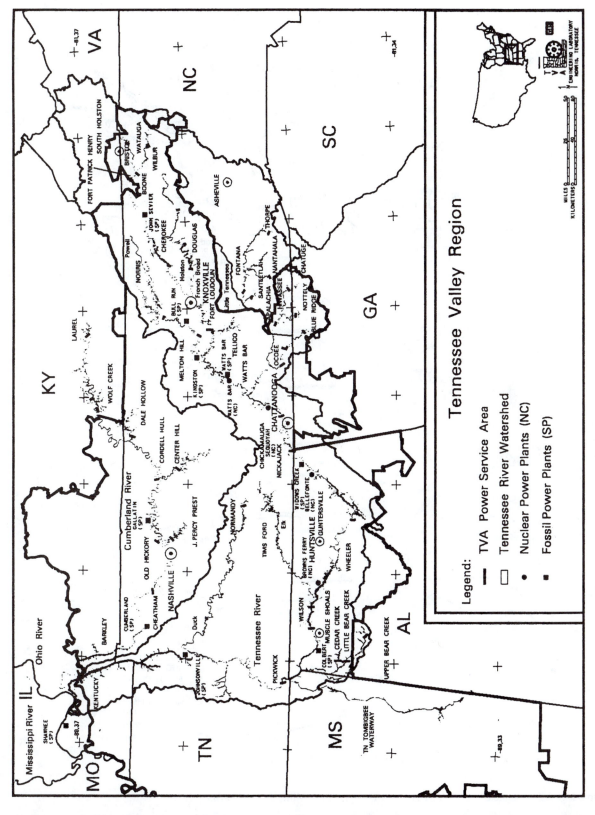

Figure 14.6.1 The Tennessee River Basin (from Miller et al. (1996)).

TENNESSEE RIVER RESERVOIR SYSTEM

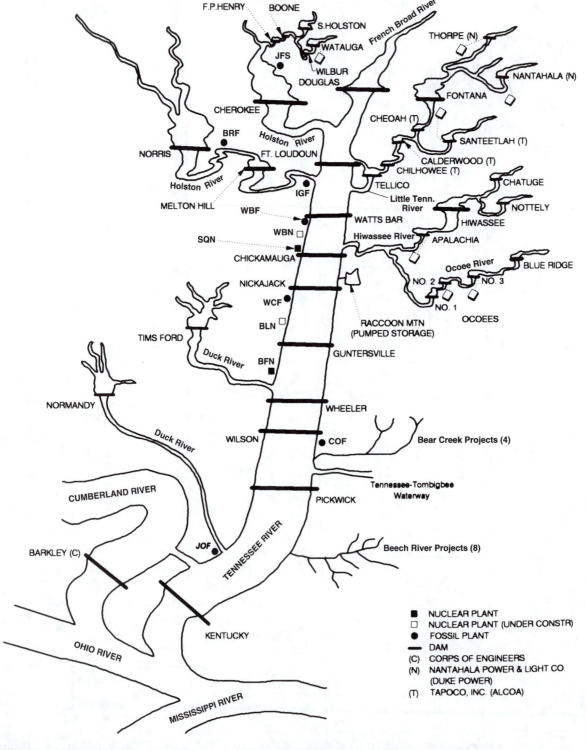

Figure 14.6.2 Schematic of the TVA water-control system (from Miller et al. (1996)).

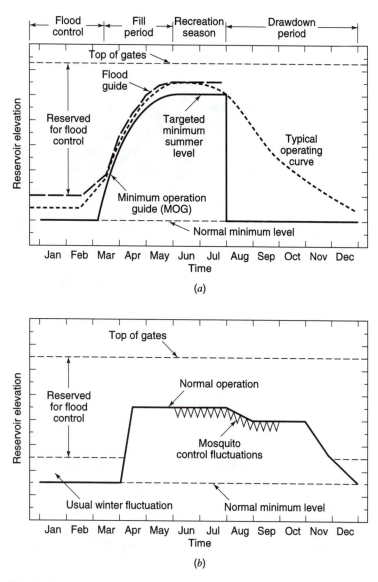

Figure 14.6.3 Typical TVA reservoir operating levels. (*a*) Tributary reservoir operations; (*b*) Mainstream reservoir operations (from Miller et al. (1996)).

During the winter flood season, reservoir levels are held below the *flood guides* (see Figure 14.6.3) to provide storage for high winter flows. During the fill period, spring rainfall is used rapidly to fill the reservoirs for the summer recreation target levels. During the fill period, an attempt is made to keep the reservoir levels between the flood guides and the *minimum operation guides* (MOGs). Summer reservoir levels are normally kept above the targeted minimum summer recreation levels until the end of July, when they are gradually lowered during the drawdown period.

Mainstream reservoir operation generally follows a similar but simpler operation rule (see Figure 14.6.3). Topography and navigation requirements allow only a few feet between winter and summer levels on the mainstream reservoir, whereas a reservoir elevation on the tributaries can fluctuate as much as 75 ft (23 m) between winter and summer. *Mosquito control fluctuations* (reservoir levels are fluctuated by 1 ft each week) are used during the late spring and summer into mainstream and reservoirs (except in Kentucky) to help control mosquitoes. The use of TVA's reservoir operating guides is unique in the United States in that most of the major reservoir

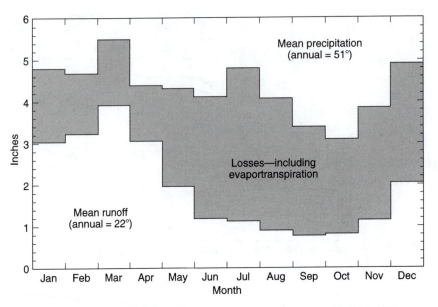

Figure 14.6.4 Tennessee River Basin mean monthly rainfall and runoff 1894–1993 (from Miller et al. (1996)).

systems operate within specified zones or pools, such as the U.S. Army Corps of Engineers (Miller et al., 1996).

The U.S. Army Corps of Engineers HEC-5 computer program (U.S. Army Corps of Engineers, 1982) models the performance of a reservoir or a reservoir system to manage excess water. The storage in each reservoir in a system is divided into zones (see Figure 11.7.1), so that inlet storage levels are defined in each zone. Table 14.6.2 summarizes the flood-control operation rules in HEC-5. Refer to the Web site (www.hec.usace.army.mil) to obtain HEC-5 and additional information.

Table 14.6.2 HEC-5 Flood-Control Operation Rules

Constraint on Release Made	Condition
Release to draw storage to top of conservation pool without exceeding channel capacity at reservoir or downstream points for which reservoir is operated	Storage is between top of conservation pool and top of flood-control pool
Release equal to or greater than minimum desired flow	Storage greater than top buffer storage
Release equal to minimum desired flow	Storage between top inactive and top of buffer pool
No release	Storage below top of inactive pool
Release required to satisfy hydropower requirement	If that release is greater than controlling desired or required flows for above conditions
Release limited to user-specified rate of change	Unless reservoir is in surcharge operation
No release that will continue to flooding downstream	If flood storage available
Release to maintain downstream flow at channel capacity	If operating for flood control
Release from reservoir at greatest level	If two or more reservoirs on parallel streams operate for common downstream point
Release to bring upper reservoir to same index level as downstream reservoir	If two reservoirs in tandem

Source: Ford and Hamilton (1996).

PROBLEMS

14.2.1 Obtain the FEMA flood hazard map for your local area and identify the 100-year floodplain, the floodway, and the floodway fringe.

14.4.1 Determine the optimal design return period for a flood-control project with damage and capital costs for the following return periods:

T	1	2	5	10	15	20	25	50	100	200
Damage cost, $ $\times 10^3$	0	40	120	280	354	426	500	600	800	1000
Capital cost, $/yr $\times 10^3$	0	6	28	46	50	54	58	80	120	160

14.4.2 The frequency-discharge-storage-damage data for existing conditions at a particular area along a river is given in Table P14.4.2. The storage-discharge data is the rating curve, the discharge-frequency data is the frequency curve, and the storage-damage data is the storage-damage curve. Plot these three relationships.

Table P14.4.2

Exceedance probability	Q (cfs)	Stage (ft)	Damage (10^6)
20	12000	30	0
10	14000	33	6
7	15000	34	10
5	16000	35	13
2	18000	38	22
1	20000	40	30
0.5	22000	42.5	40
0.2	25000	45	50
0.1	26000	46	54
0.05	28000	47	57

14.4.3 If a dike system with a capacity of 15,000 cfs were used as one alternative for the situation in problem 14.4.2, develop the damage-frequency curve for this alternative.

14.4.4 If an upstream permanent diversion that will protect up to a natural flow of 15,000 cfs is used for the situation in problem 14.4.2, develop the damage-frequency curve for this alternative.

14.4.5 If channel modifications were used for the situation in problem 14.4.2 to increase the conveyance capacity of the river up to 15,000 cfs, develop the damage-frequency curve for this alternative.

14.4.6 Referring to problems 14.4.2 and 14.4.3, determine the return period for the dike system that maximizes annual expected benefit.

14.4.7 Referring to problems 14.4.2 and 14.4.4, determine the return period for the diversion capacity with maximum expected annual benefit.

14.4.8 Referring to problems 14.4.2 and 14.4.5, determine the return period associated with the flow capacity that maximizes the expected annual net benefit in the channel modification.

14.5.1 Use the Web site for the U.S. Committee on Large Dams (www.uscold.org/~uscold/) to write a summary of the recent flood damage prevention of: (a) California's Central valley dams; (b) Oroville Dam; and (c) Truckee River dams; and (d) Missouri River dams.

14.5.2 Write a summary of the flood control aspects of the Three Gorges Dam on the Yangtze River in China.

14.6.1 Write a hsitory of the TVA starting with the Web address (www.TVA.gov), with particular emphasis on flood control.

REFERENCES

Bull, W. B., "Alluvial Fans and Pediments of Southern Arizona," in *Landscapes of Arizona*, edited by T. L. Smiley, et al., University Press of America, Lanham, pp. 229–252, 1984.

Chow, V. T., Maidment, D. R., and Mays, L. W., *Applied Hydrology*, McGraw-Hill, New York, 1988.

Corry, M. L., J. S. Jones, and D. L. Thompson, "The Design of Encroachments of Floodplains Using Risk Analysis," Hydraulic Engineering Circular No. 17, U.S. Department of Transportation, Federal Highway Administration, Washington, DC, July 1980.

Dodson, R., "Floodplain Hydraulics," *Hydraulic Design Handbook*, edited by L.W. Mays, McGraw-Hill, New York, 1999.

Eckstein, O., *Water Resources Development; The Economic of Project Evaluation*, Harvard University Press, Cambridge, MA, 1958.

Federal Interagency Floodplain Management Task Force, *Floodplain Management in the United States: An Assessment Report*, Vols. 1 and 2, FEMA, Washington DC, 1992.

Ford, D., and D. Hamilton, "Computer Models for Water-Excess Management," *Water Resources Handbook* (edited by L. W. Mays), McGraw-Hill, New York, 1996.

Grigg, N. S. and O. J. Helweg, "State-of-the-Art of Estimating Flood Damage in Urban Areas," *Water Resources Bulletin*, vol. 11, no. 2, pp. 379–390, 1975.

Grigg, N. S., *Water Resources Planning*, McGraw-Hill, New York, 1985.

Hoggan, D. H., *Computer-Assisted Floodplain Hydrology and Hydraulics*, second edition, McGraw-Hill, New York, 1997.

Interagency Floodplain Management Review Committee, Sharing the Challenge: Floodplain Management into the 21st Century, Administration Floodplain Management Task Force, Executive Office of the President, Washington, DC, June, 1994.

James, L. D. "Role of Economics in Planning Floodplain Land Use," *Journal of the Hydraulics Division, ASCE*, vol. 98, no. HY6, pp. 981–992.

Mays, L. W., and Tung, Y .K., *Hydrosystems Engineering and Management,* McGraw-Hill, New York, 1992.

Meyers, Mary F., and G. F. White, "The Challenge of the Mississippi Flood," *Environment,* vol. 35, no. 10, Washington, DC, 1993.

Miller, B. A., A. Whitelock, and R. C. Hughes, "Flood Management—The TVA Experience," *Water International,* IWRA, vol. 21, no. 3, pp. 119–130, 1996.

U.S. Army Corps of Engineers, Galveston District, Special Flood Hazard Information, Clear Creek-Brazoria, Fort Bend, Galveston, and Harris Counties, Texas, prepared for Harris Soil and Water Conservation District, Houston Texas, June 1972.

U.S. Army Corps of Engineers, *Hydrologic Engineering Methods for Water Resources Development: Reservoir System Analysis for Conservation*, vol. 9, Hydrologic Engineering Center, Davis, CA, June 1977.

U.S. Army Corps of Engineers, "HEC-5: Simulation of Flood Control and Conservation Systems, User's Manual," Hydrologic Engineering Center, Davis, CA, 1982.

U.S. Army Corps of Engineers, *HECDSS User's Guide and Utility Program Manual,* Hydrologic Engineering Center, Davis, CA,1983.

U.S. Army Corps of Engineers, *Flood Damage Analysis Package Users Manual,* Hydrologic Engineering Center, Davis, CA,1988a.

U.S. Army Corps of Engineers, "National Economic Development Procedures Manual—Urban Flood Damage," Institute for Water Resources, IWR Report 88-R-2, Ft. Belvoir, VA, 1988b.

U.S. Army Corps of Engineers, "EAD: Expected Annual Flood Damage Computation, User's Manual," CPD-30, Hydrologic Engineering Center, Davis, CA 1989.

U.S. Army Corps of Engineers, Hydrologic Engineering Center, "Federal Perspective for Flood-Damage-Reduction Studies," *Water Resources Handbook* (edited by L. W. Mays), McGraw-Hill, New York, 1996a.

U.S. Army Corps of Engineers, *Risk-Based Analysis for Flood Damage Reduction Studies*, Engineering Manual EM 1110–2–1619, Washington, DC, 1996b.

U.S. Water Resources Council, Economic and Environmental Principles and Guidelines for Water and Related Land Resources Implementation Studies, U.S. Government Printing Office, Washington, DC, 1983.

Waananen, A. O., J. T. Limerinos, W. J. Kockelman, W. E. Spangle, and M. L. Blair, "Flood-Prone Areas and Land-Use Planning-Selected Examples from the San Fransisco Bay Region", U.S. Geological Survey Professional Paper 942, California 1977.

White, G. F., "Human Adjustment to Floods: A Geographical Approach to the Flood Problem in the United States," Research Paper no. 29, University of Chicago, 1945.

Chapter 15

Stormwater Control: Storm Sewers and Detention

15.1 STORMWATER MANAGEMENT

Stormwater management is knowledge used to understand, control, and utilize waters in their different forms within the hydrologic cycle (Wanielista and Yousef, 1993). The goal of this chapter is to provide an introduction to the various concepts and design procedures involved in stormwater management. The overall key component of stormwater management is the drainage system. Urbonas and Roesner (1993) point out the following vital functions of a drainage system:

1. It removes stormwater from the streets and permits the transportation arteries to function during bad weather; when this is done efficiently, the life expectancy of street pavement is extended.
2. The drainage system controls the rate and velocity of runoff along gutters and other surfaces in a manner that reduces the hazard to local residents and the potential for damage to pavement.
3. The drainage system conveys runoff to natural or manmade major drainage ways.
4. The system can be designed to control the mass of pollutants arriving at receiving waters.
5. Major open drainage ways and detention facilities offer opportunities for multiple use such as recreation, parks, and wildlife preserves.

Storm drainage criteria are the foundation for developing stormwater control. Table 15.1.1 provides a checklist for developing storm drainage criteria. These criteria should set limits on development, provide guidance and methods of design, provide details of key components of drainage and flood control systems, and ensure longevity, safety, aesthetics, and maintainability of the system served (Urbonas and Roesner, 1993).

Table 15.1.1 Checklist for Developing Local Storm Drainage Criteria

Governing legislation and statements of policy and procedure
   Legal basis for criteria
   Define what constitutes the drainage system
   Benefits of the drainage system
   Policy for dedication of right-of-way
   Compatible multipurpose uses
   Review and approval procedures
   Procedures for obtaining variances or waivers of criteria

Table 15.1.1   Checklist for Developing Local Storm Drainage Criteria (*continued*)

Initial and major drainage system provisions
    Definitions of initial system and major system
    Where should a separate formal major drainageway begin?
    Allowable flow capacities in streets for initial and major storms
    Maximum and minimum velocities in pipes and channels
    Maximum flow depths in channels and freeboard requirements
Data required for design, such as:
    Watershed boundaries
    Local rainfall and runoff data
    History of flooding in the area
    Defined regulatory flood plains and floodways
    Existing and projected land use for project site
    Existing and planned future land uses upstream
    Existing and planned drainage systems off-site
    Tabulation of previous studies affecting site
    Conflicts with existing utilities
    Design storms, intensity-duration-frequency data
    Hydrologic methods and/or models
    Storm sewer design criteria, including materials
    Manhole details and spacing, inlet details and spacing, types of inlets, trenching, bedding,
        backfill, etc.
    Street flow calculations and limitations
Details of major system components such as channels, drop structures, erosion checks, transitions,
    major culverts and pipes, bridges, bends, energy dissipators, riprap, sediment transport
Detention requirements
    When and where to use detention
    Design storms
    Hydrologic sizing criteria and/or procedures
    Safety, aesthetics, and maintainability criteria
    Multipurpose uses and design details for each
Water quality criteria
    Goals and objectives
    Minimum capture volumes
    Required or acceptable best management practices (BMP)
    Technical design criteria for each BMP
Special considerations
    Right-of-way dedication requirements
    Use of irrigation ditches
    Flood proofing and when it is accepted
    Any other items reflecting local needs
    List of technical references

Source: Urbonas and Roesner (1993).

15.2   STORM SYSTEMS

15.2.1   Information Needs and Design Criteria

To begin the design process of a storm sewer system, one must collect a considerable amount of information. A condensed checklist of information needs for storm sewer design is presented in Table 15.2.1. There are many sources for this information, ranging from various local government agencies to federal agencies to pipe and pump manufacturers. The designer must also obtain future development information for areas surrounding the site of interest.

Table 15.2.1   Condensed Checklist of Information Needs for Storm Sewer Design

- Local storm drainage criteria and design standards
- Maps, preferably topographic, of the subbasin in which the new system is to be located
- Detailed topographic map of the design area
- Locations, sizes, and types of existing storm sewers and channels located upstream and downstream of design area
- Locations, depths, and types of all existing and proposed utilities
- Layout of design area including existing and planned street patterns and profiles, types of street cross-sections, street intersection elevations, grades of any irrigation and drainage ditches, and elevations of all other items that may post physical constraints to the new system
- Soil borings, soil mechanical properties, and soil chemistry to help select appropriate pipe materials and strength classes
- Seasonal water table levels
- Intensity-duration-frequency and design storm data for the locally required design return periods
- Pipe vendor information for the types of storm sewer pipe materials accepted by local jurisdiction

Source: Urbonas and Roesner (1993).

Design criteria vary from one city to another, but for the most part the following are a fairly standard set of assumptions and constraints used in the design of storm sewers (American Society of Civil Engineers, 1969, 1992):

(a) For small systems, free-surface flow exists for the design discharges; that is, the sewer system is designed for "gravity flow" so that pumping stations and pressurized sewers are not considered.

(b) The sewers are commercially available circular pipes.

(c) The design diameter is the smallest commercially available pipe that has flow capacity equal to or greater than the design discharge and satisfies all the appropriate constraints.

(d) Storm sewers must be placed at a depth that will not be susceptible to frost, will drain basements, and will allow sufficient cushioning to prevent breakage due to ground surface loading. Therefore, minimum cover depths must be specified.

(e) The sewers are joined at junctions such that the crown elevation of the upstream sewer is no lower than that of the downstream sewer.

(f) To prevent or reduce excessive deposition of solid material in the sewers, a minimum permissible flow velocity at design discharge or at barely full-pipe gravity flow is specified.

(g) To prevent the occurrence of scour and other undesirable effects of high-velocity flow, a maximum permissible flow velocity is also specified. Maximum velocities in sewers are important mainly because of the possibilities of excessive erosion on the sewer inverts.

(h) At any junction or manhole, the downstream sewer cannot be smaller than any of the upstream sewers at that junction.

(i) The sewer system is a dendritic network converging towards downstream without closed loops.

Table 15.2.2 lists the more important typical technical items and limitations to consider.

Table 15.2.2   Technical Items and Limitations to Consider in Storm Sewer Design

Velocity:	
Minimum design velocity	2–3 ft/s (0.6–0.9 m/s)
Maximum design velocity	
Rigid pipe	15–21 ft/s (4.6–6.4 m/s)
Flexible pipe	10–15 ft/s (3.0–4.6 m/s)
Maximum manhole spacing:	
(function of pipe size)	400–600 ft (122–183 m)

Table 15.2.2   Technical Items and Limitations to Consider in Storm Sewer Design (*continued*)

Minimum size of pipe	12–24 in (0.3–0.6 m)
Vertical alignment at manholes:	
Different size pipe	Match crown of pipe or 80 to 85% depth lines
Same size pipe	Minimum of 0.1–0.2 ft (0.03–0.06 m) in invert drop
Minimum depth of soil cover	12–24 in (0.3–0.6 m)
Final hydraulic design	Check design for surcharge and junction losses by using backwater analysis
Location of inlets	In street where the allowable gutter flow capacity is exceeded

Source: Urbonas and Roesner (1993).

15.2.2   Rational Method Design

From an engineering viewpoint the design can be divided into two main aspects: runoff prediction and pipe sizing. The rational method, which can be traced back to the mid-nineteenth century, is still probably the most popular method used for the design of storm sewers (Yen and Akan, 1999). Although criticisms have been raised of its adequacy, and several other more advanced methods have been proposed, the rational method, because of its simplicity, is still in continued use for sewer design when high accuracy of runoff rate is not essential.

Using the rational method, the storm runoff peak is estimated by the rational formula

$$Q = KCiA \qquad (15.2.1)$$

where the peak runoff rate Q is in ft^3/s (m^3/s), K is 1.0 in U.S. customary units (0.28 for SI units), C is the runoff coefficient (Table 15.2.3), i is the average rainfall intensity in in/hr (mm/hr) from intensity-duration frequency relationships for a specific return period and duration t_c in min, and A is the area of the tributary drainage area in acres (km^2). The duration is taken as the time of concentration t_c of the drainage area.

Table 15.2.3   Runoff Coefficients for Use in the Rational Method

	Return Period (years)						
Character of Surface	2	5	10	25	50	100	500
Developed							
Asphaltic	0.73	0.77	0.81	0.86	0.90	0.95	1.00
Concrete/roof	0.75	0.80	0.83	0.88	0.92	0.97	1.00
Grass areas (lawns, parks, etc.)							
Poor condition (grass cover less than 50% of the area)							
Flat, 0–2%	0.32	0.34	0.37	0.40	0.44	0.47	0.58
Average, 2–7%	0.37	0.40	0.43	0.46	0.49	0.53	0.61
Steep, over 7%	0.40	0.43	0.45	0.49	0.52	0.55	0.62
Fair condition (grass cover 50% to 75% of the area)							
Flat, 0–2%	0.25	0.28	0.30	0.34	0.37	0.41	0.53
Average, 2–7%	0.33	0.36	0.38	0.42	0.45	0.49	0.58
Steep, over 7%	0.37	0.40	0.42	0.46	0.49	0.53	0.60
Good condition (grass cover larger than 75% of the area)							
Flat, 0–2%	0.21	0.23	0.25	0.29	0.32	0.36	0.49
Average, 2–7%	0.29	0.32	0.35	0.39	0.42	0.46	0.56
Steep, over 7%	0.34	0.37	0.40	0.44	0.47	0.51	0.58

Table 15.2.3 Runoff Coefficients for Use in the Rational Method (*continued*)

	Return Period (years)						
Character of Surface	2	5	10	25	50	100	500
Undeveloped							
Cultivated land							
Flat, 0–2%	0.31	0.34	0.36	0.40	0.43	0.47	0.57
Average, 2–7%	0.35	0.38	0.41	0.44	0.48	0.51	0.60
Steep, over 7%	0.39	0.42	0.44	0.48	0.51	0.54	0.61
Pasture/range							
Flat, 0–2%	0.25	0.28	0.30	0.34	0.37	0.41	0.53
Average, 2–7%	0.33	0.36	0.38	0.42	0.45	0.49	0.58
Steep, over 7%	0.37	0.40	0.42	0.46	0.49	0.53	0.60
Forest/woodlands							
Flat, 0–2%	0.20	0.25	0.28	0.31	0.35	0.39	0.48
Average, 2–7%	0.31	0.34	0.26	0.40	0.43	0.47	0.56
Steep, over 7%	0.35	0.39	0.41	0.45	0.48	0.52	0.58

Note: The values in the table are the standards used by the City of Austin, Texas.

Source: Chow, Maidment, and Mays (1988).

In urban areas, the drainage area usually consists of subareas or subcatchments of substantially different surface characteristics. As a result, a composite analysis is required that must take into account the various surface characteristics. The areas of the subcatchments are denoted by A_j and the runoff coefficients for each subcatchment are denoted by C_j. Then the peak runoff is computed using the following form of the rational formula:

$$Q = Ki\sum_{j=1}^{m} C_j A_j \tag{15.2.2}$$

where m is the number of subcatchments drained by a sewer.

The *rainfall intensity i* is the average rainfall rate considered for a particular drainage basin or subbasin. The intensity is selected on the basis of design rainfall duration and design frequency of occurrence. The design duration is equal to the time of concentration for the drainage area under consideration. The frequency of occurrence is a statistical variable that is established by design standards or chosen by the engineer as a design parameter.

The *time of concentration* t_c used in the rational method is the time associated with the peak runoff from the watershed to the point of interest. Runoff from a watershed usually reaches a peak at the time when the entire watershed is contributing; in this case, the time of concentration is the time for a drop of water to flow from the remotest point in the watershed to the point of interest. Runoff may reach a peak prior to the time the entire watershed is contributing. A trial-and-error procedure can be used to determine the critical time of concentration. The time of concentration to any point in a storm drainage system is the sum of the inlet time t_0 and the flow time t_f in the upstream sewers connected to the catchment, that is,

$$t_c = t_0 + t_f \tag{15.2.3}$$

where the flow time is

$$t_f = \sum \frac{L_j}{V_j} \tag{15.2.4}$$

where L_j is the length of the jth pipe along the flow path in ft (m) and V_j is the average flow velocity in the pipe in ft/s (m/s). The inlet time t_0 is the longest time of overland flow of water in a catchment to reach the storm sewer inlet draining the catchment.

In the rational method each sewer is designed individually and independently (except for the computation of sewer flow time) and the corresponding rainfall intensity i is computed repeatedly for the area drained by the sewer. For a given sewer, all the different areas drained by this sewer have the same i. Thus, as the design progresses towards the downstream sewers, the drainage area increases and usually the time of concentration increases accordingly. This increasing t_c in turn gives a decreasing i that should be applied to the entire area drained by the sewer.

Inlet times, or times of concentration for the case of no upstream sewers, can be computed using a number of methods, some of which are presented in Table 15.2.4. The longest time of concentration among the times for the various flow routes in the drainage area is the critical time of concentration used.

Table 15.2.4   Summary of Time of Concentration Formulas

Method and Date	Formula for t_c (min)	Remarks
Kirpich (1940)	$t_c = 0.0078L^{0.77}S - 0.385$ L = length of channel/ditch from headwater to outlet, ft S = average watershed slope, ft/ft	Developed from SCS data for seven rural basins in Tennessee with well-defined channel and steep slopes (3% to 10%); for overland flow on concrete or asphalt surfaces multiply t_c by 0.4; for concrete channels multiply by 0.2; no adjustments for overland flow on bare soil or flow in roadside ditches.
California Culverts Practice (1942)	$t_c = 60(11.9L^3/H)^{0.385}$ L = length of longest watercourse, mi H = elevation difference between divide and outlet, ft	Essentially the Kirpich formula; developed from small mountainous basins in California (U.S. Bureau of Reclamation, 1973, 1987).
Izzard (1946)	$t_c = \dfrac{41.025(0.0007i + c)L^{0.33}}{S^{0.333}i^{0.667}}$ i = rainfall intensity, in/h c = retardance coefficient L = length of flow path, ft S = slope of flow path, ft/ft	Developed in laboratory experiments by Bureau of Public Roads for overland flow on roadway and turf surfaces; values of the retardance coefficient range from 0.0070 for very smooth pavement to 0.012 for concrete pavement to 0.06 for dense turf; solution requires iteration; product i times L should be < 500.
Federal Aviation Administration (1970)	$t_c = 1.8(1.1 - C)L^{0.50}/S^{0.333}$ C = rational method runoff coefficient L = length of overland flow, ft S = surface slope, %	Developed from airfield drainage data assembled by the Corps of Engineers; method is intended for use on airfield drainage problems, but has been used frequently for overland flow in urban basins.
Kinematic wave formulas (Morgali and Linsley (1965); Aron and Erborge (1973))	$t_c = \dfrac{0.94L^{0.6}n^{0.6}}{(i^{0.4}S^{0.3})}$ L = length of overland flow, ft n = Manning roughness coefficient i = rainfall intensity in/h S = average overland slope ft/ft	Overland flow equation developed from kinematic wave analysis of surface runoff from developed surfaces; method requires iteration since both i (rainfall intensity) and t_c are unknown; superposition of intensity–duration–frequency curve gives direct graphical solution for t_c.
SCS lag equation (U.S. Soil Conservation Service (1975))	$t_c = \dfrac{100L^{0.8}[(1000/CN) - 9]^{0.7}}{1900S^{0.5}}$ L = hydraulic length of watershed (longest flow path), ft CN = SCS runoff curve number S = average watershed slope, %	Equation developed by SCS from agricultural watershed data; it has been adapted to small urban basins under 2000 acres; found generally good where area is completely paves; for mixed areas it tens to overestimate; adjustment factors are applied to correct for channel improvement and impervious area; the equation assumes that t_c = 1.67 × basin lag.

Table 15.2.4   Summary of Time of Concentration Formulas (*continued*)

Method and Date	Formula for t_c (min)	Remarks
SCS average velocity charts (U.S. Soil Conservation Service 1975, 1986)	$$t_c = \frac{1}{60}\sum \frac{L}{V}$$ L = length of flow path, ft V = average velocity in feet per second for various surfaces found using Figure 8.8.2	Overland flow charts in U.S. Soil Conservation Service (1986) show average velocity as function of watercourse slope and surface cover.

Source: Kibler (1982).

EXAMPLE 15.2.1

The computational procedure in the rational method is illustrated through an example design of sewers to drain a 20-ac area along Goodwin Avenue in Urbana, Illinois, as shown in Figure 15.2.1. The physical characteristics of the drainage basin are given in Table 15.2.5. The catchments are identified by the manholes they drain directly into. The sewer pipes are identified by the number of the upstream manhole of each pipe. The Manning's roughness factor n is 0.014 for all the sewers in the example (adapted from Yen, 1978).

Table 15.2.5   Characteristics of Catchments of Goodwin Avenue Drainage Basin

(1) Catchment	(2) Ground Elevation at Manhole (ft)	(3) Area A (ac)	(4) Runoff Coefficient C	(5) Inlet Time t_o (min)
1.1	731.08	2.20	0.65	11.0
1.2	725.48	1.20	0.80	9.2
2.1	724.27	3.90	0.70	13.7
2.2	723.10	0.45	0.80	5.2
3.1	722.48	0.70	0.70	8.7
3.2	723.45	0.60	0.85	5.9
3.3	721.89	1.70	0.65	11.8
4.1	720.86	2.00	0.75	9.5
4.2	719.85	0.65	0.85	6.2
5.1	721.19	1.25	0.70	10.3
5.2	719.10	0.70	0.65	11.8
5.3	722.00	1.70	0.55	17.6
6.1	718.14	0.60	0.75	7.3
7.1	715.39	2.30	0.70	14.5

Source: Yen (1978).

SOLUTION

Table 15.2.6 shows the computations for the design of 12 sewer pipes, namely, all the pipes upstream of sewer 6.1. The rainfall intensity-duration relationship is developed using National Weather Service report HYDRO-35 (see Chapter 7 or Frederick et al., 1977) and plotted in Figure 15.2.2 for the design return period of two years. The entries in Table 15.2.6 are explained as follows:

Columns 1, 2, and 3: The sewer number and its length and slope are predetermined quantities.

Column 4: Total area drained by a sewer is equal to the sum of the areas of the subcatchments drained by the sewer, e.g., for sewer 3.1, the area 8.45 acres is equal to the area drained by sewer 2.1 (7.30 ac in column 4) plus the area drained by sewer 2.2 (0.45 ac) plus the incremental area given in column 6 (0.70 ac for subcatchment 3.1).

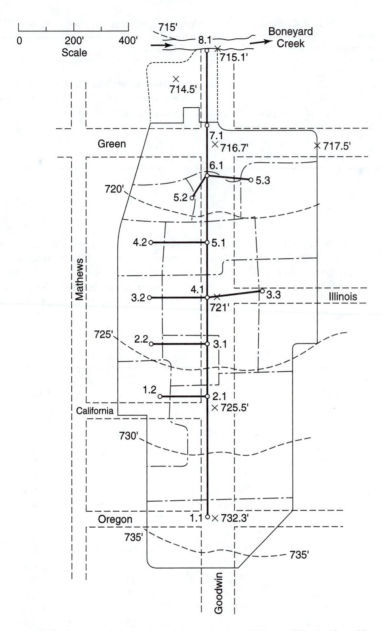

Figure 15.2.1 Goodwin Avenue drainage basin at Urbana, Illinois (from Yen (1978)).

Column 5: The identification number of the incremental subcatchments that drain directly through manhole or junction into the sewer being considered.

Column 6: Size of the incremental subcatchment identified in column 5 (Table 15.2.5).

Column 7: Value of runoff coefficient for each subcatchment (Table 15.2.5).

Column 8: Product of C and the corresponding subcatchment area.

Column 9: Summation of CA for all the areas drained by the sewer which is equal to the sum of contributing values in column 9 and the values in column 8 for that sewer, e. g., for sewer 3.1, 5.97 = 5.12 + 0.36 + (0.49).

Column 10: Values of inlet time (Table 15.2.5) for the subcatchment drained (computed using methods in Table 15.2.4), i.e., the overland flow inlet time if the upstream subcatchment is no more than one

Table 15.2.6 Design of Sewers by the Rational Method

(1) Sewer	(2) Length L	(3) Slope S (ft)	(4) Total Area Drained (ac)	(5) Catchment	(6) Increment Area (ac)	(7) C	(8) CA	(9) ΣCA	(10) Inlet Time (min)	(11) Upstream Sewer Flow Time (min)	(12) t_c (min)	(13) t_d (min)	(14) i (in/hr)	(15) Design Discharge Q_p (cfs)	(16) Computed Diameter D_r (ft)	(17) Pipe Size Used D_n (ft)	(18) Flow Velocity (fps)	(19) Sewer Flow Time min
1.1	390	0.0200	2.20	1.1	2.20	0.65	1.43	1.43	11.0	–	11.0	11.0	4.00	5.72	1.08	1.25	4.6	1.42
1.2	183	0.0041	1.20	1.2	1.20	0.80	0.96	0.96	9.2	–	9.2	9.2	4.30	4.13	1.28	1.50	2.3	1.31
2.1	177	0.0245	7.30	2.1	3.90	0.70	2.73	5.12	13.7	–	13.7	13.7	3.68	18.8	1.62	1.75	7.8	0.38
				1.1					11.0	1.4	12.4							
				1.2					9.2	1.3	10.5							
2.2	200	0.0180	0.45	2.2	0.45	0.80	0.36	0.36	5.2	–	5.2	5.2	5.30	1.91	0.73	0.83	3.5	0.95
3.1	156	0.0104	8.45	3.1	0.70	0.70	0.49	5.97	8.7	–	8.7	14.1	3.63	21.6	2.00	2.00	6.9	0.39
				2.1					13.7	0.4	14.1							
				2.2					5.2	1.0	6.2							
3.2	210	0.0175	0.60	3.2	0.60	0.85	0.51	0.51	5.9	–	5.9	5.9	5.07	2.59	0.82	0.83	4.7	0.74
3.3	130	0.0300	1.70	3.3	1.70	0.65	1.11	1.11	11.8	–	11.8	11.8	3.90	4.32	0.90	1.00	5.5	0.39
4.1	181	0.0041	12.75	4.1	2.00	0.75	1.50	9.09	9.5	–	9.5	14.5	3.60	32.7	2.79	3.00	4.6	0.65
				3.1					14.1	0.4	14.5							
				3.3					11.8	0.4	12.2							
4.2	200	0.0026	0.65	4.2	0.65	0.85	0.55	0.55	6.2	–	6.2	6.2	4.98	2.75	1.20	1.25	2.2	1.49
5.1	230	0.0028	14.65	5.1	1.25	0.70	0.88	10.52	10.3	–	10.3	15.2	3.50	36.8	3.13	3.50	3.8	1.00
				4.1					14.5	0.7	15.2							
5.2	70	0.0250	0.70	5.2	0.70	0.65	0.46	0.46	11.8	–	11.8	11.8	3.90	1.79	0.67	0.67	5.1	0.23
5.3	130	0.0060	1.70	5.3	1.70	0.55	0.94	0.94	17.6	–	17.6	17.6	3.30	3.10	1.07	1.25	2.5	0.86

Source: Yen (1978).

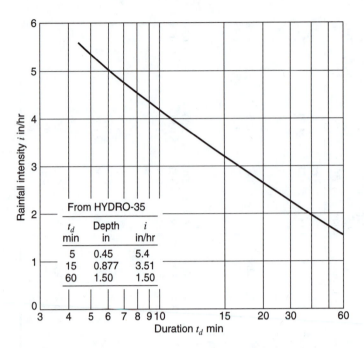

Figure 15.2.2 Variation of rainfall intensity with duration at Urbana, Illinois (from Yen (1978)).

sewer away from the sewer being designed (e.g., in designing sewer 3.1, 5.2 min for subcatchment 2.2 and 8.7 min for subcatchment 3.1); otherwise it is the total flow time to the entrance of the immediate upstream sewer (e.g., in designing sewer 3.1, 13.7 min for sewer 2.1).

Column 11: The sewer flow time of the immediate upstream sewer as given in column 19.

Column 12: The time of concentration t_c for each of the possible critical flow paths; t_c = inlet time (column 10) + sewer flow time (column 11) for each flow path.

Column 13: The rainfall duration t_d is assumed equal to the longest of the different times of concentration of different flow paths to arrive at the entrance of the sewer being considered; e.g., for sewer 3.1, t_d is equal to 14.1 min for sewer 2.1, which is longer than from sewer 2.2 (6.2 min) or directly from subcatchment 3.1 (8.7 min).

Column 14: The rainfall intensity i for the duration given in column 13 is based on HYDRO-35 for the two-year design return period (see Figure 15.2.2).

Column 15: Design discharge is computed by using equation (15.2.2), i.e., the product of columns 9 and 14.

Column 16: Required sewer diameter in feet, as computed using Manning's formula (equation 15.2.7); with $n = 0.014$, Q is given in column 15 and S_0 in column 3.

Column 17: The nearest commercial nominal pipe size that is not smaller than the computed size is adopted.

Column 18: Flow velocity computed by using $V = 4Q_p/(\pi D^2)$, i.e., column 13 multiplied by $4/\pi$ and divided by the square of column 17.

Column 19: Sewer flow time is computed as equal to L/V, i.e., column 2 divided by column 18 and converted into minutes.

This example demonstrates that in the rational method each sewer is designed individually and independently (except for the computation of sewer flow time) and the corresponding rainfall intensity i is computed repeatedly for the area drained by the sewer. For a given sewer, all the different areas drained by this sewer have the same i. Thus, as the design progresses towards downstream sewers, the drainage area increases and usually the time of concentration increases accordingly. This increasing t_c in turn gives a decreasing i, which should be applied to the entire area drained by the sewer. Failure to realize this variation of i is the most common mistake made in using the rational method for sewer design.

The size of a particular pipe is based upon computing the smallest available commercial pipe that can handle the peak flow rate determined using the rational formula (15.2.2). Manning's equation (equation (5.1.23) or (5.1.25)) has been popular in the United States for sizing pipes:

$$Q = \frac{m}{n} S_f^{1/2} A R^{2/3} \tag{15.2.5}$$

where m is 1.486 for U.S. customary units (1 for SI units), S_f is the friction slope, A is the inside cross-sectional area of the pipe $\pi D^2/4$ in ft^2 (m^2), R is the hydraulic radius, $R = A/P = D/4$ in ft (m), P is the wetted perimeter (πD) in ft (m), and K is the inside pipe diameter in ft (m). By substituting in the bed slope S_0 for the friction slope (assuming uniform flow) and $A = \pi D^2/4$ and $R = D/4$ (assuming that the pipe is flowing full under gravity, not pressurized), Manning's equation becomes

$$Q = \frac{m}{n} S_0 \left(\frac{\pi D^2}{4} \right) \left(\frac{D}{4} \right)^{2/3} = m \left(\frac{0.311}{n} \right) S_0^{1/2} D^{8/3} \tag{15.2.6}$$

Equation (15.2.6) can be solved for the diameter

$$D = \left(\frac{m_D Q n}{\sqrt{S_0}} \right)^{3/8} \tag{15.2.7}$$

where m_D is 2.16 for U.S. customary units (3.21 for SI units). Q is determined using the rational formula, and D is rounded up to the next commercial size pipe. The Darcy–Weisbach equation (4.3.13) can also be used to size pipes,

$$Q = A \left(\frac{8g}{f} R S_f \right)^{1/2} \tag{15.2.8a}$$

Equation (15.2.8a) can be solved for D using $S_f = S_0$ as

$$D = \left(\frac{0.811 f Q^2}{g S_0} \right)^{1/5} \tag{15.2.8b}$$

which is valid for any dimensionally consistent set of units.

15.2.3   Hydraulic Analysis of Designs

To analyze the hydraulic effectiveness of storm sewer design, it is necessary to analyze the hydraulic gradient. The hydraulic gradient can be used to determine if design flows can be accommodated without causing flooding at various locations or causing flows to exit the system at locations where this is not acceptable. Such analysis can be done manually or by computer. This section first discusses the form losses, then the hydraulic gradient calculations, and finally hydrograph routing.

15.2.3.1   Form Losses

During the propagation of flows through storm sewers, both open-channel flow and pressurized pipe flow can occur, depending upon the magnitude of the flows. Consequently the form loss equations for both types of flow are presented here.

Transition Losses (open-channel flow)

Contraction losses for open-channel flow are expressed as

$$H_c = 0.1\left(\frac{V_2^2}{2g} - \frac{V_1^2}{2g}\right) \text{ for } V_2 > V_1 \tag{15.2.9}$$

where V_1 is the upstream velocity and V_2 is the downstream velocity. Expansion losses are expressed as

$$H_e = 0.2\left(\frac{V_1^2}{2g} - \frac{V_2^2}{2g}\right) \text{ for } V_1 > V_2 \tag{15.2.10}$$

Simple size transitions through manholes with straight-through flow can be analyzed with the above two equations.

Transition losses (pressurized flow)

Contraction losses for pressurized flow are expressed as

$$H_c = K\left(\frac{V_2^2}{2g}\right)\left[1 - \frac{A_2}{A_1}\right]^2 \tag{15.2.11}$$

where $K = 0.5$ for a sudden contraction, $K = 0.1$ for a well-designed transition, A_1 is the cross-sectional area of flow at the beginning of the transition, and A_2 is the cross-sectional area of flow at the end of the transition. Expansion losses for pressurized flow are expressed as

$$H_e = K\left[\frac{(V_1 - V_2)^2}{2g}\right] \tag{15.2.12}$$

where $K = 1.0$ for a sudden expansion and $K = 0.2$ for a well-designed transition. These K values for the contractions and expansions are for approximation. For detailed analysis, Tables 15.2.7–15.2.10 can be used in conjunction with the following form of the headloss equation:

$$H = K\left(\frac{V^2}{2g}\right) \tag{15.2.13}$$

Exit losses can be computed with the following equation:

$$H_{\text{ext}} = K_e\left(\frac{V^2}{2g}\right) \tag{15.2.14}$$

Table 15.2.7 Values of K_2 for Determining Loss of Head Due to Sudden Enlargement in Pipes, from the Formula $H_2 = K_2(V_1^2/2g)$

$\dfrac{D_2}{D_1}$	Velocity V_1 in feet per second												
	2	3	4	5	6	7	8	10	12	15	20	30	40
1.2	.11	.10	.10	.10	.10	.10	.10	.09	.09	.09	.09	.09	.08
1.4	.26	.26	.25	.24	.24	.24	.24	.23	.23	.22	.22	.21	.20
1.6	.40	.39	.38	.37	.37	.36	.36	.35	.35	.34	.33	.32	.32
1.8	.51	.49	.48	.47	.47	.46	.46	.45	.44	.43	.42	.41	.40
2.0	.60	.58	.56	.55	.55	.54	.53	.52	.52	.51	.50	.48	.47
2.5	.74	.72	.70	.69	.68	.67	.66	.65	.64	.63	.62	.60	.58
3.0	.83	.80	.78	.77	.76	.75	.74	.73	.72	.70	.69	.67	.65
4.0	.92	.89	.87	.85	.84	.83	.82	.80	.79	.78	.76	.74	.72
5.0	.96	.93	.91	.89	.88	.87	.86	.84	.83	.82	.80	.77	.75
10.0	1.00	.99	.96	.95	.93	.92	.91	.89	.88	.86	.84	.82	.80
∞	1.00	1.00	.98	.96	.95	.94	.93	.91	.90	.88	.86	.83	.81

D_2/D_1 = ratio of larger pipe to smaller pipe; V_1 = velocity in smaller pipe
Source: American Iron and Steel Institute (1995).

Table 15.2.8 Values of K_2 for Determining Loss of Head Due to Gradual Enlargement in Pipes from the Formula $H_2 = K_2(V_1^2/2g)$

$\dfrac{D_2}{D_1}$	Angle of Cone													
	2°	4°	6°	8°	10°	15°	20°	25°	30°	35°	40°	45°	50°	60°
1.1	.01	.01	.01	.02	.03	.05	.10	.13	.16	.18	.19	.20	.21	.23
1.2	.02	.02	.02	.03	.04	.09	.16	.21	.25	.29	.31	.33	.35	.37
1.4	.02	.03	.03	.04	.06	.12	.23	.30	.36	.41	.44	.47	.50	.53
1.6	.03	.03	.04	.05	.07	.14	.26	.35	.42	.47	.51	.54	.57	.61
1.8	.03	.04	.04	.05	.07	.15	.28	.37	.44	.50	.54	.58	.61	.65
2.0	.03	.04	.04	.05	.07	.16	.29	.38	.46	.52	.56	.60	.63	.68
2.5	.03	.04	.04	.05	.08	.16	.30	.39	.48	.54	.58	.62	.65	.70
3.0	.03	.04	.04	.05	.08	.16	.31	.40	.48	.55	.59	.63	.66	.71
∞	.03	.04	.05	.06	.08	.16	.31	.40	.49	.56	.60	.64	.67	.72

D_2/D_1 = ratio of diameter of larger pipe to diameter of smaller pipe. Angle of cone is twice the angle between the axis of the cone and its side.

Source: American Iron and Steel Institute (1995).

Table 15.2.9 Values of K_3 for Determining Loss of Head Due to Sudden Contraction from the Formula $H_2 = K_2(V_1^2/2g)$

$\dfrac{D_2}{D_1}$	Velocity V_2 in feet per second												
	2	3	4	5	6	7	8	10	12	15	20	30	40
1.1	.03	.04	.04	.04	.04	.04	.04	.04	.04	.04	.05	.05	.06
1.2	.07	.07	.07	.07	.07	.07	.07	.08	.08	.08	.09	.10	.11
1.4	.17	.17	.17	.17	.17	.17	.17	.18	.18	.18	.18	.19	.20
1.6	.26	.26	.26	.26	.26	.26	.26	.26	.26	.25	.25	.25	.24
1.8	.34	.34	.34	.34	.34	.34	.33	.33	.32	.32	.31	.29	.27
2.0	.38	.38	.37	.37	.37	.37	.36	.36	.35	.34	.33	.31	.29
2.2	.40	.40	.40	.39	.39	.39	.39	.38	.37	.37	.35	.33	.30
2.5	.42	.42	.42	.41	.41	.41	.40	.40	.39	.38	.37	.34	.31
3.0	.44	.44	.44	.43	.43	.43	.42	.42	.41	.40	.39	.36	.33
4.0	.47	.46	.46	.46	.45	.45	.45	.44	.43	.42	.41	.37	.34
5.0	.48	.48	.47	.47	.47	.46	.46	.45	.45	.44	.42	.38	.35
10.0	.49	.48	.48	.48	.48	.47	.47	.46	.46	.45	.43	.40	.36
∞	.49	.49	.48	.48	.48	.47	.47	.47	.46	.45.	44	.41	.38

D_2/D_1 = ratio of larger pipe to smaller diameter; V_2 = velocity in smaller pipe.

Source: American Iron and Steel Institute (1995).

Table 15.2.10 Entrance Loss Coefficients for Corrugated Steel Pipe or Pipe-Arch

Inlet End of Culvert	Coefficient K_2
Projecting from fill (no headwall)	0.9
Headwall, or headwall and wingwalls square-edge	0.5
Mitered (beveled) to conform to fill slope	0.7
End-section conforming to fill slope	0.5
Headwall, rounded edge	0.2
Beveled ring	0.25

Source: American Iron and Steel Institute (1995).

Manhole losses

In many cases manhole losses can comprise a significant percentage of the overall losses in a storm sewer system. The losses that occur at storm sewer junctions are dependent upon the flow characteristics, junction geometry, and relative sewer diameters. For a *straight-through manhole* with no change in pipe sizes, the losses can be expressed as

$$H_m = 0.05 \frac{V^2}{2g} \tag{15.2.15}$$

Losses at *terminal manholes* can be estimated using

$$H_m = \frac{V^2}{2g} \tag{15.2.16}$$

For *junction manholes* with one or more incoming laterals, the total manhole loss (pressure change) can be estimated using the following equation form:

$$H_m = K \frac{V^2}{2g} \tag{15.2.17}$$

where Figure 15.2.3 shows manhole junction types and nomenclature. Values of K for various types of manhole configurations can be found in Figures 15.2.4–15.2.8.

Bend losses

Bend losses in storm sewers can be estimated using

$$H_b = K_b \frac{V^2}{2g} \tag{15.2.18}$$

where

$$K_b = 0.25 \sqrt{\frac{\Phi}{90}} \tag{15.2.19}$$

for curved sewer segments where the angle of deflection Φ is less than 40°. For greater angles of deflection and for bends in manholes, the loss coefficient can be obtained from Figure 15.2.9.

EXAMPLE 15.2.2

Approximate the sudden expansion loss for a 400-mm sewer pipe connecting to a 450-mm sewer pipe for a design discharge of 0.3 m³/s assuming full-pipe flow.

SOLUTION

First compute the velocity of flow in each sewer pipe:

$$V_1 = \frac{Q}{A_1} = \frac{0.3 \text{ m}^3/s}{\left[\pi \left(400/1000\right)^2\right]/4} = 2.39 \text{ m/s}$$

$$V_2 = \frac{Q}{A_2} = \frac{0.3 \text{ m}^3/s}{\left[\pi \left(450/1000\right)^2\right]/4} = 1.89 \text{ m/s}$$

The expansion loss is then determined using equation (15.2.12) with $K = 1.0$:

$$H_e = (1) \left[\frac{(2.39 - 1.89)^2}{2(9.81)} \right] = 0.0127 \text{ m}$$

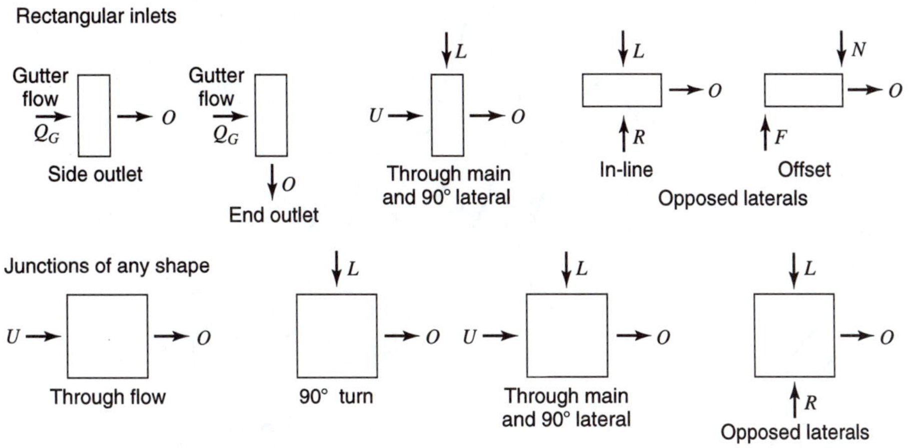

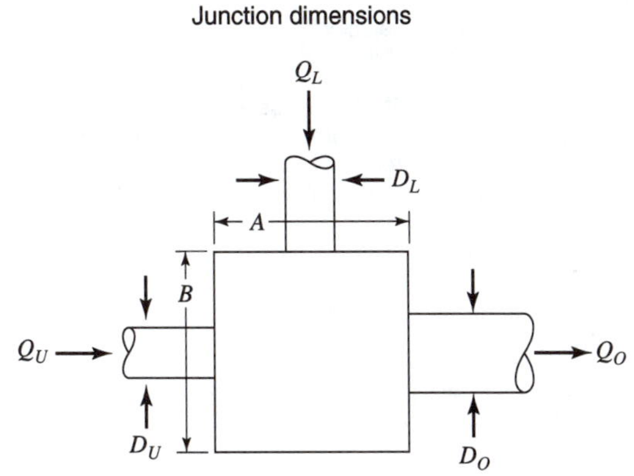

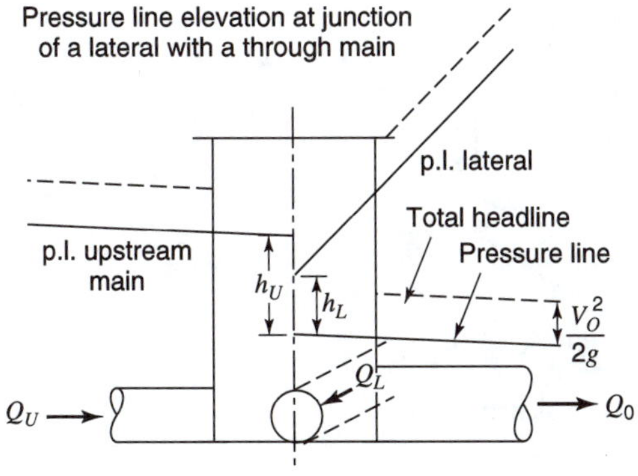

Rectangular inlets

Gutter flow Q_G Side outlet — O

Gutter flow Q_G End outlet — O

Through main and 90° lateral — $U \rightarrow \rightarrow O$, $\downarrow L$

In-line — $\downarrow L$, $\rightarrow O$, $\uparrow R$

Offset — $\downarrow N$, $\rightarrow O$, $\uparrow F$

Opposed laterals

Junctions of any shape

Through flow — $U \rightarrow \rightarrow O$

90° turn — $\downarrow L$, $\rightarrow O$

Through main and 90° lateral — $U \rightarrow \rightarrow O$, $\downarrow L$

Opposed laterals — $\downarrow L$, $\rightarrow O$, $\uparrow R$

Junction dimensions

Pressure line elevation at junction of a lateral with a through main

Nomenclature

Q	rate of flow
D	diameter of pipe
A	dimension of junction in direction of outfall pipe
B	dimension of junction at right angles to outfall pipe
d	depth of water in inlet
S	slope of pipe
S_f	friction slope
Q_G	flow into inlet through top grate
$D_O Q_O$	diameter and flow in outfall
$D_U Q_U$	diameter and flow in upstream main
$D_L Q_L$	diameter and flow in left lateral
$D_R Q_R$	diameter and flow in right lateral
$D_N Q_N$	diameter and flow in near lateral[1]
$D_F Q_F$	diameter and flow in far lateral[1]
$D_{hv} Q_{hv}$	diameter and flow in lateral w/higher-velocity flow[1]
$D_{iv} Q_{iv}$	diameter and flow in lateral w/lower-velocity flow[1]
	Pressure change coefficients for inlet water depth and an upstream pipe pressure relative to the outfall pipe pressure
K_S	water depth with all flow through grate
K_U	upstream main pressure
K_R or K_L	lateral pipe pressure
K_N	near lateral pipe pressure[3]
K_F	far lateral pipe pressure[3]
$\overline{K}_U, \overline{K}_L$	pressure coefficient at $Q_L = Q_O$
M_U, M_L	multiplier for $\overline{K}_U$ or $\overline{K}_L$ to obtain K_U or K_L

(1) Of oppoed laterals; designation with reference to position relative to outfall end of junction.
(2) For in-line opposed laterals only.
(3) Offset opposed laterals.

$$h_U = K_U \frac{V_O^2}{2g} \qquad h_L = K_L \frac{V_O^2}{2g}$$

Figure 15.2.3 Manhole junction types and nomenclature (from Sangster et al. (1958)).

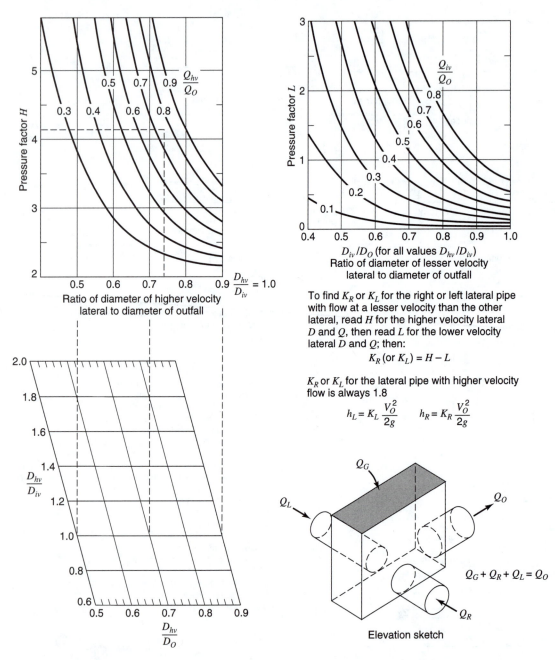

To find K_R or K_L for the right or left lateral pipe with flow at a lesser velocity than the other lateral, read H for the higher velocity lateral D and Q, then read L for the lower velocity lateral D and Q; then:

$$K_R \text{ (or } K_L) = H - L$$

K_R or K_L for the lateral pipe with higher velocity flow is always 1.8

$$h_L = K_L \frac{V_O^2}{2g} \qquad h_R = K_R \frac{V_O^2}{2g}$$

$$Q_G + Q_R + Q_L = Q_O$$

Elevation sketch

Figure 15.2.4 Rectangular manhole with in-line opposed lateral pipes each at 90° to outfall (with or without grate flow) (from Sangster et al. (1958)).

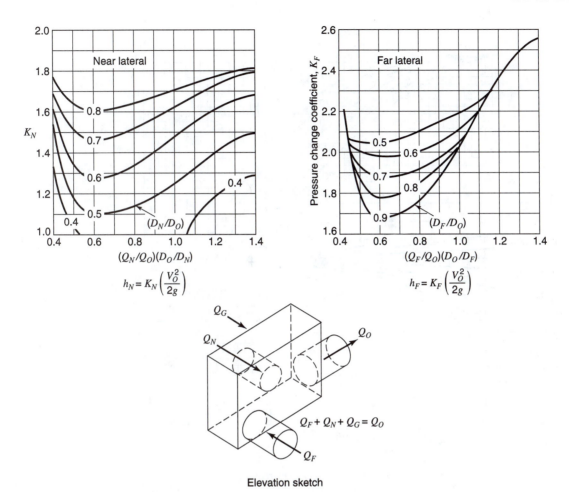

Figure 15.2.5 Rectangular manhole with offset opposed lateral pipes—each at 90° to outfall (with or without inlet flow) (from Sangster et al. (1958)).

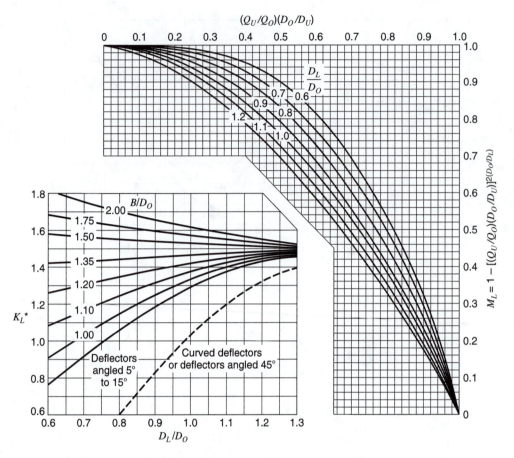

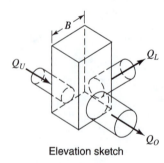

Elevation sketch

To find K_L for the for the lateral pipe, first read $\overline{K}_L^*$ from the lower graph. Next determine M_L.
Then

$$K_L = K_L^* \times M_L$$

Dashed curve for curved or 45° angled deflectors applies only to manholes without upstream in-line pipe.

Use this chart for round manholes also.

For rounded entrance to outfall pipe, reduce chart values of K_L^* by 0.2 for combining flow.

For $(Q_U/Q_O)^* \times (D_O/D_U) > 1$ use

$$h_L = K_L \frac{V_O^2}{2g} \text{ from Figure 15.2.8}$$

For $D_L = D_O < 0.6$ use

$$h_L = K_L \frac{V_O^2}{2g} \text{ from Figure 15.2.8}$$

Curved

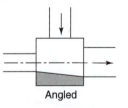

Angled

Plan of deflectors

Figure 15.2.6 Manhole at 90° deflection or on through pipeline at junction of 90° lateral pipe (lateral coefficient) (from Sangster et al. (1958)).

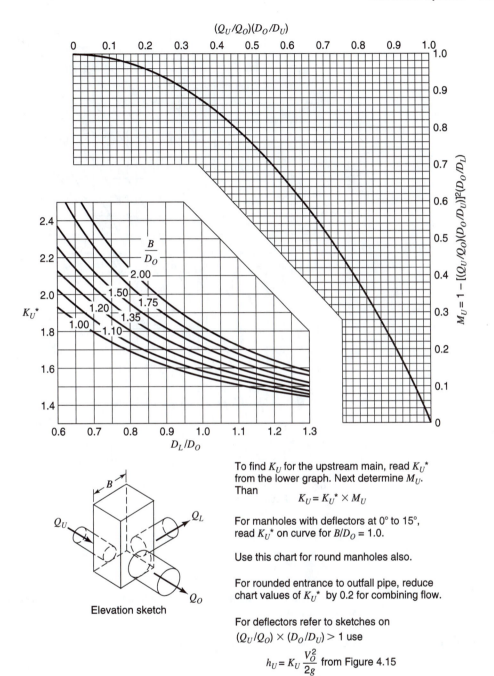

To find K_U for the upstream main, read K_U^* from the lower graph. Next determine M_U. Than

$$K_U = K_U^* \times M_U$$

For manholes with deflectors at 0° to 15°, read K_U^* on curve for $B/D_O = 1.0$.

Use this chart for round manholes also.

For rounded entrance to outfall pipe, reduce chart values of K_U^* by 0.2 for combining flow.

For deflectors refer to sketches on
$(Q_U/Q_O) \times (D_O/D_U) > 1$ use

$$h_U = K_U \frac{V_O^2}{2g} \text{ from Figure 4.15}$$

For $D_L/D_O < 0.6$ use

$$h_U = K_U \frac{V_O^2}{2g} \text{ from Figure 4.15}$$

Figure 15.2.7 Manhole on through pipeline at junction of a 90° lateral pipe (in-line pipe coefficient) (from Sangster et al. (1958)).

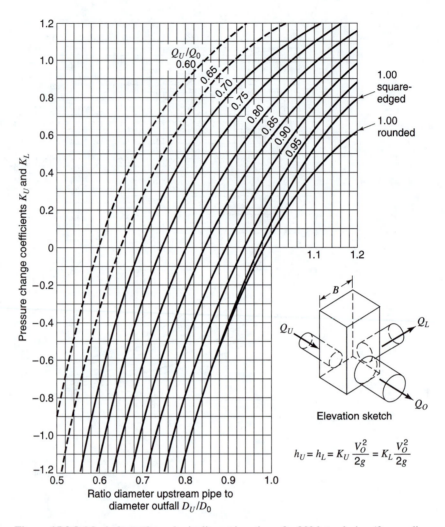

Figure 15.2.8 Manhole on through pipeline at junction of a 90° lateral pipe (for conditions outside the range of Figures 15.2.6 and 15.2.7 (from Sangster et al. (1958)).

EXAMPLE 15.2.3

Compute the bend loss for a 30° bend in a 400-mm sewer pipe with a discharge of 0.3 m³/s assuming full-pipe flow.

SOLUTION

First compute the flow velocity in the sewer pipe:

$$V = \frac{Q}{A} = \frac{0.3}{\left[\pi \left(400/1000\right)^2\right]/4} = 2.39 \text{ m/s}$$

Next compute the bend loss coefficient using equation (15.2.19):

$$K_b = 0.25\sqrt{\frac{\Phi}{90}} = 0.25\sqrt{\frac{30}{90}} = 0.144$$

Use equation (15.2.18) to compute the bend loss:

$$H_b = K_b \frac{V^2}{2g} = 0.144\frac{(2.39)^2}{2(9.81)} = 0.0419 \text{ m}$$

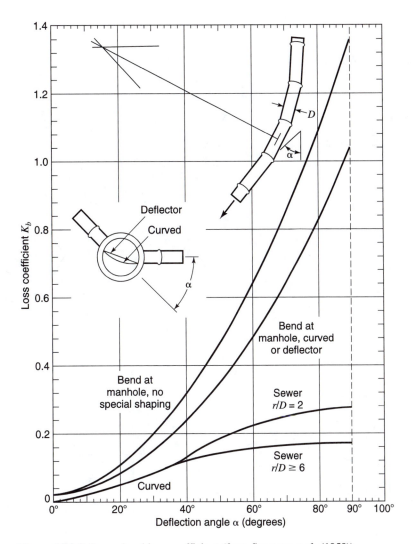

Figure 15.2.9 Sewer bend loss coefficient (from Sangster et al. (1958)).

EXAMPLE 15.2.4

The hypothetical storm sewer layout shown in Figure 15.2.10 includes an existing portion and an extension of the existing system. The objective for this example is to analyze the hydraulics of manhole number 4 (MH-4). Refer to Figure 15.2.11 for details of the manhole. We have

Top of manhole elevation	476.00 ft
Bottom of manhole elevation	470.15 ft
Manhole diameter	48.0 in
Lateral flow, Q_L	25.0 cfs
Upstream in-line flow, Q_u	46.0 cfs
Outfall flow, Q_0	71.0 cfs
Diameter of lateral line, D_L	30.0 in
Diameter of upstream in-line, D_u	42.0 in
Diameter of outfall line, D_0	48.0 in
Elevation of outfall pipe pressure line at MH = 6	475.08 ft

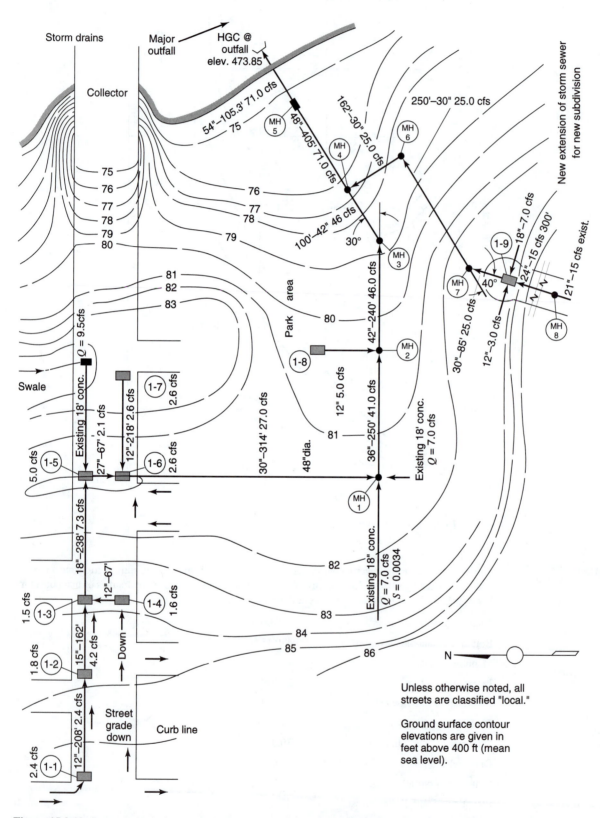

Figure 15.2.10 Storm drain design example (from Flood Control District of Maricopa County, 1992).

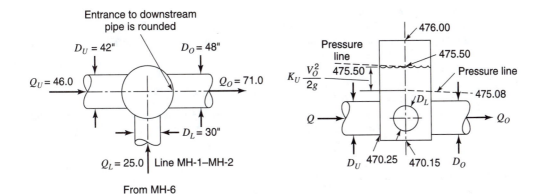

Figure 15.2.11 Storm drain design example for manhole no. 4 (from Flood Control District of Maricopa County, vol. II, 1992).

SOLUTION

1. The outfall pressure line elevation at manhole is given as 475.08 ft.

2. The velocity head at the outfall is

$$\frac{V_0^2}{2g} = \frac{1}{2g}\left(\frac{Q}{A}\right)^2 = \frac{1}{2g}\left(\frac{71}{\pi\,4^2/4}\right)^2 = 0.50 \text{ ft} \quad (\text{Note } D = 48 \text{ in} = 4 \text{ ft})$$

3. Compute the ratios

$$\frac{Q_u}{Q_o} = \frac{46}{71} = 0.65, \quad \frac{D_u}{D_0} = \frac{42}{48} = 0.88, \quad \frac{D_L}{D_0} = \frac{30}{48} = 0.63$$

4. Compute $B/D_0 = 48\,/48 = 1.0$ (where B is the manhole diameter).

5. $\left(\dfrac{Q_u}{Q_o}\right) \times \left(\dfrac{D_0}{D_u}\right) = 0.65 \times \dfrac{1}{0.88} = 0.74$

Consider the lateral pipe:

6. Using Figure 15.2.6, $K_L^* = 0.95$ for $D_L/D_0 = 0.63$ and $B/D_0 = 1.0$. For a round-edged manhole $K_L^* = K_L^* - 0.2 = 0.95 - 0.20 = 0.75$, where 0.2 is obtained from the table of reductions for K_L^* for manholes with a rounded entrance (see Table 15.2.11). When $V_0^2/2g < 1.0$ it is usually not economical to use a rounded entrance from the manhole to the outlet pipe; therefore, keep $K_L^* = 0.95$ for a square-edged entrance.

Table 15.2.11 Reductions for K_L^* for a Manhole with Rounded Entrance Reductions for K_L

B/D_0	D_L/D_0			
	0.6	0.8	1.0	1.2
1.75	0.4	0.3	0.2	0.0
1.33	0.3	0.2	0.1	0.0
1.10	0.2	0.1	0.0	0/.0

Source: Flood Control District of Maricopa County, vol. II, (1992).

7. Determine m_L using $(Q_u/Q_0) \times (D_u/D_0) = 0.74$; $m_L = 0.61$ from Figure 15.2.6.
8. $K_L = m_L \times K_L^* = 0.61 \times 0.95 = 0.58$ (for square-edged entrance).
9. Lateral pipe pressure change $= K_L(V_0^2/2g) = 0.58 \times 0.50 = 0.29$ ft.
10. Lateral pipe pressure $= 475.08 + 0.29 = 475.37$ ft.

Now consider the upstream in-line pipe:

11. From Figure 15.2.7, $K_u^* = 1.86$.
12. Because the velocity head is less than 1.0 ft/s, a rounded entrance to the outfall pipe will not be appropriate and a square-edged entrance will be used.
13. From Figure 15.2.7, $m_u = 0.45$.
14. $K_u = m_u \times K_u^* = 0.45 \times 1.86 = 0.84$.
15. $h_u = K_u \times (V_0^2/2g) = 0.84 \times 0.50 = 0.42$.
16. The in-line upstream pressure elevation is $475.08 + 0.42 = 475.50$, which is also the water surface elevation, as shown in Figure 15.2.11.

15.2.3.2   Hydraulic Gradient Calculations

Any storm sewer design must be analyzed to determine if the design flows can be accommodated without causing flows to exit the system and creating flooding conditions. Figure 15.2.12 illustrates the difference between an improper design and a proper design. Note the energy and hydraulic grade lines for the improper design as opposed to the proper design.

If the hydraulic grade line is above the pipe crown at the next upstream manhole, pressure flow calculations are indicated; if it is below the pipe crown, then open-channel flow calculations should be used at the upstream manhole. The process is repeated throughout the storm drain system. If all HGL elevations are acceptable, then the hydraulic design is adequate. If the HGL exceeds an inlet elevation, then adjustments to the trial design must be made to lower the water surface elevation. Computer programs such as HYDRA (FHWA, 1993) are recommended for the design of storm drains and include a hydraulic grade-line analysis and a pressure flow simulation.

15.2.3.3   Hydrograph Routing for Design

Hydrograph design methods consider design hydrographs as input to the upstream end of sewers and use some form of routing to propagate the inflow hydrograph to the downstream end of the sewer. The routed hydrograph is added to the surface runoff hydrograph to the manhole at the downstream junction, and the routed hydrograph for each sewer is added also. The combined hydrographs for all upstream connecting pipes plus the hydrograph for the surface runoff represents the design inflow hydrograph to the next (adjacent) downstream sewer pipe. The pipe size and sewer slope are selected by solving for the commercial size pipe that can handle the peak discharge of the inflow hydrograph and maintain a gravity flow.

A simple and rather effective hydrograph design method rather effective method is the *hydrograph time lag method* (Yen, 1978), which is a hydrologic (lumped) routing method. The inflow hydrograph of a sewer is shifted without distortion by the sewer flow time t_f to produce the sewer outflow hydrograph. The outflow hydrographs of the upstream sewers at a manhole are added, at the corresponding times, to the direct manhole inflow hydrograph to produce the inflow hydrograph for the downstream sewer in accordance with the continuity relationship

$$\sum Q_{ij} + Q_j - Q_o = \frac{ds}{dt} \qquad (15.2.20)$$

in which Q_{ij} is the inflow from the *i*th upstream sewer into the junction *j*, Q_0 is the outflow from the junction into the downstream sewer, Q_j is the direct inflow into the manhole or junction, and *S* is the water stored in the junction structure or manhole. For point type junctions where there is no storage, *dS/dt* is 0.

The sewer flow time t_f that is used to shift the hydrograph is estimated by

$$t_f = \frac{L}{V} \qquad (15.2.21)$$

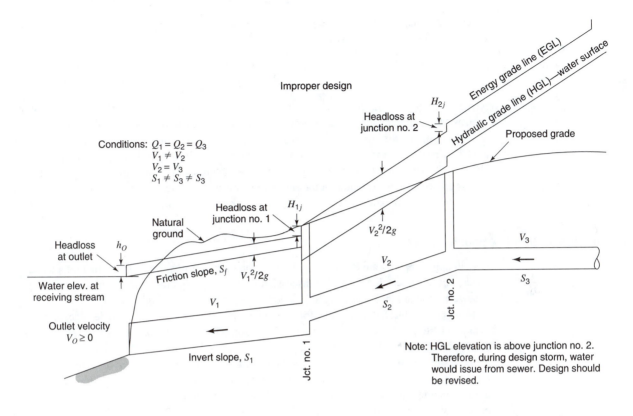

Improper design

Conditions: $Q_1 = Q_2 = Q_3$
$V_1 \neq V_2$
$V_2 = V_3$
$S_1 \neq S_3 \neq S_3$

Energy grade line (EGL)

Hydraulic grade line (HGL)—water surface

H_{2j}

Headloss at junction no. 2

Proposed grade

Headloss at junction no. 1

H_{1j}

Natural ground

$V_2^2/2g$

V_3

Headloss at outlet

h_O

Friction slope, S_f

$V_1^2/2g$

V_2

Water elev. at receiving stream

V_1

S_2

S_3

Jct. no. 2

Outlet velocity $V_O \geq 0$

Invert slope, S_1

Jct. no. 1

Note: HGL elevation is above junction no. 2. Therefore, during design storm, water would issue from sewer. Design should be revised.

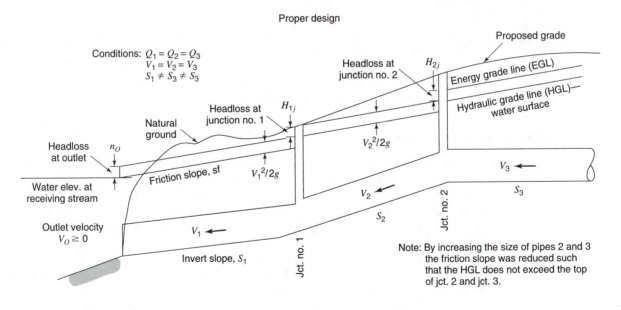

Proper design

Conditions: $Q_1 = Q_2 = Q_3$
$V_1 = V_2 = V_3$
$S_1 \neq S_3 \neq S_3$

Proposed grade

Headloss at junction no. 2

H_{2j}

Energy grade line (EGL)

Hydraulic grade line (HGL)—water surface

Headloss at junction no. 1

H_{1j}

Natural ground

$V_2^2/2g$

V_3

Headloss at outlet

n_O

Friction slope, sf

$V_1^2/2g$

V_2

Water elev. at receiving stream

S_2

S_3

Jct. no. 2

Outlet velocity $V_O \geq 0$

V_1

Invert slope, S_1

Jct. no. 1

Note: By increasing the size of pipes 2 and 3 the friction slope was reduced such that the HGL does not exceed the top of jct. 2 and jct. 3.

Figure 15.2.12 Use of energy losses in developing a storm drain system: energy and hydraulic grade lines for storm sewer under constant discharge (from AASHTO (1991)).

in which L is the length of the sewer and V is a sewer flow velocity. The velocity could be computed, assuming a full-pipe flow, using

$$V = \frac{4Q_p}{\pi D^2} \qquad (15.2.22)$$

where Q_p is the peak discharge and D is the pipe diameter. Also steady uniform flow equations such as Manning's equation or the Darcy-Weisbach equation can be used to compute the velocity.

In this method, the continuity relationship of the flow within the sewer is not directly considered. The routing of the sewer flow is done by shifting the inflow hydrograph by t_f and no consideration is given to the unsteady and nonuniform nature of the sewer flow. Shifting of hydrographs approximately accounts for the sewer flow translation time but offers no wave attenuation. However, the computational procedure through interpolation introduces numerical attenuation.

15.2.4 Storm Sewer Appurtenances

Certain appurtenances are essential to the functioning of storm sewers. These include manholes, bends, inlets, and catch basins. The discussion here briefly explains some of the common appurtenances; for additional details refer to American Society of Civil Engineers (1992).

Manholes are located at the junctions of sewer pipes and at changes of grade, size, and alignment, with street intersections being typical locations. Typical manholes for small sewers (sewer diameters less than 2 ft) are shown in Figure 15.2.13 and for intermediate sized sewers (sewer diameters greater than 2 ft) are shown in Figure 15.2.14. Manholes provide convenient access to the sewer for observation and maintenance operation. They should be designed to cause a minimum of interference with the hydraulics of the sewer. Manholes should be spaced 400 ft (120 m) or less for sewers 15 in (375 mm) or less and 500 ft (150 m) for sewers 18 to 30 in (460 to 760 mm) (Great Lakes–Upper Mississippi Board of State Sanitary Engineers, 1978). For larger sewers, spacing of up to a maximum of 600 ft (180 m) can be used.

Drop manholes (see Figure 15.2.15) are provided for sewers entering a manhole at an elevation of 24 in (0.6 m) or more above the manhole invert. These manholes are constructed with either an internal or external drop connection. For structural reasons, external connections are preferred.

Inlets are structures where stormwater enters the sewer system. Section 16.1 provides a detailed description of and design procedures for inlets.

15.2.5 Risk-Based Design of Storm Sewers

Proposed water-excess management solutions are subject—like are most solutions to engineering problems—to an element of uncertainty. The uncertainty inherent in storm sewer design derives from both hydraulic and hydrologic aspects of the problem. Recommended references on risk-based design include Ang and Tang (1975, 1984), Chow et al. (1988), Harr (1987), Kapur and Lamberson (1977), Kececioglu (1991), Mays and Tung (1992), and Yen (1986). Also see Sections 10.6 and 10.7.

The key question is the ability of the proposed sewer design to accommodate the surface runoff generated by a storm. Although a factor of safety SF is inherent in the choice of a design frequency, the relationship between the sewer capacity Q_c and the storm runoff Q_L can also be explicitly considered: that is, $Q_c = SF \times Q_L$. Using risk/reliability analysis, a probability of failure $P(Q_L > Q_c)$ can be calculated for selected frequencies and safety factors. The corresponding risks and safety factors for each return period (recurrence interval) can be plotted to derive the risk-safety factor relationship for each return period. The procedure is as follows:

1. Select the return period T.

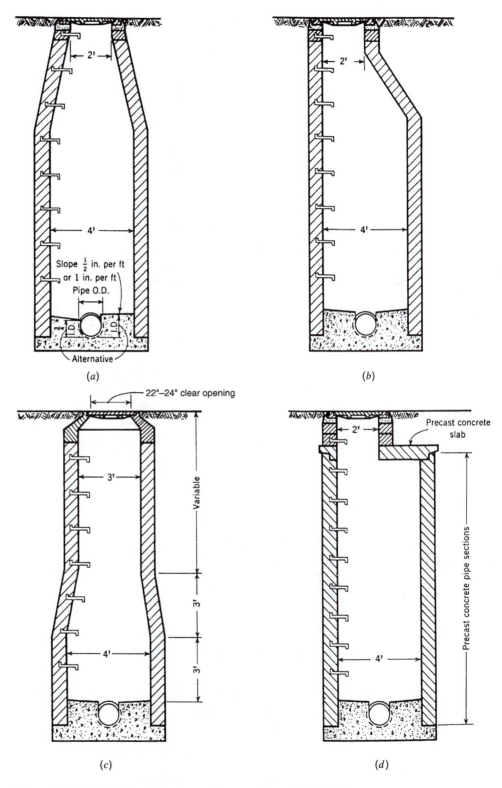

Figure 15.2.13 Typical manholes for small sewers (from ASCE (1992)).

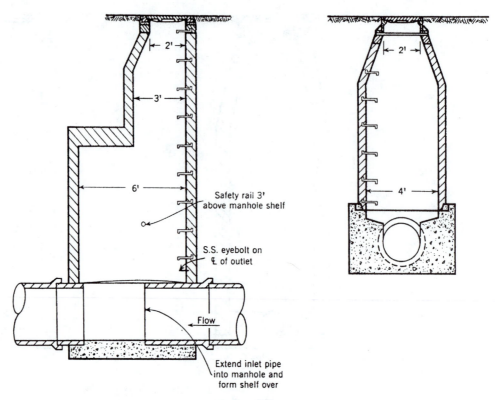

Figure 15.2.14 Two manholes for intermediate-sized sewers (from ASCE (1992)).

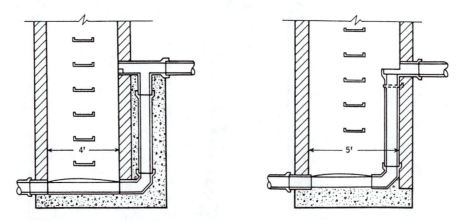

Figure 15.2.15 Drop manholes (from ASCE (1992)).

2. Use a rainfall-runoff model (such as the rational method) and perform an uncertainty analysis to compute the mean loading on the sewer $\bar{Q}_L$ (the mean surface runoff) and its coefficient of variation Ω_{QL}, where

$$\bar{Q}_L = \overline{C}\,\overline{i}\,\overline{A} \tag{15.2.23}$$

where $\overline{C}$ is the mean runoff coefficient, $\overline{A}$ is the mean basin area in acres, $\overline{i}$ is the mean rainfall intensity in in/hr, and (see derivation in Section 10.6)

$$\Omega_{QL}^2 = \Omega_C^2 + \Omega_i^2 + \Omega_A^2 \tag{15.2.24}$$

where Ω_c, Ω_i, and Ω_A are the coefficients of variation of C, i, A, respectively.

3. Select a pipe diameter D and compute $\bar{Q}_c$ and Ω_{Qc}, where $\bar{Q}_c$ is the mean capacity of the sewer and Ω_{Qc} is the coefficient of variation of Ω_c. The mean capacity can be obtained from a modified form of the Manning equation:

$$\bar{Q}_c = \frac{0.463}{\bar{n}} D^{8/3} \bar{S}_0^{1/2}$$ (15.2.25)

in which $\bar{n}$ is the mean Manning roughness coefficient, $\bar{S}_0$ is the mean sewer slope, and D is the sewer diameter. The coefficient of variation Ω_{Qc} is computed using (see derivation in Section 10.6)

$$\Omega_{Qc}^2 = \Omega_n^2 + \frac{1}{4}\Omega_{S_0}^2 + \left(\frac{8}{3}\right)^2 \Omega_D^2$$ (15.2.26)

in which Ω_n, Ω_D, and Ω_{S_0} are the coefficients of variation of n, d, and S_0, respectively.

4. Compute the risk and safety factor (see Section 10.7).
5. Repeat Step 3 for other diameters.
6. Repeat Step 2 for other rainfall duration.
7. Return to Step 1 for each return period to be considered.

EXAMPLE 15.2.5

Determine the coefficient of variation of the loading and the capacity for the following parameters. Assume a uniform distribution to define the uncertainty of each parameter.

Parameter	Mode	Range
C	0.75	0.70–0.80
i	7.5 in/hr	7.2–7.8 in/hr
A	12 acres	11.9–12.1 acres
n	0.015	0.0145–0.0155
D	5 ft	4.96–5.04 ft
S_0	0.001 ft/ft	0.0009–0.0011 ft/ft

SOLUTION

The loading Q_L is estimated using the rational equation (15.2.23), the capacity Q_c is estimated using Manning's equation (15.2.25), and the coefficients of variation of Q_L and Q_C are determined using equations (15.2.24) and (15.2.26), respectively. All the random parameters are assumed to follow a uniform distribution, so the mean and variance of each parameter are calculated using mean $= (a + b)/2$ and variance $= (b - a)^2/12$, in which a and b are the lower and upper bounds, respectively. Hence,

$$\Omega = \frac{\sqrt{\dfrac{(b-a)^2}{12}}}{(b+a)/2} = \left(\frac{b-a}{b+a}\right)\frac{1}{\sqrt{3}}$$

Based on the above formula, the means, variance, and coefficients of each parameter are calculated in the following table:

Parameter	Range	Mean	Variance	Ω
C	0.70–0.80	0.75	8.33×10^{-4}	3.85×10^{-2}
i (in/hr)	7.2–7.8	7.5	3.00×10^{-2}	2.31×10^{-2}
A (ac)	11.9–12.1	12.0	3.33×10^{-3}	4.81×10^{-3}
n	0.0145–0.0155	0.0150	8.33×10^{-8}	1.92×10^{-2}
D (ft)	4.96–5.04	5.00	5.33×10^{-4}	4.62×10^{-3}
S_0	0.0009–0.0011	0.0010	3.33×10^{-9}	5.77×10^{-2}

Now, the coefficients of Q_L and Q_C can be calculated as

$$\Omega_{Q_L} = \sqrt{(3.85 \times 10^{-2})^2 + (2.31 \times 10^{-2})^2 + (4.81 \times 10^{-3})^2} = 4.52 \times 10^{-2}$$

$$\Omega_{Q_C} = \sqrt{(1.92 \times 10^{-2})^2 + \left(\frac{8}{3} \times 4.62 \times 10^{-3}\right)^{-2} + (\frac{1}{2} \times 5.77 \times 10^{-2})^{-2}} = 3.68 \times 10^{-2}$$

EXAMPLE 15.2.6

Using the results of Example 15.2.5, determine the risk of the loading exceeding the capacity of the sewer pipe. Assume the use of a safety margin ($SM = Q_C - Q_L$) that is normally distributed.

SOLUTION

The risk is defined (see Section 10.7) as Risk = $P[Q_L > Q_C] = P[SM < 0]$, in which $SM = Q_C - Q_L$. The mean and variance of SM are, respectively, $\mu_{SM} = \mu_{Qc} - \mu_{QL}$ and $\sigma_{SM}^2 = \sigma_{Q_C}^2$. From example 15.2.5 we have

$$\mu_{Q_L} = \overline{CiA} = (0.75)(7.5)(12) = 67.50 \text{ cfs}$$

$$\sigma_{Q_L} = \mu_{Q_L}\Omega_{Q_L} = (67.50)(4.52 \times 10^{-2}) = 3.05 \text{ cfs}$$

$$\mu_{Q_C} = \frac{0.463}{\overline{n}}\left(\overline{D}\right)^{8/3}\left(\overline{S}_0\right)^{1/2} = \frac{0.464}{0.015}(5)^{8/3}(0.001)^{1/2} = 71.51 \text{ cfs}$$

$$\sigma_{Q_C} = \mu_{Q_C}\Omega_{Q_C} = (71.51)(3.68 \times 10^{-2}) = 2.63 \text{ cfs}$$

Therefore, $\mu_{SM} = 71.51 - 67.50 = 4.01$ and $\sigma_{SM}^2 = 3.05^2 + 2.63^2 = 16.22$. The risk then is calculated as

$$\text{Risk} = P[SM < 0] = P\left(Z < \frac{0 - \mu_{SM}}{\sigma_{SM}}\right)$$

$$= P\left(Z < \frac{-4.01}{\sqrt{16.22}}\right) = P(Z < -1.00) = 0.159$$

15.3   STORMWATER DRAINAGE CHANNELS

Stormwater-drainage channels (or flood-control channels) must behave in a stable, predictable manner to ensure that a known flow capacity will be available for a design storm event. In most cases, the design goal is a noneroding channel boundary, although, in certain cases, a dynamic channel is desired (Cotton, 1999).

Because most soils erode under a concentrated flow, either temporary or permanent channel linings are needed to achieve channel stability.

Channel linings can be classified in two broad categories: rigid or flexible. *Rigid linings* include channel pavements of concrete or asphaltic concrete and a variety of precast interlocking blocks and articulated mats. *Flexible linings* include such materials as loose stone (riprap), vegetation, manufactured mats of lightweight materials fabrics, or combinations of these materials. Rigid linings are capable of high conveyance and high-velocity flow. Rigid-lined channels are used to reduce the amount of land required for a surface drainage system. The selection of lining is a function of the design context related to the consequences of flooding, the availability of land and environmental needs (Cotton, 1999). Figures 15.3.1 and 15.3.2 show two constructed channels in Arizona, each with and without flow.

15.3.1   Rigid-Lined Channels

Rigid-lined channels are nonerodible channel sides typically lined with concrete grouted riprap, stone masonry, or asphalt. The steps in the design of such a channel are as follows:

Figure 15.3.1 Views of constructed reach of Cave Creek below Cave Buttes Dam, Phoenix, Arizona. The constructed reach is straight, and cross-sections are trapezoidal in shape. The bottom and sides of the channel are composed of rounded cobbles imbedded in a matrix of cement (approximate mean diameter of the rock projections was 80 mm, about half of which seemed to be exposed to flow). Roughness elements are constant throughout the reach. The channel gradient increases from about 0.002 ft/ft at cross-section 1 to about 0.010 ft/ft at cross-section 8. The stream is ephemeral, and flow is regulated by Cave Buttes Dam (Phillips and Ingersoll (1998)).

Figure 15.3.2 Views of Indian Bend Wash in Scottsdale, Arizona. The constructed channel is uniform and cross-sections are trapezoidal in shape (channel bottom is firm earth with seasonal growth of grasses and small brush) (Phillips and Ingersoll (1998)).

1. Select the Manning roughness coefficient, n (see Table 15.3.1), the channel side slope z, and the channel bottom slope S_0. The bottom slope is based upon the topography and other considerations such as alignment.

Table 15.3.1 Manning's Roughness Coefficients

| | | Manning's n | | |
| | | Depth Ranges | | |
Lining Category	Lining Type	0–0.5 ft (0–15 cm)	0.5–2.0 ft (15–60 cm)	>2.0 ft (>60 cm)
Rigid	Concrete	0.015	0.013	0.013
	Grouted riprap	0.040	0.030	0.028
	Stone masonry	0.042	0.032	0.030
	Soil cement	0.025	0.022	0.020
	Asphalt	0.018	0.016	0.016
Unlined	Bare soil	0.023	0.020	0.020
	Rock cut	0.045	0.035	0.025
Temporary	Woven paper net	0.016	0.015	0.015
	Jute net	0.028	0.022	0.019
	Fiberglass roving	0.028	0.021	0.019
	Straw with net	0.065	0.033	0.025
	Curled wood mat	0.066	0.035	0.028
	Synthetic mat	0.036	0.025	0.021
Gravel riprap	1 in (2.5 cm) D_{50}	0.044	0.033	0.030
	2 in (5 cm) D_{50}	0.066	0.041	0.034
Rock riprap	6 in (15 cm) D_{50}	0.104	0.069	0.035
	12 in (30 cm) D_{50}	—	0.078	0.040

Source: Chen and Cotton (1988).

2. Compute the uniform flow section factor (see Chapter 5)

$$AR^{2/3} = \frac{Qn}{K_a S^{1/2}} \qquad (15.3.1)$$

in which A is the cross-sectional area of flow, ft^2 (m^2), R is the hydraulic radius in ft (m), Q is the design discharge in ft^3/s (m^3/s), and $K_a = 1.49$ for U.S. customary units ($K_a = 1.0$ for SI units).

3. Determine the channel dimensions and flow depth for the uniform flow section factor computed in step 2. Choose the expression for the uniform flow section factor $AR^{2/3}$, as a function of depth, in Table 15.3.2 and solve for the depth using the value of $AR^{2/3}$ from equation (15.3.1). For a trapezoidal channel,

$$\left[\frac{\left(B_w + zy\right)^5 y^5}{\left(B_w + 2y\sqrt{1+z^2}\right)^2} \right]^{1/3} = AR^{2/3} \qquad (15.3.2)$$

Table 15.3.2   Geometric Elements of Channel Cross-Sections

Cross-Section	Area A	Wetted Perimeter P	Hydraulic Radius R	Top Width B
Rectangle	by	$b+2y$	$\dfrac{by}{b+2y}$	b
Trapezoid	$(b+zy)y$	$b+2y\sqrt{1+z^2}$	$\dfrac{(b+zy)y}{b+2y\sqrt{1+z^2}}$	$b+2zy$
Triangle	zy^2	$2y\sqrt{1+z^2}$	$\dfrac{zy}{2\sqrt{1+z^2}}$	$2zy$
Circle	$\dfrac{1}{8}(\theta-\sin\theta)D_0^2$	$\dfrac{1}{2}\theta D_0$	$\dfrac{1}{4}\left(1-\dfrac{\sin\theta}{\theta}\right)D_0$	$(\sin\tfrac{1}{2}\theta)D_0$ or $2\sqrt{y(D_0-y)}$
Parabola $y=cx$	$\dfrac{2}{3}By$	$B+\dfrac{8}{3}\dfrac{y^2}{B}$ [†]	$\dfrac{2B^2y}{3B^2+8y^2}$ [†]	$\dfrac{3}{2}\dfrac{A}{y}$
Power $y=cx^{1/m}$	$\dfrac{By}{(m+1)}$	$\dfrac{2y}{m}\sqrt{\dfrac{m^2}{c^{2m}}y^{2m-2}+1}+\sum$ [§]	$\dfrac{A}{P}$	$2\left(\dfrac{y}{c}\right)^m$
Round-cornered rectangle $(y>r)$	$\left(\dfrac{\pi}{2}-2\right)r^2+(b+2r)y$	$(\pi-2)r+b+2y$	$\dfrac{(\pi/2-2)r^2+(b+2r)y}{(\pi-2)r+b+2y}$	$b+2r$
Round-bottomed triangle	$\dfrac{B^2}{4z}-\dfrac{r^2}{z}(1-z\cot^{-1}z)$	$\dfrac{B}{z}\sqrt{1+z^2}-\dfrac{2r}{z}(1-z\cot^{-1}z)$	$\dfrac{A}{P}$	$2\left[z(y-r)+r\sqrt{1+z^2}\right]$

[*]For the section factor Z_c, the energy correction factor α or momentum correction factor β are assumed equal to unity. Otherwise, $Z_c = Q/\sqrt{g/\alpha}$ or $Z_c = Q/\sqrt{g/\beta}$.

[†]Satisfactory approximation for the interval $0 < x < 1$, where $x = 4y/b$. When $x > 1$, use the exact expression

$$P = (B/2)\left[\sqrt{1+x^2}+1/x\ \ln\left(x+\sqrt{1+x^2}\right)\right].$$

[‡]For trapezoid, approximate Y_c valid for $0.1 < Q/b^{2.5} < 0.4$; when $Q/b^{2.5} < 0.1$, use rectangular formula.

$$[§]\sum = \left(1-\dfrac{1}{m}\right)y\sum_{k=0}^{\infty}\dfrac{\left(\dfrac{1}{2}\right)_k\left(\dfrac{1}{2m-2}\right)_k\left[-\dfrac{m}{c^m}y^{m-1}\right]^{2k}}{\left(1+\dfrac{1}{2m-2}\right)_k\,k!},\ \text{where}\ (w)_k = w(w+1)\ldots(w+k-1),\ k = 1, 2,\ldots,(w)_{k=0} = 1.$$

Source: Yen (1996).

Hydraulic Mean Depth D	Uniform Flow Section Factor $AR^{2/3} = \dfrac{Qn}{K_n S^{05}}$	Critical Flow Section Factor* $Z_c = Q/\sqrt{g} = A_c \sqrt{D_c}$	Critical Depth y_c
y	$\left[\dfrac{b^5 y^5}{(b+2y)^2}\right]^{1/3}$	$by^{1.5}$	$\left(\dfrac{Z_c}{b}\right)^{2/3}$
$\dfrac{(b+zy)y}{b+2zy}$	$\left[\dfrac{(b+zy)^5 y^5}{(b+2y\sqrt{1+z^2})^2}\right]^{1/3}$	$\dfrac{[(b+zy)y]^{1.5}}{\sqrt{b+2zy}}$	$0.81\left(\dfrac{Z_c^4}{z^{1.5}b^{2.5}}\right)^{0.135} - \dfrac{b}{30z}\,‡$
$\dfrac{y}{2}$	$\left[\dfrac{z^5 y^8}{4(1+z^2)}\right]^{1/3}$	$\dfrac{\sqrt{2}}{2}zy^{2.5}$	$\left(\dfrac{\sqrt{2}Z_c}{z}\right)^{0.4}$
$\dfrac{1}{8}\left(\dfrac{\theta - \sin\theta}{\sin\frac{1}{2}\theta}\right)D_0$	$\dfrac{1}{16}\left[\dfrac{(\theta-\sin\theta)^5 D_0^3}{2\theta^2}\right]^{1/3}$	$\dfrac{\sqrt{2}}{32}\left(\dfrac{(\theta-\sin\theta)^{1.5}}{(\sin\frac{1}{2}\theta)^{0.5}}\right)D_0^{2.5}$	$\dfrac{1.01}{D_0^{0.26}}Z_c^{0.5\,‡}$ for $0.02 \le y_c/D_0 \le 0.85$
$\dfrac{2}{3}y$	$\dfrac{2}{3}\left[\dfrac{4B^7 y^5}{(3B^2+8y^2)^2}\right]^{1/3\,†}$	$\dfrac{2\sqrt{6}}{9}By^{1.5}$	$0.958(c^{0.5}\,Z_c)^{0.5}$
$\dfrac{y}{(m+1)}$	$\left[\dfrac{A^5}{P^2}\right]^{1/3}$	$\dfrac{By^{1.5}}{(m+1)^{1.5}}$	$\left[\dfrac{c^m}{2}(m+1)^{1.5}Z_c\right]^{1/(m+1.5)}$
$\dfrac{(\pi/2-2)r^2}{b+2r}+y$	$\left[\dfrac{A^5}{P^2}\right]^{1/3}$	$\dfrac{[(\pi/2-2)r^2+(b+2r)y]^{1.5}}{\sqrt{b+2r}}$	
$\dfrac{A}{B}$	$\left[\dfrac{A^5}{P^2}\right]^{1/3}$	$A\sqrt{\dfrac{A}{B}}$	

where B_w is the bottom width. By assuming several values of B_w and z, a number of combinations of section dimensions can be obtained. Final dimensions should be based upon hydraulic efficiency and practicability. If the best hydraulically efficient section is required, select the expression for $AR^{2/3}$ from Table 15.3.3 and solve for the depth using the value of $AR^{2/3}$ from equation (15.3.1). For a trapezoidal channel,

$$\sqrt{3}\left(\dfrac{y^8}{4}\right)^{1/3} = AR^{2/3} \tag{15.3.3}$$

4. Check the minimum velocity to see if the water carries the silt.
5. Add an appropriate freeboard to the depth of the channel section.

Table 15.3.3   Best Hydraulically Efficient Sections Without Freeboard

Cross-Section	Area A	Wetted Perimeter P	Hydraulic Radius R	Top Width B	Hydraulic Depth D	$AR^{2/3}$
Trapezoid, half of a hexagon	$\sqrt{3}y^2$	$2\sqrt{3}y$	$\dfrac{1}{2}y$	$\dfrac{4}{3}\sqrt{3}y$	$\dfrac{3}{4}y$	$\sqrt{3}\left(\dfrac{y^8}{4}\right)^{1/3}$
Rectangle, half of a square	$2y^2$	$4y$	$\dfrac{1}{2}y$	$2y$	y	$(2y^8)^{1/3}$
Triangle, half of a square	y^2	$2\sqrt{2}y$	$\dfrac{1}{4}\sqrt{2}y$	$2y$	$\dfrac{1}{2}y$	$\dfrac{1}{2}y^{8/3}$
Semicircle	$\dfrac{\pi}{2}y^2$	πy	$\dfrac{1}{2}y$	$2y$	$\dfrac{\pi}{4}y$	$\dfrac{\pi}{2}(2y^8)^{1/3}$
Parabola, $B = 2\sqrt{2}y$	$\dfrac{4}{3}\sqrt{2}y^2$	$\dfrac{8}{3}\sqrt{2}y$	$\dfrac{1}{2}y$	$2\sqrt{2}y$	$\dfrac{2}{3}y$	$\dfrac{2\sqrt{2}}{3}(2y^8)^{1/3}$
Hydrostatic catenary	$1.39586y^2$	$2.9836y$	$0.46784y$	$1.917532y$	$0.72795y$	$0.84122y^{8/3}$

Source: Yen (1996).

EXAMPLE 15.3.1   Design a nonerodible trapezoidal channel to carry a discharge of 11.33 m³/s.

SOLUTION

1. The Manning's $n = 0.025$ slope, $S_0 = 0.0016$.
2. The uniform flow section factor is computed using equation (15.3.1):

$$AR^{2/3} = \frac{Qn}{K_a S^{1/2}} = \frac{11.33 \times 0.025}{(0.0016)^{1/2}} = 7.08$$

3. Equation (15.3.3) is used for the best hydraulic section:

$$\sqrt{3}\left(\frac{y^8}{4}\right)^{1/3} = AR^{2/3} = 7.08, \text{ solving } y = 2.02 \text{ m.}$$

Because the best hydraulic section is half of a hexagon, the side slopes are $z = \tan 30° = 0.577$, z horizontal to 1 vertical. The area for a best hydraulic section is $A = \sqrt{3}y^2 = \sqrt{3}(2.02) = 7.07\text{m}^2 = (B_w + zy)y$, so $[B_w + (0.577)(2.02)]2.02 = 7.07$. Solving, the bottom width is 2.33 m.
4. The velocity is $Q/A = 11.33/7.07 = 1.60$ m/s, which is greater than the minimum permissible velocity for inducing silt deposition (if any exists).
5. A required freeboard, e.g., 1 m, can be added, so the total channel depth would be 3.02 m.

15.3.2   Flexible-Lined Channels

Flexible-lined channels include rock riprap and vegetable linings and are considered flexible because they can conform to change in channel slope. Flexible linings have the following advantages for stormwater conveyance: they (1) permit infiltration and exfiltration; (2) filter out contaminants; (3) provide greater energy dissipation; (4) allow flow conditions that provide better habitat opportunities for local flora and fauna, and (5) are less expensive. For a given design flow, channel geometry, and slope, the following design procedure can be used to select the appropriate flexible lining (Kouwen et al., 1969; Bathurst et al., 1981; Cotton, 1999; Wang and Shen, 1985; Chen and Cotton, 1988):

1. Choose a flexible lining from Table 15.3.4 and note its permissible shear stress τ_p.

Table 15.3.4 Permissible Shear Stresses for Lining Materials

Lining Category	Lining Type	Permissible Shear Stress, τ_p	
		lb/ft²	kg/m²
Temporary	Woven paper net	0.15	0.73
	Jute net	0.45	2.20
	Fiberglass roving		
	Single	0.60	2.93
	Double	0.85	4.15
	Straw with net	1.45	7.08
	Curled wood mat	1.55	7.57
	Synthetic mat	2.00	9.76
Vegetative	Class A	3.70	18.06
	Class B	2.10	10.25
	Class C	1.00	4.88
	Class D	0.60	2.93
	Class E	0.35	1.71
Gravel riprap	1-in	0.33	1.61
	2-in	0.67	3.22
Rock riprap	6-in	2.00	9.76
	12-in	4.00	19.52

Source: Chen and Cotton (1988).

2. Assume an appropriate flow depth y (for vegetative lining only; for nonvegetative lining, assume the range of flow depth and go to step 3).
3. Use Table 15.3.1 for nonvegetative lining to find the Manning n. For vegetative lining use Table 15.3.5 to determine the appropriate retardant class. The Manning n for vegetative lining is given by the general equation

$$n = k_1/(a_c + k_2) \tag{15.3.4}$$

in which $k_1 = R^{1/6}$, where R is the hydraulic radius, ft; $k_2 = 19.97 \log(R^{1.4}S_0^{0.4})$ where S_0 is the channel longitudinal slope, ft/ft; $a_c = 15.8, 23.0, 30.2, 34.6,$ and 37.7 for retardance classes A, B, C, D, E respectively.

Table 15.3.5 Classification of Vegetal Covers by Degree of Retardance

Retardance Class	Cover	Condition
A	Weeping lovegrass	Excellent stand, tall (average 30 in [76 cm])
	Yellow bluestem *Ischaemum*	Excellent stand, tall (average 36 in [91 cm])
B	Kudzu	Very dense growth, uncut
	Bermuda grass	Good stand, tall [average 12 in (30 cm)]
	Native grass mixture	Good stand, unmowed
	(little bluestem, bluestem, blue gamma,	
	and other long and short midwest grasses)	
	Weeping lovegrasses	Good stand, (average 24 in [61 cm])
	Lespedeza sericea	Good stand, not woody, tall (average 19 in [48 cm])

Table 15.3.5   Classification of Vegetal Covers by Degree of Retardance (*continued*)

	Alfalfa	Good stand, uncut (average 11 in [28 cm])
	Weeping lovegrass	Good stand, unmowed (average 13 in [28 cm])
	Kudzu	Dense growth, uncut
	Blue gamma	Good stand, uncut (average 13 in [28 cm])
C	Crabgrass	Fair stand, uncut (10 to 48 in [25 to 120 cm])
	Bermuda grass	Good stand, mowed (average 6 in [15 cm])
	Common lespedeza	Good stand, uncut (average 11 in [28 cm])
	Grass-legume mixture—summer (orchard grass, redtop, Italian ryegrass, and common lespedeza)	Good stand, uncut (6 to 8 in [15 to 20 cm])
	Centipedegrass	Very dense cover (average 6 in [15 cm])
	Kentucky bluegrass	Good stand, headed (6 to12 in [15 to 30 cm])
D	Bermuda grass	Good stand, cut 2.5-in height (6 cm)
	Common lespedeza	Excellent stand, uncut (average 4.5 in [11 cm])
	Buffalo grass	Good stand, uncut (3 to 6 in [8 to 15 cm])
	Grass-legume mixture—fall, spring (orchard grass, redtop, Italian ryegrass, and common lespedeza)	Good stand, uncut (4 to 5 in [10 to 13 cm])
	Lespedeza sericea	After cutting to 2-in height (5 cm) Very good stand before cutting
E	Bermuda grass	Good stand, cut to 1.5-in height (4 cm)
	Bermuda grass	Burned stubble

Source: U.S. Soil Conservation Service (1954).

4. Calculate the computed flow depth y_{comp} from the Manning equation using the value of n from step 3.

5. For vegetative lining, compare y and y_{comp}; if they are not close enough replace y based on y_{comp} for a new y and go to step 3. For nonvegetative lining, go to step 6.

6. Compute shear stress for the design condition by

$$\tau_{des} = \gamma R S_0 \tag{15.3.5}$$

If $\tau_{des} < \tau_p$ (Table 15.3.4) the lining is acceptable. Otherwise go to step 1 and choose a different lining.

EXAMPLE 15.3.2

Design a flexible-lined trapezoidal channel for a slope of $S_0 = 0.0016$ (same flow rate of 11.33 m³/s and slope as example 15.3.1). Use a nonvegetative lining.

SOLUTION

1. A gravel riprap (2.5 cm) D_{50} is chosen.
2. A flow depth of greater than 60 cm is assumed.
3. From Table 15.3.1, the Manning's roughness factor is $n = 0.03$.
4. Compute the flow depth using Manning's equation:

$$AR^{2/3} = \frac{Qn}{K_a S^{1/2}} = \frac{11.33 \times 0.03}{(0.0016)^{1/2}} = 8.50$$

Assume a bottom width of 6 m and $z = 2$. So

$$AR^{2/3} = 8.50 = \left[\frac{\left(B_w + zy\right)^5 y^5}{\left(B_w + 2y\sqrt{1+z^2}\right)^2} \right]^{1/3}$$

Solving yields y_{comp} = 1.14 m. Skip step 5 for nonvegetative lining and go to step 6.

6. The shear stress is computed using equation (15.3.5) with $R \approx y$,
$$\tau_{des} = \gamma y S_0 = (9810)(1.14 \text{ m})(0.0016) = 17.89 \text{ N/m}^2 \ (1.73 \text{ kg/m}^2)$$
The allowable shear stress is τ = 1.61 kg/m^2 (from Table 15.3.4).

Because $\tau_{des} > \tau_{allowable}$, a 5 cm gravel riprap with approximately double the allowable shear stress only increases Manning's n from 0.03 to 0.034. So, returning to step 4, we find

$$AR^{2/3} = \frac{Qn}{K_a S^{1/2}} = \frac{11.33 \times 0.034}{(0.0016)^{1/2}} = 9.63$$

$$= \left[\frac{(B_w + zy)^5 y^5)}{(B_w + 2y\sqrt{1+z^2})^2} \right]^{1/3} = 9.63$$

Solving yields $y \approx 1.22$ m. The shear stress is $\tau_{des} = 1.95$ kg/m^2 and the permissible shear stress is 3.22 kg/m^2 (Table 15.3.4).

7. Freeboard can be added.

EXAMPLE 15.3.3

Design a gravel riprap triangle-shaped channel to carry a discharge of 11.33 m^3/s at a slope of 0.0016.

SOLUTION

1. Select a D_{50} = 2.5 cm gravel riprap.
2. Assume $y >$ 60 cm.
3. From Table 15.3.1, n = 0.034.
4. Compute flow depth:

$$AR^{2/3} = \frac{Qn}{S^{1/2}} = \frac{11.33 \times 0.034}{(0.0016)^{1/2}} = 9.63$$

Using the best hydraulic section $AR^{2/3} = 1/2y^{8/3} = 9.63$; then y = 3.03 m.

5. The assumed and computed depths for the selected Manning's n are OK.
6. The design shear stress using $R = (\sqrt{2}/4)y$ is $\tau_{des} = \gamma R S_0 = 9810(\sqrt{2}/4)(3.03)(0.0016) = 16.81$ N/m^2 = 1.71 Kg/m$^2 < \tau_{allowable}$ = 3.22 (Table 15.3.4).
7. A freeboard can be added.

15.4 STORMWATER DETENTION

15.4.1 Why Detention? Effects of Urbanization

Urban stormwater management systems typically include detention and retention facilities to mitigate the negative impacts of urbanization on stormwater drainage. The effects of urbanization on stormwater runoff include increased total volumes of runoff and peak flow rates, as depicted in Figure 15.4.1. In general, major changes in flow rates in urban watershed are the result of (Chow et al., 1988):

1. The increase in the volume of water available for runoff because of the increased impervious cover provided by parking lots, streets, and roofs, which reduce the amount of infiltration;

2. Changes in hydraulic efficiency associated with artificial channels, curbing, gutters, and storm drainage collection systems, which increase the velocity of flow and the magnitude of flood peaks.

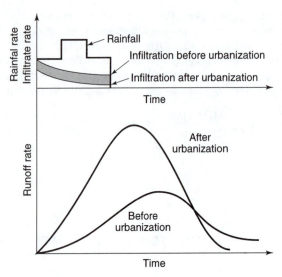

Figure 15.4.1 Effect of urbanization on stormwater runoff.

Stahre and Urbonas (1990) present the classification of storage facilities shown in Figure 15.4.2. The major classification is *source control* or *downstream control*. Source control involves the use of smaller facilities located near the source, allowing better use of the downstream conveyance system. Downstream control uses storage facilities that are larger and consequently at fewer locations, such as at watershed outlets. As Figure 15.4.2 shows, source control consists of local disposal, inlet control, and on-site detention. *Local disposal* is the use of infiltration or

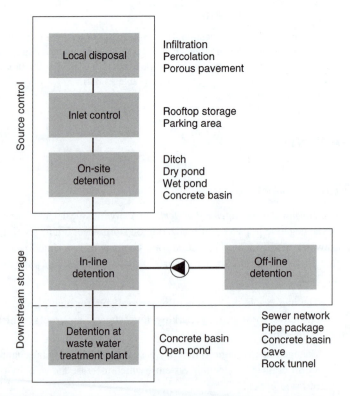

Figure 15.4.2 Classification of storage facilities (from Stahre and Urbonas (1990)).

percolation. *Inlet control* entails detaining stormwater where the precipitation occurs (such as rooftops and parking lots). *On-site detention* typically refers to detaining stormwater from larger areas than the previous two and includes swales, ditches, dry ponds, wet ponds, concrete basins (which are typically underground), and underground piping. Wet ponds have a permanent water pool as opposed to dry ponds.

Downstream storage includes *in-line detention*, *off-line detention*, and *detention at wastewater treatment plants*. In-line detention refers to detention storage in sewer lines, tunnels, storage vaults, pipes, surface ponds, or other facilities that are connected in-line with a stormwater conveyance network. Off-line storage facilities are not in line with the stormwater conveyance system.

Detention as described in this chapter is of two major types: (1) underground or subsurface systems, and (2) surface systems. Most of this section discusses surface detention: section 15.4.3 discusses various methods for sizing detention ponds, section 15.4.4 discusses various types of detention, section 15.4.5 discusses infiltration methods, and section 15.4.6 discusses water quality aspects.

Additional references on stormwater management include Overton and Meadows (1974), Meadows (1976), Stahre and Urbonas (1990), Loganathan et al. (1996), and Whipple et al. (1983).

15.4.2 Types of Surface Detention

Surface detention, for purposes of this discussion, refers to *extended detention basins* (or *dry detention basins*) and *retention ponds* (or *wet detention ponds*). Dry detention ponds empty after a storm, whereas retention ponds retain the water much longer above a permanent pool of water. Dry detention is the most widely used technique in the United States and many other countries. Figure 15.4.3 illustrates an extended detention basin. Water enters the basin, is impounded behind the embankment, and is slowly discharged through a perforated riser outlet. The coarse aggregate around the perforated riser minimizes clogging by debris. Typically once a required water-quality volume is filled, the remaining inflow is diverted around the basin or the pond overflows through a primary spillway. A large part of the sediment from the stormwater settles in the basin. Refer to Loganathan et al. (1996), Loganathan et al. (1985; 1989), Segaua and Loganathan (1992), and Wanielista and Yousef (1993).

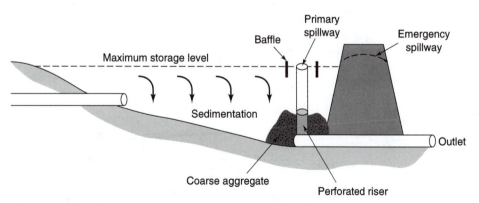

- Efficiency: Poor for detention times under 12 hours
    Good for detention times greater than 24 hours
- Function: Settle pollutants out; soluble pollutants pass through
- Maintenance is moderate if properly designed
- Improper design can make facilities an eyesore and a mosquito-breeding mudhole
- Newer designs are incorporating a shallow marsh around outlet
    Result: Better removal efficiency and no mosquito nuisance
- Regional detention facilities serving 100–200 acres can be aesthetically developed
    Result: Lower maintenance costs

Figure 15.4.3 Design of an extended detention basin (from Urbonas and Roesner (1993)).

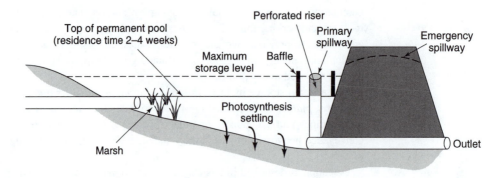

- Efficiency: Excellent if properly designed.
  Can be poor if bottom goes anoxic.
- Function: Removes pollutants by settling dissolved pollutants biochemically.
- Maintenance: Relatively free after first year except for major cleanout at
  about 10 years.
- Aesthetic design can make pond an asset to community.
  Excellent as a regional facility.

Figure 15.4.4 Design of a retention pond (from Urbonas and Roesner (1993)).

Figure 15.4.4 illustrates a retention pond, which is basically a lake that can be designed to remove pollutants. The figure illustrates the basic treatment processes that occur in the retention pond. Pollutants are removed by settling. Nutrients are removed by phytoplankton growth in the water column and by shallow marsh plants around the pond perimeter.

A multipurpose detention basin for quantity and quality is illustrated in Figure 15.4.5. The outlet works are staged so that the water-quality design volume is released very slowly. The other stages (see figure) provide storage and outlet peak discharges for erosion and flood control. Whipple et al. (1987) discuss the implementation of dual purpose stormwater detention programs.

Outlet works for extended detention and retention ponds differ because of the different operating functions of each. Figures 15.4.6 and 15.4.7 illustrate outlet works for dry (extended) detention ponds, which allow the entire storage volume to drain. The structure in Figure 15.4.6 has a fixed gate control and the structure in Figure 15.4.7 has a fixed orifice control. Figure 15.4.8 illustrates an outlet for a retention pond in which the water level drops to a permanent pool level that is controlled by positioning the openings.

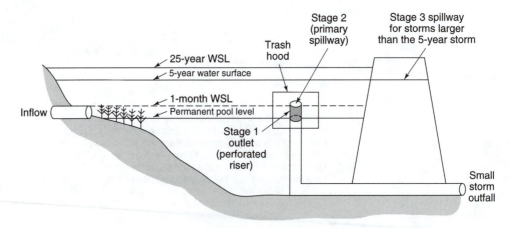

Figure 15.4.5 Conceptual design of a multipurpose pond (from Urbonas and Roesner (1993)).

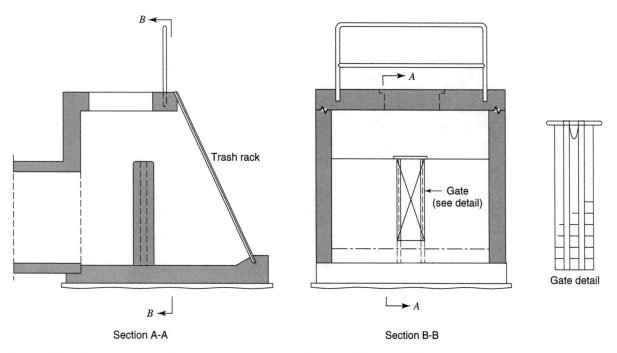

Figure 15.4.6 Outlet structure with a fixed gate control (from Stahre and Urbonas (1990)).

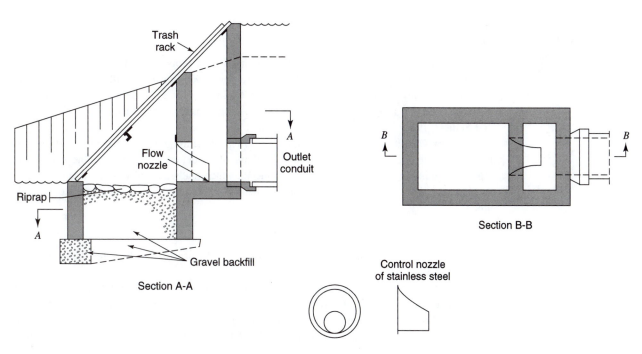

Figure 15.4.7 Outlet structure with a fixed orifice control (from Stahre and Urbonas (1990)).

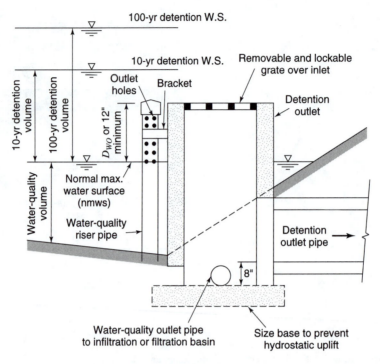

Figure 15.4.8 Outlet riser for a pond with a permanent pool (from Stahre and Urbonas (1990)).

15.4.3   Sizing Detention

Several simplified methods for sizing detention have been proposed in the literature (Abt and Grigg, 1978; Akan, 1989, 1990, 1993; Aron and Kibler, 1990; Donahue et al., 1981; Kessler and Diskin, 1991; Mays and Tung, 1992; McEnroe, 1992; and Wycoff and Singh, 1976). More sophisticated procedures using optimization have also been proposed (Bennett and Mays, 1985; Mays and Bedient, 1982; Nix and Heaney, 1988; Taur et al., 1987).

The stormwater detained during and after a storm is a function of the runoff volume, the detention basin outlet(s), and the available storage volume of the detention basin. The objective of sizing the pond is to determine the storage volume V_s, which mathematically is the time integral of the difference between the detention basin inflow and outflow hydrographs; i.e., the storage volume is

$$V_s = \int (Q_{in} - Q_{out})\, dt \tag{15.4.1}$$

where Q_{in} is the inflow rate and Q_{out} is the outflow rate. Figure 15.4.9 illustrates this integration showing the V_{max}.

Simple methods based upon regression equations, the rational method, or the modified rational method can be used to estimate preliminary sizes. This is followed up by iteratively routing one or more hydrographs through the preliminary sized pond to refine the size and outlet structures. The classic detention sizing procedure consists of the following steps (Urbonas and Roesner, 1993):

1. Estimate the preliminary storage volume V_s using a simplified procedure (see sections 15.4.3.1 and 15.4.3.2).

2. Use site topography to prepare a preliminary layout of a detention basin that has the desired volume and outlet configuration.

3. Determine stage-storage-outflow characteristics of the trial pond size.

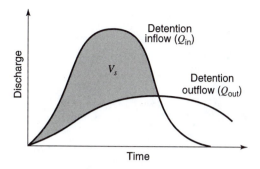

Figure 15.4.9 Detention storage volume V_s.

4. Perform routing of input hydrographs through the pond. Steps 3 and 4 can be accomplished using computer models.
5. If the trial pond (size and outlet configuration) does not satisfy desired goals and design criteria, resize the basin and or reconfigure the outlet(s) and repeat steps 3–5 until design goals and design criteria are satisfied (optimized).

The inflow hydrographs can be generated using any of a number of rainfall-runoff models (also see Akan, 1993; Chow et al., 1988; Kibler, 1982; and Urbonas and Roesner, 1993).

The American Association of State Highway Transportation Officials (AASHTO) (1991) recommended using triangular-shaped inflow and outflow hydrographs (see Figure 15.4.10) to determine preliminary estimates of storage volume V_s. The required storage volume is simply the cross-hatched area shown in Figure 15.4.10, which is computed using

$$V_s = 0.5t_b(Q_p - Q_A) \qquad (15.4.2)$$

Any consistent units may be used in equation (15.4.2). The time to peak inflow hydrograph t_p is half of the total time base of this hydrograph.

Abt and Grigg (1978) considered a triangular inflow hydrograph and a trapezoidal outflow hydrograph to develop the following relationship to estimate the required storage volume V_s using consistent units:

$$\frac{V_s}{V_r} = \left(1 - \frac{Q_A}{Q_P}\right)^2 \qquad (15.4.3)$$

where V_r is the runoff volume, Q_A is the allowable peak outflow rate, and Q_P is the peak inflow rate. This procedure assumes that the rising limbs of the inflow and outflow hydrographs coincide up to the allowable peak outflow rate (see Figure 15.4.11).

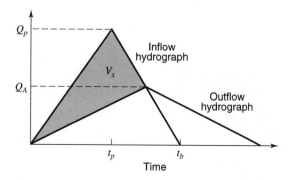

Figure 15.4.10 Inflow and outflow hydrographs for AASHTO (1991) procedure.

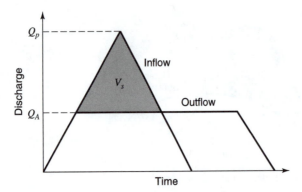

Figure 15.4.11 Inflow and outflow hydrographs for procedure by Abt and Grigg (1978).

Aron and Kibler (1990) developed an approximate method considering trapezoidal inflow hydrographs. They assumed (1) that the peak outflow hydrograph falls on the recession limb of the inflow hydrograph and (2) that the rising limb of the outflow hydrograph can be approximated by a straight line (see Figure 15.4.12). The volume of storage is computed using

$$V_s = Q_P t_D - Q_A \left(\frac{t_D + t_C}{2} \right) \tag{15.4.4}$$

where t_D is the storm duration and t_C is the time of concentration of the watershed. The design-storm duration is the one that maximizes the detention storage volume V_s for a given return period. This method uses a trial-and-error procedure to find the storm duration using the local intensity-duration-frequency (IDF) relationships. The rational formula ($Q_P = CiA$) is used to compute the peak flow rate Q_P.

AASHTO (1991) recommended an alternate estimate of storage volume using the regression equation developed by Wycoff and Singh (1986) as

$$\frac{V_s}{V_r} = \frac{1.29 \left(1 - \frac{Q_A}{Q_P} \right)^{0.153}}{\left(t_b / t_p \right)^{0.411}} \tag{15.4.5}$$

where V_s is the volume of storage in inches, V_r is the volume of runoff in inches, Q_A is the peak outflow in cfs, Q_P is the peak inflow in cfs, t_b is the time base of the inflow hydrograph in hours (determined as the time from the beginning of rise to a point on the recession limb where the flow is 5 percent of the peak), and t_p is the time to peak of the inflow hydrograph in hour. A preliminary estimate of the potential peak flow reduction for a selected storage volume is

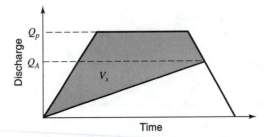

Figure 15.4.12 Inflow and outflow hydrographs for procedure by Aron and Kibler (1990).

$$\frac{Q_A}{Q_P} = 1 - 0.712 \left(\frac{V_s}{V_r}\right)^{1.328} \left(\frac{t_b}{t_p}\right)^{0.546} \tag{15.4.6}$$

EXAMPLE 15.4.1

The peak runoff rate for a 10-year storm of 133 ft³/s is to be limited to a peak of 40 ft³/s through the use of a detention basin. Determine a preliminary estimate of storage using the AASHTO (1991) method, equation (15.4.5); t_p = 30 minutes.

SOLUTION

First, by definition t_b = time base of inflow hydrograph in hours determined as the time from the beginning to a point on the recession limb where the flow is 5 percent of the peak. So

$$t_b = 60 - 30 \left[\frac{0.05(133)}{133}\right] = 58.5 \text{ min} = 0.98 \text{ hr}$$

Using equation (15.4.5) yields

$$\frac{V_s}{V_r} = 1.29 \frac{\left(1 - \frac{40}{133}\right)^{0.153}}{\left(\frac{0.98}{0.5}\right)^{0.411}} = \frac{1.29(0.95)}{1.32} = 0.93$$

$$V_r = \frac{1}{2}(60 \text{ min})\left(133 \frac{\text{ft}^3}{\text{s}}\right) \times \frac{60 \text{ sec}}{\text{min}} - \frac{1}{2}(60 \text{ min})\left(40 \frac{\text{ft}^3}{\text{s}}\right)\left(\frac{60 \text{ sec}}{\text{min}}\right)$$

$$= 239{,}400 - 72{,}000 = 167{,}400 \text{ ft}^3 = 3.84 \text{ ac-ft}$$

The volume of storage is

$$V_s = V_r(0.93) = 3.84 \,(0.93) = 3.57 \text{ ac-ft}$$

EXAMPLE 15.4.2

Solve example 15.4.1 using the Abt and Grigg (1978) method.

SOLUTION

Using equation (15.4.3) yields

$$\frac{V_s}{V_r} = \left(1 - \frac{Q_A}{Q_P}\right)^2 = (1 - 0.3)^2$$

$$V_s = 0.49V_r = 0.49(3.84) = 1.88 \text{ ac-ft}$$

The procedure adopted by the Federal Aviation Agency (FAA) (1966) is a simple mass-balance technique that is intensity-duration-frequency (IDF) based. The procedure assumes that rainfall volume accumulates with time and is a time integral of the desired IDF curve. The rainfall volume-duration curve is transformed into a runoff volume-duration curve using

$$V_{in} = K_u CiAt_D \tag{15.4.7}$$

where V_{in} is the cumulative runoff volume, ft³(m³), K_u is 1.0 (for U.S. customary units) or 0.28 (for SI units), C is the runoff coefficient, i is the storm intensity from the IDF curve at time t_D in in/h (mm/h), A is the tributary area in acres (km²), and t_D is the storm duration in seconds. The cumulative volume leaving the detention basin, V_{out}, is estimated by

$$V_{out} = kQ_{out}t_D \tag{15.4.8}$$

where V_{out} is in ft³(m³), Q_{out} is the maximum outflow rate, ft³/s (m³/s), and k is an outflow adjustment coefficient from Figure 15.4.13 ($Q_{pin} = CiA$). The FAA procedure assumes a constant outflow rate Q_{out}, which is the rate of discharge when the detention basin is full. Because discharge

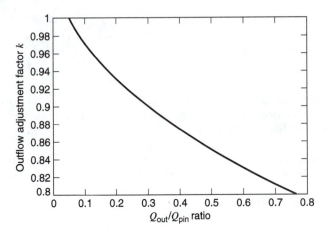

Figure 15.4.13 Outflow adjustment factor versus outflow rate/inflow peak ratio (from Urbonas and Roesner (1993)).

increases with depth of water, the outflow adjustment factor is used. The required detention volume is computed using

$$V_{\text{req}} = \max (V_{\text{in}} - V_{\text{out}}) \qquad (15.4.9)$$

which states that the required storage volume is the maximum difference between the cumulative inflow and the cumulative outflow volume.

15.4.3.1 Modified Rational Method

The *modified rational method* can be used to determine the preliminary design, which is the detention pond volume requirement for contributing drainage areas of 30 acres or less (Chow et al., 1988). For larger contributing areas, a more detailed rainfall-runoff analysis with a detention basin flow routing procedure should be used. The modified rational method is an extension of the rational method to develop hydrographs for storage design, rather than only peak discharges for storm sewer design. The shape of hydrographs produced by the modified rational method is either triangular or trapezoidal, constructed by setting the duration of the rising and recession limbs equal to the time of concentration t_c and computing the peak discharge assuming various durations. Figure 15.4.14a illustrates hydrographs drawn using the modified rational method.

An allowable discharge Q_A from a proposed detention basin can be the requirement that the peak discharge from the pond be equal to the peak of the runoff hydrograph for predeveloped conditions. The required detention storage V_s for each rainfall duration can be approximated as the cumulative volume of inflow minus the outflow, as shown in Figure 15.4.14b.

The assumptions of the modified rational method include:

1. The same assumptions as the rational method
2. The period of rainfall intensity averaging is equal to the duration of the storm
3. Because the outflow hydrograph is either triangular or trapezoidal, the effective contributing drainage area increases linearly with respect to time

An equation for the *critical storm duration*, that is, the storm duration that provides the largest storage volume, can be determined for small watersheds based upon the modified rational method. Consider a rainfall intensity-duration equation of the general form

$$i = \frac{a}{(t_D + b)^c} \qquad (15.4.10)$$

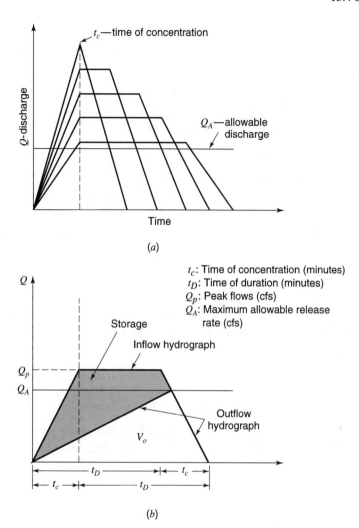

Figure 15.4.14 Modified rational method hydrographs. (*a*)
Hydrographs for different durations; (*b*) Storage requirement.

where i is the average rainfall intensity (in/hr) for the specific duration and return period, t_D is
storm duration in minutes, and a, b, and c are coefficients for a specific return period and location.
Consider the trapezoidal-shaped inflow hydrograph and outflow hydrograph in Figure 15.4.15.

Using the rational formula, the peak discharge can be expressed in terms of the storm duration:

$$Q_p = C_p iA = C_p \left(\frac{a}{(t_D + b)^c} \right) A \tag{15.4.11}$$

The inflow hydrograph volume V_i in ft^3 is expressed as

$$V_i = 60\,(0.5)\,Q_p[(t_D - t_c) + (t_D + t_c)] \tag{15.4.12}$$

where t_c is the time of concentration for proposed conditions. The outflow hydrograph volume V_0
in ft^3 is expressed as

$$V_0 = 60(0.5)Q_A(t_D + t_c) \tag{15.4.13}$$

where Q_A is the allowable peak flow release in ft^3. The storage volume V_s in ft^3 is computed using
the above expressions for V_i and V_0:

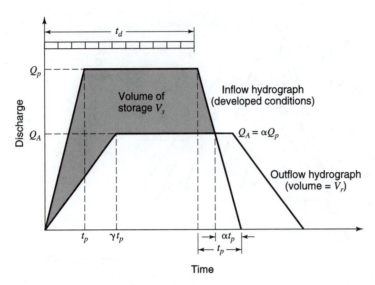

Figure 15.4.15 Inflow and outflow hydrographs for detention design. The outflow hydrograph is based on the inflow hydrograph for predeveloped conditions or on other more restrictive outflow criteria (from Donahue et al. (1981)).

$$V_s = V_i - V_0 = 60(0.5)Q_p[(t_D - t_c) + (t_D + t_c)] - 60\,(0.5)\,Q_A(t_D + t_c)$$

$$= 60\,Q_p\,t_D - 30Q_A(t_D + t_c) \tag{15.4.14}$$

The duration for the maximum detention is determined by differentiating equation (15.4.14) with respect to t_D and setting the derivative equal to zero:

$$\frac{dV_s}{dt_D} = 0 = 60t_D + \frac{dQ_p}{dt_D} = 0 = 60t_D + 60Q_p - 30Q_A$$

$$= 60t_D C_p A \frac{di}{dt_D} + 60C_p iA - 30Q_A \tag{15.4.15}$$

where

$$\frac{di}{dt_D} = \frac{d}{dt_D}\left[\frac{a}{(t_D + b)^c}\right] = \frac{-ac}{(t_D + b)^{c+1}} \tag{15.4.16}$$

so

$$\frac{dV_s}{dt_D} = 0 = 60C_p A(-ac)\frac{t_D}{(t_D + b)^{c+1}} + 60C_p\left(\frac{a}{(t_D + b)^c}\right)A - 30Q_A \tag{15.4.17}$$

Simplifying results in

$$\frac{a[t_D(1-c)+b]}{(t_D + b)^{c+1}} - \frac{Q_A}{2C_p A} = 0 \tag{15.4.18}$$

t_D in equation (15.4.18) can be solved by using Newton's iteration technique (Appendix A) where the iterative equation is

$$t_{D_{i+1}} = t_{D_i} - \frac{F(t_{D_i})}{F'(t_{D_i})} \tag{15.4.19}$$

where

$$F(t_D) = \frac{a[t_D(1-c)+b]}{(t_D+b)^{c+1}} - \frac{Q_A}{2C_pA} \tag{15.4.20}$$

and

$$F'(t_{Di}) = \frac{d[F(t_{D_i})]}{dt_D} = -\frac{a[t_D(1-c)+b](c+1)}{(t_D+b)^{c+2}} + \frac{a(1-c)}{(t_D+b)^{c+1}} \tag{15.4.21}$$

EXAMPLE 15.4.3

Determine the critical duration t_D and maximum detention storage for a 31.39-acre fully developed watershed with a runoff coefficient of $C_p = 0.95$. The allowable discharge is the predevelopment discharge of $Q_A = 59.08$ cfs. The time of concentration for proposed conditions is 21.2 minutes. The applicable rainfall intensity duration relationship is

$$i = \frac{97.86}{(t_D+16.4)^{0.76}}$$

SOLUTION

The critical storm duration t_D can be obtained by solving equation (15.4.18) using Newton's iteration as stated in equation (15.4.19). From equation (15.4.20) we find

$$F(t_D) = \frac{a[t_D(1-c)+b]}{(t_D+b)^{c+1}} - \frac{Q_A}{2C_pA}$$

$$= \frac{a[t_D(1-0.76)+16.4]}{(t_D+16.4)^{1.76}} - \frac{59.08}{2(0.95)(31.39)}$$

$$= \frac{97.86[0.24t_D+16.4]}{(t_D+16.4)^{1.76}} - 0.99 = \frac{23.49t_D+1604.90}{(t_D+16.4)^{1.76}} - 0.99$$

and from equation (15.4.21) we find

$$F'(t_D) = -\frac{a[t_D(1-c)+b](1+c)}{(t_D+b)^{c+2}} + \frac{a(1-c)}{(t_D+b)^{c+1}}$$

$$= -\frac{97.86[t_D(1-0.76)+16.4](1+0.76)}{(t_D+16.4)^{2.76}} + \frac{1-0.76}{(t_D+16.4)^{1.76}}$$

$$= -\frac{97.86(1.76)[0.24t_D+16.4]}{(t_D+16.4)^{2.76}} + \frac{97.86(0.24)}{(t_D+16.4)^{1.76}}$$

$$= \frac{41.34t_D+2824.63}{(t_D+16.4)^{1.76}} + \frac{23.49}{(t_D+16.4)^{1.76}}$$

With Newton's algorithm, the value of t_D converges to 92.0 minutes.

Next use equation (15.4.11):

$$Q_p = C_p\left[\frac{a}{(t_D+b)^c}\right]A = 0.95\left(\frac{97.86}{(92.0+16.4)^{0.76}}\right)(31.39) = 82.89 \text{ cfs}$$

Then use equation (15.4.14) to compute the detention storage:

$$V_s = 60Q_pt_D - 30Q_A(t_D+t_c) = 60(82.89)(92.0) - 30(59.08)(92.0+21.2)$$

$$= 256917 \text{ cfs} = 5.90 \text{ acre-ft}$$

Figure 15.4.15 is a representation of inflow and outflow hydrographs for a detention basin design. In this figure, α is the ratio of the peak discharge before development Q_A (or peak discharge from the detention basin that is allowable), and the peak discharge after development, Q_p:

$$\alpha = \frac{Q_A}{Q_p} \tag{15.4.22}$$

The ratio of the times to peak in the two hydrographs is γ. V_r is the volume of runoff after development. The volume of storage V_s needed in the basin is the accumulated volume of inflow minus outflow during the period when the inflow rate exceeds the outflow rate, shown shaded in the figure.

Using the geometry of the trapezoidal hydrographs, the ratio of the volume of storage to the volume of runoff V_s/V_r can be determined (Donahue et al., 1981) as:

$$\frac{V_s}{V_r} = 1 - \alpha\left[1 + \frac{t_p}{t_D}\left(1 - \frac{\gamma + \alpha}{2}\right)\right] \tag{15.4.23}$$

where t_D is the duration of the precipitation and t_p is the time to peak of the inflow hydrograph.

Consider a rainfall intensity-duration relationship of the form

$$i = \frac{a}{t_D + b} \tag{15.4.24}$$

where i is rainfall intensity and a and b are coefficients. The volume of runoff after development is equal to the volume under the inflow hydrograph:

$$V_r = Q_p t_D \tag{15.4.25}$$

The volume of storage is determined by substituting (15.4.25) into (15.4.23) and rearranging to get

$$V_s = Q_p t_D\left\{1 - \alpha\left[1 + \frac{t_p}{t_D}\left(1 - \frac{\gamma + \alpha}{2}\right)\right]\right\} \tag{15.4.26}$$

$$= t_D Q_p - Q_A t_D - Q_A t_p + \frac{\gamma Q_A t_p}{2} + \frac{Q_A^2 t_p}{2}\frac{1}{Q_p} \tag{15.4.27}$$

where α has been replaced by Q_A/Q_p.

The duration that results in the maximum detention is determined by substituting $Q_p = CiA = CAa/(t_D + b)$, then differentiating (15.4.27) with respect to t_D and setting the derivative equal to zero:

$$\frac{dV_s}{dt_D} = 0 = t_D\frac{dQ_p}{dt_D} + Q_p - Q_A + \frac{Q_A^2 t_p}{2}\left[\frac{d(1/Q_p)}{dt_D}\right] = \frac{bCAa}{(t_D + b)^2} - Q_A + \frac{Q_A^2 t_p}{2CAa}$$

where it is assumed that Q_A, t_p, and γ are constants. Solving for t_D,

$$t_D = \left(\frac{bCAa}{Q_A - \dfrac{Q_A^2 t_p}{2CAa}}\right)^{1/2} - b \tag{15.4.28}$$

The time to peak t_p is set equal to the time of concentration.

EXAMPLE 15.4.4

Determine the maximum storage for a detention pond on a 25-acre watershed for which the developed runoff coefficient is 0.8; the time of concentration before development is 25 min and after development is 15 min. The allowable discharge is 25 cfs, and $a = 96.6$ and $b = 13.9$.

SOLUTION

The maximum storage is determined using equation (15.4.27) with allowable discharge $Q_A = 25$ cfs, $t_p = 15$ min (developed condition), $C = 0.8$ (developed condition), $a = 96.6$, and $b = 13.9$. First determine the critical duration t_D using equation (15.4.28).

$$t_D = \left[\frac{(13.9)(0.8)(25)(96.6)}{25 - \left((25)^2(15)/2(0.8)(25)(96.6)\right)} \right]^{1/2} - 13.9 = 20.59 \text{ min}$$

The peak discharge is $Q_P = C\,i\,A$. Then using equation (15.4.24) and $i = a/(t_D + b)$, with $t_D = 20.59$ min, we get

$$Q_P = CA\left(\frac{a}{t_D + b}\right) = 0.8(25)\left(\frac{96.6}{20.59 + 13.9}\right) = 56.01 \text{ cfs}$$

The maximum storage is then

$$V_s = t_D Q_P - Q_A t_D - Q_A t_P + \frac{\gamma Q_A t_P}{2} + \frac{Q_A^2 t_P}{2}\frac{1}{Q_P}$$

$$V_s = (20.59)(56.01) - (25)(20.59) - (25)(15) + \frac{(25/15)(25)(15)}{2} + \frac{(25)^2(15)}{2}\frac{1}{56.01}$$

$$= 659.81 \text{ cfs(min} \times 60 \text{ s/min} = 39{,}588 \text{ ft}^3$$

15.4.3.2 Hydrograph Design Methods

A simple design procedure for sizing detention basins is now outlined that is useful in practice.

1. Determine the watershed characteristics and location of the detention basin.
2. Determine the design inflow hydrograph to the detention basin using a rainfall-runoff model.
3. Determine the detention storage-discharge relationship.
   a. Determine the storage-elevation relationship.
   b. Determine the discharge-elevation relationship for the discharge structure (culvert, spillway, etc.).
   c. Using the above relationships, develop the storage-discharge relationship.
4. Perform the computations described in Chapter 9 or section 15.4.4 for routing the inflow hydrograph through the detention basin using hydrologic reservoir routing.
5. Once the routing computations are completed, the reduced peak can be checked to see that the reduction is adequate and also to check the delay of the peak outflow.
6. Steps 3(b) through 5 of this procedure can be repeated for various discharge structures.

15.4.4 Detention Basin Routing

The hydrograph design method presented in section 15.4.3 requires routing of a design inflow hydrograph through the detention/retention basin. The level pool routing can be accomplished using the procedure in Chapter 9. An alternative procedure is presented in this section that does not require development of the storage outflow function. This method, presented in Chow et al. (1988), is based upon the Runge–Kutta method (Carnahan et al., 1969). A third-order scheme which breaks the time interval into three time increments and computes the water surface elevation and reservoir discharge for each increment.

Continuity is expressed as

$$\frac{dS}{dt} = I(t) - Q(H) \tag{15.4.29}$$

where S is the storage volume in the detention pond, $I(t)$ is the inflow into the pond as a function of time, t, and $Q(H)$ is the discharge from the pond as a function of the head or elevation H in the pond. The change in storage dS due to a change in elevation dH is

$$dS = A(H)dH \tag{15.4.30}$$

where $A(H)$ is the water surface area at elevation H. Substitution of this expression for dS into equation (15.4.29) and rearranging gives

$$\frac{dH}{dt} = \frac{I(t) - Q(H)}{A(H)} \tag{15.4.31}$$

which is an implicit differential equation.

Equation (15.4.31) is solved at each time step using three approximations of ΔH, ΔH_1, ΔH_2, and ΔH_3 at times t_j, $t_j + \Delta t/3$, and $t_j + 2\Delta t/3$, respectively. These approximations of ΔH are

$$\Delta H_1 = \left[\frac{I(t_j) - Q(H_j)}{A(H_j)}\right]\Delta t \tag{15.4.32}$$

$$\Delta H_2 = \left[\frac{I\left(t_j + \dfrac{\Delta t}{3}\right) - Q\left(H_j + \dfrac{\Delta H_1}{3}\right)}{A\left(H_j + \dfrac{\Delta H_1}{3}\right)}\right]\Delta t \tag{15.4.33}$$

$$\Delta H_3 = \left[\frac{I\left(t_j + \dfrac{2\Delta t}{3}\right) - Q\left(H_j + \dfrac{2\Delta H_2}{3}\right)}{A\left(H_j + \dfrac{2\Delta H_2}{3}\right)}\right]\Delta t \tag{15.4.34}$$

The value of H_{j+1} at time $t_{j+1} = t_j + \Delta t$ is

$$H_{j+1} = H_j + \Delta H \tag{15.4.35}$$

where

$$\Delta H = \frac{\Delta H_1}{4} + \frac{3\Delta H_3}{4} \tag{15.4.36}$$

This procedure requires the relationship of $Q(H)$ and $A(H)$ and the design detention pond inflow hydrograph.

EXAMPLE 15.4.5

Consider a 2-acre detention basin with vertical walls. The triangular inflow hydrograph increases linearly from zero to a peak of 540 cfs at 60 min and then decreases linearly to a zero discharge at 180 min. Route the inflow hydrograph through the detention basin using the head-discharge curve. Assuming the basin is initially empty, use the third-order Runge-Kutta method, with a 20-min time interval to determine the maximum depth.

Elevation (H, ft)	0.0	0.5	1.0	1.5	2.0	2.5	3.0
Discharge (Q, ft^3/s)	0.0	3.0	8.0	17	30	43	60

Elevation (H, ft)	3.5	4.0	4.5	5.0	5.5	6.0	6.5
Discharge (Q, ft^3/s)	78	97	117	137	156	173	190

Elevation (H, ft)	7.0	8.0	9.0	10	11	12
Discharge (Q, ft^3/s)	205	231	253	275	323	340

SOLUTION

The function $A(H)$ relating the water surface area to the reservoir elevation is simply $A(H) = 2(43,560)$ ft^2 = 87,120 ft^2. For all values of H, the reservoir has a base area of 2 acres and vertical sides. The routing procedure begins with determination of $I(0)$, $I(0 + 20/3)$, and $I[0 + (2/3) \times 20]$, which are determined by linear interpolation of values between 0 and 540/3 = 180 cfs, so they are respectively 0, 60, and 120 cfs.

Next ΔH_1 is computed using equation (15.4.32) with $\Delta t = 20$ min = 1200 s, $A = 87,120$ ft^2, and $I(0) = 0$ cfs. The reservoir is initially empty, so $H_1 = 0$ and $Q(H_1) = 0$, and thus $\Delta H_1 = [(0 - 0)/87120] \times 1200 = 0$. For the next $I(0+20/3) = 60$ ft^2/s, $\Delta H_2 = [(60 - 0)/87120] \times 1200 = 0.826$ ft, and $\Delta H_3 = [(120 - 3.507)/87120)](1200) = 1.605$ ft, so

$$H_2 = H_1 + \frac{\Delta H_1}{4} + \frac{3}{4}\Delta H_3 = 0 + \frac{0}{4} + \frac{3}{4}(1.605) = 1.204 \text{ ft}$$

and by linear interpolation $Q(1.204) = 11.672$ cfs.

In the next iteration, $\Delta H_1 = [(180 - 11.66)/87,120](1200) = 2.319$

$$Q\left(1.204 + \frac{2.319}{3}\right) = 29.402 \text{ cfs}$$

$\Delta H_2 = [(240 - 29.402)/87120]1200 = 2.901$ etc.

The routing computations are summarized in Table 15.4.1. The maximum depth in the pond is 12 ft.

Table 15.4.1 Routing Computation for Example 15.4.8

Time (min)	Inflow (cfs)	ΔH_1 (ft)	ΔH_2 (ft)	ΔH_3 (ft)	ΔH (ft)	Depth (H) (ft)	Outflow (cfs)
0	0	—	—	—	—	0	0
20	180	0	0.826	1.605	1.204	1.204	11.66
40	360	2.319	2.901	3.138	2.933	4.137	102.48
60	540	3.547	3.731	4.171	4.015	8.152	234.34
80	450	4.210	3.372	2.559	2.972	11.124	325.11
100	360	1.720	1.173	0.711	0.963	12.087	341.48
120	270	0.255	−0.178	−0.543	−0.344	11.743	335.63
140	180	−0.904	−1.247	−1.498	−1.350	10.393	293.91
160	90	−1.568	−1.682	−1.914	−1.828	8.565	243.43
180	0	−2.113	−2.306	−2.420	−2.343	6.222	180.55

15.4.5 Subsurface Disposal of Stormwater

Subsurface practices can be categorized as follows:
Infiltration practices:

- Swales and filter strips
- Porous pavement
- Infiltration trenches (Figure 15.4.16)
- Infiltration basins (Figure 15.4.17)
- Recharge wells (Figures 15.4.18–15.4.20)
- Underground storage (Figures 15.4.21–15.4.22)

15.4.5.1 Infiltration Practices

First the various types of infiltration practices are discussed. *Swales* are shallow vegetated trenches with nearly flat longitudinal slopes and mild side slopes. *Filter strips* are strips of land that stormwater must flow across before entering a conveyance system. These practices allow some of the runoff to infiltrate into the soil and filter the flow, removing some of the suspended solids and other pollutants attached to the solids. They also have the effect of reducing the directly connected impervious area and reducing the runoff velocity. They can be used for stormwater runoff from streets, parking lots, and roofs.

Wanielista et al. (1988) used mass balance of input and output water in swale systems to develop the following estimate of the length of a swale necessary to infiltrate all the input and rainfall excess from a specific storm event for a trapezoidal cross-sectional shape:

$$L = \frac{K \overline{Q}^{5/8} S^{3/16}}{n^{3/8} f} \tag{15.4.37}$$

where L is the length of the swale in ft (m), $\overline{Q}$ is the average runoff flow rate, ft³/s (m³/s), S is the longitudinal slope, ft/ft (m/m), n is the Manning roughness coefficient for overland flow (Tables 15.4.2 and 15.4.3), f is the infiltration rate, in/h (cm/h), and K is a constant that is a function of the side slope parameter Z (1 vertical/Z horizontal), as listed in Table 15.4.4.

Swales should be as flat as possible to maximize infiltration and to minimize resuspension of solids caused by high-flow velocities. Table 15.4.2 lists maximum or permissible velocities to reduce erosion or resuspension. The swale volume V_{swale}, for situations in which the available land is not long enough to infiltrate all the runoff, can be estimated using

$$V_{swale} = V_r - V_f \tag{15.4.38}$$

where V_{swale} is the volume of the swale (m³), V_r is the volume of runoff (m³), and V_f is the volume of infiltration (m³). V_f can be derived using

$$V_f = \overline{Q} t_r \tag{15.4.39}$$

where $\overline{Q}$ is the average infiltration flow rate in ft³/s (m³/s) (see equation (15.4.38)) and t_r is the runoff hydrograph time, seconds. With $\overline{Q} = [(Ln^{3/8}f)/(KS^{3/16})]^{8/5}$ from equation (15.4.37), the volume of swale is

$$V_{swale} = V_r - \left[\frac{Ln^{3/8} f}{KS^{3/10}} \right]^{8/5} t_r \tag{15.4.40}$$

Table 15.4.2 Maximum Permissible Design Velocities to Prevent Erosion and Manning's n for Swales

Cover	Manning's n for vR:[a]			Slope Range (%)	Maximum Permissible Velocity (ft/s)
	0.1	1.0	10		
Tufcote, Midland, and coastal Bermuda grass	0.25	0.150	0.045	0.0–5.0	6.0
				5.1–10.0	5.0
				Over 10.0	4.0
Reed canary grass	0.40	0.250	0.070	0.0–5.0	5.0
Kentucky 31 tall fescue	0.40	0.250	0.070	5.1–10.0	4.0
Kentucky bluegrass (mowed)	0.10	0.055	0.030	Over 10.0	3.0
Red fescue and Argentine Bahia	0.10	0.055	0.030	0.0–5.0	2.5
Annuals[b] and ryegrass	0.10	0.050	0.030	0.0–5.0	2.5

[a]Product of velocity and hydraulic radius (ft²/s).
[b]Annuals—use only as temporary protection until permanent vegetation is established.

Source: As presented in Wanielista and Yousef (1992).

Table 15.4.3 Overland Flow Manning's n Values for Shallow Flow $(vR < 1.0)^a$

	Recommended Value	Range of Values
Concrete	0.011	0.01–0.013
Asphalt	0.012	0.01–0.015
Bare sand	0.010	0.010–0.016
Graveled surface	0.012	0.012–0.030
Bare clay-loam (eroded)	0.012	0.012–0.033
Fallow (no residue)	0.05	0.006–0.16
Plow	0.06	0.02–0.10
Range (natural)	0.13	0.01–0.32
Range (clipped)	0.08	0.02–0.24
Grass (bluegrass sod)	0.45	0.39–0.63
Short grass prairie	0.15	0.10–0.20
Dense grass	0.24	0.17–0.30
Bermuda grass	0.41	0.30–0.48

[a]These values were determined specifically for overland flow conditions and are not appropriate for conventional open-channel flow calculations.

Source: As presented in Wanielista and Yousef (1992).

Table 15.4.4 Swale Length Formula Constant

Z (Side Slope (1 vertical/Z horizontal)	K (SI Units) (i = cm/h, Q = m³/s)	K (U.S. Units) (i = in/h, Q = ft³/s)
1	98,100	13,650
2	85,400	11,900
3	71,200	9,900
4	61,200	8,500
5	54,000	7,500
6	48,500	6,750
7	44,300	6,150
8	40,850	5,680
9	38,000	5,255
10	35,760	4,955

Source: Wanielista and Yousef (1992).

EXAMPLE 15.4.6

Determine the length of a swale needed to infiltrate an average runoff flow rate of 0.003 m³/s. The trapezoidal-shaped swale has a slope of 0.02, a Manning $n = 0.05$, an infiltration rate of 10 cm/h, and side slope of 1 vertical to $Z = 5$ horizontal.

SOLUTION

Equation (15.4.37) is used to determine the required length of the swale with $K = 54,000$ from Table 15.4.4 for $Z = 5$:

$$L = \frac{K \overline{Q}^{5/8} S^{3/16}}{n^{3/8} f} = \frac{54,000(0.003)^{5/8}(0.02)^{3/16}}{(0.05)^{3/8}(10)} = 211 \text{ m}$$

EXAMPLE 15.4.7   For the situation in example 15.4.9, only 100 m of swale was needed. How much storage volume is required for a runoff time of 120 min?

SOLUTION   Equation (15.4.40) is solved for V_{swale} with the volume of runoff $V_r = (0.003)(60)(120) = 21.6$ m³:

$$V_{swale} = V_r - \left[\frac{Ln^{3/8}f}{KS^{3/10}} \right]^{8/5} t_r = 21.6 - \left[\frac{100(0.05)^{3/8}(10)}{54,000(0.02)^{3/16}} \right]^{8/5} (60)(120)$$

$$= 21.6 \text{ m}^3 - 6.5 \text{ m}^3$$

$V_{swale} = 15.1$ m³ of storage required

Porous pavement and *modular pavement* (modular porous block pavement) can be used in parking areas to help reduce the amount of land needed for runoff quality control.

Percolation (or *infiltration*) *trenches* include both open surface type and underground (covered) trenches. Figure 15.4.16 illustrates infiltration trenches for perforated storm sewers and parking lot drainage. The perforated pipe allows distribution of stormwater along the entire length of the trench.

Perforated pipes allow the collection of sediment before it enters the aggregate backfill. Trenches are particularly suited for rights-of-way, parking lots, easements, and other areas with limited space. Their advantages are that they can be placed in narrow bands and in complex alignments. Prevention of excessive silt from entering the aggregate backfill and thus clogging the system is a major concern in design and construction. Sediment traps, filtration manholes, deep catchbasins, synthetic fibercloths, and the installation of filter bags in catch basins has proven effective (American Iron and Steel Institute, 1995).

Infiltration basins are retention facilities in which captured runoff is infiltrated into the ground. They are essentially depressions of varying size, either natural or excavated, into which stormwater is conveyed and allowed to infiltrate. Figure 15.4.17 illustrates an infiltration basin that serves the dual function of infiltration and storage. Infiltration basins are typically used in parks and urban open spaces, in highway rights-of-way, and in open spaces in freeway interchange loops. Infiltration basins are susceptible to clogging and sedimentation and can require large land areas. Standing water in these basins can create problems of security and insect breeding.

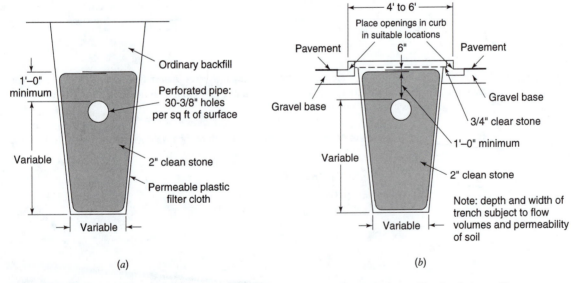

(a)   (b)

Figure 15.4.16 (a) Typical trench for perforated storm sewer; (b) Typical trench for parking lot drainage (from American Iron and Steel Institute (1995)).

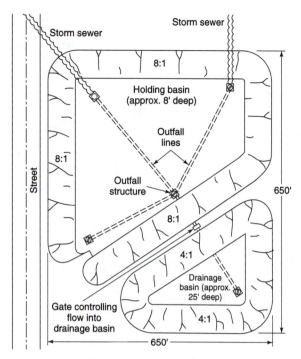

Figure 15.4.17 Infiltration basin (from American Iron and Steel Institute (1995)).

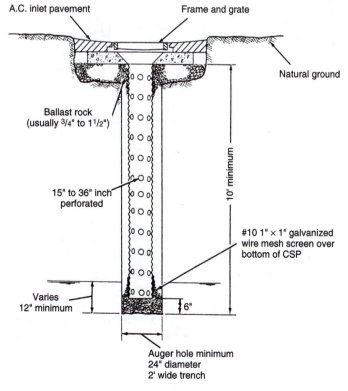

Figure 15.4.18 Recharge well (from American Iron and Steel Institute (1995)).

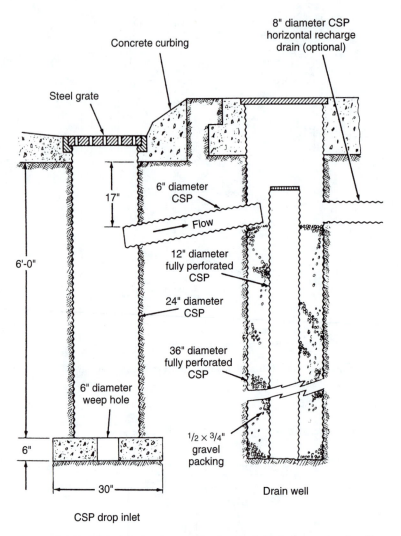

Figure 15.4.19 Typical design for combination catch basin for sand and sediment and recharge well. Catch basin would be periodically cleaned, and recharge well jetted through lower pipe to flush silt and restore permeability (from American Iron and Steel Institute (1995)).

15.4.5.2 Recharge Wells

Recharge wells can be used to dispose of stormwater directly into the subsurface. Figure 15.4.18 illustrates a recharge well. Recharge wells can be used to remove standing water in areas that are difficult to drain. They can also be used in conjunction with infiltration basins to penetrate impermeable strata. Another use is as a bottomless catchbasin in conventional minor system design. Typically, recharge wells are used for small areas and can be combined with catchbasins as illustrated in Figure 15.4.19. Figure 15.4.20 illustrates the use of a filter manhole in conjunction with a recharge well, in order to prevent excess silt entering the recharge well and causing clogging.

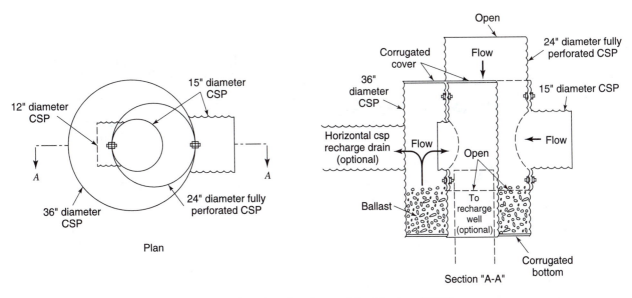

Figure 15.4.20 Typical CSP "Filter Manhole" (from American Iron and Steel Institute (1995)).

15.4.5.3 *Underground Storage*

Underground storage can be effective where surface ponds are not permitted or feasible. These storage tanks can be either *in-line*, in which the storage is incorporated directly into the sewer system, or *off-line*, in which stormwater is collected before it enters the sewer system and then discharged to either the sewer system or an open water course at a controlled rate. When the capacity of an in-line system is exceeded, surcharging in the sewer can occur. Figure 15.4.21 illustrates an off-line underground stormwater detention tank with an inlet control system.

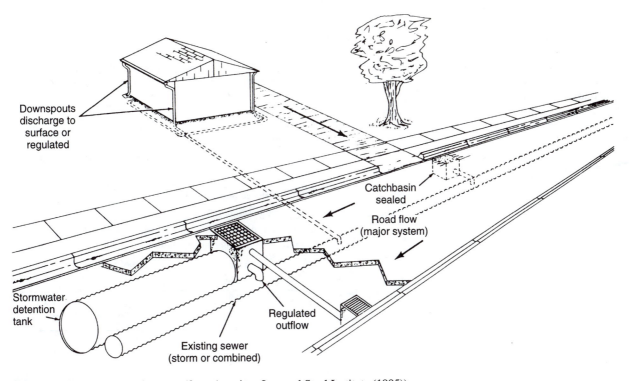

Figure 15.4.21 Inlet control system (from American Iron and Steel Institute (1995)).

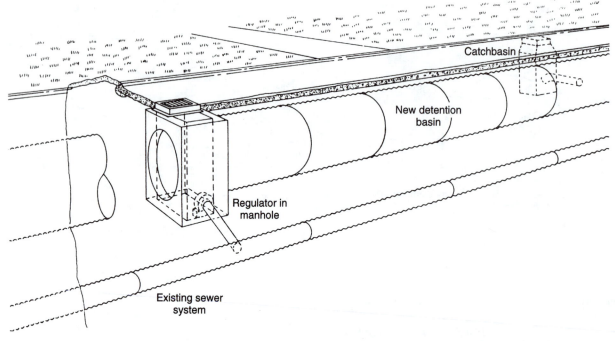

Figure 15.4.22 Typical installation of regulator for underground storage (from American Iron and Steel Institute (1995)).

Figure 15.4.22 illustrates a typical installation of a regulator for underground storage. Flow regulators at inlets to storm sewers are effective in preventing storm sewer surcharging. The simplest form of a flow regulator is an orifice for which the opening has been sized for a given discharge at the maximum head. Regulators in Figure 15.4.22 are designed to handle a discharge that the sewer can handle without excessive surcharging.

PROBLEMS

15.2.1 Determine the pipe diameters for the storm sewer system in Figure P15.2.1a, which is located in Phoenix, Arizona. The rainfall-intensity-duration frequency relationship for the Phoenix metro area is given in Figure P15.2.1b. Characteristics of the catchments are listed in Table P15.2.1. Use a return period of two years ($n = 0.014$).

Table P15.2.1 Catchment Characteristics, Problem 15.2.1

Catchment	Ground Elevation (m)	Area (km²)	Runoff Coefficient	Inlet Time (min)
1.1	300	0.01	0.60	25
1.2	298	0.008	0.75	20
2.1	296	0.005	0.80	15
3.1	294.5			

15.2.2 Rework problem 15.2.1 using a 10-year return period.

15.2.3 The simple storm sewer system below is to be designed using the following data. Assume the use of a 10-year frequency rainfall. The pipe is concrete with $n = 0.014$.

Manhole	Drainage Area (acre)	Time of Conc. (min)	C
1	2	15	0.5
2	3	20	0.8

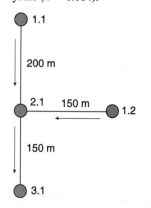

Figure P15.2.1 (*a*)

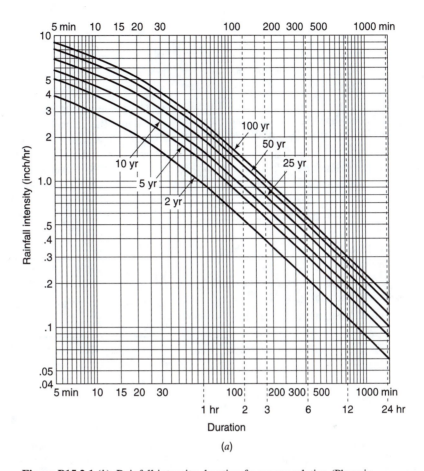

Figure P15.2.1 (*b*) Rainfall intensity-duration-frequency relation (Phoenix metro area) (from Flood Control District of Maricopa County (1992)).

Pipe	Slope (ft/ft)	Length (ft)
1–2	0.005	1000
2–3	0.006	800

15.2.4 Approximate the sudden expansion loss for a 600-mm sewer pipe connecting to a 700-mm sewer pipe for a design discharge of 0.5 m³/s.

15.2.5 Compute the bend loss for a 45° bend in a 500-mm sewer pipe with a discharge of 0.45 m³/s, assuming full-pipe flow. Assume $r/D = 2$.

15.2.6 Rework example 15.2.4 with $Q_L = 30$ cfs, $Q_u = 50$ cfs, and $Q_0 = 80$ cfs. The outfall pressure line elevation is 475.7 ft.

15.2.7 Determine the coefficient of variation of the loading and the capacity for the following parameters. Assume a uniform distribution to define the uncertainty of each parameters.

Parameter	Mode	Range
C	0.80	0.75–0.85
i	5.0 in/hr	4.5–5.5
A	10 acres	9.8–10.2
n	0.015	0.014–0.016
d	5 ft	4.98–5.02
S_0	0.0005	0.0004–0.0006

15.2.8 Rework example 15.2.5 using a triangular distribution to define the uncertainty of each parameter.

15.2.9 Using the results of problem 15.2.7, determine the risk of loading exceeding the capacity of the sewer pipe. Assume the use of a safety margin that is normally distributed.

15.2.10 Rework example 15.2.6 using a safety factor approach that is normally distributed; SF $= Q_c/Q_L$.

15.3.1 Design a nonerodible trapezoidal channel to carry a discharge of 6 m³/s. Use a Manning's $n = 0.025$ and a slope $S_0 = 0.0005$. Consider the best hydraulic section.

15.3.2 Design a concrete-lined trapezoidal channel to carry a discharge of 8 m³/s. A slope of $S_0 = 0.0001$ is to be used. Consider the best hydraulic section.

15.3.3 Design a concrete-lined rectangular channel to carry 25 ft³/s. A slope of $S_0 = 0.0001$ is to be used. Consider the best hydraulic section.

15.3.4 Design a gravel riprap-lined trapezoidal channel to carry a discharge of 11.33 m³/s. Use a slope $S_0 = 0.0016$ and consider the best hydraulic section.

15.3.5 Design a gravel riprap-lined triangular channel to carry a discharge of 11.33 m³/s. Use a slope $S_0 = 0.0016$ and consider the best hydraulic section.

15.4.1 Rework example 15.4.1 with the flow peak limited to 30 ft³/s.

15.4.2 Solve problem 15.4.1 using the Abt and Grigg (1978) method.

15.4.3 Solve example 15.4.6 using a runoff coefficient of $C_p = 0.85$ for a 15.24-acre watershed with $Q_A = 32.17$ cfs.

15.4.4 Solve example 15.4.6 using a developed runoff coefficient of 0.80.

15.4.5 Solve example 15.4.7 using a developed runoff coefficient of 0.85.

15.4.6 Solve example 15.4.7 using a developed runoff coefficient of 0.95.

15.4.7 Solve example 15.4.8 using the level pool routing procedure.

15.4.8 Solve example 15.4.8 using a time interval of 30 minutes.

15.4.9 Consider a 4047 m² (0.4047 ha) detention basin with vertical walls. The triangular inflow hydrograph increases linearly from zero to a peak of 10.2 m³/s at 60 min and then decreases linearly to zero at 150 min. The basin is initially empty and the discharge-elevation relationship is:

Elevation (H, m)	0.0	0.152	0.305	0.457	0.610	0.762	0.914
Discharge (Q, m³/s)	0.0	0.085	0.230	0.482	0.850	1.220	1.700

Elevation (H, m)	1.067	1.219	1.524	1.830	2.134	2.438	2.743	3.048
Discharge (Q, m³/s)	2.209	2.750	3.880	4.900	5.806	6.542	7.165	7.788

Use the Runge–Kutta method with a routing interval of 20 min to determine the detention basin discharge at the end of 20 min.

REFERENCES

Abt, S. R., and N. S. Grigg, "An Approximate Method for Sizing Retention Reservoirs," *Water Resources Bulletin*, AWRA, vol. 14, no. 4, pp. 956–965, 1978.

Akan, A. O., "Detention Pond Sizing for Multiple Return Periods," *Journal of Hydraulic Engineering*, vol. 115, no. 5, pp. 650–664, 1989.

Akan, A. O., "Single-Outlet Detention-Pond Analysis and Design," *Journal of Irrigation and Drainage Engineering*, vol. 116, no. 4, pp. 527–536, 1990.

Akan, A. O., *Urban Stormwater Hydrology*, Technomic Publishing, Lancaster, PA, 1993.

American Association of State Highway and Transportation Officials (AASHTO), *Model Drainage Manual*, Washington, DC, 1991.

American Iron and Steel Institute, *Modern Sewer Design*, third edition, Washington, DC, 1995.

American Society of Civil Engineers and Water Pollution Control Federation, *Design and Construction of Urban Stormwater Management Systems*, ASCE Manual and Report of Engineering Practice No.77, WEF Manual of Practice FD-20, New York, 1992.

American Society of Civil Engineering (ASCE) and Water Pollution Control Federation (WPCF), *Design and Construction of Sanitary and Storm Sewers*, ASCE Manual of Practice no. 37 and WPCF Manual of Practice no. 9, 1969.

Ang, A. H.-S., and W. H. Tang, *Probability Concepts in Engineering Planning and Design, I: Basic Principles*, John Wiley, New York, 1975.

Ang, A. H.-S., and W. H. Tang, *Probability Concepts in Engineering Planning and Design, II: Decision, Risk, and Reliability*, John Wiley, New York, 1984.

Aron, G., and C. E. Egborge, "A Practical Feasibility Study of Flood Peak Abatement in Urban Areas," report, U.S. Army Corps of Engineers, Sacramento District, Sacramento, CA, March 1973.

Aron, G., and D. F. Kibler, "Pond Sizing for Rational Formula Hydrographs," *Water Resources Bulletin*, AWRA, vol. 26, no. 2, pp. 255–258, 1990.

Bennett, M. S., and L. W. Mays, "Optimal Design of Detention and Drainage Channel Systems," *Journal of the Water Resources Planning and Management Division, ASCE*, vol. 111, no. 1, pp. 99–112.

Bathurst, J. C., R. M. Li, and D. B.Simons, "Resistance Equation for Large-Scale Roughness," *Journal of the Hydraulics Division, ASCE*, vol. 107, no. HY12, pp.1593–1613, 1981.

Carnahan, B., H. A. Luther, and J. O. Wilkes, *Applied Numerical Methods*, John Wiley, New York, 1969.

Chen, Y. H., and G. K. Cotton, "Design of Roadside Channels with Flexible Linings," Hydraulic Engineering Circular 15, FHWA-IP-87-7, Federal Highway Administration, McLean, VA, 1988.

Chow, V. T., D. R. Maidment, and L. W. Mays, *Applied Hydrology*, McGraw-Hill, New York, 1988.

Cotton, G. K., Hydraulic Design of Flood Control Channels, *Hydraulic Design Handbook*, edited by L. W. Mays, McGraw-Hill, New York, 1999.

Donahue, J. R., R. H. McCuen, and T. R. Bondelid, "Comparison of Detention Basin Planning and Design Models," *Journal of the Water Resource Planning and Management Division, ASCE*, vol. 107, no. 2, pp. 385–400, 1981.

Federal Aviation Administration, Department of Transportation, circular on airport drainage, report A/C 050-5320-5B, Washington, DC, 1970.

Federal Aviation Agency (FAA), *Airport Drainage*, Washington, DC, 1966.

Federal Highway Administration (FHWA), "HYDRAIN—Integrated Drainage Design Computer System. Vol. III. HYDRA—Storm Drains," Structure Division, Federal Highway Administration, Washington, DC, 1993.

Flood Control District of Maricopa County, Drainage Design Manual for Maricopa County, Arizona, vol. I–Hydrology, Phoenix, Arizona, 1992.

Flood Control District of Maricopa County, Drainage Design Manual for Maricopa County, Arizona, vol. II–Hydraulics, Phoenix, Arizona, 1992.

Frederick, R. H., V. A. Myers, and E. P. Auciello, "Five to 60-minute Precipitation Frequency for the Eastern and Central United States," NOAA Technical Memo NWS HYDRO-35, National Weather Service, Silver Spring, MD, June 1977.

Great Lakes–Upper Mississippi River Board of State Sanitary Engineers, Ten-State Standards, "Recommended Standards for Sewage Works," Health Education Service, Albany, NY, 1978.

Harr, M. E., *Reliability-Based Design in Civil Engineering*, McGraw-Hill, New York, 1987.

Izzard, C. F., Hydraulics of Runoff from Developed Surfaces, *Proc. Highway Research Board*, vol. 26, pp. 129–146, 1946.

Kapur, K. C., and L. R. Lamberson, *Reliability in Engineering Design*, John Wiley, New York, 1977.

Kececioglu, D., *Reliability Engineering Handbook*, vols. 1 and 2, Prentice-Hall, Englewood Cliffs, NJ, 1991.

Kessler, A., and M. H. Diskin, "The Efficiency Function of Detention Reservoirs in Urban Drainage Systems," *Water Resources Research*, American Geophysical Union, vol. 27, no. 3, pp. 253–258, 1991.

Kibler, D. F., *Urban Stormwater Hydrology*, Water Research Monograph 7, American Geophysical Union, Washington, DC, 1982.

Kirpich, Z. P., "Time of Concentration of Small Agricultural Watersheds," *Civ. Eng.*, vol. 10, no. 6, p. 362, 1940.

Kouwen, N., T. E. Unny, and H. M. Hill, "Flow Retardance in Vegetated Channels," *Journal of the Irrigation Division, ASCE*, vol. 95, no. IR2, pp. 329–342, 1969.

Lakatos, D. F., and R. H. Kropp, "Stormwater Detention—Downstream Effects on Peak Flow Rates," *Stormwater Detention Facilities*, American Society of Civil Engineering, New York, 1982.

Loganathan, G. V., D. F. Kibler, and T. J. Grizzard, "Urban Stormwater Management," *Water Resources Handbook*, edited by L.W. Mays, McGraw-Hill, New York, 1996.

Loganathan, G. V., J. W. Delleur, and R. I. Segana, "Planning Detention Storage for Stormwater Management," *Journal of Water Resources Planning and Management Division, ASCE*, vol. 111, no. 4, pp. 382–398, 1985.

Loganathan, G. V., E. W. Watkins, and D. F. Kibler, "Sizing Stormwater Detention Basins for Pollutant Removal," *Journal of Environmental Engineering, ASCE*, vol. 120, no.6, pp. 1380–1399, 1989.

Mays, L. W., and P. B. Bedient, "Model for Optimal Size and Location of Detention Basins," *Journal of the Water Resources Planning and Management Division, ASCE*, vol. 108, no. WR3, pp. 220–285, 1982.

Mays, L. W., and Y. K. Tung, *Hydrosystems Engineering and Management*, McGraw-Hill, New York, 1992.

McEnroe, B. M., "Preliminary Sizing of Detention Reservoirs to Reduce Peak Discharges," *Journal of Hydraulic Engineering, ASCE*, vol. 118, no. 11, pp. 1450–1549, 1992.

Morgali, J. R., and R. K. Linsley, "Computer Analysis of Overland Flow," *J. Hyd. Div., Am. Soc. Civ. Eng.*, vol. 91, no. HY3, pp. 81–100, May 1965.

National Highway Institute, *Urban Drainage Design*, Participant Notebook, NHI Cause No. 13027, Publication No. FHWA HI-89-035, Federal Highway Administration, U.S. Department of Transportation, Washington, DC, 1993.

Nix, S. J., and J. P. Heaney, "Optimization of stormwater Storage—Release Strategies," *Water Resources Research*, vol. 24, no. 11, pp. 1831–1838, 1988.

Overton, D. E., and M. E. Meadows, *Stormwater Modeling*, Academic Press, New York, 1976.

Phillips, J. V., and T. L. Ingersoll, "Verification of Roughness Coefficients for Selected Natural and Constructed Stream Channels in Arizona," U.S. Geological Survey Professional Paper 1584, U.S. Geological Survey, Denver, CO, 1998.

Poertner, H. G., "Practices in Detention of Urban stormwater Runoff," American Public Works Association, Special Report No. 43, 1974.

Sangster, W. M., H. M. Wood, E. T. Smerdon, and H. G. Bossy, "Pressure Changes at Storm Drain Junction," Engineering Series Bulletin, No. 41, Engineering Experiment Station, University of Missouri, Columbia, MO, 1958.

Segaua, R. I., and G. V. Loganathan, "Stormwater Detention Storage Design Under Random Pollutant Loading," *Journal of Water Resources Planning and Management, ASCE*, vol. 118, no. 5, pp. 475–491, 1992.

Stahre, P., and B. R. Urbonas, *Stormwater Detention for Drainage, Water Quality and CSO Management*, Prentice Hall, Englewood Cliffs, NJ, 1990.

Taur, C. K., G. Toth, G. E. Oswald, and L. W. Mays, "Austin Detention Basin Optimization Model," *Journal of Hydraulic Engineering, ASCE*, vol. 113, no. 7, pp. 860–878, 1987.

University of Colorado, *HYDRO POND User's Manual*, Department of Civil Engineering, University of Colorado at Denver, 1991.

U.S. Bureau of Reclamation (USBR), *Design of Small Dams*, second Edition, U.S. Government Printing Office, Denver, CO, 1973.

U.S. Bureau of Reclamation (USBR), *Design of Small Dams*, third Edition, U.S. Government Printing Office, Denver, CO, 1987.

U.S. Soil Conservation Service, "Urban Hydrology for Small Watersheds, Tech. Release 55," Washington, DC, 1975 (updated, 1986).

U.S. Soil Conservation Service (SCS), *Handbook of Channel Design for Soil and Water Conservation, SCS-TP-61*, Stillwater, OK., 1954.

Urbonas, B. R., and L. A. Roesner, "Hydrologic Design of Urban Drainage and Flood Control," *Handbook of Hydrology* (edited by D. R. Maidment), McGraw-Hill, New York, 1993.

Wang, S. Y., and H. W. Shen, "Incipient Sediment Motion and Riprap Design," *Journal of the Hydraulics Division, ASCE*, vol. 111, no. 3, pp. 521–538, 1985.

Wanielista, M. P., and Y. A. Yousef, *Stormwater Management*, Wiley Interscience, New York, 1993.

Whipple, W., N. S. Grigg, T. Grizzard, C. W. Randall, R. P. Shubinski, and L. S. Tucker, *Stormwater Management in Urbanizing Areas*, Prentice-Hall, Englewood Cliffs, NJ, 1983.

Whipple, W., R. Kropp, and S. Burke, "Implementing Dual Purpose Stormwater Detention Program," *Journal of Water Resources Planning and Management, ASCE*, vol. 113, no. 6, pp. 779–792, 1987.

Wycoff, R. L., and V. P. Singh, "Preliminary Hydrologic Design of Small Flood Detention Reservoirs," *Water Resources Bulletin*, AWRA, vol. 12, no. 2, pp. 337–349, 1976.

Yen, B. C., and A. O. Akan, "Hydraulic Design of Urban Drainage Systems," *Hydraulic Design Handbook*, edited by L. W. Mays, McGraw-Hill, New York, 1999.

Yen, B. C., "Hydraulics for Excess Water Management," *Water Resources Handbook*, edited by L.W. Mays, McGraw-Hill, New York, 1996.

Yen, B. C., ed., Storm Sewer System Design, Department of Civil Engineering, University of Illinois at Urbana-Champaign, 1978.

Yen, B. C., ed., *Stochastic and Risk Analysis in Hydraulic Engineering*, Water Resources Publications, Littleton, CO, 1986.

Chapter **16**

Stormwater Control: Street and Highway Drainage and Culverts

16.1 DRAINAGE OF STREET AND HIGHWAY PAVEMENTS

This section discusses the removal of stormwater from street and highway pavement surfaces and median areas. The removal of stormwater from streets is accomplished by collecting overland flow in gutters and intercepting the gutter flow at inlets to storm sewers. The design objective for a drainage system (for a curbed highway pavement section) is to collect runoff in the gutter and convey it to inlets in order to provide reasonable safety for traffic and pedestrians at reasonable cost. This section is based on the Federal Highway Administration's Hydraulic Engineering Circular No. 12 (Johnson and Chang, 1984). Other references are Young et al. (1993) and Young and Stein (1999).

16.1.1 Design Considerations

In the design of highway pavement drainage, two of the more significant variables are the frequency of the runoff event for design and the spread of water (*design spread*) on the pavement during the design event. The following summarizes the major considerations in the selection of design frequency and design spread:

- Classification of the highway
- Design speed of the highway
- Projected traffic volumes
- Rainfall intensities
- Capital costs
- Hazards and nuisances to pedestrian traffic
- Relative elevation of the highway and surrounding terrain

For low-volume local roads when traffic volumes and speeds are low, a two-year recurrence interval and a spread of one-half or more of the traffic lane are minimum design standards. High-speed, high-volume roads, such as freeways, are designed to minimize or eliminate spread on the traffic lanes for the design event. A 10-year frequency and a spread limited to the shoulders is common. Federal Highway Administration policy requires a 50-year frequency for underpasses and depressed sections on interstate highways. The American Association of State Highway and Transportation Officials (AASHTO, 1991) provides the design storm guidelines listed in Table 16.1.1.

Table 16.1.1 Design Storm Selection Guidelines

Roadway Classification	Exceedance Probability	Return Period
Rural principal arterial system	(2%)	(50-year)
Rural minor arterial system	(4%–2%)	(25–50-year)
Rural collector system, major	(4%)	(25-year)
Rural collector system, minor	(10%)	(10-year)
Rural local road system	(20%–10%)	(5–10-year)
Urban principal arterial system	(4%–2%)	(25–50-year)
Urban minor arterial street system	(4%)	(25-year)
Urban collector street system	(10%)	(10-year)
Urban local street system	(20%–10%)	(5–10-year)

Note: Federal law requires interstate highways to be provided with protection from the 2% flood event, and facilities such as underpasses, depressed roadways, etc., where no overflow relief is available should be designed for the 2% event.

Source: AASHTO (1991).

The rational method (discussed in Section 15.2.2) is the most commonly used method for determining discharges for highway pavement drainage. The rational formula (equation (15.2.1)) is

$$Q = KCiA \qquad (15.2.1)$$

where Q is the peak discharge, ft^3/s (m^3/s), K is 1.0 for U.S. customary units and 0.00275 for SI units, C is the dimensionless runoff coefficient, i is the average rainfall intensity, in/hr (mm/hr) for a duration equal to the time of concentration and for a design recurrence interval, and A is the drainage area, acres (hectares) (1 hectare = 0.4047 acres, 1 acre = 4047 m^2, and 1 acre = 0.004047 km^2).

The time of concentration for inlets consists of the overland flow time and the gutter flow time. If overland flow is channeled upstream of the inlet where flow enters the gutter, then the channel flow time must also be considered. Using the work by Ragan (1971), Johnson and Chang (1984) recommend the use of the following *kinematic wave equation*

$$t_c = \frac{KL^{0.6}n^{0.6}}{i^{0.4}S^{0.3}} \qquad (16.1.1)$$

where t_c is the time of concentration in seconds; L is the overland flow length, ft (m); n is Manning's roughness coefficient, i is the rainfall intensity, in/hr (m/hr); S is the average slope of the overland flow area; and $K = 56$ (26.285). To determine the flow time in a gutter, the average velocity in the gutter must be determined.

Young et al. (1993) developed an alternative method for selecting rainfall intensity that is not dependent on rainfall frequency. Instead this method uses incipient hydroplaning that causes drivers to slow down as opposed to traffic reaction due to gutter flooding. This method selects values of vehicle speed, tire tread depth, pavement texture, and tire pressure and calculates the thickness of the sheet flow film at incipient hydroplaning.

EXAMPLE 16.1.1   Use the kinematic wave formula to determine the time of concentration for a small drainage area that has a length of overland flow of 150 ft (45.72 m) and an average overland slope of 0.02 ft/ft (m/m) for a rainfall rate of 3.6 in/hr (0.091 m/hr). The drainage area is a turf ($n = 0.4$).

SOLUTION   Using the kinematic wave formula (equation 16.1.1), we find

$$t_c = \frac{KL^{0.6}n^{0.6}}{i^{0.4}S^{0.3}} = \frac{56(150)^{0.6}(0.4)^{0.6}}{(3.6)^{0.4}(0.02)^{0.3}} = 1265.5 \text{ s} = 21 \text{ min}$$

Alternatively, the time of concentration using the kinematic wave equation in SI units is

$$t_c = \frac{KL^{0.6}n^{0.6}}{i^{0.4}S^{0.3}} = \frac{26.285(45.72)^{0.6}(0.4)^{0.6}}{(0.091)^{0.4}(0.02)^{0.3}} = 1267.9 \text{ s} = 21 \text{ min}$$

EXAMPLE 16.1.2	Determine the time of concentration for an overland flow length of 100 m, on a turf surface ($n = 0.4$) with an average slope of 0.02. Use a rainfall rate of 10 cm/hr.

SOLUTION

Using equation (16.1.1) with $K = 26.285$, we get

$$t_c = \frac{KL^{0.6}n^{0.6}}{i^{0.4}S^{0.3}} = (26.285)\frac{(100)^{0.6}(0.4)^{0.6}}{(0.1)^{0.4}(0.02)^{0.3}} = \frac{(26.285)(15.85)(0.577)}{(0.398)(0.309)} = 1953.8 \text{ s} = 32.56 \text{ min}$$

Alternatively, in U.S. customary units,

$$t_c = \frac{56(328)^{0.6}(0.4)^{0.6}}{(3.98)^{0.4}(0.02)^{0.3}} = 1953.8 \text{ s} = 32.56 \text{ min}$$

16.1.2 Flow in Gutters

A *pavement gutter* conveys water during a storm event by collecting overland flow along its length and concentrating the flow as channel flow. Referring to Figure 16.1.1a, the elemental gutter flow dQ through an elemental cross-section dx of the gutter is

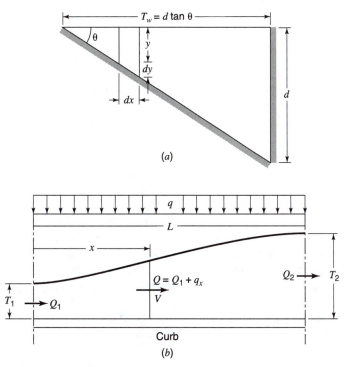

Figure 16.1.1 Pavement gutter flow. (*a*) Flow in triangular gutter; (*b*) Conceptual sketch of spatially varied gutter flow (from Johnson and Chang (1984)).

$$dQ = Vdx\left(y + \frac{dy}{2}\right) \tag{16.1.2}$$

where V is the velocity in the elemental area and y is the flow depth. Using Manning's equation, we find

$$V = \frac{1.49}{n}\left(\frac{ydx + \frac{1}{2}dydx}{\sqrt{dx^2 + dy^2}}\right)^{2/3} S^{1/2} \tag{16.1.3}$$

where n is Manning's roughness factor, S is the longitudinal slope of the gutter and the term $()^{2/3}$ is the hydraulic radius. The slope of the overland flow plane (or cross-slope) is S_x, so that $\sqrt{dx^2 + dy^2} = dx\sqrt{1 + S_x^2}$ where, $S_x = (dy/dx) = (d/T_w)$, so that

$$V = \frac{1.49}{n}\left(\frac{ydx + \frac{1}{2}dydx}{dx\sqrt{1 + S_x^2}}\right)^{2/3} S^{1/2} \tag{16.1.4}$$

For small values of S_x, equation (16.1.4) can be approximated by

$$V = \frac{1.49}{n}\left(y + \frac{dy}{2}\right)^{2/3} S^{1/2} \tag{16.1.5}$$

Substituting equation (16.1.5) into equation (16.1.2) and assuming $(y + (dy/2))^{5/3} = y^{5/3}$ results in

$$dQ = \frac{1.49}{n}y^{5/3}\frac{dy}{S_x}S^{1/2} \tag{16.1.6}$$

By integrating equation (16.1.6) across the cross-section of the gutter as y goes from 0 to d, the gutter flow can be expressed as

$$Q = \frac{0.56}{n}\frac{S^{1/2}}{S_x}d^{8/3} \tag{16.1.7}$$

The *spread* of water measured laterally from the curb is $T_w = d/\tan\theta = d/S_x$. Substituting $d = S_x T_w$ into equation (16.1.7), we can express the gutter flow in terms of the spread as

$$Q = \frac{0.56}{n}S^{1/2}S_x^{5/3}T_w^{8/3} \tag{16.1.8a}$$

or more generally as

$$Q = \left(\frac{\phi}{n}\right)S^{1/2}S_x^{5/3}T_w^{8/3} \tag{16.1.8b}$$

where ϕ is 0.56 (U.S. customary units) or 0.375 for SI units; Q is the discharge, ft^3/s (m^3/s); T_w is the top width of flow (spread), ft (m); S_x is the cross-slope, ft/ft (m/m); and S is the longitudinal slope, ft/ft (m/m). Equation (16.1.8) neglects the resistance of the curb face, which is negligible if the cross-slope is 10 percent or less (Johnson and Chang, 1984).

Figure 16.1.1b is a conceptual sketch of spatially varied gutter flow. As previously mentioned, the gutter flow time can be estimated using the average velocity. An assumption is that the flow rate in the gutter varies uniformly from Q_1 at the beginning of the section to Q_2 at the inlet (refer to Figure 16.1.1b). Equation (16.1.8a) can be expressed as

$$Q = K_1 T_w^{8/3} \tag{16.1.9}$$

where

$$K_1 = \frac{0.56}{n} S^{1/2} S_x^{5/3} \tag{16.1.10a}$$

in U.S. customary units, or

$$K_1 = \frac{0.375}{n} S^{1/2} S_x^{5/3} \tag{16.1.10b}$$

in SI units. The velocity can be expressed as

$$V = K_2 T_W^{2/3} = \frac{1.12}{n} S^{1/2} S_x^{2/3} T_W^{2/3} \tag{16.1.11a}$$

in U.S. customary units, or

$$V = \frac{0.75}{n} S^{1/2} S_x^{2/3} T_W^{2/3} \tag{16.1.11b}$$

in SI units, where

$$K_2 = \frac{1.12}{n} S^{1/2} S_x^{2/3} \tag{16.1.12a}$$

in U.S. customary units, or

$$K_2 = \frac{0.75}{n} S^{1/2} S_x^{2/3} \tag{16.1.12b}$$

in SI units.

From equation (16.1.9), $T_w^{8/3} = (Q/K_1)$, so the velocity is

$$V = \frac{dx}{dt} = \frac{K_2}{K_1^{0.25}} Q^{0.25} \tag{16.1.13}$$

Rearranging, we get

$$\frac{dx}{Q^{0.25}} = \frac{K_2}{K_1^{0.25}} dt \tag{16.1.14}$$

The flow rate downstream of section 1 is $Q = Q_1 + q_x$ and $dQ = qdx$, which can be substituted into equation (16.1.14). Integration of this equation results in

$$t = \frac{4}{3} \left(Q_2^{0.75} - Q_1^{0.75} \right) \frac{K_1^{0.25}}{K_2 q} \tag{16.1.15}$$

The average velocity $\bar{V}$ is then L/t, where L is the length shown in Figure 16.1.1b and

$$\bar{V} = \frac{L}{t} = \frac{3K_2 q}{4K_1^{0.25}} \left(\frac{L}{Q_2^{0.75} - Q_1^{0.75}} \right) \tag{16.1.16}$$

Substituting $L = (Q_2 - Q_1)/q$ and equation (16.1.9) into equation (16.1.16) yields

$$\bar{V} = \frac{3}{4} K_2 \frac{T_2^{8/3} - T_1^{8/3}}{(T_2^2 - T_1^2)} \tag{16.1.17}$$

where T_1 is the spread at the upstream end and T_2 is the spread at the downstream end of reach of gutter.

To determine the spread at the average velocity T_w using the average velocity, let V from equation (16.1.11) equal the average velocity in equation (16.1.17); solving for T_w/T_2 gives

$$\frac{T_w}{T_2} = 0.65 \left[\frac{1 - \left(\frac{T_1}{T_2}\right)^{8/3}}{1 - \left(\frac{T_1}{T_2}\right)^2} \right]^{1.5} \tag{16.1.18}$$

This equation is used to develop the numerical relationship between T_1/T_2 and T_w/T_2.

EXAMPLE 16.1.3

(a) Use equation (16.1.18) to develop the numerical relationship of T_w/T_2 for the average velocity in a reach of triangular gutter.

(b) Compute the time of flow in a gutter for the situation in which the spread T_1 at the upstream end is 3 ft, which results from the bypass flow from the upstream inlet, the design spread T_2 at the second inlet is 10 ft, the gutter slope is $S = 0.03$, and the cross-slope is $S_x = 0.02$. Manning's $n = 0.016$ and the inlet spacing is 200 ft.

SOLUTION

(a) Solving equation (16.1.18) results in the numerical relationship shown in Table 16.1.2.

Table 16.1.2 Relationship between T_1/T_2 and T_w/T_2

T_1/T_2	0	0.1	0.2	0.3	0.4	0.5	0.6	0.7	0.8	0.9	1.0
T_w/T_2	0.65	0.66	0.68	0.70	0.74	0.77	0.82	0.86	0.90	0.95	1.0

(b) From the above table, $T_1/T_2 = 3/10 = 0.3$, $T_w/T_2 = 0.7$; then $T_w = T_2(T_w/T_2) = 10(0.7) = 7$ ft. Next compute the average velocity V using equation (16.1.11):

$$V = K_2 T_w^{2/3} = \frac{1.12}{n} S^{1/2} S_x^{2/3} T_w^{2/3} = \frac{1.12}{0.016}(0.03)^{1/2}(0.02)^{2/3}(7)^{2/3} = 3.27 \text{ ft/s}$$

The time of flow is $t = L/V = 200/3.27 = 61.2$ sec $= 1.0$ min.

EXAMPLE 16.1.4

Derive the equation for the total discharge in a composite gutter section (with compound slopes) as shown in Figure 16.1.2. The distance a is referred to as the gutter depression.

SOLUTION

The depth of flow at the break in cross-section d_2 is

$$d_2 = (T_w - W)S_x$$

and the depth of flow at the curb is

$$d_1 = T_w S_x + a$$

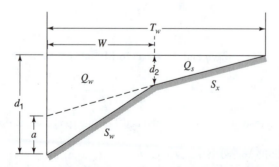

Figure 16.1.2 Flow in a composite gutter section.

The discharge for the outside section of gutter Q_s using equation (16.1.7) is

$$Q_s = \frac{0.56}{n} \frac{S^{1/2}}{S_x} d_2^{8/3} = \frac{0.56}{n} \frac{S^{1/2}\left[(T_w - W)S_x\right]^{8/3}}{S_x}$$

and the discharge in the inside section of gutter (next to the curb) is

$$Q_w = \frac{0.56}{n} \frac{S^{1/2}}{S_w} d_1^{8/3} - \frac{0.56}{n} \frac{S^{1/2}}{S_w} d_2^{8/3} = \frac{0.56}{n} \frac{S^{1/2}}{S_w} \left(d_1^{8/3} - d_2^{8/3}\right)$$

$$= \frac{0.56}{n} \frac{S^{1/2}}{S_w} \left[(T_w S_x + a)^{8/3} - \left[(T_w - W)S_x\right]^{8/3} \right]$$

The total gutter flow (Figure 16.1.2) is

$$Q = Q_w + Q_s$$

EXAMPLE 16.1.5

Using the discharge relationships developed in example 16.1.4, determine the total gutter flow for a design spread of $T_w = 8$ ft, a cross-slope of $S_x = 0.025$, and $W = 2.0$ ft with a 2-in gutter depression. The gutter slope is $S = 0.01$ and Manning's n is 0.016.

SOLUTION

First compute S_w by geometry, $S_w = (d_1 - d_2)/W$, so

$$S_w = \frac{T_w S_x + a - (T_w - W)S_x}{W} = \frac{a}{W} + S_x$$

Then $S_w = \left[(2/12)/2\right] + 0.025 = 0.108$. The total discharge is

$$Q = Q_w + Q_s$$

$$Q_s = \frac{0.56}{n} \frac{S^{1/2}\left[(T_w - W)S_x\right]^{8/3}}{S_x} = \frac{0.56(0.01)^{1/2}\left[(8-2)0.025\right]^{8/3}}{0.016(0.025)} = 0.89 \text{ cfs}$$

$$Q_w = \frac{0.56}{n} \frac{S^{1/2}}{S_w} \left\{ (T_w S_x + a)^{8/3} - \left[(T_w - W)S_x\right]^{8/3} \right\}$$

$$= \frac{0.56(0.01)^{1/2}}{(0.016)(0.108)} \left\{ \left[(8)(0.025) + (2/12)\right]^{8/3} - \left[(8-2)(0.025)\right]^{8/3} \right\}$$

$$= 32.41(0.0689 - 0.0064) = 2.03 \text{ cfs}$$

The total gutter flow is then

$$Q = Q_w + Q_s = 2.03 + 0.89 = 2.92 \text{ cfs}$$

16.1.3   Pavement Drainage Inlets

Figure 16.1.3 illustrates the four types of pavement inlets: *grate inlets*, *curb-opening inlets*, *combination inlets*, and *slotted drain inlets*. A grate inlet consists of an opening in the gutter covered by one or more grates that are parallel with the flow. A curb-opening inlet is a vertical opening in the curb, covered by a top slab. Combination inlets typically consist of both a curb opening inlet and a grate inlet placed in a side-by-side configuration. The curb-opening may be located in part upstream of the grate. Slotted drain inlets consist of a pipe cut along the longitudinal axis with a grate of spacer bars to form slot openings.

Johnson and Chang (1984) present several types of grates that have been hydraulically tested. Only a few are presented here for comparison:

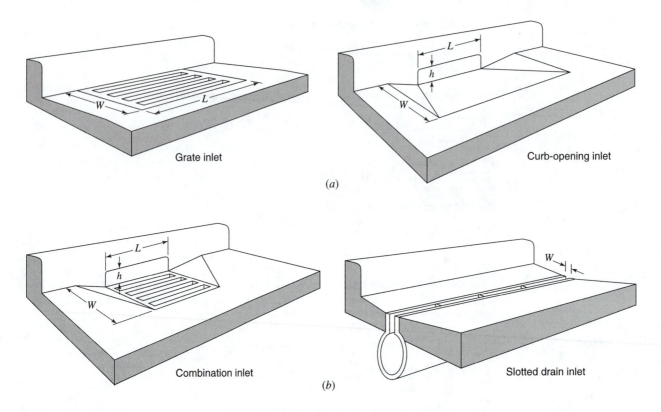

Figure 16.1.3 (*a*) Perspective views of grate and curb-opening inlets; (*b*) Perspective views of combination inlet and slotted drain inlet (from Johnson and Chang (1984)).

- P-1-7/8, which is a *bar grate* with bar spacing 1-7/8 in on center (Figure 16.1.4)
- P-1-7/8-4, which is a *parallel bar grate* with bar spacing 1-7/8 in on center and 3/8 in diameter lateral rods spaced at 4 in on center (Figure 16.1.4)
- CV-3-1/4-4-1/4, which is a *curved vane grate* with 3-3/4 in longitudinal bars and 4-1/4 in transverse bar spacing on center (Figure 16.1.5)
- 30-3-1/4-4, which is a 30° *tilt-bar grate* with 3-1/4 in and 4 in on center longitudinal and lateral bar spacing, respectively (Figure 16.1.6)
- *Reticuline*, which is a honeycomb pattern of lateral bars and longitudinal bearing bars (Figure 16.1.7)

16.1.4 Interception Capacity and Efficiency of Inlets on Grade

The *interception capacity* Q_i of an inlet is the flow intercepted by an inlet under a given set of conditions. On a uniform grade the interception capacity is dependent on the overland flow, namely the cross-slope, S_x. To a lesser extent the capacity is dependent on the roughness, the gutter longitudinal slope, total gutter flow, and inlet configuration (length, width, and crossbar arrangement for grate inlets and length of inlet for curb-opening and slotted drain inlets).

The *efficiency of an inlet* is the percent of total flow that the inlet will intercept under a given set of conditions (Johnson and Chang, 1984). Efficiency is dependent upon cross-slope S_x, longitudinal slope S, interception capacity Q_i, total gutter flow Q, and minimally on pavement slope. Efficiency is expressed as

$$E = Q_i / Q \qquad (16.1.19)$$

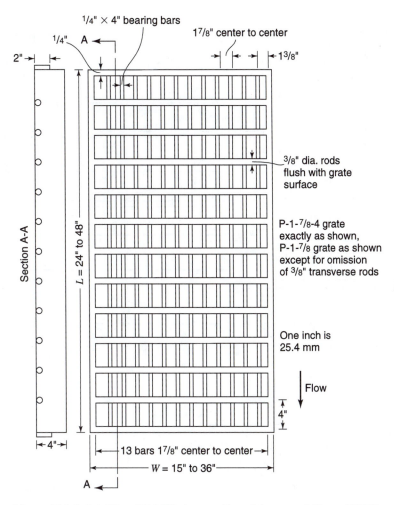

Figure 16.1.4 P-1-7/8 and P-1-7/8-4 grates (from Johnson and Chang (1984)).

The flow that is not intercepted is the *carryover* or *bypass flow* Q_b, so that

$$Q_b = Q - Q_i \qquad\qquad (16.1.20)$$

Interception capacity increases with increasing flow rate and, in general, inlet efficiency decreases with increasing flow rates.

Depth of water next to curbs is the major factor in interception capacity for both gutter inlets and curb-opening inlets. Interception capacity for grate inlets depends on the amount of water flowing over the grate, size and configuration of the grate, and velocity of flow in the gutter. Efficiency of a grate is dependent on the same factors and, in addition, on the total flow in the gutter.

Curb-opening inlet interception capacity is mostly dependent on flow depth at the curb and the curb-opening length. The use of a gutter depression at the curb opening or a depressed gutter increases the effective depth and consequently increases the inlet interception capacity and efficiency. Slotted drain inlets function as weirs, similar to curb-opening inlets. Efficiency is dependent on flow depth, inlet length, and total gutter flow. The inlet capacity of combination inlets is similar to that of grate inlets.

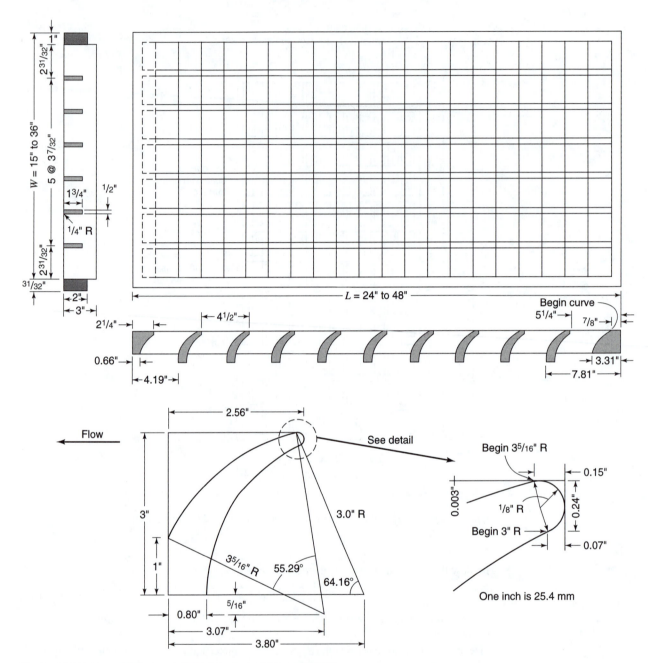

Figure 16.1.5 Curved vane grate (from Johnson and Chang (1984)).

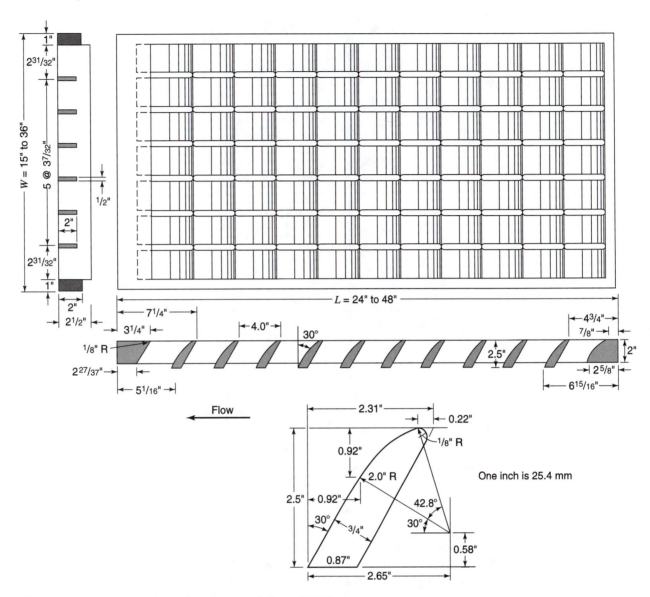

Figure 16.1.6 30° Tilt-bar grate (from Johnson and Chang (1984)).

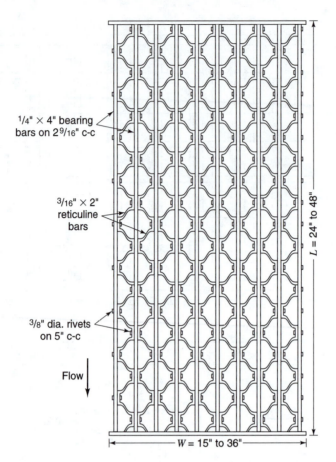

Figure 16.1.7 Reticuline grate (from Johnson and Chang (1984)).

16.1.4.1 Grate Inlets on Grade

This subsection addresses the hydraulic analysis of grates on a uniform grade. *Frontal flow* is the portion of flow that passes over the upstream side of a grate. *Splash-over* is the portion of the frontal flow at a grate that splashes over the grate and is not intercepted. This occurs when the flow velocity in the gutter exceeds a threshold value V_0. *Side flow* is the portion of gutter flow that goes around the grate when the spread T_W is larger than the grate width W. *Capture efficiency* of side flow depends on the cross-slope S_x, the grate length L, and the gutter flow velocity V.

The *frontal-flow ratio* E_0 for a straight cross-slope is

$$E_0 = \frac{Q_w}{Q} = 1 - \left(1 - \frac{W}{T_w}\right)^{8/3} \tag{16.1.21}$$

where Q_w and Q have units of ft³/s (m³/s) and W and T_w have units of ft (m). A *side-flow ratio* is expressed as

$$E_s = \frac{Q_s}{Q} = 1 - \frac{Q_w}{Q} = 1 - E_0 \tag{16.1.22}$$

where Q_s is the side flow, ft³/s (m³/s). If the gutter-flow velocity exceeds the threshold value, the *frontal-flow interception efficiency* R_f is

$$R_f = 1 - 0.09\,(V - V_0) \tag{16.1.23}$$

where the gutter-flow velocity is $V = Q/A_G$, A_G is the area of the gutter flow, and V_0 is the threshold gutter-flow velocity at which frontal flow skips over the grate without being captured. Figure 16.1.8 can be used to compute R_f for various types of grate inlets.

The *side-flow interception efficiency* R_s is the ratio of side flow intercepted to the total side flow, expressed as (Johnson and Chang, 1984)

$$R_s = \frac{1}{\left[1 + \dfrac{0.15V^{1.8}}{S_x L^{2.3}}\right]} \qquad (16.1.24)$$

The *interception efficiency of a grate inlet E* is defined as

$$E = R_f E_0 + R_s E_s \qquad (16.1.25)$$

or

$$E = R_f E_0 + R_s (1 - E_0) \qquad (16.1.26)$$

The term $R_f E_0$ is the ratio of intercepted frontal flow to total gutter flow and $R_s E_s$ is the ratio of intercepted side flow to total side flow. $R_s E_s$ is insignificant for high velocities and short grates. The *interception capacity of a grate inlet* (Q_i) is

$$Q_i = EQ = Q\,[R_f E_0 + R_s (1 - E_0)] \qquad (16.1.27)$$

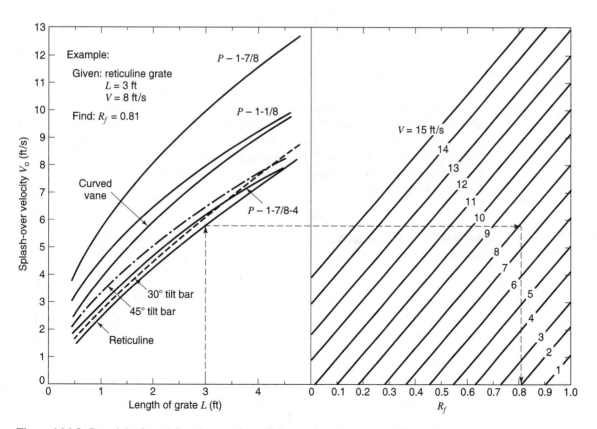

Figure 16.1.8 Grate inlet frontal flow-interception efficiency (from Johnson and Chang (1984)).

EXAMPLE 16.1.6

Using the data in example 16.1.5, find the interception capacity of a bar grate (P $-$ 1-7/8), $W = 2$ ft, $T_w = 8$ ft, $S = 0.01$, $S_x = 0.025$, and $Q = 2.92$ cfs. The grate is 2 ft long and 2 ft wide.

SOLUTION

Equation (16.1.26) is used to compute the interception capacity $Q_i = Q\,[R_f E_0 + R_s\,(1 - E_0)]$. Figure 16.1.8 is used to determine frontal-flow interception efficiency, R_f, for which the velocity Q/A is needed. The depths are $d_1 = T_w S_x + a = (8)(0.025) + (2/12) = 0.37$ ft and $d_2 = (T_w - W)S_x = (8 - 2)(0.025) = 0.15$ ft. The area of flow can be expressed as

$$A = \frac{1}{2}\left(d_1 + d_2\right)W + \frac{1}{2}d_1\left(T_w - W\right) = \frac{1}{2}(0.37 + 0.15)(2) + \frac{1}{2}(0.15)(8 - 2) = 0.97 \text{ ft}^2$$

The velocity is then $V = Q/A = 2.92/0.97 = 3.01$ ft/s. From Figure 16.1.8 with the grate length of 2 ft and the velocity of 3.01 ft/s, $R_f = 1.0$.

The side-flow interception efficiency R_s is computed using equation (16.1.24):

$$R_s = \frac{1}{\left[1 + \dfrac{0.15V^{1/8}}{S_x L^{2.3}}\right]} \frac{1}{\left[1 + \dfrac{0.15(3.01)^{1.8}}{(0.025)(2)^{2.3}}\right]} = 0.10$$

The frontal-flow ratio is computed using equation (16.1.21), $E_0 = Q_w/Q$. For this problem the grate width is 2 ft and W (the distance from curb to the break in cross-section, change in slopes) is also 2 ft. So the Q_w of 2.03 cfs from example 16.1.6 can be used; then

$$E_0 = \frac{Q_w}{Q} = \frac{2.03}{2.92} = 0.70$$

The interception capacity of the 2 $\times$ 2 ft bar grate inlet is then computed using equation (16.1.27) as

$$Q_i = 2.92\,[1 \times 0.70 + 0.10\,(1 - 0.70)] = 2.13 \text{ cfs}$$

16.1.4.2   Curb-Opening Inlets on Grade

These types of inlets are effective at locations where the flow at the curb is sufficient for the inlet to perform efficiently. They are relatively free of clogging and offer little interference to traffic. Curb-opening inlets do not function well on steep slopes because of the difficulty of directing flow into the inlet. Curb-opening inlets are much longer than grate inlets. This subsection provides the hydraulic analysis for curb-opening inlets on uniform grade.

The *full-interception curb-opening inlet length* L_T is the length to intercept 100 percent of the gutter flow, expressed as (Johnson and Chang, 1984)

$$L_T = \Phi Q^{0.42} S^{0.3}(n S_x)^{-0.6} \tag{16.1.28}$$

where Φ is 0.6 for U.S. customary units and 0.82 for SI units. The efficiency of curb-opening inlet lengths L that are shorter than L_T is

$$E = 1 - \left(1 - \frac{L}{L_T}\right)^{1.8} \tag{16.1.29}$$

For depressed curb-opening inlets or curb openings in depressed gutter sections, the length of inlet required for total interception can be determined using the *equivalent cross-slope* S_e, expressed as

$$S_e = S_x + S_w' E_0 \tag{16.1.30}$$

where S_w' is the cross-slope of the gutter measured from the cross-slope of the pavement S_x, so that $S_w' = a/12W$ where a is the gutter depression in inches and E_0 is the ratio of the flow in the depressed section to the total gutter flow Q_w/Q.

The curb-opening length required for total interception can be significantly reduced if the cross-slope (S_x) or equivalent cross-slope (S_e) is increased. The equivalent cross-slope can be increased by use of a continuously depressed gutter section or by a locally depressed gutter section. Equation (16.1.28) can be expressed using the equivalent cross-slope as

$$L_T = \Phi Q^{0.42} S^{0.3} (nS_e)^{-0.6} \tag{16.1.31}$$

This equation is applicable with either straight or compound cross-slopes.

EXAMPLE 16.1.7

Using the data from Example 16.1.6, determine the interception capacity of a 6-ft curb-opening inlet.

SOLUTION

Compute the full-interception curb-opening inlet length using equation (16.1.28):

$$L_T = \Phi Q^{0.42} S^{0.3} (nS_x)^{-0.6} = 0.6(2.92)^{0.42}(0.01)^{0.3}[0.016 \times 0.025]^{-0.6}$$

$$= 25.84 \text{ ft}$$

The efficiency is computed using equation (16.1.29):

$$E = 1 - \left(1 - \frac{L}{L_T}\right)^{1.8} = 1 - \left(1 - \frac{6.0}{25.84}\right)^{1.8} = 0.38 \text{ (or 38\%)}$$

The interception capacity is computed using $Q_i = EQ = 0.38(2.92) = 1.11$ cfs.

16.1.4.3 Slotted Drain Inlets on Grade

Slotted drain inlets have a flow interception that is similar to curb-opening inlets because each acts as a side weir in which the flow is subjected to lateral acceleration due to the cross-slope of the pavement. For slot widths greater than 1.75 in, the length required for total interception can be computed by equation (16.1.28). Similarly, equation (16.1.29) is applicable for determining the efficiency. The use of these equations is identical to that for curb-opening inlets.

16.1.4.4 Combination Inlets on Grade

The interception capacity for combination inlets on grade (with the curb opening and grate placed side by side) is not appreciably greater than that of just a grate inlet. Interception capacity is therefore determined by neglecting the curb opening. In the case of a combination inlet with the curb opening upstream of the grate, the interception capacity is the sum of the two inlets, with the exception that the frontal flow and the interception capacity of the grate are reduced by the curb-opening interception.

EXAMPLE 16.1.8

Determine the interception capacity of a combination curb-opening grate inlet in a triangular gutter section. The curb opening is 12 ft long with 10 ft of the curb opening upstream of the grate. The grate is a 2 ft by 2 ft recticuline grate placed alongside the downstream 2 ft of curb opening, $Q = 8$ ft^3/s, $S = 0.01$, $S_x = 0.025$, and $n = 0.016$.

SOLUTION

From equation (16.1.28), the full-interception curb-opening inlet length is

$$L_T = \Phi Q^{0.42} S^{0.3} (nS_x)^{-0.6} = 0.6(8)^{0.42}(0.01)^{0.3}[0.016 \times 0.025]^{-0.6}$$

$$= 39.47 \text{ ft}$$

The efficiency of the curb-opening inlet (upstream of the grate) is computed using equation (16.1.29):

$$E = 1 - \left(1 - \frac{L}{L_T}\right)^{1.8} = 1 - \left(1 - \frac{10}{39.47}\right)^{1.8} = 0.41$$

The interception capacity of the curb-opening inlet is then

$$Q_i = EQ = (0.41)(8) = 3.28 \text{ cfs}$$

To compute the interception capacity of the grate inlet, equation (16.1.27) is used:

$$Q_i = Q\,[R_f E_0 + R_s\,(1 - E_0)].$$

The flow at the grate is then $Q = 8 - 3.28 = 4.72$ cfs. Using this flow rate, the spread T_w can be computed with equation (16.1.8a):

$$Q = \frac{0.56}{n}\,S^{1/2} S_x^{5/3} T_w^{8/3}$$

$$4.72 = \frac{0.56}{0.016}(0.01)^{1/2}(0.025)^{5/3} T_w^{8/3}$$

$$T_w = 11.22 \text{ ft}$$

Next the velocity can be computed for use in determining R_f from Figure 16.1.8, so

$$V = Q/A = \frac{Q}{\left[\dfrac{1}{2} T_w^2 S_x\right]} = \frac{4.72}{\dfrac{1}{2}(11.22)^2(0.025)} = 3.00 \text{ ft/s}$$

From Figure 16.1.8, $R_f = 1.0$. The side-flow interception efficiency is computed using equation (16.1.24):

$$R_s = \frac{1}{\left[1 + \dfrac{0.15 V^{1.8}}{S_x L^{2.3}}\right]} = \frac{1}{\left[1 + \dfrac{0.15(3.00)^{1.8}}{0.025(2)^{2.3}}\right]} = 0.10$$

The frontal-flow ratio is computed using equation (16.1.21):

$$E_0 = \frac{Q_w}{Q} = 1 - \left(1 - \frac{W}{T_w}\right)^{8/3} = 1 - \left(1 - \frac{2}{11.22}\right)^{8/3} = 0.41$$

The interception capacity is then

$$Q_i = Q[R_f E_0 + R_s(1 - E_0)] = 4.72\,[1 \times 0.41 + 0.1(1 - 0.41)] = 2.21 \text{ cfs}$$

The total interception is $Q_i = Q_{\text{grate inlet}} + Q_{\text{curb open}} = 2.21 + 3.28 = 5.49$ cfs.

16.1.5   Interception Capacity and Efficiency of Inlets in Sag Locations

Inlets that are placed in sag locations operate as weirs under low heads and as orifices for higher heads. The transition between weir flow and orifice flow cannot be accurately defined, as the flow may fluctuate back and forth between these two controls. All runoff that enters sags must flow through the inlet. As a consequence, the efficiency of inlets in sags in passing debris is somewhat critical. Combination inlets and curb-opening inlets are recommended for sag locations, as grate inlets have clogging tendencies.

16.1.5.1   Grate Inlets in a Sag Location

The capacity of grate inlets Q_i under weir control is

$$Q_i = C_W P d^{1.5} \tag{16.1.32}$$

where C_W is the weir coefficient, 3.0 for U.S. customary units (1.66 for SI units), P is the grate perimeter disregarding bars and the curb side in ft (m), and d is the depth of water over the inlet in ft (m). The capacity of grate inlets under orifice control is

$$Q_i = C_0 A (2gd)^{0.5} \tag{16.1.33}$$

where C_0 is 0.67, A is the clear opening area of the grate, ft^2 (m^2), and g is the acceleration due to gravity, 32.16 ft/s^2 (9.81 m/s^2). Figure 16.1.9 provides a design solution for equations (16.1.32) and (16.1.33).

EXAMPLE 16.1.9

Consider a symmetrical sag vertical curve (with a curb) with equal bypass from inlets upgrade of the low point. Determine the grate size for a design Q of 6 ft^3/s and the curb depth. Allow for 50 percent clogging of the grate. The design spread is $T_w = 12$ ft, $S = 0.01$, $S_x = 0.025$ ft, and $n = 0.015$. What happens when the flow rate is 8 ft^3/s?

SOLUTION

According to Figure 16.1.9, a grate must have a perimeter of 12 ft, using $d = T_w S_x = 12(0.025) = 0.3$ ft and $Q = 6$ ft^3/s. Assuming 50 percent clogging by debris, the effective perimeter is reduced by 50 percent. Assume the use of a grate would meet the perimeter requirement with a double 2 ft $\times$ 5 ft grate.

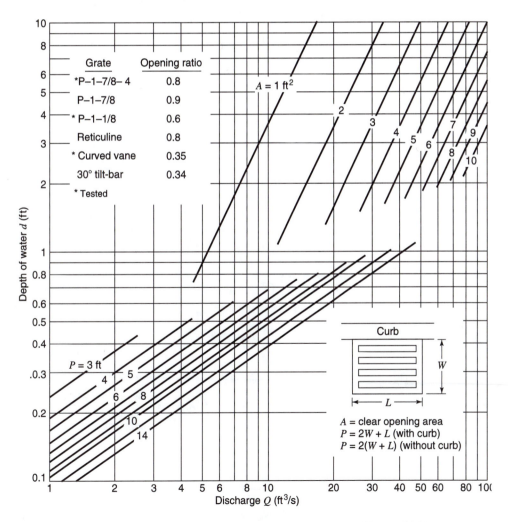

Figure 16.1.9 Grate inlet capacity in sump conditions (from Johnson and Chang (1984)).

Assuming 50 percent clogging so that the effective width of the gate is 1 ft, then the perimeter of the grate is $P = 1.0 + 1.0 + 10 = 12$ ft rather than 14 ft.

For a flow rate of 8 ft³/s and perimeter of 12 ft, the depth is 0.37 ft. The spread can be computed using $S_x = (d/T_w)$, so $T_w = (d/S_x) = 0.37/0.025 = 14.8$ ft at the flow rate of 8 ft³/s. The ponding will extend $14.8 - 12 = 2.8$ ft into the traffic lane.

16.1.5.2   Curb-Opening Inlets in a Sag Location

Curb-opening inlets in a sag operate as weirs to depths equal to the curb opening height and as orifices at depths greater than 1.4 times the opening height h. The transition of control occurs between 1.0 and 1.4 times the opening height h. The weir control equation for interception capacity of a depressed curb-opening inlet (Figure 16.1.10) is

$$Q_I = C_w(L + 1.8W)d^{1.5} \text{ (for } d \leq h + a/12, \text{ or } d \leq h + a \text{ for SI units)} \qquad (16.1.34)$$

where C_W is 2.3 (or 1.25 for SI units), L is the curb-opening length in ft (m), W is the lateral width of the depression in ft (m), a is the depth of the depression in (m), and d is the depth at curb opening measured from the normal cross-slope $d = T_w S_x$ in ft (m). As indicated by equation (16.1.34), the limitation for use of this weir equation is that depths at the curb be less than or equal to the height of the opening plus the depth of the depression. For curb-opening inlets without any depression ($W = 0$), the weir equation is simply

$$Q_i = C_w L d^{1.5} \quad \text{(for } d \leq h) \qquad (16.1.35)$$

where $C_w = 3.0$ (1.66 in SI units).

The interception capacity for orifice control ($d > 1.4h$) for both depressed and undepressed vertical curb-opening inlets is

$$Q = C_0 hL(2gd_0)^{0.5} \qquad (16.1.36a)$$

$$= C_0 A\left[2g\left(d_i - \frac{h}{2}\right)\right]^{0.5} \qquad (16.1.36b)$$

where C_0 is the weir coefficient equal to 0.67, h is the curb-opening inlet height in ft (m) (see Figure 16.1.11a), d_0 is the effective head on the center of the orifice throat in ft (m) (see Figure 16.1.11a), A is the clear area of the opening in ft² (m²), h is the height of the curb-opening orifice ($TS_x + a/12$), and d_i is the depth at the lip of the curb opening in ft (m), including any gutter

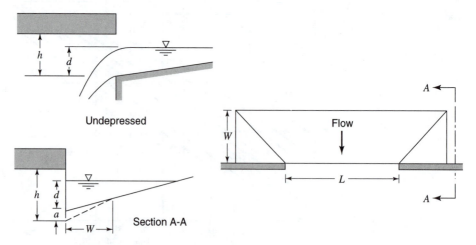

Figure 16.1.10 Depressed curb openings showing weir control in sump locations (from Johnson and Chang (1984)).

depression. The height h in equation (16.1.36) assumes a vertical orifice opening (see Figure 16.1.11). Other orifice throat locations change the effective depth on the orifice and the dimension $d_i - h/2$. For other than vertical faces, the orifice equation is

$$Q = C_0 hL(2gd_0)^{0.5} \qquad (16.1.37)$$

where h is the orifice throat width in ft (m) and d_0 is the effective head on the center of the orifice throat in ft (m); see Figures 16.1.11b and c.

EXAMPLE 16.1.10

A curb-opening inlet in a sump location is to have a curb opening length of $L = 5$ ft and a height of 5 in. (a) Determine the interception capacity for an undepressed curb opening for which $S_x = 0.05$ and $T_w = 8$ ft. (b) Determine the interception capacity for a depressed curb opening with $S_x = 0.05$, $a = 2$ in, $W = 2$ ft and $T_w = 8$ ft.

SOLUTION

(a) The depth is $d = T_w S_x = (8)(0.05) = 0.4$ ft, for which $d < h = 5/12 = 0.42$ ft, so the flow has weir control (see Figure 16.1.10). The interception capacity is computed using equation (16.1.35), $Q_i = C_w Ld^{1.5} = 3.0(5)(0.4)^{1.5} = 3.8$ ft³/s.

(b) The depth is $d = T_w S_x = (8)(0.05) = 0.4$ ft, in which $d \leq h + a/12 = 7/12$ ft, so weir flow controls (see Figure 16.1.10) and equation (16.1.34) $Q_i = C_w (L + 1.8W)d^{1.5}$ is applicable: $Q_i = 2.3 (5 + 1.8 \times 2) (0.4)^{1.5} = 5.0$ ft³/s. According to Johnson and Chang (1984), at $d = 0.4$ ft the depressed curb-opening inlet has about 30 percent more capacity than an inlet without depression. In a design situation, the flow rate would be known and the depth of the curb must be determined.

16.1.5.3 Slotted Drain Inlets in a Sag Location

These types of inlets (see Figure 16.1.3) operate as weirs to depths of about 0.2 ft (0.06 m) dependent upon the slot width (opening at top) and length, and operate as orifices for depths greater than about 0.4 ft (0.12 m). Transition flow occurs between these depths. The orifice control equation is

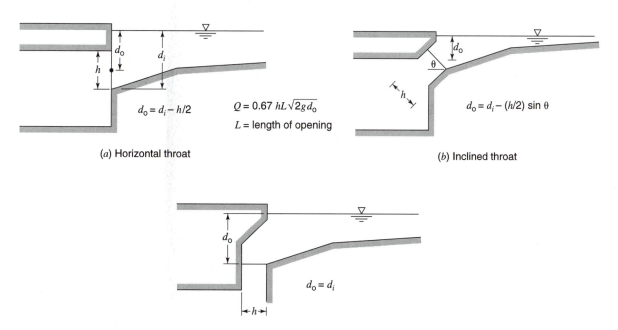

$d_0 = d_i - h/2$

$Q = 0.67\, hL\sqrt{2gd_0}$

$L = $ length of opening

(a) Horizontal throat

$d_0 = d_i - (h/2)\sin\theta$

(b) Inclined throat

$d_0 = d_i$

(c) Vertical throat

Figure 16.1.11 Curb-opening inlets (from Johnson and Chang (1984)).

$$Q_i = 0.8LW(2gd)^{0.5} \tag{16.1.38}$$

where W is the slot width in ft (m), L is the slot length in ft (m), and d is the water depth at the slot in ft (m) for $d \geq 0.4$ ft. The interception capacity of slotted inlets for depths in the range $0.2 \leq d \leq 0.4$ ft (0.12 m) can also be approximated using the above orifice equation.

EXAMPLE 16.1.11   Determine the length of a slotted inlet that is required to limit the maximum depth at the curb to 0.3 ft for a discharge of 5 ft³/s. Assuming no clogging, use a slot width of 1.75 in.

SOLUTION   Using equation (16.1.38), we find

$$Q_i = 0.8LW(2gd)^{0.5}$$

$$5 = 0.8\frac{1.75}{12}L[2(32.2)(0.3)]^{0.5}$$

$$L = 9.75 \text{ ft} \approx 10 \text{ ft}$$

16.1.5.4   Combination Inlets in a Sag Location

For situations where hazardous ponding can occur, combination inlets in sags are advised (Johnson and Chang, 1984). Equation (16.1.32) can be used for weir flow calculations, and for complete clogging of the grate, equations (16.1.33) to (16.1.35) can be utilized. For orifice flow, equations (16.1.33) and (16.1.37) can be combined:

$$Q_i = 0.67A(2gd)^{0.5} + 0.67hL(2gd_0)^{0.5} \tag{16.1.39}$$

EXAMPLE 16.1.12   Consider a combination inlet in a sag location. The grate is a P-1-7/8, that is 2×4 ft with a curb opening of $L = 4$ ft and $h = 4$ in. The cross-slope $S_x = 0.03$ and the discharge is $Q = 5$ ft³/s. Determine the depth at the curb and the spread for a grate with no clogging. What would be the depth and spread if the grate were 100 percent clogged?

SOLUTION   Using Figure 16.1.9 with $P = 2 + 2 + 4 = 8$ ft and $Q = 5$ ft³/s, we find $d = 0.36$ ft. Alternatively, using $Q = C_w Pd^{1.5}$, we get $5 = (3.0)(8.0)(d)^{1.5}$ then $d = 0.35$ ft. The spread is $T_w = d/S_x = 0.36/0.03 = 12$ ft. If the grate were 100 percent clogged, then the problem would be that of a curb-opening inlet in a sump location, so that $L = 4.0$ ft, $A = hL = (4/12)4 = 1.33$ ft². Then by equation (16.1.37),

$$Q = C_0 hL(2gd_0)^{0.5} \quad \text{or} \quad 5 = 0.67(1.33)\sqrt{2(32.2)}d_0^{0.5}$$

$$d_0 = 0.5 \text{ ft}$$

$$T_w = 0.5/0.03 = 16.7 \text{ ft}$$

Interception by the curb only is a transition stage between weir and orifice flow controls.

16.1.6   Inlet Locations

Johnson and Chang (1984) recommend that inlets be located at all low points in the gutter grade, at median breaks, intersections, crosswalks, and on side streets at intersections where drainage flows onto highway pavements. Inlet locations are often dictated by street-geometrical conditions rather than the spread-of-water computations. Sheet flow on pavements is susceptible to icing so that water should be prevented from flowing across the pavement. Therefore gutter flow should be intercepted where pavement surfaces are warped, such as at cross-slope reversals and ramps. Inlets should also be placed upgrade of bridges in order to prevent flow onto bridge decks and downgrade of bridges to intercept bridge drainage. Roadside channels should be used to intercept runoff from areas draining onto pavements. Inlets should be used where open channels are not practicable.

The design spread on the pavement is the criterion for spacing inlets on continuous grades. It is possible to establish the maximum design spacing for inlets on a continuous grade for a given design if the drainage area consists only of pavement or has reasonably uniform runoff characteristics and is rectangular in shape. The assumption is that the time of concentration is the same for all inlets.

For sag locations (see Figure 16.1.12) where significant ponding can occur, such as at underpasses and sag vertical curves in depressed locations, *flanking inlets* should be placed on each side of the inlet and at the low point. Table 16.1.3 lists the required spacing for various depths at curb criteria and vertical curve lengths defined by $K = L/A$ where L is the length of the vertical curve and A is the algebraic difference in approach grade.

Table 16.1.3   Distance to Flanking Inlets in Sag Vertical Curve Locations Using Depth-at-Curb Criteria

Speed	20	25	30	35	40	45	50	55	60		65	70
d\K	20	30	40	50	70	90	110	130	160	167	180	220
0.1	20	24	28	32	37	42	47	51	57	58	60	66
0.2	28	35	40	45	53	60	66	72	80	82	85	94
0.3	35	42	49	55	65	73	81	88	98	100	104	115
0.4	40	49	57	63	75	85	94	102	113	116	120	133
0.5	45	55	63	71	84	95	105	114	126	129	134	148
0.6	49	60	69	77	92	104	115	125	139	142	147	162
0.7	53	65	75	84	99	112	124	135	150	153	159	176
0.8	57	69	80	89	106	120	133	144	160	163	170	188

$x = (200dK)^{0.5}$, where x = distance from the low point; drainage maximum $K = 167$.

Source: Johnson and Chang (1984).

EXAMPLE 16.1.13

Consider a 32 ft pavement width on a continuous grade, $n = 0.016$, $S_x = 0.025$, $S = 0.01$, and $T_w = 8$ ft. The design rainfall intensity is 10 in/hr and $C = 0.8$. Determine the maximum design inlet spacing for a 2 ft by 2 ft curved vane grate.

SOLUTION

The discharge is expressed in terms of the maximum distance L_x where $Q = CiA$ and $A = (32 L_x/43560)$, so $Q = [0.8 \times 10 \times 32 L_x]/43560 = 0.0059L_x$ (or 0.0059 ft³/s/ft). The gutter flow can be determined using equation (16.1.8a):

$$Q = \frac{0.56}{n} S^{1/2} S_x^{5/3} T_w^{8/3} \frac{0.56}{0.016}(0.01)^{1/2}(0.025)^{5/3}(8)^{8/3} = 1.92 \text{ ft}^3\text{/s}$$

The first inlet can then be placed at L_x where $1.92 = 0.0059L_x$; then $L_x = 325$ ft.

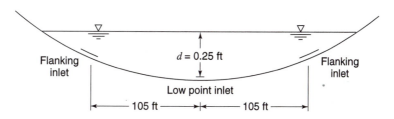

Figure 16.1.12 Example for sag locations.

Next the interception capacity of the grate inlet is determined using equation (16.1.26):

$$Q_i = Q\,[R_f E_0 + R_s(1 - E_0)]$$

where $Q = 1.92$ ft³/s. Figure 16.1.8 is used to compute R_f. Velocity can be computed using equation (16.1.11):

$$V = \frac{1.12}{0.016}(0.01)^{1/2}(0.025)^{2/3}(8)^{2/3} = 2.40 \text{ ft/s}$$

From Figure 16.1.8 for a curved vane grate, $R_f = 1.0$. Using equation (16.1.24) yields

$$R_s = \frac{1}{\left[1 + \dfrac{0.15 V^{1.8}}{S_x L^{2.3}}\right]}\,\frac{1}{\left[1 + \dfrac{0.15(2.40)^{1.8}}{0.025(2)^{2.3}}\right]} = 0.15$$

The frontal-flow ratio E_0 is computed using equation (16.1.21), $E_0 = Q_w/Q$ where $Q_w = AV$ where $V = 2.40$ ft/s, and

$$A = \frac{1}{2}\Big[(T_w S_x) + (T_w - W)S_x\Big]W = \frac{1}{2}\Big[2T_w S_x - W S_x\Big]W$$

$$= \frac{S_x W}{2}(2T_w - W) = \frac{0.025 \times 2}{2}(2 \times 8 - 2) = 0.35$$

$Q_W = 0.35\,(2.40) = 0.84$, $E_0 = 0.84/1.92 = 0.44$

The interception capacity is then

$$Q_i = 1.92\,[1 \times 0.44 + 0.15\,(1 - 0.44)] = 1.0 \text{ cfs}$$

The intervening drainage area between inlets should be sufficient to generate runoff equal to the interception capacity of the inlet; i.e., for each subsequent inlet, the interception is 1 cfs. Therefore, using $Q = 0.0059\,L_x$, we have

$$L_x = Q/0.0059 = 1/0.0059 = 169 \text{ ft}$$

The bypass runoff at the first inlet, and consequently at all other subsequent inlets, is

$$Q_b = Q - Q_i = 1.92 - 1 = 0.92 \text{ cfs}$$

The result is that the initial inlet can be placed at 325 ft from the crest and subsequent inlets at 169-ft intervals.

EXAMPLE 16.1.14

Using the data from example 16.1.13, determine the maximum design inlet spacing for a 10-ft slotted inlet. Compare the efficiency for this inlet design and the one computed in example 16.1.13.

SOLUTION

From example 16.1.13, the gutter flow $Q = 1.92$ ft³/s, so that the full-interception curb-opening inlet length L_T is computed using equation (16.1.28) as

$$L_T = 0.6 Q^{0.42} S^{0.3}(n S_x)^{-0.6}$$

$$L_T = 0.6 \times 1.92^{0.42} \times 0.01^{0.3}(0.016 \times 0.025)^{-0.6} = 21.7 \text{ ft}$$

The efficiency is computed using equation (16.1.29) as

$$E = 1 - \left(1 - \frac{L}{L_T}\right)^{1.8} = 1 - \left(1 - \frac{10}{21.7}\right)^{1.8} = 0.67$$

The interception capacity is $Q_i = EQ = 0.67(1.92) = 1.29$ ft³/s, and $Q_b = 1.92 - 1.29 = 0.63$ ft³/s. The spacing for the 10 ft slotted inlet is then computed using $Q = 0.0059 L_x$, so that $L_x = 1.29/0.0059 = 219$ ft. In example 16.1.13 the efficiency for the grate inlet design is $E = R_f E_0 + R_s(1 - E_0) = [1 \times 0.44 + 0.15(1 - 0.44)] = 0.52$ or 52 percent. For this example, the efficiency of the 10-ft slotted inlet design is 0.67 or 67 percent. This illustrates the effects of the relative efficiencies of the different inlet configurations for the chosen design condition.

EXAMPLE 16.1.15

Consider a sag vertical curve at an underpass on a four-lane divided highway facility. Determine the location of flanking inlets that will function in relief of the inlet at the low point when the depth at the curb exceeds the design depth. The design spread is not to exceed the shoulder width of 10 ft. The cross-slope is $S_x = 0.025$ and $K = 220$ (speed = 70 mph).

SOLUTION

The depth at the curb at the design spread of 10 ft is $d = T_w S_x = 10(0.025) = 0.25$ ft. Spacing to the flanking inlet is found from Table 16.1.3 as $(94 + 115)/2 = 104.5$; use 105-ft distance in Figure 16.1.12.

16.1.7   Median, Embankment, and Bridge Inlets

Median inlets are sometimes necessary to remove water in medians that could cause erosion. Figure 16.1.13 illustrates a median drop inlet. These drop inlets should be flush with the ditch bottom and constructed of traffic safety grates. Johnson and Chang (1984) present design charts and examples for median inlets.

Embankment inlets are commonly needed for collecting runoff from pavements to prevent erosion or to intercept water upgrade or downgrade of bridges. Figure 16.1.14 illustrates an embankment inlet and the catch basin with a down drain.

Bridge deck inlets have the same principles of inlet interception as pavement inlets. Figure 16.1.15 illustrates a grate inlet, which is about the maximum-size inlet for many bridge decks.

EXAMPLE 16.1.16

Determine the interception capacity of a median drop inlet ($W = 2$ ft, $L = 2$ ft) in a median ditch (see Figure 16.1.16 for definition of parameters) with $B = 2$ ft, $z = 6$, $n = 0.03$, $Q = 10$ ft³/s, $S_x = 1/6 = 0.17$, $S = 0.03$. The grate is a P-1-7/8 grate (2 ft × 2 ft).

SOLUTION

First determine the depth of flow in the median using Manning's equation for $B = 2$ ft, $Z = 6$, $Q = 10$ ft³/s. Manning's equation can be solved for the depth of flow using Newton's method, as described in Appendix A, to yield $d = 0.61$ ft. The velocity is then computed to be 5.69 ft/s.

The frontal-flow ratio can be calculated using the following formula:

$$E_0 = \frac{W}{B + dz} = \frac{2}{2 + 0.61 \times 6} = 0.35$$

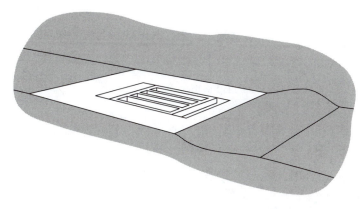

Figure 16.1.13 Median drop inlet (from Johnson and Chang (1984)).

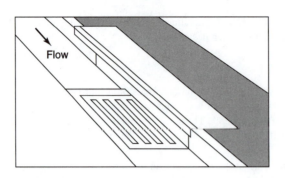

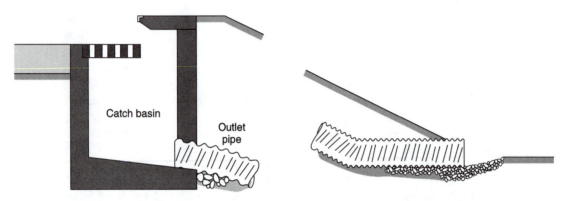

Figure16.1.14 Embankment inlet and downdrain (from Johnson and Chang (1984)).

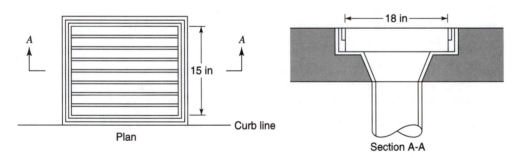

Figure 16.1.15 Bridge inlet (from Johnson and Chang (1984)).

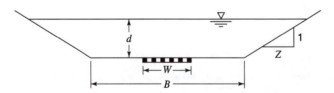

Figure 16.1.16 Median inlet in trapezoidal ditches.

$R_f = 1.0$ from Figure 16.1.8 (for $L = 2$ ft and $V = 5.69$ ft/s) and R_s is computed using equation (16.1.24) as

$$R_s = \frac{1}{1 + \dfrac{0.15(5.69)^{1.8}}{(0.17)(2)^{2.3}}} = 0.20$$

The interception efficiency is then

$$E = R_f E_0 + R_s(1 - E_0) = [1 \times 0.35 + 0.20(1 - 0.35)] = 0.48$$

The interception capacity is $Q_i = EQ = 0.48 \times 10 = 4.8$ ft^3/s, and $Q_b = 10 - 4.8 = 5.2$ ft^3/s. A berm could be placed downstream of the grate inlet for total interception of flow in the ditch.

16.2   HYDRAULIC DESIGN OF CULVERTS

Culverts are hydraulically short closed conduits that convey streamflow through a road embankment or some other type of flow obstruction. The flow in culverts may be full flow over all its length or partly full, resulting in pressurized flow and/or open-channel flow. The characteristics of flow in culverts are very complicated because the flow is controlled by many variables, including inlet geometry, slope, size, flow rate, roughness, and approach and tailwater conditions.

Culverts have numerous cross-sectional shapes, including circular, box (rectangular), elliptical, pipe arch, and arch. Shape selection is typically based upon cost of construction, limitation on upstream water surface elevation, roadway embankment height, and hydraulic performance. Culverts are also made of numerous materials, depending upon structural strength, hydraulic roughness, durability, and corrosion and abrasion resistance. Concrete, corrugated aluminum, and corrugated steel are the three most common.

Various types of inlets are also used for culverts, including both prefabricated and constructed-in-place inlets. Some of the commonly used inlets are illustrated in Figure 16.2.1. Inlet design is important because the hydraulic capacity of a culvert may be improved by the appropriate inlet selection. Natural channels are usually much wider than the culvert barrel, so that the inlet is a flow contraction and can be the primary flow control.

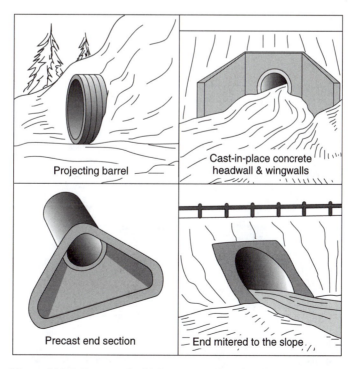

Figure 16.2.1 Four standard inlet types (schematic) (from Normann et al. (1985)).

16.2.1 Culvert Hydraulics

16.2.1.1 Types of Control

There are two basic types of flow control in culverts: *inlet control* and *outlet control*. Culverts with inlet control have high-velocity shallow flow that is supercritical, as shown in Figure 16.2.2. The control section is at the upstream end (inlet) of the culvert barrel. Culverts with outlet control have lower velocity, deeper flow that is subcritical as shown in Figure 16.2.3. The control section is at the downstream end (outlet) of the culvert barrel. Tailwater depths are either critical depth or higher.

Figure 16.2.2 illustrates four different examples of inlet control that depend on the submergence of the inlet and outlet ends of the culvert. In Figure 16.2.2a, neither end of the culvert is submerged. Flow passes through critical depth just downstream of the culvert entrance with supercritical flow in the culvert barrel. Partly full flow occurs throughout the length of the culvert, approaching normal depth at the outlet.

In Figure 16.2.2b, the outlet is submerged and the inlet is unsubmerged. The flow just downstream of the inlet is supercritical and a hydraulic jump occurs in the culvert barrel. In Figure 16.2.2c, the inlet is submerged and the outlet is unsubmerged. Supercritical flow occurs throughout the length of the culvert barrel, with critical depth occurring just downstream of the culvert entrance. Flow approaches normal depth at the downstream end. This flow condition is typical of design conditions. Figure 16.2.2d shows an unusual condition in which submergence occurs at both ends of the culvert with a hydraulic jump occurring in the culvert barrel. Note the median inlet, which provides ventilation of the culvert barrel.

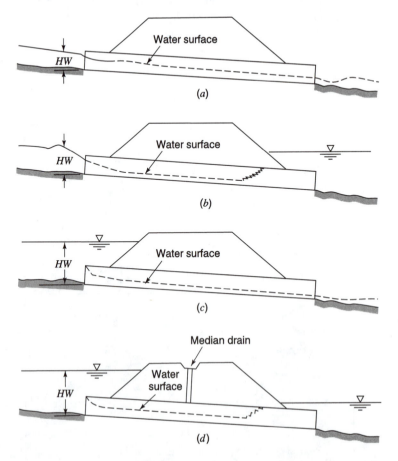

Figure 16.2.2 Types of inlet control. (*a*) Outlet submerged; (*b*) Outlet submerged, inlet unsubmerged; (*c*) Inlet submerged; (*d*) Outlet submerged (from Normann et al. (1985)).

Figure 16.2.3 illustrates five flow conditions for outlet control. Subcritical flow occurs for the partly full flow conditions. Figure 16.2.3a is the classic condition with both the inlet and the outlet submerged, with pressurized flow throughout the culvert. In Figure 16.2.3b, the outlet is submerged and the inlet is unsubmerged. In Figure 16.2.3c, the entrance is submerged enough that full flow occurs throughout the culvert length but the exit is unsubmerged. Figure 16.2.3d is a typical condition in which the entrance is submerged by the headwater and the outlet end flows freely with a low tailwater. The culvert barrel flows partly full part of the length with subcritical flow and passes through critical just upstream of the outlet. Figure 16.2.3e is another typical condition in which neither the inlet nor the outlet is submerged. The flow is subcritical and partly full throughout the length of the culvert barrel.

16.2.1.2   Inlet-Control Design Equations

A culvert under inlet-control conditions performs as an orifice when the inlet is submerged and as a weir when it is unsubmerged. The (*submerged*) *orifice discharge equation* is computed using

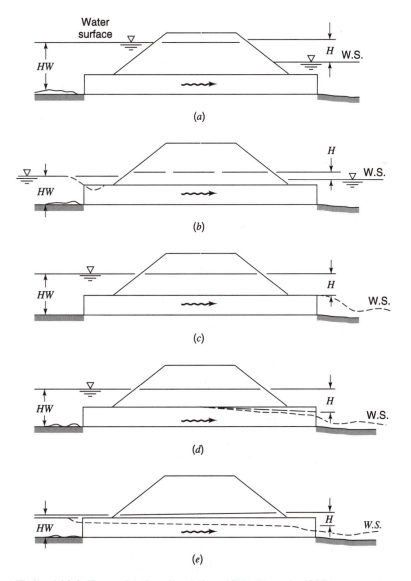

Figure 16.2.3 Types of outlet control (from Normann et al. (1985)).

$$\left[\frac{HW}{D}\right] = C\left[\frac{Q}{AD^{0.5}}\right]^2 + Y + Z \quad \text{for} \quad \left[\frac{Q}{AD^{0.5}}\right] \geq 4.0 \tag{16.2.1}$$

where HW is the headwater depth above the inlet control section invert (ft), D is the interior height of the culvert barrel (ft), Q is the discharge (ft^3/s), A is the full cross-sectional area of the culvert barrel in (ft^2), S_0 is the culvert barrel slope (ft/ft), C and Y are constants from Table 16.2.1, and Z is the slope correction factor where $Z = -0.5S_0$ in general and $Z = +0.7S_0$ for mitered inlets.

The *(unsubmerged) weir discharge equation* is (Form 1):

$$\left[\frac{HW}{D}\right] = \left[\frac{H_c}{D}\right] + K\left[\frac{Q}{AD^{0.5}}\right]^M + Z \quad \text{for} \quad \left[\frac{Q}{AD^{0.5}}\right] \leq 3.5 \tag{16.2.2}$$

where H_c is the specific head at critical depth ($H_c = d_c + V_c^2/2g$) (ft), d_c is the critical depth (ft), V_c is the critical velocity (ft/s), and K and M are constants in Table 16.2.1. A simpler equation to use for the unsubmerged condition is (Form 2):

$$\left[\frac{HW}{D}\right] = K\left[\frac{Q}{AD^{0.5}}\right]^M + Z \quad \text{for} \quad \left[\frac{Q}{AD^{0.5}}\right] \leq 3.5 \tag{16.2.3}$$

Form 2 is easier to apply and is the only documented form for some of the design inlet control nomographs in Normann et al. (1985).

Equations (16.2.1) to (16.2.3) are implemented by assuming a culvert diameter D and using it on the right-hand side of these equations and solving for [HW/D]. The headwater depth is then obtained by multiplying $D[HW/D]$. Typical inlet-control nomographs are presented in Figures 16.2.4 and 16.2.5.

Table 16.2.1   Constants for Inlet Control Design Equations

Chart[a] No.	Shape and Material	Monograph Scale	Inlet EdgeDescription	Equation Form[b]	Unsubmerged		Submerged	
					K	M	C	Y
1	Circular concrete	1	Square edge w/headwall	1	0.0098	2.0	0.0398	0.67
		2	Groove and w/headwall		.0078	2.0	.0292	.74
		3	Groove and projecting		.0045	2.0	.0317	.69
2	Circular CMP	1	Headwall	1	.0078	2.0	.0379	.69
		2	Mitered to slope		.0210	1.33	.0463	.75
		3	Projecting		.0340	1.50	.0553	.54
3	Circular	A	Beveled ring, 45° bevels	1	.0018	2.50	.0300	.74
		B	Beveled ring, 33.7° bevels		.0018	2.50	.0243	.83
8	Rectangular box	1	30° to 75° wingwall flares		.026	1.0	.0385	.81
		2	90° and 15° wingwall flares	1	.061	0.75	.0400	.80
		3	0° wingwall flares		.061	0.75	.0423	.82
9	Rectangular box	1	45° wingwall flare d = .0430	2	.510	.667	.0309	.80
		2	18° to 33.7° wingwall flare d = .0830		.486	.667	.0249	.83
10	Rectangular box	1	90° headwall w/3/4" chamfers	2	.515	.667	.0375	.79
		2	90° headwall w/45° bevels		.495	.667	.0314	.82
		3	90° headwall w/33.7° bevels		.486	.667	.0252	.865
11	Rectangular box	1	3/4" chamfers; 45° skewed headwall	2	.522	.667	.0402	.73
		2	3/4" chamfers; 30° skewed headwall		.533	.667	.0425	.705
		3	3/4" chamfers; 15° skewed headwall		.545	.667	.04505	.68
			45° bevels; 10°–45° skewed headwall		.498	.667	.0327	.75

Table 16.2.1 Constants for Inlet Control Design Equations (*continued*)

Chart[a] No.	Shape and Material	Monograph Scale	Inlet EdgeDescription	Equation Form[b]	Unsubmerged K	Unsubmerged M	Submerged C	Submerged Y
12	Rectangular box 3/4" chamfers	1	45° non offset wingwall flares	2	.497	.667	.0339	.805
		2	18.4° non offset wingwall flares		.493	.667	.0361	.806
		3	18.4° non offset wingwall flares 30° skewed barrel		.495	.667	.0386	.71
13	Rectangular box Top bevels	1	45° wingwall flares — offset	2	.495	.667	.0302	.835
		2	33.7° wingwall flares — offset		.493	.667	.0252	.881
		3	18.4° wingwall flares — offset		.497	.667	.0227	.887
16–19	CM boxes	1	90° headwall	1	.0083	2.0	.0379	.69
		2	Thick wall projecting		.0145	1.75	.0419	.64
		3	Thin wall projecting		.0340	1.5	.0496	.57
29	Horizontal ellipse concrete	1	Square edge with headwall	1	.0100	2.0	.0398	.67
		2	Groove end with headwall		.0018	2.5	.0292	.74
		3	Groove end projecting		.0045	2.0	.0317	.69
30	Vertical ellipse concrete	1	Square edge with headwall	1	.0100	2.0	.0398	.67
		2	Groove end with headwall		.0018	2.5	.0292	.74
		3	Groove end projecting		.0095	2.0	.0317	.69
34	Pipe arch 18" corner radius CM	1	90° headwall	1	.0083	2.0	.0379	.69
		2	Mitered to slope		.0300	1.0	.0463	.75
		3	Projecting		.0340	1.5	.0496	.57
35	Pipe arch 18" corner radius CM	1	Projecting	1	.0296	1.5	.0487	.55
		2	No. bevels		.0087	2.0	.0361	.66
		3	33.7° bevels		.0030	2.0	.0264	.75
36	Pipe arch 31" corner radius CM	1	Projecting	1	.0296	1.5	.0487	.55
			No. bevels		.0087	2.0	.0361	.66
			33.7° bevels		.0030	2.0	.0264	.75
40–42	Arch CM	1	90° headwall	1	.0083	2.0	.0379	.69
		2	Mitered to slope		.0300	1.0	.0463	.75
		3	Thin wall projecting		.0340	1.5	.0496	.57
55	Circular	1	Smooth tapered inlet throat	2	.534	.555	.0196	.89
		2	Rough tapered inlet throat		.519	.64	.0289	.90
56	Elliptical Inlet face	1	Tapered inlet—beveled edges	2	.536	.622	.0368	.83
		2	Tapered inlet—square edges		.5035	.719	.0478	.80
		3	Tapered inlet—thin edge projecting		.547	.80	.0598	.75
57	Rectangular	1	Tapered inlet throat	2	.475	.667	.0179	.97
58	Rectangular concrete	1	Side tapered—less favorable edges	2	.56	.667	.0466	.85
		2	Side tapered—more favorable edges		.56	.667	.3978	.87
59	Rectangular concrete	1	Slope tapered—less favorable edges	2	.50	.667	.0466	.65
			Slope tapered—more favorable edges		.50	.667	.0378	.71

[a]Chart number in Normann et al. (1985)
[b]Form 1 is equation (16.2.2)
  Form 2 is equation (16.2.3)

Source: Normann et al. (1985).

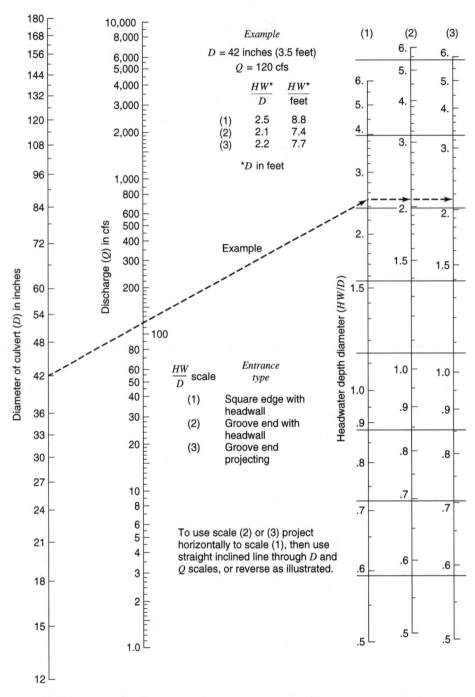

Figure 16.2.4 Headwater depth for concrete pipe culverts with inlet control (from Normann et al. (1985)).

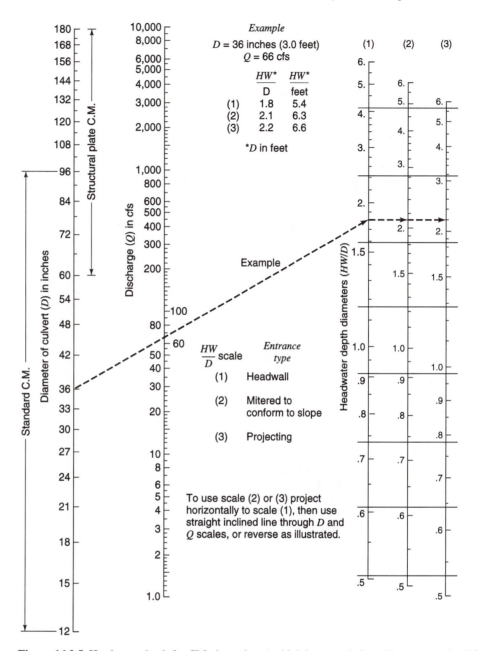

Figure 16.2.5 Headwater depth for CM pipe culverts with inlet control (from Normann et al. (1985)).

16.2.1.3 Outlet-Control Design Equations

A culvert under outlet-control conditions has either subcritical flow or full-culvert flow, so that outlet-control flow conditions can be calculated using an energy balance. For the condition of full-culvert flow, considering entrance loss H_e, friction loss (using Manning's equation) H_f, and exit loss H_0, the total headloss H is

$$H = H_0 + H_e + H_f \qquad (16.2.4a)$$

and in U.S. customary units is

$$H = \left[1 + K_e + \left(\frac{29n^2 L}{R^{1.33}}\right)\right]\frac{V^2}{2g} \tag{16.2.4b}$$

or in SI units is

$$H = \left[1 + K_e + \left(\frac{20n^2 L}{R^{1.33}}\right)\right]\frac{V^2}{2g} \tag{16.2.4c}$$

where K_e is the entrance loss coefficient, n is Manning's roughness coefficient, R is the hydraulic radius of the full-culvert barrel in ft (m), V is the velocity in ft/s (m/s), and L is the culvert length in ft (m). Other losses such as bend losses H_b, junction losses H_j, and grate losses H_g can also be added to equation (16.2.4). Table 16.2.2 lists common values of Manning's n values for culverts. Table 16.2.3 lists entrance loss coefficients for outlet control, full or part full flow.

Table 16.2.2   Manning n Values for Culverts*

Type of Conduit	Wall Description	Manning n
Concrete pipe	Smooth walls	0.010–0.013
Concrete boxes	Smooth walls	0.012–0.015
Corrugated metal pipes and boxes, annular or helical pipe (Manning n varies with barrel size)	2 2/3" by 1/2" corrugations	0.022–0.027
	6" by 1" corrugations	0.022–0.025
	5" by 1" corrugations	0.025–0.026
	3" by 1" corrugations	0.027–0.028
	6" by 2" structural plate corrugations	0.033–0.035
	9" by 2 1/2" structural plate corrugations	0.033–0.037
Corrugated metal pipes, helical corrugations, full circular flow	2 2/3" by 1/2" corrugations	0.012–0.024
Spiral rib metal pipe	Smooth walls	0.012–0.013

*Note: The values indicated in this table are recommended Manning n design values. Actual field values for older existing pipelines may vary depending on the effects of abrasion, corrosion, deflection, and joint conditions. Concrete pipe with poor joints and deteriorated walls may have n values of 0.014 to 0.018. Corrugated metal pipe with joint and wall problems may also have higher n values, and in addition may experience shape changes which could adversely affect the general hydraulic characteristics of the pipeline.

Source: Normann et al. (1985).

Table 16.2.3   Entrance Loss Coefficients for Outlet Control, Full or Partly Full $H_e = k_e \, [V^2/2g]$

Type of Structure and Design of Entrance	Coefficient k_e
Pipe, concrete	
Mitered to conform to fill slope	0.7
*End section conforming to fill slope	0.5
Projecting from fill, sq. cut end	0.5
Headwall or headwall and wingwalls	
Square-edge	0.5
Rounded (radius = 1/12D)	0.2
Socket end of pipe (groove-end)	0.2
Projecting from fill, socket end (groove-end)	0.2
Beveled edges, 33.7° or 45° bevels	0.2
Side- or slope-tapered inlet	0.2

Table 16.2.3 Entrance Loss Coefficients for Outlet Control, Full or Partly Full $H_e = k_e \, [V^2/2g]$
(*continued*)

Type of Structure and Design of Entrance	Coefficient k_e
Pipe or pipe-arch, corrugated metal	
Projecting from fill (no headwall)	0.9
Mitered to conform to fill slope, paved or unpaved slope	0.7
Headwall or headwall and wingwalls square-edge	0.5
*End section conforming to fill slope	0.5
Beveled edges, 33.7° or 45° bevels	0.2
Side- or slope-tapered inlet	0.2
Box, reinforced concrete	
Wingwalls parallel (extension of sides)	
Square-edged at crown	0.7
Wingwalls at 10° to 25° or 30° to 75° to barrel	
Square-edged at crown	0.5
Headwall parallel to embankment (no wingwalls)	
Square-edged on 3 edges	0.5
Rounded on 3 edges to radius of 1/12 barrel	
dimension, or beveled edges on 3 sides	0.2
Wingwalls at 30° to 75° to barrel	
Crown edge rounded to radius of 1/12 barrel	
dimension, or beveled top edge	0.2
Side- or slope-tapered inlet	0.2

*Note: "End section conforming to fill slope," made of either metal or concrete, are the sections commonly available from manufacturers. From limited hydraulic tests, they are equivalent in operation to a headwall in both *inlet* and *outlet* control. Some end sections, incorporating a *closed* taper in their design, have superior hydraulic performance. These latter sections can be designed using the information given for the beveled inlet.

Source: Normann et al. (1985).

Figure 16.2.6 illustrates the energy and hydraulic grade lines for full flow in a culvert. Equating the total energy at section 1 (upstream) and section 2 (downstream) gives

$$HW_0 = \frac{V_u^2}{2g} = TW + \frac{V_d^2}{2g} + H_f + H_e + H_0 \tag{16.2.5}$$

where HW_0 is the headwater depth above the outlet invert and TW is the tailwater depth above the outlet invert. Neglecting the approach velocity head and the downstream velocity head (Figure 16.2.6), equation (16.2.5) reduces to

$$HW_0 = TW + H_f + H_e + H_0 \tag{16.2.6}$$

For full flow $TW \geq D$; however, for partly full flow, the headloss should be computed from a backwater analysis. An empirical equation for the head loss H for this condition is

$$H = HW_0 - h_0 \tag{16.2.7}$$

where $h_0 = \max \, [TW, (D + d_c)/2]$.

The outlet-controlled headwater depth can be computed by first determining the tailwater depth from backwater computations where TW is measured above the outlet invert. By using equation (16.2.4) for full-flow conditions the headloss H is obtained. With equation (16.2.7) the *required outlet-controlled headwater elevation H* is obtained as

$$HW = H + h_0 - LS_0 \tag{16.2.8}$$

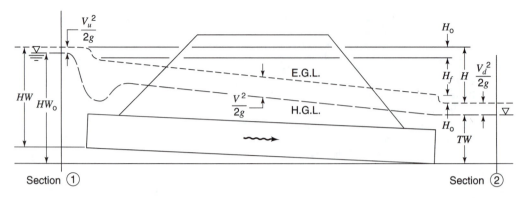

Figure 16.2.6 Full-flow energy and hydraulic grade lines (from Normann et al. (1985)).

Sample outlet-control nomographs are shown in Figures 16.2.7 and 16.2.8. Using the value of H from these nomographs, equation (16.2.8) can be implemented to compute HW_0. For Manning's n value different from that of the outlet nomograph, a modified length L_1 is used as the length scale:

$$L_1 = L\left(\frac{n_1}{n}\right)^2 \tag{16.2.9}$$

where L is the actual culvert length, n_1 is the desired Manning's n, and n is the Manning n from the chart.

The larger of the headwater elevation, obtained from the inlet- and outlet-control calculation, is adopted as the design headwater elevation. If a design headwater elevation exceeds the permissible headwater elevation, a new culvert configuration is selected and the process is repeated. Under outlet-control conditions a larger barrel is necessary since inlet improvement may have only limited effect. In the case of very large culverts, the use of multiple culverts may be required with the new design discharge taken as the ratio of the original discharge to the number of culverts. Figure 16.2.9 illustrates computation of the outlet velocity under inlet control and outlet control.

EXAMPLE 16.2.1

Analyze a 6 ft × 5 ft square-edged reinforced concrete box culvert (designed for outlet control) for a roadway crossing to pass a 50-year discharge of 300 ft³/s with the following site conditions (adapted from Normann et al., 1985):

Shoulder elevation = 113.5 ft

Stream bed elevation at culvert face = 100.0 ft

Natural stream slope = 2%

Tailwater depth = 4.0 ft

Approximate culvert length = 250 ft

Maximum allowable upstream water surface (head) elevation = 110 ft (based on adjacent structures)

The inlet is not to be depressed (no fall). Refer to Figure 16.2.10 for further details.

SOLUTION

Consider an outlet control and determine the headwater elevation (EL_{h0}) in steps 1–8.

Step 1 The tailwater depth is specified as 4.0 ft, which is obtained from backwater computations or from normal depth calculations.

Step 2 The critical depth is computed as $d_c = \sqrt[3]{\dfrac{(300/6)^2}{32.2}} = 4.27$ ft.

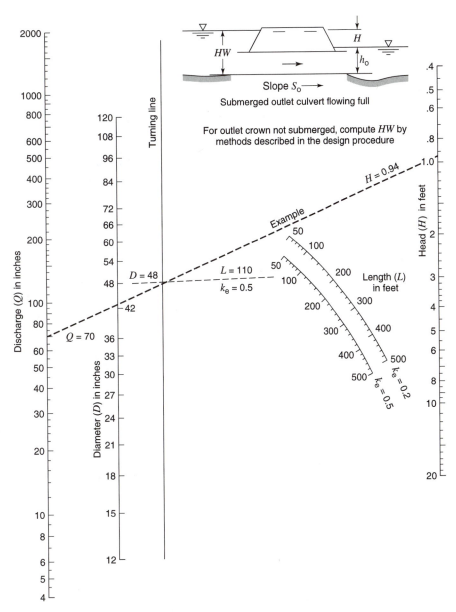

Figure 16.2.7 Head for concrete pipe culverts flowing full, $n = 0.012$ (from Normann et al. (1985)).

Step 3 $\dfrac{d_c + D}{2} = \dfrac{4.27 + 5.0}{2} = 4.64$ ft.

Step 4 $h_0 = TW$ or $(d_c + D)/2$, whichever is larger. For this problem $h_0 = 4.64$ ft.

Step 5 Use Table 16.2.3 to obtain the entrance loss coefficient. For the square-edged entrance, $K_e = 0.5$.

Step 6 Determine headlosses through the culvert barrel; use equation (16.2.4):

$$H = \left[1 + K_e + \left(\frac{29n^2 L}{R^{1.33}}\right)\right]\frac{V^2}{2g}$$

where $A = 6 \times 5 = 30$ ft^2, $V = 300/30 = 10$ ft/s, $R = A/P = 30/(6 + 6 + 5 + 5) = 1.36$ ft. For

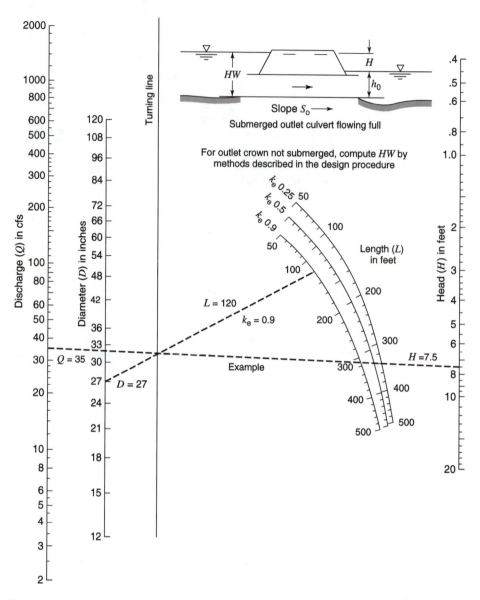

Figure 16.2.8 Head for standard CM pipe culverts flowing full, $n = 0.024$ (from Normann et al. (1985)).

concrete box culvert, take $n = 0.012$. So, $H = \left[1 + 0.5 + \left(\dfrac{29(0.012)^2(250)}{1.36^{1.33}}\right)\right]\dfrac{10^2}{2(32.2)} = 3.41 \text{ ft}$

Because $TW < D$ there is only partly full flow at the exit. The headlosses would be more accurately computed from a backwater analysis.

Step 8 Determine the required outlet control head water elevation (EL_{h0}), where $EL_{h0} = EL_0 + H + h_0$, where EL_0 is the invert elevation at the outlet:

$EL_0 = EL_i - S_0 L = 100 - 0.02(250) = 95 \text{ ft}$

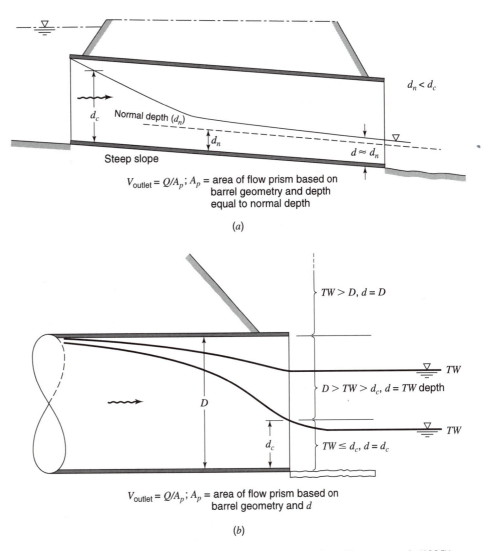

Figure 16.2.9 Outlet velocity for (a) inlet and (b) outlet control (from Normann et al. (1985)).

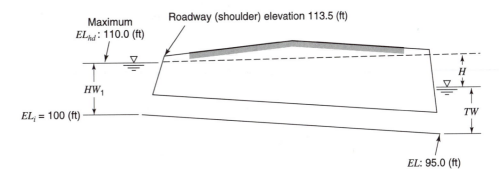

Figure 16.2.10 Details of culvert for Example 16.2.1.

Then $EL_{h0} = 95 + 3.41 + 4.63 = 103$ ft. Also the approach velocity head ($V_u^2/2g$) and the downstream velocity head can be considered in the calculation of EL_{h0} by adding $V_d^2/2g$ and subtracting $V_u^2/2g$ from the right-hand side of the above equation for EL_{h0}. Consider inlet control and determine headwater elevation, EL_{hi}.

Step 9 The design headwater elevation is now computed as $EL_{hi} = HW_i + EL_i$, so HW_i must be computed using equation (16.2.1), where

$$\frac{HW}{D} = C\left[\frac{Q}{AD^{0.5}}\right]^2 + Y + Z$$

and C and Y are obtained from Table 16.2.1 as $C = 0.0423$ and $Y = 0.82$ for a rectangular box culvert with $0°$ wing wall flares. $Z = -0.5S_0 = -0.5(0.02) = -0.01$

$$\frac{HW}{D} = 0.0423\left[\frac{300}{30(5)^{0.5}}\right]^2 + 0.82 - 0.01 = 1.66$$

To check,

$$\left[\frac{Q^2}{AD^{0.5}}\right] = 4.47 > 4$$

$$HW_i = D\left[\frac{HW}{D}\right] = 5(1.66) = 8.28 \text{ ft}$$

$$EL_{hi} = 8.28 + 100 = 108.28 \text{ ft}$$

The design headwater elevation of 108.28 ft exceeds the outlet-control headwater elevation of 103 ft. Also, the headwater elevation is less than the roadway shoulder elevation of 113.5 ft.

This design is OK; however, a smaller culvert could be considered. In fact, for this problem a 5 ft × 5 ft reinforced concrete culvert with either a square-edged entrance or a 45° beveled-edge entrance will work, as shown by Normann et al. (1985).

16.2.2   Culvert Design

The hydrologic analysis for culverts involves estimation of the design flow rate based upon the climatological and watershed characteristics. Chapters 7 through 9 and 15 cover the various methods used. The previous section described the use of the hydraulic equations and nomographs for the design of culverts under inlet and outlet conditions. This section concentrates on the use of performance curves for the design process. *Performance curves* are relationships of the flow rate versus the headwater depth or elevation for different culvert designs, including the inlet configuration. Both inlet and outlet performance curves are developed.

An overall performance curve can be developed using the following procedure (Norman et al., 1985):

1. Select a range of flow rates and determine the corresponding headwater elevation for the culvert. The flow rate should cover a range of flows of interest above and below the design discharge. Both inlet and outlet control headwater are computed.

2. Combine the inlet- and outlet-control performance curves into a single curve.

3. For roadway overtopping (culvert headwater elevation > roadway crest), compute the equivalent upstream water depth above the roadway crest for each flow rate using the weir equation

$$Q_0 = C_d L(HW_r)^{1.5} \tag{16.2.10}$$

where Q_0 is the overtopping flow rate in ft³/s (m³/s), C_d is the discharge coefficient ($C_d = k_t C_r$, see Figure 16.2.11), L is the length of roadway crest overtopped in ft (m), and HW_r is the

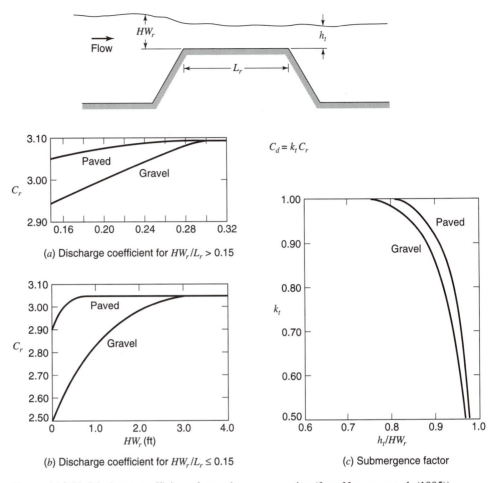

Figure 16.2.11 Discharge coefficients for roadway overtopping (from Normann et al. (1985)).

upstream depth measured from the roadway crest to the water surface upstream of the weir drawdown in ft (m).

4. Add the culvert flow and roadway overtopping flow for the corresponding headwater elevations to obtain the overall culvert performance curve. Figure 16.2.12 shows a culvert performance curve with roadway overtopping, showing the outlet-control portion and the inlet-control portion.

Tuncok and Mays (1999) provide a brief review of various computer models for culverts including HYDRAIN (www.fhwa.dot.gov) by the Federal Highway Administration CAP (http://water.usgs.gov/software/) by the U.S. Geological Survey.

EXAMPLE 16.2.2

The objective is to develop the performance curve for an existing 7-ft by 7-ft and 200-ft long concrete box culvert on a 5 percent slope that was designed for a 50-year flood of 600 ft³/s at a design headwater elevation of 114 ft (refer to Figure 16.2.12 for further details). The roadway is a 40 ft wide gravel road that can be approximated as a broad-crested weir with centerline elevation of 116 ft. The culvert inlet invert elevation is 100 ft. The tailwater depth-discharge relationship is:

Q (ft³/s)	400	600	800	1000	1200
TW (ft)	2.6	3.1	3.8	4.1	4.5

(modified from Normann et al. (1985)).

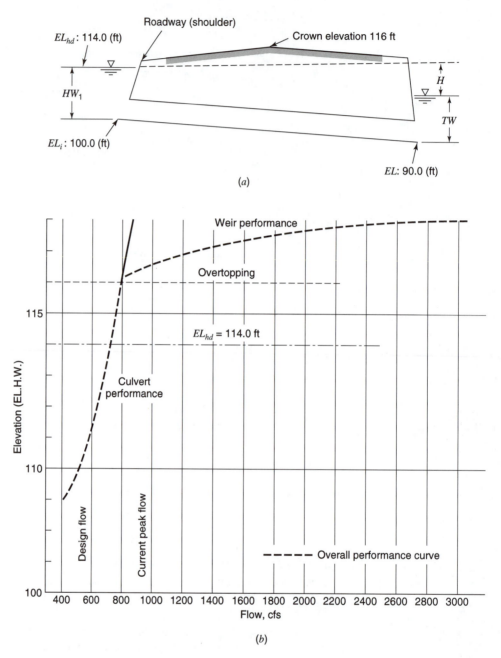

Figure 16.2.12 (*a*) Culvert profile; (*b*) Performance curve (from Johnson and Chang (1984)).

The flow width on the roadway for various elevations are:

Elevation (ft)	116.0	116.5	117.0	117.5	118.0	119.0
Flow width (ft)	0	100	150	200	250	300

SOLUTION

The same basic type of calculations performed in example 16.2.1 are followed in Table 16.2.4 for various discharges ranging from 400 ft^3/s to 1000 ft^3/s. From Figure 16.2.11, we find

$$C_d = 2.70 @ HW_r = 0.57$$

$$Q = CL(HW_r)^{1.5}$$

$$C_d = 2.92 @ HW_r = 1.5$$

$$K_t = 1$$

The performance curve computations are given in Table 16.2.5. The resulting performance curve is shown in Figure 16.2.12b.

Table 16.2.4 Discharge-headwater Computations for Culvert Flow—Example 16.2.2

Total Flow	Flow Per Barrel	Headwater Calculations										Control Head Elevation
		Inlet Control			Outlet Control							
Q (cfs)	Q/M	HW_i/D	MW_i	EL_{hi}	TW	d_c	$d_c + D$	h_0	k_e	H	EL_{h_0}	
400	57.1	1.15	8.1	108.1	2.6	4.6	5.8	5.8	0.5	1.95	97.8	108.1
600	85.7	1.65	11.6	111.6	3.1	6.1	6.6	6.6	0.5	4.4	101.0	111.6
700	100.0	1.95	13.7	113.7	3.5	6.8	6.9	6.9	0.5	6.0	102.9	113.7
800	114.3	2.35	16.5	116.5	3.8	>7	7.0	7.0	0.5	7.9	104.9	116.5
850	121.9	2.55	17.9	117.9	3.9	>7	7.0	7.0	0.5	9.0	106.0	117.9
1000	142.6	3.21	22.5	122.5	4.1	>7	7.0	7.0	0.5	1.26	109.6	122.5

Table 16.2.5 Performance Curve Computations—Example 16.2.2

Q_c Culvert flow	EL_h	H_0	Q_0 Overtopping flow	Q Total flow
400	108.1	—	—	400
600	111.6	—	—	600
700	113.7	—	—	700
800	116.5	0.5	191	991
850	117.5	1.5	1073	1923

PROBLEMS

16.1.1 Determine the time of concentration for an overland flow length of 100 m on a turf surface ($n = 0.4$) with an average slope of 0.02 for a design frequency of 10 years in Phoenix, AZ (see Figure P15.2.1).

16.1.2 Determine the time of concentration for an overland flow length of 200 m on a bare sand ($n = 0.01$) with an average slope of 0.003 m/m for a design frequency of 10 years in Phoenix, AZ (see Figure P15.2.1).

16.1.3 Determine the time of concentration for an overland flow length of 200 ft on an area ($n = 0.10$) in Colorado Springs, CO, for a design frequency of 25 years. The rainfall intensity-duration relationship for a 25-year frequency is as given below. Take the average slope of the area as 0.005 ft/ft.

Duration (min)	5	10	15	30	60
Rainfall intensity (in/hr)	7.3	5.7	4.8	3.3	2.1

16.1.4 Determine the time of concentration for an overland flow length of 400 ft on an area of bluegrass sod ($n = 0.45$) in Charlotte, NC, for a 5-year recurrence interval. The 5-year rainfall intensity-duration relationship (i in in/hr and t_D in min) is

$$i = \frac{57}{(t_D + 12)^{0.77}}$$

Take the average slope of the overland flow area as 0.010 ft/ft.

16.1.5 Determine the peak runoff from 500 ft of pavement (32 ft wide) that drains toward a gutter (for a 10-year frequency storm) in Phoenix, AZ. The pavement slope is 0.005, $n = 0.016$, and $C = 0.9$.

16.1.6 Rework problem 16.1.5 for a 25-year storm.

16.1.7 Determine the runoff from 600 ft of pavement (32 ft wide) that drains toward a gutter for a 25-year frequency storm in

Colorado Springs, CO. The pavement slope is 0.02, $n = 0.016$, and $C = 0.9$. See problem 16.1.3 for rainfall information.

16.1.8 Compute the time of flow in a gutter for the situation in which the spread T_1 at the upstream end is 1 ft, which results from the bypass flow from the upstream inlet. The design spread T_2 at the second inlet is 5 ft, the gutter slope is $S = 0.025$, and the cross slope is $S_x = 0.02$. Manning's $n = 0.016$ and the inlet spacing is 200 ft.

16.1.9 Rework problem 16.1.8 with $T_1 = 0.5$ m, $T_2 = 1.5$ m, and the inlet spacing equals 100 m. All other information is the same as in problem 16.1.8.

16.1.10 Compute the flow rate in a gutter section with a design spread of $T_w = 8$ ft and a cross-slope of $S_x = 0.025$. The gutter slope $S = 0.01$ and Manning's n is 0.016.

16.1.11 Compute the flow rate in a gutter section with a design spread of 2.45 m and cross-slope of $S_x = 0.025$. The gutter slope $S = 0.01$ and Manning's n is 0.016.

16.1.12 Compute the flow rate in a composite gutter section with a design spread of $T_w = 8$ ft, a cross-slope of $S_x = 0.03$, $W = 2.0$ ft with a 2-in gutter depression. The gutter slope is $S = 0.005$ and Manning's n is 0.016.

16.1.13 Compute the flow rate in a composite gutter section with a design spread of $T_w = 2.45$ m, a cross-slope of $S_x = 0.03$, $W = 0.60$ m with a 5-cm gutter depression. The gutter slope is $S = 0.005$ and Manning's n is 0.016.

16.1.14 Using the data for the composite gutter section in problem 16.1.12, find the interception capacity of a bar grate (P-1-7/8) that is 2 ft long and 2 ft wide.

16.1.15 Rework problem 16.1.14 using a parallel bar grate that is 2 ft long and 2 ft wide.

16.1.16 Rework problem 16.1.14 using a curved vane grate (CV-3-1/4-4-1/4) that is 2 ft long and 2 ft wide.

16.1.17 Using the data in problem 16.1.13, find the interception capacity of a bar grate (P-1-7/8) that is 0.6 m long and 0.6 m wide.

16.1.18 Rework problem 16.1.17 using a curved vane grate (CV-3-1/4-4-1/4) that is 0.6 m long and 0.6 m wide.

16.1.19 Using the data in problem 16.1.14, determine the interception capacity for an 8-ft curb opening.

16.1.20 Using the data in problem 16.1.16, determine the interception capacity for a 1.83 m curb opening.

16.1.21 Rework Example 16.1.6 for a 5 ft curb-opening inlet.

16.1.22 Determine the interception capacity for a 3.048 m curb-opening inlet given $S_x = 0.03$, $n = 0.016$, $S = 0.035$, and $T_w = 2.44$ m.

16.1.23 Solve problem 16.1.22 for a depressed 3.048 m curb-opening inlet with $a = 5.08$ cm and $W = 0.61$ m.

16.1.24 Determine the interception capacity for a 10-ft curb-opening inlet given $S_x = 0.03$, $n = 0.016$, $S = 0.035$, $T_w = 8$ ft.

16.1.25 Solve problem 16.1.24 for a depressed 10 ft curb-opening inlet with $a = 2$ in and $W = 2$ ft.

16.1.26 Determine the interception capacity of a combination curb opening-grate inlet on grade. The curb opening is 10 ft long and the grate is a 2 ft by 2 ft reticuline grate placed along side the downstream 2 ft of the curb opening ($Q = 7$ ft³/s, $S = 0.04$, $S_x = 0.03$, and $n = 0.016$).

16.1.27 Compare the interception capacity of the combination inlet described in problem 16.1.26 with the interception capacity of just using the curb-opening inlet and with the interception capacity of just using the grate inlet.

16.1.28 Determine the grate size for a symmetrical sag vertical curve (with a curb) with equal bypass from inlets upgrade of the low point. Consider 50 percent clogging of the grate. The design discharge is 8 ft³/s and $Q_b = 3.6$ ft³/s. Determine the depth at the curb for a check discharge of 11 ft³/s ($Q_b = 4.4$ ft³/s). Check the spread at $s = 0.003$ on approaches to the low point ($T_w = 10$ ft, $S_x = 0.03$).

16.1.29 Compute the interception capacity of a 6-ft long and 5-in high curb opening inlet in a sag location.

(a) First consider an undepressed curb opening ($S_x = 0.05$ and $T_w = 8$ ft).

(b) Next consider a depressed curb opening ($S_x = 0.05$, $a = 2$ in, $W = 2$ ft, and $T_w = 8$ ft).

16.1.30 Determine the length of a slotted inlet in a sump location that is required to limit the maximum depth at a curb to 0.4 ft for a discharge of 3.0 ft³/s. Assume no clogging and use a slot width of 1.75 in.

16.1.31 Compute the interception capacity of a combination inlet in sag location. The grate is a P-1-7/8 that is 2 ft × 6 ft with a curb opening of 12 ft and $h = 4$ in. The cross-slope $S_x = 0.03$ and the discharge is 10 ft³/s. What is the depth at the curb and the spread? Assume no clogging.

16.1.32 Consider a 26 ft pavement width on a continuous grade, $n = 0.016$, $S_x = 0.03$, $S = 0.03$, $T_w = 8$ ft, $i = 10.7$ in/hr, and $C = 0.8$. Determine the maximum inlet spacing for a 2-ft by 2-ft curved vane inlet.

16.1.33 Consider a 7.92-m pavement width on a continuous grade, $n = 0.016$, $S_x = 0.03$, $S = 0.03$, $T_w = 2.45$ m, $i = 27.2$ cm/hr, and $C = 0.8$. Determine the maximum inlet spacing for a 0.61-m by 0.61-m curved vane inlet.

16.1.34 Using the data from problem 16.1.32, determine the maximum design inlet spacing for a 10-ft curb opening depressed 2 in from the normal cross-slope in a 2-ft wide gutter.

16.1.35 Using the data from problem 16.1.33, determine the maximum design inlet spacing for a 3.048 m curb opening depressed 5.08 cm from the normal cross-slope in a 0.61-m wide gutter.

16.1.36 Using the data from problem 16.1.32, determine the maximum inlet spacing using a 15 ft slotted inlet.

16.1.37 Using the data from problem 16.1.33, determine the maximum inlet spacing using a 4.57 m slotted inlet.

16.1.38 Consider a sag vertical curve at an underpass on a four-lane divided highway facility. The spread at design Q is not to exceed the shoulder width of 10 ft, the cross-slope is $S_x = 0.05$, and $K = 130$. (a) Determine the location of flanking inlets if they are to function in relief of the inlet at the low point when depth at the curb exceeds design depth. (b) Determine the location of flanking inlets when the depth at the curb is 0.2 ft less than the depth at design spread.

16.1.39 Consider a 2 ft by 2 ft P-1-7/8 grate that is to be placed in a flanking inlet location in a sag vertical curve that is 250 ft downgrade from the most downstream curved vane inlet in problem 16.1.32 ($Q_b = 2.37$ ft³/s, $S_x = 0.03$, $T_w = 8$ ft, $n = 0.016$, $i = 10.7$ in/hr). The slope on the curve at the inlet is $S = 0.006$. Determine the spread at the flanking inlet and at $S = 0.003$.

16.1.40 Rework example 16.1.16 to determine the interception capacity of a larger median drop inlet ($W = 2$ ft, $L = 4$ ft).

16.2.1 Rework example 16.2.1 using a 5 ft × 5 ft culvert with a square-edged entrance.

16.2.2 Rework example 16.2.1 using a 5 ft × 5 ft culvert with a 45° beveled-edged entrance.

16.2.3 Rework example 16.2.1 for a discharge of 200 ft³/s.

16.2.4 A culvert is to be designed for a new roadway crossing for a 25-year peak discharge of 200 ft³/s. Use a circular corrugated metal pipe culvert with standard 2-2/3 by 1/2 corrugations and beveled edges. The site conditions include: elevation at culvert face = 100 ft; natural stream bed slope = 1%; tailwater for 25-year flood = 3.5 ft; approximate culvert length = 200 ft; shoulder elevation = 110 ft. Base the design headwater on the shoulder elevation with a 2-ft freeboard (elevation of 108 ft). Set the inlet invert at the natural stream bed elevation (no fall). Analyze the design.

16.2.5 Rework problem 16.2.4 using a concrete pipe with a grooved end.

16.2.6 Rework example 16.2.2 to develop the performance curve for an 8 ft by 7 ft concrete box culvert with a square-edged entrance. All other conditions are the same as in example 16.2.2.

16.2.7 Determine if a 1.5 m × 1.5 m square-edged reinforced concrete box culvert is adequate for a roadway crossing to pass a discharge of 8.50 m³/s. The inlet is not depressed (no fall). The site conditions are as follows:

Shoulder elevation = 34.6 m

Stream bed elevation at culvert face = 30.5 m

Natural stream slope = 2%

Tailwater depth = 1.2 m

Approximate culvert length = 76.2 m

Upstream water surface (head) elevation = 33.5 m

REFERENCES

American Association of State Highway and Transportation Officials (AASHTO), *Model Drainage Manual*, Washington, DC, 1991.

Johnson, F. L., and F. M Chang, *Drainage of Highway Pavements*, Hydraulic Engineering Circular, No. 12, Federal Highway Administration, McLean, VA, 1984.

Normann, J. M., R. J. Houghtalen, and W. J. Johnson, *Hydraulic Design of Highway Culverts*, Hydraulic Design Series No. 5, Report No. FHWA-IP-85-15, Federal Highway Administration, U.S. Department of Transportation, McLean, VA, 1985.

Ragan, R. M., "A Nomograph Based on Kinematic Wave Theory for Determining Time of Concentration for Overland Flow," Report No. 44, prepared by Civil Engineering Department, University of Maryland at College Park, Maryland State Highway Administration and Federal Highway Administration, December 1971.

Tuncok, I. K. and L. W. Mays, "Hydraulic Design of Culverts and Highway Structures," *Hydraulic Design Handbook*, edited by L. W. Mays, McGraw-Hill, New York, 1999.

Young, G. K., Jr., and S. M. Stein, "Hydraulic Design of Drainage for Highways," *Hydraulic Design Handbook*, edited by L. W. Mays, McGraw-Hill, New York, 1999.

Young, G. K., Jr., S. M. Walker, and F. Chang, *Design of Bridge Deck Drainage*, Hydraulic Engineering Circular No. 21, FHWA-SA-92-010, Federal Highway Administration, U.S. Department of Transportation, Washington DC, 1993.

Chapter 17

Design of Spillways and Energy Dissipation for Flood Control Storage and Conveyance Systems

This chapter is an introduction to the design of hydraulic structures (spillways and energy dissipators) for flood-control storage systems (reservoirs). A brief introduction to the development of the probable maximum flood is followed by a section on dams including types of dams, hazard classification of dams, spillway capacity criteria, safety of existing dams, and hydraulic analysis of dam failure. The major emphasis of this chapter is on the design of spillways, in particular overflow spillways, and the design of hydraulic jump type stilling basins and energy dissipators. The major portion of the discussion follows the U.S. Bureau of Reclamation (1987) criteria and presents the design procedures of that publication.

17.1 HYDROLOGIC CONSIDERATIONS

The purpose of this section is briefly to define the steps in determining the probable maximum flood (PMF) for the design and analysis of dams and their appurtenances. Estimated limiting value are defined in Chapter 7. The determination of the PMF involves two basic steps: (1) to synthesize a hydrograph of inflow into the reservoir and (2) to model or simulate the movement of the flood through the reservoir and past the dam. The various steps in such an analysis, all of which have been discussed in the National Research Council's (NRC's) report, *Safety of Existing Dams: Evaluation and Improvement*, Chapter 4 (1983), are generally as follows (NRC, 1985):

1. Dividing drainage area into subareas, if necessary
2. Deriving runoff model
3. Determining probable maximum precipitation (PMP) using criteria contained in the NOAA Hydrometeorological Report series (see Section 7.2.5)

4. Arranging PMP increments into a logical storm rainfall pattern called the probable maximum storm (PMS)

5. Estimating for each time interval the losses from rainfall due to such actions as surface detention and infiltration within the watershed

6. Deducting losses from rainfall to estimate rainfall excess values for each time interval

7. Applying rainfall excess values to a runoff model of each subarea of the basin

8. Adding to storm runoff hydrograph allowances for base flow of stream runoff from prior storms to obtain the synthesized flood hydrograph for each subarea

9. Routing of the flood for each subarea to points of interest

10. Routing the inflow through the reservoir storage, outlets, and spillways to obtain estimates of storage elevations, discharges at the dam, tailwater elevations, etc., that describe the passage of the flood through the reservoir. (This is essentially a process of accounting for volumes of water in inflow, storage, and outflow through the flood period. If there are several reservoirs in the watershed, the reservoir routing is repeated from the uppermost to the most downstream reservoir, in turn.)

17.2   DAMS

17.2.1   Type of Dams

Figure 17.2.1 illustrates the basic types of dams, classified on the basis of the type and materials of construction. This figure illustrates gravity dams, arch dams, buttress dams, and embankment dams. Dams can also be classified according to use or hydraulic design.

The most common type of dam is embankment *earthfill dams*. Their construction is principally from required excavation using the available materials from the construction. Most new earthfill dams are roll fill type dams, which can be further classified as homogenous, zoned, or diaphragm (U.S. Bureau of Reclamation, 1987). *Homogenous dams* are composed of only one kind of material, exclusive of the slope protection. The material used must be impervious enough to provide an adequate water barrier and the slope must be relatively flat for stability. Figure 17.2.2 illustrates the seepage through a completely homogenous dam. It is more common today to build modified homogenous sections in which pervious materials are placed to control steeper slopes. Three methods are used: rockfill toe, horizontal drainage blanket, and inclined filter drain with a horizontal drainage blanket, as illustrated in Figure 17.2.3. Pipe drains are also used for drainage on small dams in conjunction with a horizontal drainage blanket or a pervious zone.

For *diaphragm-type earthfill dams*, the embankment is constructed of pervious materials (sand, gravel, or rock). A thin diaphragm of impermeable material is used to form a water barrier. The diaphragm may vary from a blanket on the upstream face to a central vertical core. Diaphragms may consist of earth, portland cement concrete, bituminous concrete, or other materials.

Zoned embankment-type earthfill dams have a central impervious core that is flanked by a zone of materials considerably more pervious, called *shells*. These shells enclose, support, and protect the imperious core. Figure 17.2.4 illustrates the size range of impervious cores used in zoned embankments.

The other type of embankment dam is the *rockfill dam*, which consists of rock of all sizes to provide stability and an impervious core membrane. Membranes include an upstream facing of impervious soil, a concrete slab, asphaltic concrete paving, steel plates, other impervious soil (U.S. Bureau of Reclamation, 1987). Figure 17.2.5 illustrates the typical maximum section of an earth core rockfill dam using a central core. Figure 17.2.6 illustrates a decked rockfill dam that has an asphaltic concrete membrane on the upstream face.

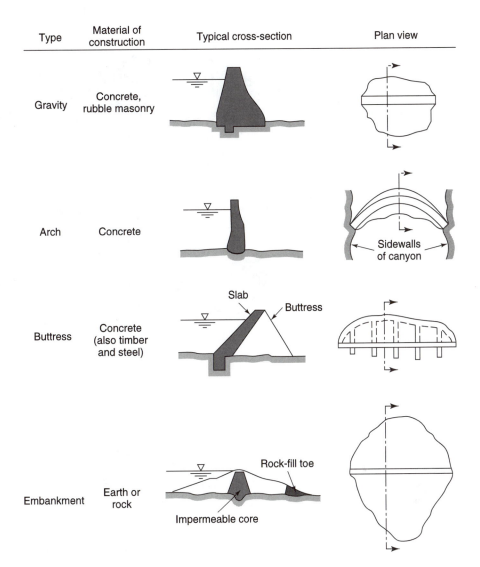

Figure 17.2.1 Basic types of dams (from Linsley et al. (1992)).

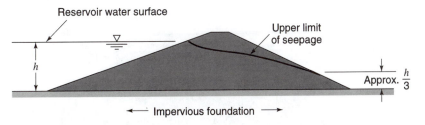

Figure 17.2.2 Seepage through a completely homogeneous dam (from U.S. Bureau of Reclamation (1987)).

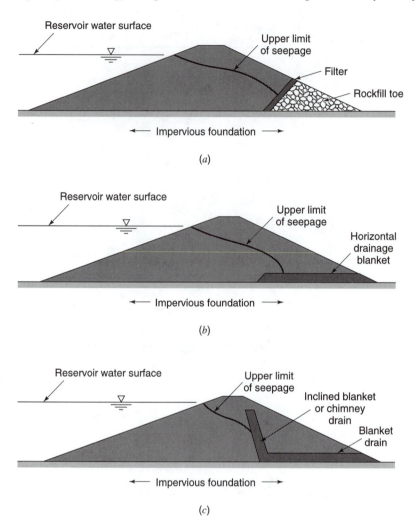

Figure 17.2.3 Seepage through modified homogeneous dams: (*a*) With rockfill toe; (*b*) With horizontal drainage blanket; (*c*) With chimney drain (from U.S. Bureau of Reclamation (1987)).

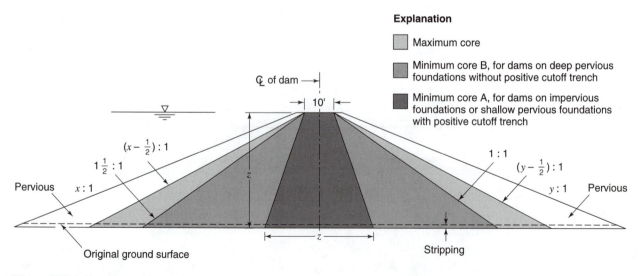

Figure 17.2.4 Size range of impervious cores used in zoned embankments (from U.S. Bureau of Reclamation (1987)).

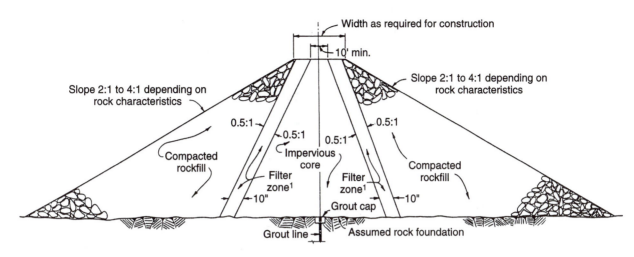

Figure 17.2.5 Typical maximum section of an earth-core rockfill dam using a central core (from U.S. Bureau of Reclamation (1987)).

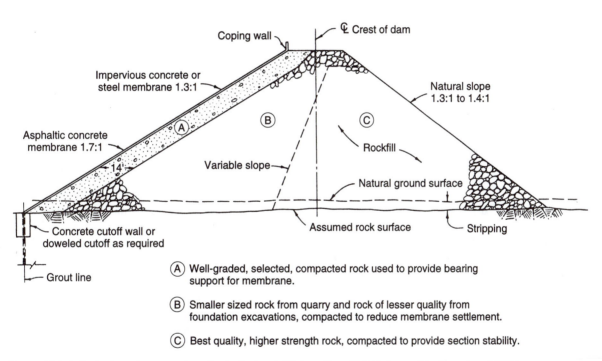

Ⓐ Well-graded, selected, compacted rock used to provide bearing support for membrane.

Ⓑ Smaller sized rock from quarry and rock of lesser quality from foundation excavations, compacted to reduce membrane settlement.

Ⓒ Best quality, higher strength rock, compacted to provide section stability.

Figure 17.2.6 Typical maximum section of a decked rockfill dam (from U.S. Bureau of Reclamation (1987)).

Concrete arch dams are used where the ratio of the width between abutments to the height is not great and where there is solid rock capable of resisting arch thrust at the abutment (foundation). Arch dams can be either single- or multiple-arch dams. *Concrete buttress dams* consist of flat deck and multiple arch structures.

17.2.2 Hazard Classification of Dams

Many systems have been developed for the hazards classification of dams relating to hydrologic and seismic events (NRC, 1985). Most of these systems utilize dam height, volume of water impounded, and probable effects of dam failure due to the downstream. The classifications used by the U.S. Army Corps of Engineers (1982), as shown in Table 17.2.1, have been adopted by several federal and state agencies has been adapted.

Table 17.2.1 Terms for Classifying Hazard Potentials

Category	Impoundment (ac-ft)	Height of Dam (ft)
Size of dam[a]		
Small	50 to 1,000	25 to 40
Intermediate	1,000 to 50,000	40 to 100
Large	Over 50,000	over 100

Category	Loss of Life (Extent of Development)	Economic Loss
Hazard potential classification		
Low	None expected (no permanent structures for human habitation)	Minimal (undeveloped to occasional structures or agriculture)
Significant	Few (no urban developments and no more than a small number of inhabitable structures)	Appreciable (notable agriculture, industry, or structures)
High	More than few	Excessive (extensive community, industry, or structures)

[a]Criterion that places project in largest category governs.

Source: National Research Council (1985).

The U.S. Army Corps of Engineers' (1991) policy for *dam safety evaluation* is that "a dam failure must not present a hazard to human life...." Dam safety evaluation must: (1) formulate any reservoir plan to comply with this safety requirement and (2) evaluate the impact of catastrophic failure of any proposed reservoir plan to confirm that their performance constraint is satisfied. The four design standards, which depend on the type of dam and risk to life, are described in Table 17.2.2. Section 17.2.4 discusses the method for failure evaluation.

Table 17.2.2 Functional Design Standards for New Dams Designed by U.S. Army Corps of Engineers

Standard 1: Design dam and spillway large enough to assure that the dam will not be overtopped by floods up to probable maximum categories.

Standard 2: Design the dam and appurtenances so that the structure can be overtopped without failing and, insofar as practicable, without suffering serious damage.

Standard 3: Design the dam and appurtenances in such a manner as to assure that breaching of the structure from overtopping would occur at a relatively gradual rate, such that the rate and magnitude of increases in flood stages downstream would be within acceptable limits, and that damage to the dam itself would be located where it could be most economically repaired.

Standard 4: Keep the dam low enough and storage impoundment small enough that no serious hazard would exist downstream in the event of breaching, and so that repairs to the dam would be relatively inexpensive and simple to accomplish.

17.2.3 Spillway Capacity Criteria

Table 17.2.3 compares the various spillway capacity criteria used in the United States as of 1985. These are based on the classification used for the National Dam Inspection Program. This table illustrates the diversity of criteria of various federal, state, and local government agencies, professional societies, and privately owned firms in the United States The spillway requirements for large high-hazard dams are fairly consistent, but there are fairly widespread differences in criteria for other classes of dams. The Institution of Civil Engineers (1978) in the United Kingdom developed the standards presented in Table 17.2.4.

Table 17.2.3 Comparison of Indicated Spillway Capacity Criteria in Use or Proposed by Federal Agencies

Federal Agencies	Hazard Class High — Size of Dam			Hazard Class Significant — Size of Dam			Hazard Class Low — Size of Dam		
	Large	Intermediate	Small	Large	Intermediate	Small	Large	Intermediate	Small
Ad Hoc ICODS of FCCSET	PMF	PMF	PMF	PMF	PMF	PMF	*	*	*
Bureau of Reclamation	PMF	PMF	PMF	*	*	*	*	*	*
FERC	PMF	PMF	PMF	PMF	PMF	PMF	*	*	*
Forest Service	(See Corps criteria for National Dam Inspection Program)								
ICODS	PMF	PMF	PMF	*	*	*	*	*	*
National Weather Service	(Does not establish criteria for dams)								
SCS	PMP	PMP	PMP	$(P_{100}+0.4(PMP-P_{100}))$			*	*	*
TVA	PMF	PMF	PMF	(TVA Max. Prob. Fld.)			*	*	*
Corps of Engineers (Corps Projects)	PMF	PMF	PMF	*	*	*	*	*	*
Corps of Engineers (National Inspection Program)	PMF	PMF	1/2 PMF to PMF	PMF	1/2 PMF to PMF	100 yr to 1/2 PMF	1/2 PMF to PMF	100 yr to 1/2 PMF	50 yr to 100 yr
Nuclear Regulatory Commission	PMF	PMF	PMF	(See Corps criteria for National Dam Inspection program).					

Source: National Research Council (1985).

Table 17.2.4 Reservoir Flood and Wave Standards by Dam Category

Category	Initial reservoir condition is tolerable	Dam Design Flood Inflow			Concurrent wind speed and minimum wave surcharge allowance
		General standard is warranted	Minimum standard if rare overtopping is tolerable	Alternative standard if economic study is warranted	
(a) Reservoir where a breach will endanger lives in a community	Spilling long-term average daily inflow	Probable maximum flood (PMF)	0.5 PMF or 10,000-year flood (take larger)	Not applicable	Winter: maximum hourly wind once in 10 years Summer; average
(b) Reservoirs where a breach: (i) may endanger lives not in a community;	Just full (i.e., no spill)	0.5 PMF or 10,000-year flood (take larger)	0.3 PMF or 1000-year flood (take larger)	Flood with probability that minimizes spillway plus damage cost, inflow	annual maximum hourly wind Wave surcharge allowance not less than 0.6 m

Table 17.2.4   Reservoir Flood and Wave Standards by Dam Category (*continued*)

| Category | Initial reservoir condition is tolerable | Dam Design Flood Inflow | | | Concurrent wind speed and minimum wave surcharge allowance |
		General standard is warranted	Minimum standard if rare overtopping is tolerable	Alternative standard if economic study is warranted	
(ii) will result in extensive damage				not to be less than minimum standard but may exceed general standard	
(c) Reservoirs where a breach will pose negligible risk to life and cause limited damage	Just fill (i.e., no spill)	0.3 PMF or 1000-year flood (take larger)	0.2 PMF or 150-year flood (take larger)		Average annual maximum hourly wind; wave surcharge allowance not less than 0.4 m
(d) Special cases where no loss of life can be foreseen as a result of a breach and very limited additional flood damage will be caused	Spilling long-term average daily flow	0.2 PMF or 150-year flood	Not applicable	Not applicable	Average annual maximum hourly wind; wave surcharge allowance not less than 0.3 m

Where reservoir control procedure requires, and discharge capacities permit, operation at or below specified levels defined throughout the year, these may be adopted providing they are specified in the certificates or reports for the dam. Where a proportion of PMF is specified it is intended that the PMF hydrograph should be computed and then all ordinates be multiplied by 0.5, 0.3, or 0.2 as indicated.

Source: Institution of Civil Engineers (1978).

17.2.4   Safety of Existing Dams

A National Research Council (1983) report presented evaluation matrices of various types of dams and appurtenances. Table 17.2.5 is the evaluation matrix for embankment dams and Table 12.2.6 is the evaluation matrix for appurtenant structures, which are other structures around a dam that are necessary for operation. Table 17.2.5 for embankment dams lists possible defects, possible indicators, possible causes, effects, and potential remedial measures. Table 17.2.6 for appurtenances lists defects, causes, effects, and remedies.

Table 17.2.5   Evaluation Matrix of Embankment Dams

Defect	Possible Indicators	Possible Causes	Effects	Potential Remedial Measures

(A) Embankment mass movement (slope failure)	Slumps on embankment face
	Longitudinal cracks
	Arcuate cracks
	Hummocky (irregular) slope
	Bulge in slope
	Sag in crest
	Bent tree trunks
	Misaligned guard rails or similar structures
		Inadequate strength
		Slopes too steep
		Phreatic surface too high
		Cracking due to differential settlement
		Earthquake
		Rapid drawdown of reservoir or tailwater
		Large trees on dam overturned
		Spillway or surface drainage discharge eroding embankment
		Temporary saturation due to rain storms, snowmelt, or high tailwater
		Decaying organic material in embankment
			Possible massive failure of dam
			Damage to spillway or outlet works, resulting in dam failure
				Determine specific cause(s) by test borings, strength tests, and piezometers. Based on test results, design appropriate remedies. Some alternatives are:
				    Free-draining downstream buttress
				    Flatten slopes
				    Lower the phreatic surface (upstream barrier, internal slurry wall or membrane cutoff, grouting)
				    Remove and replace weak soils
				    Control surface erosion with riprap or other means
				    Realign-relocate appurtenant structures as required

Table 17.2.5 Evaluation Matrix of Embankment Dams (*continued*)

Defect	Possible Indicators	Possible Causes	Effects	Potential Remedial Measures
				Permanent partial reduction in pool level
				In some cases total draining and breaching are required for safety or are more economical

(B) Embankment excessive seepage	Seepage carrying soil fines
	Sinkholes on embankment face
	Boils
	Concentrated seepage
	Unusual wetness on embankment slope
	Unusually soft or quick embankment slope
	Marsh-type vegetation on embankment slope
		Lack of appropriate internal drainage
		Inadequate core or cutoff material
		Inappropriate embankment
		Layering of relatively permeable zones in embankment
		Clogging of drains or filters
		Burrows caused by muskrats, beavers, groundhogs, foxes, moles, chipmunks
		Surface erosion gullies intersecting seepage zone
		Temporary saturation due to rain storms, snowmelt
		Seepage into, out of, or along conduits and drains
			Dam failure by internal erosion
			Structural failure due to uplift of appurtenant structures
			Loss of storage
				Distinguishing unsafe seepage from normal seepage requires considerable judgment. Amount of change in the rate of seepage is an important factor. Many require installation of piezometers to help determine seriousness. Highly concentrated seepage *or* evidence of internal erosion or mass movement definitely requires treatment. If it appears that seepage line is high enough to threaten mass stability, consider steps under mass movement above. *If mass movement is not indicated, a filtered drain in the area(s) of concern is usually most appropriate.* Other alternatives:
				  Upstream seepage barrier (blanket)
				  Install seepage cutoff beneath crest, such as slurry wall, thin membrane wall, grouting
				  Filtered relief wells
				  Fill gullies with filtered drain, riprap, prevent further erosion
				  Remove trees, replace soil
				  Trap and remove animals
				  In some cases total draining and breaching is the most economical safe action

Source: National Research Council (1985).

Table 17.2.6 Evaluation Matrix of Appurtenant Structures

Type of Defect	Causes	Effects	Remedies
Defective spillways	Insufficient analysis	Overtopping[a] Erosion or washout on downstream side	Reevaluate spillway capacity using present-day hydrologic techniques
	Design error		Using watershed model simulation and prototype studies in design

	New criteria established	Erosion along and around spillway chute	Institute major repairs:
			  Increase spillway capacity
			  Construction of auxiliary or emergency

	Major or unpredicted events	Breach	Alternate methods[d]
			  Revise reservoir operating procedures
			  Restrict reservoir utilization

Table 17.2.6 Evaluation Matrix of Appurtenant Structures (*continued*)

Type of Defect	Causes	Effects	Remedies
			Require attendance of dam personnel during flood events
			Establish well-defined emergency procedure
Obstruction to spillways and outlet works	Excess trash[b] burden	Overtopping	Install log booms or trash racks based on use of reservoir, anticipated trash burden, etc.
		Erosion	Perform maintenance as required to remove excess trash buildup
		Damage to trash racks	
Defective gates and hoists	Mechanical breakdown	Upsets normal operation characteristics of dam	Perform regular maintenance on mechanical equipment
	Inadequate gate seals	Vibrations	Check bottom gate seals for damage
		Fatigue cracking	Provide for sharp clean flow breakoff
	Cavitation around gate guides	Damage to gate frames and operating shaft	Repair cavitated areas[c] with steel liners; check that all gate frames are securely mounted
	Differential foundation settlement	Gates becoming inoperable	Repair foundation
	Trash and debris	Gate frames crack	
		Vibration	Install trash tracks
		Trash can knock gates from frames	
	Galvanic corrosion and/or mineral deposits	Corrode moveable parts; makes gates inoperable	Provide cathodic protection
			Exercise gate to prevent formation of deposits
	Poor design and/or inadequate operational procedures	Vibration	Revise operating procedures
		Unbalanced flow (can cause other problems to occur, such as buckling of steel liners and concrete erosion)	Provide adequate air vents
Defective conduits	Surface irregularities (offset joints, voids, transverse grooves, roughness)	Cavitation erosion	Grinding surfaces to smoothness that will prevent cavitation erosion
		Piping	Air vents at irregularities
			Require close construction tolerances
			Provide aeration grooves to draw air into flowing water
	Sealing in conduit	Unsteady flow conditions	Perform prototype studies to modify
		Structural vibrations	Adequate air vents
	Unsymmetrical flow	Cavitation	Repair concrete
			Install guide vanes
		Erosion in stilling basins	Baffle blocks at terminal structure
			Adequate air vents
	Settlement of foundations	Joint separation	Stabilize foundations
		Structural cracking	Replace joint collars
		Piping	Replace joint seals
	Corrosion	Piping of embankment material through holes	Replace or repair conduit
Defective drainage system	Inadequate design	Uncontrolled seepage	Investigate and modify
			Install new or improve existing drain field
	Improper installation	Piping	Provide relief wells
		Boils	Reduce reservoir pool level
	Inadequate filter layer	Saturated conditions	Improve filter layer
		Seepage of fines from foundation	
	Mineral deposition	Clogging	Ream drains. Drill supplemental drains

Table 17.2.6 Evaluation Matrix of Appurtenant Structures (*continued*)

Type of Defect	Causes	Effects	Remedies
Erosion	Inadequate design of spill-ways and stilling basins	Fluctuating positive to negative or uplift pressures can develop on spillways and stilling basins (can cause cracking of concrete slabs in stilling basins and subsequent removal of embankment material); this fluctuation of pressure can demolish a spillway or stilling basin	Increase thickness of concrete slabs Impose tailwater elevation that will force hydraulic jump Provide floor drain openings in locations to avoid subjecting them to fluctuating pressures
	Structural cracks in concrete slabs of spillways and stilling basins	Water seepage through slab and eroding of embankment materials	Pressure grout cracks in slab Replace with thicker slab
		Development of voids under slab	Evaluate effectiveness of energy dissipators and replace if necessary
		Loss of slab support	
		Breakup of slab	Fill voids under concrete slabs Anchor invert
	Unsymmetrical operation of outlet gates	Unsymmetrical loading of spillway	Operate gates symmetrically
		Scour actions in discharge area	Repair with erosion resistant aggregate and high-strength concrete
	Excessive discharges Abrasive objects in stilling basin (rocks, construction debris, etc.)	Abrasion and cavitation erosion of concrete in spillway and stilling basins	Repair with special concretes and steel plates Line dissipators with steel plates
		Damage to chute blocks and enegy dissipators	Install riprap
		Breakup of slabs and destruction of spillway	

[a]Overtopping is more critical on earth or rockfilled dams. Concrete dams can stand a limited amount of overtopping.
[b]Large trash, such as logs, etc., can damage spillways, stilling basins, and energy-dissipating blocks as it is carried over the spillway.
[c]New techniques for repair; polymer-impregnated concrete has been used to repair cavitation in concrete tunnels and stilling basins.
[d]New technique for repair; for spillway repair, rollcrete has been used as an alternative repair method.

Source: National Research Council (1983).

17.2.5 Hydraulic Analysis of Dam Failures

Dam safety studies typically require an evaluation of the downstream propagation of a flood hydrograph (wave). This evaluation requires determination of the movement of the flood wave as a function of time. Such analysis needs to provide peak water surface elevations, peak discharges, and the timing of these peak elevations and discharges at various locations downstream of the dam.

Forecasting downstream flash floods due to dam failures is an application of flood routing that has received considerable attention. The most widely used dam-breach model in the United States has been the U.S. National Weather Service DAMBRK/88 model (Fread, 1977, 1980, 1981). This model consists of three functional parts: (1) temporal and geometric description of the dam breach; (2) computation of the breach outflow hydrograph; and (3) routing the breach outflow hydrograph downstream.

The FLDWAV model (Fread, 1985) is a synthesis of DWOPER (see Chapter 9) and DAMBRK/88, with additional modeling capabilities not available in either of the other models. FLDWAV is a generalized dynamic wave model for one-dimensional unsteady flows in a single or branched waterway.

These unsteady flow models are used to develop the *flood inundation boundary*, which is the areal extent of flooding. *Dam failure inundation maps*, are used to depict the areal extent of flood-

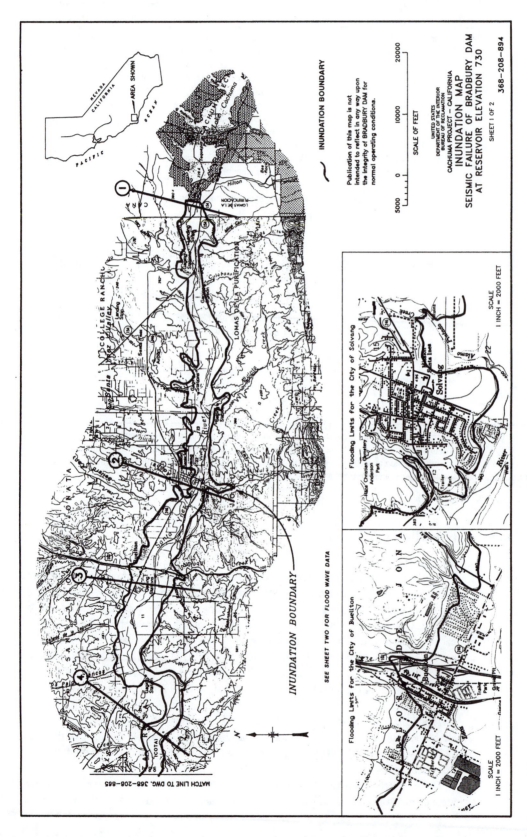

Figure 17.2.7 Inundation map of the Bradbury Dam in California (from Graham and Yang (1996)).

ing, as shown in Figure 17.2.7 for the Bradbury Dam in California. There inundation maps are used for several purposes, including the following (Graham and Yang, 1996):

- Emergency action and evacuation planning
- Determining the population at risk
- Hazard classification of dams
- An aid in selection of risk reduction alternatives

The U.S. Bureau of Reclamation began a dam-mapping program in 1980.

17.2.6 Examples of Dams and Spillways

Figures 17.2.8–17.2.12 show the Roosevelt Dam (Figure 17.2.8), Horse Mesa Dam (Figure 17.2.9), Stewart Mountain Dam (Figures 17.2.10 and 17.2.11), and Granite Reef Dam (Figure 17.2.12), all on the Salt River in Arizona. These figures show releases being made during January 1993. Figures 17.2.13a and b show the intake tower and the side-channel spillways at the Hoover Dam on the Colorado River in Arizona. Figures 17.2.14 a–e show the Lake Vallicito Dam and parts of the spillway. Figures 17.2.15a and b show the Atatürk Dam and spillway channel located on the Euphrates River in Turkey. Figure 17.2.16 shows the Navajo Dam on the San Juan River in New Mexico showing the ungated overflow spillway. Figure 17.2.17 shows an overflow spillway on the Santa Cruz Lake Dam on the Rio Frijoles in New Mexico. Figure 17.2.18 shows the inlet (baffle apron spillway) structure and outlet structure for a detention basin in Phoenix, Arizona. Figure 17.2.19 shows a small detention basin dam and the outlet structure (culvert spillway) for a detention basin in Scottsdale, Arizona. Figure 17.2.20 shows a small dam and culvert spillway under construction in northern New Mexico. Figure 17.2.21 shows the outlet structure under construction for the New Waddel Dam on the Agua Fria River in Arizona. Also refer back to Figure 13.1.5.

Figure 17.2.8 Roosevelt Dam on the Salt River in Arizona showing releases in January 1993. (Courtesy of the U.S. Bureau of Reclamation, photograph by J. Madrigal, Jr. (1993).)

Figure 17.2.9 Horse Mesa Dam on the Salt River in Arizona showing releases in January 1993. (Courtesy of the U.S. Bureau of Reclamation, photograph by J. Madrigal, Jr. (1993).)

Figure 17.2.10 Stewart Mountain Dam on the Salt River in Arizona showing releases in January 1993. (Courtesy of the U.S. Bureau of Reclamation, photograph by J. Madrigal, Jr. (1993).)

(a)

(b)

Figure 17.2.11 Stewart Mountain Dam on the Salt River in Arizona. (a) and (b) Radial gates on the spillways (Photographs by L. W. Mays).

Figure 17.2.12 Granite Reef Dam on the Salt River in Arizona showing releases in January 1993. (Courtesy of the U.S. Bureau of Reclamation, photograph by J. Madrigal, Jr. (1993).)

(*a*)

Figure 17.2.13 Hoover Dam on the Colorado River. (*a*) Side channel spillway with dam in background.

(*b*)

Figure 17.2.13 Hoover Dam on the Colorado River. (*b*) Intake towers to 30 ft diameter steel penstocks (Photographs by L. W. Mays).

(*a*)

Figure 17.2.14 Lake Vallicito Dam in Colorado. (*a*) Downstream side of dam.

(*b*)

(*c*)

Figure 17.2.14 Lake Vallicito Dam in Colorado (*continued*). (*b*) Overflow spillway with radial gates; (*c*) Radial gates.

(*d*)

(*e*)

Figure 17.2.14 Lake Vallicito Dam in Colorado (*continued*). (*d*) Motors and mechanisms to lift radial gates; (*e*) Spillway chute and downstream spillway channel (Photographs by L. W. Mays).

(*a*)

(*b*)

Figure 17.2.15 Atatürk Dam on Euphrates River in Turkey. (*a*) Dam with overflow spillway showing spillway chute; (*b*) Spillway chute and downstream channel. Also refer to Figures 11.4.8 and 13.2.2 (Photographs by L. W. Mays).

Figure 17.2.16 Navajo Dam on San Juan River in New Mexico showing ungated overflow spillway (Photograph by L. W. Mays).

Figure 17.2.17 Spillway on Santa Cruz Lake Dam on the Rio Frijoles in New Mexico (Photograph by L. W. Mays).

(*a*)

(*b*)

Figure 17.2.18 Inlet and outlet structures for a detention basin in Phoenix, Arizona. (*a*) Inlet structure (baffle apron spillway); (*b*) Outlet structure (Photographs by L. W. Mays).

Figure 17.2.19 Culvert spillway for a detention basin in Scottsdale, Arizona (Photograph by L. W. Mays).

Figure 17.2.20 Culvert spillway under construction in northern New Mexico (Photograph by L. W. Mays).

(*a*)

(*b*)

Figure 17.2.21 Photos of outlet structure at New Waddell Dam. (*a*) Structures under construction; (*b*) Prior to filling of Lake Pleasant. (Courtesy Central Arizona Project.)

17.3   SPILLWAYS

17.3.1   Functions of Spillways

The functions of spillways are "for storage and detention dams to release surplus water or flood water that cannot be contained in the allotted storage space, and for diversion dams to bypass flows exceeding those turned into the diversion system" (U.S. Bureau of Reclamation, 1987). Safe spillways are extremely important and cannot be over emphasized. Many dam failures have been caused by improperly designed spillways or by spillways of insufficient capacity. Also, spillways must be hydraulically and structurally adequate. They must be located to prevent the erosion and undermining of the downstream toe of a dam. To determine the best combination of storage capacity and spillway capacity for selected inflow design floods, hydrologic, hydraulic, design, cost, and environmental function must be considered. Spillways can be thought of as safety valves for dams.

Components of spillways include the following:

- *Entrance channels* convey water from the reservoir to the control structure
- *Control structures* regulate the outflow from the reservoir
- *Discharge channels* convey flow released through the control structure to the stream bed below the dam

- *Terminal structures* provide energy dissipation of the flow to prevent erosion and scour in the downstream stream bed
- *Outlet channels* convey the spillway flow from the terminal structure to the river channel below the dam

Spillways are typically classified according to features that pertain to control, to the discharge channel, or some other features. They are often referred to as *controlled* or *uncontrolled spillways* depending on whether they are gated or ungated, respectively. Spillway types include (U.S. Bureau of Reclamation, 1987):

- Overfall spillways (Figures 17.2.8, 17.2.9, and 17.2.17)
- Ogee (overflow) spillways (Figures 17.2.15a and 17.2.16)
- Labyrinth spillways (see U.S. Bureau of Reclamation, 1987)
- Spillway chutes (Figures 17.2.14e and 17.2.15b)
- Conduit and channel spillways (Figure 17.2.13a)
- Drop inlet (shaft or morning glory) spillways (Figures 17.2.13b and 17.2.21)
- Baffle apron drop spillways (Figure 17.2.18)
- Culvert spillways (Figure 17.2.19)
- Siphon spillways

Additional discussion of the above types of spillways is in Coleman et. al. (1999).

17.3.2 Overflow and Free-Overfall (Straight Drop) Spillways

Free overfall (straight drop) spillways allow the flow to drop freely from the crest (see Figure 17.3.1). These types of spillways are characterized by the following (U.S. Bureau of Reclamation, 1987):

- Suited to a thin arch or crest that has a nearly vertical downstream face.
- Flows may be free discharge or may be supported along a narrow section of the crest.
- In many cases the crest is extended in the form of an overhanging lip to direct small discharge away from the face of the overfall section.
- The underside of the nappe is ventilated to prevent a pulsating and fluctuating jet.
- A deep plunge pool will develop at the base of the overfall as a result of scour if artificial protection is not provided.
- A hydraulic jump can form on flat aprons if the tailwater has sufficient depth.
- The major hydraulic problems with free overfall spillways are the characteristics of the control and the dissipation of flow in the downstream basin.
- Flow in the downstream basin can be dissipated by three basic approaches (U.S. Bureau of Reclamation, 1987):
  - by a hydraulic jump (see Figure 17.3.1)
  - by impact and turbulence induced by impact blocks (see Figure 17.3.1)
  - by a slotted grating dissipator installed immediately downstream from the control.

The hydraulic control of free-overfall spillways can be sharp-crested to provide a fully contracted vertical jet, broad-crested to cause a fully suppressed jet, or even shaped to increase crest efficiency. The discharge for these types of spillways is of the form (see Section 5.7)

$$Q = CLH_e^{3/2} \tag{17.3.1}$$

where Q is the discharge, C is the discharge coefficient, L is the effective length of the crest, and H_e is the actual head (total energy head) on the crest including the approach velocity head:

$$H_e = H + \frac{V_a^2}{2g} \tag{17.3.2}$$

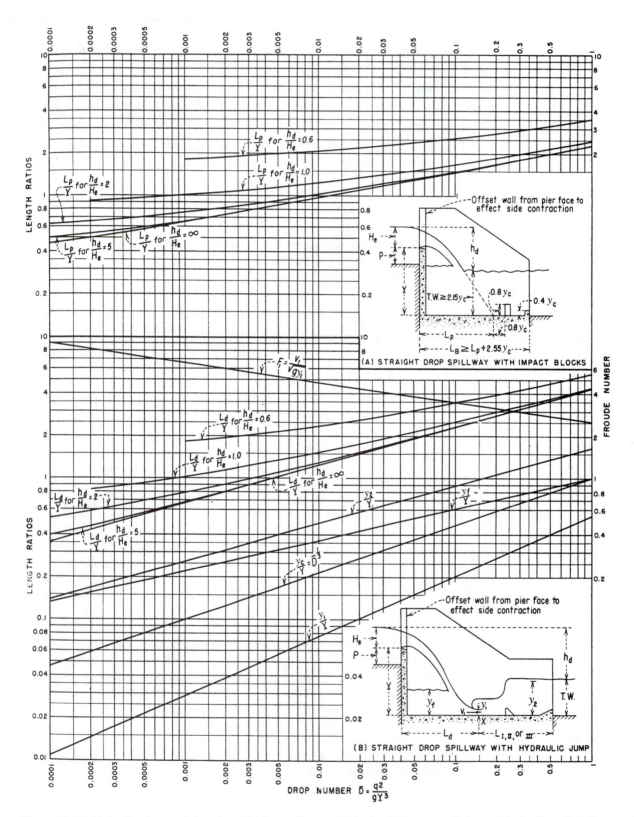

Figure 17.3.1 Hydraulic characteristics of straight drop spillways with hydraulic jump or with impact blocks (from U.S. Bureau of Reclamation (1987)).

where H is the head due to depth of water above the spillway crest and $V_a^2/2g$ is the approach velocity head.

When crest pier and abutment are shaped to cause side contraction of the flow, the effective crest length L is less than the net crest length.

The effective length of the crest is determined using

$$L = L' - 2\left(NK_p + K_a\right)H_e \tag{17.3.3}$$

where

L' = net length of the crest, N = number of piers, K_a = abutment contraction coefficient (approximately 0.2), and K_p = pier contraction coefficient:

- For square-nosed piers with corners rounded on a radius equal to about 0.1 of the pier thickness: $K_p = 0.02$
- For round-nosed piers: $K_p = 0.01$
- For pointed-nose piers: $K_p = 0.0$

Overflow spillways can be gated or ungated and provide for flow over an arch or arch-buttress dam, wherein the flow free-falls some distance before entering a plunge-pool energy dissipator in the tail race (see Figures 17.2.8, 17.2.9, and 17.2.17).

17.3.3   Ogee (Overflow) Spillways

Ogee (overflow) spillways have a control weir that is ogee-shaped (S-shaped in profile). These spillways can be gated (Figure 17.2.10, 17.2.11, and 17.2.12) or ungated (Figures 17.2.12 and 17.2.16) and they normally provide for flow over a gravity dam section. Flow remains in contact with the spillway surface. The upper part of the ogee spillway conforms closely to the profile of the lower nappe of a ventilated sheet falling from a sharp-crested weir (see Figure 17.3.2). This shape results in a pressure distribution on the crest that is near atmospheric for the design discharge. Discharges less than the design discharge produce pressures on the spillway face that are above atmospheric, whereas discharges greater than the design flow cause subatmospheric pressures. These subatmospheric pressures have the hydraulic effect of increasing the discharge-passing capability of the spillway, but they can also lead to cavitation for high heads, which can cause vibration and surface erosion.

Crest shapes have been studied extensively by the U. S. Bureau of Reclamation over the years. Figure 17.3.2 illustrates the suggested spillway shape for a vertical upstream face as well as values of K and n for different upstream inclinations and velocities of approach. H_0 is the design head and h_a is the approach velocity head, $h_a = V_a^2/2g$. The shape equation for the portion downstream of the apex of the crest is

$$\frac{y}{H_0} = -K\left(\frac{x}{H_0}\right)^n \tag{17.3.4}$$

where H_0 is the design head and n and K are functions of h_a/H_0, as given in Figure 17.3.2.

The discharge over an ogee crest is described by equation (17.3.1). The discharge coefficient is influenced by a number of factors, including:

- depth of approach (Figure 17.3.3)
- heads different from design heads (Figure 17.3.4)
- upstream face slope (Figure 17.3.5)
- downstream apron interference (Figure 17.3.6)
- downstream submergence (Figure 17.3.7)

Figure 17.3.3 presents values of discharge coefficients (C_0); $C = C_0$ for the situation when $H_e = H_0$, or $H_e/H_0 = 1$ (which is for the ideal nappe shape). The discharge coefficient varies with the values of P/H, where P is the height of the spillway crest above the channel bed.

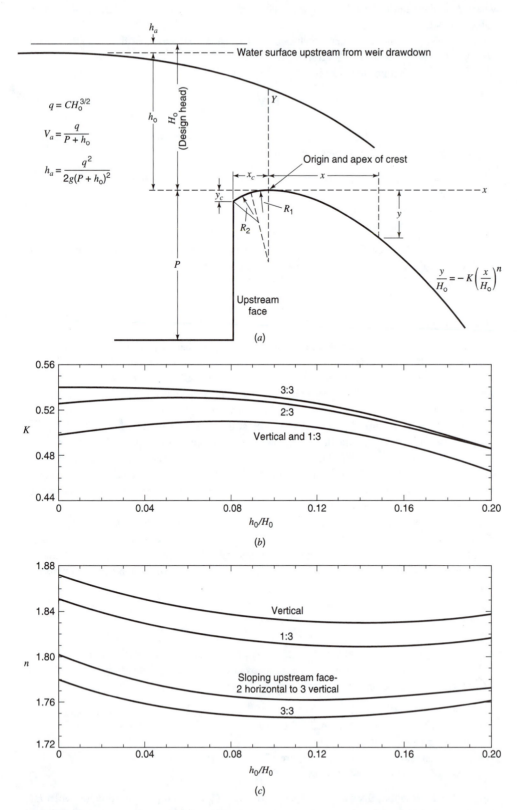

Figure 17.3.2 Factors for definition of nappe-shaped crest profiles. (*a*) Elements of nappe-shaped crest profiles; (*b*) Values of K; (*c*) Values of n (from U.S. Bureau of Reclamation (1987)).

The effect of heads different from the design head on the discharge coefficient are presented in Figure 17.3.4. This figure shows the variation of the coefficient as related to values of H_e/H_0, where H_e is the actual head considered. The effect of the upstream face shape on the discharge coefficient is illustrated in Figure 17.3.5, the effect of downstream apron interference on the discharge coefficient is illustrated in Figure 17.3.6, and the effect of downstream submergence (tail-water effects) is illustrated in Figure 17.3.7.

Gate-controlled ogee crest discharge (Figures 17.2.10 and 17.2.11) is similar to discharge through an orifice, given as

$$Q = CDL\sqrt{2gH} \tag{17.3.5}$$

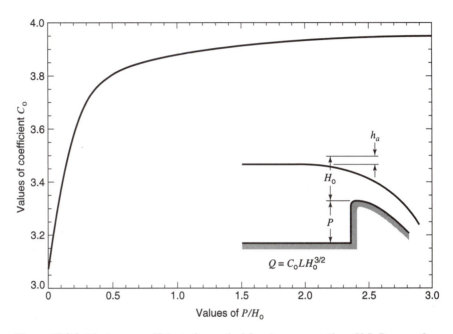

Figure 17.3.3 Discharge coefficients for vertical-faced ogee crest (from U.S. Bureau of Reclamation (1987)).

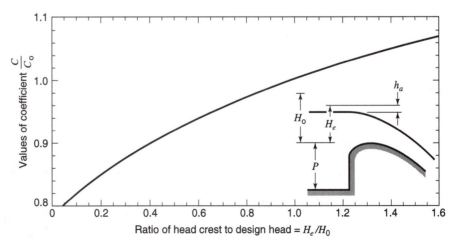

Figure 17.3.4 Discharge coefficients for other than the design head (from U.S. Bureau of Reclamation (1987)).

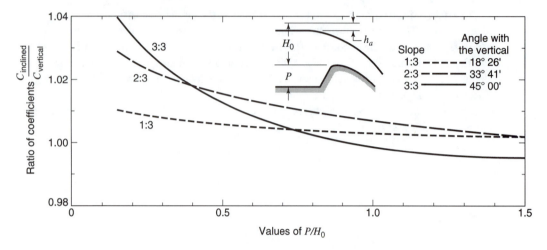

Figure 17.3.5 Discharge coefficients for ogee-shaped crest with sloping upstream face (from U.S. Bureau of Reclamation (1987)).

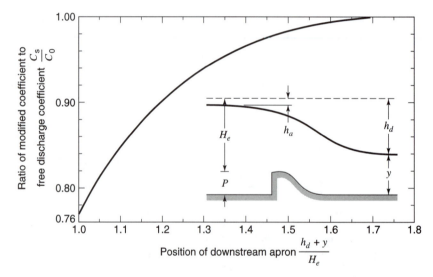

Figure 17.3.6 Ratio of discharge coefficients resulting from apron effects (from U.S. Bureau of Reclamation (1987)).

where C is the discharge coefficient, which is dependent upon the characteristics of the flow lines approaching and leaving the orifice, which are in turn dependent upon the crest shape and the type of gate (see Figure 17.3.8), H is the head to the center of the gate opening (including the velocity head of approach) (see Figure 17.3.8), D is the shortest distance from the gate lip to the crest curve, and L is the crest width. Figure 17.3.8 presents the variation of discharge coefficients for different angles θ. This figure can be used for leaf gates or radial gates located at the crest or downstream of the crest. θ is the angle formed by the tangent to the gate lip and the tangent to the crest curve at the nearest point of the crest curve for radial gates (U.S. Bureau of Reclamation, 1987).

Subatmospheric pressure on both uncontrolled and gate-controlled ogee crests must be given consideration in the design of these structures. For an uncontrolled ogee crest, Figure 17.3.9a illustrates the approximate force diagram of subatmospheric pressure when the design head used to define the crest shape is 75 percent of the maximum head. For a gate-controlled ogee crest, Figure 17.3.9b shows that subatmospheric pressures are equal to about one-tenth of the design head for small gate openings and the ogee is shaped to an ideal nappe profile for maximum head.

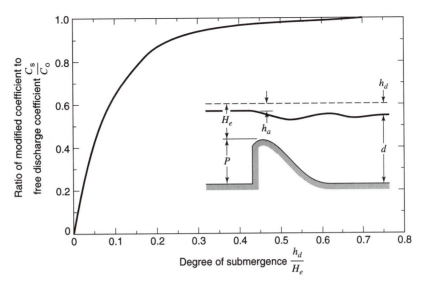

Figure 17.3.7 Ratio of discharge coefficients caused by tailwater effects (from U.S. Bureau of Reclamation (1987)).

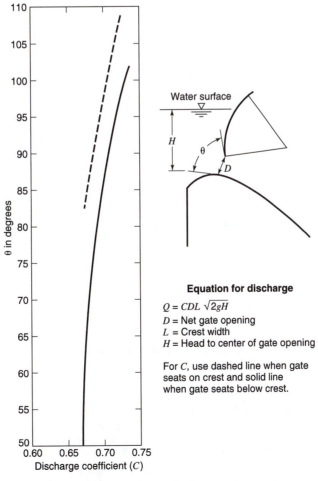

Equation for discharge

$Q = CDL\sqrt{2gH}$

D = Net gate opening
L = Crest width
H = Head to center of gate opening

For C, use dashed line when gate seats on crest and solid line when gate seats below crest.

Figure 17.3.8 Discharge coefficient for flow under gates (from U.S. Bureau of Reclamation (1987)).

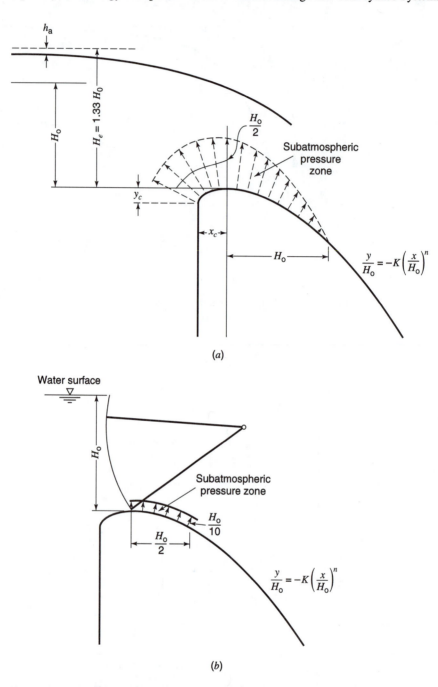

Figure 17.3.9 Subatmospheric crest pressures. (*a*) For $H_0/H_e = 0.75$; (*b*) For undershot gate flow (from U.S. Bureau of Reclamation (1987)).

EXAMPLE 17.3.1

Design an uncontrolled overflow ogee crest for a spillway to discharge 20,000 cfs. The upstream face of the crest is sloped 3:3 and a bridge is to span the crest. Bridge piers 18 in-wide (pier contraction $K_p = 0.05$) with rounded noses are to be provided. The bridge spans (center to center of piers) are not to exceed 30 ft. The maximum expected head $H_e = 10$ ft. Neglect velocity of approach head. The distance from the spillway crest to the reservoir bottom at the dam is 18 ft. The abutment coefficient $K_a = 0.10$. Your design should be based upon the design head being equal to the maximum expected head. What will be the discharges for different expected heads of up to 16 ft?

SOLUTION

The design head H_0 is selected as the maximum expected head H_e, so that $H_e/H_0 = 1.0$. The discharge coefficient C_0 is obtained from Figure 17.3.3 for $P/H_0 = 18/10 = 1.8$ (assume that the ogee is formed to the ideal nappe shape for $H_e/H_0 = 1.0$), $C_0 = 3.925$. The upstream face of the crest is sloped 3:3; therefore from Figure 17.3.5, $C_{inclined}/C_{vertical} = 0.992$ (by extrapolation). The discharge coefficient is $C = C_0 (C_{inclined}/C_{vertical}) = (3.925)(0.992) = 3.894$.

The effective crest length is defined by equation (17.3.3): $L = L' - 2(NK_p + K_a)H_e$. Using the discharge equation (17.3.1), $Q = CLH_e^{3/2}$, the effective crest length L can be determined as

$$L = \frac{Q}{CH_e^{3/2}} = \frac{20,000}{(3.894)(10^{3/2})} = 162.4 \text{ ft}$$

The net length of the crest is $L' = L + 2(NK_p + K_a)H_e$ where $K_p = 0.05$ and $K_a = 0.1$. Assume four piers; then $L' = 162.4 + [2(4)(0.05) + 2(0.1)](10) = 168.4$ ft. The five span lengths not accounting for pier widths are $168.4/(4 + 1) = 33.7$ ft, which is greater than the allowable 30 ft in the problem statement. Assume five piers; then $L' = 162.4 + [2(5)(0.05) + 2(0.1)](10) = 169.4$ ft. The span length not accounting for pier widths is $169.4/(5 + 1) = 28.2$ ft < 30 ft.

Next determine the discharge-head relationship. Because the heads now considered are different from H_0, a correction is made to C using Figure 17.3.4:

$$C = \left(\frac{C}{C_0}\right)(C_0)\left(\frac{C_{inclined}}{C_{vertical}}\right)$$

The effective crest length is $L = 169.4 - [2(5)(0.05) + 2(0.1)]H_e = 169.4 - 0.7H_e$.

The shape of the ogee crest is computed using equation (17.3.4) as $y/H_0 = -K(x/H_0)^n$, where $K = 0.54$ for $h_a/H_0 = 0$ using the 3:3 curve in Figure 17.3.2b and $n = 1.78$ using the 3:3 curve for $h_a/H_0 = 0$ in Figure 17.3.2c. Then $y/H_0 = -0.54(x/H_0)^{1.78}$. This equation is used to shape the ogee crest as shown in Figure 17.3.2. The discharges for different expected heads of up to 16 ft are given in Table 17.3.1.

Table 17.3.1 Discharge Computations for Example 17.3.1

H_c (ft)	$\dfrac{H_c}{H_0}$	$\dfrac{C}{C_0}$	C	Q (cfs)
1	0.1	0.820	3.193	519
2	0.2	0.852	3.317	1524
4	0.4	0.900	3.504	4552
6	0.6	0.940	3.660	8736
8	0.8	0.972	3.785	13909
10	1.0	1.000	3.894	20000
12	1.2	1.025	3.991	26,943
14	1.4	1.050	4.088	34,777
16	1.6	1.070	4.166	43,300

$$\frac{C_i}{C_u} \times C_0 = 0.922 \times 3.925 = 3.894$$

For $H_c = 1 \quad C = 0.820 \times 3.894 = 3.193$

EXAMPLE 17.3.2

A gate-controlled ogee spillway has been designed that uses Tainter gates for control. Determine the head-discharge curves for two different openings of the Tainter gate, for $D = 2$ ft and 6 ft. The heads (including the approach velocity) to the bottom of the orifice range from 4 to 13 ft. The upstream face of the crest is vertical with $P = 10$ ft and designed for $H_0 = 8$ ft, $L = 50$ ft ($\theta = 90°$). The gate seats on the crest.

SOLUTION

For $D = 2$ ft use equation (17.3.5) for all depths, where $C = 0.68$ from Figure 17.3.8, so $Q = CDL\sqrt{2gH} = (0.68)(2)(50)\sqrt{2gH} = 68\sqrt{2gH}$ where H is the head to the center of the gate opening. For $D = 6$ ft, use equation (17.3.5) for $H > 6$ ft and use equation (17.3.1) for $H \leq 6$ ft. For $H > 6$ ft, $Q = (0.68)(6)(50)\sqrt{2gH} = 204\sqrt{2gH}$. For $H \leq 6$ ft, $Q = CLH^{3/2}$, where $C_0 = 3.91$ from Figure 17.3.3 for $P/H_0 = 10/8 = 1.25$. For heads other than the design heads, Figure 17.3.4 is used to determine C/C_0, e.g., $C/C_0 = 0.92$ for $H_e/H_0 = 0.5$ then $C = C_0(H_e/H_0) = (3.91)(0.5) = 3.60$. The computations are summarized in Table 17.3.2.

Table 17.3.2   Discharge Computation for Example 17.3.2

Total Head	H	D	C Fig. (17.3.8)	Q(cfs) Eq. (17.3.5)	C_0 Fig. (17.3.3)	H/H_0	C/C_0 Fig. (17.3.4)	C	Q(cfs) Eq. (17.3.1)
5	4	2	0.68	1091					
7	6	2	0.68	1337					
9	8	2	0.68	1543					
11	10	2	0.68	1726					
13	12	2	0.68	1890					
4	4	6			3.91	0.5	0.92	3.60	1,440
6	6	6			3.91	0.75	0.96	3.75	2,756
10	7	6	0.68	4,331					
11	8	6	0.68	4,630					
13	10	6	0.68	5,177					

17.3.4   Side Channel Spillways

As shown in Figures 17.2.13a and 17.3.10, side channel spillways have a control weir that is alongside and approximately parallel to the upper part of the spillway discharge channel. Discharge over the crest falls into a side channel or into a conduit or tunnel. Discharge characteristics are similar to an uncontrolled overflow spillway and depend on the weir crest profile. The methods described in the previous section can be used to determine the overflow spillway discharge, and methods for water surface profile determination (Chapter 15) can be used to analyze the flow profile in the side channel.

Side channel flow characteristics are illustrated in Figure 17.3.11. When the bottom of the side channel trough is selected so that the depth below the hydraulic gradient is greater than the minimum specific energy depth, the flow will be either subcritical or supercritical depending on the critical slope or downstream control. For slopes greater than critical without downstream control, supercritical flow occurs throughout the channel (see profile B' in Figure 17.3.11a). If there is a downstream control in the channel, the flow is supercritical (see profile A' in Figure 17.3.11a). The hydraulic effects for these two profiles are illustrated in Figure 17.3.11b.

EXAMPLE 17.3.3

Design a side channel spillway 200 ft long to discharge a maximum of 4000 ft^3/s. The spillway crest is at an elevation of 100 ft; assume a crest coefficient of 3.6.

SOLUTION

The discharge per unit length is $q = 4000/200 = 20$ ft^3/s/ft. The design head is then computed using equation (17.3.1) as $H_0 = (q/C)^{2/3} = (20/3.6)^{2/3} = 3.1$ ft. Consider a trapezoidal section for the side channel trough with 0.5:1 side slopes and a bottom width of 20 ft. Consider a bottom slope of 0.01 with the upstream end of the side channel spillway having a bottom elevation of 80 ft.

A control section is placed downstream of the side channel trough with a 20 ft transition from the 0.5:1 side slope of the trough section to a rectangular section 20 ft wide at the control with no change in the bottom elevation at the control. Refer to Figure 17.3.12.

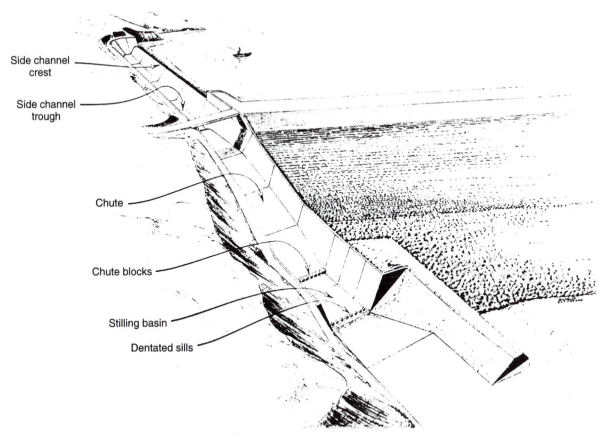

Figure 17.3.10 Typical side channel and chute spillway arrangement (from U.S. Bureau of Reclamation (1987)).

The critical depth for the flow at the control is $y_c = \sqrt[3]{q^2/g} = \sqrt[3]{(4000/20)^2/32.2} = 10.75$ ft and the velocity is $V_c = q/y_c = 200/10.75 = 18.60$ ft/s. The velocity head is $V_c^2/2g = (18.6)^2/64.4 = 5.37$ ft. The transition losses are approximated as 0.2 times the difference of the velocity heads between the ends of the transition. The energy equation can be written for the ends of the transition as (see Figure 17.3.12)

$$\frac{V^2}{2g} + y = \frac{V_c^2}{2g} + y_c + 0.2\left(\frac{V_c^2}{2g} - \frac{V^2}{2g}\right)$$

Substituting the known values, we find

$$\frac{V^2}{2g} + y = 5.37 + 10.75 + 0.2\left(5.37 - \frac{V^2}{2g}\right)$$

$$1.2\frac{V^2}{2g} + y = 17.19$$

$$V = Q/A = Q/((b+zy)y) = 4000/\left[(20+0.5y)y\right]$$

$$\frac{1.2}{2(32.2)}\left[\frac{4000}{\left[(20+0.5y)y\right]}\right]^2 + y = 17.19$$

Solving yields $y \approx 15.6$ ft; then the flow area is $A = [20+0.5(15.6)]15.6 = 434$ ft^2, $V = 4000/434 = 9.2$ ft/s, $V^2/2g = 1.31$ ft. The headloss through the transition is $0.2(5.37 - 1.31) = 0.81$ ft. The maximum bottom elevation of the side channel spillway could be placed so that there is a tolerable crest

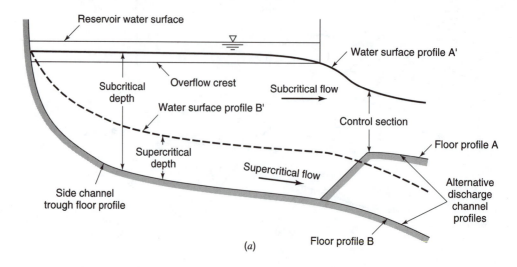

(a)

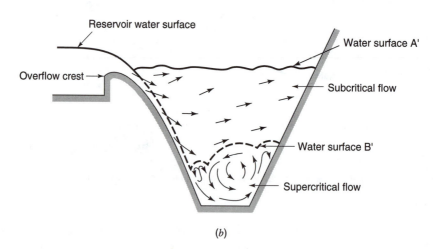

(b)

Figure 17.3.11 Side channel flow characteristics. (a) Profile; (b) Cross-section (from U.S. Bureau of Reclamation (1987)).

submergence of 2.0 ft. Next a water surface profile must be performed to relate the elevation to the reservoir water level.

17.3.5   Drop Inlet (Shaft or Morning Glory) Spillways

Drop inlet spillways are characterized as spillways in which the discharge enters over a horizontal lip, drops through a vertical or sloping shaft, and then discharges through a horizontal or nearly horizontal conduit or tunnel. These structures have three basic components: a control weir, a vertical transition, and a closed discharge channel. Figure 17.3.13 illustrates a typical installation of a shaft spillway for a small dam. Figure 17.3.14 illustrates a morning-glory spillway, which has a funnel-shaped inlet.

There are three types of flow control in drop inlet spillways: crest control, orifice control, and full-pipe flow control. The control shifts according to the relative discharge capacities of the weir, transition, and the conduit or tunnel. The flow condition and discharge characteristics of morning glory spillways are illustrated in Figure 17.3.14. For low flows the flow rate is governed by the

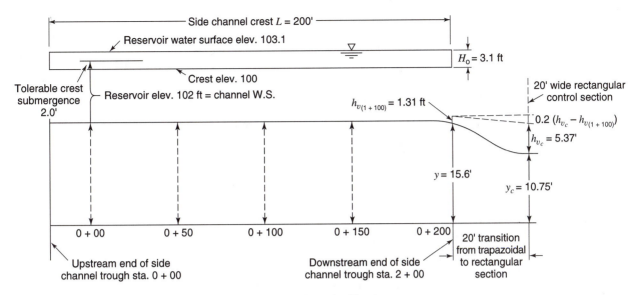

Figure 17.3.12 Example 17.3.3—Hydraulic design for side channel spillway.

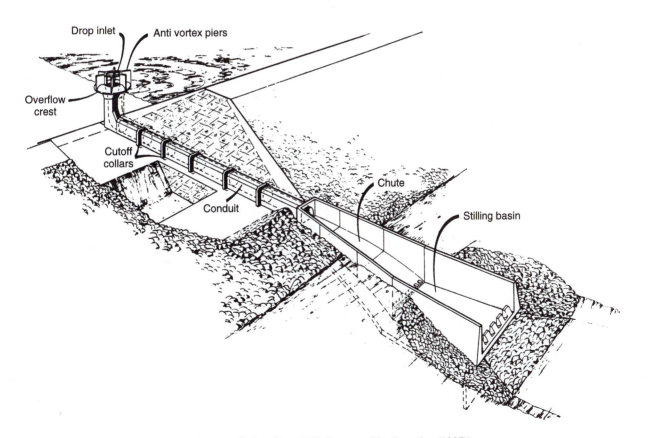

Figure 17.3.13 Drop inlet spillway for a small dam (from U.S. Bureau of Reclamation (1987)).

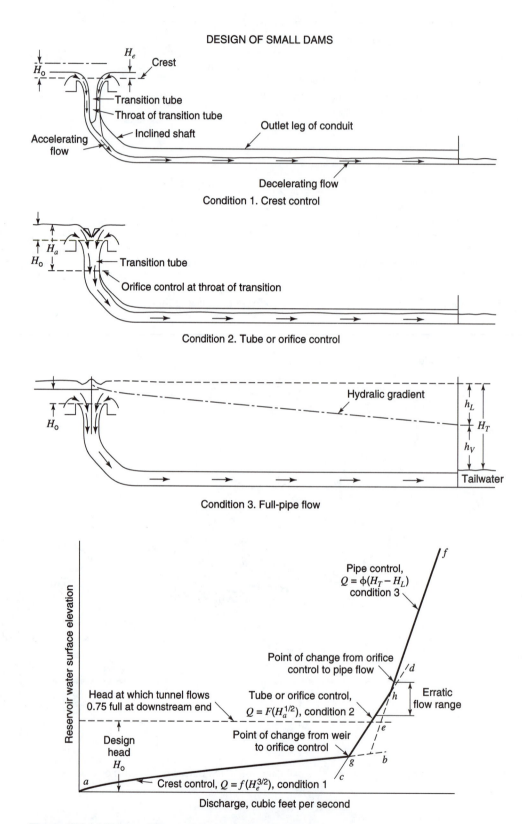

Figure 17.3.14 Nature of flow and discharge characteristics of a morning-glory spillway (from U.S. Bureau of Reclamation (1987)).

relationship of weir flow $Q = f(H_e^{3/2})$, i.e., equation (17.3.1). A continuous volume of air persists throughout the spillway components.

Figure 17.3.15a illustrates the elements of a nappe shaped profile for a circular weir. The head on the weir is measured from the crest and the length L is the circumference of the weir crest ($2\pi R_s$). The weir discharge is then

$$Q = C_0\left(2\pi R_s\right)H_0^{3/2} \tag{17.3.6}$$

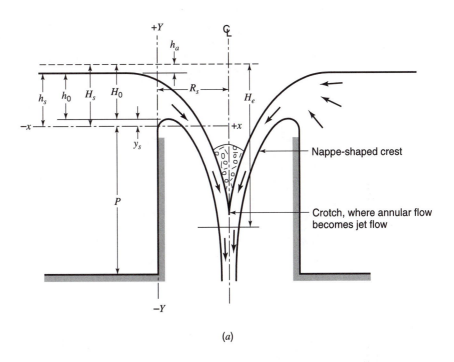

(a)

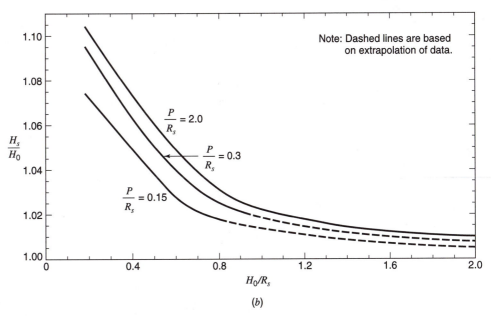

(b)

Figure 17.3.15 Circular sharp-crested weirs: (a) Elements of nappe-shaped profile for circular weir; (b) Relationship of H_s/H_0 to H_0/R_s for circular sharp-crested weirs (from U.S. Bureau of Reclamation (1987)).

The discharge coefficients for a circular crest differ from that for a straight crest because of the effect of submergence and back pressure incident to the joining of the converging flow (U.S. Bureau of Reclamation, 1987). Figure 17.3.16 presents the relationship of the circular crest coefficient C_0 to H_0/R_s for different approach depths. Figure 17.3.17 presents the circular crest coefficients for other than design head.

Weir control governs when free flow prevails for values of H_0/R_s less than approximately 0.45. As H_0/R_s increases above 0.45, the weir becomes partially submerged, and as H_0/R_s approaches 1.0 the entrance becomes submerged when orifice flow controls $Q = f(H_a^{1/2})$. When the downstream pipe becomes full, pipe flow prevails so that relatively large increases in head produce small increases in discharge. These spillways are commonly designed so that the flow depth in the downstream tunnel is less than or equal to 75 percent of area, allowing air to enter from the downstream end and preventing subatmospheric pressure in the tunnel.

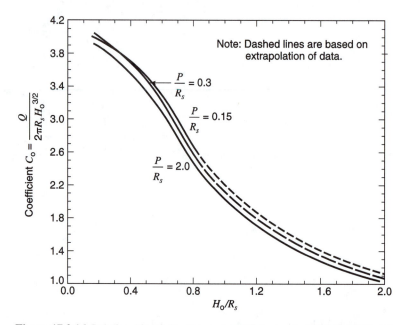

Figure 17.3.16 Relationship of circular crest coefficient c_0 to H_0/R_s for different approach depths (aerated nappe) (from U.S. Bureau of Reclamation (1987)).

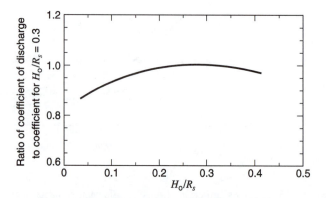

Figure 17.3.17 Circular crest discharge coefficient for other than design head (from U.S. Bureau of Reclamation (1987)).

Given a design discharge and maximum expected head, the following procedure can be used to size a morning-glory spillway.

Step 0. Select the crest elevation and outlet invert elevation.

Step 1. Determine the minimum radius of the crest so that subatmospheric pressure can be tolerated.

(a) Assume R_s, determine H_0/R_s, and obtain C_0 from Figure 17.3.15 for P/R_s.

(b) Compute the discharge $Q = C_0(2\pi R_s)H_0^{3/2}$ and compare with the required discharge.

Step 2. The shape of the crest can be computed using tables in the U.S. Bureau of Reclamation (1987), such as in Table 17.3.3. Use Figure 17.3.14 to obtain values of H_s/H_0 in order to compute H_s/R_s used on the tables.

Table 17.3.3 Coordinates of Lower Nappe Surface Different Values of $\dfrac{H_s}{R}$ when $\dfrac{P}{R} = 2$

[Negligible approach velocity and aerated nappe]

$\dfrac{H_s}{R}$	0.00	0.10*	0.20	0.25	0.30	0.35	0.40	0.45	0.50	0.60	0.80	1.00	1.20	1.50	2.00
$\dfrac{X}{H_p}$							$\dfrac{Y}{H_s}$ for portion of the surface above the weir crest								
0.000	0.000	0.000	0.000	0.000	0.000	0.000	0.000	0.000	0.000	0.000	0.000	0.000	0.000	0.000	0.000
.010	.0150	.0145	.0133	.0130	.0128	.0125	.0122	.0119	.0116	.0112	.0104	.0095	.0086	.0077	.0070
.020	.0280	.0265	.0250	.0243	.0236	.0231	.0225	.0220	.0213	.0202	.0180	.0159	.0140	.0115	.0090
.030	.0395	.0365	.0350	.0337	.0327	.0317	.0308	.0299	.0289	.0270	.0231	.0198	.0168	.0126	.0085
.040	.0490	.0450	.0435	.0417	.0403	.0389	.0377	.0363	.0351	.0324	.0268	.0220	.0176	.0117	.0050
.050	.0575	.0535	.0506	.0487	.0471	.0454	.0436	.0420	.4002	.0368	.0292	.0226	.0168	.0092	
.060	.0650	.0605	.0570	.0550	.0531	.0510	.0489	.0470	.0448	.0404	.0305	.0220	.0147	.0053	
.070	.0710	.0665	.0627	.0605	.0584	.0560	.0537	.0514	.0487	.0432	.0308	.0201	.0114	.0001	
.080	.0765	.0710	.0677	.0655	.0630	.0603	.0578	.0550	.0521	.0455	.0301	.0172	.0070		
.090	.0820	.0765	.0722	.0696	.0670	.0640	.0613	.0581	.0549	.0471	.0287	.0135	.0018		
.100	.0860	.0810	.0762	.0734	.0705	.0672	.0642	.0606	.0570	.0482	.0264	.0089			
.120	.0940	.0880	.0826	.0790	.0758	.0720	.0683	.0640	.0596	.0483	.0195				
.140	.1000	.0935	.0872	.0829	.0792	.0750	.0705	.0654	.0599	.0460	.0101				
.160	.1045	.0980	.0905	.0855	.0812	.0765	.0710	.0651	.0585	.0418					
.180	.1080	.1010	.0927	.0872	.0820	.0766	.0705	.0637	.0559	.0361					
.200	.1105	.1025	.0938	.0877	.0819	.0756	.0688	.0611	.0521	.0292					
.250	.1120	.1035	.0926	.0850	.0773	.0683	.0596	.0495	.0380	.0068					
.300	.1105	.1000	.0850	.0764	.0668	.0559	.0446	.0327	.0174						
.350	.1060	.0930	.0750	.0650	.0540	.0410	.0280	.0125							
.400	.0970	.0830	.0620	.0500	.0365	.0220	.0060								
.450	.0845	.0700	.0450	.0310	.0170	.000									
.500	.0700	.0520	.0250	.0100											
.550	.0520	.0320	.0020												
.600	.0320	.0080													
.650	.0090														

$\dfrac{Y}{H_s}$	0.00	0.10*	0.20	0.25	0.30	0.35	0.40	0.45	0.50	0.60	0.80	1.00	1.20	1.50	2.00
							$\dfrac{X}{H_s}$ for portion of the surface above the weir crest								
0.000	0.668	0.615	0.554	0.520	0.487	0.450	0.413	0.376	0.334	0.262	0.158	0.116	0.093	0.070	0.048
−.020	.705	.652	.592	.560	.526	.488	.452	.414	.369	.293	.185	.145	.120	.096	.074
−.040	.742	.688	.627	.596	.563	.524	.487	.448	.400	.320	0.212	.165	.140	.115	.088
−.060	.777	.720	.660	.630	.596	.557	.519	.478	.428	.342	.232	.182	.155	.129	.100
−.080	.808	.752	.692	.662	.628	.589	.549	.506	.454	.363	.250	.197	.169	.140	.110
−.100	.838	.784	.722	.692	.657	.618	.577	.532	.478	.381	.266	.210	.180	.150	.118
−.150	.913	.857	.793	.762	.725	.686	.641	.589	.531	.423	.299	.238	.204	.170	.132
−.200	.978	.925	.860	.826	.790	.745	.698	.640	.575	.459	.326	.260	.224	.184	.144
−.250	1.040	.985	.919	.883	.847	.801	.750	.683	.613	.490	.348	.280	.239	.195	.153

Table 17.3.3　Coordinates of Lower Nappe Surface Different Values of $\dfrac{H_s}{R}$ when $\dfrac{P}{R} = 2$ (*continued*)

[Negligible approach velocity and aerated nappe]

$\dfrac{H_s}{R}$	0.00	0.10*	0.20	0.25	0.30	0.35	0.40	0.45	0.50	0.60	0.80	1.00	1.20	1.50	2.00
$\dfrac{Y}{H_s}$							$\dfrac{X}{H_s}$ for portion of the surface above the weir crest								
−.300	1.100	1.043	.976	.941	.900	.852	.797	.722	.648	.518	.368	.296	.251	.206	.160
−.400	1.207	1.150	1.079	1.041	1.000	.944	.880	.791	.705	.562	.400	.322	.271	.220	.168
−.500	1.308	1.246	1.172	1.131	1.087	1.027	.951	.849	.753	.598	.427	.342	.287	.232	.173
−.600	1.397	1.335	1.250	1.215	1.167	1.102	1.012	.898	.793	.627	.449	.359	.300	.240	.179
−.800	1.563	1.500	1.422	1.369	1.312	1.231	1.112	.974	.854	.673	.482	.384	.320	.253	.184
−1.000	1.713	1.646	1.564	1.508	1.440	1.337	1.189	1.030	.899	.710	.508	.402	.332	.260	.188
−1.200	1.846	1.780	1.691	1.635	1.553	1.422	1.248	1.074	.933	.739	.528	.417	.340	.266	
−1.400	1.970	1.903	1.808	1.748	1.653	1.492	1.293	1.108	.963	.760	.542	.423	.344		
−1.600	2.085	2.020	1.918	1.855	1.742	1.548	1.330	1.133	.988	.780	.553	.430			
−1.800	2.196	2.130	2.024	1.957	1.821	1.591	1.358	1.158	1.008	.797	.563	.433			
−2.000	2.302	2.234	2.126	2.053	1.891	1.630	1.381	1.180	1.025	.810	.572				
−2.500	2.557	2.475	2.354	2.266	2.027	1.701	1.430	1.221	1.059	.838	.588				
−3.000	2.778	2.700	2.559	2.428	2.119	1.748	1.468	1.252	1.086	.853					
−3.500		2.916	2.749	2.541	2.171	1.777	1.489	1.267	1.102						
−4.000		3.114	2.914	2.620	2.201	1.796	1.500	1.280							
−4.500		3.306	3.053	2.682	2.220	1.806	1.509								
−5.000		3.488	3.178	2.734	2.227	1.811									
−5.500		3.653	3.294	2.779	2.229										
−6.000		3.820	3.405	2.812	2.232										
$\dfrac{H_s}{R}$	0.00	0.10	0.20	0.25	0.30	0.35	0.40	0.45	0.50	0.60	0.80	1.00	1.20	1.50	2.00

*The tabulation for $H_s/R = 0.10$ was obtained by interpolation between $H_s/R = 0$ and 0.20.

Source: U.S. Bureau of Reclamation (1987)

Step 3. Determine transition shape for the design discharge and maximum expected head for the crest elevation. Use the following equation (U.S. Bureau of Reclamation, 1987):

$$R = 0.204 \frac{Q^{1/2}}{H_a^{1/4}} \tag{17.3.7}$$

where H_a is the distance between the water surface and the elevation being considered and R is the transition shape radius. Vary H_a to determine the relationship between throat radius and elevation. This equation assumes that the total losses through the transition are 0.1 H_a.

Step 4.　Determine the minimum conduit diameter that can pass the flow from the transition section to the conduit portal without the conduit flowing more than 75 percent at the downstream end.

(a) Select a trial diameter and throat diameter and find the corresponding throat location.

(b) Compute the length from the transition throat to the outlet portal.

(c) Approximate friction losses by assuming the conduit flows 75 percent full for its entire length, using

$$V = q/(0.75A), \quad S_f = \left(\frac{V_n}{1.49(D/2)^{2/3}} \right)^2, \quad h_{Lf} = L \cdot S_f$$

(d) Use the energy equation from the throat to the outlet to check elevation of invert at the outlet portal required to pass the design discharge through the selected conduit diameter. You may have to increase the conduit and repeat 4a through 4d.

EXAMPLE 17.3.4

Design an ungated drop inlet (morning glory) spillway that will operate under a maximum surcharge head of $H_0 = 10$ ft that can limit the outflow to 1500 ft³/s. Determine (a) the minimum radius of the overflow crest, (b) the transition shaft radius, (c) the minimum uniform conduit diameter considering a 75 percent full flow in order to allow air to pass up the conduit from the downstream portal in order to prevent sub-atmospheric pressures in the conduit, and (d) develop the discharge-water surface elevation relationship. The horizontal length of the conduit is 250 ft. The crest elevation is 100 ft and outlet invert elevation is 60 ft. This example design must minimize the overflow crest radius because the intake is formed as a tower away from the abutment. Also, subatmospheric pressure along the crest can be tolerated; now the conduit portion of the spillway must not flow more than 75 percent full at the downstream end.

SOLUTION

Step 1. Assume $R_s = 5.5$ ft, then $H_0/R_s = 10/5.5 = 1.82$; assume $P/R_s \geq 2$, then $C_0 = 1.14$ (Figure 17.3.16). The discharge is computed using equation (17.3.6): $Q = 1.14(2\pi)(5.5)(10)^{3/2} = 1246 < 1500$ ft³/s.

Next try $R_s = 6.0$ ft, then $H_0/R_s = 1.67$; assume $P/R_s \geq 2$; then $C_0 = 1.23$ and $Q = 1466$ ft³/s. Next try $R_s = 6.5$ ft; then $H_0/R_s = 1.54$, and assume $P/R_s \geq 2$; then $C_0 = 1.31$ and $Q = 1692$ ft³/s. Select $R_s = 6.0$ ft as 1465 ft³/s is closer to the 1500 ft³/s.

Step 2. The shape of the crest is computed using Table 17.3.3 with $H_0/R_s = 1.67$. From Figure 17.3.15b, $H_s/H_0 = 1.01$ for $H_0/R_s = 1.67$ and $P/R_s = 2$. Therefore, $H_s = 1.01H_0 = 1.01(10) = 10.1$ ft. From Figure 17.3.15a, $y_s = H_s - H_0 = 10.1 - 10 = 0.10$ ft.

The X-Y coordinate of the crest profile for the portion above the weir crest is computed as given in the table below for $H_s/R_s = 10.1/6.0 = 1.68$:

X/H_s	Y/H_s*	$X = \left(\dfrac{X}{H_s}\right)H_s$	$Y = \left(\dfrac{Y}{H_s}\right)H_s$
		(ft)	(ft)
0.000	0.0000	0.000	0.000
0.010	0.0074	0.101	0.075
0.020	0.0106	0.202	0.107
0.030	0.0111	0.303	0.112
0.040	0.0093	0.404	0.094

$X_s = 0.303$ ft and $Y_s = 0.112$ ft are taken. The X-Y coordinates of the crest profile below the weir crest were calculated similarly:

Y/H_s	X/H_s*	$Y = \left(\dfrac{Y}{H_s}\right)H_s$	$X = \left(\dfrac{X}{H_s}\right)H_s$
		(ft)	(ft)
0.000	0.062	0.000	0.626
−0.020	0.087	−0.202	0.879
−0.040	0.105	−0.404	1.061
−0.060	0.119	−0.606	1.202
−0.080	0.129	−0.808	1.303
−0.100	0.138	−1.010	1.394
−0.150	0.156	−1.515	1.576
−0.200	0.170	−2.020	1.717
−0.250	0.181	−2.525	1.828
−0.300	0.189	−3.030	1.909
−0.400	0.201	−4.040	2.030
−0.500	0.211	−5.050	2.131
−0.600	0.218	−6.060	2.202
−0.800	0.228	−8.080	2.303
−1.000	0.234	−10.100	2.363

*Values were linearly interpolated between $H_s/R_s = 1.50$ and 2.00.

Step 3. Determine the transition shape. The transition shape radius is computed using equation (17.3.7):

$$R = (0.204)\frac{Q^{1/2}}{H_a^{1/4}} = 0.204\frac{(1500)^{1/2}}{H_a^{1/4}} = \frac{7.9}{H_a^{1/4}}$$

where H_a is the distance below the water surface elevation (110 ft) and the elevation being considered. The shape radius is computed for different elevations.

Elevation (ft)	100	99	98	97	96	95
H_a (ft)	10	11	12	13	14	15
R (ft)	4.44	4.34	4.24	4.16	4.08	4.01

Elevation (ft)	94	93	92	91	90
H_a (ft)	16	17	18	19	20
R (ft)	3.95	3.89	3.84	3.78	3.74

Step 4. Determine minimum conduit diameter.

(a) Consider a conduit diameter of 8 ft (or radius of 4 ft). This corresponds to $H_a = [0.204(1500)^{1/2}/4]^4 = 15.22$ ft. The throat location is $110 - 15.22 = 94.78$ ft elevation.

(b) Assume conduit length of 250 ft from throat transition to the outlet portal.

(c) Friction losses are approximated assuming the conduit flows 75 percent full for its entire length, so the velocity is $V = Q/(0.75A) = 1500/[0.75\pi(4)^2] = 39.8$ ft/s and $V^2/2g = 24.6$ ft. Using Manning's equation, $S_f = 0.08$ and $h_{Lf} = 20$ ft (with $n = 0.014$). For 75 percent full, $d/D = 0.702$, so $d = 0.702(8) = 5.62$ ft.

(d) Apply the energy equation to check outlet invert elevation.

Invert elevation = throat elevation + velocity head at throat − velocity head in the conduit flowing 75 percent full − friction losses in conduit − depth at outlet

$$= 94.78 + \left(\frac{1}{1.1}(110 - 94.78) = 13.84\right) - 24.6 - 20 - 5.62$$

$$= 58.4 \text{ ft}$$

This invert elevation is very close to the assumed invert elevation of 60 ft. A water surface profile through the conduit should be computed.

17.3.6 Baffled Chute Spillways

Figures 17.2.18a and 17.3.18 illustrates a baffled-chute spillway, which is used to lower water from one level to another in the case where a stilling basin is not desired. The baffle pier dissipates energy as water flows down the chute in order to decrease flow velocities of water entering the downstream channel. These spillways have the following advantages: (a) being economical, (b) low terminal velocities even with large elevation drops, (c) no requirements for initial tailwater depth, and (d) no effect on spillway operation due to downstream degradation. Figure 17.3.19 presents recommended baffle pier heights and allowable velocities for baffled-chute spillways. Baffled-chute spillways are normally constructed at slopes of 2:1 or flatter extending below the channel outlet floor, and design capacities have varied from less than 10 to over 80 cfs/ft of width (U.S. Bureau of Reclamation, 1987). USBR model studies have shown that a baffled-chute spillway design can be based on a discharge of about two-thirds the maximum expected discharge.

The typical hydraulic design procedure for a baffled chute spillway is given below (adapted from U.S. Bureau of Reclamation, 1987):

1. Determine maximum expected discharge.
2. Determine unit design discharge $q = Q/W$, where W is the chute width.
3. Determine entrance velocity, ideally as $V_1 = \sqrt[3]{gq} - 5$, which is curve D in Figure 17.3.19.

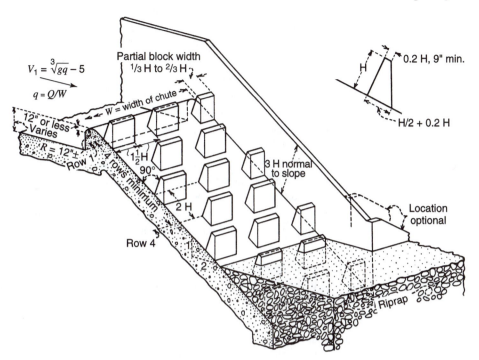

Figure 17.3.18 Basic proportions of a baffled chute spillway (from U.S. Bureau of Reclamation (1987)).

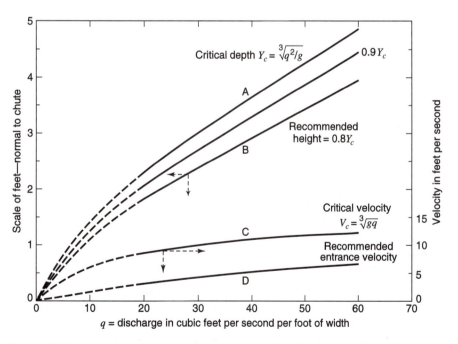

Figure 17.3.19 Recommended baffle pier heights and allowable velocities for baffled chute spillways (from U.S. Bureau of Reclamation (1987)).

4. Select a vertical offset between the approach channel flow and the chute to establish a desirable uniform entrance velocity V_1 (see Figure 17.3.18).

5. Select the baffle pier height H as $0.8d_c$ or $0.9d_c$, where d_c is the critical depth $d_c = \sqrt[3]{q^2/g}$. Refer to curve A in Figure 17.3.19.

6. Select baffle pier width and space equal to about $1.5H$, but not less than H. Figure 17.3.18 gives suggested cross-sectional dimensions of baffle piers.

7. Determine spacing between the rows of baffles as H divided by the slope. For a 2:1 slope, the row spacing is $2H$.

8. Baffle piers are typically placed with the upstream face normal to the chute flow surface. Also, vertical-faced piers can be used; these tend to produce more splash and less bed scour, but differences are minor.

9. Select at least four rows of baffle piers to establish full control of the flow. There should be at least one row of buried baffle piers with an additional row to protect against degradation.

10. Determine chute training wall heights as $3H$.

11. Place riprap consisting of 6- to 12-inch stones at the downstream ends of the training walls to prevent eddies. This riprap should be extended appreciably into the flow area.

EXAMPLE 17.3.5   Determine the dimensions for a baffled chute spillway for a maximum expected discharge of 5000 ft³/s. The spillway slope is 2:1 and the water is lowered vertically 30 ft.

SOLUTION

1. The maximum expected discharge is 5000 ft³/s.
2. Select a width of 80 ft so that the unit discharge is $q = 5000/80 = 62.5$ ft³/s/ft.
3. The entrance velocity is $V = \sqrt[3]{gq} - 5 = \sqrt[3]{32.2(62.5)} - 5 = 7.63$ ft/s. Also, from Figure 17.3.19, $V = 7.6$ ft/s.
4. A vertical offset of 1 ft is arbitrarily selected (12 inches or less is recommended).
5. The critical depth is $y_c = \sqrt[3]{q^2/g} = \sqrt[3]{62.5^2/32.2} = 4.95$ ft. Choose $H = 0.8\,y_c = 0.8(4.95) = 3.96$, and use $H = 4.0$ ft.
6. Referring to Figure 17.3.18, pier width is $1.5H = 6$ ft; bottom length is $H/2 + 0.2H = 2.8$ ft.
7. The longitudinal spacing of the baffle rows is $H /$ slope $= H / 0.5 = 2H = 8$ ft.
8. Use an upstream baffle face that is normal to the chute spillway.
9. At a 2:1 slope, the horizontal chute length is approximately $2(30) = 60$ ft so that the minimum chute length is $\sqrt{30^2 + 60^2} = 67$ ft. The chute length can be made 80 ft to allow for at least one buried baffle row; then 10 rows of baffles are needed.
10. The chute training wall height is $3H = 3(4) = 12$ ft.
11. Select a riprap ($D_{50} = 6$ to 12 inches) and extend approximately 30 ft into the flow area.

17.3.7 Culvert Spillways

A culvert spillway (see Figure 17.2.19) usually consists of a simple culvert placed through a dam or along an abutment, usually on a uniform grade, with a vertical or inclined entrance. These spillways are circular, square, rectangular, or some other cast-in-place shapes. These spillways can discharge freely or can flow into an open channel. Culvert spillways are for elevation drops of less than 25 feet. As illustrated in Figure 17.3.20, these spillways have a variety of flow conditions depending on whether or not the inlet is submerged and whether the slope is mild or steep. The flow operation is either partial flow or full flow depending upon the condition. The hydraulics of culverts has been discussed in Chapter 15. The same design procedures discussed in Chapter 15 may be followed to analyze culvert spillway hydraulics.

Part full flow—inlet not submerged

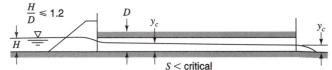

① Mild slope. Subcritical flow, control at critical depth at outlet.

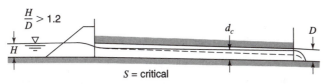

② Critical slope. Subcritical flow, control at critical depth at outlet or some greater depth if backwater exists.

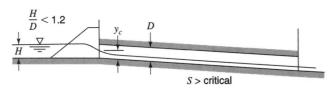

③ Steep slope. Superficial flow, control at critical depth at inlet.

Part full flow—inlet submerged

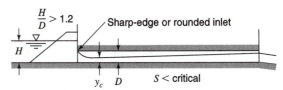

④ Mild slope. Supercritical flow, orifice flow control at inlet.

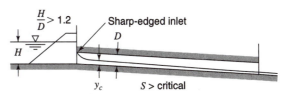

⑤ Steep slope. Supercritical flow, orifice flow control at inlet.

Full flow—inlet submerged

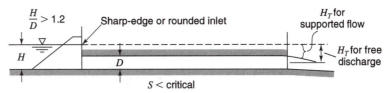

⑥ Mild slope. Control at outlet. Effective head ($H_T - \Sigma$ losses).

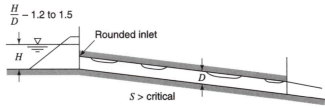

⑦ Steep slope. Supercritical pulsating slug flow, control switching between inlet and some section within the conduit.

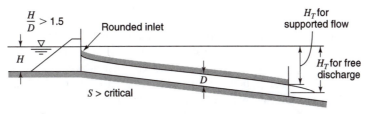

⑧ Steep slope. Control at outlet. Effective head ($H_T - \Sigma$ losses).

Figure 17.3.20 Typical flow conditions for culvert spillways on mild and steep slopes (from U.S. Bureau of Reclamation (1987)).

17.4   HYDRAULIC-JUMP TYPE STILLING BASINS AND ENERGY DISSIPATORS

The purpose of this section is to describe the various stilling basins and energy dissipators that are used to dissipate the energy of flow from the spillway before the discharge is returned to the downstream environment. Additional material is in Mays (1999) and Wei and Lindell (1999).

17.4.1   Types of Hydraulic Jump Basins

Using the U.S. Bureau of Reclamation (1987) classification, there are five basic hydraulic-jump type basins that are briefly described in this section:

Basin I   Horizontal aprons
Basin II   Stilling basins for high dam and earth dam spillways and large canal structures
Basin III   Short stilling basins for canal structures, small outlet works, and small spillways
Basin IV   Stilling basins and wave suppressors for canal structures, outlet works, and diversion dams
Basin V   Stilling basins with sloping aprons

Many of the basins are based upon developing a hydraulic jump for the purpose of dissipating energy before the flow enters the downstream channel. The U.S. Bureau of Reclamation (Peterka, 1984) performed a comprehensive series of tests to determine the properties of hydraulic jumps. Figure 17.4.1 illustrates the characteristic form of the hydraulic jump related to the Froude number. For a Froude number of 1.0 for the incoming flow, the flow is critical and a jump cannot form. For Froude numbers between 1.0 to about 1.7, the incoming flow is slightly below critical depth and a jump still cannot form. For Froude numbers above 1.7, the incoming flows can produce hydraulic jumps with the characteristics shown in Figure 17.4.1.

Characteristics of the hydraulic jump and notation are shown in Figure 17.4.2. From Section 5.5, the expression for conjugate depths of a hydraulic jump can be expressed using equation (5.5.2) as

$$y_2 = -\frac{y_1}{2} + \sqrt{\frac{2V_1^2 y_1}{g} + \frac{y_1^2}{4}} \tag{17.4.1}$$

or as

$$y_1 = -\frac{y_2}{2} + \sqrt{\frac{2V_2^2 y_2}{g} + \frac{y_2^2}{4}} \tag{17.4.2}$$

Substituting the Froude number

$$F_{r_1} = \frac{V_1}{\sqrt{gy_1}} \tag{17.4.3}$$

into equation (17.4.1) results in (given in Chapter 5):

$$\frac{y_2}{y_1} = \frac{1}{2}\left(\sqrt{1 + 8F_{r_1}^2} - 1\right) \tag{5.5.4}$$

The main concern in the hydraulic design of a hydraulic-jump stilling basin is the determination of the basin width and elevation to form a stable hydraulic jump in the basin. This requires that the jump is neither swept out of the basin nor drowned as the discharge over the spillway varies. If the water depth in the stream below the basin (the tailwater depth), is lower than the conjugate depth of the jump in the basin, the jump will be swept out of the basin. This is likely to lead to erosion of the unprotected stream bed unless the basin is long enough to contain the jump. On the other hand, if the tailwater depth is higher than the conjugate jump depth, the jump moves upstream until the conjugate and tailwater depths become equal. In the process, the jump may be

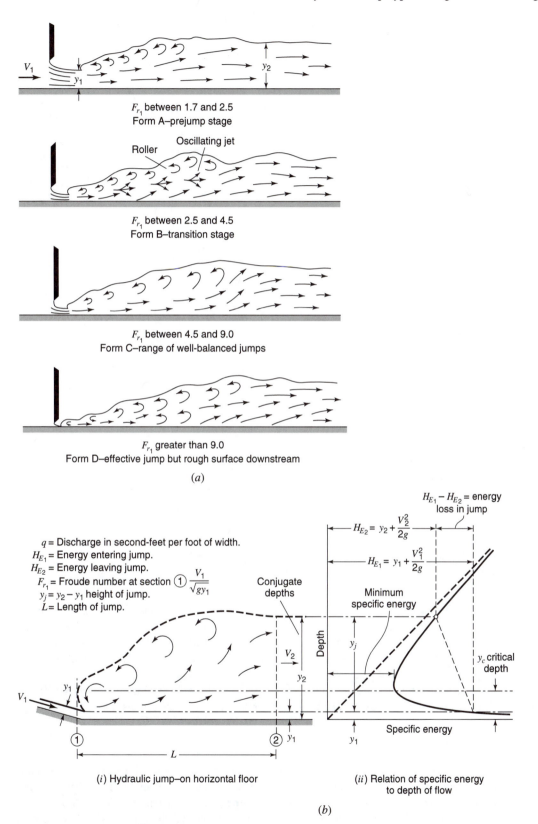

Figure 17.4.1 Hydraulic jumps on horizontal apron. (*a*) Characteristic forms of hydraulic jump related to the Froude number; (*b*) Hydraulic jump symbols and characteristics (from U.S. Bureau of Reclamation (1987)).

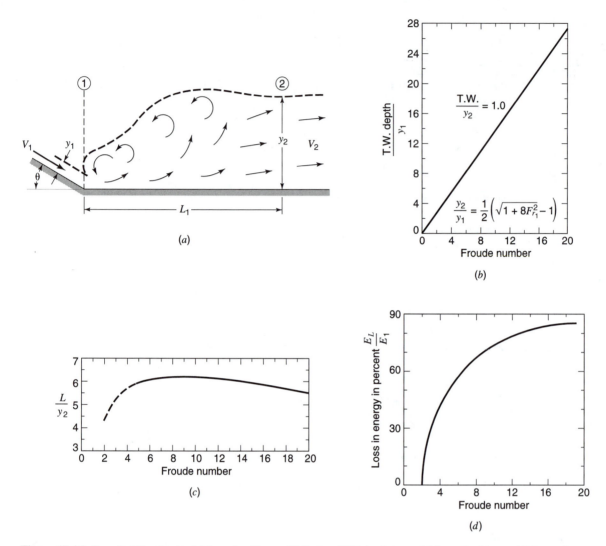

Figure 17.4.2 Type-I stilling basin. (*a*) Length of jump; (*b*) Ratio of *TW* depth to y_1; (*c*) Length of jump; (*d*) Loss of energy in jump (from U.S. Bureau of Reclamation (1987)).

drowned, losing its efficiency as an energy dissipator. Although both situations are undesirable, the first can be far more serious. Therefore, the design objective is to select the basin width and elevation to match the tailwater and conjugate depths at all discharges and to contain the jump in the basin. However, this may be difficult to achieve due to limitations imposed on selection of the proper basin width and/or elevation by the topography and economy. The general practice is to sacrifice jump efficiency in order to prevent bed scour (or to avoid extremely long basins). This is done by matching the tailwater and conjugate depths only for the design discharge.

17.4.2 Basin I

For Froude numbers less than 1.7, no special stilling basin is required. Channel lengths must extend beyond the point where the depth starts to change to not less than $4y_2$. These basins do not require baffle or dissipation devices. These basins are referred to as type-I basins (U.S. Bureau of Reclamation, 1987). For Froude numbers between 1.7 and 2.5 the type-I basin also applies. Characteristics of the type-I basins are shown in Figure 17.4.2.

17.4.3 Basin II

Basins that have been used with high dam and earth dam spillways and large canal structures are type-II basins (see Figure 17.4.3). These basins contain chute blocks at the upstream end and a dentated sill near the downstream end. Baffle piers are not needed because of the relatively high velocity entering the jump. These basins are for Froude numbers above 4.5 or velocities above 50 ft/s. Relationships illustrating stilling basins proportional to minimum tailwater depths and lengths of jump, as a function of Froude number, are presented in Figure 17.4.3.

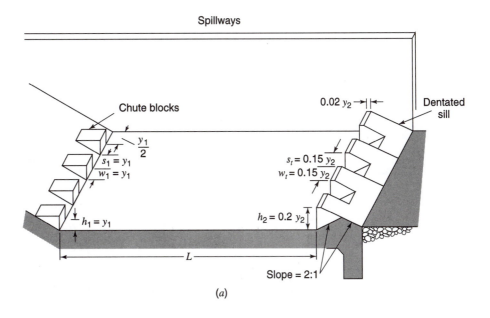

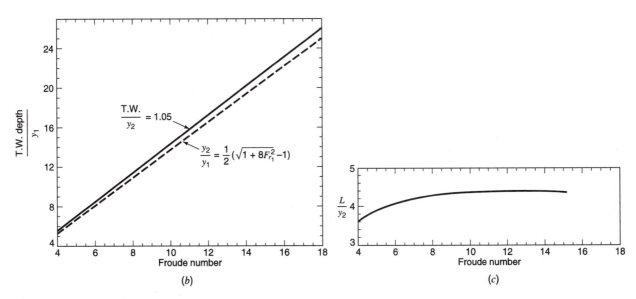

Figure 17.4.3 Stilling basin characteristics for Froude numbers above 4.5. (*a*) Type-II basin dimensions; (*b*) Minimum tailwater depths; (*c*) Length of jump (from U.S. Bureau of Reclamation (1987)).

17.4.4 Basin III

Type-III basins are shorter basins than the type-II with a simpler end sill and with baffle piers downstream of the chute blocks (see Figure 17.4.4). The incoming velocity for the type-III basin must be limited to prevent the possibility of low pressures on the baffle piers that can result in cavitation. The type-III basin length is about 60 percent of the type-II basin. Type-III basins are used on small spillways, outlet works, and small canal structures where V_1 does not exceed 50 to 60 ft/sec and the Froude number $F_1 > 4.5$.

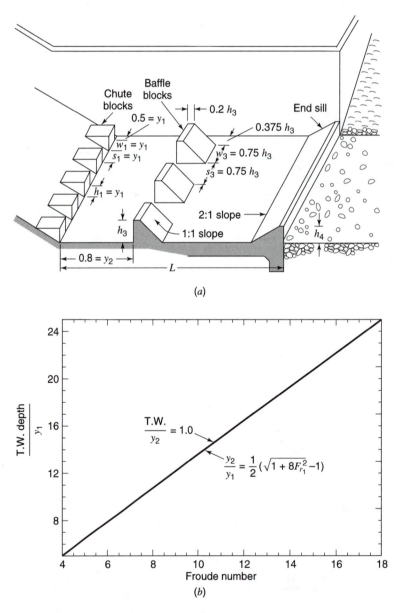

Figure 17.4.4 Stilling basin characteristics for Froude numbers above 4.5 where incoming velocity $V_1 \leq 60$ ft/s. (*a*) Type-III basin dimensions; (*b*) Minimum tailwater depths;

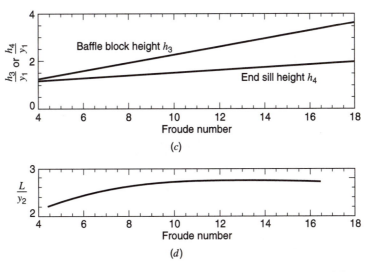

Figure 17.4.4 Stilling basin characteristics for Froude numbers above 4.5 where incoming velocity $V_1 \leq 60$ ft/s. (*continued*) (*c*) Height of baffle blocks and end sill; (*d*) Length of jump (from U.S. Bureau of Reclamation (1987)).

17.4.5   Basin IV

Type-IV basins are used where the Froude number is in the range of 2.5 to 4.5, which is typical of canal structures and occasionally of low dams (small spillways), small outlet works, and diversion dams. In this case the hydraulic jump is not fully developed and the main concern is the waves created in the unstable hydraulic jump. These basins reduce excessive waves created in imperfect jumps. Figure 17.4.5 illustrates the characteristics of this basin along with an alternate design and wave suppressors that may also be used in place of the type-III basin.

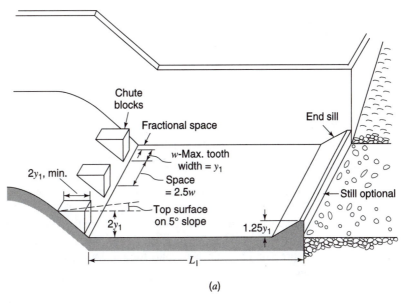

Figure 17.4.5 Stilling basin characteristics for Froude numbers between 2.5 and 4.5. (*a*) Type-IV basin dimensions.

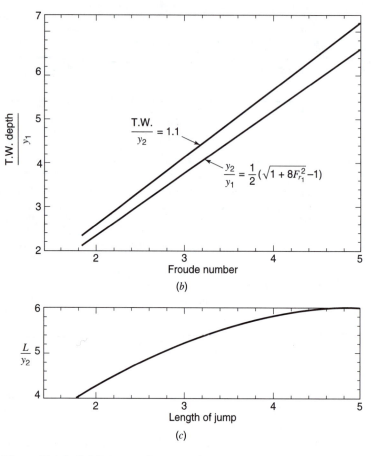

Figure 17.4.5 (*b*) Minimum tailwater depths; (*c*) Length of jump.

Figure 17.4.6 shows an alternative low Froude number stilling basin. The type-IV basin has large deflector blocks that are similar to but larger than chute blocks, and an optional solid end sill. The design shown in Figure 17.4.6 does not have chute blocks, but does have large baffle piers and a dentated end sill.

17.4.6   Basin V

Type-V basins (Figure 17.4.7) are stilling basins with sloping aprons, which are for use where structural economics make the sloping apron more desirable. They are usually used on high dam spillways. Sloping aprons need a greater tailwater depth than the horizontal (type-I) basins. Four cases are illustrated in Figure 17.4.7. Case A is a jump on the horizontal apron. Case B is also a horizontal apron but the toe of the jump forms on the slope and the jump ends on the horizontal apron. For case C the toe of the jump is on the slope and the end of the jump is at the junction of the slope and the horizontal apron. For case D the entire jump forms on the slope. Case B is the one usually encountered in sloping apron design (Peterka, 1978).

The expression for conjugate depths for a hydraulic jump on a sloping apron was derived by Kindsvater (1944):

$$\frac{y_2}{y_1} = \frac{1}{2\cos\phi}\left[\sqrt{\frac{8F_{r_1}^2\cos^3\phi}{1-2K\tan\phi}+1}-1\right] \qquad (17.4.4)$$

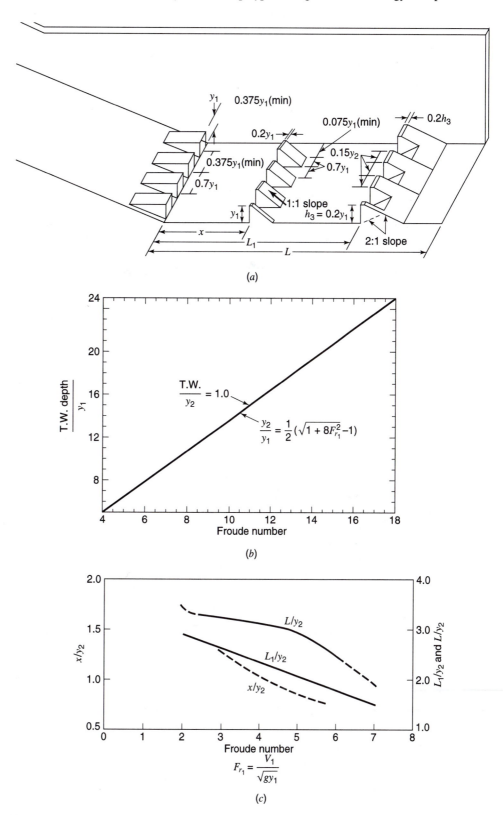

Figure 17.4.6 Characteristics for alternative low Froude number stilling basins. (*a*) Dimensions for alternative low Froude number basin; (*b*) Minimum tailwater depths; (*c*) Length of jump (from U.S. Bureau of Reclamation (1987)).

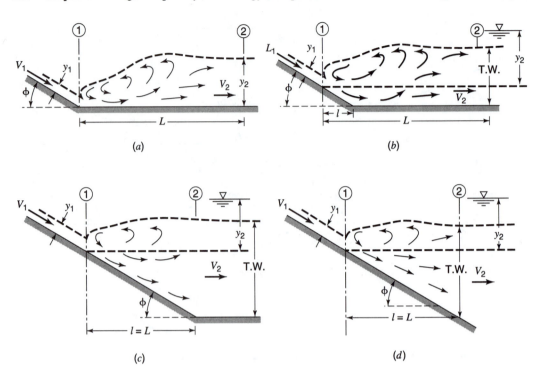

Figure 17.4.7 Sloping aprons (Basin V) (from Peterka (1978)).

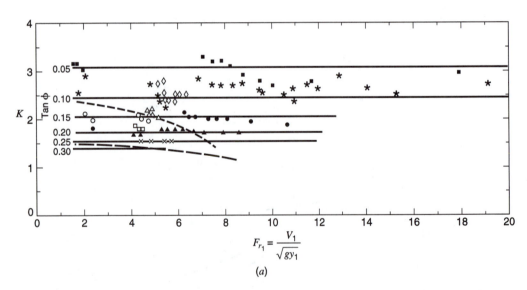

Figure 17.4.8 Shape factor, K, in jump formula. K as function of F_{r_1} (Basin V, Case D) (from Peterka (1978)).

where K is a dimensionless shape factor that varies mainly with the slope of the apron and minimally with the Froude number (see Peterka, 1984). Figure 17.4.8 illustrates the variation of K with $\tan\phi$, assuming that K is independent of F_{r_1}.

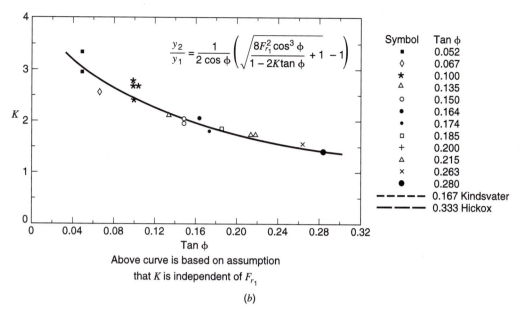

Above curve is based on assumption
that K is independent of F_{r_1}

(b)

Figure 17.4.8 Shape factor, K, in jump formula. K as independent of F_{r_1} (Basin V, Case D) (from Peterka (1978)).

17.4.7   Tailwater Considerations for Stilling Basin Design

For the various terminal structures previously discussed, the tailwater conditions are a major concern in the design process. For hydraulic-jump type basins, the basin floor level must be selected to provide a hydraulic jump elevation that conforms to tailwater elevation for various discharges in order to prevent sweepout of the jump from the basin. *A tailwater rating curve*, which is the stage-discharge relationship of the natural stream or river below a dam, must be determined using the water surface profile (backwater) determination method, as discussed in Chapter 5. Tailwater rating curves are dependent upon the natural conditions of the downstream river characteristics and are not normally changed by the spillway release characteristics.

The objective is to compare an elevation- (conjugate depth-) discharge relationship with the tailwater rating curve determined from a backwater analysis. The conjugate depth- (elevation-) discharge relationship is developed for a certain spillway basin design including the basin width and basin floor elevation. Consider conjugate depth curve 1 in Figure 17.4.9. The stilling basin elevation has been set so that at maximum spillway design capacity the tailwater rating curve and the conjugate-depth curve intersect. For smaller discharges the tailwater elevation is greater than the conjugate depth (elevation), which results in an excess tailwater that can cause a drowned jump. Drowned jumps unfortunately do not achieve the jump type dissipation. Curve 2 has a stilling basin elevation higher than curve 1, causing the conjugate depth (elevation) to be greater than the tailwater rating curve for the entire range of discharges. Curve 3 has a stilling basin elevation lower than that for the basin of curve 1, causing the tailwater elevation to be greater than the conjugate depth (elevation) for the entire range of discharges.

EXAMPLE 17.4.1

Estimate the length of the concrete apron ($S_0 = 0.001$) for a type-I stilling basin downstream from an overflow spillway (Figure 17.4.10). The spillway crest is 50 ft long and consider a discharge of 4000 cfs. Manning's roughness factor $n = 0.025$. Should a type-I basin be used for this situation, based upon the information available?

SOLUTION

The objective is to determine $\Delta x + L_j$ where L_j is the length of the jump. Flow rate is $q = 4000/50 = 80$ ft^3/s/unit width of spillway crest. Use the energy equation from the spillway crest to the toe (or bottom of spillway) and ignore energy loss, which is conservative:

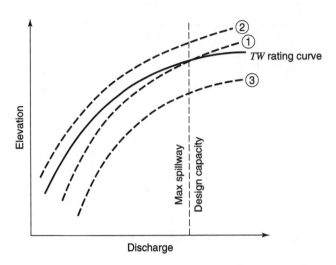

Figure 17.4.9 Relationships of conjugate depth curves to tailwater rating curve.

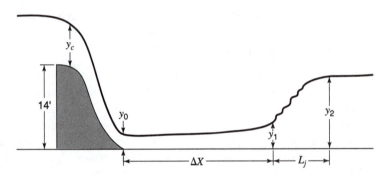

Figure 17.4.10 Example 17.4.1.

$$E_c + 14 = y_0 + \frac{V^2}{2g}$$

where the specific energy at critical flow is determined using

$$E_c = \frac{3}{2}\left(\frac{q^2}{g}\right)^{1/3} = \frac{3}{2}\left(\frac{6400}{32.2}\right)^{1/3} = 8.75 \text{ ft}$$

The energy equation is now solved for y_0:

$$8.75 + 14 = y_0 + \left(\frac{80}{y_0}\right)^2 \frac{1}{2(32.2)}$$

$$22.75 = y_0 + \frac{99.38}{y_0^2}$$

from which $y_0 = 2.20$ ft. Next use Manning's equation to solve for y_2, which assumes normal depth occurs after the jump:

$$Q = \frac{1.49}{n} AR^{2/3} S_0^{1/2} = \frac{1.49}{0.025}(50y_2)\left(\frac{50y_2}{50 + 2y_2}\right)^{2/3}(0.001)^{1/2} = 4000 \text{ cfs}$$

Solving by Newton's method (Appendix A) yields $y_2 = 10.98$ ft. Next compute the depth y_1 using

$$\frac{y_2}{y_1} = \frac{1}{2}\left[-1 + \sqrt{1 + \frac{8q^2}{gy_1^3}}\right] \Rightarrow \frac{10.98}{y_1} = \frac{1}{2}\left[-1 + \sqrt{1 + \frac{8(80)^2}{gy_1^3}}\right]$$

Solving, $y_1 = 2.66$ ft. Δx can now be determined using the energy equation between depths y_0 and y_1:

$$\Delta x = \frac{\left(y_0 + \dfrac{V_0^2}{2g}\right) - \left(y_1 + \dfrac{V_1^2}{2g}\right)}{S_f - S_0}$$

$V_0 = q/y_0 = 80/2.20 = 36.4$ ft/s
$V_1 = q/y_1 = 80/2.66 = 30.1$ ft/s
$V_{ave} = 33.3$ ft/s

The friction slope is now determined using Manning's equation:

$$S_f = \frac{n^2 V_{ave}^2}{2.22 R_{ave}^{4/3}}$$

Assuming the stilling basin is the same width as the spillway crest, the hydraulic radius R_{ave} is:

$$R_{ave} = \frac{1}{2}\left[\frac{50 \times 2.20}{50 + 2(2.20)} + \frac{50 \times 2.66}{50 + 2(2.66)}\right] = 2.21 \text{ ft}$$

The friction slope is then

$$S_f = \frac{(0.025)^2 (33.3)^2}{2.22(2.21)^{4/3}} = 0.1087$$

and

$$\Delta x = \frac{\left(2.20 + \dfrac{(36.4)^2}{29}\right) - \left(2.66 + \dfrac{(30.1)^2}{29}\right)}{0.1087 - 0.001} = \frac{22.8 - 16.7}{0.1077} = 56.6 \text{ ft}$$

To compute the length of the jump L_j Figure 17.4.2c is used. The Froude number F_{r_1} is

$$F_{r_1} = \frac{V_1}{\sqrt{gy_1}} = \frac{30.1}{\sqrt{32.2(2.66)}} = 3.25$$

From Figure 17.4.2c, for $F_{r_1} = 3.25$, $L_j/y_2 = 5.4$, so $L_j = 5.4y_2 = 5.4 (10.98) = 59.3$ ft. The length of stilling basin should be at least $(\Delta x + L_j) = 116$ ft to prevent the jump from leaving the basin. Actually, for this example a type-III stilling basin is more appropriate because of the Froude number, $F_{r_0} = 4.3$.

EXAMPLE 17.4.2

Select the type of stilling basin considering the crest length = 100 ft and the discharge = 15000 cfs for elevations of AA equal to 0, 20, and 40 ft (refer to Figure 17.4.11).

SOLUTION

Assume a rectangular stilling basin is built.

$$q = \frac{Q}{L} = \frac{15000}{100} = 150 \text{ cfs/ft}$$

The energy equation from spillway crest to just before the jump is

$$Z_c + E_c = Z_1 + y_1 + \frac{V_1^2}{2g}$$

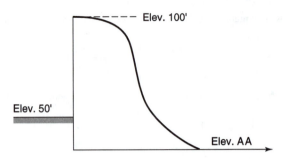

Figure 17.4.11 Example 17.4.2.

$$100 + E_c = y_1 + \frac{V_1^2}{2g}$$

For elevation of AA $= 0$ ft, $Z_1 = 0$,

$$E_c = \frac{3}{2}\left(\frac{q^2}{g}\right)^{1/3} = \frac{3}{2}\left(\frac{150^2}{32.2}\right)^{1/3} = 13.31$$

$$V_1 = \frac{150}{y_1}$$

Thus

$$100 + 13.31 = y_1 + \frac{\left(\dfrac{150}{y_1}\right)^2}{2(32.2)}$$

which simplifies to $y_1^3 - 113.31y_1^2 + 349.38 = 0$, $y_1 = 1.77$ ft.

The Froude number at section 1 is $F_{r_1} = V_1/\sqrt{gy_1} = 150/y_1\sqrt{gy_1} = 150/\left(1.77\sqrt{32.2 \times 1.77}\right) = 11.23$.

Since $F_{r_1} = 11.23 > 4.5$, the type of basin can be type II or III. As a check, the incoming velocity is $V_1 = 150/1.77 = 84.7$ ft/s > 50 ft/s, so a type-II basin is recommended.

For elevation of AA $= 20$ ft,

$$Z_c + E_c = Z_1 + y_1 + \frac{V_1^2}{2g}$$

$$100 + 13.31 = 20 + y_1 + \frac{150^2}{2(32.2)y_1^2}$$

$$y_1^3 - 93.31y_1^2 + 349.38 = 0$$

$$y_1 = 1.96 \text{ ft}$$

The Froude number is $F_{r_1} = \dfrac{150}{1.96\sqrt{32.2 \times 1.96}} = 9.63 > 4.5$, $V_1 = \dfrac{150}{1.96} = 76.5 > 50$ ft/s, so use a type-II basin.

For elevation of AA $= 40$ ft,

$$100 + 13.31 = 40 + y_1 + \frac{150^2}{2(32.2)y_1^2}$$

$$y_1^3 - 73.31y_1^2 + 349.38 = 0$$
$$y_1 = 2.22 \text{ ft}$$

The Froude number is $F_{r_1} = \dfrac{150}{2.22\sqrt{32.2 \times 2.22}} = 7.99 > 4.5,$

$$V_1 = \frac{150}{2.22} = 67.57 > 50 \text{ ft/s}$$

so use type-II basin.

EXAMPLE 17.4.3

The objective of this example is to determine the appropriate stilling basin floor elevation in order to prevent the sweepout of a hydraulic jump from the basin. The crest length is 76.6 ft and the discharge is 20,000 ft³/s (refer to Figure 17.4.12).

SOLUTION

First assume a basin floor elevation of 16.7 ft; then $Z_c = 50 - 16.7 = 33.3$ ft,

$q = 20{,}000/76.6 = 261$ ft³/sec/ft of width; $E_c = \dfrac{3}{2}\sqrt[3]{q^2/g} = \dfrac{3}{2}\sqrt[3]{(261)^2/(32.2)} = 19.3$ ft.

The energy equation between the crest and depth y_1 before the jump is used to compute y_1 (ignoring losses):

$$E_c + Z_c = y_1 + \frac{V_1^2}{2g} = y_1 + \frac{q^2}{2gy_1^2}$$

$$19.3 + 33.3 = y_1 + \frac{(261)^2}{2(32.2)y_1^2}$$

$$y_1^3 - 52.6y_1^2 + 1057.8 = 0$$

Solving yields $y_1 = 4.7$ ft; then $V_1 = 261/4.7 = 55.5$ ft/s and the Froude number is $F_{r_1} = V_1/\sqrt{gy_1} = 55.5/\sqrt{32.2(4.7)} = 4.53$. Because $F_{r_1} > 4.5$ and $V_1 > 50$ ft/s, a type-II stilling basin would be recommended. Using Figure 17.4.3b $TW/y_1 = 6.2$ for $F_{r_1} = 4.53$ and $TW/y_2 = 1.05$ then $TW = 6.2y_1 = 6.2(4.7) = 29.1$ ft and $y_2 = TW/1.05 = 27.7$ ft. Other stilling basin floor elevations should be looked at with the objective of selecting the best.

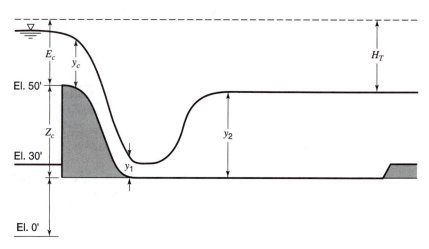

Figure 17.4.12 Example 17.4.3.

EXAMPLE 17.4.4

The objective of this problem is to determine the floor elevation for the stilling basin shown in Figure 17.4.13. The results of a survey at a point on a stream at which the basin is to be built are given in Figure 17.4.14, showing the relationships of hydraulic radius and cross-sectional area of flow as a function of elevation. Manning's n is 0.030 and the average slope of the stream bed is 0.00375. The design discharge for the ogee-type spillway is to be 20,000 cfs and the spillway crest length is 100 ft at elevation 3260 ft. The approach channel floor elevation is to be at 3210 ft. The crest is to be shaped using a design head that is 85 percent of the maximum head. Note that the maximum head corresponds to design discharge. A hydraulic-jump stilling basin having the same width as the spillway crest is to be provided. Determine a suitable floor elevation for the hydraulic-jump stilling basin. Ignore the effect of abutments and piers in your design. Present a graphical comparison of computed water surface elevations in the stilling basin with tailwater elevations for all discharges up to the maximum of design discharge.

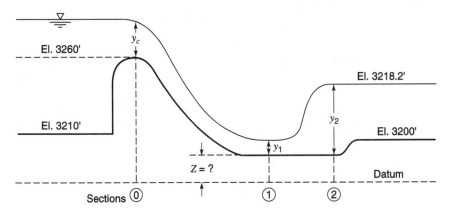

Figure 17.4.13 Example 17.4.4.

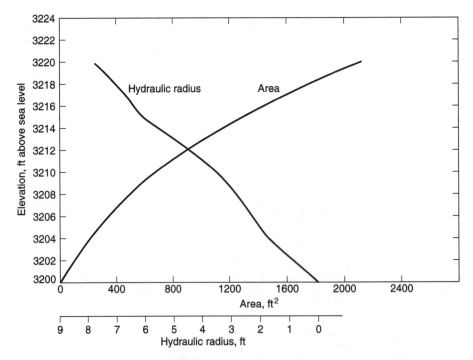

Figure 17.4.14 Information to construct tailwater rating curve for Example 17.4.4.

SOLUTION

To determine the flow elevation, the energy equation between cross-sections 0 and 1 along with the hydraulic jump momentum balance equation are used. The energy between cross-sections 0 and 1 (neglecting losses) is

$$3260 + E_c = Z + E_1$$

$$3260 + \frac{3}{2}\sqrt[3]{q^2/g} = Z + y_1 + \frac{V_1^2}{2g}$$

$$3260 + \frac{3}{2}\frac{\left(\dfrac{20{,}000}{100}\right)^{2/3}}{32.2^{1/3}} = Z + y_1 + \frac{(20{,}000)^2}{2(32.2)(100y_1)^2}$$

Solving yields $Z = 3276.1 - y_1 - \dfrac{621.1}{y_1^2}$. Using equation (5.5.4), we find

$$\frac{y_2}{y_1} = \frac{1}{2}\left(\sqrt{1 + 8F_{r_1}^2} - 1\right) = \frac{1}{2}\left(\sqrt{1 + \frac{8q^2}{gy_1^3}} - 1\right)$$

$$y_2 = \frac{y_1}{2}\left(\sqrt{1 + \frac{9937.9}{y_1^3}} - 1\right)$$

From geometry, $y_2 + Z = 3218.2$; then rearranging and substituting the above equation for y_2 gives us

$$Z = 3218.2 - y_2 = 3218.2 - \frac{y_1}{2}\left(\sqrt{1 + \frac{9937.9}{y_1^3}} - 1\right)$$

Now the above energy and momentum equations can be combined:

$$3276.1 - y_1 - \frac{621.1}{y_1^2} = 3218.2 - \frac{y_1}{2}\left(\sqrt{1 + \frac{9937.9}{y_1^3}} - 1\right)$$

$$\left(\frac{1242.2}{y_1^3} - \frac{115.8}{y_1} + 3\right)^2 - \left(1 + \frac{9937.9}{y_1^3}\right) = 0$$

which can be solved using Newton's method (Appendix A) to obtain $y_1 = 2.7$, and

$$Z = 3276.1 - 2.7 - \frac{621.1}{(2.7)^2} = 3188.2 \text{ ft}$$

The tailwater rating curve is developed using area–elevation and hydraulic radius–elevation relationships in Figure 17.4.14. Table 17.4.1 presents the tailwater rating curve computations where $K = \dfrac{1.49}{n}AR^{2/3}$.

Table 17.4.1   Tailwater Rating Curve (Figure 17.4.15)

El. (ft)	A (ft²)	R	K	Q (cfs)
3200	0	0	—	—
3204	200	1.65	13870	849
3207	400	2.5	36595	2241
3210	660	3.4	74119	4535
3213	1010	4.9	144716	8862
3215	1270	6.0	208275	12754
3217	1580	6.6	270112	16908
3220	2120	7.7	410577	25143

$S = 0.00375,\ n = 0.03,\ K = \dfrac{1.49}{n}AR^{2/3},\ Q = K\sqrt{S}.$

The conjugate water depths in the basin for various discharge rates may be determined by equating the energies at cross-sections 0 and 1 (neglecting any losses and setting $Z = 3188.2$ ft):

$$(50 + 21.8) + E_c = y_1 + \frac{V_1^2}{2g}$$

or

$$71.8 + \frac{3}{2}\sqrt[3]{\frac{q^2}{g}} = y_1 + \frac{q^2}{2gy_1^2}$$

to compute y_1 for various values of q. The momentum equation (equation (5.5.4)) is then used to compute y_2. The resulting water surface level is $y_2 + Z$. Table 17.4.2 provides the computations and the results are plotted on Figure 17.4.15 along with the tailwater rating curve. For discharges less than $Q_d = 20,000$ cfs, the tailwater depths are higher than the conjugate depths. Therefore the jump will move upstream and probably be drowned at the toe. This is a safe practice when we are concerned primarily with preventing the jump from sweeping downstream on the unprotected stream bed. The disadvantage would be the loss in efficiency, since the submerged jump does not dissipate as much energy.

Table 17.4.2   Discharge–Conjugate Depth Curves

Q (cfs)	q	y_c	E_c	$71.8 + E_c$	y_1	V_1	$F_{r_1}^2$	y_2	El. ($y_2 + Z$)
25,000	250	12.47	18.71	90.51	3.34	74.9	52.2	32.5	3220.7
20,000	200	10.75	16.13	87.93	2.70	74.1	63.2	29.0	3217.2
15,000	150	8.87	13.31	85.11	2.05	73.2	81.2	25.1	3213.3
10,000	100	6.77	10.16	81.96	1.39	71.9	115.5	20.4	3203.6
5,000	50	4.27	6.40	78.20	0.71	70.4	216.8	14.4	3202.6

EXAMPLE 17.4.5

Design a free-overfall spillway and hydraulic-jump basin to discharge 500 ft³/s with a drop of 12 ft. The tailwater elevation is 108 ft and the approach channel length is 20 ft. The approach floor is level with the spillway crest, which is at elevation 120 ft. Select an energy dissipator (adapted from U.S. Bureau of Reclamation, 1987).

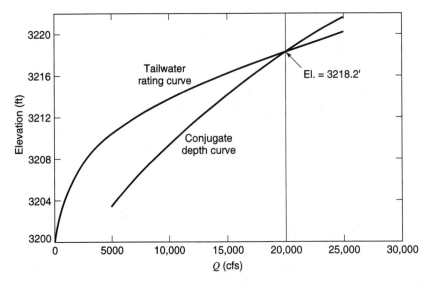

Figure 17.4.15 Tailwater rating and conjugate depth curves for Example 17.4.4.

SOLUTION

Step 1. Select an effective length of the spillway crest as 15 ft, and use $C = 3.0$.

Step 2. The unit discharge is $q = 500/15 = 33.3$ ft^3/s, so the design head is

$$H_e = (q/C)^{2/3} = (33.3/3.0)^{2/3} = 5 \text{ ft.}$$

The reservoir water surface level is then $120 + 5 = 125$ ft, and the drop in elevation from the reservoir level to the tailwater level is 17 ft.

Step 3. An offset of 0.5 ft is provided along each side of the weir to effect side contraction for aerating the underside of the sheet (U.S. Bureau of Reclamation, 1987). The offset is assumed square-cornered. The net crest length is then from equation (17.3.3), adding the offset, 2(0.5)

$$L' = L + 2K_a H_e + 2(0.5) = 15 + 2(0.2)(5) + 1.0 = 18 \text{ ft}$$

Step 4. Select apron floor elevation. Referring to the notation in Figure 17.3.1, $h_d = 125 - 108 = 17$ ft. Neglecting losses and applying the energy equation at cross-sections before and after the jump, we find

$$h_d + y_2 = y_1 + V_1^2/2g$$

$$h_d = \frac{2q^2}{gy_2^2 \left[\sqrt{1 + \dfrac{8q^2}{gy_2^3}} - 1\right]^2} + \frac{y_2}{2}\left(\sqrt{1 + \frac{8q^2}{gy_2^3}} - 1\right) - y_2$$

Using $h_d = 17$ ft and $q = 33.3$ ft^3/s/ft, we find $y_2 = 8.8$ ft. The apron floor elevation would then be $108 - 8.8 = 99.2$ ft.

Step 5. The drop distance is $Y = 120 - 99.2 = 20.8$ ft, and the drop number is

$$\bar{D} = \frac{q^2}{gY^3} = \frac{(33.3)^2}{32.2(20.8)^3} = 0.0038$$

Using Figure 17.3.1 with $\bar{D} = 0.0038$ yields, the length ratio, $y_2/Y = 0.375$ and then $y_2 = 0.375(20.8) = 7.8$ ft. The elevation of the apron is adjusted to $108 - 7.8 = 100.2$ ft. The adjusted value of Y is $Y = 120 - 100.2 = 19.8$ ft, and $\bar{D} = 0.0044$.

Step 6. From Figure 17.3.1, $\bar{D} = 0.0044$ and $h_d/H_e = 17/5 = 3.4$, $L_d/Y = 1.02$ and $L_d = 20.2$ ft, $y_1/Y = 0.054$ and $y_1 = 0.054(20.8) = 1.1$ ft, and $F_{r_1} = 5.3$.

Step 7. For $F_{r_1} = 5.3$, $y_1 = 1.1$ ft, and $y_2 = 7.8$ ft, the Type-III basin in Figure 17.4.4 is used. From Figure 17.4.4d, for $F_r = 5.3$, $L/y_2 = 2.37$ and $L = 2.37(7.8) = 18.5$ ft. The length of the basin measured from the vertical crest is $L_d + L = 20.2 + 18.5 = 38.7$ ft.

Step 8. Baffle block heights are approximately $1.5y_1 = 1.5(1.1) = 1.65$ ft (20 in), 14 in wide and spaced at about 28 in center to center.

EXAMPLE 17.4.6

For Example 17.4.5, use an impact block basin instead of a type-III basin.

SOLUTION

Step 1. The critical depth is $y_c = \sqrt[3]{(33.3)^2/32.2} = 3.3$ ft.

Step 2. Using Figure 17.3.1 for $\bar{D} = 0.0044$ and $h_d/H_e = 3.4$, $L_p/Y = 0.85$ and L_p 0.85(19.8) = 17.0 ft.

Step 3. Minimum length of the basin is $L_B = L_p + 2.55y_c = 17.0 + 2.55(3.3) = 25.4$ ft or 26 ft.

Step 4. Minimum tailwater depth is $2.15y_c = 7.1$ ft (from Figure 17.3.1a), which places the floor elevation at 100.9 ft.

Step 5. Distance from the vertical crest to the baffle blocks is $L + 0.8\,y_c = 17 + 0.8(3.3) = 19.6$ ft or 20 ft.

Step 6. Baffle block height is $0.8y_c = 0.8(3.3)2.6$ ft ≈ 3 ft and 18 inches wide spaced at 3-ft center to center. The end sill is $0.4y_c = 1.5$ ft high.

Comparing this design with example 17.4.5 shows that the impact basin is considerably smaller than the hydraulic-jump basin. Impact basins should be limited to drop distances of 20 ft or less.

17.5 OTHER TYPES OF STILLING BASINS

The previous basins (I–V) are hydraulic-jump type basins, which require tailwater, whereas the type-VI basin is an *impact-type energy dissipator* (see Figure 17.5.1). This basin is contained in a boxlike structure and does not require tailwater for successful performance, though it will improve performance. Use of this basin is limited to entrance velocities that do not exceed 30 ft/sec and Froude numbers of 1.5 to 7.0, (Peterka, 1978). Dimensions are determined using the table in Figure 17.5.1 with the maximum expected discharge. Multiple units can be used side by side quite effectively and economically.

In cases where the tailwater depth is too large for the formation of a hydraulic jump, *submerged bucket dissipators* (deflectors)(type VII basins) can be used to dissipate high-energy flows. These submerged buckets include solid buckets and slotted buckets (see Figures 17.5.1 and 17.5.2). The solid buckets do not have the slots shown in Figure 17.5.1.

The hydraulic action of these two types of buckets is illustrated in Figure 17.5.2. Hydraulic behavior involves the formation of two rollers, a bucket roller and a ground roller. The *bucket roller* is formed at the surface, moves counter clockwise and is contained in the curved bucket. The *ground roller* moves clockwise and is downstream of the bucket. Energy is dissipated by these two rollers.

Figure 17.5.1 provides various relationships that can be used in the design of these basins. To determine the minimum bucket radius for a given Froude number, the bucket radius dimensionless ratio $R/(y_1 + V_1^2/2g)$, where R is the bucket radius, must be plotted as a function of the Froude number using maximum capacity discharges.

The purpose of *flip-bucket type basin*s is to throw the water downstream to minimize riverbed damage. These basins are not substitutes for an energy dissipation because the bucket is not able to dissipate energy. A free jet of water falling vertically into water in a riverbed will create a plunge pool whose depth is related to the discharge, height of the fall, depth of tail water, and the bed material. For more horizontal jets with an angle of impingement less than 25 degrees above the horizontal, the jet rides and skips across the surface, causing waves and eddies in the basin. Figure 17.5.3 illustrates a plunge-basin energy dissipator.

Hollow-jet valve stilling basins (type VIII basins)(Figure 17.5.1) are used to dissipate hydraulic energy at the downstream end of outlet works control structures. These basins are usually constructed within or adjacent to powerhouse structures.

STILLING BASIN FOR PIPE ON OPEN CHANNEL OUTLET (BASIN VI)

Use on pipe or open channel outlets. Sizes and discharges from table v_1 should not exceed 30 feet per second. No tailwater required.

Froude number usually 15 to 7 but not important.

May substitute for basin IV.

Energy loss greater than in comparable jump.

SLOTTED AND SOLID BUCKET FOR HIGH, MEDIUM AND LOW DAM SPILLWAYS (BASIN VII)

For use on spillways, drops, chutes, etc. where crest flow is not submerged. V_1 and F_{r_2} have no limits. Data may also be used to design a solid-type roller bucket.

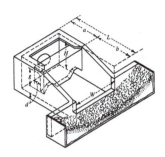

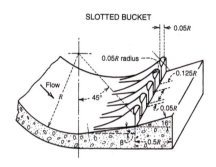

SLOTTED BUCKET

BASIC DIMENSIONS

| Pipe | | | Feet and inches | | | | | | | |
Dia. in.	Area sq ft.	Q	W	H	L	a	b	c	d	g
18	177	21	5-5	4-3	7-4	3-3	4-1	2-4	0-11	2-1
24	314	36	6-9	5-3	9-0	3-11	5-1	2-10	1-2	2-6
30	491	59	8-0	6-3	10-8	4-7	6-1	3-4	1-4	3-0
36	707	65	9-3	7-3	12-4	5-3	7-1	3-10	1-7	3-6
42	962	115	10-6	6-0	14-0	6-0	8-0	4-5	1-9	3-11
48	1257	151	11-9	9-0	15-6	6-9	6-11	4-11	2-0	4-5
54	1590	191	13-0	9-9	17-4	7-4	10-0	5-5	2-2	4-11
60	1963	234	14-3	10-9	19-0	8-0	11-0	5-11	2-5	5-4
72	2827	339	16-6	12-3	22-0	9-3	12-9	6-11	2-9	6-2

MINIMUM ALLOWABLE BUCKET RADIUS

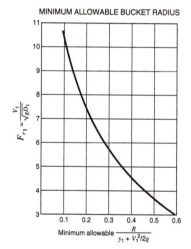

DISCHARGE LIMITS

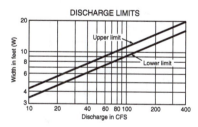

COMPARISON OF ENERGY LOSSES

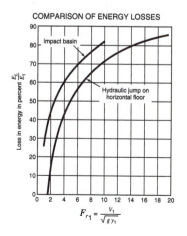

TAILWATER SWEEPOUT DEPTH

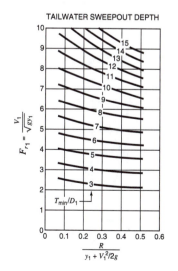

Figure 17.5.1 Summary of characteristics of stilling basin types VI, VII, and VIII (from Peterka (1978)).

SLOTTED AND SOLID BUCKET FOR HIGH, MEDIUM AND LOW DAM SPILLWAYS (CONTINUED) (BASIN VII)

For use on spillways, drops, chutes, etc. where crest flow is not submerged. V_1 and F_{r2} have no limits. Data may also be used to design a solid-type roller bucket.

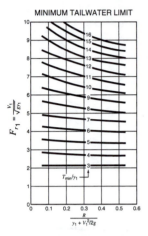

MINIMUM TAILWATER LIMIT

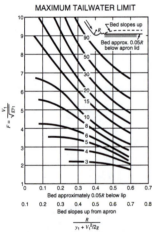

MAXIMUM TAILWATER LIMIT

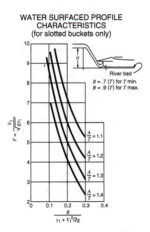

WATER SURFACED PROFILE CHARACTERISTICS
(for slotted buckets only)

STILLING BASIN FOR HIGH HEAD OUTLET WORKS UTILIZING HOLLOW JET VALVE CONTROL (BASIN VIII)

This stilling basin, about 50% shorter than a conventional basin, is used to dissipate hydraulic energy at the downstream end of an outlet works control structure. To reduce cost and save space, the stilling basin is usually constructed within or adjacent to a powerhouse structure.

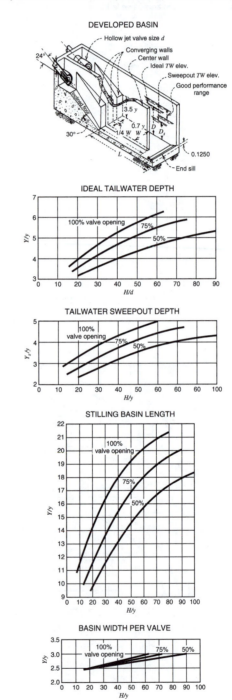

Figure 17.5.1 Summary of characteristics of stilling basin types VI, VII, and VIII (Depths D_1 and D_2 are y_1 and y_2)(from Peterka (1978)).

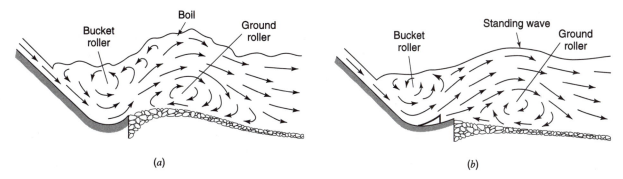

Figure 17.5.2 Hydraulic action of solid and slotted buckets (from U.S. Bureau of Reclamation (1987)).

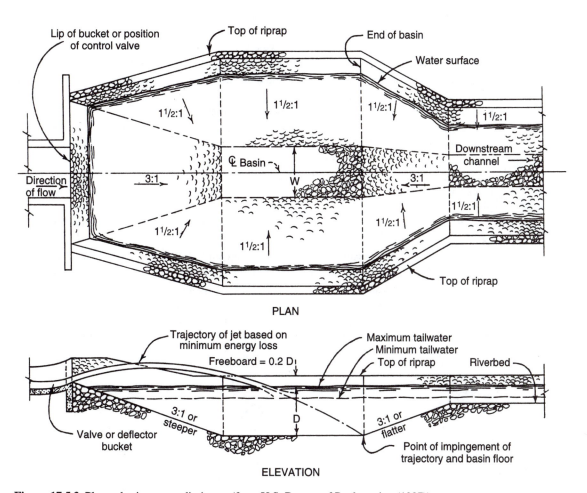

Figure 17.5.3 Plunge basin energy dissipator (from U.S. Bureau of Reclamation (1987)).

17.6   GATES AND VALVES

The two major types of gates are spillway crest gates and outlet works (or conduit) gates. Each of these is briefly described in this section along with diagrams illustrating their construction.

17.6.1   Spillway Crest Gates

Spillway crest gates are used to provide additional storage above a spillway crest. These gates are movable and include several types: (1) flash boards, stop logs, and needles; (2) radial (tainter) gates (Figure 17.2.11 and 17.2.14c); (3) flap gates; (4) drum gates; (5) vertical-lift gates; (6) bear-trap gates; and (7) rolling gates. Zipparro and Hasen (1993) provide detailed descriptions of these gates.

Flash boards are located on top of fixed crests to provide extra depth of storage and can be easily removed at times of flooding by deliberate failure or by some simple tripping device. *Stop logs* are horizontal timbers spanning between piers that can be raised by a hoist. *Needles* are timbers set on end side by side, with the lower ends supported on the spillway crest and the upper ends supported by a runway or bridge.

Tainter gates (radial gates) are the most widely used type of crest gate for large installations. These gates typically consist of a plate formed to a segment of a cylinder that is supported by a steel framework. The steel framework is pivoted on turnings that are supported in the downstream portion of the piers. Figure 17.6.1 illustrates three types of conventional radial gates. Radial gates have the advantages of being simple, reliable, and inexpensive.

Flap gates are basically flat or curved leaves hinged at bearings along the lower edge. The position of the leaf is controlled by hoisting attachments that pull or push or by hydraulic (or screw stem) hoists that push at selected locations under the gate.

Drum gates are acute circular sections formed by skin plates attached to an internal bracing and hinged at the center of curvature, which may be either upstream or downstream, as illustrated in Figure 17.6.2.

Vertical-lift gates are rectangular timber or steel gates that are supported by vertical guides in which the gates are moved vertically.

Bear-trap gates are basically two leaves, the upstream leaf and the downstream leaf, as illustrated in Figure 17.6.3. When the gate is lowered, the leaves are horizontal with one leaf lying on top of the other. These gates are well suited for surface regulation and for passing drift and ice.

Rolling (or roller) gates are steel cylinders that span between piers that have an inclined rack, along which the gate is raised or lowered by a chain (hoist cable). The cylinder has gear teeth so that when pulled, the gate rolls up the rack. Figure 17.6.4 illustrates three types of rolling gates.

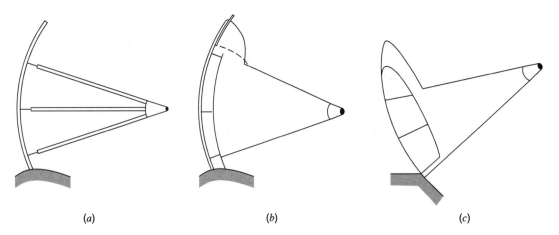

(a)   (b)   (c)

Figure 17.6.1 Types of radial gate: (*a*) Standard gate; (*b*) Gate with flap; (*c*) Submersible gate (from Zipparro and Hasen, 1993).

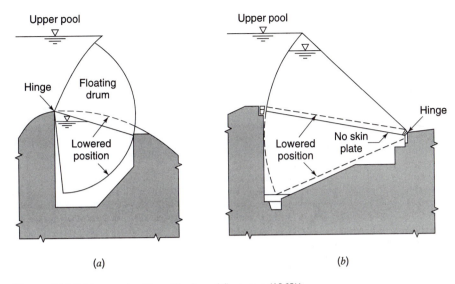

Figure 17.6.2 Drum gates (from Davis and Sorensen (1969)).

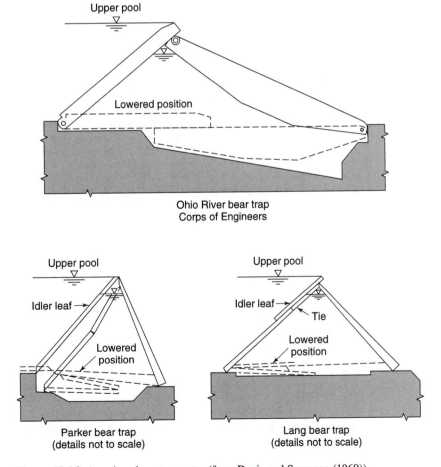

Figure 17.6.3 American bear-trap gates (from Davis and Sorensen (1969)).

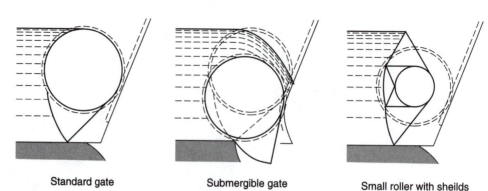

Standard gate   Submergible gate   Small roller with sheilds

Figure 17.6.4 Types of rolling gates (from Davis and Sorensen (1969)).

17.6.2 Gates for Outlet Works

Gates for outlet works are closure devices in which a leaf or closure member is moved across a fluidway from an external position to control the flow of water. These types of gates include *conduit sliding gates* and *jet-flow gates*, which are designed for throttling conditions to regulate flows. Other types of gates are *rising-follower gates*, *ring seal gates*, and *bulkhead gates*, which are operated fully opened or closed and are never used for throttling or regulating flow. Figure 17.6.5 illustrates typical intake gating arrangements. *Guard gates* operate fully open or closed and function as secondary devices in case a primary closure device is inoperable. *Bulkhead gates* are typically installed at the entrance for purposes of drawing water from fluidways for inspection and maintenance.

Slide gates are primarily used for the control of discharges from outlet conduits in dams, for both guard and regulating services. Slide gates such as a tandem slide-gate installation can be placed on a slope so that flow can be downward into a stilling basin.

Ring-follower gates consist of a leaf, which has a bulkhead portion that blocks the fluidway when the gate is closed, and a follower portion with a circular opening that aligns concentrically with the fluidway when the gate is open.

A *ring-seal gate* is a type of ring-follower gate having roller trains and wheels to reduce friction and a movable, hydraulically activated ring seal. These gates have the feature of reduced friction and hoist capacity over the sliding type ring-follower gates.

Jet-flow gates consist of a flat-bottomed leaf, a body, a bonnet, and a bonnet cover on which the operating hoist is mounted.

17.6.3 Valves for Outlet Works

Valves are closure devices in which the closure member remains fixed axially with respect to the fluidway and can be rotated or moved longitudinally to control the discharge (Davis and Sorensen, 1969). The types of valves include needle valves or tube valves, cone dispersion (Howell-Buger) valves, and hollow-jet valves.

Needle valves and *tube valves* have a bulb-shaped fixed steel jacket. The valve closes against the casing in the downstream direction. The open valve produces a solid circular jet. These can be used in submerged conditions.

Cone dispersion valves are installed at the end of outlets discharging into the atmosphere. Figure 17.6.6 illustrates a cone dispersion valve. These valves have been used for reaches up to 250 m (Novak et al.,1996). The *hollow-jet valve*, illustrated in Figure 17.6.7, closes in the upstream direction.

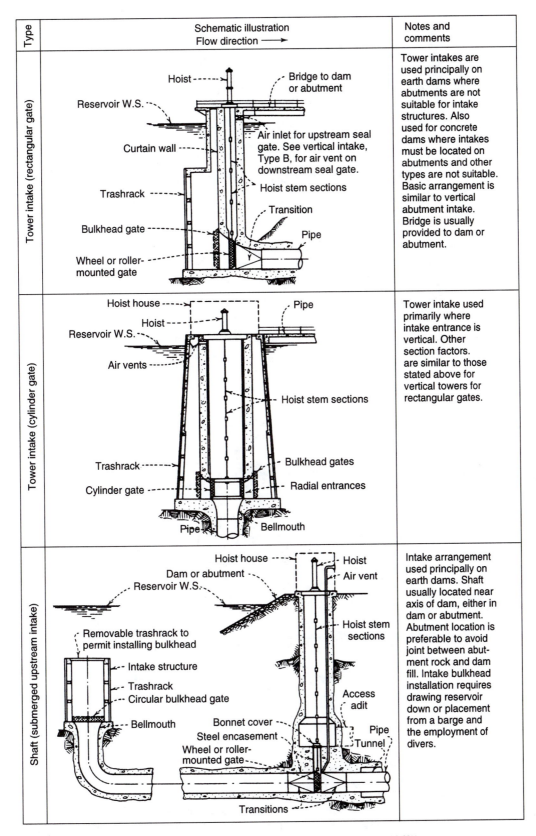

Type	Schematic illustration Flow direction ⟶	Notes and comments
Tower intake (rectangular gate)	Hoist Reservoir W.S. Bridge to dam or abutment Air inlet for upstream seal gate. See vertical intake, Type B, for air vent on downstream seal gate. Curtain wall Hoist stem sections Trashrack Transition Bulkhead gate Pipe Wheel or roller-mounted gate	Tower intakes are used principally on earth dams where abutments are not suitable for intake structures. Also used for concrete dams where intakes must be located on abutments and other types are not suitable. Basic arrangement is similar to vertical abutment intake. Bridge is usually provided to dam or abutment.
Tower intake (cylinder gate)	Hoist house Pipe Hoist Reservoir W.S. Air vents Hoist stem sections Trashrack Bulkhead gates Cylinder gate Radial entrances Pipe Bellmouth	Tower intake used primarily where intake entrance is vertical. Other section factors. are similar to those stated above for vertical towers for rectangular gates.
Shaft (submerged upstream intake)	Hoist house Dam or abutment Reservoir W.S. Hoist Air vent Hoist stem sections Removable trashrack to permit installing bulkhead Intake structure Trashrack Circular bulkhead gate Access adit Bellmouth Bonnet cover Steel encasement Pipe Tunnel Wheel or roller-mounted gate Transitions	Intake arrangement used principally on earth dams. Shaft usually located near axis of dam, either in dam or abutment. Abutment location is preferable to avoid joint between abutment rock and dam fill. Intake bulkhead installation requires drawing reservoir down or placement from a barge and the employment of divers.

Figure 17.6.5 Typical intake gating arrangements (from Davis and Sorensen (1969)).

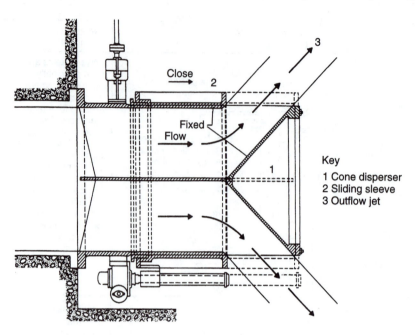

Figure 17.6.6 Cone dispersion valve (after Smith (1978))(as presented in Novak et al. (1996)).

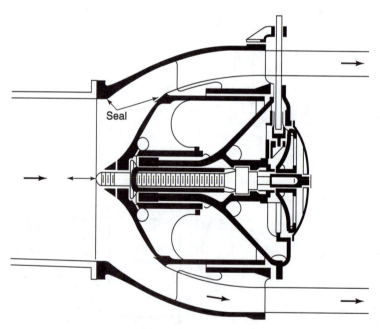

Figure 17.6.7 Hollow-jet valve (after Smith (1978))(as presented in Novak et al. (1996)).

17.7 OUTLET WORKS

The basic function of any outlet is to provide a means of releasing water from a reservoir for desired purposes. Figure 17.7.1 illustrates typical outlet arrangements for the two basic types of outlets: *power outlets* and *outlet works*. Outlet works regulate or release water impounded by a dam. Outlet works release incoming flows at a retarded rate, or divert incoming flows into canals or pipelines, or release stored water for downstream needs, for reservoir evacuation (emptying), or for a combination of multiple-purposes (U.S. Bureau of Reclamation, 1987). They can be classified according to purpose, physical and structural arrangement, or hydraulic operation. Outlet works may be used in lieu of service spillways combined with an auxiliary or secondary spillway. They may also be used as a flood control regulator to lower or empty a reservoir.

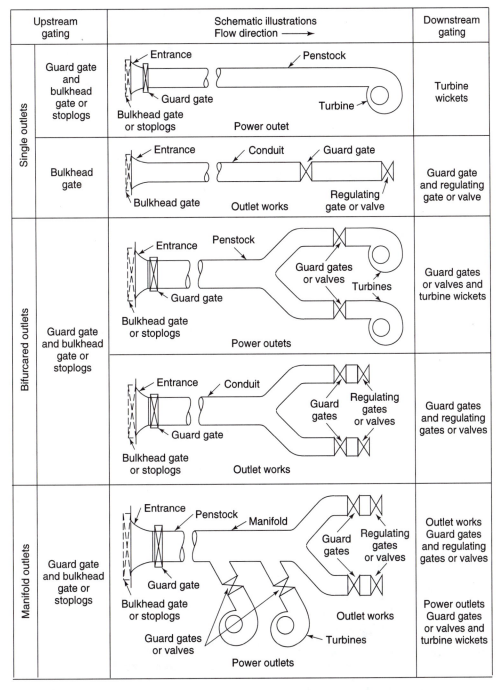

Figure 17.7.1 Schematic of typical outlet arrangements (from Davis and Sorensen (1969)).

PROBLEMS

17.3.1 Design an uncontrolled overflow ogee crest for a spillway to discharge 45,000 cfs. The upstream face of the crest is vertical and a bridge is to span the crest. Bridge piers 24 in-wide (pier contraction coefficient = 0.05) with rounded noses are to be provided. The abutment coefficient is 0.10. The bridge spans (center to center of piers) are not to exceed 25 ft. The maximum expected head is 10 ft. Neglect velocity of approach. The design should be based upon economic considerations such that the design head is no less than 75 percent of the maximum head. The distance from the spillway crest to the reservoir bottom at the dam is 40 ft.

17.3.2 A gate-controlled ogee spillway has been designed that uses Tainter gates for control. You have been consulted to determine the head-discharge curves for two different openings of the Tainter gate, for $D = 5$ and 10 ft. The total heads (including the approach velocity) to the bottom of the orifice range from 4 to 18 ft. The upstream face of the crest is vertical with $P = 20$ ft and designed for $H_0 = 12$ ft, $L = 100$ ft. The gate seats on the crest ($\theta = 90°$).

17.3.3 An ogee overflow section is formed to the ideal nappe shape and is 100 ft long. The design head $H_0 = 8$ ft and height of crest above channel bottom = 12 ft. (a) Find the flow rate at the design head. (b) If, under unusual flood conditions, the head becomes 12 ft, find the flow rate. (c) If the head drops to 4.0 ft, find the flow rate. Take the upstream face of the overflow section as vertical and assume no downstream apron interfernce and downstream submergence.

17.3.4. Rework example 17.3.4 for a maximum surcharge head of 5 ft.

17.3.5. Rework example 17.3.4 for a maximum discharge of 2500 cfs

17.3.6. Determine the head on the crest-discharge curve for a morning-glory spillway with a crest radius of 7.0 ft, considering both crest control and throat control. The morning-glory spillway was designed for a maximum surcharge head of 10 ft and to limit the outflow to 2000 cfs. Assume a coefficient of 3.75 for $H_e/R_s = 0.3$.

Crest elevation = 100 ft

Maximum water surface elevation = 110 ft

Throat radius = 4.5 ft

Circular conduit diameter = 9.0 ft

Consider heads (on the crest) of 2, 3, 4, and 6 ft. Which of the elevations have crest control and which ones have throat control?

17.3.7. Rework example 17.3.5 with a discharge of 2500 ft³/s.

17.3.8. Design a baffle chute spillway for a maximum expected discharge of 5000 ft³/s.

17.4.1 Select the type of stilling basin for the following ogee spillway section (refer to Figure P17.4.1), designed for $H_0 = 10.00$ ft: crest length of spillway = 100 ft, slope = 0.001. What length L of stilling basin selected will be required? Assume $n = 0.025$.

17.4.2 Plot the discharge as a function of head H_1 for flow through a Tainter gate mounted atop an ogee spillway, for a gate opening of 5 ft and heads H_1 ranging from 2.0 feet to 10.0 feet. The ogee spillway is designed for a head $H_0 = 7.0$ feet and the height of the crest above the channel bottom is 15.0 feet.

17.4.3 Rework example 17.4.1 using a discharge of 2000 ft³/s.

17.4.4 Rework example 17.4.2 using a discharge of 7500 ft³/s.

17.4.5 Rework example 17.4.3 using a discharge of 10,000 ft³/s.

17.4.6 Rework example 17.4.4 using a discharge of 30,000 ft³/s.

17.4.7 Rework example 17.4.5 using a spillway crest length of 20 ft.

17.4.8 Design an impact basin for the spillway designed in problem 17.4.5.

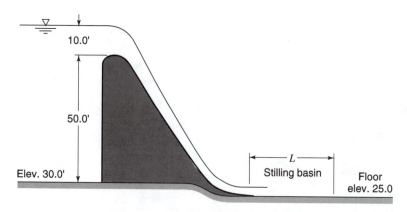

Figure P17.4.1 Problem 17.4.1.

REFERENCES

Chow, V. T., D. R. Maidment, and L. W. Mays, *Applied Hydrology*, McGraw-Hill, New York, 1988.

Coleman, H. W., C. Y. Wei, and J. E. Lindell, "Hydraulic Design of Spillways," *Hydraulic Design Handbook*, edited by L.W. Mays, McGraw-Hill, New York, 1999.

Davis, C. V., and K. E. Sorensen, *Handbook of Applied Hydraulics*, 3rd edition, McGraw-Hill, New York, 1969.

Erbiste, P. C. F., "Developments in Hydraulic Gates," *Hydropower and Dams*, 1(1), pp. 51–56, 1994.

Fread, D. L., "The Development and Testing of Dam-Break Flood Forecasting Model," *Proceedings, Dam-break Flood Modeling Workshop*, U.S. Water Resources Council, Washington, DC, pp. 164–197, 1977.

Fread, D. L., "Capabilities of NWS Model to Forecast Flash Floods Caused by Dam Failures," *Preprint Volume, Second Conference on Flash Floods*, March 18–20, Am. Meteorol. Soc., Boston, MA, pp. 171–178, 1980.

Fread, D. L., "Some Limitations of Dam-Breach Flood Routing Models," Preprint, Amer. Soc. Civ. Eng. Fall Convention, St. Louis, MO, October 1981.

Fread, D. L., "Channel Routing," *Hydrological Forecasting*, edited by M. G. Anderson and T. P. Burt, John Wiley and Sons, New York, pp. 437–503, 1985.

Graham, W. J., and C. T. Yang, "Dam Safety and Nonstructural Damage Reduction Measures," *Water International*, vol. 21, no. 3, pp. 138–143, September 1996.

Hager, W. H., "Discharge Characteristics of Gated Standard Spillways," *Water Power and Dam Construction*, 40(1), pp. 15–26.

Institution of Civil Engineers, *Floods and Reservoir Safety: An Engineering Guide*, London, 1978.

Kindsvater, C. E., "The Hydraulic-Jump in Sloping Channels," *Transactions ASCE*, vol. 109, p. 1107, 1944.

Lemperiere, F., "Overspill Fence Gates," *Water Power and Dam Construction*, 44, pp. 47–48, 1992.

Linsley, R. K., J. B. Franzini, D. L. Freyberg, and G. Tchobanoglous, *Water Resources Engineering*, 4th edition, McGraw-Hill, New York, 1992.

Mays, L. W. (editor-in-chief), *Hydraulic Design Handbook*, McGraw-Hill, New York, 1999.

National Research Council, Committee on Safety Criteria of Existing Dams, *Safety of Existing Dams: Evaluation and Improvement*, National Academy Press, Washington, DC, 1983.

National Research Council, Committee on Safety Criteria for Dams, *Safety of Dams: Flood and Earthquake Criteria*, National Academy Press, Washington, DC, 1985.

Novak, P., A. I. B. Moftat, C. Nalluri, and R. Narayanan, *Hydraulic Structures*, 2nd edition, E. & F. N. Spon, London, 1996.

Peterka, A. J., *Hydraulic Design of Spillways and Energy Dissipators*, U.S. Bureau of Reclamation, 1978.

Smith, C. D. *Hydraulic Structures*, University of Saskatchewan, Saskatchewan, Canada, 1978.

Thomas, H. H., *The Engineering of Large Dams*, Wiley, London, 1976.

U.S. Army Corps of Engineers, Waterways Experiment Station, *Hydraulic Design Criteria*, Washington, DC, 1959.

U.S. Army Corps of Engineers, *Report of the Chief of Engineers to the Secretary of the Army on the National Program of Inspection of Non-Federal Dams*, Washington, DC, 1982.

U.S. Army Corps of Engineers, *Inflow Design Floods for Dams and Reservoirs*, Engineering Regulation 1110-8-2, 1991.

U.S. Bureau of Reclamation, *Design of Small Dams*, 3rd edition, U.S. Government Printing Office, Washington, DC, 1987.

Wei, C. Y. and J. E. Lindell, "Hydraulic Design of Stilling Basins and Energy Dissipators," *Hydraulic Design Handbook*, edited by L.W. Mays, McGraw-Hill, New York, 1999.

Zipparro, V. J., and H. Hasen, *Davis' Handbook of Applied Hydraulics*, 4th edition, McGraw-Hill, New York, 1993.

Appendix A

Newton–Raphson Method

FINDING THE ROOT FOR A SINGLE NONLINEAR EQUATION

One of the most widely used methods for determining the root of a function ($f(x) = 0$) is the *Newton–Raphson equation*. Referring to Figure A.1, using an initial guess of the root as x_i, a tangent can be extended from point $[x_i, f(x_i)]$ to the x-axis. The point (value of x) where the tangent crosses the x-axis, x_{i+1}, usually represents an improved estimate of the root. The first derivative at $[x_i, f(x_i)]$ is the gradient or slope, $f'(x_i)$, expressed as

$$\left(\frac{df}{dy}\right)_i = f'(x_i) = \frac{f(x_i) - 0}{x_i - x_{i+1}} \tag{A.1}$$

which can be rearranged to yield the Newton–Raphson equation.

$$x_{i+1} = x_i - \frac{f(x_i)}{f'(x_i)} \tag{A.2}$$

The above derivation is simply a geometrical interpretation. See Chapra and Canale (1988) for a more rigorous mathematical derivation based on the Taylor series.

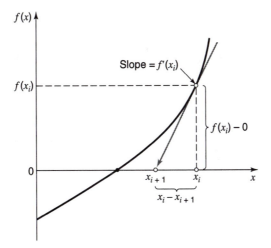

Figure A.1 Graphical depiction of the Newton–Raphson method. A tangent to the function of x_i [that is, $f'(x_i)$] is extrapolated down to the x axis to provide an estimate of the root at x_{i+1}.

APPLICATION TO SOLVE MANNING'S EQUATION FOR NORMAL DEPTH

Given a discharge Q, the normal depth is computed using Manning's equation (see Chapter 5).

$$Q = \frac{1.49}{n} AR^{2/3} S_0^{1/2}$$

where the cross-sectional area of flow A and the hydraulic radius R are both functions of the depth y. Starting with an initial guess of y_i, for the normal depth, the resulting discharge is Q_i:

$$Q_i = \frac{1.49}{n} A_i R_i^{2/3} S_0^{2/3} \tag{A.3}$$

To solve this problem for y given Q using the Newton–Raphson method, the nonlinear function to determine the root of is

$$f(y_i) = Q_i - Q \tag{A.4}$$

The gradient, $f'(y_i)$ is

$$\frac{df}{dy_i} = \frac{dQ_i}{dy_i} \tag{A.5}$$

because Q is a constant. Hence, assuming Manning's n is constant,

$$f'(y_i) = \frac{d}{dy}\left[\frac{1.49}{n} S_0^{1/2} A_i R_i^{2/3} \right]$$

$$= \frac{1.49}{n} S_0^{1/2} \left(\frac{2AR^{-1/3}}{3} \frac{dR}{dy} + R^{2/3} \frac{dA}{dy} \right)_i$$

$$= \frac{1.49}{n} S_0^{1/2} A_i R_i^{2/3} \left(\frac{2}{3R} \frac{dR}{dy} + \frac{1}{A} \frac{dA}{dy} \right)_i$$

$$= Q_i \left(\frac{2}{3R} \frac{dR}{dy} + \frac{1}{A} \frac{dA}{dy} \right)_i \tag{A.6}$$

where the subscript i outside the parentheses indicates that the contents are evaluated for $y = y_i$. The Newton–Raphson equation (A.2) in terms of the normal depth is

$$y_{i+1} = y_i - \frac{f(y_i)}{f'(y_i)} \tag{A.7}$$

Substituting equation (A.4) and (A.6) into equation (A.7) gives the Newton–Raphson equation for solving Manning's equation:

$$y_{i+1} = y_i - \frac{Q_i - Q}{Q_i \left(\dfrac{2}{3R} \dfrac{dR}{dy} + \dfrac{1}{A} \dfrac{dA}{dy} \right)_i} \tag{A.8}$$

For a rectangular channel, $A = B_w y$ and $R = \dfrac{B_w y}{(B_w + 2y)}$ where B_w is the channel width; equation (A.8) becomes

$$y_{i+1} = y_i - \frac{1 - Q/Q_i}{\left(\dfrac{5B_w + 6y_i}{3y_i(B_w + 2y_i)} \right)} \tag{A.9}$$

Values for the *channel shape function* $\left(\dfrac{2}{3R}\dfrac{dR}{dy} + \dfrac{1}{A}\dfrac{dA}{dy} \right)$ for other cross sections are given in Table 5.2.1.

FINDING THE ROOTS OF A SYSTEM OF NONLINEAR EQUATIONS

The above discussion focused on the determination of the roots of a single equation. A related problem is to find the roots of simultaneous equations,

$$f_1(x_1, x_2, ..., x_n) = 0$$
$$f_2(x_1, x_2, ..., x_n) = 0$$

$$\cdot \qquad \cdot$$
$$\cdot \qquad \cdot$$
$$\cdot \qquad \cdot$$

$$f_n(x_1, x_2, ..., x_n) = 0$$

(A.10)

The solution of this system consists of a set of x's that simultaneously result in all the equations equaling zero.

The Newton–Raphson method is an iterative technique for solving a system of nonlinear algebraic equations. It uses the same idea as presented above for the determination of the roots of a single equation, except that here the solution is for a vector of variables rather than for a single variable. Consider a system of equations denoted in vector form as

$$f(\mathbf{x}) = \mathbf{0} \tag{A.11}$$

where $\mathbf{x} = (x_1, x_2, ..., x_n)$ is the vector of unknown quantities and for iteration k, $x_k = \left(x_1^k, x_2^k, ..., x_n^k \right)$. The nonlinear system can be linearized to

$$f(\mathbf{x}^{k+1}) \approx f(\mathbf{x}^k) + \mathbf{J}(\mathbf{x}^k)(\mathbf{x}^{k+1} - \mathbf{x}^k) \tag{A.12}$$

where $\mathbf{J}(\mathbf{x}^k)$ is the *Jacobian*, which is a coefficient matrix made up of the first partial derivatives of $f(\mathbf{x})$ evaluated at $\mathbf{x}^k$. The right-hand side of equation (A.12) is the nonlinear vector function of $\overline{\mathbf{x}}^k$. Basically, an iterative procedure is used to determine $\mathbf{x}^{k+1}$ that forces the residual error $f(\overline{\mathbf{x}}^{k+1})$ in equation (A.12) to zero. This can be accomplished by setting $f(\overline{\mathbf{x}}^{k+1}) = 0$ and rearranging equation (A.12) to reread

$$\mathbf{J}(\mathbf{x}^k)(\mathbf{x}^{k+1} - \mathbf{x}^k) = -f(\mathbf{x}^k) \tag{A.13}$$

This system is solved for $(\mathbf{x}^{k+1} - \mathbf{x}^k) = \Delta\mathbf{x}^k$, and the improved estimate of the solution, $\mathbf{x}^{k+1}$, is determined knowing $\Delta\mathbf{x}^k$. The process is repeated until $(\mathbf{x}^{k+1} - \mathbf{x}^k)$ is smaller than some specified tolerance.

Let the initial estimate for the roots be $x_1^0, x_2^0, ..., x_n^0$ where the superscript indicates the number of the iteration—0 for the initial estimate, 1 for values obtained after one iteration, and so on. We can expand and rearrange equation (A.13) as

$$
\begin{bmatrix}
\dfrac{\partial f_1}{\partial x_1} \dfrac{\partial f_1}{\partial x_2} & \cdots & \dfrac{\partial f_1}{\partial x_i} & \cdots & \dfrac{\partial f_1}{\partial x_n} \\[2ex]
\dfrac{\partial f_2}{\partial x_1} \dfrac{\partial f_2}{\partial x_2} & \cdots & \dfrac{\partial f_2}{\partial x_i} & \cdots & \dfrac{\partial f_2}{\partial x_n} \\[2ex]
\cdots\cdots\cdots\cdots\cdots\cdots\cdots\cdots\cdots \\
\cdots\cdots\cdots\cdots\cdots\cdots\cdots\cdots\cdots \\
\dfrac{\partial f_i}{\partial x_1} \dfrac{\partial f_i}{\partial x_2} & \cdots & \dfrac{\partial f_i}{\partial x_i} & \cdots & \dfrac{\partial f_i}{\partial x_n} \\[2ex]
\cdots\cdots\cdots\cdots\cdots\cdots\cdots\cdots\cdots \\
\cdots\cdots\cdots\cdots\cdots\cdots\cdots\cdots\cdots \\
\dfrac{\partial f_n}{\partial x_1} \dfrac{\partial f_n}{\partial x_2} & \cdots & \dfrac{\partial f_n}{\partial x_i} & \cdots & \dfrac{\partial f_n}{\partial x_n}
\end{bmatrix}^{0}
\begin{bmatrix}
\Delta x_1 \\ \Delta x_2 \\ \cdot \\ \cdot \\ \Delta x_i \\ \cdot \\ \cdot \\ \cdot \\ \Delta x_n
\end{bmatrix}
=
\begin{bmatrix}
f_1 \\ f_2 \\ \cdot \\ \cdot \\ f_i \\ \cdot \\ \cdot \\ \cdot \\ f_n
\end{bmatrix}^{0}
\tag{A.14}
$$

REFERENCE

Chapra, S. C. and R. P. Canale, *Numerical Methods for Engineers*, Second Edition, McGraw-Hill, New York, 1988.

Index